The Digital Signal Processing Handbook, Second Edition

Digital Signal Processing Fundamentals
Video, Speech, and Audio Signal Processing and Associated Standards
Wireless, Networking, Radar, Sensor Array Processing, and Nonlinear Signal Processing

CRC Press
Taylor & Francis Group
6000 Broken Sound Parkway NW, Suite 300
Boca Raton, FL 33487-2742

Printed in the United States of America on acid-free paper
10 9 8 7 6 5 4 3 2 1

International Standard Book Number: 978-1-4200-4608-3 (Hardback)

Library of Congress Cataloging-in-Publication Data

Video, speech, and audio signal processing and associated standards / Vijay K. Madisetti.
p. cm.
"Second edition of the DSP Handbook has been divided into three parts."
Includes bibliographical references and index.
ISBN 978-1-4200-4608-3 (alk. paper)
1. Signal processing--Digital techniques--Standards. 2. Digital video--Standards. 3. Image processing--Digital techniques--Standards. 4. Speech processing systems--Standards 5. Sound--Recording and reproducing--Digital techniques--Standards I. Madisetti, V. (Vijay) II. Digital signal processing handbook. III. Title.

TK5102.9.V493 2009
621.382'2--dc22 2009022594

Visit the Taylor & Francis Web site at
http://www.taylorandfrancis.com

and the CRC Press Web site at
http://www.crcpress.com

The Digital Signal Processing Handbook

SECOND EDITION

Video, Speech, and Audio Signal Processing and Associated Standards

EDITOR-IN-CHIEF

Vijay K. Madisetti

CRC Press
Taylor & Francis Group
Boca Raton London New York

CRC Press is an imprint of the
Taylor & Francis Group, an **informa** business

The Electrical Engineering Handbook Series

Series Editor
Richard C. Dorf
University of California, Davis

Titles Included in the Series

The Handbook of Ad Hoc Wireless Networks, Mohammad Ilyas
The Avionics Handbook, Second Edition, Cary R. Spitzer
The Biomedical Engineering Handbook, Third Edition, Joseph D. Bronzino
The Circuits and Filters Handbook, Second Edition, Wai-Kai Chen
The Communications Handbook, Second Edition, Jerry Gibson
The Computer Engineering Handbook, Vojin G. Oklobdzija
The Control Handbook, William S. Levine
The CRC Handbook of Engineering Tables, Richard C. Dorf
The Digital Avionics Handbook, Second Edition Cary R. Spitzer
The Digital Signal Processing Handbook, Second Edition, Vijay K. Madisetti
The Electrical Engineering Handbook, Second Edition, Richard C. Dorf
The Electric Power Engineering Handbook, Second Edition, Leonard L. Grigsby
The Electronics Handbook, Second Edition, Jerry C. Whitaker
The Engineering Handbook, Third Edition, Richard C. Dorf
The Handbook of Formulas and Tables for Signal Processing, Alexander D. Poularikas
The Handbook of Nanoscience, Engineering, and Technology, Second Edition William A. Goddard, III, Donald W. Brenner, Sergey E. Lyshevski, and Gerald J. Iafrate
The Handbook of Optical Communication Networks, Mohammad Ilyas and Hussein T. Mouftah
The Industrial Electronics Handbook, J. David Irwin
The Measurement, Instrumentation, and Sensors Handbook, John G. Webster
The Mechanical Systems Design Handbook, Osita D.I. Nwokah and Yidirim Hurmuzlu
The Mechatronics Handbook, Second Edition, Robert H. Bishop
The Mobile Communications Handbook, Second Edition, Jerry D. Gibson
The Ocean Engineering Handbook, Ferial El-Hawary
The RF and Microwave Handbook, Second Edition, Mike Golio
The Technology Management Handbook, Richard C. Dorf
The Transforms and Applications Handbook, Second Edition, Alexander D. Poularikas
The VLSI Handbook, Second Edition, Wai-Kai Chen

Contents

PART III Image and Video Processing

Preface

Digital signal processing (DSP) is concerned with the theoretical and practical aspects of representing information-bearing signals in a digital form and with using computers, special-purpose hardware and software, or similar platforms to extract information, process it, or transform it in useful ways. Areas where DSP has made a significant impact include telecommunications, wireless and mobile communications, multimedia applications, user interfaces, medical technology, digital entertainment, radar and sonar, seismic signal processing, and remote sensing, to name just a few.

Given the widespread use of DSP, a need developed for an authoritative reference, written by the top experts in the world, that would provide information on both theoretical and practical aspects in a manner that was suitable for a broad audience—ranging from professionals in electrical engineering, computer science, and related engineering and scientific professions to managers involved in technical marketing, and to graduate students and scholars in the field. Given the abundance of basic and introductory texts on DSP, it was important to focus on topics that were useful to engineers and scholars without overemphasizing those topics that were already widely accessible. In short, the DSP handbook was created to be relevant to the needs of the engineering community.

A task of this magnitude could only be possible through the cooperation of some of the foremost DSP researchers and practitioners. That collaboration, over 10 years ago, produced the first edition of the successful DSP handbook that contained a comprehensive range of DSP topics presented with a clarity of vision and a depth of coverage to inform, educate, and guide the reader. Indeed, many of the chapters, written by leaders in their field, have guided readers through a unique vision and perception garnered by the authors through years of experience.

The second edition of the DSP handbook consists of *Digital Signal Processing Fundamentals*; *Video, Speech, and Audio Signal Processing and Associated Standards*; and *Wireless, Networking, Radar, Sensor Array Processing, and Nonlinear Signal Processing* to ensure that each part is dealt with in adequate detail and that each part is then able to develop its own individual identity and role in terms of its educational mission and audience. I expect each part to be frequently updated with chapters that reflect the changes and new developments in the technology and in the field. The distribution model for the DSP handbook also reflects the increasing need by professionals to access content in electronic form anywhere and at anytime.

Video, Speech, and Audio Signal Processing and Associated Standards, as the name implies, provides a comprehensive coverage of the basic foundations of speech, audio, image, and video processing and associated applications to broadcast, storage, search and retrieval, and communications.

This book needs to be continuously updated to include newer aspects of these technologies, and I look forward to suggestions on how this handbook can be improved to serve you better.

MATLAB® is a registered trademark of The MathWorks, Inc. For product information, please contact:

The MathWorks, Inc.
3 Apple Hill Drive
Natick, MA 01760-2098 USA
Tel: 508 647 7000
Fax: 508-647-7001
E-mail: info@mathworks.com
Web: www.mathworks.com

Editor

Vijay K. Madisetti is a professor in the School of Electrical and Computer Engineering at the Georgia Institute of Technology in Atlanta. He teaches graduate and undergraduate courses in digital signal processing and computer engineering, and leads a strong research program in digital signal processing, telecommunications, and computer engineering.

Dr. Madisetti received his BTech (Hons) in electronics and electrical communications engineering in 1984 from the Indian Institute of Technology, Kharagpur, India, and his PhD in electrical engineering and computer sciences in 1989 from the University of California at Berkeley.

He has authored or edited several books in the areas of digital signal processing, computer engineering, and software systems, and has served extensively as a consultant to industry and the government. He is a fellow of the IEEE and received the 2006 Frederick Emmons Terman Medal from the American Society of Engineering Education for his contributions to electrical engineering.

Contributors

Kenzo Akagiri
Sony Corporation
Tokyo, Japan

Osama Al-Shaykh
Packet Video
San Diego, California

Yucel Altunbasak
School of Electrical and Computer Engineering
Georgia Institute of Technology
Atlanta, Georgia

Robert L. Andersen
Dolby Laboratories, Inc.
San Francisco, California

Soo Hyun Bae
Sony U.S. Research Center
San Jose, California

Arshdeep Bahga
School of Electrical and Computer Engineering
Georgia Institute of Technology
Atlanta, Georgia

Kurt Baudendistel
Momentum Data Systems
Fountain Valley, California

Jan Biemond
Faculty of Electrical Engineering, Mathematics, and Computer Science
Delft University of Technology
Delft, the Netherlands

Ralph Braspenning
Philips Research Laboratories
Eindhoven, the Netherlands

Richard V. Cox
AT&T Research Labs
Florham Park, New Jersey

Grant A. Davidson
Dolby Laboratories, Inc.
San Francisco, California

Sean Dorward
Bell Laboratories
Lucent Technologies
Murray Hill, New Jersey

Ruggero E. H. Franich
AEA Technology
Culham Laboratory
Oxfordshire, United Kingdom

Sadaoki Furui
Department of Computer Science
Tokyo Institute of Technology
Tokyo, Japan

W. Gehrke
Philips Semiconductors
Hamburg, Germany

Jan J. Gerbrands
Department of Electrical Engineering
Delft University of Technology
Delft, the Netherlands

Gerard de Haan
Philips Research Laboratories
Eindhoven, the Netherlands

and

Information Communication Systems Group
Eindhoven University of Technology
Eindhoven, the Netherlands

Joseph L. Hall
Bell Laboratories
Lucent Technologies
Murray Hill, New Jersey

Emile A. Hendriks
Information and Communication Theory Group
Delft University of Technology
Delft, the Netherlands

Nikil Jayant
School of Electrical and Computer Engineering
Georgia Institute of Technology
Atlanta, Georgia

James D. Johnston
AT&T Research Labs
Florham Park, New Jersey

Biing-Hwang Juang
School of Electrical and Computer Engineering
Georgia Institute of Technology
Atlanta, Georgia

Masayuki Katakura
Sony Corporation
Kanagawa, Japan

M. Kohut
Sony Corporation
San Diego, California

Reginald L. Lagendijk
Information and Communication Theory Group
Delft University of Technology
Delft, the Netherlands

Vijay K. Madisetti
School of Electrical and Computer Engineering
Georgia Institute of Technology
Atlanta, Georgia

Russell M. Mersereau
School of Electrical and Computer Engineering
Georgia Institute of Technology
Atlanta, Georgia

Ralph Neff
Video and Image Processing Laboratory
University of California
Berkeley, California

Masayuki Nishiguchi
Sony Corporation
Tokyo, Japan

Peter Noll
Institute for Telecommunications
Technical University of Berlin
Berlin, Germany

Joseph Olive
Bell Laboratories
Lucent Technologies
Murray Hill, New Jersey

P. Pirsch
Laboratory for Information Technology
University of Hannover
Hannover, Germany

Schuyler R. Quackenbush
Audio Research Labs
Scotch Plains, New Jersey

and

AT&T Research Labs
Florham Park, New Jersey

Lawrence R. Rabiner
Department of Electrical and Computer Engineering
Rutgers University
New Brunswick, New Jersey

and

AT&T Research Labs
Florham Park, New Jersey

Tor A. Ramstad
Department of Electronics and Telecommunications
Norwegian University of Science and Technology
Trondheim, Norway

Stanley J. Reeves
Electrical and Computer Engineering Department
Auburn University
Auburn, Alabama

Aaron E. Rosenberg
Center for Advanced Information Processing
Rutgers University
Piscataway, New Jersey

E. Saito
Sony Corporation
Kanagawa, Japan

Juergen Schroeter
AT&T Research Labs
Florham Park, New Jersey

John Shore
Entropic Research Laboratory, Inc.
Washington, District of Columbia

Deepen Sinha
Bell Laboratories
Lucent Technologies
Murray Hill, New Jersey

M. Mohan Sondhi
Bell Laboratories
Lucent Technologies
Murray Hill, New Jersey

Richard Sproat
Bell Laboratories
Lucent Technologies
Murray Hill, New Jersey

David Taubman
Hewlett Packard
Palo Alto, California

A. Murat Tekalp
Department of Electrical and Electronics Engineering
Koç University
Istanbul, Turkey

Keisuke Toyama
Sony Corporation
Tokyo, Japan

Kyoya Tsutsui
Sony Corporation
Tokyo, Japan

Kou-Hu Tzou
Hyundai Network Systems
Seoul, Korea

Lucas J. van Vliet
Department of Imaging Science and Technology
Delft University of Technology
Delft, the Netherlands

H. Yamauchi
Sony Corporation
Kanagawa, Japan

Ian T. Young
Department of Imaging Science and Technology
Delft University of Technology
Delft, the Netherlands

Avideh Zakhor
Video and Image Processing Laboratory
University of California
Berkeley, California

I

Digital Audio Communications

Nikil Jayant
Georgia Institute of Technology

AS I PREDICTED IN THE SECTION INTRODUCTION FOR THE 1997 version of this book, digital audio communications has become nearly as prevalent as digital speech communications. In particular, new technologies for audio storage and transmission have made available music and wideband signals in a flexible variety of standard formats.

The fundamental underpinning for these technologies is audio compression based on perceptually tuned shaping of the quantization noise. Chapter 1 in this part describes aspects of psychoacoustics that have led to the general foundations of "perceptual audio coding." Succeeding chapters in this part cover established examples of "perceptual audio coders." These include MPEG standards, and coders developed by Dolby, Sony, and Bell Laboratories.

The dimensions of coder performance are quality, bit rate, delay, and complexity. The quality vs. bit rate trade-offs are particularly important.

Audio Quality

The three parameters of digital audio quality are "signal bandwidth," "fidelity," and "spatial realism."

Compact-disc (CD) signals have a bandwidth of 20–20,000 Hz, while traditional telephone speech has a bandwidth of 200–3400 Hz. Intermediate bandwidths characterize various grades of wideband speech and audio, including roughly defined ranges of quality referred to as AM radio and FM radio quality (bandwidths on the order of 7–10 and 12–15 kHz, respectively).

In the context of digital coding, fidelity refers to the level of perceptibility of quantization or to reconstruction noise. The highest level of fidelity is one where the noise is imperceptible in formal listening tests. Lower levels of fidelity are acceptable in some applications if they are not annoying, although in general it is good practice to sacrifice some bandwidth in the interest of greater fidelity, for a given bit rate in coding. Five-point scales of signal fidelity are common both in speech and audio coding.

Spatial realism is generally provided by increasing the number of coded (and reproduced) spatial channels. Common formats are 1-channel (mono), 2-channel (stereo), 5-channel (3 front, 2 rear), 5.1-channel (5-channel plus subwoofer), and 8-channel (6 front, 2 rear). For given constraints on bandwidth and fidelity, the required bit rate in coding increases as a function of the number of channels; but the increase is slower than linear, because of the presence of interchannel redundancy. The notion of perceptual coding originally developed for exploiting the perceptual irrelevancies of a single-channel audio signal extends also to the methods used in exploiting interchannel redundancy.

Bit Rate

The CD-stereo signal has a digital representation rate of 1406 kilobits per second (kbps). Current technology for perceptual audio coding, notably MP3 audio reproduces CD-stereo with near-perfect fidelity at bit rates as low as 128 kbps, depending on the input signal. CD-like reproduction is possible at bit rates as low as 64 kbps for stereo. Single-channel reproduction of FM-radio-like music is possible at 32 kbps. The single-channel reproduction of AM-radio-like music and wideband speech is possible at rates approaching 16 kbps for all but the most demanding signals. Techniques for so-called pseudo-stereo can provide additional enhancement of digital single-channel audio.

Applications of Digital Audio

The capabilities of audio compression have combined with increasingly affordable implementations on platforms for digital signal processing (DSP), native signal processing (NSP) in a computer's (native) processor, and application-specific integrated circuits (ASICs) to create revolutionary applications of digital audio. International and national standards have contributed immensely to this revolution. Some of these standards only specify the bit-stream syntax and decoder, leaving room for future, sometimes proprietary, enhancements of the encoding algorithm.

The domains of applications include "transmission" (e.g., digital audio broadcasting), "storage" (e.g., the iPod and the digital versatile disk [DVD]), and "networking" (music preview, distribution, and publishing). The networking applications, aided by significant advances in broadband access speeds, have made digital audio communications as commonplace as digital telephony.

The Future of Digital Audio

Remarkable as the capabilities and applications mentioned above are, there are even greater challenges and opportunities for the practitioners of digital audio technology. It is unlikely that we have reached or even approached the fundamental limits of performance in terms of audio quality at a given bit rate. Newer capabilities in this technology (in terms of audio fidelity, bandwidth, and spatial realism) will continue to lead to newer classes of applications in audio communications. New technologies for universal coding will create interesting new options for digital networking and seamless communication of speech and music signals. Advances in multichannel audio capture and reproduction will lead to more sophisticated and user-friendly technologies for telepresence. In the entertainment domain, the audio dimension will continue to enhance the overall quality of visual formats such as multiplayer games and 3D-television.

1

Auditory Psychophysics for Coding Applications

Joseph L. Hall
Lucent Technologies

In this chapter, we review properties of auditory perception that are relevant to the design of coders for acoustic signals. The chapter begins with a general definition of a perceptual coder, then considers what the "ideal" psychophysical model would consist of, and what use a coder could be expected to make of this model. We then present some basic definitions and concepts. The chapter continues with a review of relevant psychophysical data, including results on threshold, just-noticeable differences (JNDs), masking, and loudness. Finally, we attempt to summarize the present state of the art, the capabilities and limitations of present-day perceptual coders for audio and speech, and what areas most need work.

1.1 Introduction

A coded signal differs in some respect from the original signal. One task in designing a coder is to minimize some measure of this difference under the constraints imposed by bit rate, complexity, or cost. What is the appropriate measure of difference? The most straightforward approach is to minimize some physical measure of the difference between original and coded signal. The designer might attempt to minimize RMS difference between the original and coded waveform, or perhaps the difference between original and coded power spectra on a frame-by-frame basis. However, if the purpose of the coder is to encode acoustic signals that are eventually to be listened to* by people, these physical measures do not directly address the appropriate issue. For signals that are to be listened to by people, the "best" coder is the one that sounds the best. There is a very clear distinction between "physical" and "perceptual" measures of a signal (frequency vs. pitch, intensity vs. loudness, for example). A perceptual coder can be defined as a coder that minimizes some measure of the difference between original and coded signal so as to minimize the perceptual impact of the coding noise. We can define the best coder given a particular set of constraints as the one in which the coding noise is least objectionable.

* Perceptual coding is not limited to speech and audio. It can be applied also to image and video [16]. In this chapter we consider only coders for acoustic signals.

It follows that the designer of a perceptual coder needs some way to determine the perceptual quality of a coded signal. "Perceptual quality" is a poorly defined concept, and it will be seen that in some sense it cannot be uniquely defined. We can, however, attempt to provide a partial answer to the question of how it can be determined. We can present something of what is known about human auditory perception from psychophysical listening experiments and show how these phenomena relate to the design of a coder.

One requirement for successful design of a perceptual coder is a satisfactory model for the signal-dependent sensitivity of the auditory system. Present-day models are incomplete, but we can attempt to specify what the properties of a complete model would be. One possible specification is that, for any given waveform (the signal), it accurately predicts the loudness, as a function of pitch and of time, of any added waveform (the noise). If we had such a complete model, then we would in principle be able to build a transparent coder, defined as one in which the coded signal is indistinguishable from the original signal, or at least we would be able to determine whether or not a given coder was transparent. It is relatively simple to design a psychophysical listening experiment to determine whether the coding noise is audible, or equivalently, whether the subject can distinguish between original and coded signal. Any subject with normal hearing could be expected to give similar results to this experiment. While present-day models are far from complete, we can at least describe the properties of a complete model.

There is a second requirement that is more difficult to satisfy. This is the need to be able to determine which of two coded samples, each of which has audible coding noise, is preferable. While a satisfactory model for the signal-dependent sensitivity of the auditory system is in principle sufficient for the design of a transparent coder, the question of how to build the best nontransparent coder does not have a unique answer. Often, design constraints preclude building a transparent coder. Even the best coder built under these constraints will result in audible coding noise, and it is under some conditions impossible to specify uniquely how best to distribute this noise. One listener may prefer the more intelligible version, while another may prefer the more natural sounding version. The preferences of even a single listener might very well depend on the application. In the absence of any better criterion, we can attempt to minimize the loudness of the coding noise, but it must be understood that this is an incomplete solution.

Our purpose in this chapter is to present something of what is known about human auditory perception in a form that may be useful to the designer of a perceptual coder. We do not attempt to answer the question of how this knowledge is to be utilized, how to build a coder. Present-day perceptual coders for the most part utilize a "feedforward" paradigm: analysis of the signal to be coded produces specifications for allowable coding noise. Perhaps a more general method is a "feedback" paradigm, in which the perceptual model somehow makes possible a decision as to which of two coded signals is "better." This decision process can then be iterated to arrive at some optimum solution. It will be seen that for proper exploitation of some aspects of auditory perception the feedforward paradigm may be inadequate and the potentially more time-consuming feedback paradigm may be required. How this is to be done is part of the challenge facing the designer.

1.2 Definitions

In this section, we define some fundamental terms and concepts and clarify the distinction between physical and perceptual measures.

1.2.1 Loudness

When we increase the intensity of a stimulus its loudness increases, but that does not mean that intensity and loudness are the same thing. "Intensity" is a physical measure. We can measure the intensity of a signal with an appropriate measuring instrument, and if the measuring instrument is standardized and calibrated correctly, anyone else anywhere in the world can measure the same signal and get the same result. "Loudness" is "perceptual magnitude." It can be defined as "that attribute of auditory sensation in terms of which sounds can be ordered on a scale extending from quiet to loud" [23, p. 47]. We cannot measure it directly. All we can do is ask questions of a subject and from the responses attempt to infer something about loudness. Furthermore, we have no guarantee that a particular stimulus will be as loud for one subject as for another. The best we can do is assume that, for a particular stimulus, loudness judgments for one group of normal-hearing people will be similar to loudness judgments for another group.

There are two commonly used measures of loudness. One is "loudness level" (unit *phon*) and the other is "loudness" (unit *sone*). These two measures differ in what they describe and how they are obtained. The phon is defined as the intensity, in dB sound pressure level (SPL), of an equally loud 1 kHz tone. The sone is defined in terms of subjectively measured loudness ratios. A stimulus half as loud as a one-sone stimulus has a loudness of 0.5 sones, a stimulus 10 times as loud has a loudness of 10 sones, etc. A 1 kHz tone at 40 dB SPL is arbitrarily defined to have a loudness of one sone.

The argument can be made that loudness matching, the procedure used to obtain the phon scale, is a less subjective procedure than loudness scaling, the procedure used to obtain the sone scale. This argument would lead to the conclusion that the phon is the more objective of the two measures and that the sone is more subject to individual variability. This argument breaks down on two counts: first, for dissimilar stimuli even the supposedly straightforward loudness-matching task is subject to large and poorly understood order and bias effects that can only be described as subjective. While loudness matching of two equal-frequency tone bursts generally gives stable and repeatable results, the task becomes more difficult when the frequencies of the two tone bursts differ. Loudness matching between two dissimilar stimuli, as for example between a pure tone and a multicomponent complex signal, is even more difficult and yields less stable results. Loudness-matching experiments have to be designed carefully, and results from these experiments have to be interpreted with caution. Second, it is possible to measure loudness in sones, at least approximately, by means of a loudness-matching procedure. Fletcher [6] states that under some conditions loudness adds. Binaural presentation of a stimulus results in loudness doubling; and two equally loud stimuli, far enough apart in frequency that they do not mask each other, are twice as loud as one. If loudness additivity holds, then it follows that the sone scale can be generated by matching loudness of a test stimulus to binaural stimuli or to pairs of tones. This approach must be treated with caution. As Fletcher states, "However, this method [scaling] is related more directly to the scale we are seeking (the sone scale) than the two preceding ones (binaural or monaural loudness additivity)" [6, p. 278]. The loudness additivity approach relies on the assumption that loudness summation is perfect, and there is some more recent evidence [28,33] that loudness summation, at least for binaural vs. monaural presentation, is not perfect.

1.2.2 Pitch

The American Standards Association defines pitch as "that attribute of auditory sensation in which sounds may be ordered on a musical scale." Pitch bears much the same relationship to frequency as loudness does to intensity: frequency is an objective physical measure, while pitch is a subjective perceptual measure. Just as there is not a one-to-one relationship between intensity and loudness, so also there is not a one-to-one relationship between frequency and pitch. Under some conditions, for example, loudness can be shown to decrease with decreasing frequency with intensity held constant, and pitch can be shown to decrease with increasing intensity with frequency held constant [40, p. 409].

1.2.3 Threshold of Hearing

Since the concept of threshold is basic to much of what follows, it is worthwhile at this point to discuss it in some detail. It will be seen that thresholds are determined not only by the stimulus and the observer but also by the method of measurement. While this discussion is phrased in terms of threshold of hearing, much of what follows applies as well to differential thresholds (JNDs) discussed in Section 1.2.4.

By the simplest definition, the threshold of hearing (equivalently, auditory threshold) is the lowest intensity that the listener can hear. This definition is inadequate because we cannot directly measure the listener's perception. A first-order correction, therefore, is that the threshold of hearing is the lowest intensity that elicits from the listener the response that the sound is audible. Given this definition, we can present a stimulus to the listener and ask whether he or she can hear it. If we do this, we soon find that identical stimuli do not always elicit identical responses. In general, the probability of a positive response increases with increasing stimulus intensity and can be described by a "psychometric function" such as that shown for a hypothetical experiment in Figure 1.1. Here the stimulus intensity (in dB) appears on the abscissa and the probability $P(C)$ of a positive response appears on the ordinate. The yes–no experiment could be described by a psychometric function that ranges from zero to one, and threshold could be defined as the stimulus intensity that elicits a positive response in 50% of the trials.

A difficulty with the simple yes–no experiment is that we have no control over the subject's "criterion level." The subject may be using a strict criterion ("yes" only if the signal is definitely present) or a lax criterion ("yes" if the signal might be present). The subject can respond correctly either by a positive response in the presence of a stimulus ("hit") or by a negative response in the absence of a stimulus ("correct rejection"). Similarly, the subject can respond incorrectly either by a negative response in the presence of a stimulus ("miss") or by a positive response in the absence of a stimulus ("false alarm"). Unless the experimenter is willing to use an elaborate and time-consuming procedure that involves assigning rewards to correct responses and penalties to incorrect responses, the criterion level is uncontrolled.

The field of psychophysics that deals with this complication is called "detection theory." The field of psychophysical detection theory is highly developed [12] and a complete description is far beyond the scope of this chapter. Very briefly, the subject's response is considered to be based on an internal "decision variable," a random variable drawn from a distribution with mean and standard deviation that depend on the stimulus. If we assume that the decision variable is normally distributed with a fixed standard deviation σ and a mean that depends only on stimulus intensity, then we can define an "index of sensitivity" d' for a given stimulus intensity as the difference between m_0 (the mean in the absence of the stimulus) and m_s (the mean in the presence of the stimulus), divided by σ.

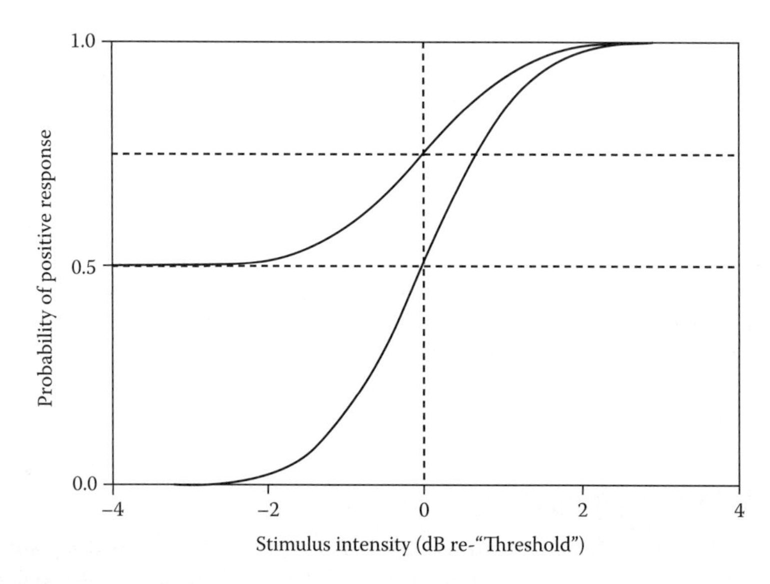

FIGURE 1.1 Idealized psychometric functions for hypothetical yes–no experiment (0–1) and for hypothetical 2FIC experiment (0.5–1).

An "ideal observer" (a hypothetical subject who does the best possible job for the task at hand) gives a positive response if and only if the decision variable exceeds an internal criterion level. An increase in criterion level decreases the probability of a false alarm and increases the probability of a miss.

A simple and satisfactory way to deal with the problem of uncontrolled criterion level is to use a "criterion-free" experimental paradigm. The simplest is perhaps the two-interval forced choice (2IFC) paradigm, in which the stimulus is presented at random in one of two observation intervals. The subject's task is to determine which of the two intervals contained the stimulus. The ideal observer selects the interval that elicits the larger decision variable, and criterion level is no longer a factor. Now, the subject has a 50% chance of choosing the correct interval even in the absence of any stimulus, so the psychometric function goes from 0.5 to 1.0 as shown in Figure 1.1. A reasonable definition of threshold is $P(C) = 0.75$, halfway between the chance level of 0.5 and 1. If the decision variable is normally distributed with a fixed standard deviation, it can be shown that this definition of threshold corresponds to a d' of 0.95.

The number of intervals can be increased beyond two. In this case, the ideal observer responds correctly if the decision variable for the interval containing the stimulus is larger than the largest of the $N-1$ decision variables for the intervals not containing the stimulus. A common practice is, for an N-interval forced choice paradigm (NIFC), to define threshold as the point halfway between the chance level of $1/N$ and one. This is a perfectly acceptable practice so long as it is recognized that the measured threshold is influenced by the number of alternatives. For a 3IFC paradigm this definition of threshold corresponds to a d' of 1.12 and for a 4IFC paradigm it corresponds to a d' of 1.24.

1.2.4 Differential Threshold

The differential threshold is conceptually similar to the auditory threshold discussed above, and many of the same comments apply. The differential threshold, or JND, is the amount by which some attribute of a signal has to change in order for the observer to be able to detect the change. A tone burst, for example, can be specified in terms of frequency, intensity, and duration, and a differential threshold for any of these three attributes can be defined and measured.

The first attempt to provide a quantitative description of differential thresholds was provided by the German physiologist E. H. Weber in the first half of the nineteenth century. According to "Weber's law," the JND ΔI is proportional to the stimulus intensity I, or $\Delta I/I = K$, where the constant of proportionality $\Delta I/I$ is known as the "Weber fraction." This was supposed to be a general description of sensitivity to changes of intensity for a variety of sensory modalities, not limited just to hearing, and it has since been applied to perception of nonintensive variables such as frequency. It was recognized at an early stage that this law breaks down at near-threshold intensities, and in the latter half of the nineteenth century the German physicist G. T. Fechner suggested the modification that is now known as the "modified Weber law." $\Delta I/(I + I_0) = K$, where I_0 is a constant. While Weber's law provides a reasonable first-order description of intensity and frequency discrimination in hearing, in general it does not hold exactly, as will be seen below.

As with the threshold of hearing, the differential threshold can be measured in different ways, and the result depends to some extent on how it is measured. The simplest method is a same-different paradigm, in which two stimuli are presented and the subject's task is to judge whether or not they are the same. This method suffers from the same drawback as the yes–no paradigm for auditory threshold: we do not have control over the subject's criterion level.

If the physical attribute being measured is simply related to some perceptual attribute, then the differential threshold can be measured by requiring the subject to judge which of two stimuli has more of that perceptual attribute. A JND for frequency, for example, could be measured by requiring the subject to judge which of two stimuli is of higher pitch; or a JND for intensity could be measured by requiring the subject to judge which of two stimuli is louder. As with the 2IFC paradigm discussed above for auditory threshold, this method removes the problem of uncontrolled criterion level.

There are more general methods that do not assume a knowledge of the relationship between the physical attribute being measured and a perceptual attribute. The most useful, perhaps, is the NIFC method: N stimuli are presented, one of which differs from the other $N - 1$ along the dimension being measured. The subject's task is to specify which one of the N stimuli is different from the other $N - 1$.

Note that there is a close parallel between the differential threshold and the auditory threshold described in Section 1.2.4. The auditory threshold can be regarded as a special case of the JND for intensity, where the question is by how much the intensity has to differ from zero in order to be detectable.

1.2.5 Masked Threshold

The "masked threshold" of a signal is defined as the threshold of that signal (the "probe") in the presence of another signal (the "masker"). A related term is "masking," which is the elevation of threshold of the probe by the masker: it is the difference between masked and absolute threshold. More generally, the reduction of loudness of a suprathreshold signal is also referred to as masking. It will be seen that masking can appear in many forms, depending on spectral and temporal relationships between probe and masker.

Many of the comments that applied to measurement of absolute and differential thresholds also apply to measurement of masked threshold. The simplest method is to present masker plus probe and ask the subject whether or not the probe is present. Once again there is a problem with criterion level. Another method is to present stimuli in two intervals and ask the subject which one contains the probe. This method can give useful results but can, under some conditions, give misleading results. Suppose, for example, that the probe and masker are both pure tones at 1 kHz, but that the two signals are 180° out of phase. As the intensity of the probe is increased from zero, the intensity of the composite signal will first decrease, then increase. The two signals, masker alone and masker plus probe, may be easily distinguishable, but in the absence of additional information the subject has no way of telling which is which.

A more robust method for measuring masked threshold is the NIFC method described above, in which the subject specifies which of the N stimuli differs from the other $N - 1$. Subjective percepts in masking experiments can be quite complex and can differ from one observer to another. In the NIFC method, the observer has the freedom to base judgments on whatever attribute is most easily detected, and it is not necessary to instruct the observer what to listen for.

Note that the differential threshold for intensity can be regarded as a special case of the masked threshold in which the probe is an intensity-scaled version of the masker.

A note on terminology: suppose two signals, $x_1(t)$ and $[x_1(t) + x_2(t)]$ are just distinguishable. If $x_2(t)$ is a scaled version of $x_1(t)$, then we are dealing with intensity discrimination. If $x_1(t)$ and $x_2(t)$ are two different signals, then we are dealing with masking, with $x_1(t)$ the masker and $x_2(t)$ the probe. In either case, the difference can be described in several ways. These ways include (1) the intensity increment between $x_1(t)$ and $[x_1(t) + x_2(t)]$, ΔI; (2) the intensity increment relative to $x_1(t)$, $\Delta I/I$; (3) the intensity ratio between $x_1(t)$ and $[x_1(t) + x_2(t)]$, $(I + \Delta I)/I$; (4) the intensity increment in dB, $10 \times \log_{10}(\Delta I/I)$; and (5) the intensity ratio in dB, $10 \times \log_{10}[(I + \Delta I)/I]$. These ways are equivalent in that they show the same information, although for a particular application one way may be preferable to another for presentation purposes. Another measure that is often used, particularly in the design of perceptual coders, is the intensity of the probe $x_2(t)$. This measure is subject to misinterpretation and must be used with caution. Depending on the coherence between $x_1(t)$ and $x_2(t)$, a given probe intensity can result in a wide range of intensity increments ΔI. The resulting ambiguity has been responsible for some confusion.

1.2.6 Critical Bands and Peripheral Auditory Filters

The concepts of "critical bands" and "peripheral auditory" filters are central to much of the auditory modeling work that is used in present-day perceptual coders. Scharf, in a classic review article [33], defines the empirical critical bandwidth as "that bandwidth at which subjective responses rather abruptly change." Simply put, for some psychophysical tasks the auditory system behaves as if it consisted of a bank of band-pass filters (the critical bands) followed by energy detectors. Examples of critical-band behavior that are particularly relevant for the designer of a coder include the relationship between bandwidth and loudness (Figure 1.5) and the relationship between bandwidth and masking (Figure 1.10). Another example of critical-band behavior is phase sensitivity: in experiments measuring the detectability of amplitude and of frequency modulation, the auditory system appears to be sensitive to the relative phase of the components of a complex sound only so long as the components are within a critical band [9,45].

The concept of the critical band was introduced more than a half-century ago by Fletcher [6], and since that time it has been studied extensively. Fletcher's pioneering contribution is ably documented by Allen [1], and Scharf's 1970 review article [33] gives references to some later work. More recently, Moore and his co-workers have made extensive measurements of peripheral auditory filters [24].

The value of critical bandwidths has been the subject of some discussion, because of questions of definition and method of measurement. Figure 1.2 [31, Fig. 1] shows critical bandwidth as a function of frequency for Scharf's empirical definition (the bandwidth at which subjective responses undergo some sort of change). Results from several experiments are superimposed here, and they are in substantial agreement with each other. Moore and Glasberg [26] argue that the bandwidths shown in Figure 1.2 are determined not only by the bandwidth of peripheral auditory filters but also by changes in processing efficiency. By their argument, the bandwidth of peripheral auditory filters is somewhat smaller than the values shown in Figure 1.2 at frequencies above 1 kHz and substantially smaller, by as much as an octave, at lower frequencies.

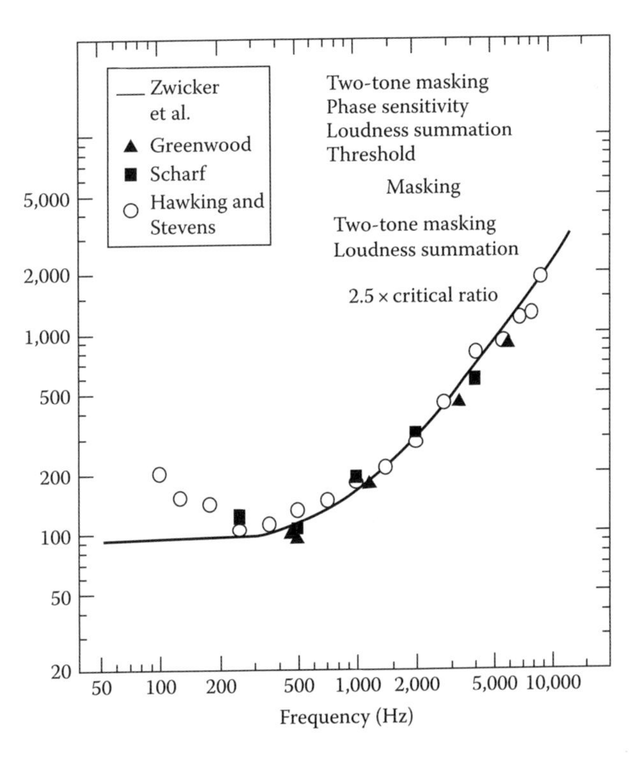

FIGURE 1.2 Empirical critical bandwidth. (From Scharf, B., Critical bands, in *Foundations of Modern Auditory Theory*, Vol. 1, Chap. 5, Tobias, J.V., Ed., Academic Press, New York, 1970. With permission.)

1.3 Summary of Relevant Psychophysical Data

In Section 1.2, we introduced some basic concepts and definitions. In this section, we review some relevant psychophysical results. There are several excellent books and book chapters that have been written on this subject, and we have neither the space nor the inclination to duplicate material found in these other sources. Our attempt here is to make the reader aware of some relevant results and to refer him or her to sources where more extensive treatments may be found.

1.3.1 Loudness

1.3.1.1 Loudness Level and Frequency

For pure tones, loudness depends on both intensity and frequency. Figure 1.3 (modified from [37, p. 124]) shows loudness level contours. The curves are labeled in phons and, in parentheses, sones. These curves have been remeasured many times since, with some variation in the results, but the basic conclusions remain unchanged. The most sensitive region is around 2–3 kHz. The low-frequency slope of the loudness level contours is flatter at high loudness levels than at low. It follows that loudness level grows more rapidly with intensity at low frequencies than at high. The 38- and 48-phon contours are (by definition) separated by 10 dB at 1 kHz, but they are only about 5 dB apart at 100 Hz.

Figure 1.3 also shows contours that specify the dynamic range of hearing. Tones below the 8-phon contour are inaudible, and tones above the dotted line are uncomfortable. The dynamic range of hearing, the distance between these two contours, is greatest around 2–3 kHz and decreases at lower and higher frequencies. In practice, the useful dynamic range is substantially less. We know today that

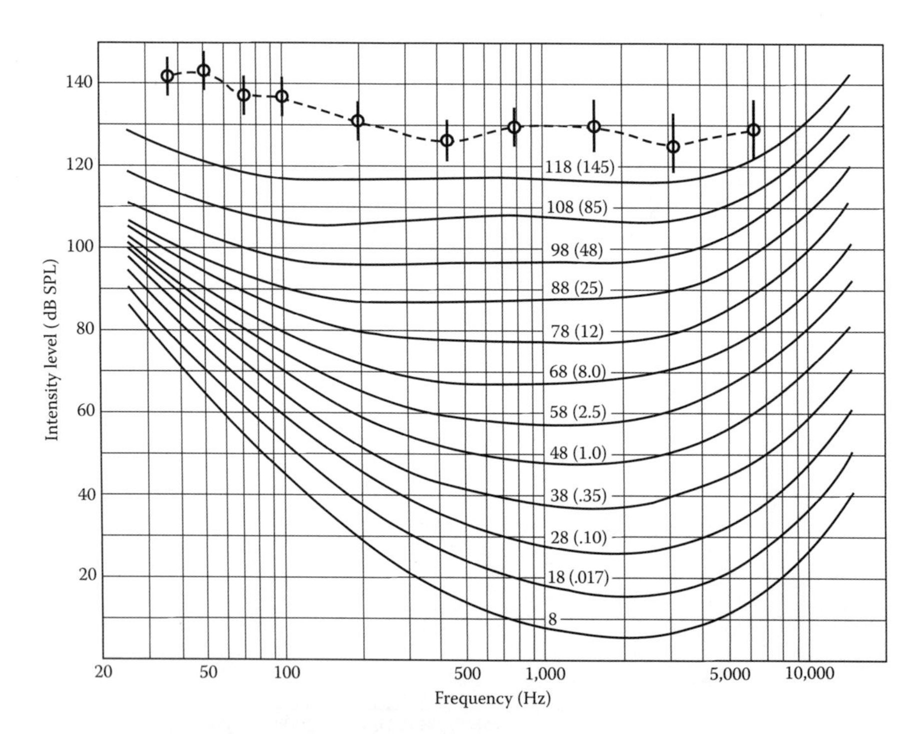

FIGURE 1.3 Loudness level contours. Parameters: phons (sones). The bottom curve (8 phons) is at the threshold of hearing. The dotted line shows Wegel's 1932 results for "threshold of feeling." This line is many dB above levels that are known today to produce permanent damage to the auditory system. (Modified from Stevens, S.S. and Davis, H.W., *Hearing*, John Wiley & Sons, New York, 1938.)

extended exposure to sounds at much lower levels than the dotted line in Figure 1.3 can result in temporary or permanent damage to the ear. It has been suggested that extended exposure to sounds as low as 70–75 dB(A) may produce permanent high-frequency threshold shifts in some individuals [39].

1.3.1.2 Loudness and Intensity

Figure 1.4 (modified from [32, Fig. 5]) shows "loudness growth functions," the relationship between stimulus intensity in dB SPL and loudness in sones, for tones of different frequencies. As can be seen in Figure 1.4, the loudness growth function depends on frequency. Above about 40 dB SPL for a 1 kHz tone the relationship is approximately described by the power law $L(I) = (I/I_0)^{1/3}$, so that if the intensity I is increased by 9 dB the loudness L is approximately doubled.* The relationship between loudness and intensity has been modeled extensively [1,6,46].

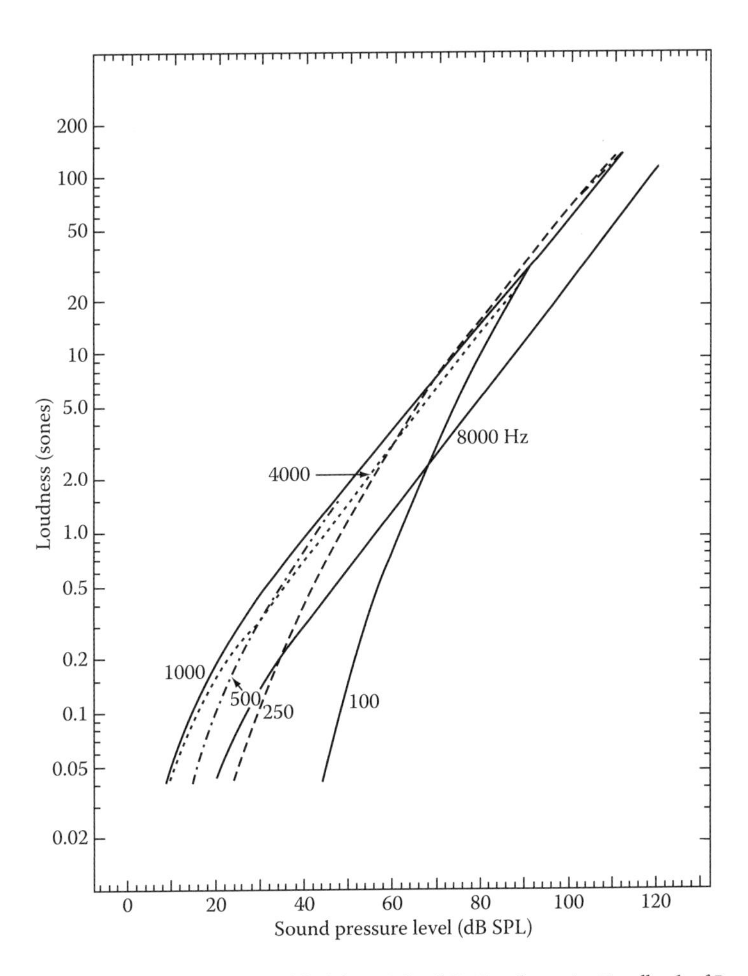

FIGURE 1.4 Loudness growth functions. (Modified from Scharf, B., Loudness, in *Handbook of Perception*, Vol. IV, Hearing, Chap. 6, Carterette, E.C. and Friedman M.P., Eds., Academic Press, New York, 1978.)

* This power-law relationship between physical and perceptual measures of a stimulus was studied in great detail by S.S. Stevens. This relationship is now commonly referred to as Stevens. This relationship is now commonly referred to as Steven's Law. Stevens measured exponents for many sensory modalities, ranging from a low of 0.33 for loudness and brightness to a high of 3.5 for electric shock produced by a 60 Hz electric current delivered to the skin.

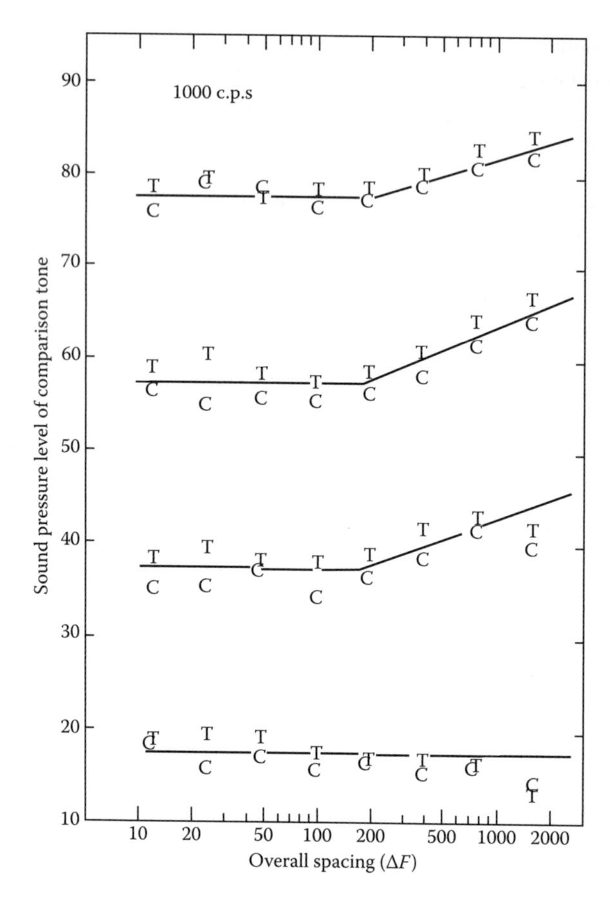

FIGURE 1.5 Loudness vs. bandwidth of tone complex. (From Zwicker, E. et al., *J. Acoust. Soc. Am.*, 29, 548, 1957. With permission.)

1.3.1.3 Loudness and Bandwidth

The loudness of a complex sound of fixed intensity, whether a tone complex or a band of noise, depends on its bandwidth, as is shown in Figure 1.5 [48, Fig. 3]. For sounds well above threshold, the loudness remains more or less constant so long as the bandwidth is less than a critical band. If the bandwidth is greater than a critical band, the loudness increases with increasing bandwidth. Near threshold the trend is reversed, and the loudness decreases with increasing bandwidth.*

These phenomena have been modeled successfully by utilizing the loudness growth functions shown in Figure 1.4 in a model that calculates total loudness by summing the specific loudness per critical band [49]. The loudness growth function is very steep near threshold, so that dividing the total energy of the signal into two or more critical bands results in a reduction of total loudness. The loudness growth function well above threshold is less steep, so that dividing the total energy of the signal into two or more critical bands results in an increase of total loudness.

1.3.1.4 Loudness and Duration

Everything we have talked about so far applies to steady-state, long-duration stimuli. These results are reasonably well understood and can be modeled reasonably well by present-day models. However,

* These data were obtained by comparing the loudness of a single 1 kHz tone and the loudness for a four-tone complex of the specified bandwidth centered at 1 kHz. The systematic difference between results when the tone was adjusted ("T" symbol) and when the complex was adjusted ("C" symbol) is an example of the bias effects mentioned in Section 1.2.

there is a host of psychophysical data having to do with aspects of temporal structure of the signal that are less well understood and less well modeled. The subject of temporal dynamics of auditory perception is an area where there is a great deal of room for improvement in models for perceptual auditory coders. One example of this subject is the relationship between loudness and duration discussed here. Other examples appear in a later section on temporal aspects of masking.

There is general agreement that, for fixed intensity, loudness increases with duration up to stimulus durations of a few hundred milliseconds. (Other factors, usually discussed under the terms adaptation or fatigue, come into play for longer durations of many seconds or minutes. We will not discuss these factors here.) The duration below which loudness increases with increasing duration is sometimes referred to as "the critical duration." Scharf [32] provides an excellent summary of studies of the relationship between loudness and duration. In his survey, he cites values of critical duration ranging from 10 to over 500 ms. About half the studies in Scharf's survey show that the total energy (intensity × duration) stays constant below the critical duration for constant loudness, while the remaining studies are about evenly split between total energy increasing and total energy decreasing with increasing duration.

One possible explanation for this confused state of affairs is the inherent difficulty of making loudness matches between dissimilar stimuli, discussed in Section 1.2.1. Two stimuli of different durations differ by more than "loudness," and depending on a variety of poorly understood experimental or individual factors what appears to be the same experiment may yield different results in different laboratories or with different subjects.

Some support for this explanation comes from the fact that studies of threshold intensity as a function of duration are generally in better agreement with each other than studies of loudness as a function of duration. As discussed in Section 1.2.3 measurements of auditory threshold depend to some extent on the method of measurement, but it is still possible to establish an internally consistent criterion-free measure. The exact results depend to some extent on signal frequency, but there is reasonable agreement among various studies that total energy at threshold remains approximately constant between about 10 and 100 ms. (See [41] for a survey of studies of threshold intensity as a function of duration.)

1.3.2 Differential Thresholds

1.3.2.1 Frequency

Figure 1.6 shows frequency JND as a function of frequency and intensity as measured in the most recent comprehensive study [43]. The frequency JND generally increases with increasing frequency and decreases with increasing intensity, ranging from about 1 Hz at low frequency and moderate intensity to more than 100 Hz at high frequency and low intensity.

The results shown in Figure 1.6 are in basic agreement with results from most other studies of frequency JND's with the exception of the earliest comprehensive study, by Shower and Biddulph [43, p. 180]. Shower and Biddulph [35] found a more gradual increase of frequency JND with frequency. As we have noted above, the results obtained in experiments of this nature are strongly influenced by details of the method of measurement. Shower and Biddulph measured detectability of frequency modulation of a pure tone; most other experimenters measured the ability of subjects to correctly identify whether one tone burst was of higher or lower frequency than another. Why this difference in procedure should produce this difference in results, or even whether this difference in procedure is solely responsible for the difference in results, is unclear.

The Weber fraction $\Delta f/f$, where Δf is the frequency JND, is smallest at mid frequencies, in the region from 500 Hz to 2 kHz. It increases somewhat at lower frequencies, and it increases very sharply at high frequencies above about 4 kHz. Wier et al. [43] in their Figure 39.1, reproduced here as our Figure 1.6, plotted log Δf against $\sqrt{f}$. They found that this choice of axes resulted in the closest fit to a straight line. It is not clear that this choice of axes has any theoretical basis; it appears simply to be a choice that

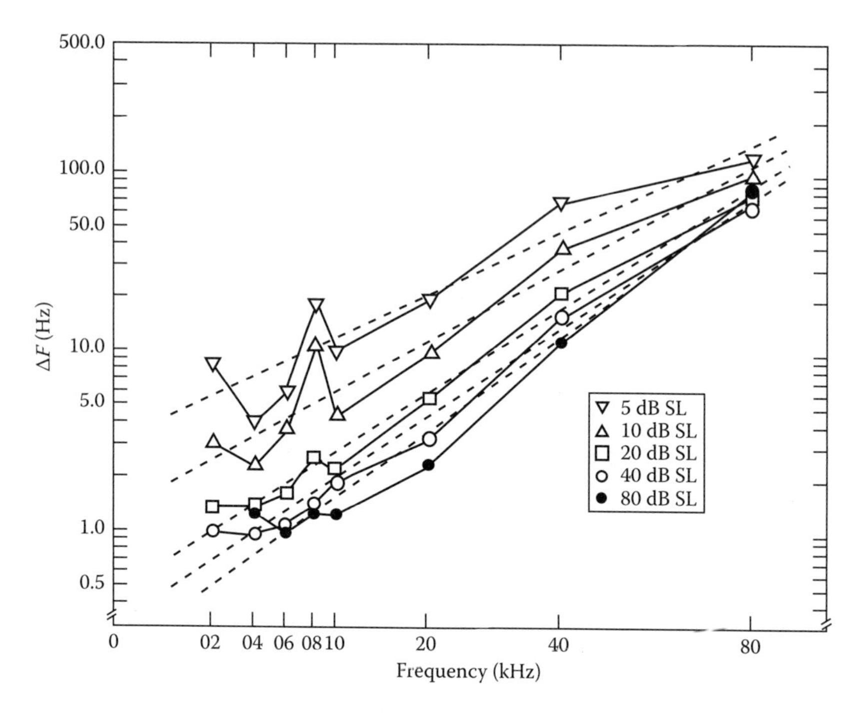

FIGURE 1.6 Frequency JND as a function of frequency and intensity. (Modified from Wier, C.C. et al., *J. Acoust. Soc. Am.*, 61, 178, 1977. With permission.)

happens to work well. There have been extensive attempts to model frequency selectivity. These studies suggest that the auditory system uses the timing of individual nerve impulses at low frequencies, but that at high frequencies above a few kHz this timing information is no longer available and the auditory system relies exclusively on place information from the mechanically tuned inner ear.

Rosenblith and Stevens [30] provide an interesting example of the interaction between method of measurement and observed result. They compared frequency JNDs using two methods. One was an "AX" method, in which the subject judged whether the second of a pair of tone bursts was of higher or lower frequency than the first of the pair. The other was an "ABX" method, in which the subject judged whether the third of three tone bursts, at the same frequency as one of the first two tone bursts, was more similar to the first or to the second burst. They found that frequency JNDs measured using the AX method were approximately half the size of frequency JNDs measured using the ABX method, and they concluded that "... it would be rather imprudent to postulate a "true" DL (difference limen), or to infer the behavior of the peripheral organ from the size of a DL measured under a given set of conditions." They discussed their results in terms of information theory, an active topic at the time, and were unable to reach any definite conclusion. An analysis of their results in terms of detection theory, which at that time was in its infancy, predicts their results almost exactly.*

1.3.2.2 Intensity

The Weber fraction $\Delta I/I$ for pure tones is not constant but decreases slightly as stimulus intensity increases. This change has been termed the "near miss to Weber's law" In most studies, the Weber

* Assume RV's A, B and X are drawn independently from normal distributions with means m_A, m_B, m_C, respectively, and equal stard deviations σ, while the relevant decision variable in the AX experiment has mean $m_A - m_X$ and standard deviation $\sqrt{2} \times \sigma$, while the relevant decision variable in the ABX experiment has mean $m_A - m_B$ and standard deviation $\sqrt{6} \times \sigma$, a value almost twice as large.

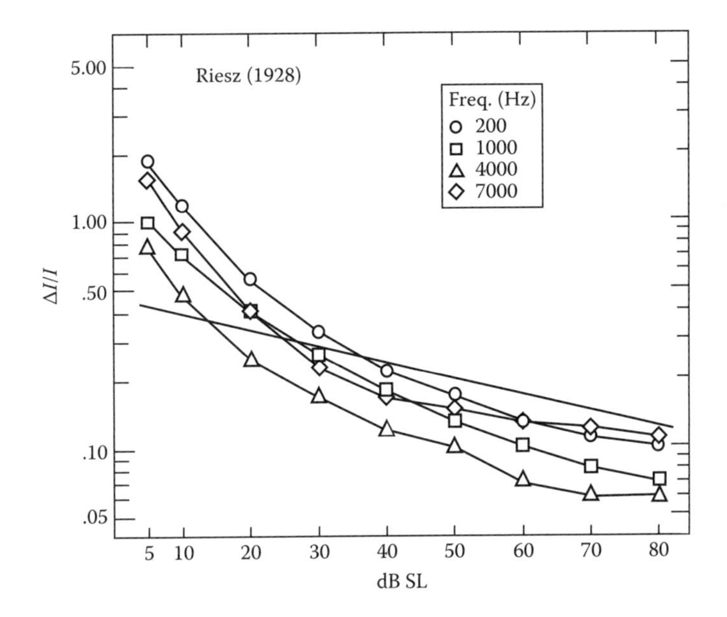

FIGURE 1.7 Summary of intensity JNDs for pure tones. Jesteadt et al. [18] found that the Weber fraction $\Delta I/I$ was independent of frequency (straight line). Riesz [29], using a different procedure, found a dependence (connected points). (From Jesteadt, W. et al., *J. Acoust. Soc. Am.*, 61, 169, 1977. With permission.)

fraction has been found to be independent of frequency. An exception is Riesz's study [29], in which the Weber fraction was at a minimum at approximately 2 kHz and increased at higher and lower frequencies.

Typical results are summarized in Figure 1.7 [18, Fig. 4]. The solid straight line is a good fit to Jesteadt's intensity JND data at frequencies from 200 Hz to 8 kHz. The Weber fraction decreases from about 0.44 at 5 dB SL (decibels above threshold) to about 0.12 at 80 dB SL. These results are in substantial agreement with most other studies with the exception of Riesz's study. Riesz's data are shown in Figure 1.7 as the curves identified by symbols. There is a larger change of intensity JND with intensity, and the intensity JND depends on frequency.

There is an interesting parallel between the results for intensity JND and the results for frequency JND. In both cases, results from most studies are in agreement with the exception of one study: Shower and Biddulph for frequency JND, and Riesz for intensity JND. In both cases, most studies measured the ability of subjects to correctly identify the difference between two tone bursts. Both of the outlying studies measured, instead, the ability of subjects to identify modulation of a tone: Shower and Biddulph used frequency modulation and Riesz used amplitude modulation. It appears that a modulated continuous tone may give different results than a pair of tone bursts. Whether this is a real effect, and, if it is, whether it is due to stimulus artifact or to properties of the auditory system, is unclear. The subject merits further investigation.

The Weber fraction for wideband noise appears to be independent of intensity. Miller [21] measured detectability of intensity increments in wideband noise and found that the Weber fraction $\Delta I/I$ was approximately constant at 0.099 above 30 dB SL. It increased below 30 dB SL, which led Miller to revive Fechner's modification of Weber's law as discussed above in Section 1.2.4.

1.3.3 Masking

No aspect of auditory psychophysics is more relevant to the design of perceptual auditory coders than masking, since the basic objective is to use the masking properties of speech to hide the coding noise. It will be seen that while we can use present-day knowledge of masking to great advantage, there is still much to be learned about properties of masking if we are to fully exploit it. Since some of the major unresolved problems in modeling masking are related to the relative bandwidth of masker and probe, our approach here is to present masking in terms of this relative bandwidth.

1.3.3.1 Tone Probe, Tone Masker

At one time, perhaps because of the demonstrated power of the Fourier transform in the analysis of linear time-invariant systems, the sine wave was considered to be the "natural" signal to be used in studies of human hearing. Much of the earliest work on masking dealt with the masking of one tone by another [42]. Typical results are shown in Figure 1.8 [3, Fig. 1]. Similar results appear in Wegel and Lane [42]. The abscissa is probe frequency and the ordinate is masking in dB, the elevation of masked over absolute threshold (15 dB SPL for 400 Hz tone). Three curves are shown, for 400 Hz maskers at 40, 60, and 80 dB SPL.

Masking is greatest for probe frequencies slightly above or below the masker frequency of 400 Hz. Maximum probe-to-masker ratios are −19 dB for an 80 dB SPL masker (probe intensity elevated 46 dB above the absolute threshold of 15 dB SPL), −15 dB for a 60 dB SPL masker, and −14 dB for a 40 dB SPL masker.

Masking decreases as probe frequency gets closer to 400 Hz. The probe frequencies closest to 400 Hz are 397 and 403 Hz, and at these frequencies the threshold probe-to-masker ratio is −26 dB for an 80 dB SPL masker, −23 dB for a 60 dB SPL masker, and −21 dB for a 40 dB SPL masker.

Masking also decreases as probe frequency gets further away from masker frequency. For the 40 dB SPL masker this selectivity is nearly symmetric in log frequency, but as the masker intensity increases the masking becomes more and more asymmetric so that the 400 Hz masker produces much more masking at higher frequencies than at lower.

The irregularities seen near probe frequencies of 400, 800, and 1200 Hz are the result of interactions between masker and probe. When masker and probe frequencies are close, beating results. Even when their

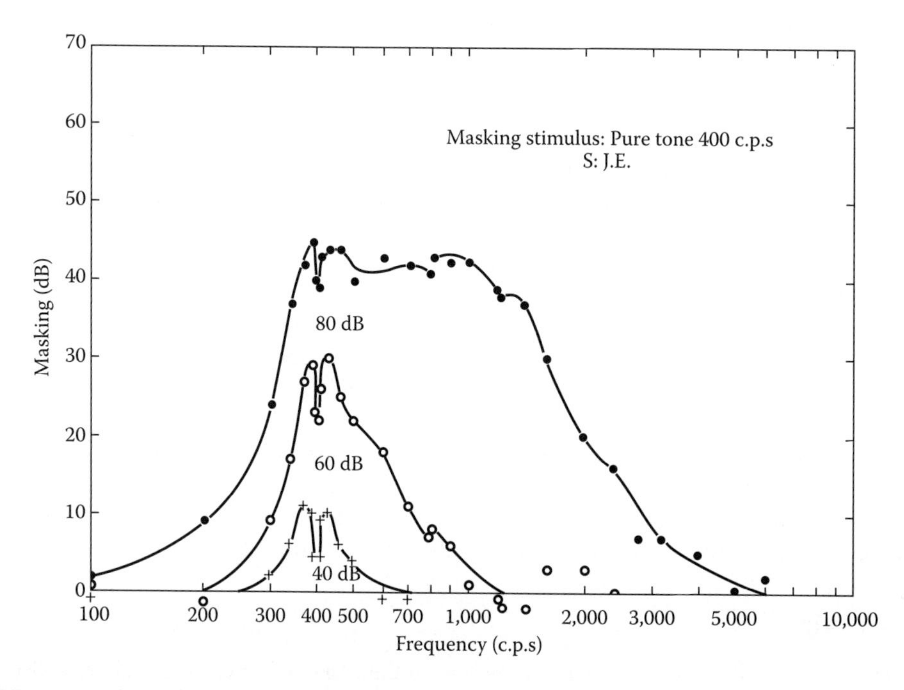

FIGURE 1.8 Masking of tones by a 400 Hz tone at 40, 60, and 80 dB SPL. (From Egan, J.P. and Hake, H.W., *J. Acoust. Soc. Am.*, 22, 622, 1950. With permission.)

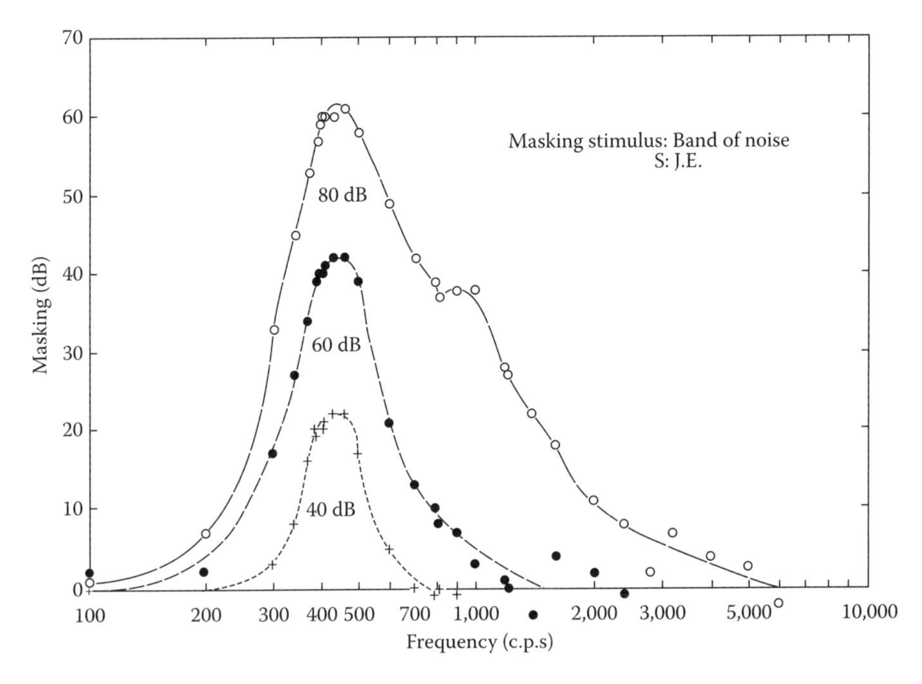

FIGURE 1.9 Masking of tones by a 90 Hz wide band of noise centered at 410 Hz at 40, 60, and 80 dB SPL. (From Egan, J.P. and Hake, H.W., *J. Acoust. Soc. Am.*, 22, 622, 1950. With permission.)

frequencies are far apart, nonlinear effects in the auditory system result in complex interactions. These irregularities provided incentive to use narrow bands of noise, rather than pure tones, as maskers.

1.3.3.2 Tone Probe, Noise Masker

Fletcher and Munson [8] were among the first to use bands of noise as maskers. Figure 1.9 [3, Fig. 2] shows typical results. The conditions are similar to those for Figure 1.8 except that now the masker is a band of noise 90 Hz wide centered at 410 Hz. The maximum probe-to-masker ratios occur for probe frequencies slightly above the center frequency of the masker, and they are much greater than they were for the tone maskers shown in Figure 1.8. Maximum probe-to-masker ratios are −4 dB for an 80 dB SPL masker and −3 dB for 60 and 40 dB SPL maskers. The frequency selectivity and upward spread of masking seen in Figure 1.8 appear in Figure 1.9 as well, but the irregularities seen at harmonics of the masker frequency are greatly reduced.

An important effect that occurs in connection with masking of a tone probe by a band of noise is the relationship between masker bandwidth and amount of masking. This relationship can be presented in many ways, but the results can be described to a reasonable degree of accuracy by saying that noise energy within a narrow band of frequencies surrounding the probe contributes to masking while noise energy outside this band of frequencies does not. This is one manifestation of the "critical band" described in Section 1.2.6.

Figure 1.10 [2, Fig. 6] shows results from a series of experiments designed to determine the widths of critical bands. We are most concerned here with the closed symbols and the associated solid and dotted straight lines. These show an expanded and elaborated repeat of a test Fletcher reported in 1940 to measure the width of critical bands, and the results shown here are similar to Fletcher's results [7, Fig. 124]. The closed symbols show threshold level of probe signals at frequencies ranging from 500 Hz to 8 kHz in dB relative to the intensity of a masking band of noise centered at the frequency of the test signal and with the bandwidth shown on the abscissa. The intensity of the masking noise is 60 dB SPL per 1/3 octave. Note that for narrowband maskers the probe-to-masker ratio is nearly

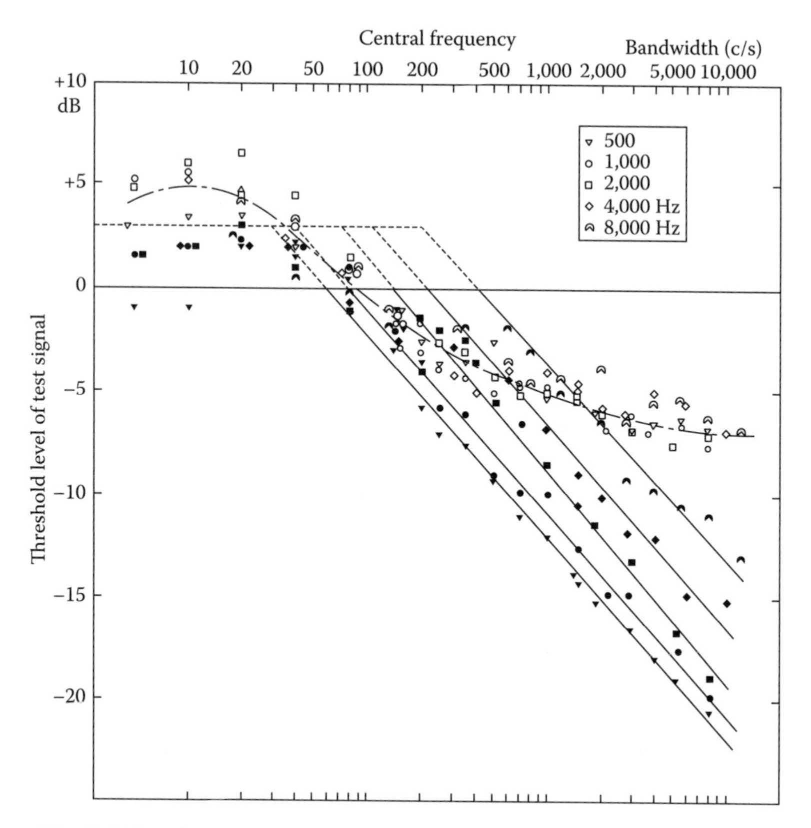

FIGURE 1.10 Threshold level of probe signals from 500 Hz to 8 kHz relative to overall level of noise masker at bandwidth shown on the abscissa. (Modified from Bos, C.E. and de Boer, E., *J. Acoust. Soc. Am.*, 39, 708, 1966.)

independent of bandwidth, while for wideband maskers the probe-to-masker ratio decreases at approximately 3 dB per doubling of bandwidth. This result indicates that above a certain bandwidth, approximated in Figure 1.10 as the intersection of the asymptotic narrowband horizontal line and the asymptotic wideband sloping lines, noise energy outside of this band does not contribute to masking.

The results shown in Figure 1.10 are from only one of many studies of masking of pure tones by noise bands of varying bandwidths that lead to similar conclusions. The list includes Feldtkeller and Zwicker [5] and Greenwood [13]. Scharf [33] provides additional references.

1.3.3.3 Noise Probe, Tone or Noise Masker

Masking of bands of noise, either by tone or noise maskers, has received relatively little attention. This is unfortunate for the designer who is concerned with masking wideband coding noise. Masking of noise by tones is touched on in Zwicker [47], but the earliest study that gives actual data points appears to be Hellman [15]. The threshold probe-to-masker ratios for a noise probe approximately one critical band wide were −21 dB for a 60 dB SPL masker and −28 dB for a 90 dB SPL masker. Threshold probe-to-masker ratios for an octave-band probe were −55 dB for 1 kHz maskers at 80 and 100 dB SPL. A 1 kHz masker at 90 dB SPL produced practically no masking of a wideband probe.

Hall [34] measured threshold intensity for noise bursts one-half, one, and two critical bands wide with various center frequencies in the presence of 80 dB SPL pure-tone maskers ranging from an octave below to an octave above the center frequency. Figure 1.11 shows results for a critical-band 1 kHz probe. The threshold probe-to-masker ratio for a 1 kHz masker is −24 dB, in agreement with Hellman's results, and Figure 1.11 shows the same upward spread of masking that appears in Figures 1.8 and 1.9. (Note that

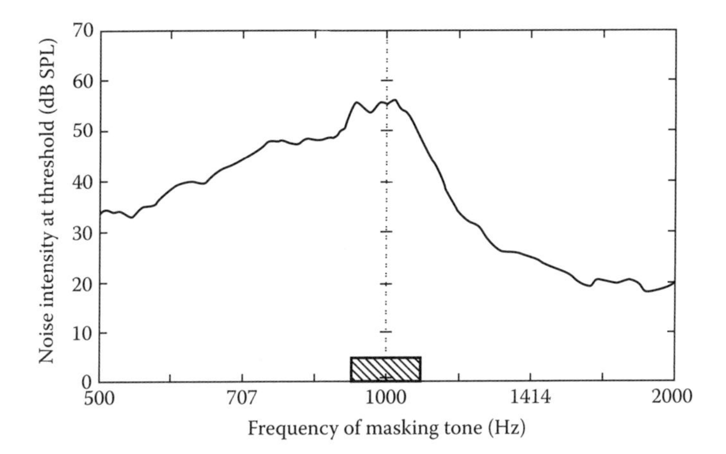

FIGURE 1.11 Threshold intensity for a 923–1083 Hz band of noise masked by an 80-dB SPL tone at the frequency shown on the abscissa. (From Schroeder, M.R. et al., *J. Acoust. Soc. Am.*, 66, 1647, 1979. With permission.)

in Figures 1.8 and 1.9 the masker is fixed and the abscissa is probe frequency, while in Figure 1.11 the probe is fixed and the abscissa is masker frequency.) A tone below 1 kHz produces more masking than a tone above 1 kHz.

Masking of noise by noise is confounded by the question of phase relationships between probe and masker. If masker and probe are identical in bandwidth and phase, then as we saw in Section 1.2.5 the masked threshold becomes identical to the differential threshold. Miller's [21] Weber fraction $\Delta I/I$ of 0.099 for intensity discrimination of wideband noise, phrased in terms of intensity of the just-detectable increment, leads to a probe-to-masker ratio of −26.3 dB.

More recently, Hall [14] measured threshold intensity for various combinations of probe and masker bandwidths. These experiments differ from earlier experiments in that phase relationships between probe and masker were controlled: all stimuli were generated by adding together equal-amplitude random phase sinusoidal components, and components common to probe and masker had identical phase. Results for one subject are shown in Figure 1.12. Masker bandwidth appears on the abscissa, and the parameter is probe bandwidth: $A \Rightarrow 0$ Hz, $B \Rightarrow 4$ Hz, $C \Rightarrow 16$ Hz, and $D \Rightarrow 64$ Hz. All stimuli were centered at 1 kHz and the overall intensity of the masker was 70 dB SPL.

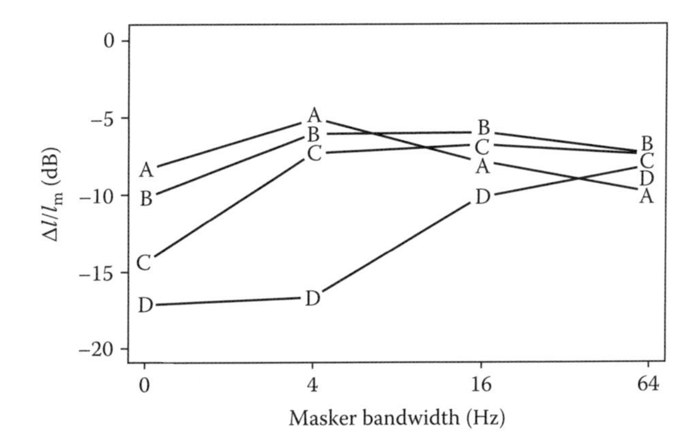

FIGURE 1.12 Intensity increment between masker alone and masker plus just-detectable probe. Probe bandwidth 0 Hz (A); 4 Hz (B); 16 Hz (C); 64 Hz (D). Frequency components common to probe and masker have identical phase. (From Hall, J.L., *J. Acoust. Soc. Am.*, 101, 1023, 1997. With permission.)

Figure 1.12 differs from Figures 1.8 through 1.11 in that the vertical scale shows intensity increment between masker alone and masker plus just-detectable probe rather than intensity of the just-detectable probe, and the results look quite different. For all probe bandwidths shown, the intensity increment varies only slightly so long as the masker is at least as wide as the probe. The intensity increment decreases when the probe is wider than the masker.

1.3.3.4 Asymmetry of Masking

Inspection of Figures 1.8 through 1.11 reveals large variation of threshold probe-to-masker intensity ratios depending on the relative bandwidth of probe and masker. Tone maskers produce threshold probe-to-masker ratios of −14 to −26 dB for tone probes, depending on the intensity of the masker and the frequency of the probe (Figure 1.8), and threshold probe-to-masker ratios of −21 to −28 dB for critical-band noise probes ([15]; also Figure 1.11). On the other hand, a tone masked by a band of noise is audible only at much higher probe-to-masker ratios, in the neighborhood of 0 dB (Figures 1.9 and 1.10). This "asymmetry of masking" (the term is due to Hellman [15]) is of central importance in the design of perceptual coders because of the different masking properties of noise-like and tone-like portions of the coded signal [19]. Current perceptual models do not handle this asymmetry well, so it is a subject we must examine closely.

The logical conclusion to be drawn from the numbers in the preceding paragraph at first appears to be that a band of noise is a better masker than a tone, for both noise and tone probes. In fact, the correct conclusion may be completely different. It can be argued that so long as the masker bandwidth is at least as wide as the probe bandwidth, tones or bands of noise are equally effective maskers and the psychophysical data can be described satisfactorily by current energy-based perceptual models, properly applied. It is only when the bandwidth of the probe is greater than the bandwidth of the masker that energy-based models break down, and some criterion other than average energy must be applied.

Figure 1.13 shows Egan and Hake's results for 80 dB SPL tone and noise maskers superimposed on each other. Results for the tone masker are shown as a solid curve and results for the noise masker are

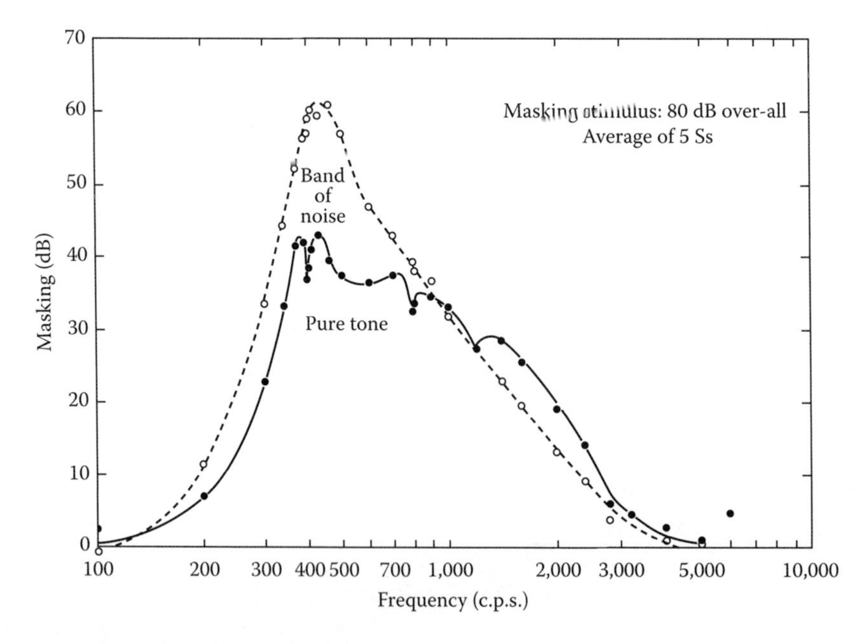

FIGURE 1.13 Masking produced by a 400 Hz masker at 80 dB SPL and a 90 Hz wide band of noise centered at 410 Hz. (From Egan, J.P. and Hake, H.W., *J. Acoust. Soc. Am.*, 22, 622, 1950. With permission.)

shown as a dashed curve. (These curves are not identical to the corresponding curves in Figures 1.8 and 1.9: They are average results from five subjects, while Figures 1.8 and 1.9 were for a single subject.) The maximum amount of masking produced by the band of noise is 61 dB, while the tone masker produces only 37 dB of masking for a 397 Hz probe.

The difference between tone and noise maskers may be more apparent than real, and for masking of a tone the auditory system may be similarly affected by tone and noise maskers. What is plotted in Figure 1.13 is the elevation in threshold intensity of the probe tone by the masker, but the discrimination the subject makes is in fact between masker alone and masker plus probe. As was discussed above in Section 1.5, since coherence between tone probe and masker depends on the bandwidth of the masker, a probe tone of a given intensity can produce a much greater change in intensity of probe plus masker for a tone masker than for a noise masker.

The stimulus in the Egan and Hake experiment with a 400 Hz masker and a 397 Hz probe is identical to the stimulus Riesz used to measure intensity JND (see Section 1.2). As Egan and Hake observe "... When the frequency of the masked stimulus is 397 or 403 c.p.s. [Hz], the amount of masking is evidently determined by the value of the differential threshold for intensity at 400 c.p.s." [3, p. 624]. Specifically, for the results shown in Figure 1.13, the threshold intensity of a 397 Hz tone is 52 dB SPL. This leads to a Weber fraction $\Delta I/I$ (power at envelope maximum minus power at envelope minimum, divided by power at envelope minimum) of 0.17, which is only slightly higher than values obtained by Riesz and by Jesteadt et al. shown in Figure 1.7.

The situation with noise masker is more difficult to analyze because of the random nature of the masker. The effective intensity increment between masker alone and masker plus probe depends on the phase relationship between probe and 400 Hz component of the masker, which are uncontrolled in the Egan and Hake experiment, and also on the effective time constant and bandwidth of the analyzing auditory filter, which are unknown. However, for the experiment shown in Figure 1.12 the maskers were computer-generated repeatable stimuli, so that the intensity of masker plus probe could be computed. The results shown in Figure 1.12 lead to a Weber fraction $\Delta I/I$ of 0.15 for tone masked by tone and 0.10 for tone masked by 64 Hz wide noise. Results are similar for noise masked by noise, so long as the masker is at least as wide as the probe. Weber fractions for the 64 Hz wide masker in Figure 1.12 range from 0.18 for a 4 Hz wide probe to 0.15 for a 64 Hz wide probe.

Our understanding of the factors leading to the results shown in Figure 1.12 is obviously very limited, but these results appear to be consistent with the view that to a first-order approximation the relevant variable for masking is the Weber fraction $\Delta I/I$, the intensity of masker plus probe relative to the intensity of the masker, so long as the masker is at least as wide as the probe. This is true for both tone and noise maskers. Because of changes in coherence between probe and masker as masker bandwidth changes, the corresponding probe intensity at threshold can be much lower for a tone masker than for a probe masker, as is shown in Figure 1.13.

The asymmetry that Hellman was primarily concerned with in her 1972 paper is the striking difference between the threshold of a band of noise masked by a tone and of a tone masked by a band of noise. It appears that this is a completely different effect than the asymmetry shown in Figure 1.13 and one that cannot be accounted for by current energy-based models of masking. The difference between the -5 and $+5$ dB threshold probe-to-masker ratios seen in Figures 1.9 and 1.10 for tones masked by noise and the -21 and -28 dB threshold probe-to-masker ratios for noise masked by tone reported by Hellman and seen in Figure 1.11 is due in part to the random nature of the noise masker and to the change in coherence between masker and probe that we have already discussed. Even when these factors are controlled, as in Figure 1.12, decrease of masker bandwidth for a 64 Hz wide band of noise results in a decrease of threshold intensity increment. (The situation is complicated by the possibility of off-frequency listening. As we have already seen, neither a tone nor a noise masker masks remote frequencies effectively. The 64 Hz band is narrow enough that off-frequency listening is not a factor.) These and similar results lead to the conclusion that present-day models operating on signal power are inadequate and that some envelope-based measure, such as the envelope maximum or ratio of envelope maximum to minimum, must be considered [10,11,38].

1.3.3.5 Temporal Aspects of Masking

Up until now, we have discussed masking effects with simultaneous masker and probe. In order to be able to deal effectively with a dynamically varying signal such as speech, we need to consider nonsimultaneous masking as well. When the probe follows the masker, the effect is referred to as "forward masking." When the masker follows the probe, it is referred to as "backward masking." Effects have also been measured with a brief probe near the beginning or the end of a longer-duration masker. These effects have been referred to as "forward" or "backward fringe masking," respectively [44, p. 162].

The various kinds of simultaneous and nonsimultaneous masking are nicely illustrated in Figure 1.14 [4, Fig. 1]. The masker was wideband noise at an overall level of 70 dB SPL and the probe was a brief 1.9 kHz tone burst. The masker was on continuously except for a silent interval of 25, 50, 200, or 500 ms. beginning at the 0 ms point on the abscissa. The four sets of data points show thresholds for probes presented at various times relative to the gap for the four gap durations. Probe thresholds in silence and in continuous noise are indicated on the ordinate by the symbols "Q" and "CN."

Forward masking appears as the gradual drop of probe threshold over a duration of more than 100 ms following the cessation of the masker. Backward masking appears as the abrupt increase of masking, over a duration of a few tens of ms, immediately before the reintroduction of the masker. Forward fringe masking appears as the more than 10 dB overshoot of masking immediately following the reintroduction of the masker, and backward fringe masking appears as the smaller overshoot

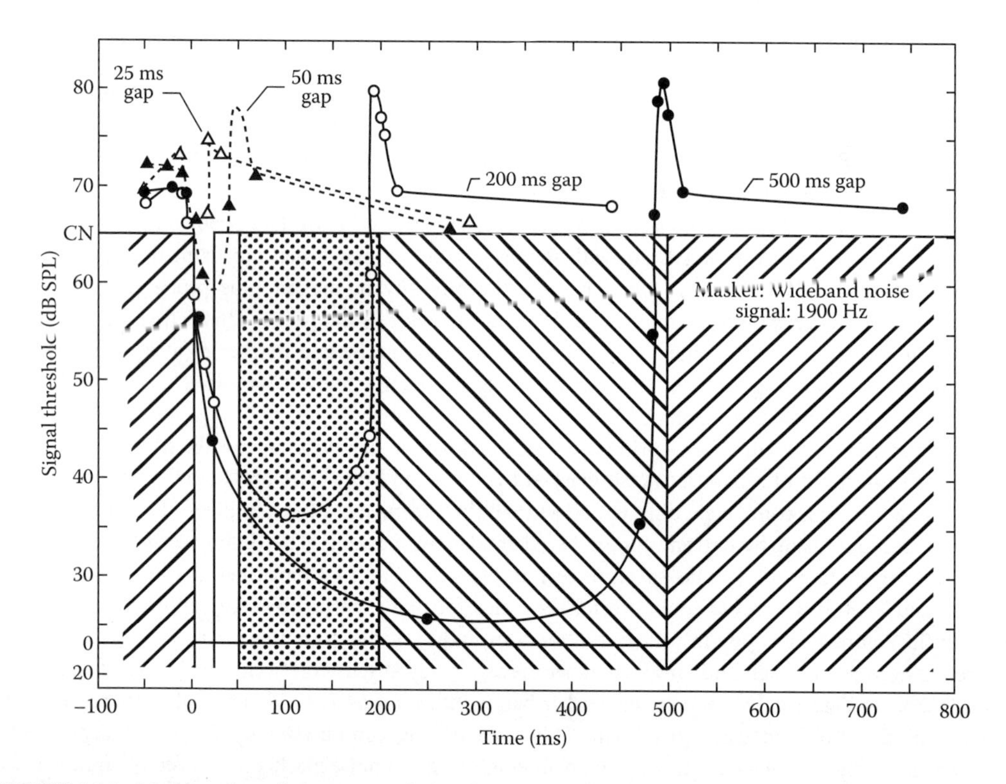

FIGURE 1.14 Masking of tone by ongoing wide-band noise with silent interval of 25, 50, 200, or 500 ms. This figure shows simultaneous, forward, backward, forward fringe, and backward fringe masking. (From Elliott, L.L., *J. Acoust. Soc. Am.*, 45, 1277, 1969. With permission.)

immediately preceding the cessation of the masker. Backward masking is an important effect for the designer of coders for acoustic signals because of its relationship to audibility of preecho. It is a puzzling effect, because it is caused by a masker that begins only after the probe has been presented. Stimulus-related electrophysiological events can be recorded in the cortex several tens of ms after presentation of the stimulus, so there may be some physiological basis for backward masking. It is an unstable effect, and there is some evidence that backward masking decreases with practice [20,23, p. 119].

Forward masking is a more robust effect, and it has been studied extensively. It is a complex function of stimulus parameters, and we do not have a comprehensive model that predicts amount of forward masking as a function of frequency, intensity, and time course of masker and of probe. The following two examples illustrate some of its complexity.

Figure 1.15 [17, Fig. 1] is from a study of the effects of masker frequency and intensity on forward masking. Masker and probe were of the same frequency. The left and right columns show the same data, plotted on the left against probe delay with masker intensity as a parameter and plotted on the right

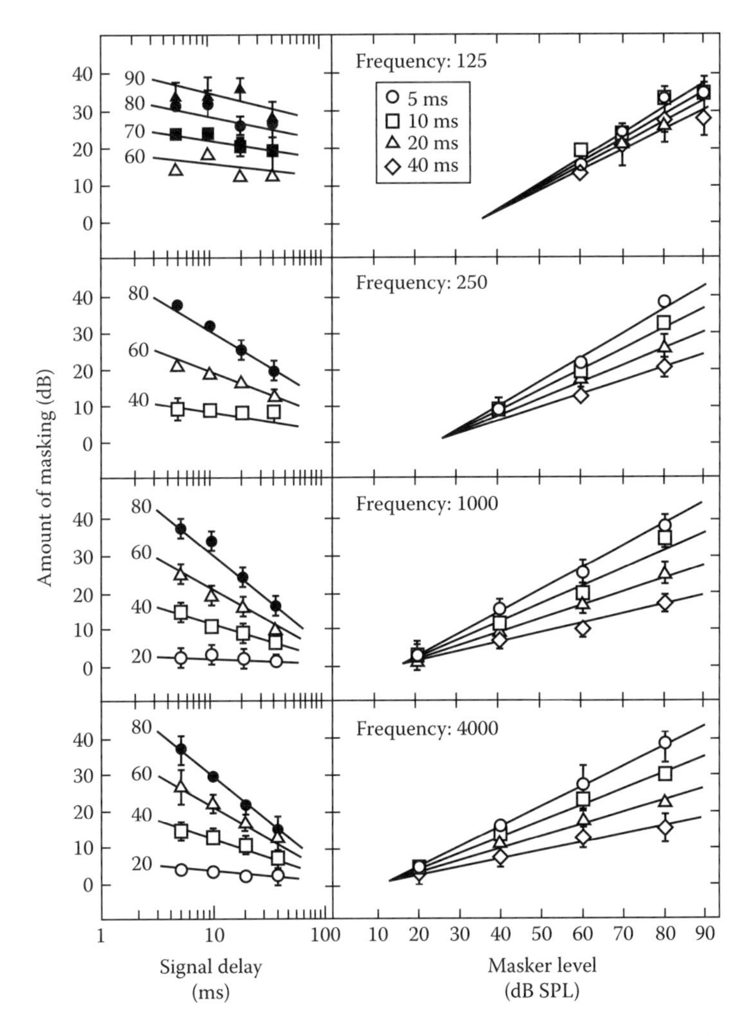

FIGURE 1.15 Forward masking with identical masker and probe frequencies, as a function of frequency, delay, and masker level. (From Jesteadt, W. et al., *J. Acoust. Soc. Am.*, 71, 950, 1982. With permission.)

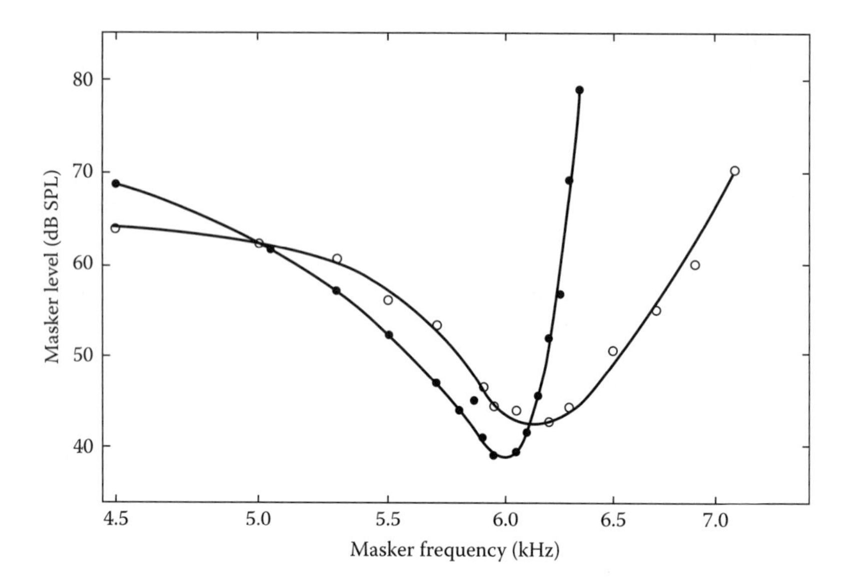

FIGURE 1.16 Simultaneous (open symbols) and forward (closed symbols) masking of a 6 kHz probe tone at 36 dB SPL. Masker frequency appears on the abscissa, and masker intensity just sufficient to mask the probe appears on the abscissa. (Modified from Moore, B.C.J., *J. Acoust. Soc. Am.*, 63, 524, 1978.)

against masker intensity with probe delay as a parameter. The amount of masking depends in an orderly way on masker frequency, masker intensity, and probe delay. Jesteadt et al. were able to fit these data with a single equation with three free constants. This equation, with minor modification, was later found to give a satisfactory fit to data obtained with forward masking by wideband noise [25].

Striking effects can be observed when probe and masker frequencies differ. Figure 1.16 (modified from [22, Fig. 8]) superimposes simultaneous (open symbols) and forward (filled symbols) masking curves for a 6 kHz probe at 36 dB SPL, 10 dB above the absolute threshold of 26 dB SPL. Rather than showing the amount of masking for a fixed masker, Figure 1.16 shows masker level, as a function of masker frequency, sufficient to just mask the probe. It is clear that simultaneous and forward masking differ, and that the difference depends on the relative frequency of masker and probe. Results such as those shown in Figure 1.16 are of interest to the field of auditory physiology because of similarities between forward masking results and frequency selectivity of primary auditory neurons.

1.4 Conclusions

Notwithstanding the successes obtained to date with perceptual coders for speech and audio [16,19,27,36], there is still a great deal of room for further advancement. The most widely applied perceptual models today apply an energy-based criterion to some critical-band transformation of the signal and arrive at a prediction of acceptable coding noise. These models are essentially refinements of models first described by Fletcher and his coworkers and further developed by Zwicker and others [34]. These models do a good job describing masking and loudness for steady-state bands of noise, but they are less satisfactory for other signals. We can identify two areas in which there seem to be great room for improvement. One of these areas presents a challenge jointly to the designer of coders and to the auditory psychophysicist, and the other area presents a challenge primarily to the auditory psychophysicist.

One area for additional research has to do with asymmetry of masking. Noise is a more effective masker than tones, and this difference is not handled well by present-day perceptual models. Present-day coders first compute a measure of tonality of the signal and then use this measure empirically to obtain an estimate of masking. This empirical approach has been applied successfully to a variety of signals, but it is possible that an approach that is less empirical and more based on a comprehensive model of auditory perception would be more robust.

As discussed in Section 1.3.3.4, there is evidence that there are two separate factors contributing to this asymmetry of masking. The difference between noise and tone maskers for narrowband coding noise appears to result from problems with signal definition rather than a difference in processing by the auditory system, and it may be that an effective way of dealing with it will result not from an improved understanding of auditory perception but rather from changes in the coder. A feedforward prediction of acceptable coding noise based on the energy of the signal does not take into account phase relationships between signal and noise. What may be required is a feedback, analysis-by-synthesis approach, in which a direct comparison is made between the original signal and the proposed coded signal. This approach would require a more complex encoder but leave the decoder complexity unchanged [27]. The difference between narrowband and wideband coding noise, on the other hand, appears to call for a basic change in models of auditory perception. For largely historical reasons, the idea of signal energy as a perceptual measure is deeply ingrained in present-day perceptual models. There is increasing realization that under some conditions signal energy is not the relevant measure but that some envelope-based measure may be required.

A second area in which additional research may prove fruitful is in the area of temporal aspects of masking. As is discussed in Section 1.3.3.5, the situation with time-varying signal and noise is more complex than the steady-state situation. There is an extensive body of psychophysical data on various aspects of nonsimultaneous masking, but we are still lacking a satisfactory comprehensive perceptual model. As is the case with asymmetry of masking, present-day coders deal with this problem at an empirical level, in some cases very effectively. However, as with asymmetry of masking, an approach based on fundamental properties of auditory perception would perhaps be better able to deal with a wide variety of signals.

References

1. Allen, J.B., Harvey Fletcher's role in the creation of communication acoustics, *J. Acoust. Soc. Am.*, 99: 1825–1839, 1996.
2. Bos, C.E. and de Boer, E., Masking and discrimination, *J. Acoust. Soc. Am.*, 39: 708–715, 1966.
3. Egan, J.P. and Hake, H.W., On the masking pattern of a simple auditory stimulus, *J. Acoust. Soc. Am.*, 22: 622–630, 1950.
4. Elliott, L.L., Masking of tones before, during, and after brief silent periods in noise, *J. Acoust. Soc. Am.*, 45: 1277–1279, 1969.
5. Feldtkeller, R. and Zwicker, E., *Das Ohr als Nachrichtenempfänger*, S. Hirzel, Stuttgart, Germany, 1956.
6. Fletcher, H., Loudness, masking, and their relation to the hearing process and the problem of noise measurement, *J. Acoust. Soc. Am.*, 9: 275–293, 1938.
7. Fletcher, H., *Speech and Hearing in Communication, ASA Edition*, Allen, J.B., Ed., American Institute of Physics, New York, 1995.
8. Fletcher, H. and Munson, W.A., Relation between loudness and masking, *J. Acoust. Soc. Am.*, 9: 1–10, 1937.
9. Goldstein, J.L., Auditory spectral filtering and monaural phase perception, *J. Acoust. Soc. Am.*, 41: 458–479, 1967.
10. Goldstein, J.L., Comparison of peak and energy detection for auditory masking of tones by narrow-band noise, *J. Acoust. Soc. Am.*, 98(A): 2907, 1995.
11. Goldstein, J.L. and Hall, J.L., Peak detection for auditory sound discrimination, *J. Acoust. Soc. Am.*, 97(A): 3330, 1995.
12. Green, D.M. and Swets, J.A., *Signal Detection Theory and Psychophysics*, John Wiley & Sons, New York, 1966.

13. Greenwood, D.D., Auditory masking and the critical band, *J. Acoust. Soc. Am.*, 33: 484–502, 1961.
14. Hall, J.L., Asymmetry of masking revisited: Generalization of masker and probe bandwidth, *J. Acoust. Soc. Am.*, 101: 1023–1033, 1997.
15. Hellman, R.P., Asymmetry of masking between noise and tone, *Percept. Psychophs.*, 11: 241–246, 1972.
16. Jayant, N., Johnston, J., and Safranek, R., Signal compression based on models of human perception, *Proc. IEEE*, 81: 1385–1422, 1993.
17. Jesteadt, W., Bacon, S.P., and Lehman, J.R., Forward masking as a function of frequency, masker level, and signal delay, *J. Acoust. Soc. Am.*, 71: 950–962, 1982.
18. Jesteadt, W., Wier, C.C., and Green, D.M., Intensity discrimination as a function of frequency and sensation level, *J. Acoust. Soc. Am.*, 61: 169–177, 1977.
19. Johnston, J.D., Audio coding with filter banks, in *Subband and Wavelet Transforms, Design and Applications*, Chap. 9, Akansu, A.N. and Smith, M.J.T., Eds., Kluwer Academic, Boston, MA, 1966a.
20. Johnston, J.D., Personal communication, 1996b.
21. Miller, G.A., Sensitivity to changes in the intensity of white noise and its relation to masking and loudness, *J. Acoust. Soc. Am.*, 19: 609–619, 1947.
22. Moore, B.C.J., Psychophysical tuning curves measured in simultaneous and forward masking, *J. Acoust. Soc. Am.*, 63: 524–532, 1978.
23. Moore, B.C.J., *An Introduction to the Psychology of Hearing,* Academic Press, London, 1989.
24. Moore, B.C.J., *Frequency Selectivity in Hearing,* Academic Press, London, 1986.
25. Moore, B.C.J. and Glasberg, B.R., Growth of forward masking for sinusoidal and noise maskers as a function of signal delay: Implications for suppression in noise, *J. Acoust. Soc. Am.*, 73: 1249–1259, 1983a.
26. Moore, B.C.J. and Glasberg, B.R., Suggested formulae for calculating auditory-filter bandwidths and excitation patterns, *J. Acoust. Soc. Am.*, 74: 750–757, 1983b.
27. Quackenbush, S.R. and Noll, P., MPEG/Audio coding standards in *Video, Speech, and Audio Signal Processing and Associated Standards,* Chap. 4, Vijay K. Madisetti, Ed., Taylor & Francis, Boca Raton, Florida, 2009.
28. Reynolds, G.S. and Stevens, S.S., Binaural summation of loudness, *J. Acoust. Soc. Am.*, 32: 1337–1344, 1960.
29. Riesz, R.R., Differential intensity sensitivity of the ear for pure tones, *Phys. Rev.*, 31: 867–875, 1928.
30. Rosenblith, W.A. and Stevens, K.N., On the DL for frequency, *J. Acoust. Soc. Am.*, 25: 980–985, 1953.
31. Scharf, B., Critical bands, in *Foundations of Modern Auditory Theory*, Vol. 1, Chap. 5, Tobias, J.V., Ed., Academic Press, New York, 1970.
32. Scharf, B., Loudness, in *Handbook of Perception*, Vol. IV, Hearing, Chap. 6, Carterette, E.C. and Friedman, M.P., Eds., Academic Press, New York, 1978.
33. Scharf, B. and Fishkin, D., Binaural summation of loudness: Reconsidered, *J. Exp. Psychol.,* 86: 374–379, 1970.
34. Schroeder, M.R., Atal, B.S., and Hall, J.L., Optimizing digital speech coders by exploiting masking properties of the human ear, *J. Acoust. Soc. Am.*, 66: 1647–1652, 1979.
35. Shower, E.G. and Biddulph, R., Differential pitch sensitivity of the ear, *J. Acoust. Soc. Am.*, 3: 275–287, 1931.
36. Sinha, D., Johnston, J.D., Dorward, S., and Quackenbush, S.R., The perceptual audio coder (PAC) in *Video, Speech, and Audio Signal Processing and Associated Standards*, Chap. 4, Vijay K. Madisetti, Ed., Taylor & Francis, Boca Raton, Florida, 2009.
37. Stevens, S.S. and Davis, H.W., *Hearing,* John Wiley & Sons, New York, 1938.
38. Strickland, E.A. and Viemeister, N.F., Cues for discrimination of envelopes, *J. Acoust. Soc. Am.*, 99: 3638–3646, 1996.

39. Von Gierke, H.E. and Ward, W.D., Criteria for noise and vibration exposure, in *Handbook of Acoustical Measurements and Noise Control*, 3rd ed., Chap. 26, Harris, C.M., Ed., McGraw-Hill, New York, 1991.
40. Ward, W.D., Musical perception, in *Foundations of Modern Auditory Theory*, Vol. 1, Chap. 11, Tobias, J.V., Ed., Academic Press, New York, 1970.
41. Watson, C.S. and Gengel, R.W., Signal duration and signal frequency in relation to auditory sensitivity, *J. Acoust. Soc. Am.*, 46: 989–997, 1969.
42. Wegel, R.L. and Lane, C.E., The auditory masking of one pure tone by another and its probable relation to the dynamics of the inner ear, *Phys. Rev.*, 23: 266–285, 1924.
43. Wier, C.C., Jesteadt, W., and Green, D.M., Frequency discrimination as a function of frequency and sensation level, *J. Acoust. Soc. Am.*, 61: 178–184, 1977.
44. Yost, W.A., *Fundamentals of Hearing, an Introduction*, 3rd ed., Academic Press, New York, 1994.
45. Zwicker, E., Die Grenzen der Hörbarkeit der Amplitudenmodulation und der Frequenzmodulation eines Tones, *Acustica*, 2: 125–133, 1952.
46. Zwicker, E., Über psychologische und methodische Grundlagen der Lautheit, *Acustica*, 8: 237–258, 1958.
47. Zwicker, E., Über die Lautheit von ungedrosselten und gedrosselten Schallen, *Acustica*, 13: 194–211, 1963.
48. Zwicker, E., Flottorp, G., and Stevens, S.S., Critical bandwidth in loudness summation, *J. Acoust. Soc. Am.*, 29: 548–557, 1957.
49. Zwicker, E. and Scharf, B., A model of loudness summation, *Psychol. Rev.*, 16: 3–26, 1965.

2

MPEG Digital Audio Coding Standards

Schuyler R. Quackenbush
Audio Research Labs
and
AT&T Research Labs

Peter Noll
Technical University of Berlin

2.1 Introduction

2.1.1 PCM Bit Rates

Typical audio signal classes are telephone speech, wideband speech, and wideband audio, all of which differ in bandwidth, dynamic range and in listener expectation of offered quality. The quality of telephone-bandwidth speech is acceptable for telephony and for some videotelephony and videoconferencing services. Higher bandwidths (7 kHz for wideband speech) may be necessary to improve the intelligibility and naturalness of speech. Wideband (high fidelity) audio representation including multichannel audio is expected in entertainment applications such as recorded music, motion picture sound tracks and digital audio broadcast (DAB), providing bandwidths of at least 15 kHz but more typically 18–20 kHz.

The conventional digital format for these signals is PCM, with sampling rates and amplitude resolutions (PCM bits per sample) as given in Table 2.1.

The compact disc (CD) is today's de facto standard of digital audio representation. On a CD with its 44.1 kHz sampling rate the resulting stereo bit rate is $2 \times 44.1 \times 16 \times 1000 \equiv 1.41$ Mbps (see Table 2.2). However, the CD needs a significant overhead for a run-length-limited line code, which maps 8

TABLE 2.1 Basic Parameters for Three Classes of Acoustic Signals

	Frequency Range (Hz)	Sampling Rate (kHz)	PCM Bits per Sample	PCM Bit Rate (kbps)
Telephone speech	300–3,400[a]	8	8	64
Wideband speech	50–7,000	16	8	128
Wideband audio (stereo)	10–20,000	48[b]	2×16	2×768

[a] Bandwidth in Europe; 200–3200 Hz in the United States.
[b] Other sampling rates: 44.1 kHz, 32 kHz.

TABLE 2.2 CD Bit Rates

Storage Device	Audio Rate (Mbps)	Overhead (Mbps)	Total Bit Rate (Mbps)
CD	1.41	2.91	4.32

Note: Stereophonic signals, sampled at 44.1 kHz.

information bits into 14 bits, for synchronization and for error correction, resulting in a 49 bit representation of each 16 bit audio sample. Hence, the total stereo bit rate is $1.41 \times 49/16 = 4.32$ Mbps.

For intermediate storage during the production chain and for archiving of audio signals, sampling rates of at least 2×44.1 kHz and amplitude resolutions of up to 24 bit per sample are typical. In the production chain light compression (e.g., 2 : 1) should be used, however archival application should employ lossless compression.

2.1.2 Bit Rate Reduction

Although high bit rate channels and networks have become the norm, low bit rate coding of audio signals is still of vital importance in many applications, such as terrestrial and satellite DAB. The main motivations for low bit rate coding are the need to minimize transmission cost and maximize throughput over channels of limited capacity such as mobile radio channels, terrestrial radio channels or satellite channels.

Basic requirements in the design of low bit rate audio coders are first, to retain a high quality of the reconstructed signal with robustness to variations in spectra and levels. In the case of stereophonic and multichannel signals spatial integrity is an additional dimension of quality. Second, robustness against random and bursty channel bit errors and packet losses is required. Third, low complexity and low power consumption are highly relevant. For example, in broadcast and playback applications, the complexity and power consumption of audio decoders used must be low, whereas constraints on encoder complexity are much more relaxed. Additional network-related requirements may be low encoder/decoder delays, robustness against errors introduced by cascading codecs, and a graceful degradation of quality with increasing bit error rates in mobile radio and broadcast applications. Finally, in professional applications, the coded bit streams must allow editing, fading, mixing, and dynamic range compression [1].

We have seen rapid progress in bit rate compression techniques for speech and audio signals [2–8]. Linear prediction, subband coding (SBC), transform coding (TC), as well as various forms of vector quantization and entropy coding techniques have been used to design efficient coding algorithms which can achieve substantially more compression than was thought possible only a few years ago. Recent results in speech and audio coding indicate that an excellent coding quality can be obtained with bit rates of 1.5 bit per sample. Expectations over the next decade are that these rates will continue to be reduced. Such reductions shall be based mainly on employing sophisticated forms of adaptive noise shaping and parametric coding which are controlled by psychoacoustic criteria. In storage and packet-network-based applications additional savings are possible by employing variable-rate coding with its potential to offer a time-independent constant-quality performance.

Compressed digital audio representations can be made less sensitive to channel impairments than analog ones if source and channel coding are implemented appropriately. Bandwidth expansion has often been mentioned as a disadvantage of digital coding and transmission, but with today's data compression and multilevel signaling techniques, channel bandwidths can be reduced actually, compared with analog systems. In broadcast systems, the reduced bandwidth requirements, together with the error robustness of the coding algorithms, will allow an efficient use of newly available radio spectrum.

2.1.3 MPEG Standardization Activities

Of particular importance for digital audio is the standardization work within the International Organization for Standardization (ISO/IEC), intended to provide international standards for audiovisual coding. ISO has set up a Working Group WG 11 to develop such standards for a wide range of communications-based and storage-based applications. This group is called MPEG, an acronym for Moving Pictures Experts Group.

MPEG's initial effort was the MPEG Phase 1 (MPEG-1) coding standards IS 11172 supporting bit rates of around 1.2 Mbps for video (with video quality comparable to that of analog video cassette recorders) and 256 kbps for two-channel audio (with audio quality comparable to that of CDs) [9].

The MPEG-2 standard IS 13818 specifies standards for high-quality video (including high-definition TV [HDTV]) in bit rate ranges from 3 to 15 Mbps and above. It also specifies standards for mono, stereo, and multichannel audio in bit rate ranges from 32 to 320 kbps [10,11].

The MPEG-4 standard IS 14496 specifies standards for audiovisual coding for applications ranging from mobile access low complexity multimedia terminals to high quality multichannel sound systems. MPEG-4 supports interactivity, providing a high degree of flexibility and extensibility [12].

MPEG-D standard IS 23003 specifies new audio coding technology. Its first standard is IS 23003-1 MPEG Surround, which specifies coding of multichannel signals in a way that is compatible with legacy stereo receivers and which exploits what is known about the human perception of a sound stage [13].

MPEG-1, MPEG-2, MPEG-4, and MPEG-D standardization work will be described in Sections 2.3 through 2.6. Web information about MPEG is available at a number of addresses. The official MPEG Web site offers crash courses in MPEG and ISO, an overview of current activities, MPEG requirements, workplans, and information about documents and standards [14]. Links lead to collections of frequently asked questions, listings of MPEG, multimedia, or digital video related products, MPEG/Audio resources, software, audio test bitstreams, etc.

2.2 Key Technologies in Audio Coding

Some of the first proposals to reduce wideband audio coding rates have built upon speech coding technology. Differences between audio and speech signals are manifold; however, audio coding implies higher sampling rates, better amplitude resolution, higher dynamic range, larger variations in power density spectra, stereophonic and multichannel audio signal presentations, and, finally, higher listener expectation of quality. Indeed, the CD with its 16 bit per sample PCM format sets a benchmark for consumer's expectation of quality.

Speech and audio coding are similar in that in both cases quality is based on the properties of human auditory perception. On the other hand, speech can be coded very efficiently because a speech production model is available, whereas nothing similar exists for audio signals.

Modest reductions in audio bit rates have been obtained by instantaneous companding (e.g., a conversion of uniform 14 bit PCM into a 11 bit nonuniform PCM presentation) or by forward-adaptive PCM (block companding) as employed in various forms of near-instantaneously companded audio multiplex (NICAM) coding [ITU-R, Rec. 660]. For example, the British Broadcasting Corporation (BBC) has used the NICAM 728 coding format for digital transmission of sound in several European broadcast television networks; it uses 32 kHz sampling with 14 bit initial quantization followed by a compression to a 10 bit format on the basis of 1 ms blocks resulting in a total stereo bit rate of 728 kbps [15]. Such adaptive PCM schemes can solve the problem of providing a sufficient dynamic range for audio coding but they are not efficient compression schemes because they do not exploit statistical dependencies between samples and do not sufficiently remove signal irrelevancies.

In recent audio coding algorithms five key technologies play an important role: perceptual coding, frequency domain coding, window switching, dynamic bit allocation, and parametric coding. These will be covered next.

2.2.1 Auditory Masking and Perceptual Coding

2.2.1.1 Auditory Masking

The inner ear performs short-term critical band analyses where frequency-to-place transformations occur along the basilar membrane. The power spectra are not represented on a linear frequency scale but in frequency bands called critical bands. The auditory system can roughly be described as a band-pass filterbank, consisting of heavily overlapping band-pass filters with bandwidths in the order of 50–100 Hz for signals below 500 Hz and up to 5000 Hz for signals above 10 kHz. The human auditory system typically is modeled using 25 critical bands covering frequencies of up to 20 kHz.

Simultaneous masking is a frequency domain phenomenon where a low-level signal (the maskee) can be made inaudible (masked) by a simultaneously occurring stronger signal (the masker), if masker and maskee are close enough to each other in frequency [16,17]. Such masking is greatest in the critical band in which the masker is located, and it is effective to a lesser degree in neighboring bands. A masking threshold can be measured below which the low-level signal will not be audible. This masked signal can consist of low-level signal elements or noise resulting from quantizing higher-level signal elements. The masking threshold varies in time in response to the instantaneous signal statistics. In the context of source coding it is also known as the threshold of just noticeable distortion (JND) [18]. It depends on the sound pressure level (SPL), the frequency of the masker, and on characteristics of masker and maskee. Take the example of the masking threshold for the narrowband masker with a SPL of 60 dB, as shown in Figure 2.1: the four maskees located near 1 kHz will be masked as long as their individual SPLs are below the masking threshold. The slope of the masking threshold is steeper towards lower frequencies, i.e., higher frequencies are more easily masked. It should be noted that the distance between masker and

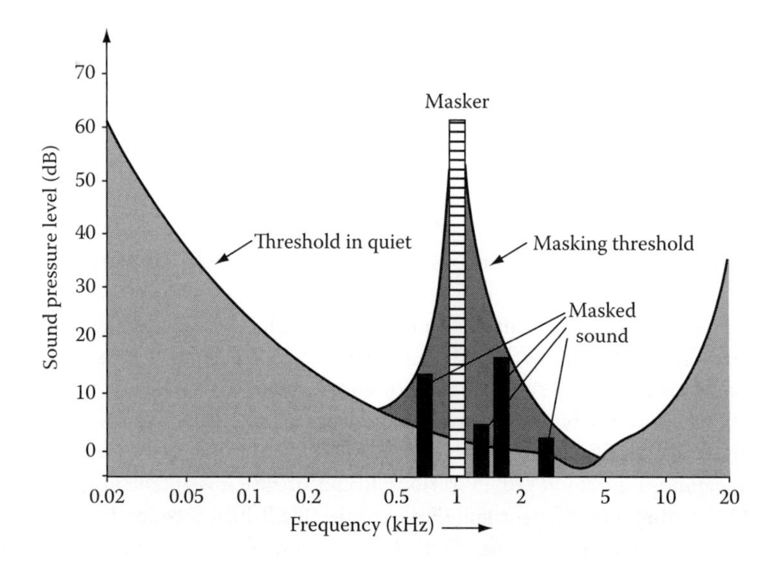

FIGURE 2.1 Threshold in quiet and masking threshold. Acoustical events in the shaded areas will not be audible.

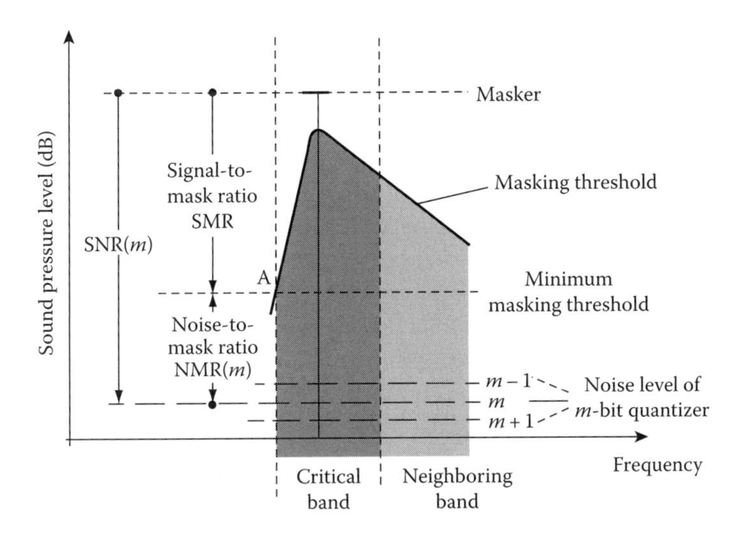

FIGURE 2.2 Masking threshold and SMR. Acoustical events in the shaded areas will not be audible.

masking threshold is smaller in noise-masking-tone experiments than in tone-masking-noise experiments, i.e., noise is a better masker than a tone. In MPEG coders both thresholds play a role in computing the masking threshold.

Without a masker, a signal is inaudible if its SPL is below the threshold in quiet which expresses the lower limit of human audibility. This depends on frequency and covers a dynamic range of more than 60 dB as shown in the lower curve of Figure 2.1.

The qualitative sketch of Figure 2.2 gives a few more details about the masking threshold: in a given critical band, tones below this threshold (darker area) are masked. The distance between the level of the masker and the masking threshold is called signal-to-mask ratio (SMR). Its maximum value is at the left border of the critical band (point A in Figure 2.2), its minimum value occurs in the frequency range of the masker and is around 6 dB in noise-masks-tone experiments. Assume an m bit quantization of an audio signal. Within a critical band the quantization noise will not be audible as long as its signal-to-noise ratio (SNR) is higher than its SMR. Noise and signal contributions outside the particular critical band will also be masked, although to a lesser degree, if their SPL is below the masking threshold.

Defining SNR(m) as the SNR resulting from an m bit quantization, the perceivable distortion in a given subband is measured by the noise-to-mask ratio (NMR)

$$\mathrm{NMR}(m) = \mathrm{SMR} - \mathrm{SNR}(m) \quad \text{(in dB)}.$$

The NMR(m) describes the difference in dB between the SMR and the SNR to be expected from an m bit quantization. The NMR value is also the difference (in dB) between the level of quantization noise and the level where a distortion may just become audible in a given subband. Within a critical band, coding noise will not be audible as long as NMR(m) is negative.

We have just described masking by only one masker. If the source signal consists of many simultaneous maskers, each has its own masking threshold, and a global masking threshold can be computed that describes the threshold of JNDs as a function of frequency.

In addition to simultaneous masking, the time domain phenomenon of "temporal masking" plays an important role in human auditory perception. It may occur when two sounds appear within a small interval of time. Depending on the individual SPLs, the stronger sound may mask the weaker one, even if the maskee precedes the masker (Figure 2.3)!

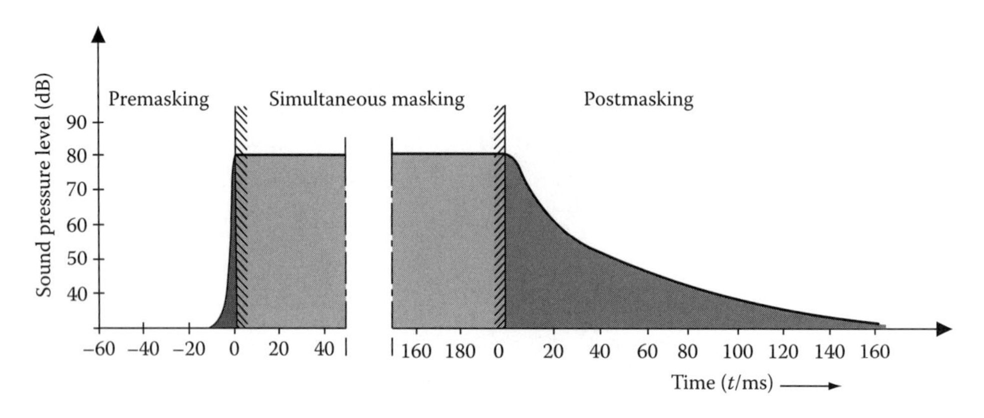

FIGURE 2.3 Temporal masking. Acoustical events in the shaded areas will not be audible.

Temporal masking can help to mask preechoes caused by the spreading of a relatively high level of quantization error over the entire coding block due to coding a signal onset. The duration within which premasking applies is significantly less than one-tenth of that of the postmasking which is in the order of 50–200 ms. Both pre- and postmasking are exploited in MPEG/Audio coding algorithms.

2.2.1.2 Perceptual Coding

Digital coding at high bit rates is typically waveform-preserving, i.e., the amplitude-vs.-time waveform of the decoded signal approximates that of the input signal. The difference signal between input and output waveform is then the basic error criterion of coder design. Waveform coding principles are covered in detail in Ref. [2]. At lower bit rates, issues concerning the production and perception of audio signals have to be taken into account when designing a coder. Error criteria have to be such that the output signal is perceived by the human receiver as being most nearly identical to the input signal, rather than it being that the output signal is most nearly identical, as in an SNR sense, to the input signal. Basically, an efficient source coding algorithm will (1) remove redundant components of the source signal by exploiting correlations between its samples and (2) remove irrelevant components by exploiting knowledge of the human auditory system. Irrelevancy manifests itself as unnecessary amplitude or frequency resolution—portions of the source signal that are masked and hence do not need to be coded or transmitted.

The dependence of human auditory perception on frequency and the accompanying tolerance of errors can (and should) directly influence encoder designs; noise-shaping techniques can emphasize coding noise in frequency bands where that noise perceptually is not important. To this end, the noise shaping must be dynamically adapted to the actual short-term input spectrum in accordance with the SMR. This can be done in different ways, although frequency weighting based on linear filtering, as is typical in speech coding, cannot make full use of the results from psychoacoustics. Therefore, in wide-band audio coding, noise-shaping parameters are dynamically controlled in a more efficient way to exploit simultaneous masking and temporal masking.

Figure 2.4 depicts the structure of a perception-based coder that exploits auditory masking. The encoding process is controlled by the SMR vs. frequency curve from which the needed amplitude resolution (and hence the bit allocation and rate) in each frequency band is derived. The SMR is typically determined from a high resolution, say, a 1024 point fast Fourier transform (FFT)-based spectral analysis of the audio block to be coded. Principally, any coding scheme can be used that can be dynamically controlled by such perceptual information. Frequency domain coders (see Section 2.2.2) are of particular interest because they offer a direct method for noise shaping. If the frequency resolution of these coders is high enough, the SMR can be derived directly from the subband samples or transform coefficients without running a FFT-based spectral analysis in parallel [19].

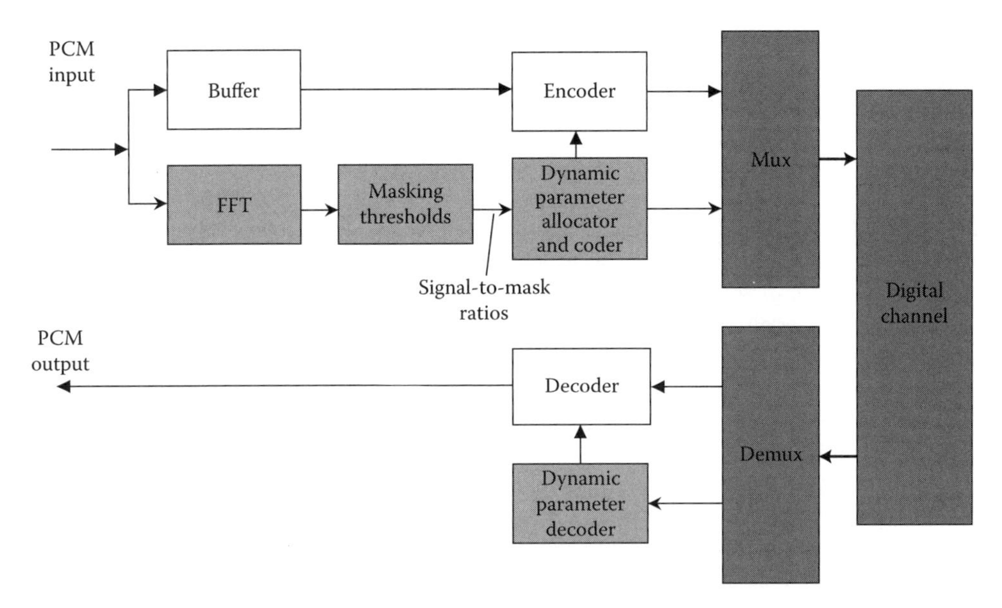

FIGURE 2.4 Block diagram of perception-based coders.

If the necessary bit rate for a complete masking of distortion is available, the coding scheme will be perceptually transparent, i.e., the decoded signal is then subjectively indistinguishable from the source signal.

Practical designs should not compress to the limits of JND because postprocessing of the acoustic signal by the end user and multiple encoding/decoding processes in transmission links have to be considered. Moreover, our current knowledge about auditory masking is very limited. Generalizations of masking results, derived for simple and stationary maskers and for limited bandwidths, may be appropriate for most source signals, but may fail for others. Therefore, as an additional requirement, we need a sufficient safety margin in practical designs of such perception-based coders. It should be noted that the MPEG/Audio coding standard is open for better encoder-located psychoacoustic models because such models are not normative elements of the standard (see Section 2.3).

2.2.2 Frequency Domain Coding

As one example of dynamic noise-shaping, quantization noise feedback can be used in predictive schemes [20,21]. However, frequency domain coders with dynamic allocations of bits (and hence of quantization noise contributions) to subbands or transform coefficients offer an easier and more accurate way to control the quantization noise [2,19].

In all frequency domain coders, redundancy (the nonflat short-term spectral characteristics of the source signal) and irrelevancy (signals below the psychoacoustical thresholds) are exploited to reduce the transmitted data rate with respect to PCM. This is achieved by splitting the source spectrum into frequency bands to generate nearly uncorrelated spectral components, and by quantizing these separately. Two coding categories exist, TC and SBC. The differentiation between these two categories is mainly due to historical reasons. Both use an analysis filterbank in the encoder to decompose the input signal into subsampled spectral components. The spectral components are called subband samples if the filterbank has low frequency resolution, otherwise they are called spectral lines or transform coefficients. These spectral components are recombined in the decoder via synthesis filterbanks.

In SBC, the source signal is fed into an analysis filterbank consisting of M band-pass filters which are contiguous in frequency so that the set of subband signals can be recombined additively to produce the original signal or a close version thereof. Each filter output is critically decimated (i.e., sampled at twice the nominal bandwidth) by a factor equal to M, the number of band-pass filters. This decimation results in an aggregate number of subband samples that equals that in the source signal. In the receiver, the sampling rate of each subband is increased to that of the source signal by filling in the appropriate number of zero samples. Interpolated subband signals appear at the band-pass outputs of the synthesis filterbank. The sampling processes may introduce aliasing distortion due to the overlapping nature of the subbands. If perfect filters, such as two-band quadrature mirror filters or polyphase filters, are applied, aliasing terms will cancel and the sum of the band-pass outputs equals the source signal in the absence of quantization [22–25]. With quantization, aliasing components will not cancel ideally; nevertheless, the errors will be inaudible in MPEG/Audio coding if a sufficient number of bits is used. However, these errors may reduce the original dynamic range of 20 bits to around 18 bits [26].

In TC, a block of input samples is linearly transformed via a discrete transform into a set of near-uncorrelated transform coefficients. These coefficients are then quantized and transmitted in digital form to the decoder. In the decoder, an inverse transform maps the signal back into the time domain. In the absence of quantization errors, the synthesis yields exact reconstruction. We have already mentioned that the decoder-based inverse transform can be viewed as the synthesis filterbank, the impulse responses of its band-pass filters equal the basis sequences of the transform. The impulse responses of the analysis filterbank are just the time-reversed versions thereof. The finite lengths of these impulse responses may cause so-called block boundary effects. State-of-the-art transform coders employ a modified discrete cosine transform (MDCT) filterbank as proposed by Princen and Bradley [24]. The MDCT is typically based on a 50% overlap between successive analysis blocks. Without quantization they are free from block boundary effects and in the presence of quantization, block boundary effects are deemphasized due to the overlap of the filter impulse responses in reconstruction.

Hybrid filterbanks, i.e., combinations of discrete transform and filterbank implementations, have frequently been used in speech and audio coding [27,28]. One of the advantages is that different frequency resolutions can be provided at different frequencies in a flexible way and with low complexity. A high spectral resolution can be obtained in an efficient way by using a cascade of a filterbank (with its short delays) and a linear MDCT transform that splits each subband sequence further in frequency content to achieve a high frequency resolution. MPEG-1/Audio coders use a subband approach in layers I and II, and a hybrid filterbank in layer III.

2.2.3 Window Switching

A crucial issue in frequency domain coding of audio signals is the potential for preechoes. Consider the case in which a silent period is followed by a percussive sound, such as from castanets or triangles, within the same coding block. Such an onset ("attack") will cause comparably large instantaneous quantization errors. In TC, the inverse transform in the decoding process will distribute such errors over the entire block; similarly, in SBC, the decoder band-pass filters will spread such errors. In both mappings preechoes can become distinctively audible, especially at low bit rates with comparably high levels of quantization error. Preechoes can be masked by the time domain effect of premasking if the time spread is of short length (in the order of a few milliseconds). Therefore, they can be reduced or avoided by using blocks of short lengths. However, a larger percentage of the total bit rate is typically required for the transmission of side information if the blocks are shorter. A solution to this problem is to switch between block sizes of different lengths as proposed by Edler (window switching) [29], in which typical block sizes are between $N = 64$ and $N = 1024$. The small blocks are only used to control preecho artifacts during nonstationary periods of the signal, otherwise the coder uses long blocks. It is clear that the block size selection has to be based on an analysis of the characteristics of the actual audio coding block. Figure 2.5 demonstrates the effect in TC: if the block size is $N = 1024$ (Figure 2.5b) preechoes are clearly (visible and) audible whereas a block size of 256 will reduce these effects because they are limited to the block where the signal attack and the corresponding quantization errors occur (Figure 2.5c), in which case premasking renders the coding noise inaudible.

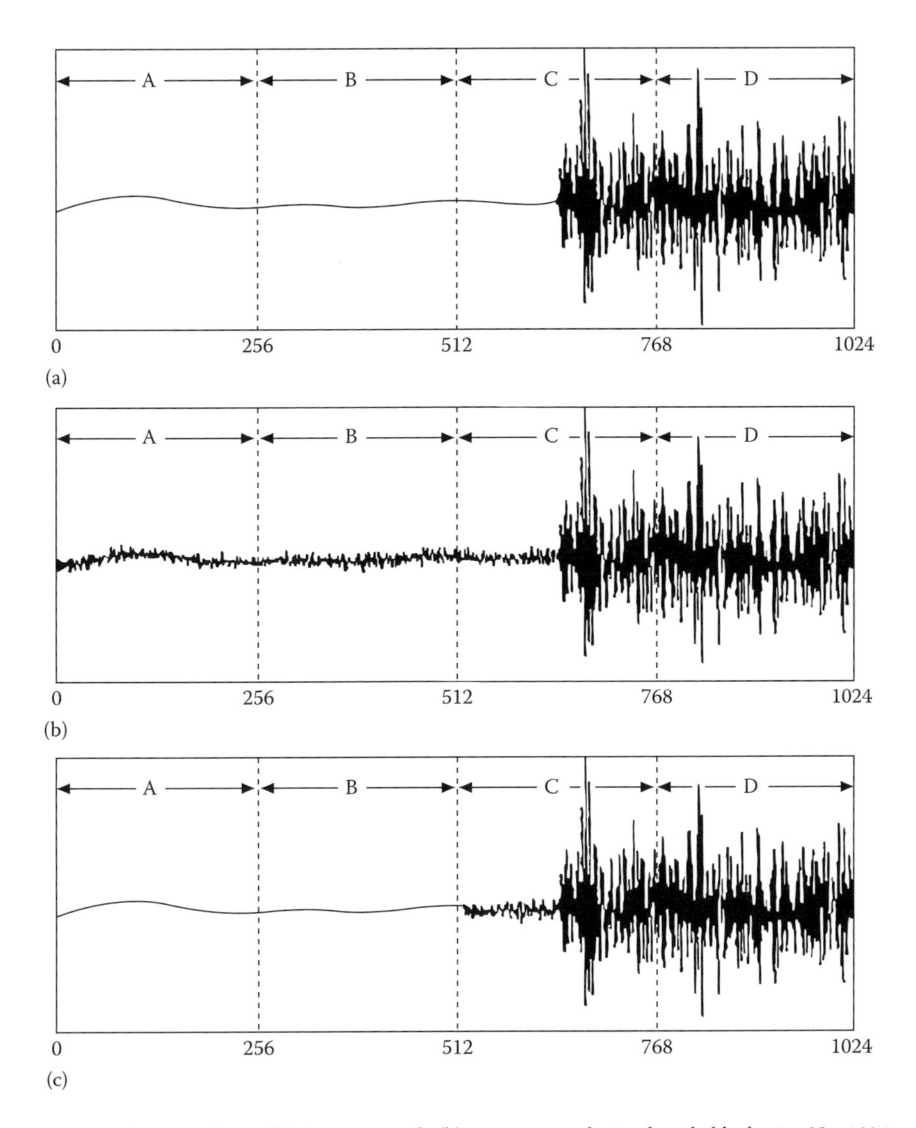

FIGURE 2.5 Window switching. (a) Source signal, (b) reconstructed signal with block size $N = 1024$, and (c) reconstructed signal with block size, $N = 256$. (From Iwadare, M. et al., *IEEE J. Sel. Areas Commun.*, 10, 138, 1992. With permission.)

2.2.4 Dynamic Bit Allocation

Frequency domain coding significantly gains in performance if the number of bits assigned to each of the quantizers of the transform coefficients is adapted to short-term spectrum of the audio coding block on a block-by-block basis. In the mid-1970s, Zelinski and Noll introduced "dynamic bit allocation" and demonstrated significant SNR-based and subjective improvements with their adaptive transform coding (ATC, see Figure 2.6 [19,30]). They proposed a DCT mapping and a dynamic bit allocation algorithm which used the DCT transform coefficients to compute a DCT-based short-term spectral envelope. Parameters of this spectrum were coded and transmitted. From these parameters, the short-term spectrum was estimated using linear interpolation in the log-domain. This estimate was then used to calculate the optimum number of bits for each transform coefficient, both in the encoder and decoder.

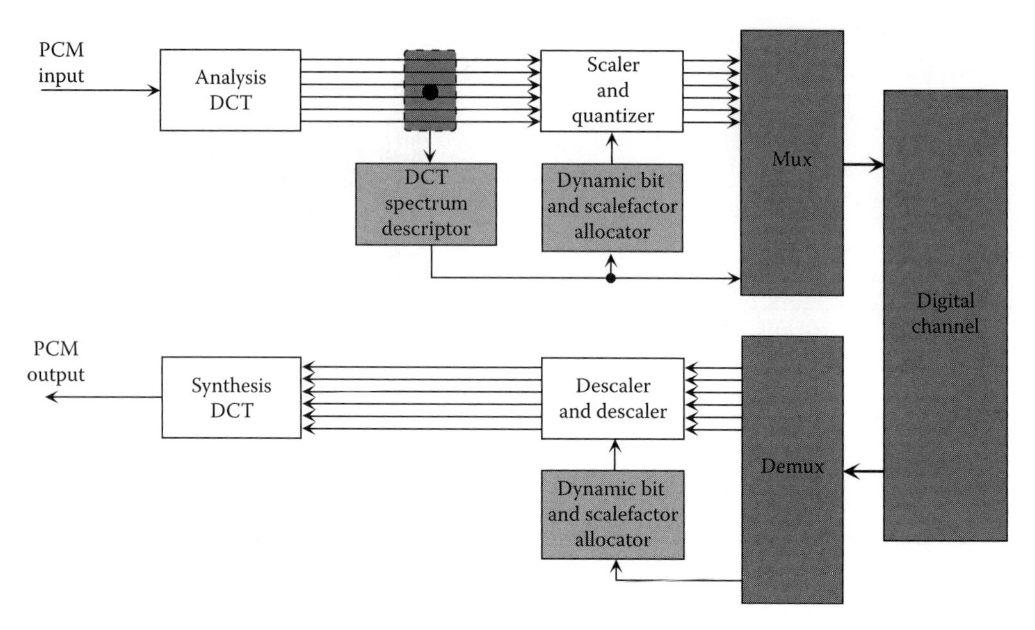

FIGURE 2.6 Conventional ATC.

That ATC had a number of shortcomings, such as block boundary effects, preechoes, marginal exploitation of masking, and insufficient quality at low bit rates. Despite these shortcomings, we find many of the features of that ATC in more modern frequency domain coders.

MPEG/Audio coding algorithms, described in detail in the next section, make use of the above key technologies.

2.2.5 Parametric Coding

The classic perceptual coder described in Section 2.2.1 shapes the quantization noise so that it is below the masking threshold. This is a very successful method when there is sufficient bit rate to insure that the NMR (in dB) is always negative. However, this is not always possible when operating over channels where more throughput, or higher compression, is needed. The most straightforward strategy is to reduce the signal bandwidth, reducing the "bits per Hz" from the perspective of frequency domain coding, and thereby keeping the NMR negative.

Recent MPEG technology has explored two additional mechanisms, beyond noise shaping in the frequency domain, to exploit perception and the human auditory system. Both make use of parametric representations of the signal for coding efficiency and exploit areas of human perception that permit less precise representations of the signal being coded.

The first is exemplified by spectral band replication (SBR), the key technology in high efficiency AAC (HE-AAC) (Section 2.5.2). SBR exploits the fact that the human auditory system is less sensitive to spectral structure at high frequencies as compared to low frequencies. HE-AAC divides the signal spectrum in two, codes the low band using a classic perceptual coder and codes the high band in a parametric fashion based on the lower spectral components.

The second is exemplified by MPEG Surround (Section 2.6.1), whose key technology is collapsing a multichannel signal into a stereo (or even mono) downmix that is transmitted to the decoder. However, additionally, side information is transmitted that is a parametric representation of how to upmix the stereo signal to the desired multichannel signal. This exploits the fact that the human auditory system is less sensitive to the sound stage of a signal in the presence of spatial masking. In other words, if the signal to code is a solo musical instrument, then its position in the sound stage must be carefully controlled. However, if the signal to code is multiple instruments, one of which dominates the other, then placement of the other instruments in the sound stage is far less critical.

The most recent MPEG technologies exploit all of these aspects of the human auditory system to provide the maximum coding efficiency. The following table lists MPEG technologies and the aspects of human perception that are exploited to deliver coding gain. Note that MPEG Surround does not specify a method for coding the downmix channels, so that table indicates the "core coder" used for the downmix.

MPEG Technology	Coding Strategies
AAC	Perceptually shaped noise in the frequency domain
HE-AAC	Perceptually shaped noise in the frequency domain plus parametric coding of the spectrum
MPEG Surround and AAC as core coder	Perceptually shaped noise in the frequency domain plus parametric coding of the sound stage
MPEG Surround and HE-AAC as core coder	Perceptually shaped noise in the frequency domain plus parametric coding of the spectrum plus parametric coding of the sound stage

2.3 MPEG-1/Audio Coding

The MPEG-1/Audio coding standard [8,31–33] is a very widely adopted standard in many application areas including consumer electronics, professional audio processing, telecommunications, broadcasting and Internet streaming and downloading [34]. The standard combines features of "MUSICAM" and "ASPEC coding algorithms" [35,36]. The development of the MPEG-1/Audio standard is described in Refs. [33,37]. The MPEG-1/Audio standard has subjective quality that is perceptually equivalent to CD quality (16 bit PCM) at stereo bit rates given in Table 2.3 for many types of music. Because of its high dynamic range, MPEG-1/Audio has the potential to exceed the quality of a CD [34,38].

2.3.1 The Basics

2.3.1.1 Structure

The basic structure follows that of perception-based coders (see Figure 2.4). In the first step, the audio signal is converted into spectral components via an analysis filterbank; layers I and II make use of a subband filterbank, layer III employs a hybrid filterbank. Each spectral component is quantized and coded with the goal to keep the quantization noise below the masking threshold. The number of bits for each subband and a scalefactor are determined on a block-by-block basis, each block having 12 (layer I) or 36 (layers II and III) subband samples (see Section 2.2). The number of quantizer bits is obtained from a dynamic bit allocation algorithm (layers I and II) that is controlled by a psychoacoustic model (see below). The subband codewords, scalefactors, and bit allocation information are multiplexed into one

TABLE 2.3 Approximate MPEG-1 Bit Rates for Transparent Representations of Audio Signals and Corresponding Compression Factors (Compared to CD Bit Rate)

MPEG-1 Audio Coding	Approximate Stereo Bit Rates for Transparent Quality (kbps)	Compression Factor
Layer I	384	4
Layer II	192	8
Layer III	128[a]	12

[a] Average bit rate; variable bit rate coding assumed.

bitstream, together with a header and optional ancillary data. In the decoder, the synthesis filterbank reconstructs a block of 32 audio output samples from the demultiplexed bitstream.

MPEG-1/Audio supports sampling rates of 32, 44.1, and 48 kHz and bit rates ranging between 32 kbps (mono) for all layers and 448, 384, or 320 kbps (stereo) for layers I, II and III, respectively. Lower sampling rates (16, 22.05, and 24 kHz) have been defined in MPEG-2 for better audio quality at bit rates at, or below, 64 kbps per channel [10]. The corresponding maximum audio bandwidths are 7.5, 10.3, and 11.25 kHz. The syntax, semantics, and coding techniques of MPEG-1 are maintained except for a small number of additional parameters.

2.3.1.2 Layers and Operating Modes

The standard consists of three layers I, II, and III of increasing complexity, delay, and subjective performance. From a hardware and software standpoint, the higher layers incorporate the main building blocks of the lower layers (Figure 2.7). A standard full MPEG-1/Audio decoder is able to decode bit streams of all three layers. The standard also supports MPEG-1/Audio layer X decoders (X = I, II, or III). Usually, a layer II decoder will be able to decode bitstreams of layers I and II, a layer III decoder will be able to decode bitstreams of all three layers.

2.3.1.3 Stereo Redundancy Coding

MPEG-1/Audio supports four modes: mono, stereo, dual mono (two unrelated channels, such as in bilingual programs), and joint stereo. In joint stereo mode, layers I and II exploit interchannel dependencies to reduce the overall bit rate by using an irrelevancy reducing technique called intensity stereo. It is known that above 2 kHz and within each critical band, the human auditory system bases its perception of stereo imaging more on the temporal envelope of the audio than on its temporal fine structure. Therefore, the MPEG audio compression algorithm supports a stereo redundancy coding mode called intensity stereo coding which reduces the total bit rate without violating the spatial integrity of the stereophonic signal.

In intensity stereo mode, the encoder codes some upper-frequency subband outputs with a single sum signal L + R (or some linear combination thereof) instead of sending independent left (L) and right (R) subband signals. The decoder reconstructs the left and right channels based only on the single L + R signal and on independent left and right channel scalefactors. Hence, the spectral shape of the left and right outputs is the same within each intensity-coded subband but the magnitudes are different [39]. The optional joint stereo mode will only be effective if the required bit rate exceeds the available bit rate, and it will only be applied to subbands corresponding to frequencies of around 2 kHz and above.

Layer III also uses intensity stereo coding, but can additionally use mid/side (M/S) coding, in which the left and right channel signals are encoded as middle (L + R) and side (L − R) channels.

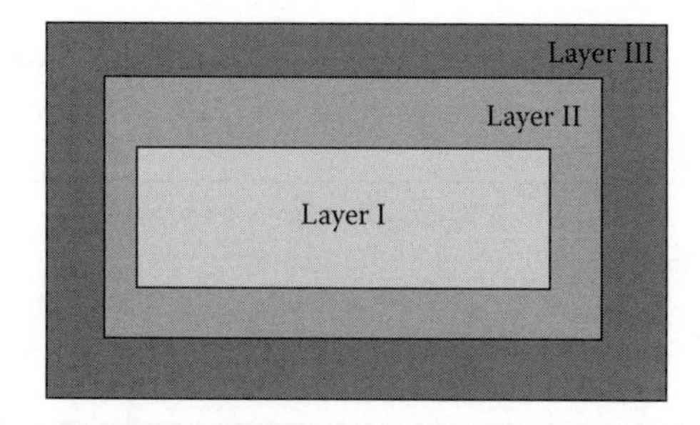

FIGURE 2.7 Hierarchy of layers I, II, and III of MPEG-1/Audio.

2.3.1.4 Psychoacoustic Models

We have already mentioned that the adaptive bit allocation algorithm is controlled by a psychoacoustic model. This model computes SMR taking into account the short-term spectrum of the audio block to be coded and knowledge about noise masking. This potentially computationally complex model is only required in the encoder, permitting the decoder to be much less complex. Such asymmetry in complexity is a desirable feature for audio playback and audio broadcasting applications.

The normative part of the standard describes the decoder and the meaning of the encoded bitstream, but the encoder is not standardized thus leaving room for an evolutionary improvement of the encoder. In particular, different psychoacoustic models can be used ranging from very simple ones (or none at all) to very complex ones based on requirements for target signal quality and target implementation complexity. Information about the short-term spectrum can be obtained in various ways, for example, as an FFT-based spectral analysis of the audio input samples or, less accurately, directly from the spectral components as in a conventional ATC [19]; see also Figure 2.6. Encoders can be optimized for given applications but, regardless of any specific encoder design choices, the standard insures that they interoperate with all MPEG-1/Audio decoders.

The informative part of the standard gives two examples of FFT-based psychoacoustic models; see also [8,33,40]. Both models identify, in different ways, tonal and nontonal spectral components and use the corresponding results of tone-masks-noise and noise-masks-tone experiments in the calculation of the global masking thresholds. Details are given in the standard, experimental results for both psychoacoustic models described in Ref. [40]. In the informative part of the standard a 512 point FFT is proposed for layer I, and a 1024 point FFT for layers II and III. In both models, the audio input samples are Hann-weighted. Model 1, which may be used for layers I and II, computes for each masker its individual masking threshold, taking into account its frequency position, power, and tonality information. The global masking threshold is obtained as the sum of all individual masking thresholds and the absolute masking threshold. The SMR is then the ratio of the maximum signal level within a given subband and the minimum value of the global masking threshold in that given subband (see Figure 2.2).

Model 2, which may be used for all layers, is more complex: tonality is assumed when a simple prediction indicates a high prediction gain, the masking thresholds are calculated in the cochlea domain, i.e., properties of the inner ear are taken into account in more detail, and, finally, in case of potential pre-echoes the global masking threshold is adjusted appropriately.

2.3.2 Layers I and II

MPEG layer I and II coders have very similar structures. The layer II coder achieves better performance, mainly because the overall scalefactor side information is reduced by exploiting redundancies between the scalefactors. Additionally, a slightly finer quantization is provided.

2.3.2.1 Filterbank

Layer I and II coders map the digital audio input into 32 subbands via equally spaced band-pass filters (Figures 2.8 and 2.9). A polyphase filter structure is used for the frequency mapping; its filters have 512 coefficients. Polyphase structures are computationally very efficient because a DCT can be used in the filtering process, and they are of moderate complexity and low delay. On the negative side, the filters are equally spaced, and therefore the frequency bands do not correspond well to the critical band partitions (see Section 2.2.1). At 48 kHz sampling rate, each band has a width of $24000/32 = 750$ Hz; hence, at low frequencies, a single subband covers a number of adjacent critical bands. The subband signals are resampled (critically decimated) to a rate of 1500 Hz. The impulse response of subband k, $h_{\mathrm{sub}(k)}(n)$, is obtained by multiplication of the impulse response of a single prototype lowpass filter, $h(n)$, by a modulating function which shifts the lowpass response to the appropriate subband frequency range:

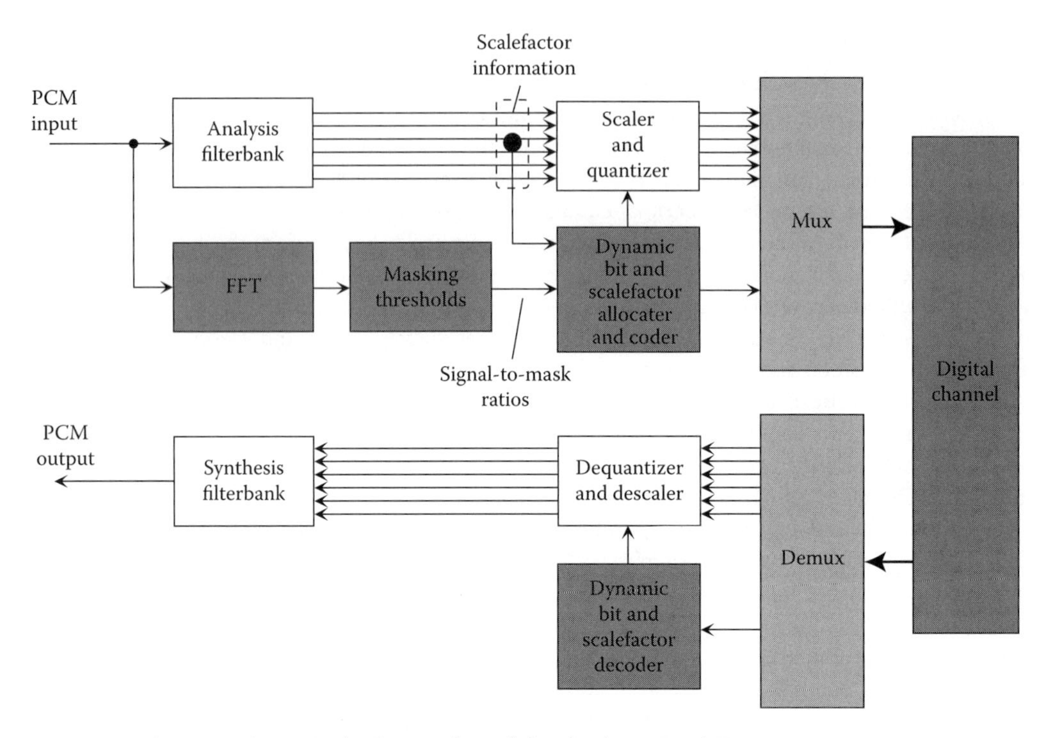

FIGURE 2.8 Structure of MPEG-1/Audio encoder and decoder, layers I and II.

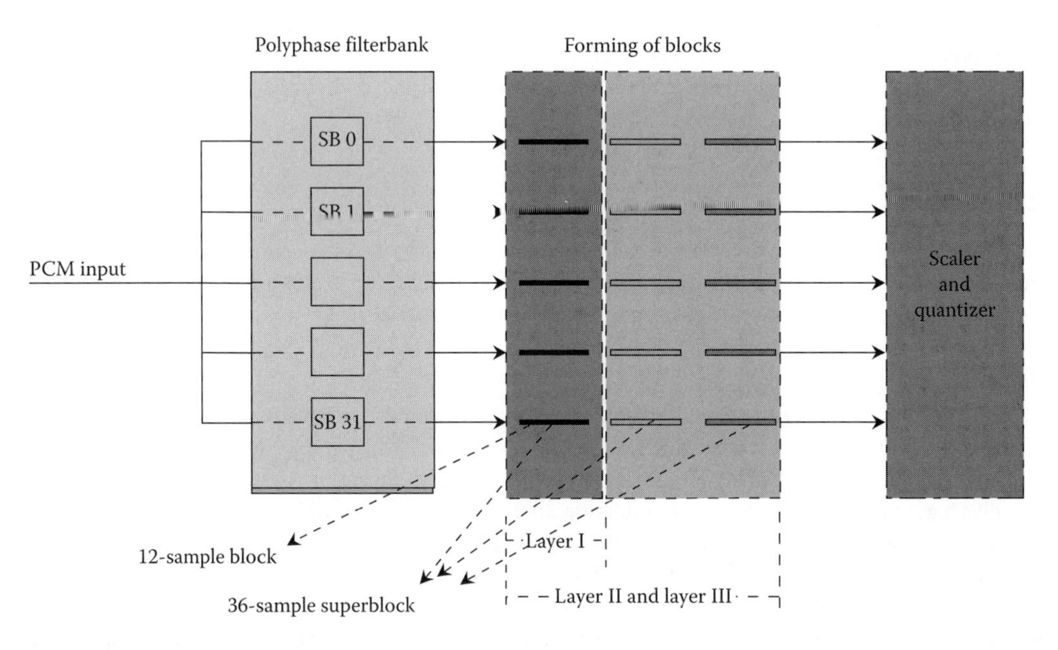

FIGURE 2.9 Block companding in MPEG-1/Audio coders.

$$h_{\text{sub}(k)}(n) = h(n)\cos\left[\frac{(2k+1)\pi n}{2M} + \varphi(k)\right]; \quad M = 32;\ k = 0, 1, \ldots, 31;\ n = 0, 1, \ldots, 511$$

The prototype low-pass filter has a 3 dB bandwidth of $750/2 = 375$ Hz, and the center frequencies are at odd multiples thereof (all values at 48 kHz sampling rate). The subsampled filter outputs exhibit a significant overlap. However, the design of the prototype filter and the inclusion of appropriate phase shifts in the cosine terms result in an aliasing cancellation at the output of the decoder synthesis filterbank. Details about the coefficients of the prototype filter and the phase shifts $\varphi(k)$ are given in the ISO/MPEG standard. Details about an efficient implementation of the filterbank can be found in Refs. [26,40] and in the standardization documents.

2.3.2.2 Quantization

The number of quantizer levels for each spectral component is obtained from a dynamic bit allocation rule that is controlled by a psychoacoustic model. The bit allocation algorithm selects one uniform midtread quantizer out of a set of available quantizers such that both the bit rate requirement and the masking requirement are met. The iterative procedure minimizes the NMR in each subband. It starts with the number of bits for the samples and scalefactors set to zero. In each iteration step, the quantizer SNR(m) is increased for the one subband quantizer producing the largest value of the NMR at the quantizer output. (The increase is obtained by allocating one more bit). For that purpose, NMR(m) = SMR − SNR(m) is calculated as the difference (in dB) between the actual quantization noise level and the minimum global masking threshold. The standard provides tables with estimates for the quantizer SNR(m) for a given m.

Block companding is used in the quantization process, i.e., blocks of decimated samples are formed and divided by a scalefactor such that the sample of largest magnitude is unity. In layer I blocks of 12 decimated and scaled samples are formed in each subband (and for the left and right channel) and there is one bit allocation for each block. At 48 kHz sampling rate, 12 subband samples correspond to 8 ms of audio. There are 32 blocks, each with 12 decimated samples, representing $32 \times 12 = 384$ audio samples.

In layer II in each subband a 36 sample superblock is formed of three consecutive blocks of 12 decimated samples corresponding to 24 ms of audio at 48 kHz sampling rate. There is one bit allocation for each 36 sample superblock. All 32 superblocks, each with 36 decimated samples, represent, altogether, $32 \times 36 = 1152$ audio samples. As in layer I, a scalefactor is computed for each 12 sample block. A redundancy reduction technique is used for the transmission of the scalefactors: depending on the significance of the changes between the three consecutive scalefactors, one, two, or all three scalefactors are transmitted, together with a 2 bit scalefactor select information. Compared with layer I, the bit rate for the scalefactors is reduced by around 50% [33]. Figure 2.9 indicates the block companding structure.

The scaled and quantized spectral subband components are transmitted to the receiver together with scalefactor, scalefactor select (layer II), and bit allocation information. Quantization with block companding provides a very large dynamic range of more than 120 dB. For example, in layer II uniform midtread quantizers are available with $3, 5, 7, 9, 15, 31, \ldots, 65{,}535$ levels for subbands of low index (low frequencies). In the mid- and high-frequency region, the number of levels is reduced significantly. For subbands of index 23–26 there are only quantizers with 3, 5, and 65,535 (!) levels available. The 16 bit quantizers prevent overload effects. Subbands of index 27–31 are not transmitted at all. In order to reduce the bit rate, the codewords of three successive subband samples resulting from quantizing with 3-, 5-, and 9-step quantizers are assigned one common codeword. The savings in bit rate is about 40% [33].

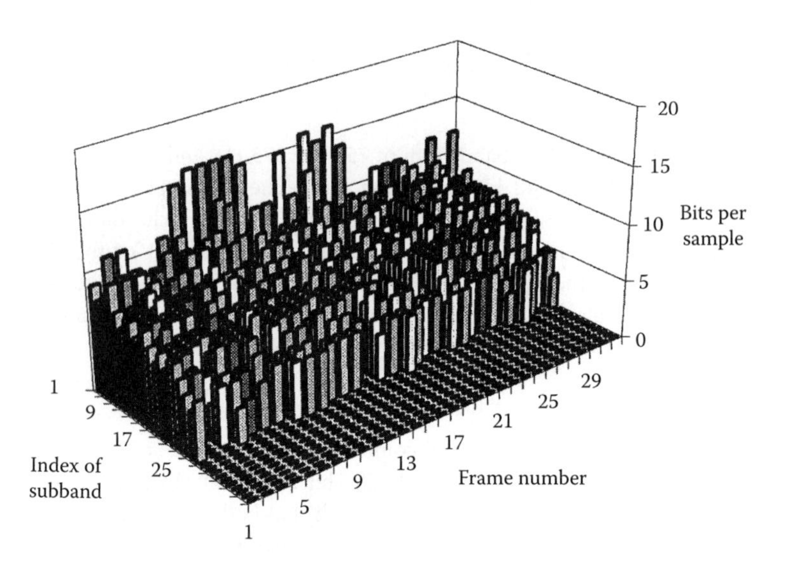

FIGURE 2.10 Time-dependence of assigned number of quantizer bits in all subbands for a layer II encoded high quality speech signal.

Figure 2.10 shows the time-dependence of the assigned number of quantizer bits in all subbands for a layer II encoded high quality speech signal. Note, for example, that quantizers with ten or more bits resolution are only employed in the lowest subbands, and that no bits have been assigned for frequencies above 18 kHz (subbands of index 24–31).

2.3.2.3 Decoding

The decoding is straightforward: the subband sequences are reconstructed on the basis of blocks of 12 subband samples taking into account the decoded scalefactor and bit allocation information. If a subband has no bits allocated to it, the samples in that subband are set to zero. Each time the subband samples of all 32 subbands have been calculated, they are applied to the synthesis filterbank, and 32 consecutive 16 bit PCM format audio samples are calculated. If available, as in bidirectional communications or in recorder systems, the encoder (analysis) filterbank can be used in a reverse mode in the decoding process.

2.3.3 Layer III

Layer III of the MPEG-1/Audio coding standard introduces many new features (see Figure 2.11), in particular a switched hybrid filterbank. In addition, it employs an analysis-by-synthesis approach, an advanced preecho control, and nonuniform quantization with entropy coding. A buffer technique, called bit reservoir, leads to further savings in bit rate. Layer III is the only layer that provides support for variable bit rate coding [41].

2.3.3.1 Switched Hybrid Filterbank

In order to achieve a frequency resolution that is a closer approximation to critical band partitions, the 32 subband signals are subdivided further in frequency content by applying to each of the subbands a 6- or 18-point MDCT block transform having 50% overlap. The MDCT windows contain, respectively, 12 or 36 subband samples. The maximum number of frequency components is $32 \times 18 = 576$ each representing a bandwidth of only $24{,}000/576 = 41.67$ Hz. Because the 18-point block transform provides better frequency resolution, it is normally applied, whereas the 6-point block transform provides better time resolution and is applied in case of expected preechoes (see Section 2.2.3). In practice, a preecho is

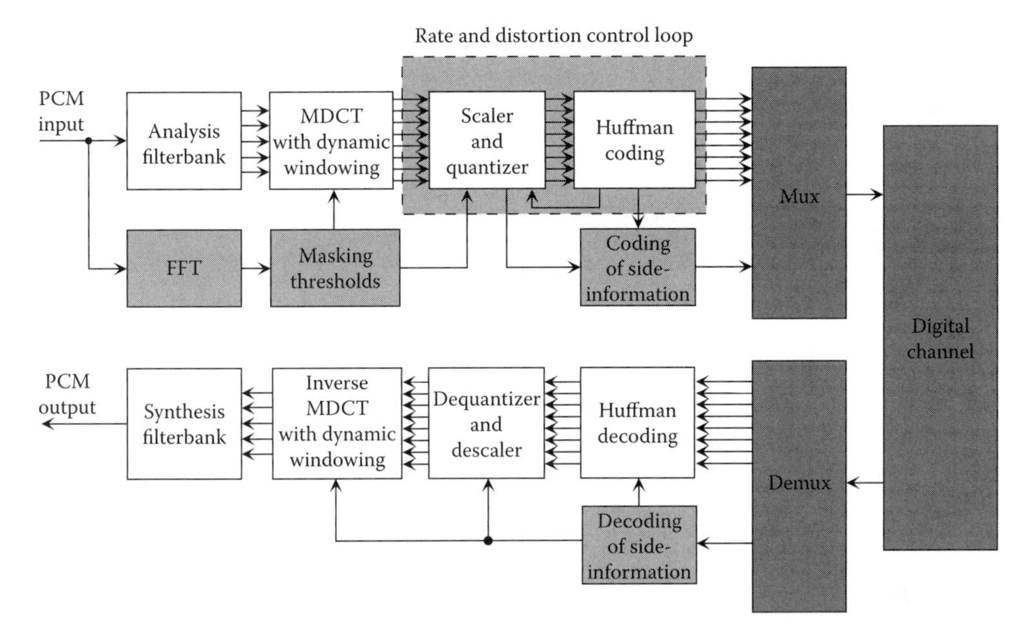

FIGURE 2.11 Structure of MPEG-1/Audio encoder and decoder, layer III.

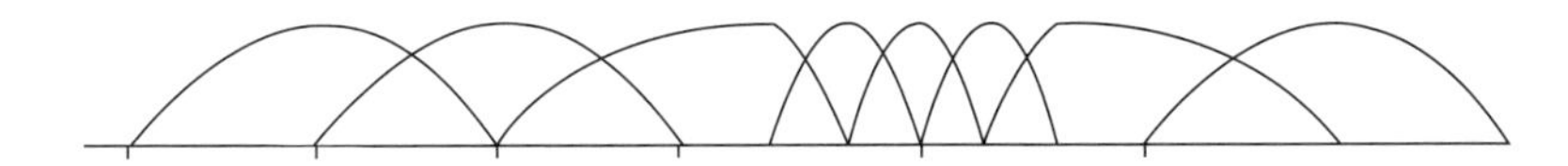

FIGURE 2.12 Typical sequence of windows in adaptive window switching.

assumed when an instantaneous demand for a high number of bits occurs. Two special MDCT windows, a start window and a stop window, are needed to facilitate transitions between short and long blocks so as to maintain the time domain alias cancellation feature of the MDCT [25,29,30]. Figure 2.12 shows a typical sequence of windows.

2.3.3.2 Quantization and Coding

The MDCT output samples are nonuniformly quantized thus providing smaller mean-squared errors while retaining SMR margins. Huffman coding, based on 32 code tables and additional run-length coding are applied to represent the quantizer indices in an efficient way. The encoder maps the variable wordlength codewords of the Huffman code tables into a constant bit rate by monitoring the state of a bit reservoir. The bit reservoir ensures that the decoder buffer neither underflows nor overflows when the bitstream is presented to the decoder at a constant rate.

In order to keep the quantization noise in all critical bands below the global masking threshold, a process known as noise allocation, an iterative analysis-by-synthesis method is employed whereby the process of scaling, quantization, and coding of spectral data is carried out within two nested iteration loops. The decoding is the inverse of the encoding process.

2.3.4 Frame and Multiplex Structure

2.3.4.1 Frame Structure

Figure 2.13 shows the frame structure of MPEG-1/Audio-coded signals, both for layer I and layer II. Each frame has a header; its first part contains 12 synchronization bits, 20 system information bits, and an

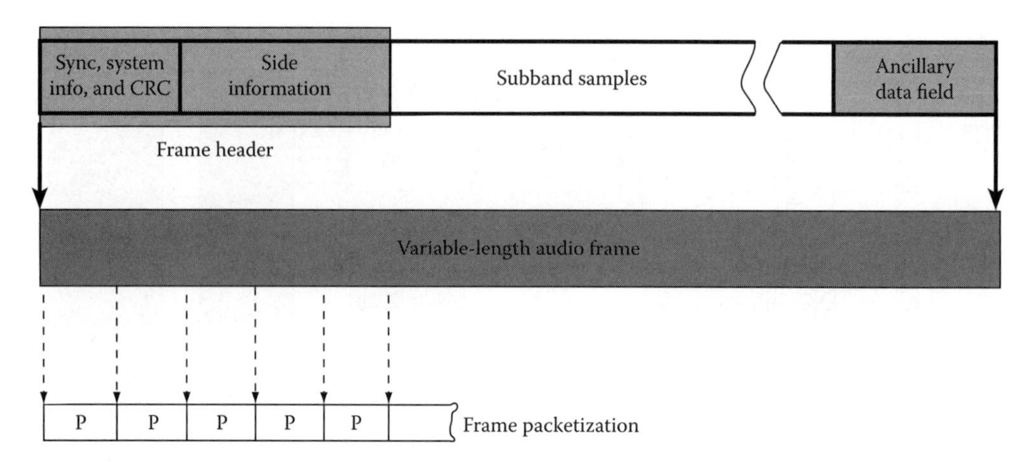

FIGURE 2.13 MPEG-1 frame structure and packetization. Layer I: 384 subband samples; layer II: 1152 subband samples; packets P: 4 byte header; 184 byte payload field (see also Figure 2.14).

optional 16 bit cyclic redundancy check code. Its second part contains side information about the bit allocation and the scalefactors (and, in layer II, scalefactor information). As main information, a frame carries a total of 32×12 subband samples (corresponding to 384 PCM audio input sample—equivalent to 8 ms at a sampling rate of 48 kHz) in layer I, and a total of 32×36 subband samples in layer II (corresponding to 1152 PCM audio input samples—equivalent to 24 ms at a sampling rate of 48 kHz). Note that the layer I and II frames are autonomous: each frame contains all information necessary for decoding. Therefore, each frame can be decoded independently from previous frames, and each defines an entry point for audio storage, audio editing and broadcasting applications. Please note that the lengths of the frames are not fixed, due to (1) the length of the main information field, which depends on bit rate and sampling frequency, (2) the side information field which varies in layer II, and (3) the ancillary data field, the length of which is not specified.

As with to layer I and II, layer III can operate over constant-rate transmission channels and has constant-rate headers. However, layer III is an instantaneously variable rate coder that adapts to the constant-rate channel by using a bit-buffer and back-pointers. Each header signals the start of another block of audio signal. However, due to the layer III syntax, the frame of data associated with that next block of audio signal may be in a prior segment of the bitstream, pointed to by the back-pointer. See Figure 2.11 and specifically the curved arrows pointing to main_data_begin.

2.3.4.2 Multiplex Structure

We have already mentioned that the systems part of the MPEG-1 coding standard IS 11172 defines a packet structure for multiplexing audio, video, and ancillary data bitstreams in one stream. The variable-length MPEG frames are broken down into packets. The packet structure uses 188 byte packets consisting of a 4 byte header followed by 184 bytes of payload (see Figure 2.15). The header includes a sync byte, a 13 bit field called packet identifier to inform the decoder about the type of data, and additional information. For example, a 1 bit payload unit start indicator indicates if the payload starts with a frame header. No predetermined mix of audio, video, and ancillary data bitstreams is required, the mix may change dynamically, and services can be provided in a very flexible way. If additional header information is required, such as for periodic synchronization of audio and video timing, a variable-length adaptation header can be used as part of the 184 byte payload field.

Although the lengths of the frames are not fixed, the interval between frame headers is constant (within a byte) via the use of padding bytes. The MPEG systems specification describes how MPEG-compressed audio and video data streams are to be multiplexed together to form a single data stream. The terminology and the fundamental principles of the systems layer are described in Ref. [42].

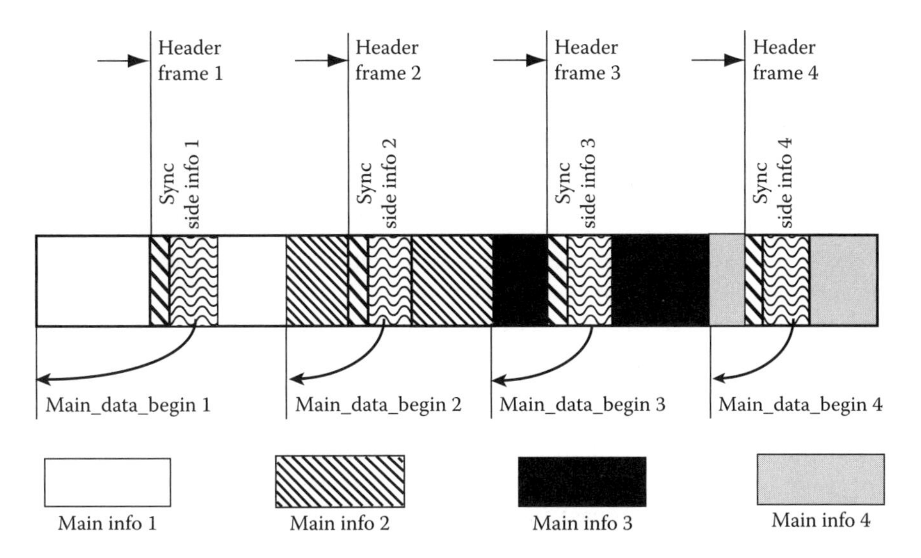

FIGURE 2.14 MPEG-1 layer III frame structure during active use of bit buffer.

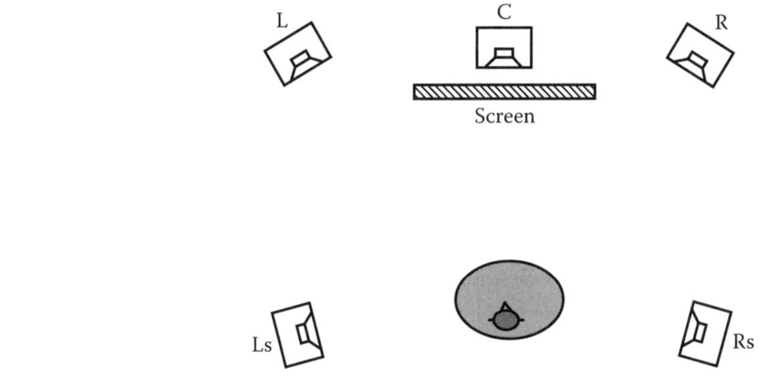

FIGURE 2.15 MPEG packet delivery.

2.3.5 Subjective Quality

The standardization process included extensive subjective tests and objective evaluations of parameters such as complexity and overall delay. The MPEG (and equivalent ITU-R) listening tests were carried out under very similar and carefully defined conditions in which approximately 60 experienced listeners, and 10 stereo test sequences were used, with listening via both loudspeakers and headphones. In order to detect the smallest impairments, the 5-point ITU-R impairment scale was used in all experiments. Details are given in Refs. [43,44]. Critical test items were chosen in the tests to evaluate the coders by their worst case (not average) performance. The subjective evaluations, which were based on triple stimulus/hidden reference/double blind test methodology, demonstrate very similar and stable evaluation results. In these tests the subject is offered three signals, A, B, and C (triple stimulus) [46]. A is always the unprocessed source signal (the reference). B and C, or C and B, are the hidden reference and the system under test, respectively. The selection is neither known to the test subjects nor to the test administrator (double blind test). The subjects must decide which of B or C is the reference and then grade the other signal.

The MPEG-1/Audio coding standard has shown an excellent performance for all layers at the rates given in Table 2.3. It should be mentioned again that the standard leaves room for encoder-based improvements by using better psychoacoustic models. Indeed, many improvements have been achieved since the first subjective results had been carried out in 1991.

2.4 MPEG-2/Audio Multichannel Coding

The logical next step MPEG Audio was the definition of a multichannel audio representation system to create a convincing, lifelike soundfield both for audio-only applications and for audiovisual systems, including video conferencing, multimedia services, and digital cinema. Multichannel systems can also provide multilingual channels and additional channels for visually impaired (a verbal description of the visual scene) and for hearing impaired (dialog with enhanced intelligibility). ITU-R has recommended a five-channel loudspeaker configuration, referred to as 3/2-stereo, with a left and a right channel (L and R), an additional center channel C, two side/rear surround channels (LS and RS) augmenting the L and R channels, see Figure 2.16 (ITU-R Rec. 775). Such a configuration offers an improved realism of auditory ambience with a stable frontal sound image and a large listening area.

Multichannel digital audio systems support p/q presentations with p front and q back channels, and also present the possibility of transmitting independent monophonic or stereophonic programs and/or a number of commentary or multilingual channels. Typical combinations of channels are shown.

1 channel:	1/0-configuration:	Centre (monophonic)
2 channels:	2/0-configuration:	Left, right (stereophonic)
3 channels:	3/0-configuration:	Left, right, centre
4 channels:	3/1-configuration:	Left, right, centre, monosurround
5 channels:	3/2-configuration:	Left, right, centre, surround left, surround right
5.1 channels:	3/2 configuration plus LFE:	Left, right, centre, surround left, surround right, low frequency effects

ITU-R Rec. 775 provides a set of downward mixing equations if the number of loudspeakers is to be reduced (downward compatibility). An additional low frequency enhancement (LFE-or subwoofer-) channel is particularly useful for HDTV applications, it can be added, optionally, to any of the configurations although it is most commonly added to the 5 channel (3/2) configuration. The LFE channel extends the low frequency content between 15 and 120 Hz in terms of both frequency and level.

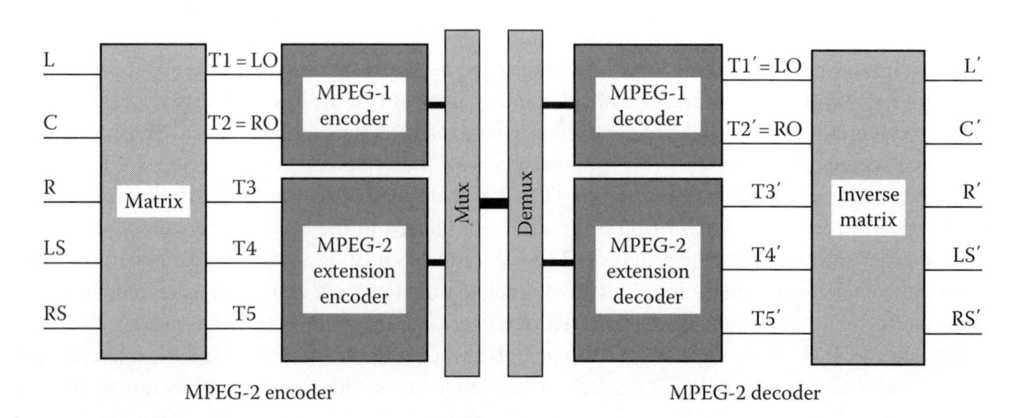

FIGURE 2.16 3/2 Multichannel loudspeaker configuration.

The LFE subwoofer can be positioned freely in the listening room to reproduce this LFE signal. (Film industry uses a similar system for their digital sound systems).*

In order to reduce the overall bit rate of multichannel audio coding systems, redundancies and irrelevancies, such as interchannel dependencies and interchannel masking effects, respectively, may be exploited.

A second phase in MPEG, MPEG-2 Audio extended the sampling rates and number of channels supported by MPEG Audio. MPEG-2 Audio is a superset of MPEG-1 Audio and supports sampling rates from 12 through 48 kHz. In addition, it supports multichannel audio programs using two distinct coding methods.

2.4.1 Backward Compatible Coding

The first method of multichannel coding is backwards-compatible such that MPEG-2 decoders play the multichannel program and legacy MPEG-1 decoders play a stereo version of the program [45,47]. The beauty of the backwards-compatible method is that the program is coded as a stereo downmix plus additional coded channels that permit reconstruction of the multichannel program in a matrix-upmix manner. Thus, the MPEG-1 decoder merely decodes the stereo program and skips the additional compressed information. MPEG-2 decoders decode the stereo program and then use the additional channels to dematrix the dowmixed stereo signal to obtain the desired multichannel signal. A typical set of matrixing equations is

$$\mathrm{LO} = \alpha(\mathrm{L} + \beta C + \delta \mathrm{LS})$$

$$\alpha = \frac{1}{1+\sqrt{2}}; \quad \beta = \delta = \sqrt{2}$$

$$\mathrm{RO} = \alpha(\mathrm{R} + \beta C + \delta \mathrm{RS})$$

The factors α, β, and δ attenuate the signals to avoid overload when calculating the compatible stereo signal (LO, RO). The signals LO and RO are transmitted in MPEG-1 format in transmission channels T1 and T2. Channels T3, T4, and T5 together form the multichannel extension signal. They have to be chosen such that the decoder can recompute the complete 3/2-stereo multichannel signal. Interchannel redundancies and masking effects are taken into account to find the best choice. A simple example is T3 = C, T4 = LS, and T5 = RS.

2.4.2 Advanced Audio Coding

The second method of multichannel coding is non-backwards-compatible. AAC supports applications that do not require compatibility with the existing MPEG-1 stereo format. AAC is compatible with the sampling rates, audio bandwidth, and channel configurations of MPEG-2/Audio, but is capable of operating at bit rates from 64 kbps up to a bit rate sufficient for highest quality audio.

MPEG-2/AAC became International Standard in April 1997 (ISO/MPEG 13818-7). The standard offers EBU broadcast quality a bit rate of 320 kbps for five channels and has many applications in both consumer and professional areas.

2.4.2.1 Profiles

In order to serve different needs, the standard offers three profiles:

- Main profile
- Low complexity profile
- Saleable-sampling-rate profile

* A 3/2-configuration with five high-quality full-range channels plus a subwoofer channel is often called a 5.1 system.

The main profile offers a backward adaptive predictor of the value in each MDCT transform bin. This provides additional compression for steady-state, tonal signals.

The low complexity profile strikes a compromise between compression efficiency and complexity and has received the widest adoption by industry and so will be described in this section. It does not include the backward-adaptive predictors or the sampling-rate-saleable preprocessing module.

The sampling-rate-scaleable profile offers a preprocessing module that allows for reconstruction at sampling rates of 6, 12, 18, and 24 kHz from a single compressed representation (bitstream). A detailed description of the MPEG-2 AAC multichannel standard can be found in the literature [48–50].

2.4.2.2 AAC Encoder Modules

The low complexity profile of the MPEG-2 AAC standard is comprised of the following modules:

- Time-to-frequency mapping (filterbank)
- Temporal noise shaping (TNS)
- Perceptual model
- Joint channel coding
- Quantization and coding
- Bit stream multiplexer

A block diagram of the AAC encoder is shown in Figure 2.17. Each of these encoder blocks is described in turn:

Filterbank: AAC uses a resolution-switching filterbank which can switch between a high frequency resolution mode of 1024 bands (for maximum statistical gain during intervals of signal stationarity), and a high time resolution mode of 128 bands (for maximum time-domain coding error control during intervals of signal nonstationarity). The MDCT is used to implement the filterbank, which has the properties of perfect reconstruction of the output (when no quantization is present), critical sampling (so that the number of new PCM samples processed by the transform exactly equals the number of time/frequency (T/F) coefficients produced), and 50% overlapping transform windows (which provide signal smoothing on reconstruction).

TNS: The TNS tool modifies the filterbank characteristics so that the combination of the two tools is better able to adapt to the T/F characteristics of the input signal. It shapes the quantization noise in the time domain by doing an open loop linear prediction in the frequency domain, so that it permits the coder to exercise control over the temporal structure of the quantization noise within a filterbank window. TNS is a new technique which has proved very effective for improving the quality of signals such as speech when coded at low bit rates.

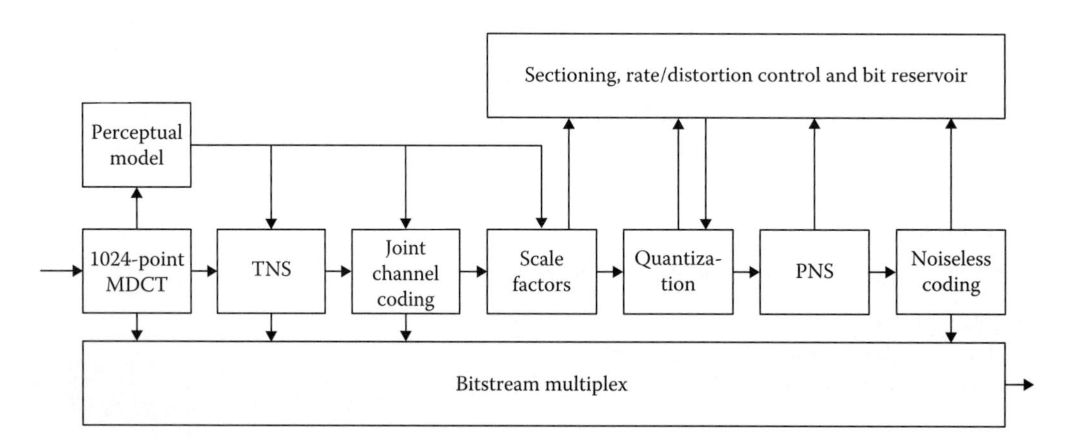

FIGURE 2.17 MPEG-4 AAC encoder block diagram.

Perceptual model: The perceptual model estimates the masking threshold, which is the level of noise that is subjectively just noticeable given the current input signal. Because models of auditory masking are primarily based on frequency domain measurements, these calculations typically are based on the short-term power spectrum of the input signal. Threshold values are adapted to the T/F resolution of the filterbank outputs. The threshold of masking is calculated relative to each frequency coefficient for each audio channel for each frame of input signal, so that it is signal-dependent in both time and frequency.

Joint channel coding: This block actually comprises three tools, intensity coding, M/S stereo coding (also known as "sum/difference coding") and coupling channel coding, all of which seek to protect the stereo or multichannel signal from noise imaging, while achieving coding gain based on exploiting the correlation between two or more channels of the input signal. M/S stereo coding, intensity stereo coding, and L/R (independent) coding can be combined by selectively applying them to different frequency regions, and by using these tools, it is possible to avoid expensive overcoding when using binaural masking level depression to correctly account for noise imaging, and very frequently to achieve a significant saving in bit rate.

Scalefactors, quantization, coding, and rate/distortion control: The spectral coefficients are coded using one quantizer per scalefactor band, which is a division of the spectrum roughly equal to one-third Bark. The psychoacoustic model specifies the quantizer step size (inverse of the scalefactor) per scalefactor band. As with MPEG-1 layer III, AAC is an instantaneously variable rate coder that similarly uses a bit reservoir. If the coded audio is to be transmitted over a constant rate channel then the rate/distortion module adjusts the step sizes and number of quantization levels so that a constant rate is achieved.

The quantization and coding processes work together. The first quantizes the spectral components and the second applies Huffman coding to vectors of quantized coefficients in order to extract additional redundancy from the nonuniform probability of the quantizer output levels. In any perceptual encoder, it is very difficult to control the noise level accurately, while at the same time achieving an "optimum quantizer" (in the minimum mean square error sense). It is, however, quite efficient to allow the quantizer to operate unconstrained, and to then remove the redundancy in the quantizer outputs through the use of entropy coding.

The noiseless coding segments the set of 1024 quantized spectral coefficients into sections, such that a single Huffman codebook is used to code each section. For reasons of coding efficiency, section boundaries can only be at scalefactor band boundaries so that for each section of the spectrum, one must transmit the length of the section, in scalefactor bands, and the Huffman codebook number used for the section. Sectioning is dynamic and typically varies from block to block, such that the number of bits needed to represent the full set of quantized spectral coefficients is minimized.

The rate/distortion tool adjusts the scalefactors such that more (or less) noise is permitted in the quantized representation of the signal which, in turn, requires fewer (or more) bits. Using this mechanism the rate/distortion control tool can adjust the number of bits used to code each audio frame and hence adjust the overall bit rate of the coder.

Bitstream multiplexer: The multiplexer (MUX) assembles the various tokens to form a coded frame, or access unit. An access unit contains all data necessary to reconstruct the corresponding time-domain signal block. The MPEG-4 system layer specifies how to carry the sequence of access units over a channel or store them in a file (using the MPEG-4 file format).

AAC has a flexible bit stream syntax for a coded frame that permits up to 48 main channels and up to 16 LFE channels to be carried in an access unit, but in a manner that does not incur any overhead for the additional channels. In this respect, it is as efficient for mono, stereo and 5.1 channel representations.

2.4.3 Subjective Quality

Extensive formal subjective, independently run at German Telekom and BBC (U.K.) under the umbrella of the MPEG-2 standardization effort, have been carried out to compare MPEG-2 AAC coders, operating, respectively, at 256 and 320 kbps, and a BC MPEG-2 layer II coder,* operating at 640 kbps [51]. Critical test items were used and the listening test employed the BS-1116 triple-stimulus hidden reference (Ref/A/B) methodology [46], a methodology appropriate for assessment of very small impairments as would be found in nearly transparent coding technology. At each test site more than 20 subjects participated in the test. All coders showed excellent performance, with a slight advantage of the 320 kbps MPEG-2 AAC coder compared with the 640 kbps MPEG-2 layer II BC coder. The performances of those coders are indistinguishable from the original in the sense of the EBU definition of indistinguishable quality [52].

2.5 MPEG-4/Audio Coding

MPEG-4/Audio is a very broad initiative that proposed not only to increase the compression efficiency of audio coding, but also to support a number of new coding mythologies and functionalities. It supports

- Synthetic speech and audio engines (below 2 kbps)
- A parametric coding scheme for low bit rate speech coding (2–4 kbps)
- An analysis-by-synthesis coding scheme for medium bit rates (6–16 kbps)
- A subband and bandwith replication scheme for higher bit rates (32–48 kbps)
- A subband/transform-based coding scheme for highest bit rates (64–128 kbps)
- Lossless coding for perfect reconstruction of the original signal (750 kbps)

Of all of these coding tools and functionalities, two coding profiles have received widespread industry adoption: MPEG-4 AAC profile and MPEG-4 HE-AAC V2 profile. Each of these technologies will be described.

2.5.1 AAC

MPEG-4 AAC builds upon MPEG-2 AAC, and in fact the basic coded representations are identical except that MPEG-4 AAC incorporates one additional tool: Perceptual noise substitution (PNS). Therefore please refer to the description in Section 2.4.2. A description of the PNS module follows:

PNS—This tool identifies segments of spectral coefficients that appear to be noise-like and codes them as random noise. It is extremely efficient in identifying, since for the segment, all that need be transmitted is a flag indicating that PNS is used and a value indicating the average power of the noise. The decoder reconstructs an estimate of the coefficients using a pseudorandom noise generator weighted by the signaled power value.

2.5.2 High Efficiency AAC

The MPEG-4 HE-AAC audio compression algorithm consists of an MPEG-4 AAC core coder augmented with the MPEG-4 SBR tool [54–56]. The encoder SBR tool is a preprocessor for the core encoder, and the decoder SBR tool is a postprocessor for the core decoder, as shown in Figure 2.18. The SBR tool essentially converts a signal at a given sampling rate and bandwidth into a signal at half the sampling rate and bandwidth, passes the low-bandwidth signal to the core codec, and codes the high-bandwidth signal using a compact parametric representation. The lowband signal is coded by the core coder, and the highband compressed data and the parametric data are transmitted over the channel. The core decoder reconstructs the lowband signal and the SBR decoder uses the parametric data to reconstruct the highband data, thus recovering the full-bandwidth signal. This combination provides a significant improvement in performance relative to that of the core coder by itself, which can be used to lower the bit rate or improve the audio quality.

* A 1995 version of this latter coder was used, therefore its test results do not reflect any subsequent enhancements.

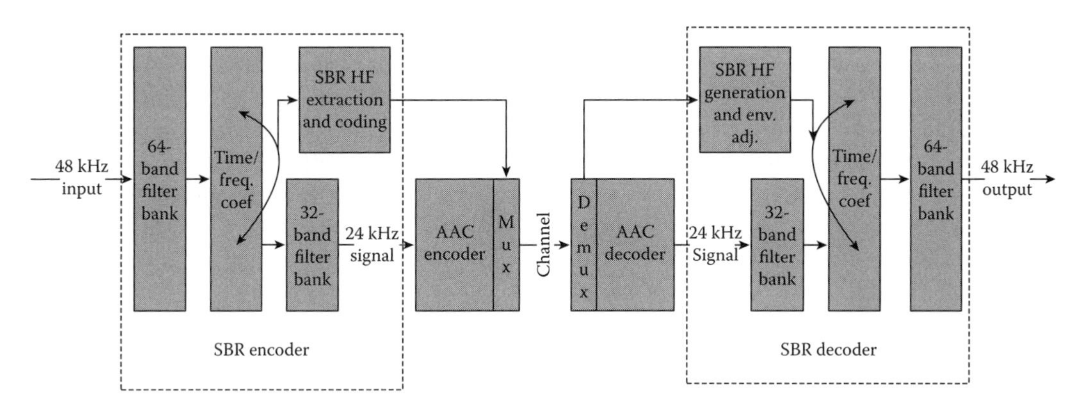

FIGURE 2.18 Block diagram of HE-AAC encoder and decoder.

2.5.2.1 SBR Principle

A perceptual audio coder, such as MPEG-4 AAC, provides coding gain by shaping the quantization noise such that it is always below the masking threshold. However, if the bit rate is not sufficiently high, the masking threshold will be violated, permitting coding artifacts to become audible. The usual solution adopted by perceptual coders in this case is to reduce the bandwidth of the coded signal, thus effectively increasing the available bits per sample to be coded. The result will sound cleaner but also duller due to the absence of high-frequency components.

The SBR tool gives perceptual coders an additional coding strategy (other than bandwidth reduction) when faced with severe bit rate restrictions. It exploits the human auditory system's reduced acuity to high-frequency spectral detail to permit it to parametrically code the high frequency region of the signal. When using the SBR tool, the lower frequency components of the signal (typically from 0 to between 5 and 13 kHz) are coded using the core codec. Since the signal bandwidth is reduced, the core coder will be able to code this signal without violating the masking threshold. The high-frequency components of the signal are reconstructed as a transposition of the low frequency components followed by an adjustment of the spectral envelope. In this way, a significant bit rate reduction is achieved while maintaining the same audio quality, or alternatively an improved audio quality is achieved while maintaining the same bit rate.

2.5.2.2 SBR Technology

Using SBR, the missing high frequency region of a low-pass filtered signal can be recovered based on the existing low pass signal and a small amount of side information, or control data. The required control data is estimated in the encoder based on the original wideband signal. The combination of SBR with a core coder (in this case MPEG-4 AAC) is a dual rate system, where the underlying AAC encoder/decoder is operated at half the sampling rate of the SBR encoder/decoder. A block diagram of the HE-AAC compression system, consisting of SBR encoder and its submodules, AAC core encoder/decoder and SBR decoder and its submodules is shown in Figure 2.18.

A major module in the SBR tool is a 64-band pseudoquadrature mirror analysis/synthesis filterbank (QMF). Each block of 2048 PCM input samples processed by the analysis filterbank results in 32 subband samples in each of 64 equal-width subbands. The SBR encoder contains a 32-band synthesis filterbank whose inputs are the lower 32 bands of the 64 subbands and whose output is simply a band limited (to one-half the input bandwidth) and half-sampling rate version of the input signal. Actual implementations may use more efficient means to accomplish this, but the illustrated means provides a clearer framework for understanding how SBR works.

The key aspect of the SBR technology is that the SBR encoder searches for the best match between the signal in the lower subbands and those in the higher subbands (indicated by the curved arrow in the Time/Freq. Coef. box in Figure 2.18), such that the high subbands can be reconstructed by transposing the low subband signals up to the high subbands. This transposition mapping is coded as SBR control data and sent over the channel. Additional control parameters are estimated in order to ensure that the high-frequency reconstruction results in a highband that is as perceptually similar as possible to the original highband. The majority of the control data is used for a spectral envelope representation. The spectral envelope information has varying time and frequency resolution such that it can control the SBR process in a perceptually relevant manner while using as small a side information rate as possible. Additionally, information on whether additional components such as noise and sinusoids are needed as part of the high-band reconstruction is coded as side information. This side information is multiplexed into the AAC bit stream (in a backward-compatible way).

In the HE-AAC decoder, the bit stream is demultiplexed, the SBR side information is routed to the SBR decoder and the AAC information is decoded by the AAC decoder to obtain the half-sampling rate signal. This signal is filtered by a 32-band QMF analysis filterbank to obtain the low 32-subbands of the desired T/F coefficients. The SBR decoder then decodes the SBR side information and maps the low-band signal up to the high subbands, adjusts its envelope, and adds additional noise and sinusoids if needed. The final step of the decoder is to reconstruct the output block of 2048 time-domain samples using a 64-band QMF synthesis filterbank whose inputs are the 32 low subbands resulting from processing the AAC decoder output and 32 high subbands resulting from the SBR reconstruction. This results in an up-sampling by a factor of two.

2.5.3 Subjective Quality

In 1998 MPEG conducted an extensive test of the quality of stereo signals coded by MPEG-2 AAC [53]. Since MPEG-4 AAC is identical to MPEG-2 AAC except for the inclusion of the PNS tool and since the MPEG-2 test was much more comprehensive than any test done for MPEG-4 AAC, it is referenced here. Critical test items were used and the listening test employed the BS-1116 triple-stimulus hidden reference (Ref/A/B) methodology [46], a methodology appropriate for assessment of very small impairments as would be found in nearly transparent coding technology. Thirty subjects participated in the test. The test concluded that for coding the stereo test items:

- AAC at 128 kbps had performance that provided "indistinguishable quality" in the EBU sense of the phrase [52]
- AAC at 128 kbps had performance that was significantly better than MPEG-1 Layer III at 128 kbps
- AAC at 96 kbps had performance comparable to that of MPEG-1 layer III at 128 kbps

MPEG conducted several tests on the quality of HE-AAC, including a final verification test [57]. There were two test sites in the test using stereo items: France Telecom and T-Systems Nova, together comprising more than 25 listeners, and 4 test sites in the test using mono items: Coding Technologies, Matsushita, NEC, and Panasonic Singapore Laboratories, together comprising 20 listeners. In all cases the MUSHRA test methodology was used [58], which is appropriate for assessment of intermediate quality coding technology.

The overall conclusions of the test were that, for the stereo items:

- The audio quality of HE-AAC at 48 kbps is in the lower range of "Excellent" on the MUSHRA scale.
- The audio quality of HE-AAC at 48 kbps is equal or better than that of MPEG-4 AAC at 60 kbps.
- The audio quality of HE-AAC at 32 kbps is equal or better than that of MPEG-4 AAC at 48 kbps.

2.6 MPEG-D/Audio Coding

MPEG-4 provided standards for audio coding with a rich set of functionalities. MPEG-D is more in the tradition of MPEG-1 and MPEG-2, in which compression is the primary goal. The first standard in MPEG-D is MPEG Surround, which is a technology for the compression of multichannel audio signals [59–61]. It exploits human perception of the auditory sound stage, and in particular preserves and codes only those components of a multichannel signal that the human auditory system requires so that the decoded signal is perceived to have the same sound stage as the original signal.

2.6.1 MPEG Surround

The MPEG Surround algorithm achieves compression by encoding an *N*-channel (e.g., 5.1 channels) audio signal as a 2-channel stereo (or mono) audio signal plus side information in the form of a secondary bit stream which contains "steering" information. The stereo (or mono) signal is encoded and the compressed representation sent over the transmission channel along with the side information, such that a spatial audio decoder can synthesize a high quality multichannel audio output signal in the receiver. Although the audio compression technology is quite flexible, such that it can take *N* input audio channels and compress those to one or two transmitted audio channels plus side information, discussed here is only a single example configuration, which is the case of coding a 5.1 channel signal as a compressed stereo signal plus some side information. This configuration is illustrated in Figure 2.19.

Referring to Figure 2.19, the MPEG Surround encoder receives a multichannel audio signal, x_1 to $x_{5.1}$, where x is a 5.1 channel (left, center, right, left surround, right surround, LFE) audio signal. Critical in the encoding process is that a "downmix" signal is derived from the multichannel input signal, and that this downmix signal is compressed and sent over the transmission channel rather than the multichannel signal itself. A well-optimized encoder is able to create a downmix that is, by itself, a faithful 2-channel stereo equivalent of the multichannel signal, and which also permits the MPEG Surround decoder to create a perceptual equivalent of the multichannel original. The downmix signal x_{t1} and x_{t2} (2-channel stereo) is compressed and sent over the transmission channel. The MPEG Surround encoding process is agnostic to the audio compression algorithm used. It could be any of a number of high-performance compression algorithms such as MPEG-1 layer III, MPEG-4 AAC or MPEG-4 HE-AAC, HDC, or even

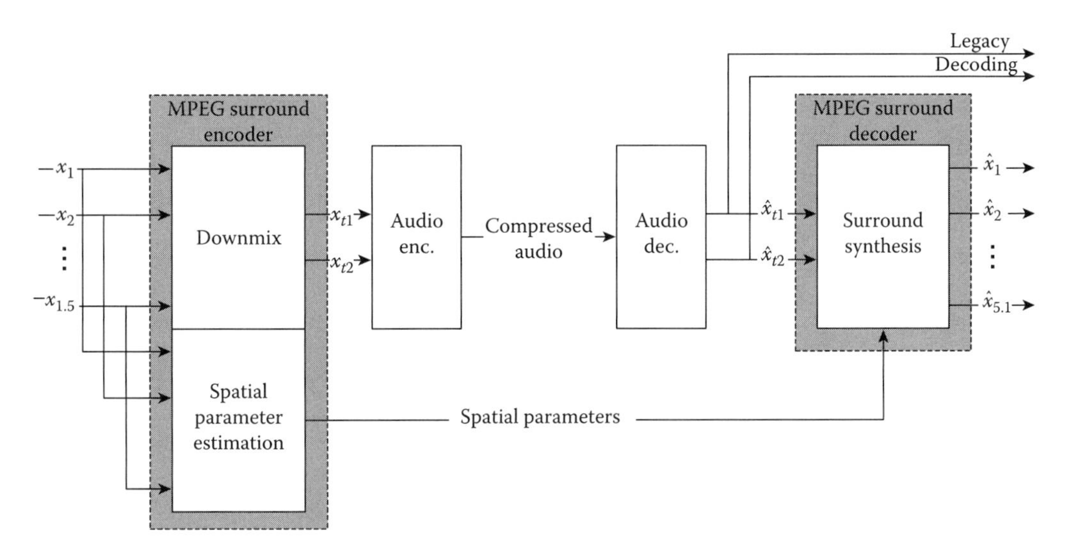

FIGURE 2.19 Principle of MPEG Surround.

PCM. The audio decoder reconstructs the downmix as $\hat{x}_{t1}$ and $\hat{x}_{t2}$. Legacy systems would stop at this point, otherwise this decoded signal plus the spatial parameter side information are sent to the MPEG Surround decoder which reconstructs the multichannel signal $\hat{x}_1$ to $\hat{x}_{5.1}$.

The heart of the encoding process is the extraction of spatial cues from the multichannel input signal that capture the salient perceptual aspects of the multichannel sound image (this is also referred to as "steering information" in that it indicates how to "steer" sounds to the various loudspeakers). Since the input to the MPEG Surround encoder is a 5.1-channel audio signal that is mixed for presentation via loudspeaker, sounds located at arbitrary points in space are perceived as phantom sound sources located between loudspeaker positions. Because of this, MPEG Surround computes parameters relating to the differences between each of the input audio channels. These cues are comprised of

- Channel level differences (CLD), representing the level differences between pairs of audio signals.
- Interchannel correlations (ICC), representing the coherence between pairs of audio signals.
- Channel prediction coefficients (CPC), able to predict an audio signal from others.
- Prediction error (or residual) signals, representing the error in the parametric modeling process relative to the original waveform.

A key feature of the MPEG Surround technique is that the transmitted downmix is an excellent stereo presentation of the multichannel signal. Hence legacy stereo decoders do not produce a "compromised" version of the signal relative to MPEG Surround decoders, but rather the very best version that stereo can render. This is vital, since stereo presentation will remain pervasive due to the number of applications for which listening is primarily via headphones (e.g., portable music players).

In the decoder, spatial cues are used to upmix the stereo transmitted signal to a 5.1-channel signal. This operation is done in the T/F domain as is shown in Figure 2.20. Here an analysis filterbank converts the input signal into two channels of T/F representation, where the upmix occurs (schematically illustrated in Figure 2.21), after which the synthesis filterbank converts the six channels of T/F data into a 5.1-channel audio signal. The filterbank must have frequency resolution comparable to that of the human auditory system, and must be oversampled so that processing in the T/F domain does not introduce aliasing distortion. MPEG Surround uses the same filterbank as is used in HE-AAC, but with division of the lowest frequency bands into additional subbands using an MDCT.

The upmix process applies mixing and decorrelation operations to regions of the stereo T/F signal to form the appropriate regions of the 5.1-channel T/F signals. This can be modeled as two matrix

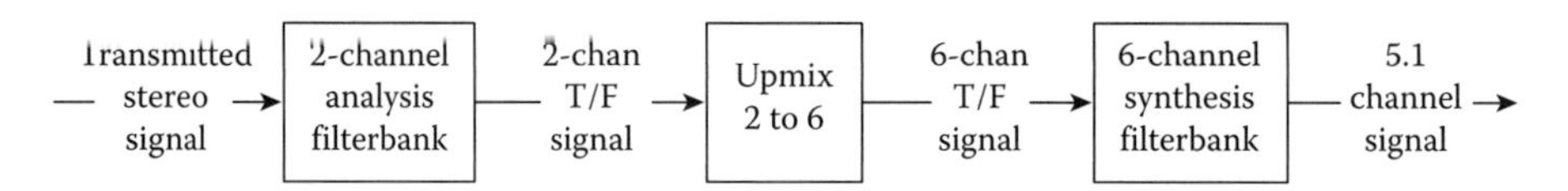

FIGURE 2.20 Block diagram of MPEG Surround spatial synthesis.

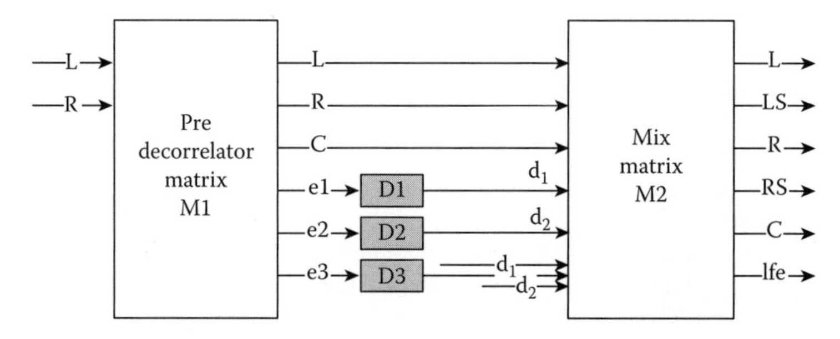

FIGURE 2.21 Block diagram of MPEG Surround upmix.

operations (M1 and M2) plus a set of decorrelation filters (D_1–D_3), all of which are time-varying. Note that in the figure the audio signals (L, R, C, LS, RS, LFE) are in the T/F domain.

In addition to encoding and decoding multichannel material via the use of side information, as just described, the MPEG Surround standard also includes operating modes that are similar to conventional matrix surround systems. It can encode a multichannel signal to a matrixed stereo signal, which it can decode back to the multichannel signal. This mode is MPEG Surround with "zero side information," and is fully interoperable with conventional matrixed surround systems. However, MPEG Surround has an additional feature in this mode: it can produce a matrix-compatible stereo signal and, if there is a side information channel, it can transmit information permitting the MPEG Surround decoder to "undo" the matrixing, apply the normal MPEG Surround multichannel decoding, and present a multichannel output that is superior to that of a matrix decoder upmix.

MPEG Surround has a unique architecture that, by its nature, can be a bridge between the distribution of stereo material and the distribution of multichannel material. The vast majority of audio decoding and playback systems are stereo, and MPEG Surround maintains compatibility with that legacy equipment. Furthermore, since MPEG Surround transmits an encoded stereo (or mono) signal plus a small amount of side information, it is compatible with most transmission channels that are currently designed to carry compressed stereo (or mono) signals. For multichannel applications requiring the lowest possible bit rate, MPEG Surround based on a single transmitted channel can result in a bit rate saving of more than 80% as compared to a discrete 5.1 multichannel transmission.

2.6.2 Subjective Quality

In 2007 MPEG conducted a test of the quality of MPEG Surround [62]. It was conducted at several test sites and comprised in total more than 40 listeners. Critical 5-channel test items were used that demonstrated an active sound stage and the MUSHRA test methodology was employed [58].

The test report showed that, when coding 5-channel signals:

- At 160 kbps, MPEG Surround in combination with the HE-AAC downmix coder achieved performance that was in the middle of the "Excellent" quality range on the MUSHA scale.
- At 64 kbps, MPEG Surround in combination with the HE-AAC downmix coder achieved performance that was at the high end of the "Good" quality range on MUSHA scale.

In addition, evaluations during the standardization process showed that MPEG Surround, when employing the residual coding tool, has the ability to scale from the high range of "Good" quality to the high range of "Excellent" quality as its bit rate increases from 160 to 320 kbps [61]. In particular:

- At 320 kbps, MPEG Surround in combination with the AAC downmix coder achieves performance that is comparable to that of AAC when coding the 5 discrete channels at 320 kbps.

2.7 Applications

MPEG/Audio compression technologies play an important role in consumer electronics, professional audio, telecommunications, broadcasting, and multimedia communication and storage. Many applications are based on delivering digital audio signals over digital broadcast and transmission systems such as satellite, terrestrial and cellular radio networks, local area networks (LAN) and wide area networks (WAN), and many also include storage of the digital audio content. A few examples of such applications are described in this section.

No MPEG audio application is as pervasive in consumer electronics as the portable music player, most often enabled by MPEG-1/Audio layer III (MP3). It would seem that every young person in every subway in every city in Asia, North America and Europe has a hardware-based MP3 portable player. These players employ low-cost, low-power hardware decoders that typically implement the MPEG-1/Audio layer III and MPEG-4/Audio AAC specifications. With the advent of remarkably inexpensive miniature hard-disk drives and solid-state memory, the level of compression is less of an issue as is the degree of interoperability, hence the success of standardized MPEG technology.

A number of DAB and digital video broadcast (DVB) services have recently been introduced. In Europe, the Eureka 147 project created a DAB system that addresses a number of technical issues in digital broadcasting [63–65]. The system uses MPEG-1/Audio layer II for coding the audio programs. The sampling rate is 48 kHz and the ancillary data field is used for program-associated data (PAD). The system has a significant bit rate overhead for error correction based on punctured convolutional codes in order to support source-adapted channel coding, i.e., an unequal error protection that is in accordance with the sensitivity to channel errors of individual bits or groups of bits in the coded representation [66]. Additionally, error concealment techniques are applied to provide a graceful degradation in case of severe errors. In the UnitedStates, systems for terrestrial and satellite based DAB have been deployed recently. Although these systems do not use an MPEG specification, they use the same technology (SBR) as is found in MPEG-4 HE-AAC. Both terrestrial and satellite systems make use of error correcting codes and simple redundant transmission channels as a means to overcome the difficulties of transmission in the FM terrestrial band (88–108 MHz) and the satellite band. This includes simulcasting analog and digital versions of the same audio program in the FM band. The European solution employs a very different system design since it is based on newly allocated channels [67].

Another DAB system, released in 2007, is DAB+. This is not backward-compatible with Eureka DAB, so that receivers that support the new standard will not be able to receive Eureka DAB broadcasts. However, DAB+ is approximately four times more efficient than Eureka DAB (which uses MPEG-1/Audio layer II) due to the use by DAB+ of the MPEG-4/Audio HE-AAC V2 profile codec. Using this MPEG technology, DAB+ can provide excellent quality with a bit rate as low as 64 kbps (see Section 2.5.3). Hence, DAB+ can deliver more choice to the consumer, in terms of broadcast material, than Eureka DAB. It is expected that reception will also be more robust in DAB+ as compared to DAB due to the use of Reed-Solomon error correction coding.

Advanced digital TV systems provide HDTV delivery to the public by terrestrial broadcasting and a variety of alternate channels (e.g., satellite, cable, and fiber) and hard media (e.g., Blu-ray disc) and offer full-motion high-resolution video and high-quality multichannel surround audio. The overall bit rate may be transmitted within the bandwidth of an analog VHF or UHF television channel. The U.S. Grand Alliance HDTV system and the European DVB system both make use of the MPEG-2 video compression system and of the MPEG-2 transport layer which uses a flexible ATM-like packet protocol with headers/descriptors for multiplexing audio and video bit streams in one stream with the necessary information to keep the streams synchronized when decoding. The systems differ in the way the audio signal is compressed: the Grand Alliance system uses the Dolby AC-3 TC technique [68–70], whereas the DVB system uses the MPEG-2/Audio algorithm.

Although multichannel audio finds a compelling application in HDTV, both in the cinema and at home, another potential application for multichannel digital audio is playback in the automobile. The automobile environment is quite suitable for enjoying multichannel music, in that many automobiles already have five or more loudspeakers and a subwoofer and, in contrast to a home environment, the automotive listener is in a known and fixed position relative to the loudspeakers. Although automobiles are a difficult environment to achieve proper spatial imaging due to a mismatched speaker and listener position, surround sound greatly improves this situation in that it provides a larger optimum listening area (i.e., "sweet spot").

This suggests that surround sound may be an excellent match to terrestrial digital radio broadcasting in that it provides, in addition to the existing benefits of delivering a digitally coded stereo signal to legacy receivers, an additional benefit to new receivers: highly discrete multichannel sound. MPEG Surround requires almost no additional resources to transmit the multichannel signal as compared to a stereo signal, and the low bit rate spatial side information could be transmitted using existing data fields in the transmission multiplex.

2.8 Conclusions

Low bit rate digital audio has application in many different fields, such as consumer electronics, professional audio processing, telecommunications, broadcasting and multimedia communication and storage. Perceptual coding, as frequency domain shaped noise, parametric spectral coding and parametric soundstage coding, has paved the way to high compression rates in audio coding. International standards ISO/MPEG-1/Audio layer II and III (MP3), MPEG-4 AAC and MPEG-4 HE-AAC have been very widely deployed in the marketplace. Software encoders and decoder and single chip implementations are available from numerous suppliers.

MPEG/Audio coders are controlled by psychoacoustic models which are not constrained by the standard. They will certainly be improved thus leading to an evolutionary improvement in coding efficiency.

Digital multichannel audio improves stereophonic images and will be of importance both for audio-only and multimedia applications. MPEG audio offers AAC, HE-AAC and MPEG Surround as coding schemes that serve different needs in presenting multichannel material.

Emerging activities of the ISO/MPEG expert group aim at proposals for coding of audio signals that employ both models of perception and models of the signal source. These new coding methods will offer higher compression rates over the entire range of audio from high fidelity audio coding to low-rate speech coding. Because the basic audio quality will be more important than compatibility with existing or upcoming standards, this activity will open the door for completely new solutions.

References

1. Bruekers, A. A. M. L. et al., Lossless coding for DVD audio, *101th Audio Engineering Society Convention*, Los Angeles, CA, Preprint 4358, 1996.
2. Jayant, N. S. and Noll, P., *Digital Coding of Waveforms: Principles and Applications to Speech and Video*, Prentice-Hall, Englewood Cliffs, NJ, 1984.
3. Spanias, A. S., Speech coding: A tutorial review, *Proc. IEEE*, 82(10), 1541–1582, Oct.1994.
4. Jayant, N. S., Johnston, J. D., and Shoham, Y., Coding of wideband speech, *Speech Commun.*, 11, 127–138, June 1992.
5. Gersho, A., Advances in speech and audio compression, *Proc. IEEE*, 82(6), 900–918, June 1994.
6. Noll, P., Wideband speech and audio coding, *IEEE Commun. Mag.*, 31(11), 34–44, Nov. 1993.
7. Noll, P., Digital audio coding for visual communications, *Proc. IEEE*, 83(6), 925–943, June 1995.
8. Brandenburg, K., Introduction to perceptual coding, in *Collected Papers on Digital Audio Bit-Rate Reduction*, N. Gilchrist and C. Grewin, Eds., pp. 23–30, Audio Engineering Society, New York, 1996.
9. ISO/IEC 11172-3:1993 Information technology—Coding of moving pictures and associated audio for digital storage media at up to about 1,5 Mbit/s—Part 3: Audio.
10. ISO/IEC 13818-3:1998, Information technology—Generic coding of moving pictures and associated audio information—Part 3: Audio.
11. ISO/IEC 13818-7:2006, Information technology—Generic coding of moving pictures and associated audio information—Part 7: Advanced audio coding (AAC).
12. ISO/IEC 14496-3:2005, Information technology—Coding of audio-visual objects—Part 3: Audio.
13. ISO/IEC 23003-1:2007, Information technology–MPEG audio technologies–Part 1: MPEG surround.
14. The MPEG home page: http://www.chiariglione.org/mpeg
15. Hathaway, G. T., A NICAM digital stereophonic encoder, in *Audiovisual Telecommunications*, Nigthingale, N. D. Ed., pp. 71–84, Chapman & Hall, New York, 1992.
16. Zwicker, E. and Feldtkeller, R., *Das Ohr als Nachrichtenempfänger*, S. Hirzel Verlag, Stuttgart, Germany, 1967.

17. Moore, B. C. J., Masking in the human auditory system, in *Collected Papers on Digital Audio Bit-Rate Reduction*, N. Gilchrist and C. Grewin, Eds., pp. 9–19, Audio Engineering Society, New York, 1996.
18. Jayant, N. S., Johnston, J. D., and Safranek, R., Signal compression based on models of human perception, *Proc. IEEE*, 81(10), 1385–1422, Oct. 1993.
19. Zelinski, R. and Noll, P., Adaptive transform coding of speech signals, *IEEE Trans. Acoust. Speech Signal Process*, ASSP-25(4), 299–309, Aug. 1977.
20. Noll, P., On predictive quantizing schemes, *Bell Syst. Tech. J.*, 57, 1499–1532, 1978.
21. Makhoul, J. and Berouti, M., Adaptive noise spectral shaping and entropy coding in predictive coding of speech. *IEEE Trans. Acoust. Speech Signal Process.*, 27(1), 63–73, Feb. 1979.
22. Esteban, D. and Galand, C., Application of quadrature mirror filters to split band voice coding schemes, *Proceedings of the International Conference on ICASSP'77*, pp. 191–195, Hartford, CT, May 1977.
23. Rothweiler, J. H., Polyphase quadrature filters, a new subband coding technique, *Proceedings of the International Conference on ICASSP'83*, pp. 1280–1283, Boston, MA, 1983.
24. Princen, J. and Bradley, A., Analysis/synthesis filterbank design based on time domain aliasing cancellation, *IEEE Trans. Acoust. Speech Signal Process.*, ASSP-34(5), 1153–1161, Oct. 1986.
25. Malvar, H. S., *Signal Processing with Lapped Transforms*, Artech House, Norwood, MA, 1992.
26. Hoogendorn, A., Digital compact cassette, *Proc. IEEE*, 82(10), 1479–1489, Oct. 1994.
27. Yeoh, F. S. and Xydeas, C. S., Split-band coding of speech signals using a transform technique, *Proc. ICC*, 3, 1183–1187, 1984.
28. Granzow, W., Noll, P., and Volmary, C., Frequency-domain coding of speech signals, (in German), *NTG-Fachbericht No. 94*, pp. 150–155, VDE-Verlag, Berlin, 1986.
29. Edler, B., Coding of audio signals with overlapping block transform and adaptive window functions, (in German), *Frequenz*, 43, 252–256, 1989.
30. Zelinski, R. and Noll, P., Adaptive Blockquantisierung von Sprachsignalen, *Technical Report No. 181*, Heinrich-Hertz-Institut für Nachrichtentechnik, Berlin, 1975.
31. van der Waal, R. G., Brandenburg, K., and Stoll, G., Current and future standardization of high-quality digital audio coding in MPEG, *Proceedings of the IEEE ASSP Workshop on Applications of Signal Processing to Audio and Acoustics*, pp. 43–46, New Paltz, NY, Oct. 1993.
32. Noll, P. and Pan, D., ISO/MPEG audio coding, *Int. J. High Speed Electron. Syst.*, 8(1), 69–118, 1997.
33. Brandenburg, K. and Stoll, G., The ISO/MPEG-audio codec: A generic standard for coding of high quality digital audio, *J. Audio Eng. Soc.*, 42(10), 780–792, Oct. 1994.
34. van de Kerkhof, L. M. and Cugnini, A. G., The ISO/MPEG audio coding standard, *Widescreen Review*, 1994.
35. Dehery, Y. F., Stoll, G., and Kerkhof, L. V. D., MUSICAM source coding for digital sound, *17th International Television Symposium*, Montreux, Switzerland, Record 612–617, June 1991.
36. Brandenburg, K., Herre, J., Johnston, J. D., Mahieux, Y., and Schroeder, E.F., ASPEC: Adaptive spectral perceptual entropy coding of high quality music signals, *90th Audio Engineering Society Convention*, Paris, Preprint 3011, 1991.
37. Musmann, H. G., The ISO audio coding standard, *Proceedings of the IEEE Globecom*, San Diego, CA, Dec. 1990.
38. van der Waal, R. G., Oomen, A. W. J., and Griffiths, F. A., Performance comparison of CD, noise-shaped CD and DCC, *Proceedings of the 96th Audio Engineering Society Convention*, Amsterdam, the Netherlands, Preprint 3845, 1994.
39. Herre, J., Brandenburg, K., and Lederer, D., Intensity stereo coding, *96th Audio Engineering Society Convention*, Amsterdam, the Netherlands, Preprint 3799, 1994.
40. Pan, D., A tutorial on MPEG/audio compression, *IEEE Trans. Multimed.*, 2(2), 60–74, 1995.
41. Brandenburg, K. et al., Variable data-rate recording on a PC using MPEG-audio layer III, *95th Audio Engineering Society Convention*, New York, 1993.

42. Sarginson, P. A., MPEG-2: Overview of the system layer, BBC Research and Development Report, BBC RD 1996/2, 1996.
43. Ryden, T., Grewin, C., and Bergman, S., The SR report on the MPEG audio subjective listening tests in Stockholm April/May 1991, ISO/IEC JTC1/SC29/WG 11: Doc.-No. MPEG 91/010, May 1991.
44. Fuchs, H., Report on the MPEG/audio subjective listening tests in Hannover, ISO/IEC JTC1/SC29/WG 11: Doc.-No. MPEG 91/331, Nov. 1991.
45. Stoll, G. et al., Extension of ISO/MPEG-audio layer II to multi-channel coding: The future standard for broadcasting, telecommunication, and multimedia application, *94th Audio Engineering Society Convention*, Berlin, Preprint 3550, 1993.
46. ITU-R Recommendation BS. 1116-1, Methods for the subjective assessment of small impairments in audio systems including multichannel sound systems, Geneva, Switzerland, 1997.
47. Grill, B. et al., Improved MPEG-2 audio multi-channel encoding, *96th Audio Engineering Society Convention*, Amsterdam, the Netherlands, Preprint 3865, 1994.
48. Bosi, M. et al., ISO/IEC MPEG-2 advanced audio coding, 101*st Audio Engineering Society Convention*, Los Angeles, CA, Preprint 4382, 1996.
49. Bosi, M. et al., ISO/IEC MPEG-2 advanced audio coding, *J. Audio Eng. Soc.*, 45(10), 789–814, Oct. 1997.
50. Johnston J. D. et al., NBC-audio-stereo and multichannel coding methods, *96th Audio Engineering Society Convention*, Los Angeles, CA, Preprint 4383, 1996.
51. ISO/IEC JTC1/SC29/WG11/N1419, Report on the formal subjective listening tests of MPEG-2 NBC multichannel audio coding.
52. ITU-R Document TG 10-2/3, Oct. 1991.
53. Meares, D., Watanabe, K., and Scheirer, E., Report on the MPEG-2 AAC stereo verification tests, ISO/IEC JTC1/SC29/WG11/N2006. Feb. 1998.
54. Dietz, M., Liljeryd, L., Kjörling, K., and Kunz, O., Spectral band replication, a novel approach in audio coding, *112th AES Convention Proceedings, Audio Engineering Society*, Munich, Germany, Preprint 5553, May 10–13, 2002.
55. Ehret, A., Dietz, M., and Kjorling, K., State-of-the-art audio coding for broadcasting and mobile applications, *114th AES Convention*, Amsterdam, the Netherlands, Preprint 5834, Mar. 2003.
56. Wolters, M. et al., A closer look into MPEG-4 high efficiency AAC, *115th AES Convention*, New York, Preprint 5871, Oct. 2003.
57. ISO/IEC JTC1/SC29/WG11/N6009, Report on the verification tests of MPEG-4 high efficiency AAC.
58. ITU-R Recommendation BS.1534-1, Method for the subjective assessment of intermediate quality level of coding systems: MUlti-Stimulus test with Hidden Reference and Anchor (MUSHRA), Geneva, Switzerland, 1998–2000. Available at http://ecs.itu.ch
59. Breebaart, J. et al., MPEG surround–the ISO/MPEG standard for efficient and compatible multichannel audio coding, *122nd AES Convention*, Vienna, AT, Preprint 7084, May 2007.
60. Breebaart, J. et al., Background, Concept, and architecture for the recent MPEG surround standard on multichannel audio compression, *J. Audio Eng. Soc.*, 55(5), 331–351, May 2007.
61. Breebaart, J. et al., A study of the MPEG surround quality versus bit-rate curve, *123rd Audio Engineering Society Convention*, New York, Preprint 7219, Oct. 2007.
62. ISO/IEC JTC1/SC29/WG11/N8851, Report on MPEG surround verification tests.
63. Lau, A. and Williams, W. F., Service planning for terrestrial digital audio broadcasting, EBU Technical Review, pp. 4–25, Geneva, Switzerland, 1992.
64. Plenge, G., DAB—A new sound broadcasting systems: status of the development—routes to its introduction, EBU Review, April 1991.
65. ETSI, European Telecommunication Standard, Draft prETS 300 401, Jan. 1994.

66. Weck, Ch., The error protection of DAB, *Audio Engineering Society Conference on DAB—The Future of Radio*, London, U.K., May 1995.
67. Jurgen, R. D., Broadcasting with digital audio, *IEEE Spectr.*, 33(3), 52–59, March 1996.
68. Todd, C. et al., AC-3: Flexible perceptual coding for audio transmission and storage, *96th Audio Engineering Society Convention*, Amsterdam, the Netherlands, Preprint 3796, 1994.
69. Hopkins, R., Choosing an American digital HDTV terrestrial broadcasting system, *Proc. IEEE*, 82(4), 554-563, 1994.
70. Basile, C. et al., The US HDTV standard: The grand alliance, *IEEE Spectr.*, 32(4), 36–45, Apr. 1995.

3

Dolby Digital Audio Coding Standards

Robert L. Andersen
Dolby Laboratories, Inc.

Grant A. Davidson
Dolby Laboratories, Inc.

3.1 Introduction

The evolution of digital audio delivery formats continues to bring profound changes to the consumer entertainment experience. Beginning with introduction of the compact disc (CD) in 1982, consumers have benefited from the analog to digital migration of music delivery formats. Today, the CD audio format remains a primary means for reproducing very high-quality audio. However, a variety of innovative and compelling new delivery channels for digital audio and video entertainment have become available. Many of these delivery channels are incapable of delivering stereo audio signals directly from CDs, not to mention multichannel audio programs with accompanying video. The CD format supports stereo audio playback with 16 bit per sample pulse code modulation (PCM) and a sample rate of 44.1 kHz. This results in a total bit-rate of 1.41 megabits per second (Mbps), which is prohibitively high for many existing and emerging entertainment applications. Digital audio compression is essential for reducing the CD audio bit-rate to an appropriate level for these new applications.

Perceptual audio codecs deployed in today's digital television (DTV) systems reduce the CD bit-rate by factors ranging from 7:1 to 14:1. Compression factors of 30:1 and higher are possible for state-of-the-art perceptual audio codecs combined with parametric coding. All these forms of data compression exert a cost, manifested as measurable distortion in the received audio signal. Distortion increases with the amount of compression applied, leading to a trade-off between decoded signal fidelity and transmission efficiency. Up to a point, coding distortion can be made imperceptible to human subjects. Perceptual audio codecs accomplish this by exploiting known limitations in human perception as well as statistical redundancies in the signal. Signal components that can be removed with no perceptible consequence are perceptually irrelevant. Classic techniques such as transform coding and linear prediction are employed to remove signal redundancy. In essence, the objective of data compression is to minimize bit-rate for a given level of perceived distortion, or alternatively, to minimize perceivable distortion at a specified bit-rate. Hence, the single most important descriptor of a codec is the subjective quality delivered and the manner in which it diminishes with decreasing bit-rate. A comprehensive tutorial review of perceptual audio coding principles and algorithms is presented in Ref. [1].

Digital media broadcasting is one of the most widespread applications for audio compression. Current applications in this area include television and radio broadcasting over terrestrial, satellite, and cable delivery channels. DTV has been rolled out on a wide scale in nearly all major global markets. In the United States, audio and video compression were both necessary to meet the Federal Communication Commission (FCC) requirement that one high-definition television (HDTV) channel fit within the 6 MHz spectrum allocation of a single NTSC (analog) channel. Outside North America, digital video broadcasting (DVB) is the most widely adopted suite of standards for digital television. DVB systems distribute multimedia using a variety of means, including satellite (DVB-S), cable (DVB-C), and terrestrial television (DVB-T). In Europe, the digital audio broadcasting (DAB) standard, also known as Eureka 147, is used for radio broadcasting in several countries. An enhanced version called DAB+, employing the high-efficiency AAC codec [2] to enable higher audio quality at a significantly lower bit-rate, is more broadly adopted. In Japan and Brazil, the integrated services digital broadcasting (ISDB) format is employed to deliver DTV and digital radio services.

In recent years, new multimedia delivery services have emerged that expand the availability of broadcast digital media, in particular for consumers on-the-go. Digital video broadcasting-handheld (DVB-H) has been endorsed by the European Union as the preferred technology for terrestrial DTV and radio broadcasting to mobile devices. Primary competitors to DVB-H are the two digital multimedia broadcasting standards. They operate over both terrestrial (T-DMB) and satellite (S-DMB) transmission links. In the United States, a similar standard is in development (ATSC-M/H), allowing Advanced Television Systems Committee (ATSC) DTV signals to be broadcast to mobile devices.

In addition to mobile broadcasting, digital media conveyance over packet-based broadband and narrowband computer networks is expanding rapidly. As consumer penetration of broadband Internet access continues to expand, residential Internet protocol television (IPTV) is expected to follow. IPTV offers the compelling advantage that digital media can be delivered on-demand, and customized for each consumer. Along with the increase in proliferation of mobile devices with music and video playback capability, consumers have embraced new means for distributing their digital media over wireless local area networks and personal area networks such as Bluetooth.

Driven by these market developments, the demand for more efficient audio compression technologies continues. Owners of digital media content are expanding availability of their product to the consumer, and this expansion leads to an increase in the number of companies competing as content providers. Providers strive to offer more content choices to differentiate themselves, which in turn increases demand on available transmission and storage channels. In many cases, this demand is most easily met by deploying more efficient audio and video compression technologies. Furthermore, as the broadcast market shifts from MPEG-2 video compression to the more efficient MPEG-4 AVC/ITU-T H.264 standard, new audio compression technologies are needed to provide comparable efficiency gains. Mobile, on-demand, and viewer-customized delivery services place particularly high demands upon broadcast bandwidth, motivating the need for very low bit-rate audio compression technologies.

Dolby Laboratories' first audio codec developed for digital media delivery to consumer electronic products is Dolby AC-3 (also known as "Dolby Digital") [3]. AC-3 is a perceptual audio codec capable of delivering one to 5.1 discrete audio channels for simultaneous presentation to a listener, either with or without accompanying video. AC-3 was the first multichannel surround sound codec offered to the broadcast market, and also the first audio codec to incorporate extensive metadata features for enhancing the listener experience. Such features include dialog normalization, dynamic range control, and decoder channel downmixing. The latter feature enables multichannel surround audio programs to be reproduced on home 5.1 channel systems, as well as on mono and stereo playback devices.

AC-3 was first standardized in May 1995 by the ATSC as an outgrowth of a competitive selection process. In December 1996, the U.S. FCC adopted the ATSC standard for DTV [4] which is consistent with a consensus agreement developed by a broad cross section of parties, including the broadcasting and computer industries. Since then, several backwards-compatible metadata feature enhancements have been added, such as karaoke support and a backwards-compatible alternate bit stream syntax defining new metadata parameters including extended downmixing options. In addition, AC-3 has been included in ITU-R recommendations and standardized by ETSI.

In the years following its introduction, AC-3 gained widespread adoption both in North America and South Korea as the mandated audio format for DTV, and worldwide as a mandated audio format for DVD and Blu-ray Disc. AC-3 has also been widely adopted in markets where it is only an optional audio format, such as DVB in Europe. This may be attributed in part to the large installed base of AC-3 decoders in home A/V systems. As of the time of this writing, more than two billion licensed AC-3 decoders have been deployed worldwide.

While AC-3 has proven to be an effective and versatile audio codec, in 2002 Dolby Laboratories realized that this format could no longer meet the most demanding performance requirements of emerging market applications. A nonbackwards compatible audio codec called Enhanced AC-3 (E-AC-3, also known as "Dolby Digital Plus") was developed to address this market need [5]. Since new E-AC-3 codecs were likely to coexist with AC-3 within the same digital media channel, optimized tandem coding performance was deemed to be a high design priority. In addition, E-AC-3 supports speaker configurations up to 15.1 channels, allowing five front speakers, four rear speakers, and height speakers. As presented later in this chapter, E-AC-3 includes several new features to improve coding efficiency at very low bit-rates, and supports shorter frame lengths for interactive audio applications such as video game sound. E-AC-3 has been standardized by ATSC [3] and ETSI, and is an optional audio format, alongside AC-3, in the DVB standard. E-AC-3 decoders are also recommended for primary audio applications (movie soundtrack delivery), and mandatory for secondary audio applications (audio intended to be mixed with the primary audio), in Blu-ray Disc.

The task of formally evaluating AC-3 and E-AC-3 subjective audio quality, while intensive and time consuming, is critical so that codec adopters can make informed decisions about their audio services. AC-3 was designed to meet the strict audio quality requirement for broadcast applications established in ITU-R Recommendation BS.1115 [6]. This requirement, called "broadcast quality," implies that impairments on carefully selected "worst-case" audio sequences for a particular codec are rated on average by listeners as "imperceptible" or "perceptible but not annoying." The test methodology for obtaining these subjective quality ratings is defined in ITU-R Recommendation BS.1116-1 [7], and is the most sensitive method available.

The subjective quality of AC-3 has been evaluated in two formal BS.1116 listening tests [8,9]. In the most recent test in 2001, AC-3 was shown to satisfy the requirements for ITU-R broadcast quality for stereo signals at a bit-rate of 192 kilobits per second (kbps). Similarly high performance is obtained with 5.1 channel signals at a bit-rate of 448 kbps.

The subjective quality of E-AC-3 has also been evaluated, but predominately at lower bit-rates intended to deliver digital audio with intermediate quality. In this case, the most appropriate test methodology is ITU-R Recommendation BS.1534-1 [10]. The subjective listening test results for E-AC-3 are fully presented in Refs. [11,12].

3.2 AC-3 Audio Coding

Dolby AC-3 is a perceptual audio codec capable of encoding a range of audio channel formats into a bit stream ranging from 32 to 640 kbps. The channel configurations supported by AC-3 meet the recommendations for multichannel sound reproduction contained in ITU-R BS.775-1 [13]. AC-3 is capable of encoding a range of discrete, 20 kHz bandwidth audio program formats, including one to three front channels and zero to two rear channels. An optional low-frequency effects (LFE or subwoofer) channel is also supported. The most common audio program formats are stereo (two front channels) and 5.1 channel surround sound (three front channels, two surround channels, plus the LFE channel, denoted ".1"). AC-3 is typically utilized at 160–192 kbps for stereo programs and 384–448 kbps for 5.1.

There are a diverse set of requirements for a codec intended for widespread application. While the most critical members of the audience may be anticipated to have complete six-speaker multichannel reproduction systems, most of the audience may be listening in mono or stereo, and still others will have three front channels only. Some of the audience may have matrix-based (e.g., Dolby Surround) multichannel reproduction equipment without discrete channel inputs, thus requiring a dual-channel matrix-encoded output from the AC-3 decoder. Most of the audience welcomes a restricted dynamic range reproduction, while a few in the audience will wish to experience the full dynamic range of the original signal. The visually and hearing impaired wish to be served. All of these and other diverse needs were considered early in the AC-3 design process. Solutions to these requirements have been incorporated from the beginning, leading to a self-contained and efficient system.

As an example, one of the more important listener features built-in to AC-3 is dynamic range compression. This feature allows the program provider to implement subjectively pleasing dynamic range reduction for most of the intended audience, while allowing individual members of the audience the option to experience more (or all) of the original dynamic range. At the discretion of the program originator, the encoder computes dynamic range control values and places them into the AC-3 bit stream. The compression is actually applied in the decoder, so the encoded audio has full dynamic range. It is permissible (under listener control) for the decoder to fully or partially apply the dynamic range control values. In this case, some of the dynamic range will be limited. It is also permissible (again under listener control) for the decoder to ignore the control words, and hence reproduce full-range audio. By default, AC-3 decoders will apply the compression intended by the program provider.

Other user features include decoder downmixing to fewer channels than were present in the bit stream, dialog normalization, and Dolby Surround compatibility. A complete description of these features and the rest of the ATSC Digital Audio Compression Standard is contained in Ref. [3].

AC-3 achieves high coding gain (the ratio of the encoder input bit-rate to the encoder output bit-rate) by quantizing a frequency domain representation of the audio signal. A block diagram of this process is shown in Figure 3.1. The first step in the encoding process is to transform the representation of audio from a sequence of PCM signal sample blocks into a sequence of frequency coefficient blocks. This is done in the analysis filter bank as follows. Signal sample blocks of length 512 are multiplied by a set of window coefficients and then transformed into the frequency domain. Each sample block is overlapped by 256 samples with the two adjoining blocks. Due to the overlap, every PCM input sample is represented in two adjacent transformed blocks. The frequency domain representation includes decimation by an extra factor of two so that each frequency block contains only 256 coefficients.

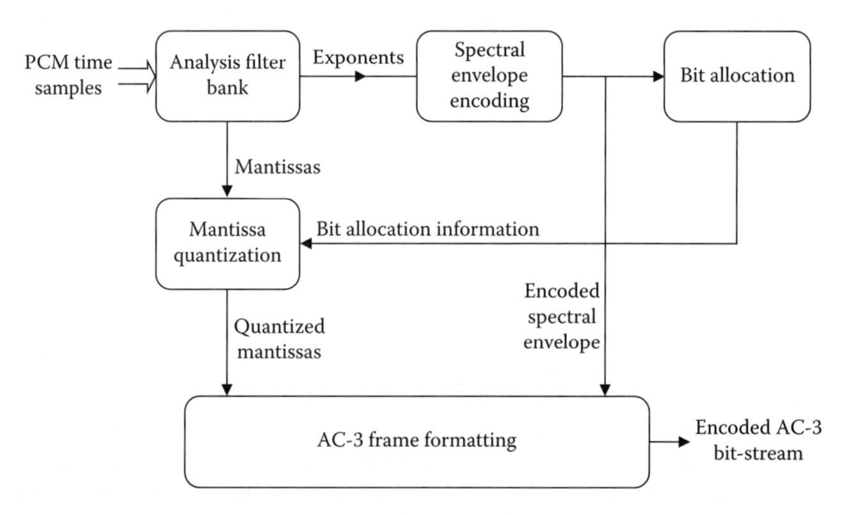

FIGURE 3.1 The AC-3 encoder.

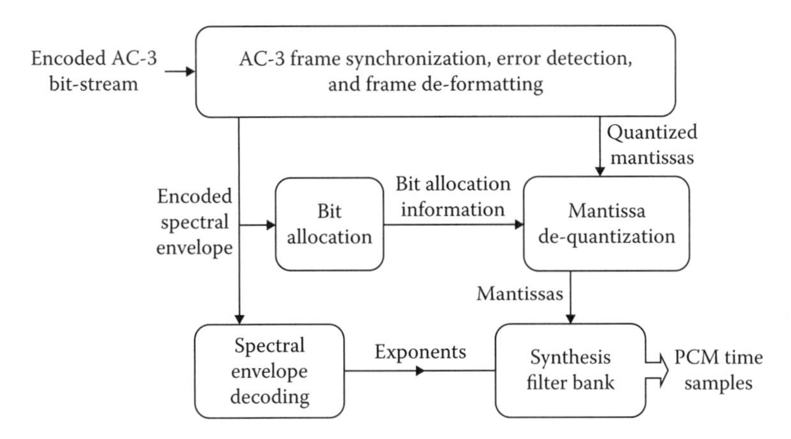

FIGURE 3.2 The AC-3 decoder.

The individual frequency coefficients are then converted into a binary exponential notation as a binary exponent and a mantissa. The set of exponents is encoded into a coarse representation of the signal spectrum which is referred to as the spectral envelope. This spectral envelope is processed by a bit allocation routine to calculate the amplitude resolution required for encoding each individual mantissa. The spectral envelope and the quantized mantissas for six audio blocks (1536 audio samples) are formatted into one AC-3 synchronization frame. The AC-3 bit stream is a sequence of consecutive AC-3 frames.

A block diagram of the decoding process, which is essentially a mirror-inverse of the encoding process, is shown in Figure 3.2. The decoder must synchronize to the encoded bit stream, check for errors, and deformat the various types of data such as the encoded spectral envelope and the quantized mantissas. The spectral envelope is decoded to reproduce the exponents. The bit allocation routine is run and the results used to unpack and dequantize the mantissas. The exponents and mantissas are transformed back into the time domain to produce decoded PCM time samples. Figures 3.1 and 3.2 present a somewhat simplified, high-level view of an AC-3 encoder and decoder.

Table 3.1 presents the different channel formats which are accommodated by AC-3. The 3 bit control variable acmod is embedded in the bit stream to convey the encoder channel configuration to the decoder. If acmod is "000," then two completely independent program channels (dual mono) are encoded into the bit stream (referenced as Ch1, Ch2). The traditional mono and stereo formats are denoted when acmod equals "001" and "010," respectively. If acmod is greater than "011," the bit stream format includes one or more surround channels. The optional LFE channel is enabled/disabled by a separate control bit called lfeon.

TABLE 3.1 AC-3 Audio Coding Modes

acmod	Audio Coding Mode	Number of Full Bandwidth Channels	Channel Array Ordering
"000"	1+1	2	Ch1, Ch2
"001"	1/0	1	C
"010"	2/0	2	L, R
"011"	3/0	3	L, C, R
"100"	2/1	3	L, R, S
"101"	3/1	4	L, C, R, S
"110"	2/2	4	L, R, SL, SR
"111"	3/2	5	L, C, R, SL, SR

TABLE 3.2 AC-3 Audio Coding Bit-Rates

frmsizecod	Nominal Bit-Rate (kbps)	frmsizecod	Nominal Bit-Rate (kbps)	frmsizecod	Nominal Bit-Rate (kbps)
0	32	14	112	28	384
2	40	16	128	30	448
4	48	18	160	32	512
6	56	20	192	34	576
8	64	22	224	36	640
10	80	24	256		
12	96	26	320		

Table 3.2 presents the different bit-rates which are accommodated by AC-3. In principle it is possible to use the bit-rates in Table 3.2 with any of the channel formats from Table 3.1. However, in high-quality applications employing the best known encoder, the typical bit-rate is 192 kbps for two channels, and 384 kbps for 5.1 channels.

3.2.1 Bit Stream Syntax

An AC-3 serial coded audio bit stream is composed of a contiguous sequence of synchronization frames. A synchronization frame is defined as the minimum-length bit stream unit which can be decoded independently of any other bit stream information. Each synchronization frame represents a time interval corresponding to 1536 samples of digital audio (for example, 32 ms at a sampling rate of 48 kHz). All of the synchronization codes, preamble, coded audio, error correction, and auxiliary information associated with this time interval is completely contained within the boundaries of one synchronization frame.

Figure 3.3 shows the various bit stream elements within each synchronization frame. The five different components are: SI (synchronization information), BSI (bit stream information), AB (audio block), AUX (auxiliary data field), and CRC (cyclic redundancy code). The SI and CRC fields are of fixed-length, while the length of the other four depends upon programming parameters such as the number of encoded audio channels, the audio coding mode, and the number of optionally conveyed listener features. The length of the AUX field is adjusted by the encoder such that the CRC element falls on the last 16 bit word of the frame. A summary of the bit stream elements and their purpose is presented in Table 3.3.

The number of bits in a synchronization frame (frame length) is a function of sampling rate and total bit-rate. In a conventional encoding scenario, these two parameters are fixed, resulting in synchronization frames of constant length. However, AC-3 also supports variable-rate audio applications, as will be discussed shortly.

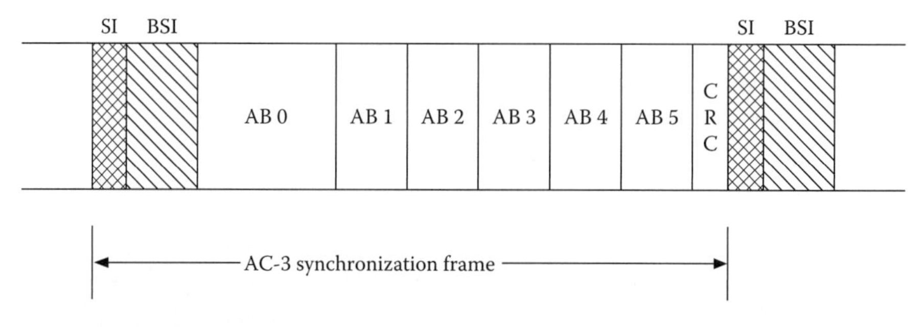

FIGURE 3.3 AC-3 synchronization frame.

TABLE 3.3 AC-3 Bit Stream Elements

Bit Stream Element	Purpose	Length (Bits)
SI	Synchronization information—Header at the beginning of each frame containing information needed to acquire and maintain bit stream synchronization	40
BSI	Bit stream information—Preamble following SI containing parameters describing the coded audio service, e.g., number of input channels (acmod), dynamic compression control word (dynrng), and program time codes (timecod1, timecod2)	Variable
AB	Audio block—Coded information pertaining to 256 quantized samples of audio from all input channels. There are six audio blocks per AC-3 synchronization frame	Variable
Aux	Auxiliary data field—Block used to convey additional information not already defined in the AC-3 bit stream syntax	Variable
CRC	Frame error detection field—Error check field containing a CRC word for error detection. An additional CRC word is located in the SI header, the use of which is optional	17

Each audio block contains coded information for 256 samples from each input channel. Within one synchronization frame, the AC-3 encoder can change the relative size of the six audio blocks depending on audio signal bit demand. This feature is particularly useful when the audio signal is nonstationary over the 1536-sample synchronization frame. Audio blocks containing signals with a high bit demand can be weighted more heavily than others in the distribution of the available bits (bit pool) for one frame. This feature provides one mechanism for local variation of bit-rate while keeping the overall bit-rate fixed.

In applications such as digital audio storage, an improvement in audio quality can often be achieved by varying the bit-rate on a long-term basis (more than one synchronization frame). This can also be implemented in AC-3 by adjusting the bit-rate of different synchronization frames on a signal-dependent basis. In regions where the audio signal is less bit-demanding (for example, during quiet passages), the frame bit-rate is reduced. As the audio signal becomes more demanding, the frame bit-rate is increased so that coding distortion remains inaudible. Frame-to-frame bit-rate changes selected by the encoder are automatically tracked by the decoder.

3.2.2 Analysis/Synthesis Filterbank

The design of an analysis/synthesis filterbank is fundamental to any frequency-domain audio codec. The frequency and time resolution of the filterbank play critical roles in determining the achievable coding gain. Of significant importance as well are the properties of critical sampling and overlap-add reconstruction. This section discusses these properties in the context of the AC-3 multichannel audio codec.

Of the many considerations involved in filterbank design, two of the most important for audio coding are the window shape and the impulse response length. The window shape affects the ability to resolve frequency components which are in close proximity, and the impulse response length affects the ability to resolve signal events which are short in time duration. For transform codecs, the impulse response length is determined by the transform block length.

A long transform length is most suitable for input signals whose spectrum remains stationary, or varies only slowly with time. A long transform length provides greater frequency resolution, and hence improved coding performance for such signals. On the other hand, a shorter transform length, possessing greater time resolution, is more effective for coding signals which change rapidly in time. The best of both cases can be obtained by dynamically adjusting the frequency/time resolution of the transform depending upon spectral and temporal characteristics of the signal being coded. This behavior is very similar to that known to occur in human hearing, and is embodied in AC-3.

The transform selected for use in AC-3 is based upon a 512-point modified discrete cosine transform (MDCT) [14]. In the encoder, the input PCM block for each successive transform is constructed by taking 256 samples from the last half of the previous audio block and concatenating 256 new samples from the current block. Each PCM block is therefore overlapped by 50% with its two neighbors. In the decoder, each inverse transform produces 512 new PCM samples, which are subsequently windowed, 50% overlapped, and added together with the previous block. This approach has the desirable property of crossfade reconstruction, which reduces waveform discontinuities (and audible distortion) at block boundaries.

3.2.2.1 Window Design

To achieve perfect-reconstruction with a unity-gain MDCT transform filterbank, the shape of the analysis and synthesis windows must satisfy two design constraints. First of all, the analysis/synthesis windows for two overlapping transform blocks must be related by

$$a_i(n + N/2)^* s_i(n + N/2) + a_{i+1}(n)^* s_{i+1}(n) = 1, \quad n = 0, \ldots, N/2 - 1 \tag{3.1}$$

where
- $a(n)$ is the analysis window
- $s(n)$ is the synthesis window
- n is the sample number
- N is the transform block length
- i is the transform block index

This is the well-known condition that the analysis/synthesis windows must add so that the result is flat [15]. The second design constraint is

$$a_i(N/2 - n - 1)^* s_i(n) - a_i(n)^* s_i(N/2 - n - 1) = 0, \quad n = 0, \ldots, N/2 - 1 \tag{3.2}$$

This constraint must be satisfied so that the time-domain alias distortion introduced by the forward transform is completely canceled during synthesis.

To design the window used in AC-3, a convolution technique was employed which guarantees that the resultant window satisfies Equation 3.1. Equation 3.2 is then satisfied by choosing the analysis and synthesis windows to be equal. The procedure consists of convolving an appropriately chosen symmetric kernel window with a rectangular window. The window obtained by taking the square root of the result satisfies Equation 3.1. Trade-offs between the width of the window main-lobe and the ultimate rejection can be made simply by choosing different kernel windows. This method provides a means for transforming a kernel window having desirable spectral analysis properties (such as in Ref. [16]) into one satisfying the MDCT window design constraints.

The window generation technique is based upon the following equation:

$$a_i(n) = s_i(n) = \sqrt{\frac{\sum_{j=L}^{M} [w(j) r(n-j)]}{\sum_{j=0}^{K} [w(j)]}} \quad \text{for } n = 0, \ldots, N - 1 \tag{3.3}$$

where

$$L = \begin{cases} 0 & 0 \le n < N - K \\ n - N + K + 1 & N - K \le n < N \end{cases}$$

$$M = \begin{cases} n & 0 \le n < K \\ K & K \le n < N \end{cases}$$

where

$w(n)$ is the kernel window of length $K+1$

$r(n)$ is a rectangular window of length $N-K$

N is the transform sample block length

K is the width of the (nonflat) transition region in the resulting window (note that K must satisfy $0 \leq K \leq N/2$)

The rectangular window is defined as

$$r(n) = \begin{cases} 0 & 0 \leq n < (N/2-K)/2 \quad \text{and} \quad (3N/2-K)/2 \leq n < N-K \\ 1 & (N/2-K)/2 \leq n < (3N/2-K)/2 \end{cases} \tag{3.4}$$

The rectangular window is defined to contain $(N/2-K)/2$ zeros, followed by $N/2$ unity samples, followed by another $(N/2-K)/2$ zeros. The AC-3 window uses $K=N/2$, implying the transition region length is one-half the total window length.

The Kaiser–Bessel window is used as the kernel in designing the AC-3 analysis/synthesis windows because of its near-optimal transition band slope and good ultimate rejection characteristic. A scalar parameter α in the Kaiser–Bessel window definition can be adjusted to vary this ratio. The AC-3 window uses $\alpha = 5$.

The selection of the Kaiser–Bessel window function and alpha factor used for the AC-3 algorithm is determined by considering the shape of masking template curves. A useful criterion is to use a filter response which is at or below the worst-case combination of all masking templates [17]. Such a filter response is advantageous in reducing the number of bits required for a given level of audio quality. When the filter response is at or below the worst-case combination of all masking templates, the number of bits assigned to transform coefficients adjacent to each tonal component is reduced.

3.2.2.2 Transform Equations

The transform employed in AC-3 is an extension of the oddly stacked TDAC (OTDAC) filter bank reported by Princen and Bradley [14]. The extension involves the capability to switch transform block length from $N=512$ to $N=256$ for audio signals with rapid amplitude changes. As originally formulated by Princen, the filter bank operates with a time-invariant block-length, and therefore has constant time/frequency resolution. An adaptive time/frequency resolution transform can be implemented by changing the time offset of the transform basis functions during short blocks. The time offset is selected to preserve critical sampling and perfect reconstruction before, during, and following transform length changes.

Prior to transforming the audio signal from time to frequency domain, the encoder performs an analysis of the spectral and/or temporal nature of the input signal and selects the appropriate block length. A one bit code per channel per audio block is embedded in the bit stream which conveys length information (blksw = 0 or 1 for 512 or 256 samples, respectively). The decoder uses this information to deformat the bit stream, reconstruct the mantissa data, and apply the appropriate inverse transform equations.

Transforming a long block (512 samples) produces 256 unique transform coefficients. Short blocks are constructed starting with 512 windowed audio samples and splitting them into two abutting subblocks of length 256. Each subblock is transformed independently, producing 128 unique nonzero transform coefficients. Hence the total number of transform coefficients produced in the short-block mode is identical to that produced in long-block mode, but with doubly improved temporal resolution. Transform coefficients from the two subblocks are interleaved together on a coefficient-by-coefficient basis. This block is quantized and transmitted identically to a single long block.

A similar, mirror image procedure is applied in the decoder. Quantized transform coefficients for the two short transforms arrive in the decoder interleaved in frequency. The decoder processes the interleaved sequences identically to long-block sequences, except during the inverse transformation as described below.

A definition of the AC-3 forward transform equation for long and short blocks is

$$X(k) = 1/\mathrm{N} \sum_{n=0}^{N-1} x(n) \cos((2\pi/N)(k + 1/2)(n + n_0)), \quad k = 0, 1, \ldots, N - 1 \tag{3.5}$$

where

n is the sample index
k is the frequency index
$x(n)$ is the windowed sequence of N audio samples
$X(k)$ is the resulting sequence of transform coefficients

The corresponding inverse transform equation for long and short blocks is

$$y(n) = \sum_{k=0}^{N-1} X(k) \cos(2\pi/N(k + 1/2)(n + n_0)), \quad n = 0, 1, \ldots, N - 1 \tag{3.6}$$

Parameter n_0 represents a time offset of the modulator basis vectors used in the transform kernel. For long blocks, and for the second of each short block pair, $n_0 = 257/2$. For the first short block, $n_0 = 1/2$.

When $x(n)$ in Equation 3.5 is real, $X(k)$ is odd-symmetric for the MDCT. Therefore, only $N/2$ unique nonzero transform coefficients are generated for each new block of N samples. Accordingly, some information is lost during the transform, which ultimately leads to an alias component in $y(n)$. However, with an appropriate choice of n_0, and in the absence of transform coefficient quantization, the aliasing is completely canceled during the window/overlap/add procedure following the inverse transform. Hence, the AC-3 filterbank has the properties of critical sampling and perfect reconstruction. A fundamental advantage of this approach is that 50% frame overlap is achieved without increasing the required bit-rate. Any nonzero overlap used with conventional transforms (such as the discrete Fourier transform [DFT] or standard DCT) precludes critical sampling, generally resulting in a higher bit-rate for the same level of subjective quality.

Several memory and computation-efficient techniques are available for implementing the AC-3 forward and inverse transforms (for example, see [18]). The most efficient ones can be derived by rewriting Equations 3.5 and 3.6 in the form of an N-point DFT and inverse DFT (IDFT), respectively, combined with two complex vector multiplications. The DFT and IDFT can be efficiently computed using an fast Fourier transform (FFT) and inverse FFT (IFFT), respectively. Two properties further reduce the fast transform length. Firstly, the input signal is real, and secondly, the N-length sequence $y(n)$ contains only $N/2$ unique samples. When these two properties are combined, the result is an $N/4$-point complex FFT or IFFT. The AC-3 decoder filter bank computation rate is about 13 multiply accumulate operations per sample per channel, including the window/overlap/add. This computation rate remains virtually unchanged during block length changes.

3.2.3 Spectral Envelope

The most basic form of audio information conveyed by an AC-3 bit stream consists of quantized frequency coefficients. The coefficients are delivered in floating-point form, whereby each consists of an exponent and a mantissa. The exponents from one audio block provide an estimate of the overall spectral content as a function of frequency. This representation is often termed a spectral envelope. This section describes spectral envelope coding strategies in AC-3, and explores an important relationship between exponent coding and mantissa bit allocation.

Due to the inherent variety of audio spectra within one frame, the AC-3 spectral envelope coding scheme contains significant degrees of freedom. In essence, the six spectral envelopes contained in one frame represent a two-dimensional signal, varying in time (block index) and frequency. AC-3 spectral envelope coding provides for variable coarseness of representation in both dimensions. In the frequency domain, either one, two, or four mantissas can be shared by one floating-point exponent. In the time dimension, any two or more consecutive audio blocks from one frame can share a common set of exponents.

The concepts of spectral envelope coding and bit allocation are closely linked in AC-3. More specifically, the effectiveness with which mantissa bits are utilized can depend greatly upon the encoder's choice of spectral envelope coding. To see this, note that the dominant contributors to the total bit-rate for a frame are the audio exponents and mantissas. Sharing exponents in either the time or frequency dimension, or both, reduces the total cost of exponent transmission for one frame. More liberal use of exponent sharing therefore frees more bits for mantissa quantization. Conversely, retransmitting exponents increases the total cost of exponent transmission for one frame relative to mantissa quantization. Furthermore, the block positions at which exponents are retransmitted can significantly alter the effectiveness of mantissa bit assignments amongst the various audio blocks. As will be seen in Section 3.6, bit assignments are derived in part from the coded spectral envelope. In summary, the encoder decisions regarding when to use frequency or time exponent sharing, and when to retransmit exponents depend upon signal conditions. Collectively, these decisions are called exponent strategy.

For short-term stationary signals, the signal spectrum remains substantially invariant from block-to-block. In this case, the AC-3 encoder transmits exponents once in audio block 0, and then typically reuses them for blocks 1–5. The resulting bit allocation would generally be identical for all six blocks, which is appropriate for these signal conditions.

For short-term nonstationary signals, the signal spectrum changes significantly from block-to-block. In this case, the AC-3 encoder transmits exponents in block 0 and typically in one or more other blocks as well. In this case, exponent retransmission produces a time trajectory of coded spectral envelopes which better matches dynamics of the original signal. Ultimately, this results in a quality improvement if the cost of exponent retransmission is less than the benefit of redistributing mantissa bits amongst blocks.

Exponent strategy decisions can be based, for example, on a cost–benefit analysis for each frame. The objective of such an analysis would be to minimize a cost–benefit ratio by considering encoding parameters such as total available bit-rate, audibility of quantization noise (noise-to-mask ratio), exponent coding mode for each audio block (reuse, D15, D25, or D45), channel coupling on/off, and reconstructed audio bandwidth.

The block(s) at which bit assignment updates occur is governed by several different parameters, but primarily by the exponent strategy fields. AC-3 bit streams contain coded exponents for up to five independent channels, and for the coupling and low frequency effects channels (when enabled). The respective exponent strategy fields are called chexpstr[ch], cplexpstr, and lfeexpstr. Bit allocation updates are triggered if the state of any one or more strategy flags is D15, D25, or D45; however, updates can be triggered in between shared exponent block boundaries as well.

Exponents are 5 bit values which indicate the number of leading zeros in the binary representation of a frequency coefficient. For the D15 exponent strategy, the unsigned integer exponent $e(i)$ represents a scale factor for the ith mantissa, equal to $2^{-e(i)}$. Frequency coefficients are normalized in the encoder by multiplying by $2^{e(i)}$, and denormalized in the decoder by multiplying by $2^{-e(i)}$. Exponent values are allowed to range from 0 (for the largest value coefficients with no leading zeroes) to 24. Exponents for coefficients which have more than 24 leading zeroes are fixed at 24, and the corresponding mantissas are allowed to have leading zeros. Exponents require five bits in order to represent all allowed values.

AC-3 exponent transmission employs differential coding, in which the exponents for a channel are differentially coded across frequency. The first exponent of a full bandwidth or LFE channel is always sent as a 4 bit absolute value, ranging from 0–15. The value indicates the number of leading zeros of the first (DC term) transform coefficient. Successive exponents (ascending in frequency) are sent as differential values which must be added to the prior exponent value in order to form the next absolute value.

TABLE 3.4 Exponent Bit-Rate for Different Exponent Strategies

	Share Interval (Number of Audio Blocks)					
Exponent Strategy	1	2	3	4	5	6
D15	2.33	1.17	0.78	0.58	0.47	0.39
D25	1.17	0.58	0.39	0.29	0.23	0.19
D45	0.58	0.29	0.19	0.15	0.12	0.10

Note: Bits per frequency coefficient.

The differential exponents are combined into groups in the audio block. The grouping is done by one of three methods, D15, D25, or D45. The number of grouped differential exponents placed in the audio block for a particular channel depends on the exponent strategy and on the frequency bandwidth information for that channel. The number of exponents in each group depends only on the exponent strategy.

Exponent strategy information for every channel is included in every AC-3 audio block. Information is never shared across frames, so block 0 will always contain a strategy indication for each channel.

The three exponent strategies provide a trade-off between bit-rate required for exponents, and their frequency resolution. The overall exponent bit-rate for a frame depends upon the exponent strategy, the number of blocks over which the exponents are shared, and the audio signal bandwidth. Table 3.4 presents the per-coefficient bit-rate required to transmit the spectral envelope for each strategy, and for each block share interval. The D15 mode provides the finest frequency resolution (one exponent per frequency coefficient), while the D45 mode consumes the lowest per-coefficient bit-rate.

The absolute exponents found in the bit stream at the beginning of the differentially coded exponent sets are sent as 4 bit values which have been limited in either range or resolution in order to save one bit. For full bandwidth and LFE channels, the initial 4 bit absolute exponent represents a value from 0 to 15. Exponent values larger than 15 are limited to a value of 15. For the coupled channel, the 5 bit absolute exponent is limited to even values, and the least significant bit is not transmitted. The resolution has been limited to valid values of 0, 2, 4, ..., 24. Each differential exponent can take on one of five values: $-2, -1, 0, +1, +2$. This allows deltas of up to ± 2 (± 12 dB) between exponents. These five values are mapped into the values 0, 1, 2, 3, 4 before being grouped, as presented in Table 3.5.

TABLE 3.5 Mapping of Differential Exponent Values

Differential Exponent	Mapped Value, M_i
+2	4
+1	3
0	2
−1	1
−2	0

In D15 mode, the above mapping is applied to each individual differential exponent for coding into the bit stream. In D25 mode, each pair of differential exponents is represented by a single mapped value in the bit stream. In this mode, the differential exponent is used once to compute an exponent which is shared between two consecutive frequency coefficients.

The D45 mode is similar to D25 mode except that quadruplets of differential exponents are represented by a single mapped value. Again, the differential exponent is used once to compute an exponent which is shared between four consecutive frequency coefficients.

For all modes, sets of three adjoining (in frequency) mapped values (M_1, M_2, and M_3) are grouped together and coded as a 7 bit unsigned integer I according to the following relation:

$$I = 25M_1 + 5M_2 + M_3. \quad (3.7)$$

Following the exponent strategy fields in the bit stream is a set of 6 bit channel bandwidth codes, chbwcod[ch]. These are only present for independent channels (not in coupling) that have new exponents in the current block. The channel bandwidth code defines the end mantissa bin number for that channel according to the following

$$\text{endmant[ch]} = ((\text{chbwcod[ch]} + 12)*3) + 37. \quad (3.8)$$

Exponent strategy for each full bandwidth channel and the LFE channel can be updated independently. LFE channel exponents are restricted only to reuse or D15 mode. If a full bandwidth channel is in coupling, exponents up to the start coupling frequency are transmitted. If the full bandwidth channel is not in coupling, exponents up to the channel bandwidth code are transmitted. If coupling is on for any full bandwidth channel, a separate and independent set of exponents is transmitted for the coupling channel. Coupling start and end frequencies are transmitted as 4 bit indices.

3.2.4 Multichannel Coding

In the context of AC-3, multichannel audio is defined as two or more full bandwidth channels which are intended for simultaneous presentation to a listener. Multichannel audio coding offers new opportunities in bit-rate reduction beyond those commonly employed in monophonic codecs. The goal of multichannel coding is to compress an audio program by exploiting redundancy between the channels and irrelevancy in the signal while preserving both sound clarity and spatial characteristics of the original program. AC-3 achieves this goal by preserving listener cues which affect perceived directionality of hearing (localization).

The motivation for multichannel audio coding is provided by an understanding of how the ear extracts directional information from an incident sound wave. Hearing research suggests that the auditory system does not evaluate every detail of the complicated interaural signal differences, but rather derives what information is needed from definite, easily recognizable attributes [19]. For example, localization of signals are generally distinguished by

1. Interaural time differences (ITD)
2. Interaural level differences (ILD)

The ITD cues are caused by the difference between the time of arrival of a sound at both ears. ILD cues are sound pressure level differences caused by a different acoustic transfer functions from the acoustic source to the two ears. Most authors agree ITD is the most important attribute of the audio signal relating to the formation of lateral displacements [19]. For tones below about 800 Hz, perceived lateral displacement is approximately linear as a function of the difference in the time of arrival for the two ears, up to an ITD of 600 μs. Full lateral displacement is obtained with an ITD of approximately 630 μs. At any given time, the auditory event corresponding to the shorter time of arrival is dominant.

The ear is able to evaluate spectral components of the ear input signals individually with respect to ITD. Lateral displacement of the auditory event attainable for pure tones is most perceptible below 800 Hz. However, for some nontonal signals above 800 Hz, such as narrowband noise, the ear is still able to detect ITDs. In this case, the interaural temporal displacement of the energy envelope of the signal is generally regarded as the criterion involved. Experiments have indicated that the ITD of the signal's fine temporal structure contributes negligibly to localization; instead, the ear evaluates only the energy envelopes. The processing occurs individually in each of a multiplicity of spectral bands. The spectrum is dissected to a degree determined by the finite spectral resolution of the inner ear. Then, the envelopes of the separate spectral components are evaluated individually. These experimental results form the basis for the use of channel coupling in AC-3.

3.2.4.1 Channel Coupling

Channel coupling is a method for reducing the bit-rate of multichannel programs by summing two or more correlated channel spectra in the encoder. Frequency coefficients for the single combined (coupled) channel are transmitted in place of the individual channel spectra, together with additional side information. The side information consists of a unique set of coupling coefficients for each channel. In the decoder, frequency coefficients for each output channel are computed by multiplying the coupled channel frequency coefficients by the coupling coefficients for that channel. Coupling coefficients are computed in the encoder in a manner which preserves the short-time energy envelope of the original signals, thereby preserving spatialization cues used by the listener.

Coupling is active only above the coupling start frequency; below this boundary, frequency coefficients are coded independently. The coupling start frequency can be changed from one audio block to the next, and coupling can be disabled if desired. Any combination of two or more full bandwidth channels can be coupled; each channel has an associated channel-in-coupling bit to indicate if it has been included in the coupling channel.

Channel coupling is intended for use only when independent channel coding at the given bit-rate and desired audio bandwidth would result in audible artifacts. As the audio bit-rate is lowered with a fixed bandwidth of 20 kHz, a point is eventually reached where audible coding errors will occur for critical signals. In these circumstances, channel coupling reduces the need for encoders to take more drastic measures to eliminate artifacts, such as lowering the audio bandwidth.

A diagram depicting the encoder coupling procedure for the case of three input channels is shown in Figure 3.4. The coupling channel is formed as the vector summation of frequency coefficients from all channels in coupling. An optional signal-dependent phase adjustment is applied to the frequency coefficients prior to summation so that phase cancellation does not occur. For each input channel in coupling, the AC-3 encoder then calculates the power of the original signal and the coupled signal. The power summation is performed individually on a number of bands. For the simplified case of Figure 3.4, there are two such bands. In a typical application the number of bands is 14, but can vary between 1 and 18. Next, the power ratio between the original signal and the coupled channel is computed for each input channel and each band. Denoted a coupling coordinate, these ratios are quantized and transmitted to the decoder.

To reconstruct the spectral coefficients corresponding to one channel's worth of transform data, quantized spectral coefficients representing the uncoupled portion of the transform block are prepended to a set of scaled coupling channel spectral coefficients. The scaled coupling channel coefficients are

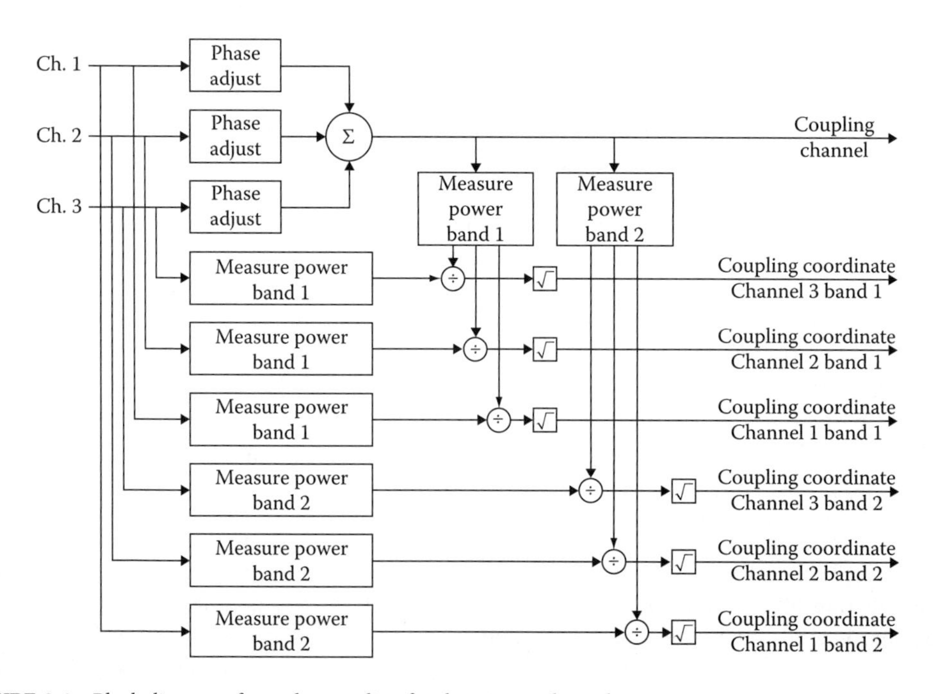

FIGURE 3.4 Block diagram of encoder coupling for three input channels.

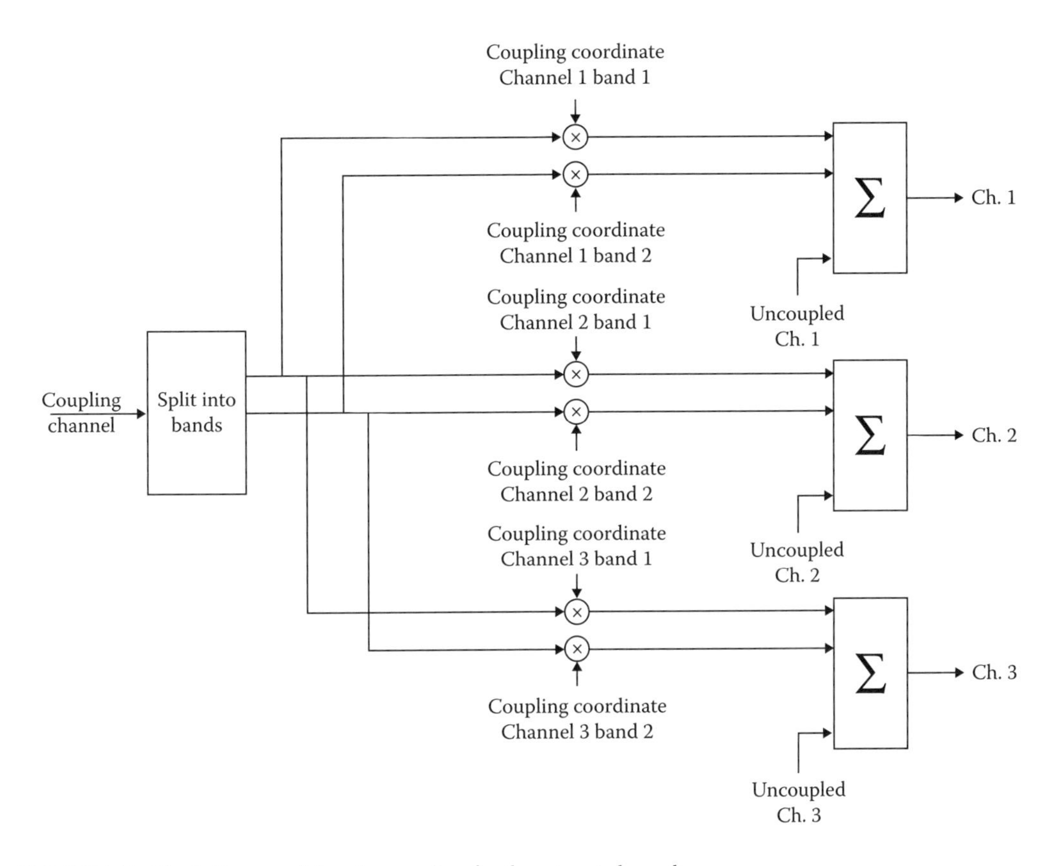

FIGURE 3.5 Block diagram of decoder coupling for three input channels.

generated for each channel by multiplying the coupling coordinates for each band by the received coupling channel coefficients, as shown in the diagram in Figure 3.5.

Coupling parameters, such as the coupling start/end frequencies and which channels are in coupling, are always transmitted in block 0. They are also optionally transmitted in blocks 1 through 5. Typically only channels with similar spectral shapes are coupled. Level differences between channels are accounted for by the coupling coefficients. It is noteworthy that if in-band spectral differences between coupled input channels are due to level only, the original input channels can still be recovered exactly in the decoder, in the absence of frequency coefficient quantization.

The coupling coefficient dynamic range is −132 to +18 dB, with step sizes varying between 0.28 and 0.53 dB. The lowest coupling start frequency is 3.42 kHz at a 48 kHz sampling frequency.

3.2.4.2 Rematrixing

Rematrixing in AC-3 is a channel combining technique in which sum and difference signals of highly correlated channels are coded rather than the original channels themselves. That is, rather than code and pack left and right (L and R) in a two channel codec, the encoder constructs:

$$\begin{aligned} \mathrm{L}' &= (\mathrm{L} + \mathrm{R})/2 \\ \mathrm{R}' &= (\mathrm{L} - \mathrm{R})/2. \end{aligned} \tag{3.9}$$

The usual quantization and data packing operations are then performed on L' and R'.

In the decoder, the original L and R signals are reconstructed using the inverse equations:

$$\begin{aligned} \mathrm{L} &= \mathrm{L}' + \mathrm{R}' \\ \mathrm{R} &= \mathrm{L}' - \mathrm{R}'. \end{aligned} \tag{3.10}$$

Clearly, if the original stereo signal were identical in both channels (i.e., two-channel mono), L′ is identical to L and R, and R′ is identically zero. Therefore, the R′ channel can be coded with very few bits, increasing accuracy in the more important L′ channel. Rematrixing is only applicable in the 2/0 encoding mode (acmod = 2).

Rematrixing is particularly important when conveying Dolby Surround encoded programs. Consider again a two channel mono source signal. A Dolby Pro Logic decoder will steer all in-phase information to the center channel, and all out-of-phase information to the surround channel. Without rematrixing, the Pro Logic decoder will receive the signals:

$$\begin{aligned} Q(\mathrm{L}) &= \mathrm{L} + n_1 \\ Q(\mathrm{R}) &= \mathrm{R} + n_2 \end{aligned} \tag{3.11}$$

where n_1 and n_2 are uncorrelated quantization noise sequences added by the process of bit-rate reduction. The Pro Logic decoder will then construct the center and surround channels as

$$\begin{aligned} C &= ((\mathrm{L} + n_1) + (\mathrm{R} + n_2))/2 \\ S &= ((\mathrm{L} + n_1) - (\mathrm{R} + n_2))/2. \end{aligned} \tag{3.12}$$

In the case of the center channel, n_1 and n_2 add, but remain masked by the dominant L + R signal. In the surround channel, however, L − R cancels to zero, and the surround speakers reproduce the difference in the quantization noise sequences $(n_1 - n_2)$.

If rematrixing is active, the left and right channels will be reproduced as

$$\begin{aligned} \mathrm{L} &= (\mathrm{L}' + n_3) + (\mathrm{R}' + n_4) \cong \mathrm{L} + n_3 \\ \mathrm{R} &= (\mathrm{L}' + n_3) - (\mathrm{R}' + n_4) \cong \mathrm{R} + n_3, \end{aligned} \tag{3.13}$$

where the approximation is made since the quantization noise $n_3 \gg n_4$. More importantly, the center and surround channels will be more faithfully reproduced as

$$\begin{aligned} C &\cong ((\mathrm{L} + n_3) + (\mathrm{R} + n_3))/2 = (\mathrm{L} + \mathrm{R})/2 + n_3 \\ S &\cong ((\mathrm{L} + n_3) - (\mathrm{R} + n_3))/2 = (\mathrm{L} - \mathrm{R})/2. \end{aligned} \tag{3.14}$$

In this case, the quantization noise in the surround channel is much lower in level.

In AC-3, rematrixing is performed independently in separate frequency bands. There are two to four contiguous bands with boundary locations dependent on coupling information. At a sampling rate of 48 kHz, bands 0 and 1 start at 1.17 and 2.30 kHz, respectively. Bands 2 and 3 start at 3.42 and 5.67 kHz. If coupling is not is use, band 3 stops at a frequency given by the channel bandwidth code. The band boundaries scale proportionally for other sampling frequencies.

Rematrixing is never used in the coupling channel. If coupling and rematrixing are simultaneously in use, the highest rematrixing band ends at the coupling start frequency.

3.2.5 Parametric Bit Allocation

The process of distributing a finite number of bits B to a block of M frequency bands so as to minimize a suitable distortion criterion is called bit allocation. The result is a bit assignment $b(k)$, $k = 0, 1, \ldots, M-1$, which defines the word length of the frequency coefficient(s) transmitted in the kth band. The bit assignment is performed subject to the constraint:

$$\sum_{k=0}^{M-1} b(k) = B. \tag{3.15}$$

B is determined from the transmission channel capacity, expressed in bps, the block length, and other parameters as well. Performance gains may be realized by allowing B to vary from block to block depending on signal characteristics.

3.2.5.1 Bit Allocation Strategies

In applications such as digital audio broadcasting and HDTV, one encoder typically distributes programs to many decoders. In these situations, it is advantageous to make the encoder as flexible as possible. If quality improvements are possible even after the decoder design is standardized, the useful life of the coding algorithm can be extended. The bit allocation strategy is a natural candidate for improvement since it plays a crucial role in determining the ultimate quality achievable by a coding algorithm.

One approach to achieving flexibility is to use a forward-adaptive bit allocation strategy, in which the bit assignment $b(k)$ for all bands is explicitly conveyed in the bit stream as side information. A second strategy is termed backward-adaptive allocation, in which $b(k)$ is computed in the encoder and then recomputed in the decoder. Since the computation is based upon quantized information which is transmitted to the decoder anyway, the side information to convey $b(k)$ is not required. The bits which are saved can be used to encode the frequency coefficients themselves.

AC-3 employs an alternative approach [20]. Called parametric bit allocation, the technique combines the advantages of forward- and backward-adaptive strategies. It employs a hearing model combining elements from Ref. [21] with new features based on more recent psychoacoustic experiments. The term "parametric" refers to the notion that the model is defined by several key variables which influence the masking curve shape and amplitude, and hence the bit assignment. A difference between AC-3 and previous codecs is that both the encoder and decoder contain the model, eliminating the need to transmit $b(k)$ explicitly. Only the essential model parameters (psychoacoustic features) are conveyed to the decoder. These parameters can be transmitted with significantly fewer bits than $b(k)$ itself. Furthermore, an improvement path is provided since the specific parameter values are selected by the encoder.

Equally significant, the parametric approach provides latitude for the encoder to adjust the time and frequency resolution of $b(k)$. Bit allocation updates are always present in block 0, and are optionally transmitted in blocks 1 through 5. The frequency resolution of $b(k)$ can be adjusted from 3.9 to 15.6 bands/kHz (one bit assignment every 94 to 375 Hz). The AC-3 encoder typically makes these adjustments in accordance with spectral and temporal changes in the audio signal itself, in a manner similar to the human ear.

Secondary strategy information which also affects block-to-block changes in bit assignment is the bit allocation information, coupling strategy, and signal-to-noise ratio (SNR) offset. This information is always transmitted in block 0, and is optionally transmitted in each subsequent audio block. The presence/absence of the information is controlled by the respective "exists" bits baie, cplstre, and snroffste. The exists bits are set to 1 in blocks 1 to 5 only when a change in strategy results in better audio quality than would be obtained by reusing parameters from the preceding block.

Signal conditions may arise in which the masking curve, and therefore $b(k)$, cannot be sufficiently optimized using the built-in parametric model. Accordingly, AC-3 encoders contain a provision for adjusting the masking curve in accordance with an independent psychoacoustic analysis. This is accomplished by transmitting additional bit stream codes, designated as deltas, which convey differences between the two masking curves.

3.2.5.2 Spreading Function Shape

In Schroeder's model for computing a masking threshold [21], one of the key variables influencing the degree of masking of one spectral component by another is the shape of the spreading function. If the spreading function is a unit impulse, the excitation function (and therefore the masking curve) will be identical in shape to the input signal spectrum. This corresponds to the case where no masking whatsoever is assumed, with the result that all frequency coefficients receive the same bit assignment, and the quantization noise spectrum will conform in shape to the input signal spectrum. As the spreading function is broadened, progressively greater degrees of masking are modeled. This yields a noise spectrum which in general contains peaks and valleys which are aligned with features of the input signal spectrum, but broadened in character. As the spreading function is flattened further, eventually a limit is reached where the noise spectrum will be white, corresponding to the minimum mean-square error bit assignment. This noise shaping behavior is identical to the concept of Refs. [22,23].

Since the spreading function strongly influences the level and extent of assumed masking, a parametric description is provided in AC-3. The parametric spreading function is one mechanism available to an AC-3 encoder for making compatible adjustments to the masking model.

The range of spreading function parameter variation was obtained by distilling a family of prototype functions from the available masking data [20]. The variation in shape of the four composite masking curves for 0.5, 1, 2, and 4 kHz tones is shown in Figure 3.6. We have approximated the envelope of upward masking for each composite curve by two linear segments. The spreading function is defined as the point-by-point maximum of the two segments across frequency. In the example shown in Figure 3.6, the composite curves can be reasonably approximated by choosing an appropriate slope and vertical offset for each linear segment. Hence, four parameters are transmitted in the bit stream to define the spreading function shape.

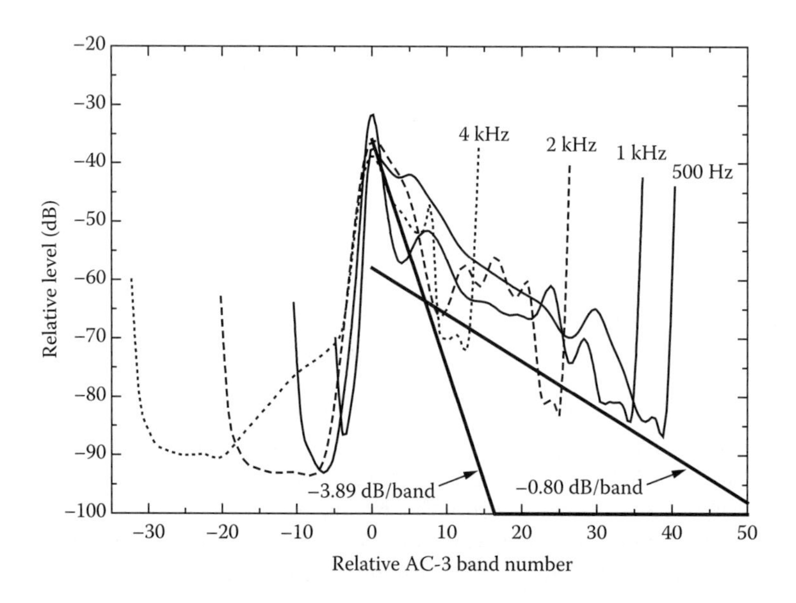

FIGURE 3.6 Comparison between 500 Hz–4 kHz masking templates and the two slope spreading function.

3.2.5.3 Algorithm Description

The AC-3 bit allocation strategy places the majority of computation in the encoder. For example, the encoder can trade off reconstructed signal bandwidth versus quantization noise power, change the degree of upward masking of one spectral component by another, modify the bit assignment as a function of acoustic signal level, and adaptively control total harmonic distortion. The bit assignment can also be adjusted according to an arbitrary second masking model, as described in Section 3.2.5.3.5. The encoder iteratively converges on an optimal solution. On the other hand, the decoder makes only one pass through the received parameters and exponent data, and is therefore considerably simpler. The bit assignment is reconstructed in the decoder using only basic two's complement operators: add, compare, arithmetic left/right shift, and table lookup.

3.2.5.3.1 Frequency Banding

The first step in the masking curve computation is to convert a block of power spectrum samples, taken at equidistant frequency intervals, into a Bark spectrum. This is accomplished by subdividing the power spectrum into multiple frequency bands and then integrating the spectrum samples within each band. The bands are nonuniform in width and derived from the critical-bandwidths defined by Zwicker [24]. The psychoacoustic basis for this procedure is that each critical band corresponds to a fixed distance along the basilar membrane, and therefore to a constant number of auditory nerve fibers.

The summation of the power spectrum samples during linear to Bark frequency conversion requires a linear summation. However, the logarithm of those quantities is most readily available in AC-3. Therefore, a log-adder is employed. The log-addition of two quantities log(a) and log(b) is computed using the relation:

$$\log(a+b) = \max(\log(a), \log(b)) + \log\left(1 + e^{-d}\right) \tag{3.16}$$

where e is the logarithm base, and

$$d = |\log(a) - \log(b)|. \tag{3.17}$$

The second term on the right side of the equation is implemented as a subtraction log(a) log(b), followed by absolute value and a table lookup. The content of the table at address d is: $\log(1+e^{-d})$. Therefore, the complete log-addition is performed with only add, compare and table lookup instructions.

3.2.5.3.2 Masking Convolution

The technique for modeling masking effects developed by Schroeder specifies a convolution. At every frequency point, masking contributions from all other spectral components are weighted and summed. The output of a linear recursive (e.g., infinite impulse response [IIR]) filter may also be viewed as a weighted summation of input samples. Therefore, the convolution of the spreading function with the linear-amplitude critical band density in Schroeder's model can be approximated by applying a time-varying linear recursive filter to the spectral components. Upward masking is modeled by filtering the frequency samples from low to high frequency. Downward masking is modeled by filtering input samples in the reverse order. The filter order and coefficients are determined from the desired spreading function. To compute the excitation function using the logarithmic-amplitude critical band density used in AC-3, the linear recursive filter is replaced with an equivalent filter which processes logarithmic spectral samples.

The conversion of the linear recursive filter to an equivalent log-domain filter is straightforward. By writing out the difference equation for an IIR filter and taking the logarithm of both sides of the equation, an expression relating the log excitation function with the log power spectrum can be derived. The multiplications and additions of the IIR filter are replaced with additions and log-additions, respectively, in the log domain filter. To implement the two-slope spreading function, two filters are connected in parallel. Each filter implements the characteristic of one of the segments, and the overall excitation value $E(k)$ at band k is computed as the larger of the two filter output samples. The log-domain equations for computing $E(k)$ are

$$\begin{aligned} x_0 &= (x_0 - d_0) \oplus (P(k) - g_0) \\ x_1 &= (x_1 - d_1) \oplus (P(k) - g_1) \\ E(k) &= \max(x_0, x_1) \end{aligned} \tag{3.18}$$

where

$P(k)$ is the log-amplitude power spectrum
d_0 and d_1 are the dB spreading function decay values for the first and second segment, respectively
g_0 and g_1 are the dB offsets of the two spreading function segments
$\oplus$ symbol denotes log-addition as defined previously

For each of 50 bands, the value of two accumulators x_0 and x_1 is computed by performing a log-addition of the previous accumulator value, decayed by d_0 or d_1, and the current power spectrum value scaled by the gain g_0 or g_1. In AC-3, log-addition is replaced by a maximum operator to reduce computation. This is also more conservative in that additive masking is not assumed.

3.2.5.3.3 Compensation for Decoder Selectivity

One basis for determining the bit assignments $b(k)$ in a perception-based allocation strategy is to compute the difference between the signal spectrum and the predicted masking curve. An implicit assumption of this technique is that quantization noise in one particular band is independent of bit assignments in neighboring bands. This is not always a reasonable assumption because the finite frequency selectivity and the high degree of overlap between bands in the decoder filter bank cause localized spreading of the error spectrum (leakage from one band into neighboring bands). The effect is predominant at low frequencies where the slope of the masking curve can equal or exceed the slope of the filter bank transition skirts. Hence, under some conditions, a basis other than the difference between the signal spectrum and masking curve is warranted.

As discussed in Ref. [20], decoder selectivity compensation has been found to improve subjective coding performance at low frequencies. Accordingly, the AC-3 masking model employs a straightforward, recursive algorithm for applying compensation from 0 to 2.3 kHz. Although the compensation is a filter bank response correction, and not a psychoacoustic effect of human hearing, it can be incorporated into the computation of the excitation curve.

3.2.5.3.4 Parameter Variation

The masking computation primarily represents the backward-adaptive portion of the bit allocation strategy. However, a number of parameters defining the masking model are transmitted in the compressed bit stream. These represent part of the forward-adaptive portion of the bit allocation strategy. As discussed earlier, the shape of the prototype spreading function is controlled by four parameters, where a pair of parameters correspond to each segment. The first linear segment slope is adjustable between −2.95 and −5.77 dB per band, with offsets ranging from −6 to −48 dB. The second segment slope can be adjusted between −0.70 and −0.98 dB per band, with offsets ranging from −49 to −63 dB. The syntax of AC-3 allows the first segment to be controlled independently for each channel. Parameters for the second segment are common to all channels. There are 512 unique spreading function shapes available to an AC-3 encoder.

3.2.5.3.5 Delta Bit Allocation

Another forward-adaptive component of the bit allocation strategy is a parametric adjustment that is optionally made to the masking curve computed by the masking model. This adjustment is conveyed to the decoder with the delta bit allocation. For each channel, the encoder can specify nearly arbitrary adjustments to the computed masking curve at a certain cost in bit-rate that otherwise would be used directly to code audio data. Delta bit allocation is used by the encoder to specify a masking curve, and hence a bit assignment, that cannot be generated by the parametric model alone. This feature is useful, for example, if future research points to masking behavior which cannot be simulated by the existing model. In this case, benefits of any new research can be added to an AC-3 encoder by augmenting the parametric model with the improved one.

Determination of the desired delta bit allocation function is straightforward. In an encoder, both the standard AC-3 masking model and the improved one are run in parallel to determine two masking curves. The desired delta bit allocation function is equal to the difference between the masking curves. It may be advantageous to only approximate the desired difference to reduce the required data expenditure. Note that an encoder should first exhaust the flexibility granted by the nondelta parameters before committing any bits to delta bit allocation. The other parameters are less expensive in terms of bit-rate as they must be transmitted periodically in any case.

The delta function is constrained to have a "stair step" shape. Each tread of the stair step corresponds to the masking level adjustment for an integral number of adjoining one-half Bark bands. Taken together, the stair steps comprise a number of nonoverlapping variable-length segments. The segments are run-length coded for efficient transmission.

3.2.6 Quantization and Coding

All mantissas are quantized to a fixed level of precision indicated by the corresponding bap. Mantissas quantized to 15 or fewer levels use symmetric quantization. Mantissas quantized to more than 15 levels use asymmetric quantization (a conventional two's complement representation).

Some quantized mantissa values are grouped together and encoded into a common codeword. In the case of the 3-level quantizer, three quantized values are grouped together and represented by a 5 bit codeword in the bit stream. In the case of the 5-level quantizer, three quantized values are grouped and represented by a 7 bit codeword. For the 11-level quantizer, two quantized values are grouped and represented by a 7 bit codeword. Groups are filled in the order that the mantissas are processed. If the number of mantissas in an exponent set does not fill an integral number of groups, the groups are shared across exponent sets. The next exponent set in the block continues filling the partial groups.

In the encoder, each frequency coefficient is normalized by applying a left-shift equal to its associated exponent (0–24). The mantissa is then quantized to a number of levels indicated by the corresponding bap.

Table 3.6 presents the assignment between bap number and number of quantizer levels. If a bap equals 0, no bits are sent for the mantissa. For more efficient bit utilization, grouping is used for bap values of 1, 2, and 4 (3-, 5-, and 11-level quantizers).

For bit allocation pointer values between 6 and 15, inclusive, asymmetric fractional two's complement quantization is used. No grouping is employed for asymmetrically quantized mantissas.

For bap values of 1–5, inclusive, the mantissas are represented by coded values. The coded values are converted to standard two's complement fractional binary words. The number of bits indicated by a mantissa's bap are extracted from the bit stream and right justified. This coded value is treated as a table index and is used to look up the quantized mantissa value. The resulting mantissa value is right shifted by the corresponding exponent to generate the transform coefficient value.

The AC-3 decoder uses random noise (dither) values instead of quantized values when the number of bits allocated to a mantissa is zero ($bap = 0$). The decoder substitution of random values for the quantized mantissas with $bap = 0$ is conditional on the value of a bit conveyed in the bit stream (dithflag). There is a separate dithflag bit for each transmitted channel. When $dithflag = 1$, the random noise value is used. When $dithflag = 0$, a true zero value is used.

TABLE 3.6 Quantizer Levels and Mantissa Bits versus Bap

Bap	Quantizer Levels	Mantissa Bits (Group Bits/Num in Group)
0	0	0
1	3	1.67 (5/3)
2	5	2.33 (7/3)
3	7	3
4	11	3.5 (7/2)
5	15	4
6	32	5
7	64	6
8	128	7
9	256	8
10	512	9
11	1,024	10
12	2,048	11
13	4,096	12
14	16,384	14
15	65,536	16

3.2.7 Error Detection

There are several ways in which the AC-3 data may determine that errors are contained within a frame of data. The decoder may be informed of that fact by the transport system which has delivered the data. Data integrity may be checked using the embedded cyclic redundancy check words (CRCs). Also, some simple consistency checks on the received data can indicate that errors are present. The decoder strategy when errors are detected is user definable. Possible responses include muting, block repeats, frame repeats, or more elaborate schemes based on waveform interpolation to "fill in" missing PCM samples. The amount of error checking performed, and the behavior in the presence of errors are not specified in the AC-3 ATSC standard, but are left to the application and implementation.

Each AC-3 frame contains two 16 bit CRC words. As discussed in Section 3.2, CRC1 is the second 16 bit word of the frame, immediately following the synchronization word. CRC2 is the last 16 bit word of the frame, immediately preceding the synchronization word of the following frame. CRC1 applies to the first 5/8 of the frame, not including the synchronization word. CRC2 provides coverage for the last 3/8 of the frame as well as for the entire frame (not including the synchronization word). Decoding of CRC word(s) allows errors to be detected.

The following generator polynomial is used to generate both of the 16 bit CRC words in the encoder:

$$x^{16} + x^{15} + x^2 + 1. \tag{3.19}$$

The CRC calculation may be implemented by one of several standard techniques. A convenient hardware implementation is a linear feedback shift register. Details of this technique are presented in Ref. [3].

3.3 Enhanced AC-3 Audio Coding

Although AC-3 had become a very popular coding format by 2002, an update was necessary to better serve the needs of both new and existing markets. Rather than design a completely new audio codec, Dolby Laboratories developed a nonbackwards compatible extension to AC-3. This extension, called Enhanced AC-3 (E-AC-3), was designed to satisfy two main goals. The first goal was to provide a significantly wider range of operation than AC-3. In order to support some of the additional features required by existing markets, it was necessary for E-AC-3 to support operation at higher bit-rates than AC-3. In order to make E-AC-3 suitable for new applications requiring low bit-rate operation, it was necessary to provide better audio quality than AC-3 at lower bit-rates. The second goal was to retain as many of the core features and technologies of AC-3 as possible while still satisfying the first goal. While it was not possible to make E-AC-3 and AC-3 completely compatible and still satisfy the first goal, the decision to retain many elements of AC-3 allowed a high degree of interoperability between the two formats. This feature facilitates computationally efficient, high-quality conversion between the two formats.

Because E-AC-3 is based to a large extent on the AC-3 technology described in Section 3.2, Sections 3.3.1 through 3.3.4 focus only on the differences between the two codecs. Specifically, Section 3.3.1 describes how the E-AC-3 synchronization frame differs from AC-3. Section 3.3.2 describes key additional features of E-AC-3, such as flexible bit-rate and framing structure, bit stream optimizations, channel and program extensions, and bit stream mixing. Section 3.3.3 describes new coding tools supported by E-AC-3, such as improved filterbank resolution and quantization techniques, enhanced channel coupling, audio bandwidth extension, and transient prenoise processing. Section 3.3.4 describes conversion of E-AC-3 bit streams to AC-3 bit streams. For a complete description of E-AC-3, see [3].

3.3.1 Synchronization Frame

In order to maintain a high degree of interoperability, the E-AC-3 synchronization frame preserves the AC-3 frame structure, with two main exceptions. The first exception is that while all E-AC-3 frames include the five basic elements that comprise an AC-3 frame (synchronization information, bit stream information, audio block, auxiliary data block, and cyclic redundancy code), E-AC-3 frames also include an audio frame (AF) element, located immediately following the BSI element and immediately preceding the audio block element. The AF element carries information that is required for decoding the individual audio blocks that make up the audio block element, but only needs to be transmitted once per frame because it does not change from one block to the next within the frame. The second exception is that while all AC-3 frames are required to carry exactly six 256-sample audio blocks, an E-AC-3 frame may carry either one, two, three, or six 256-sample audio blocks. Because both the bit-rate and the decoder latency are related to the number of audio blocks per frame, allowing frames to carry a variable number of blocks increases the flexibility of E-AC-3. Figure 3.7 shows the E-AC-3 synchronization frame, including all of the AC-3 elements and the new AF element. Because audio blocks 1–5 are optional in E-AC-3 frames, they are shaded differently than audio block 0, which is required.

One further difference between E-AC-3 and AC-3 synchronization frames is related to error checking. An AC-3 frame contains two CRC words: the first, located in the synchronization information section, covers the first 5/8th of the frame; the second, located at the end of the frame, covers the last 3/8th of the frame. An E-AC-3 frame contains only one CRC word, located at the end of the frame, which covers the

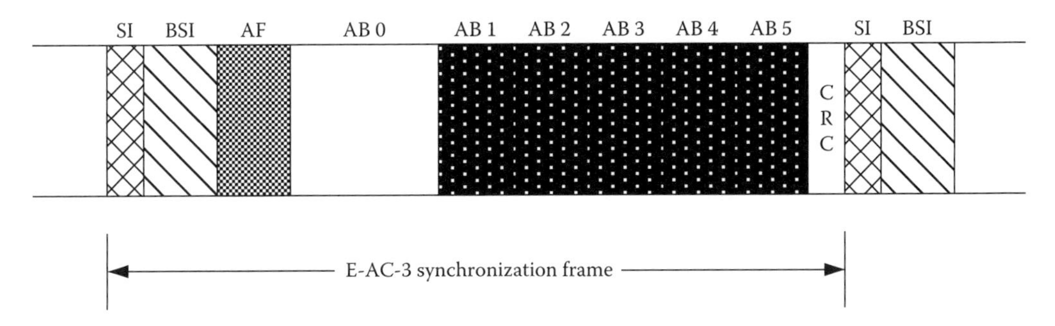

FIGURE 3.7 E-AC-3 synchronization frame.

entire frame. There are two main consequences to this approach: the first is that an error anywhere in an E-AC-3 frame will render the entire frame invalid; the second is that a decoder must receive the entire E-AC-3 frame prior to decoding any portion of it, which can increase codec latency. For applications which require low codec latency, such as audio for games, E-AC-3 encoders may reduce codec latency by choosing to carry fewer audio blocks per frame, as described above.

3.3.2 New Features

Many new bit stream parameters have been added to the E-AC-3 bit stream syntax to enable new features and increase flexibility. While it is beyond the scope of this document to describe all of these new parameters, the most important new features are discussed below.

3.3.2.1 Flexible Bit-rate and Framing Structure

In order to meet the changing needs of markets in which AC-3 is already established, and to enable expansion into new markets, E-AC-3 was designed with significantly more bit-rate flexibility than AC-3. This additional flexibility, which includes a much wider range of bit-rates and much finer bit-rate resolution, is enabled by two new bit stream parameters. The first parameter is the number of 16-bit data words that make up the frame, and varies between 1 and 2048. The second parameter is the number of audio blocks carried by the frame, and as mentioned previously may be set to either one, two, three, or six.

As the amount of storage available for compressed audio in some media formats has increased, the demand for audio coded at higher bit-rates has also increased. In order to meet this demand, E-AC-3 was designed to support bit-rates considerably higher than the maximum AC-3 bit-rate of 640 kbps. Table 3.7 presents the maximum instantaneous E-AC-3 bit-rate in kbps as a function of sample rate and number of blocks per frame. For a given combination of sample rate and number of blocks per frame, the maximum instantaneous E-AC-3 bit-rate is achieved when a frame contains 2048 16-bit words. For all combinations, the maximum instantaneous E-AC-3 bit-rate exceeds the maximum AC-3 bit-rate. The absolute maximum instantaneous E-AC-3 bit-rate of 6.144 Mbps is nearly ten times higher than the maximum AC-3 bit-rate.

To support a wide range of applications with differing target bit-rates, as well as applications for which channel capacity may vary slightly or significantly over time, it is advantageous to have very fine control over the exact operating bit-rate of the compressed audio stream. For this reason, E-AC-3 supports significantly finer bit-rate control than AC-3. Table 3.8 presents the E-AC-3 bit-rate adjustment

TABLE 3.7 E-AC-3 Maximum Bit-Rates (kbps)

		Blocks per Frame			
		1	2	3	6
Sample rate (kHz)	48	6144	3072	2048	1024
	44.1	5644	2882	1881	940
	32	4096	2048	1365	682

TABLE 3.8 E-AC-3 Bit-Rate Adjustment Resolution (kbps)

		Blocks per Frame			
		1	2	3	6
Sample rate (kHz)	48	3.000	1.500	1.000	0.500
	44.1	2.756	1.378	0.918	0.459
	32	2.000	1.000	0.666	0.333

resolution in kbps as a function of sample rate and number of blocks per frame. For a given combination of sample rate and number of blocks per frame, the bit-rate adjustment resolution represents the increase in bit-rate that results from adding one additional 16-bit data word to a frame. For all combinations, the E-AC-3 bit-rate adjustment resolution is finer than the finest AC-3 bit-rate adjustment resolution of 8 kpbs. The absolute finest E-AC-3 bit-rate resolution of 0.333 kbps is 24 times finer than the finest AC-3 bit-rate adjustment resolution.

3.3.2.2 Bit Stream Syntax Optimizations

AC-3 was designed to support a broad range of applications, and therefore the AC-3 bit stream syntax includes many adjustable parameters. However, in some applications this flexibility is not exercised, and it is common for the adjustable parameters to always take the same value. Additionally, many parameters that are sent explicitly in the bit stream can be inferred from other parameters. The result is a bit stream syntax which, while very flexible, contains many small inefficiencies. While such inefficiencies are not critical for the moderate to high bit-rate applications targeted by AC-3, they are extremely important for very low bit-rate applications, in which even small inefficiencies may degrade sound quality. Therefore, two changes were made to maximize efficiency of the E-AC-3 bit stream syntax. First, the syntax was designed so that the most common parameter settings require the least amount of data to transmit. Second, wherever possible parameters are inferred from other parameters rather than explicitly transmitted.

The main drawback of removing inefficiency from the bit stream is that it often results in reducing the flexibility available to the codec. While removing inefficiency by reducing flexibility can be very beneficial for low bit-rate applications, it may not be appropriate for all applications. For example, in some higher bit-rate applications, where efficiency is less critical, increased flexibility may be more valuable than increased efficiency. With this in mind, the E-AC-3 syntax was also designed to be scalable, so that increased flexibility could be obtained by slightly reducing the efficiency. In the most flexible mode, the E-AC-3 bit stream syntax supports all of the flexibility of the AC-3 bit stream syntax. The overall result of these bit stream optimizations is a scalable E-AC-3 bit stream syntax, which allows the user to balance flexibility and efficiency depending on the requirements of a specific application.

3.3.2.3 Substream Extensions

Individual AC-3 bit streams are constrained to carrying no more than one independent program containing no more than 5.1 channels of audio. However, some current and emerging applications require a codec that is capable of carrying multiple independent programs, or more than 5.1 channels of audio, in a single bit stream. To support these applications, E-AC-3 allows frames from multiple substreams to be time-multiplexed into a single bit stream. Substreams come in one of four types, with the type for a given frame indicated by a bit stream parameter. Independent and dependent substreams are currently the only substream types used in practice. Since the two other types of substreams are not used in practice, they are not described in this section. An independent substream consists of a complete, stand-alone audio program that is suitable to be reproduced independently of any other substreams. Every E-AC-3 bit stream is required to have at least one, and not more than eight, independent substreams. Independent substreams are numbered in ascending order in the order in which they are present in the bit stream. Every independent substream may have anywhere from zero to eight dependent substreams associated with it. Dependent substreams are intended to enhance or augment the audio carried in an independent substream, and are not intended to be reproduced as complete, stand-alone programs. Dependent substreams are numbered in ascending order in the order in which they are present in the bit stream, and must immediately follow the independent substream with which they are associated. For all independent and dependent substreams, the E-AC-3 bit stream syntax includes a parameter that indicates the number of the substream. Some specific applications of substream extensions are described in the following sections.

3.3.2.3.1 Channel Extensions

Channel extensions, which allow an E-AC-3 bit stream to carry programs containing more than 5.1 channels of audio, are the most common application of dependent substreams. Programs containing more than 5.1 channels of audio can be carried in a single bit stream using a combination of independent and dependent substreams. An independent substream carries a 5.1-channel downmix of the complete audio program, and one or more dependent substreams carry the replacement and supplemental channels necessary to reproduce the complete audio program. When a dependent substream carries replacement channels, those channels replace channels carried in the independent substream. When a dependent substream carries supplemental channels, those channels are reproduced in addition to the channels carried in the independent substream. A dependent substream may carry replacement channels, supplemental channels, or a combination of both replacement and supplemental channels.

An E-AC-3 bit stream using channel extensions to carry a program of more than 5.1 channels of audio is compatible with any E-AC-3 decoder, whether or not the decoder supports the reproduction of more than 5.1 channels of audio. Simple decoders, which output 5.1 or fewer channels, decode only the independent substream and ignore all dependent substreams. Advanced decoders, which are capable of reproducing the complete audio program, decode both the independent and dependent substreams, and combine all of the decoded channels together to reproduce the complete audio program.

Figures 3.8 through 3.10 show how channel extensions are used to carry a 7.1-channel audio program using one independent and one dependent substream. Figure 3.8 shows a block diagram of the 7.1-channel encoding and decoding process. The encoder creates a 5.1-channel downmix of the 7.1-channel program, and then uses a 5.1-channel encoder to generate independent substream 0 carrying the 5.1-channel downmix, a 4.0-channel encoder to generate dependent substream 0 carrying the four surround channels from the original 7.1-channel program, and a bit stream multiplexer to combine the independent and dependent substreams into a single E-AC-3 bit stream. The bit stream is passed through a transmission channel and received by the decoder. The decoder uses a bit stream demultiplexer to separate the independent and the dependent substreams, a 5.1-channel decoder to decode independent substream 0, and a 4.0-channel decoder to decode dependent substream 0. The decoder then replaces the left and right surround channels from independent substream 0 with the left and right surround channels from dependent substream 0, and combines the left, center, and right channels from independent substream 0 with the left and right surround replacement channels and the left and right rear surround supplemental channels from dependent substream 0 to reproduce the original 7.1-channel audio program. Figure 3.9 shows a block diagram of a simple decoder, which uses a bit stream demultiplexer to separate the independent and dependent substreams, a 5.1-channel decoder to decode independent substream 0, and ignores dependent substream 0. Figure 3.10 shows two frames of the encoded, time-multiplexed E-AC-3 bit stream.

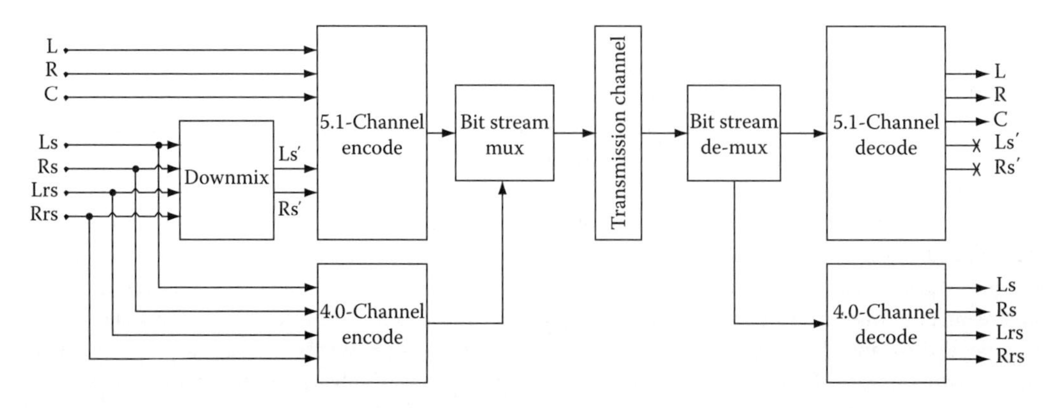

FIGURE 3.8 7.1-channel encoding/decoding system.

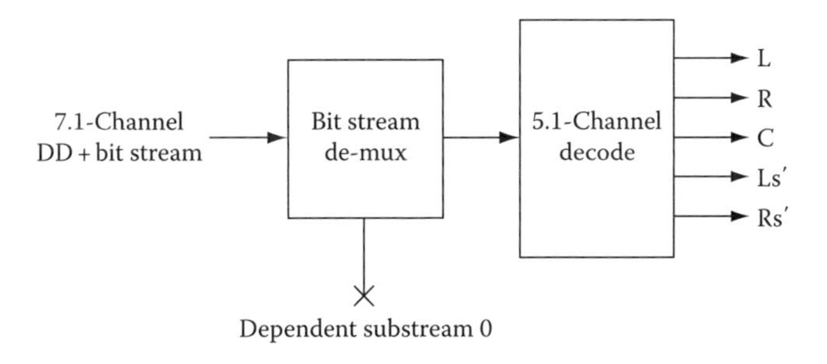

FIGURE 3.9 Simple 5.1-channel decoder.

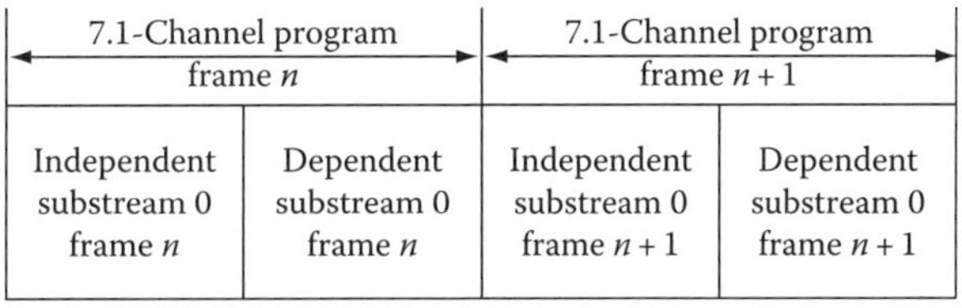

7.1-Channel program frame n		7.1-Channel program frame $n+1$	
Independent substream 0 frame n	Dependent substream 0 frame n	Independent substream 0 frame $n+1$	Dependent substream 0 frame $n+1$

FIGURE 3.10 E-AC-3 bit stream carrying 7.1-channels of audio.

3.3.2.3.2 Program Extensions

There are two main applications of independent substreams. The first application, program extensions, enables a single E-AC-3 bit stream to carry more than one independent program. A separate independent substream is used to carry each independent program. One example of the program extensions application is the carriage of different language versions of the same program material within the same bit stream. For example, independent substream 0 may contain an English language version of the program, while independent substream 1 contains a Spanish language version of the program. An E-AC-3 bit stream using program extensions to carry multiple independent programs is compatible with any E-AC-3 decoder. A simple decoder always decodes the program carried by independent substream 0 and ignores all other substreams. A more advanced decoder would allow the user to select any of the independent programs carried in the bit stream and ignore all other independent programs. In the multiple language example, the simple decoder would always reproduce the English language version carried by independent substream 0 and ignore the Spanish language version carried by independent substream 1. An advanced decoder would allow the user to choose either the English or the Spanish language version, decode the corresponding independent substream, and ignore the other independent substream. Figure 3.11 shows an example of two frames of an E-AC-3 bit stream configured to carry two independent programs.

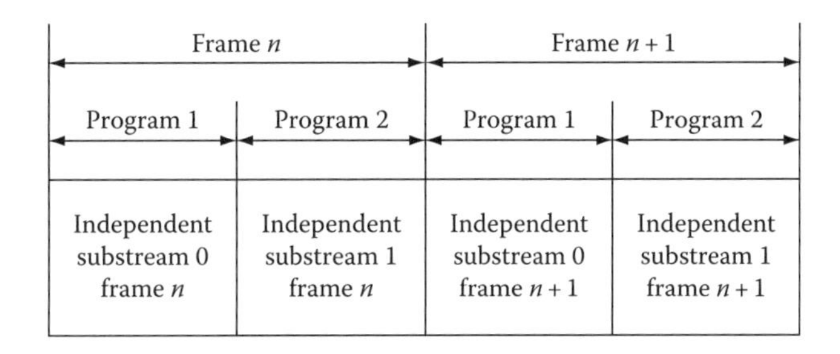

FIGURE 3.11 E-AC-3 bit stream carrying two independent programs.

3.3.2.3.3 *Bit Stream Mixing*

The second application of independent substreams is bit stream mixing. In the bit stream mixing application, a main audio program is carried by independent substream 0 and an auxiliary program is carried by independent substream 1. The substream containing the auxiliary program may also carry mixing control information. The audio contained in the auxiliary program can be reproduced as a standalone program, or be mixed with the audio contained in the main program. One example of the bit stream mixing application is the carriage of a 5.1-channel film soundtrack and a 2.0-channel commentary track in the same bit stream. An E-AC-3 bit stream using substreams to support the mixing application is compatible with any E-AC-3 decoder, whether or not the decoder supports bit stream mixing. A simple decoder always decodes the program carried by independent substream 0 and ignores all other substreams. An advanced decoder would decode both the independent substream carrying the main audio program and the independent substream carrying the auxiliary program and mixing control information, and then mix the decoded audio from the main and auxiliary programs according to the mixing control information.

Figure 3.12 shows a block diagram of the bit stream mixing encoding and decoding process for a 5.1-channel film soundtrack and an associated 2.0-channel commentary track. The mixing encoder uses a 5.1-channel encoder to generate independent substream 0 carrying the film soundtrack, a 2.0-channel encoder to generate independent substream 1 carrying the commentary track and mixing control information, and a bit stream multiplexer to combine the independent substreams into a single E-AC-3 bit stream. In some cases, it may be necessary or beneficial for the 5.1-channel and 2.0-channel encoders to share information. The format of the resulting E-AC-3 bit stream is identical to the bit stream shown in Figure 3.11. The bit stream is passed through a transmission channel and received by the mixing decoder. The mixing decoder uses a bit stream demultiplexer to separate the independent substreams, a 5.1-channel decoder to decode independent substream 0, a 2.0-channel decoder to decode independent substream 1, and a program mixer to combine the 5.1-channel film soundtrack and the 2.0-channel commentary track into a 5.1-channel signal according to the mixing control information. Note that although it is not explicitly shown, a simple decoder, like the one shown in Figure 3.9, is compatible with this application, and would decode only the independent substream carrying the 5.1-channel film soundtrack and ignore the independent substream carrying the 2.0-channel commentary track and mixing control data.

3.3.3 New Coding Tools

In order to provide improved compression efficiency, and therefore improved audio quality for low bitrate applications, several new coding tools are supported by E-AC-3. The following sections describe each of the new coding tools in detail.

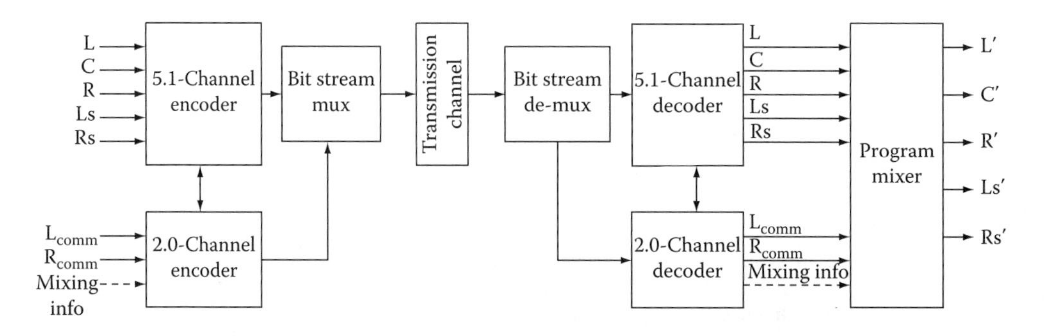

FIGURE 3.12 Bit stream mixing encoding/decoding system.

3.3.3.1 Improved Filterbank Resolution

AC-3 employs a 512-point time to frequency transform, which by current standards is considered to be relatively short. The transform length used by AC-3 was specifically chosen to balance two opposing design goals, coding efficiency and ease of implementation. It is well known in the field of audio coding that as the transform length increases, the coding efficiency of a transform codec for stationary signals also increases. Unfortunately, increasing the transform length also increases the computational complexity, memory requirements, and latency of the codec, and correspondingly, the hardware cost. While today's hardware may be able to support much longer transforms, during the 1990s when AC-3 was being developed, the increase in cost associated with supporting longer transforms was prohibitive, especially for the low-cost decoding applications commonly associated with AC-3. Therefore, the AC-3 transform length was chosen to be as long as possible without requiring prohibitively costly hardware.

Because E-AC-3 was designed for maximum interoperability with AC-3, it employs the same 512-point time to frequency transform. While this transform is very effective and enables AC-3 to provide excellent audio quality for moderate bit-rate applications, it does not provide the coding efficiency for stationary signals required to support some of the low bit-rate applications targeted by E-AC-3. Therefore, a new technique was required which could increase the coding efficiency of E-AC-3 for stationary signals without significantly impacting the interoperability with AC-3. To meet this requirement, the adaptive hybrid transform (AHT) tool was developed.

The AHT tool increases the effective length, and correspondingly the coding efficiency for stationary signals, of the standard AC-3 transform by applying, to each active transform bin, a secondary transform spanning multiple transform blocks. A standard AC-3 frame contains six audio blocks, with each block comprising up to 256 active transform bins. Stated differently, for each active transform bin, an AC-3 frame contains transform coefficients for six consecutive blocks, or time slots. For stationary signals, the transform coefficients for a given transform bin do not change significantly from one block to the next. Therefore, over the six blocks of the frame, the coefficients for a given transform bin contain a significant amount of redundancy, and coding them independently is not very efficient. Applying the secondary transform significantly reduces the redundancy and produces a more compact representation of the coefficients, which can then be coded more efficiently.

During the development of the AHT tool, several factors were considered when determining an appropriate choice for the secondary transform. Firstly, because the secondary transform is applied by the E-AC-3 encoder, and must be inverted by the E-AC-3 decoder, a perfectly invertible transform was needed. Secondly, in order to ensure that the AHT tool did not introduce dependencies between adjacent E-AC-3 frames, no overlap between adjacent secondary transform blocks was allowed. Thirdly, because overlap between adjacent secondary transform blocks was not allowed, and a perfectly invertible transform was needed, windowing of the secondary transform blocks was also not allowed. Finally, to ensure that the number of coefficients output from the secondary transform matched the number of coefficients input to the secondary transform, a real-valued transform was needed. The Type II DCT, defined in Ref. [25], satisfies all of these requirements, and was determined to be an appropriate choice for the secondary transform employed by the AHT tool.

Because the AHT tool was specifically designed to improve the coding efficiency of E-AC-3 for stationary signals, it is not appropriate to use the AHT tool to code nonstationary signals. Therefore, for each frame, the E-AC-3 encoder must decide based on input signal characteristics whether to enable the AHT tool. To reduce the amount of additional work the encoder must perform in order to make the decision, the E-AC-3 bit stream syntax places restrictions on when the AHT tool may be used. The first restriction is that the AHT tool may only be enabled for frames that carry six audio blocks. Because frames that contain fewer than six audio blocks are typically used only for higher bit-rate applications, for which improved coding efficiency is less critical, there is no need to use the AHT tool for frames that contain fewer than six audio blocks.

The second restriction imposed by the E-AC-3 bit stream syntax is that the AHT tool may only be enabled for frames which use a specific exponent time sharing strategy. The exponent time sharing strategy of a frame indicates for which blocks new exponents are transmitted, and for which blocks the exponents from the previous block are reused. In order to determine the optimal exponent time sharing strategy for a frame, the encoder performs an analysis of the input signal characteristics. For stationary signals, in which the power spectrum does not change significantly from one block to the next, it is not necessary to retransmit exponents every block. In fact, for signals that are stationary over an entire frame, exponents are typically transmitted only for the first block of the frame and then reused for all remaining blocks. By restricting the AHT tool to operate only for frames in which exponents are transmitted for the first block and reused for all remaining blocks, the E-AC-3 bit stream syntax ensures that the AHT tool is only used for appropriate signal conditions, without requiring the encoder to perform a significant amount of additional input signal analysis.

Because the signal characteristics of each channel transmitted by E-AC-3 may be completely unrelated, in some frames all channels may benefit from using the AHT tool, while in other frames only a subset or even none of the channels may benefit from using the AHT tool. Furthermore, in addition to the full bandwidth channels, it may be beneficial to use the AHT tool for the LFE or coupling channel as well. In order to support all of these possibilities, the E-AC-3 bit stream syntax allows for independent control over the use of the AHT tool for every channel, including the LFE and coupling channels. Finally, in addition to the restrictions imposed by the E-AC-3 bit stream syntax discussed above, additional restrictions are imposed on the use of the AHT tool for the coupling channel. Figure 3.13 shows a diagram of the transformation of a frame of MDCT mantissas into a frame of AHT mantissas. For each transform bin, the MDCT mantissas from blocks 0 through 5 are transformed by the AHT transform into six AHT mantissas.

3.3.3.2 Improved Quantization Techniques

AC-3 quantizes transform coefficient mantissas using conventional uniform scalar quantization. For each transform coefficient mantissa, the AC-3 bit allocation routine determines the appropriate quantizer resolution based on the pyschoacoustic masking properties of the input signal and the bit-rate requirements of the transmission channel. The quantization resolution assigned to each mantissa by the AC-3 bit allocation routine is proportional to a signal-to-mask ratio (SMR) calculated during the bit allocation routine. For mantissas with lower SMRs, symmetric quantization is used, while for mantissas with higher SMRs, asymmetric quantization is used. In general, for every 6 dB SMR increase, one additional bit of quantizer resolution is allocated. Unfortunately, this approach results in some mantissas being quantized with more resolution than is required by the psychoacoustic model. Typically this is not a significant problem for mantissas that have a very high SMR, which are relatively uncommon and considered to be

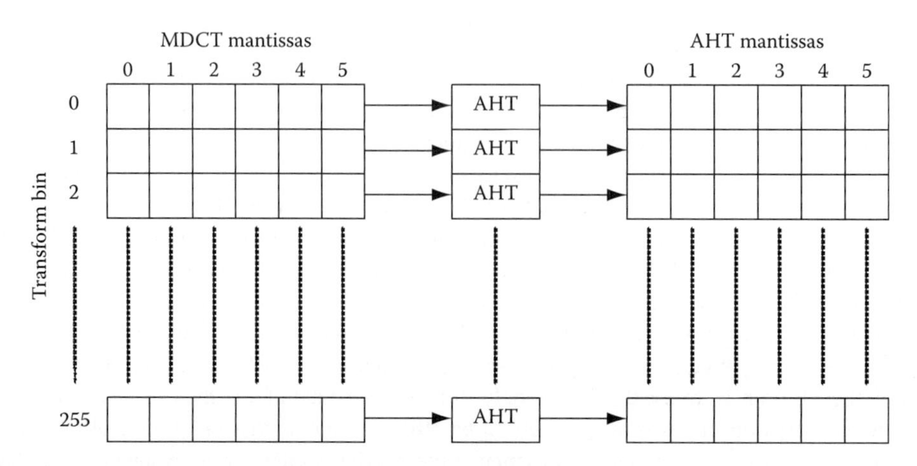

FIGURE 3.13 Transformation of MDCT mantissas to AHT mantissas.

perceptually important. However, it may be a problem for mantissas that have a very low SMR, which are relatively common, especially in low bit-rate applications.

In AC-3, any transform coefficient mantissa that has an SMR below 1.5 dB is allocated 0 bits and is not transmitted to the decoder. All transform coefficient mantissas that have SMRs between 1.5 dB and 9 dB are quantized using a 3-level symmetric quantizer, the lowest resolution quantizer available in AC-3. For transform coefficient mantissas with SMRs in the lower portion of this range, the quantizer provides more resolution than is required by the psychoacoustic masking model, which reduces the efficiency of the system. However, because this is the lowest resolution quantizer available, it must be used and nothing can be done about the reduced efficiency.

E-AC-3 determines SMR using the same method as AC-3, and is therefore subject to the same quantization efficiency issues. In order to address these issues, and take full advantage of the improved coding efficiency provided by the AHT tool, E-AC-3 includes a modified mapping of SMR to quantizer resolution for signals coded using the AHT tool. For signals not coded using the AHT tool, the mapping remains identical to AC-3. The major difference between the mapping used for AC-3 and the modified mapping used for E-AC-3 is that mantissas with SMRs ranging from 1.5 to 12 dB, which are mapped to one of two quantizers in AC-3, are mapped to one of seven quantizers in E-AC-3. Table 3.9 presents the mapping of SMR to quantizer index for mantissas with SMRs below 12 dB for both AC-3 and E-AC-3. In Table 3.9, the quantizer index 0 indicates that no bits are allocated and the mantissa is not transmitted. However, as will be discussed further in the following section, AC-3 quantizer indices 1 and 2 do not indicate the same quantizers as E-AC-3 quantizer indices 1 and 2.

3.3.3.2.1 Vector Quantization

In order to benefit from the modified SMR to quantizer mapping for mantissas with low SMRs coded using the AHT tool, E-AC-3 requires a new quantization tool. This new tool must provide the quantizer resolutions required by the psychoacoustic model for mantissas with very low SMRs at a lower bit per mantissa cost than the quantizers used in AC-3. In order to meet these requirements, the new tool must use nonuniform quantization. One very effective nonuniform quantization tool is vector quantization. In vector quantization, a multidimensional data vector is represented by a single codeword. The codeword indicates a specific vector of a predefined table of vectors. The length of the codeword depends on the number of vectors in the table, with longer tables requiring longer codewords. In general, as the number of vectors in the table increases, the quantizer resolution increases. For detailed information about vector quantization, see [26].

E-AC-3 uses six-dimensional vector quantization for transform coefficient mantissas with low SMRs coded by the AHT tool. Each vector consists of the six transform coefficient mantissas output from the secondary transform applied by the AHT tool. Because the AHT tool is used only for frames carrying six

TABLE 3.9 AC-3 and E-AC-3 SMR to Quantizer Mappings

SMR Range	AC-3 Quantizer Index	E-AC-3 Quantizer Index
<1.5 dB	0	0
[1.5 dB, 3.0 dB)	1	1
[3.0 dB, 4.5 dB)	1	2
[4.5 dB, 6.0 dB)	1	3
[6.0 dB, 7.5 dB)	1	4
[7.5 dB, 9.0 dB)	1	5
[9.0 dB, 10.5 dB)	2	6
[10.5 dB, 12.0 dB)	2	7

TABLE 3.10 Vector Quantization Codeword and Table Lengths

SMR Range	E-AC-3 Quantizer Index	Codeword Length (Bits)	Table Length
[1.5 dB, 3.0 dB]	1	2	4
[3.0 dB, 4.5 dB]	2	3	8
[4.5 dB, 6.0 dB]	3	4	16
[6.0 dB, 7.5 dB]	4	5	32
[7.5 dB, 9.0 dB]	5	7	128
[9.0 dB, 10.5 dB]	6	8	256
[10.5 dB, 12.0 dB]	7	9	512

audio blocks in which exponents are shared across all blocks, and the SMR for each mantissa is determined from its exponent, the SMR and required quantization resolution for each mantissa of a vector are the same. Therefore, a single SMR can be associated with each mantissa vector, and based on that SMR, the table length that provides the appropriate quantization resolution is selected. Table 3.10 presents the length of the vector quantization (VQ) table and codeword associated with each of the nonzero E-AC-3 quantization indices listed in Table 3.9. For mantissa vectors with a low SMR, a short table is sufficient to provide the quantization resolution required by the psychoacoustic model. As the mantissa vector SMR increases, and increased quantization resolution is needed, longer tables are required. In E-AC-3, the vector quantization tool is used only for mantissa vectors with SMRs below 12 dB. For mantissa vectors with SMRs above 12 dB, the tables required to provide sufficient quantization resolution require too much memory to be practical in low-cost decoding applications.

In the vector quantization tool used by E-AC-3, a single codeword of between 2 and 9 bits is used to represent each mantissa vector. To determine the optimal codeword for each mantissa vector, the E-AC-3 encoder first selects the appropriate table based on the mantissa vector SMR, and then searches through each vector in the table to determine the vector that best matches the mantissa vector. The optimal codeword for the mantissa vector, which is simply the table index of the best matching vector, is transmitted to the decoder. Upon receiving the codeword, the E-AC-3 decoder can restore the quantized mantissa coefficients by extracting from the appropriate table the vector indicated by the codeword. The vector quantization tool is very efficient because it can represent six mantissas by using between 2 and 9 bits. Table 3.11 presents the bits per mantissa costs of AC-3 and E-AC-3 for each of the SMR ranges that utilize vector quantization. In all cases, the quantizers used by E-AC-3 require fewer bits per mantissa than the corresponding AC-3 quantizers. This is possible because the quantization resolutions provided by the vector quantization tool better match the quantization resolutions required by the psychoacoustic model than the quantization resolutions provided by the uniform quantizers of AC-3.

TABLE 3.11 AC-3 and E-AC-3 Mantissa Transmission Costs

	AC-3		E-AC-3	
SMR Range	Quantizer Index	Bits per Mantissa	Quantizer Index	Bits per Mantissa
[1.5 dB, 3.0 dB)	1	10/6	1	2/6
[3.0 dB, 4.5 dB)	1	10/6	2	3/6
[4.5 dB, 6.0 dB)	1	10/6	3	4/6
[6.0 dB, 7.5 dB)	1	10/6	4	5/6
[7.5 dB, 9.0 dB)	1	10/6	5	7/6
[9.0 dB, 10.5 dB)	2	14/6	6	8/6
[10.5 dB, 12.0 dB)	2	14/6	7	9/6

3.3.3.2.2 Gain-Adaptive Quantization

In addition to the use of vector quantization for mantissas with low SMRs coded by the AHT tool, E-AC-3 includes another new quantization tool, referred to as gain-adaptive quantization (GAQ), designed to increase quantization efficiency without affecting quantization resolution for mantissas with higher SMRs coded by the AHT tool. The GAQ tool was originally developed for use in the Dolby E audio distribution codec and is described in detail in Refs. [27,28].

The GAQ tool reduces the cost of transmitting low-amplitude mantissas by amplifying them prior to quantization, quantizing the amplified mantissas with lower resolution, and then attenuating the dequantized mantissas to restore their original amplitude. High-amplitude mantissas are not amplified prior to quantization, and therefore must not be attenuated after dequantization. In order to prevent high-amplitude mantissas from being attenuated, a special codeword must precede each high-amplitude mantissa in the bit stream. Including the special codeword before each high-amplitude mantissa significantly increases the cost of transmitting high-amplitude mantissas. Therefore, the efficiency of the GAQ tool is directly related to the mantissa amplitude distribution.

In E-AC-3, the GAQ tool takes advantage of the skewing of the mantissa amplitude distribution towards smaller mantissas that often results from the use of the AHT tool. Using the GAQ tool for signals with such a mantissa amplitude distribution is considerably more efficient than using uniform quantization. In some cases however, the amplitude distribution of the mantissas output by the AHT tool is not significantly skewed toward smaller mantissas, and uniform quantization is more efficient than the GAQ tool. For these cases, the GAQ tool includes a fallback mode that emulates the uniform quantization used by AC-3, so that regardless of the amplitude distribution of the mantissas output from the AHT tool, the GAQ tool is always at least as efficient as, and typically more efficient than, the uniform quantization used by AC-3.

In addition to the mantissa amplitude distribution, the efficiency of the GAQ tool is also affected by the amount of amplification applied to low-amplitude mantissas prior to quantization. For example, consider a mantissa with an amplitude less than 0.5. When quantized by an m-bit uniform quantizer, the mantissa will have at least one leading sign bit. When quantized by the GAQ tool, the mantissa is amplified by a factor of two prior to quantization, quantized using an $(m - 1)$-bit uniform quantizer, and then attenuated by a factor of two, effectively reducing the cost of transmitting the mantissa by one bit without decreasing the quantization resolution. Similarly, mantissas with amplitudes less than 0.25 are amplified by a factor of four prior to quantization and then quantized with an $(m - 2)$-bit uniform quantizer to reduce the cost of transmitting the mantissa by two bits without decreasing the effective quantization resolution. Generally, for each doubling of the amplification factor, the cost of transmitting a low-amplitude mantissa is reduced by one bit without decreasing the effective quantization resolution. Unfortunately, in order for a decoder to know how much attenuation to apply to each dequantized mantissa, the gain factor applied to each mantissa prior to quantization must also be transmitted. If the cost of transmitting the gain factors for each mantissa, which is related to the number of possible gain factors, is too high, the reduction in quantized mantissa transmission cost achieved by the GAQ tool may be partially or completely offset by the gain factor transmission cost. Therefore, in order to be useful in practice, the GAQ tool must provide a good balance between the quantized mantissa transmission cost and the gain factor transmission cost.

The GAQ tool used by E-AC-3 uses two techniques to limit the gain factor transmission cost. The first technique is to limit the number of possible gain factors, and the second technique is to transmit gain factors that apply to a group of mantissas rather than a single mantissa. In E-AC-3, the GAQ tool only supports gain factors of one, two, and four, and each transmitted gain factor applies to the group of six mantissas output from the AHT tool for a given transform bin. Although these techniques lower the gain factor transmission cost, they may reduce the quantized mantissa transmission efficiency. Because each gain factor applies to a group of six mantissas, using a gain factor of one results in any low-amplitude mantissas in the group being coded less efficiently, while using a gain factor of two or four results in any high-amplitude mantissas in the group being coded less efficiently. Typically, an E-AC-3 encoder will select for each group the gain factor that results in the lowest quantized mantissa transmission cost.

To further reduce the gain factor transmission cost, four operating modes are defined, and each channel that uses the GAQ tool operates in one of the four modes for the entire frame. In the first operating mode, all gain factors are assumed to be one, and are therefore not transmitted. This mode effectively disables the GAQ tool and is nearly equivalent to the uniform quantization used by AC-3. The main difference is that this mode uses symmetric quantization rather than asymmetric quantization. This is the fallback mode discussed previously, and is used for signals in which the mantissa amplitude distribution for a channel using the AHT tool is nearly uniform.

In the second mode, only gain factors of one and two are allowed. A 1-bit codeword is transmitted for each transform bin, which indicates the gain factor for the group of six mantissas output from the AHT tool for that transform bin. For each mantissa group, the encoder selects the gain factor which minimizes the quantized mantissa transmission cost for the group. Typically, this mode is used when the mantissa amplitude distribution for a channel using the AHT tool is moderately skewed toward smaller mantissas.

The third mode is identical to the second mode, except that gain factors of one and four are allowed rather than gain factors of one and two. Again, for each mantissa group, the encoder selects the gain factor which minimizes the quantized mantissa transmission cost for the group. This mode is typically used when the mantissa amplitude distribution for a channel using the AHT tool is significantly skewed toward smaller values.

Finally, in the fourth mode, gain factors of one, two, and four are allowed. Because there are three possible gain factors, a 1-bit codeword is not sufficient to indicate the gain factor. However, using a 2-bit codeword to indicate the gain factor is inefficient because one state of the codeword is never used. Therefore, to minimize the gain factor transmission cost, gain factors are grouped into triplets and compositely coded using a single 5-bit codeword, resulting in a transmission cost of 1.67 bits per gain factor. Again, for each mantissa group, the encoder selects the gain factor which minimizes the quantized mantissa transmission cost for the group. This mode is typically used when the mantissa amplitude distribution for a channel using the AHT tool falls between moderately skewed toward smaller mantissas, where the second mode is optimal, and significantly skewed toward smaller mantissas, where the third mode is optimal.

In order to determine which operating mode to use for each channel, the encoder calculates the transmission cost for each of the four modes and selects the mode that minimizes the transmission cost. The transmission cost for the first mode depends only on the bit allocation for each transform bin, and is essentially equivalent to the transmission cost of using uniform quantization. For the other three modes, in addition to the bit allocation for each transform bin, the transmission cost also depends on the gain factors and the mantissa amplitude distribution, and may be higher, lower, or equivalent to uniform quantization. When the first mode has the lowest transmission cost, using the GAQ tool is as efficient and accurate as using uniform quantization, since it yields equivalent quantization resolution at essentially the same transmission cost. When any of the other modes have a lower transmission cost than the first mode, using the GAQ tool is as accurate as, and more efficient than, uniform quantization, since it yields equivalent quantization resolution at a lower transmission cost. Therefore, if used properly, the GAQ tool can often increase, but never reduce, the quantization efficiency for mantissas with high SMRs.

3.3.3.3 Enhanced Channel Coupling

AC-3 supports a data compression tool called channel coupling, which relies on psychoacoustic properties of the human hearing system to enable significant data reduction without introducing significant audible artifacts. Unfortunately, channel coupling is most effective at higher frequencies, and as the frequency at which channel coupling begins, commonly called the coupling frequency, is lowered, artifacts in the decoded audio become more apparent. For low bit-rate applications, like some of those targeted by E-AC-3, it is advantageous to lower the coupling frequency as much as possible in order to maximize data compression. However, if the coupling frequency is lowered too much, artifacts due to channel coupling can become unacceptable.

Two main issues lead to the degradation of the performance of channel coupling at lower frequencies. Enhanced channel coupling was designed to address those issues, enabling improved performance of channel coupling at lower frequencies, resulting in improved quality for low bit-rate applications. In order to understand the two main issues that are addressed by enhanced channel coupling, a brief review of how channel coupling operates is necessary. In channel coupling, above the coupling frequency a composite channel of frequency coefficients, commonly called the coupling channel, is generated and transmitted to the decoder in place of discrete frequency coefficients for each channel. In addition to the coupling channel frequency coefficients, scale factors for each channel, commonly called coupling coordinates, are computed and transmitted to the decoder. The decoder uses the coupling coordinates to restore the high frequency spectral envelope of each channel from the coupling channel frequency coefficients.

The first issue that leads to the degradation of the performance of channel coupling at lower frequencies is that due to phase differences in the coupled channels, significant cancellation may occur when generating the coupling channel. In some situations, the cancellation may be so significant that, due to the limited dynamic range of the coupling coordinates, the decoder is unable to faithfully restore the spectral envelope of the original signal. The result is a reduction of the energy of the decoded signal with respect to the original signal, which may result in a decrease in the perceived loudness of the decoded signal. When the coupling frequency is relatively high, the energy loss is constrained to the high frequency region of the spectrum and may not be noticeable. However, as the coupling frequency is lowered into the low to mid-frequency range, over which the ear is more sensitive, the energy loss may become more apparent, resulting in a significant decrease in the perceived loudness of the decoded signal.

Enhanced channel coupling addresses the issue of cancellation during generation of the coupling channel by modifying the method used by standard coupling for generating the coupling channel frequency coefficients. For standard coupling, the coupling channel frequency coefficients are generated by a simple summation of the frequency coefficients of each individual channel. For enhanced coupling, prior to summing the frequency coefficients for each individual channel, a phase compensation is applied to the coefficients of each channel. The phase compensation aligns the phases of the frequency coefficients of the individual channels to minimize cancellation. Once phase compensation has been applied, a simple summation of the phase-compensated frequency coefficients of the individual channels can be performed with minimal cancellation occurring.

The second issue that leads to the degradation of the performance of channel coupling at lower frequencies is that the channel coupling tool does not typically preserve the phase information of the individual channels. Above the coupling frequency, each decoded output channel is simply an amplitude scaled version of the coupling channel, and therefore all decoded output channels are reproduced with identical phase. Interchannel phase differences which existed in the original signal are not reproduced in the decoded signal. When the coupling frequency is relatively high, the loss of phase information for individual channels is typically not audible. However, as the coupling frequency is lowered into the low to mid-frequency range, over which the ear is more responsive to phase differences between the individual channels, the loss of interchannel phase differences may become more apparent, resulting in a perceived loss of dimensionality of the signal.

Enhanced channel coupling addresses the issue of loss of phase information by including additional metadata parameters in the bit stream that help the decoder model the interchannel phase relationships of the original signal. These additional metadata parameters specify angular differences between each channel and a reference channel, commonly chosen to be the first channel participating in channel coupling. To generate each output channel, the decoder applies to the coupling channel an angular rotation based on the additional metadata. The resulting output signals model the interchannel phase differences present in the original signals. By enabling the decoder to model the interchannel phase differences of the original channels, enhanced coupling can reduce the perceived loss of dimensionality which may result from using the channel coupling tool at lower frequencies.

In addition to the degradation of the performance of channel coupling at lower frequencies described above, using channel coupling can also be detrimental to two-channel matrix encoded signals. Typically, matrix decoding devices determine steering cues from interchannel phase relationships. When channel coupling is applied to a two-channel matrix encoded program, above the coupling frequency the interchannel phase relationships are not preserved, and both channels are reproduced with identical phase. Because matrix decoding systems typically reproduce signals with identical phases from the front channels, any high-frequency surround channel information present in the two-channel matrix encoded signal will be lost as a result of channel coupling. Since matrix encoded signals often contain very little high-frequency surround channel information, this is not a significant problem for moderate bit-rate applications in which the coupling frequency is relatively high. However, as the coupling frequency is lowered into the low to mid-frequency range to support low bit-rate applications, the loss of surround information above the coupling frequency from two-channel matrix encoded signals becomes much more noticeable. Because enhanced channel coupling attempts to restore the interchannel phase relationships of the original signal, it is better suited to coding two-channel matrix encoded signals than channel coupling, especially for low bit-rate applications for which the coupling frequency is in the low to mid-frequency range.

3.3.3.4 Audio Bandwidth Extension

While channel coupling, either standard or enhanced, is capable of providing significant data reduction, for some very low bit-rate applications, channel coupling may not provide sufficient data reduction to yield acceptable audio quality at the target bit-rates. Additionally, because channel coupling requires at least two channels, it cannot be used to provide data reduction for monophonic signals. Therefore, to provide additional data reduction for multichannel signals, and facilitate data reduction for monophonic signals, E-AC-3 includes an audio bandwidth extension tool, referred to as Spectral Extension (SPX), that can be used in place of, or in addtition to, channel coupling. While audio bandwidth extension was originally proposed to provide significant data reduction for low-bandwidth speech coding applications, if care is taken, it is possible to apply the bandwidth extension concepts originally proposed for speech coding to wideband audio coding.

The concept behind audio bandwidth extension is very simple. First, an encoding device generates a bandlimited version of a wideband signal. The bandlimited signal is then transmitted to a decoding device, which synthesizes the high frequency portion of the signal from the bandlimited version. In addition to the bandlimited signal, the encoder may also generate and transmit some additional data to assist the decoder in synthesizing the high frequency portion of the signal. Typically, even a relatively small amount of additional data can significantly increase the effectiveness of the bandwidth extension tool. Figure 3.14 shows the original spectral envelope of a wideband audio signal, and Figure 3.15 shows the spectral envelope of the bandlimited version of the signal generated by an encoder. Figures 3.16 through 3.18 show steps of the high frequency synthesis performed by the decoding device.

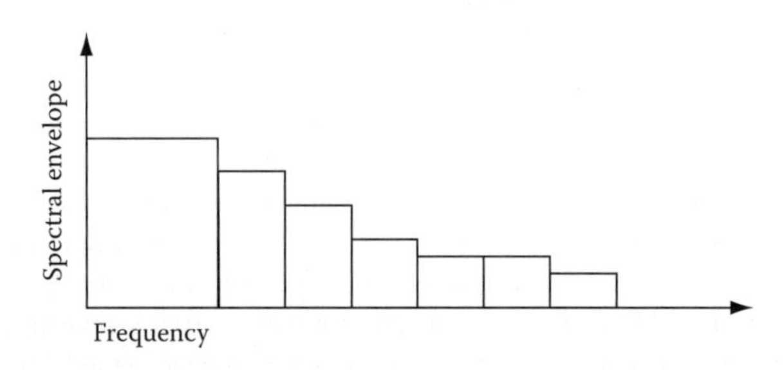

FIGURE 3.14 Spectral envelope of original signal.

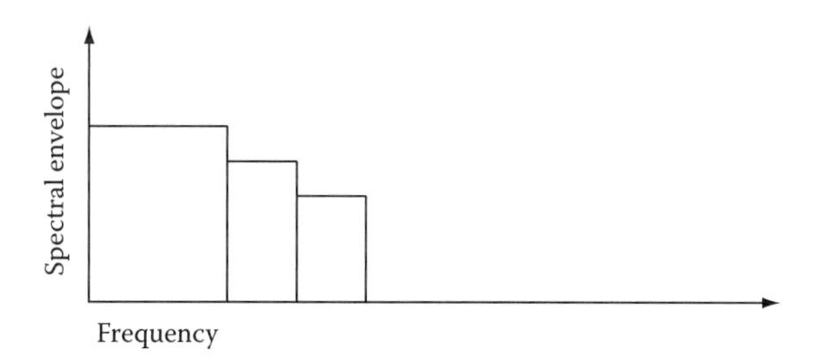

FIGURE 3.15 Spectral envelope of bandlimited signal.

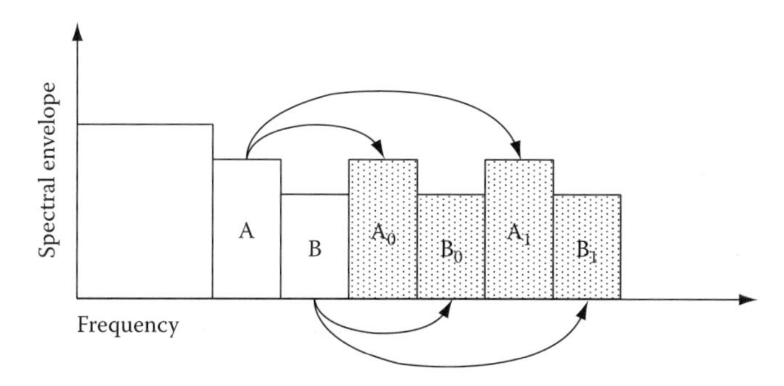

FIGURE 3.16 Synthesized high frequency signal after broadband linear translation.

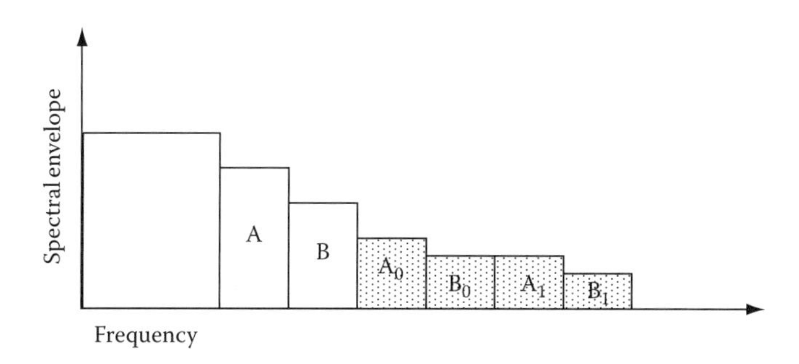

FIGURE 3.17 Synthesized high frequency signal after spectral envelope modification.

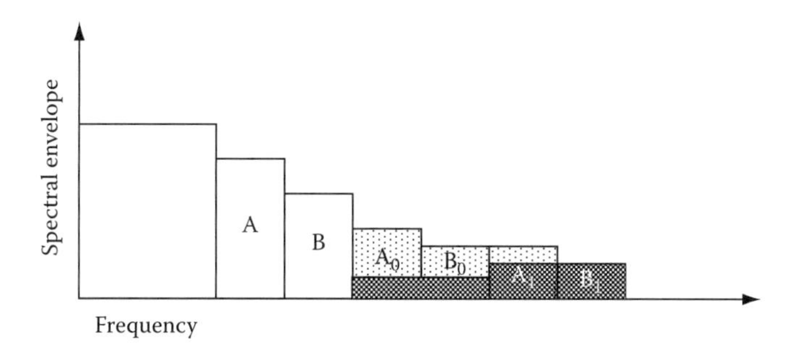

FIGURE 3.18 Synthesized high frequency signal after noise blending.

Both the amount of data reduction and the quality of the decoded audio signal resulting from using bandwidth extension depend on how much of the original signal, and how much additional data, is transmitted to the decoder. Transmitting more of the original signal improves the audio quality of the decoded signal, but decreases the amount of data reduction. Similarly, transmitting less of the original signal increases the amount of data reduction, but decreases the audio quality of the decoded signal. To a lesser extent, the same is true of the additional data. Therefore, designing and using a bandwidth extension system requires trade-offs between the amount of data reduction achieved and the audio quality of the decoded signal.

For both low bandwidth speech and wideband audio coding, many methods of bandwidth extension, varying from quite simple to extremely complex, have been proposed. One example of such an audio bandwidth extension method is spectral band replication (SBR), described in Ref. [29]. Typically, as the complexity of the bandwidth extension method increases, the quality of the synthesized signal also increases. However, if the complexity of the bandwidth extension tool is too high, the cost of implementing the technique may limit the usefulness of the method in practice, especially for relatively low-cost, mass market decoding applications. In order to constrain decoder complexity, the SPX tool in E-AC-3 uses a relatively simple, but very effective, audio bandwidth extension method.

The bandwidth extension method employed by the SPX tool uses a technique called linear translation to synthesize the high frequency components of a signal. In linear translation, a copy of the low frequency components of a signal is translated up in frequency to synthesize the high frequency components of the signal. Typically, groups of consecutive low frequency components are shifted by a constant amount to reproduce groups of consecutive high frequency components. In some cases, multiple groups of consecutive high frequency components may be synthesized from the same group of consecutive low frequency components. Figure 3.16 shows an example of linear translation in which two consecutive groups of low frequency components from the bandlimited signal shown in Figure 3.15 are copied and translated to synthesize four consecutive groups of high frequency components. Low frequency group A is shifted once to generate high frequency group A_0 and then again to generate high frequency group A_1. Similarly, low frequency group B is shifted once to generate high frequency group B_0 and again to generate high frequency group B_1. The synthesized high frequency components are shaded in Figure 3.16 to indicate that they are different from the high frequency components of the original signal shown in Figure 3.14.

Linear translation relies on the fact that the low frequency content of an audio signal is related to the high frequency content, and that a good approximation of the original high frequency components of a signal can be obtained from the low frequency components of the signal. However, because linear translation of frequency components does not maintain the exact relationships between low and high frequency harmonic components, some high frequency harmonic components may be reproduced at slightly different locations than they existed in the original signal. Fortunately, as frequency increases, humans become less sensitive to errors in the location of harmonic components, and therefore, even though the high frequency harmonic components synthesized by linear translation may not be reproduced in exactly the same location as they existed in the original signal, as long as they are reproduced sufficiently close to their original location, the synthesized signal will sound natural.

Although, as mentioned previously, the high and low frequency content of an audio signal are typically related, they are rarely identical. Therefore, after performing linear translation, the high frequency portion of the synthesized signal rarely matches the high frequency portion of the original signal, and must be modified so that it more closely approximates the high frequency portion of the original signal. Because humans typically respond mostly to the envelope and not the phase of high frequency signals, it is often sufficient to modify the synthesized signal so that the high frequency spectral envelope of the synthesized signal matches the high frequency spectral envelope of the original signal, without being concerned about matching the phase of the synthesized signal to the phase of the original signal. In order to appropriately modify the high frequency spectral envelope of the synthesized signal, the SPX tool requires information about the high frequency spectral envelope of the original signal. This information is determined by the encoder and transmitted to the decoder in the E-AC-3 bit stream. Figure 3.17 shows the result of modifying the high frequency spectral envelope of the signal in Figure 3.16 based on the information provided by the encoder about the high frequency spectral envelope of the original signal. Again, the high frequency spectral envelope of the synthesized signal is shaded to indicate that the individual high frequency spectral components of Figure 3.17 may be different from the individual high frequency spectral components of the original signal shown in Figure 3.14.

Another characteristic of audio signals that the SPX tool must account for is that they tend to become more noise-like at higher frequencies. Therefore, the high frequency signal components synthesized by linear translation may not be sufficiently noise-like, which can result in an unnatural, metallic sounding characteristic in the decoded audio signal. In order to create a more natural sounding decoded signal, the SPX tool includes a method for blending noise into the synthesized high frequency signal components in a way that increases the proportion of noise-like signal components in the synthesized high frequency signal without altering the synthesized high frequency spectral envelope. Typically, the amount of noise blended into the synthesized high frequency components by the SPX tool increases as the frequency of the synthesized component increases. However, in order to blend the correct proportion of noise-like signal components into the synthesized high frequency signal, the SPX tool requires information about the noise-like characteristics of the high frequency portion of the original signal. This information is determined by the encoder and transmitted to the decoder in the E-AC-3 bit stream. Figure 3.18 shows the result of blending noise in increasing proportion into the synthesized high frequency signal components of the signal in Figure 3.17 based on the information provided by the encoder about the noise-like characteristics of the original signal. In Figure 3.18, the darker shaded regions represent the noise components blended into the synthesized high frequency components in increasing proportion. Figure 3.18 shows the final result of the bandwidth extension process: a signal in which the spectral envelope of the original signal shown in Figure 3.14 has been faithfully restored from the bandlimited signal shown in Figure 3.15 and additional data provided by the encoder. The shaded regions of Figure 3.18 indicate that while the spectral envelope of the synthesized high frequency signal matches the spectral envelope of the original signal shown in Figure 3.14, the individual spectral components of the synthesized signal may be different from the original signal.

In order to make the SPX tool useful for a variety of applications operating at many different bit-rates, a significant amount of control over the SPX tool is built into the E-AC-3 bit stream syntax. An encoder may select, from one of eight possibilities, the frequency at which the SPX tool begins operating. This is useful because for lower bit-rate applications, it may be advantageous to begin operating the SPX tool at a relatively low frequency, while for higher bit-rate applications it may be advantageous to begin operating the SPX tool at a relatively high frequency. Also, although the frequency at which the SPX tool begins operating is typically fixed for a specific application, in some cases, to improve performance for signals with dynamically changing characteristics, an encoder may choose to dynamically alter the frequency at which the SPX tool begins operating, or completely disable the SPX tool for one, several, or all of the individual audio channels. Finally, for some applications, an encoder may choose to use the SPX tool in conjunction with the channel coupling tool. For these applications, where operating the SPX tool too low in frequency does not provide adequate audio quality, and coding channels independently up to the frequency at which the SPX tool begins operating requires too much data, using channel coupling for the mid-range frequency components provides an effective transition between independently coding low frequency components and using the SPX tool to synthesize high frequency components.

3.3.3.5 Transient Prenoise Processing

The introduction of distracting quantization noise prior to a transient event in a decoded audio signal is a well-known problem in transform coding. The problem occurs because quantization noise introduced into the signal during the encoding process spreads uniformly throughout the transform block during the decoding process. When a transform block contains a strong transient event, in which a very loud sound is preceded by silence or near silence, the quantization noise reproduced in the decoded signal preceding the transient can be audible and distracting. Two main factors, duration and level, influence the audibility of the prenoise. The transform block length determines the duration of the prenoise, and as the transform block length increases, the duration of the prenoise increases. The bit-rate at which the signal is coded determines the level of quantization noise, and therefore prenoise, that is introduced into the decoded signal. As the level of quantization noise introduced into the signal increases, the level of the prenoise increases.

Block-length switching is a common method employed by transform codecs, including both AC-3 and E-AC-3, to reduce the audiblilty of transient prenoise. Block-length switching involves shortening the transform length for signals which contain transient events, and is used to limit the duration of the prenoise to minimize audibility. Typically, prenoise that is limited to less than 5 ms in duration is masked by the transient event, while prenoise greater than 5 ms in duration may be audible. Unfortunately, for some of the very low bit-rate applications for which E-AC-3 is targeted, a significant amount of quantization noise may be added to the signal, and block-length switching alone may not be sufficient to reduce audible prenoise to an acceptable level. Additionally, because the efficiency of transform codecs is decreased when the transform block length is decreased, it may be desirable for very low bit-rate applications to disable block-length switching entirely and always use longer transform lengths. To provide improved performance and increased flexibility for low bit-rate applications, E-AC-3 includes an additional tool, referred to as transient prenoise processing (TPNP), for reducing the audibility of transient prenoise. The TPNP tool can be used in conjunction with, or as a substitute for, block-length switching.

It is common for audio signals to contain long, with respect to the transform block length, periods of silence or low-level audio immediately preceding transient events. Therefore, it is also common for a decoded signal to contain a long period of silence or low-level audio immediately preceding transient prenoise introduced during the decoding process. The TPNP tool takes advantage of this characteristic by replacing samples containing prenoise with samples from the quiet section preceding the samples containing prenoise. The TPNP tool operates on the decoded time-domain audio samples, and may be implemented as a single-ended, decoder only operation, or as a double-ended operation in which complex signal analysis is performed by an encoder, metadata is transmitted to a decoder, and the decoder performs only simple synthesis operations. Because the single-ended approach significantly increases decoder complexity and latency, it is undesirable for some of the low-cost, mass-market decoding applications targeted by E-AC-3. Therefore, the double-ended approach, which has minimal impact on decoder complexity and latency, but requires a small amount of additional bit stream metadata, is used in E-AC-3.

The analysis portion of the TPNP tool, which contains the most complicated processing, is performed by the E-AC-3 encoder. The analysis consists of three main operations: determining when transient events occur, determining the set of decoded samples which will contain prenoise, and determining the optimal set of replacement samples used to overwrite the decoded samples containing prenoise. The encoder already determines when transients occur in order to perform block-length switching, and can reuse this information for the TPNP analysis processing. The set of decoded samples which will contain prenoise extends from the first sample of the block containing the transient to the beginning of the transient. Since the location of the first sample of the block is inherently known by the encoder, determining the location of the beginning of the transient is sufficient to determine the set of decoded samples which will contain prenoise. Although the encoder already detects when a block contains a transient, it does not provide information about the location of the beginning of the transient within the block. Therefore, the encoder must perform additional analysis to determine the location of the beginning of the transient. After performing the analysis, the location of the beginning of the transient is quantized to within four samples in order to reduce the amount of data transmitted to the decoder.

To determine the optimal set of replacement samples used to overwrite the decoded samples containing prenoise, the encoder uses a technique called maximum similarity processing. First, the encoder selects a set of candidate replacement samples, referred to as the synthesis buffer, which is larger than the total number of samples to be replaced, and consists of samples that will not contain audible prenoise. Next, the encoder compares a subset of samples from the synthesis buffer to the samples which will be replaced. Then, the encoder compares a different subset of samples from the synthesis buffer to the samples which will be replaced. The encoder performs a fixed number of comparisons using different subsets, and chooses the subset which has maximum similarity to the samples being replaced. After performing the maximum similarity processing, an indication of the optimal set of replacement samples from the synthesis buffer is transmitted to the decoder.

Because the majority of the TPNP analysis processing is performed by the E-AC-3 encoder, the synthesis processing performed by the TPNP tool in the E-AC-3 decoder is very simple. First, the decoder determines from metadata transmitted in the E-AC-3 bit stream for which channels, if any, TPNP is enabled. Next, for those channels for which TPNP is enabled, the decoder determines the location of the transient from bit stream metadata. Then, the decoder determines from bit stream metadata which samples will be replaced, and which samples will be used for replacement. Finally, the decoder performs the sample replacement. The set of samples to be replaced consists of three subregions. In the first subregion, the samples to be replaced are faded out and the replacement samples are faded in. In the second subregion, the samples to be replaced are completely overwritten by the replacement samples. Finally, in the third subregion, the samples to be replaced are faded in and the replacement samples are faded out. The crossfades are performed so that any discontinuites introduced into the signal by the overwriting process are deemphasized.

Figure 3.19 shows an example signal processed with the TPNP tool. The top waveform of the figure contains a lengthy segment of prenoise and represents the decoded output of a transform codec prior to modification by the TPNP tool. The prenoise is immediately preceded by a segment of silence. The bottom waveform shows the top waveform after modification by the TPNP tool. Samples from the silent segment preceding the transient have been used to overwrite the majority of the prenoise, resulting in a final output signal with significantly reduced prenoise duration. For additional details on the TPNP tool, see [30].

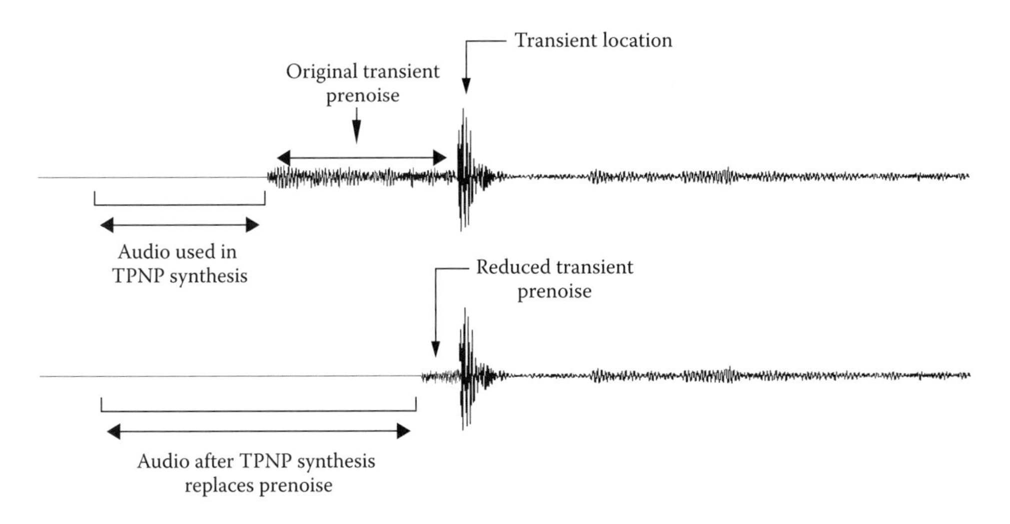

FIGURE 3.19 Audio signal processed by TPNP tool.

3.3.4 Efficient, High Quality E-AC-3 to AC-3 Conversion

As mentioned earlier, a key design goal for E-AC-3 was maintaining a high degree of interoperability with AC-3. While it was not possible to make E-AC-3 totally backwards compatible with AC-3 and still satisfy all of the other design goals, it was possible to design E-AC-3 to enable a higher quality, more efficient conversion of E-AC-3 bit streams to AC-3 bit streams than a traditional transcode. There are two main drawbacks normally associated with transcoding between two incompatible formats. The first drawback is that transcoding typically introduces a significant amount of additional distortion into the coded signal. The main source of additional distortion is tandem coding loss, which generally occurs because the frequency coefficients quantized during the first coding stage are different from the frequency coefficients quantized during the second stage, and the quantization errors introduced during each coding stage combine additively. The second drawback is that transcoding is computationally complex, and often prohibitively costly for low-cost consumer applications. Typically, transcoding requires an input signal in a first compressed audio format to be decoded to uncompressed PCM audio data, and then recompressed into an output signal in the second compressed audio format. While decoding devices are often designed for low-cost consumer applications, and therefore relatively efficient, encoding devices, especially those aimed at providing very high audio quality, are often very complex. Because transcoding devices must perform both decoding and encoding operations, they are often significantly more complex, and therefore more expensive, than ordinary decoding devices.

E-AC-3 was designed to allow conversion of E-AC-3 bit streams to AC-3 bit streams with lower tandem coding loss than a traditional transcoding device. Because tandem coding loss typically occurs when the frequency coefficients from the first coding stage differ from the frequency coefficients of the second coding stage, one way to reduce tandem coding loss is to ensure that the frequency coefficients of the first and second coding stages match. When this condition is met, the total quantization error introduced by the first and second coding stages is roughly equivalent to the larger of the quantization errors introduced by each individual coding stage, rather than the sum of the quantization errors introduced by each coding stage. In order to ensure that the frequency samples between the first and second coding stages match, two conditions must be met: first, the same time to frequency transform must be used for each coding stage, and second, the input sample alignment must be the same for each coding stage. Because E-AC-3 and AC-3 use the same time to frequency transform, provided that the conversion device maintains the input sample framing of the E-AC-3 bit stream when generating the AC-3 bit stream, the frequency coefficients of each coding stage will match, and tandem coding losses can be minimized.

Although the E-AC-3 to AC-3 conversion device attempts to minimize tandem coding loss, in many cases it cannot eliminate tandem coding loss completely. For reasons that will be explained later, in the AC-3 bit stream generated by the conversion device, all channels are coded independently. If any coding tools are used by E-AC-3, some or all of the frequency coefficients of the E-AC-3 and AC-3 bit streams may be different, and tandem coding losses may be unavoidable. For example, when the E-AC-3 bit stream uses the AHT tool, due to the secondary transform, none of the frequency coefficients of the E-AC-3 and AC-3 bit streams will match. Because it is not always possible to ensure that all frequency coefficients of the E-AC-3 and AC-3 streams match, it is advantageous to minimize the amount of quantization error introduced during the second coding stage in order to minimize tandem coding losses. Therefore, the AC-3 bit stream generated by the conversion device is always coded at 640 kbps, the maximum bit-rate supported by AC-3.

In addition to minimizing tandem coding losses, E-AC-3 was also designed to allow conversion of E-AC-3 bit streams to AC-3 bit streams with lower complexity than a traditional transcoding device. Typically, transcoding devices perform a complete decode of the compressed input signal to uncompressed PCM, followed by a complete encode of the PCM data into the desired output format. However, because of the similarities between E-AC-3 and AC-3, some complex decoding and encoding operations can be eliminated from the conversion device without impacting audio quality.

Typically, AC-3 encoding devices perform a series of PCM processing operations prior to transforming PCM input signals to the frequency domain. Some examples of these operations are input signal filtering, dynamic range control parameter calculation, and transient detection. These operations are also typically peformed by E-AC-3 encoding devices, and since the conversion device maintains the alignment between the E-AC-3 and AC-3 bit streams, the results calculated by the E-AC-3 encoder can be extracted from the E-AC-3 bit stream and reused by the conversion device. Because the conversion device does not need to perform any PCM processing operations, and both formats use a common transform, conversion between the two formats can be accomplished entirely in the transform domain, and frequency to time and time to frequency transformations, a significant source of complexity, can be completely eliminated from the conversion device.

Another source of significant complexity that can be removed from the conversion device is the use of joint channel coding tools. Although the decoding operations associated with channel coupling and rematrixing, the joint channel coding tools supported by AC-3, are relatively simple, the encoding operations associated with these tools are computationally very intensive. Therefore, in order to simplify the E-AC-3 to AC-3 conversion process, all channels are coded independently, and no joint channel encoding is performed by the conversion device. Although coding all channels independently reduces the efficiency of AC-3, because the conversion device always produces a 640 kbps output bit stream, the reduced efficiency is not a significant problem. Channel coupling, which can be used to provide significant data reduction for signals containing two or more channels, is commonly disabled entirely by AC-3 encoding devices operating at 640 kbps. Rematrixing, which can only be used for stereo signals, is unnecessary when operating at 640 kbps, which significantly exceeds the per channel bit-rate required by AC-3 to transparently code channels independently.

Although some complex operations, like block transforms and joint channel coding, can be completely eliminated from the conversion device, many others are essential to maintaining acceptable audio quality. Fortunately, because of the similarities between E-AC-3 and AC-3, some of these essential encoding operations can be moved upstream from the conversion device to the E-AC-3 encoding device without affecting the quality of the AC-3 bit stream produced by the conversion device. Because encoding devices are typically low-volume professional products, for which cost is less critical than mass-market consumer devices, shifting complexity from conversion devices to the E-AC-3 encoding devices is often beneficial.

One example of moving a significant source of encoding complexity from the conversion device to the E-AC-3 encoding device is the binary search performed to determine the optimal transform coefficient mantissa bit allocation. The mantissa bit allocation, which is dependent on the psychoacoustic masking curve, can be altered by applying a global offset to the masking curve. An encoding device must perform a binary search of all possible offsets to determine the offset that results in allocating as many mantissa bits as possible without exceeding the mantissa bit budget. Because each iteration of the binary search requires recomputing the manitssa bit allocation, the iterative search is computationally very intensive. Fortunately, all of the information required to determine the optimal offset for the AC-3 bit stream generated by the conversion device is available to the E-AC-3 encoder. The E-AC-3 encoder can perform the iterative search for the conversion device and include the optimal offset value in the E-AC-3 bit stream. Therefore, instead of performing a computationally intensive iterative search to determine the optimal masking curve offset, the conversion device can simply extract the optimal offset from the E-AC-3 bit stream, and only perform the mantissa bit allocation once.

A second example of moving a significant source of encoding complexity from the conversion device to the E-AC-3 encoding device is the exponent sharing strategy calculation. The exponent sharing strategy calculation determines how exponents are shared across both time and frequency. Typically, the exponent sharing strategy is determined by examining how the power spectrum of a signal varies with time and frequency. In order to determine the true power spectrum of the signal, both the MDCT and modified discrete sine transform (MDST) coefficients are required. However, the conversion device only has access to the quantized MDCT coefficients from the E-AC-3 bit stream. While it is possible to determine the MDST coefficients from the MDCT coefficients, doing so is computationally expensive

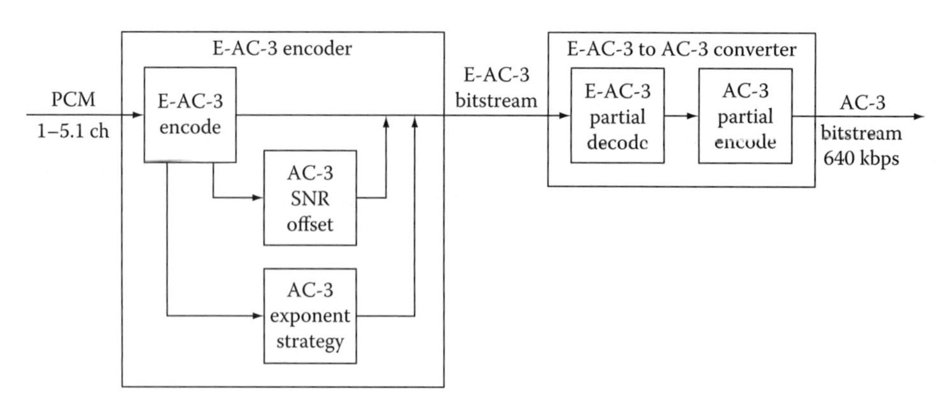

FIGURE 3.20 E-AC-3 to AC-3 conversion process.

and may introduce additional latency. Furthermore, depending on the amount of quantization error introduced into the MDCT coefficients by the E-AC-3 encoder, the resulting MDCT and MDST coefficients may be significantly different than the unquantized coefficients, and may not yield the optimal exponent sharing strategy. Fortunately, E-AC-3 encoding devices typically compute both the MDCT and MDST coefficients, and prior to MDCT quantization can compute the ideal exponent sharing strategy for the AC-3 bit stream generated by the conversion device and include it in the E-AC-3 bit stream. By moving the exponent sharing strategy calculation from the conversion device to the E-AC-3 encoding device, not only is the computational complexity of the conversion device reduced, but also the quality of the AC-3 bit stream generated by the conversion device is increased.

Figure 3.20 shows a block diagram of the E-AC-3 to AC-3 conversion process. First, an E-AC-3 encoder encodes a PCM audio signal into an E-AC-3 bit stream. The E-AC-3 encoder also determines the optimal exponent sharing strategy and masking curve offset for the AC-3 bit stream produced by the conversion device, and packs them into the E-AC-3 bit stream. The conversion device receives the E-AC-3 bit stream and performs a partial decode to convert the frequency coefficients of the E-AC-3 bit stream into a format compatible with the AC-3 bit stream. Finally, the conversion device uses the exponent sharing strategy and masking curve offset information from the E-AC-3 bit stream to encode the converted frequency coefficients into a 640 kbps AC-3 bit stream.

3.4 Conclusions

In this chapter, we presented two perceptual audio codecs, AC-3 and E-AC-3, that have played a significant role in the creation of new consumer electronics products. AC-3 is used worldwide to deliver high-quality multichannel audio for applications such as DVD and HDTV. AC-3 was the first multichannel surround sound codec offered to the broadcast market, and also the first audio codec to incorporate extensive metadata features to enhance the listening experience. E-AC-3 is deployed in next generation Blu-ray disc players and digital media broadcast systems. E-AC-3 is a nonbackwards compatible extension to AC-3, the core technology on which it is built. E-AC-3 enables new features and increases compression efficiency, while maintaining a high degree of interoperability with AC-3.

While it is insightful to reflect on the deployment history, architecture, and market applications of these two codecs, it is also of interest to consider their futures. Although AC-3 has been in use since the early 1990s, it is still very relevant in the broadcast, PC, and consumer electronics markets 15 years later, and likely will remain so for many years to come. The success of AC-3, as well as other audio codecs such as MPEG-1 Layer 3 (MP3) and MPEG-4 HE-AAC, has demonstrated that a significant market demand exists for high-quality audio data compression. Today, competing audio codecs are proliferating in the marketplace, leading to new challenges that slow the widespread adoption of new formats such as

E-AC-3. However, because of the desire to ensure audio content portability and compatibility for consumers, it is common practice for consumer electronics products to support multiple codecs. As it is likely that this practice will continue well into the future, it is reasonable to assume that, as established codecs, both AC-3 and E-AC-3 will be included in such products.

In general, the number of audio delivery applications is likely to continue expanding at a rapid pace, and in turn the demand for efficient, high-quality audio codecs is anticipated to continue. Although some researchers may reasonably argue that asymptotic limits on the performance of audio codecs have been reached, or are nearly within reach, for the foreseeable future many others will continue striving to advance the state of the art by developing new perceptual audio codecs as well as refining existing ones.

References

1. T. Painter and A. Spanias, Perceptual coding of digital audio, *Proceedings of the IEEE*, 88(4), 451–515, April 2000.
2. ISO/IEC 14496-3:2005(E), Information technology—coding of audio-visual objects—part 3: Audio, International Organization for Standardization, Geneva, Switzerland, 2005.
3. Document A/52B, Digital audio compression tandard (AC-3, E-AC-3), Advanced Television Systems Committee, Washington, D.C., June 14, 2005.
4. Document A/53, ATSC digital television standard, parts 1–6, Advanced Televison Systems Committee, Washington, D.C., January 3, 2007.
5. L.D. Fielder, et. al., Introduction to dolby digital plus, an enhancement to the dolby digital coding system, *117th Convention of the Audio Engineering Society*, San Francisco, CA, Preprint 6196, October 2004.
6. Recommendation ITU-R BS.1115, Low bit-rate audio coding, International Telecommunication Union, Geneva, Switzerland 1994.
7. Recommendation ITU-R BS.1116-1, Methods for the subjective assessment of small impairments in audio systems including multichannel sound systems, International Telecommunication Union, Geneva, Switzerland, October 1997.
8. G. Soulodre, T. Grusec, M. Lavoie, and L. Thibault, Subjective evaluation of state-of-the-art 2-channel audio codecs, *Journal of the Audio Engineering Society*, 46, 164–177, March 1998.
9. D. Grant, G. Davidson, and L. Fielder, Subjective evaluation of an audio distribution coding system, *111th Convention of the Audio Engineering Society*, San Francisco, CA, Preprint 5443, Sept. 2001.
10. Recommendation ITU-R BS.1534-1, Method for the subjective assessment of intermediate quality levels of coding systems, International Telecommunication Union, Geneva, Switzerland, January 2003.
11. L. Gaston and R. Sanders, Evaluation of HE-AAC, AC-3 and E-AC-3 Codecs, *Journal of the Audio Engineering Society*, 56, 140–155, March 2008.
12. EBU-Tech 3324, EBU evaluations of multichannel audio codecs, European Broadcasting Union, Geneva, Switzerland, September 2007.
13. Recommendation ITU-R BS.775-1, Multichannel stereophonic sound system with and without accompanying picture, International Telecommunication Union, Geneva, Switzerland 1994.
14. J. P. Princen, A. W. Johnson, and A. B. Bradley, Subband/transform coding using filter bank designs based on time domain aliasing cancellation, *IEEE International. Conference on Acoustics, Speech, and Signal Processing*, Dallas, TX, pp. 2161–2164, 1987.
15. R. E. Crochiere and L. R. Rabiner, *Multirate Digital Signal Processing*, Prentice-Hall, Inc., Englewood Cliffs, NJ, pp. 356–358, 1983.
16. F. J. Harris, On the use of windows for harmonic analysis of the discrete fourier transform, *Proceedings of the IEEE*, 66(1), 51–83, January 1978.

17. L. D. Fielder, M. Bosi, G. Davidson, M. Davis, C. Todd, and S. Vernon, AC-2 and AC-3: low-complexity transform-based audio coding, Audio Engineering Society Publication Collected Papers on *Digital Audio Bit Rate Reduction*, Neil Gilchrist and Christer Grewin, eds., San Francisco, CA, pp. 54–72, 1996.
18. D. Sevic and M. Popovic, A new efficient implementation of the oddly-stacked Princen-Bradley filter bank, *IEEE Signal Processing Letters*, 1(11), 166–168, November 1994.
19. J. Blauert, *Spacial Hearing*, The MIT Press, Cambridge, MA, 1974.
20. G. Davidson, Parametric bit allocation in a perceptual audio coder, *97th Convention of the Audio Engineering Society*, San Francisco, CA, Preprint 3921, November 1994.
21. M. Schroeder, B. Atal, and J. Hall, Optimizing digital speech coders by exploiting masking properties of the human ear, *Journal of the Acoustical Society of America*, 66(6), 1647–1652, December 1979.
22. J. Tribolet and R. Crochiere, Frequency domain coding of speech, *IEEE Transactions on Acoustics, Speech, and Signal Processing*, ASSP-27(5), 512–530, October 1979.
23. R. Crochiere and J. Tribolet, Frequency domain techniques for speech coding, *Journal of the Acoustical Society of America*, 66(6), 1642–1646, December 1979.
24. E. Zwicker, Subdivision of the audible frequency range into critical bands (Frequenzgruppen), *Journal of the Acoustical Society of America*, 33, 248, February 1961.
25. K. R. Rao and P. Yip, *Discrete Cosine Transform*, Academic Press, Boston, MA, p. 11, 1990.
26. A. Gersho and R. M. Gray, *Vector Quantization and Signal Compression*, Kluwer Academic Publisher, Boston, MA, p. 309, 1992.
27. M. Truman, G. Davidson, A. Ubale, and L. Fielder, Efficient bit allocation, quantization, and coding in an audio distribution system, *107th Convention of the Audio Engineering Society*, San Francisco, CA, Preprint 5068, September 1999.
28. L. D. Fielder and G. A. Davidson, Audio coding tools for digital television distribution, *108th Convention of the Audio Engineering Society*, San Francisco, CA, Preprint 5104, February 2000.
29. M. Dietz, L. Liljeryd, K. Kjorling, and O. Kunz, Spectral band replication, a novel approach in audio coding, *112th Convention of the Audio Engineering Society*, San Francisco, CA, Preprint 5553, May 2002.
30. B. G. Crockett, Improved transient pre-noise performance of low bit rate audio coders using time scaling synthesis, *117th Convention of the Audio Engineering Society*, San Francisco, CA, Preprint 6184, October 2004.

4

The Perceptual Audio Coder

Deepen Sinha
Lucent Technologies

James D. Johnston
AT&T Research Labs

Sean Dorward
Lucent Technologies

Schuyler R. Quackenbush
Audio Research Labs

and

AT&T Research Labs

PAC is a perceptual audio coder that is flexible in format and bit rate, and provides high-quality audio compression over a variety of formats from 16 kbps for a monophonic channel to 1024 kbps for a 5.1 format with four or six auxiliary audio channels, and provisions for an ancillary (fixed rate) and auxiliary (variable rate) side data channel. In all of its forms it provides efficient compression of high-quality audio. For stereo audio signals, it provides near compact disc (CD) quality at about 56–64 kbps, with transparent coding at bit rates approaching 128 kbps.

PAC has been tested both internally and externally by various organizations. In the 1993 ISO-MPEG-2 5-channel test, PAC demonstrated the best decoded audio signal quality available from any algorithm at 320 kbps, far outperforming all algorithms, including the layer II and layer III backward compatible algorithms. PAC is the audio coder in most of the submissions to the U.S. Digital Audio Radio (DAR) standardization project, at bit rates of 160 kbps or 128 kbps for two-channel audio compression. It has been adapted by various vendors for the delivery of high quality music over the Internet as well as ISDN links. Over the years PAC has evolved considerably. In this chapter we present an overview for the PAC algorithm including some recently introduced features such as the use of a signal-adaptive switched filterbank for efficient encoding of nonstationary signals.

4.1 Introduction

With the overwhelming success of the compact disc (CD) in the consumer audio marketplace, the public's notion of "high-quality audio" has become synonymous with "CD quality." The CD represents stereo audio at a data rate of 1.4112 Mbps (mega bits per second). Despite continued growth in the capacity of storage and transmission systems, many new audio and multimedia applications require a lower data rate.

In compression of audio material, human perception plays a key role. The reason for this is that source coding, a method used very successfully in speech signal compression, does not work nearly as well for music. Recent U.S. and international audio standards work (HDTV, DAB, MPEG-1, MPEG-2, CCIR) therefore has centered on a class of audio compression algorithms known as perceptual coders. Rather than minimizing analytic measures of distortion, such as signal-to-noise ratio (SNR), perceptual coders attempt to minimize perceived distortion. Implicit in this approach is the idea that signal fidelity perceived by humans is a better quality measure than "fidelity" computed by traditional distortion measures. Perceptual coders define CD quality to mean "listener indistinguishable from CD audio" rather than "two channel of 16-bit audio sampled at 44.1 kHz."

PAC, the perceptual audio coder [10], employs source coding techniques to remove signal redundancy and perceptual coding techniques to remove signal irrelevancy. Combined, these methods yield a high compression ratio while ensuring maximal quality in the decoded signals. The result is a high quality, high compression ratio coding algorithm for audio signals. PAC provides a 20 Hz–20 kHz signal bandwidth and codes monophonic, stereophonic, and multichannel audio. Even for the most difficult audio material it achieves approximately 10 to 1 compression while rendering the compression effects inaudible. Significantly higher level of compression, e.g., 22 to 1, is achieved with only a little loss in quality.

The PAC algorithm has its roots in a study done by Johnston [7,8] on the perceptual entropy (PE) vs. the statistical entropy of music. Exploiting the fact that the PE (the entropy of that portion of the music signal above the masking threshold) was less than the statistical entropy resulted in the perceptual transform coder (PXFM) [8,16]. This algorithm used a 2048 point real fast Fourier transform (FFT) with 1/16 overlap, which gave good frequency resolution (for redundancy removal) but had some coding loss due to the window overlap.

The next-generation algorithm was ASPEC [2], which used the modified discrete cosine transform (MDCT) filterbank [15] instead of the FFT, and a more elaborate bit allocation and buffer control mechanism as a means of generating constant-rate output. The MDCT is a critically sampled filterbank, and so does not suffer the 1/16 overlap loss that the PXFM coder did. In addition, ASPEC employed an adaptive window size of 1024 or 256 to control noise spreading resulting from quantization. However, its frequency resolution was half that of PXFM's resulting in some loss in the coding efficiency (c.f., Section 4.3).

PAC as first proposed in Ref. [10] is a third-generation algorithm learning from ASPEC and PXFM-Stereo [9]. In its current form, it uses a long transform window size of 2048 for better redundancy removal together with window switching for noise spreading control. It adds composite stereo coding in a flexible and easily controlled form, and introduces improvements in noiseless compression and threshold calculation methods as well. Additional threshold calculations are made for stereo signals to eliminate the problem of binaural noise unmasking.

PAC supports encoders of varying complexity and quality. Broadly speaking, PAC consists of a core codec augmented by various enhancements. The full capability algorithm is sometimes also referred to as enhanced PAC (or EPAC). EPAC is easily configurable to (de)activate some or all of the enhancements depending on the computational budget. It also provides a built-in scheduling mechanism so that some of the enhancements are automatically turned on or off based on averaged short term computational requirement.

One of the major enhancements in the EPAC codec is geared towards improving the quality at lower bit rates of signals with sharp attacks (e.g., castanets, triangles, drums, etc.). Distortion of attacks is a particularly noticeable artifact at lower bit rates. In EPAC, a signal adaptive switched filterbank which switches between a MDCT and a wavelet transform is employed for analysis and synthesis [18]. Wavelet transform offer natural advantages for the encoding of transient signals and the switched filterbank scheme allows EPAC to merge this advantage with the advantages of MDCT for stationary audio segments.

Real-time PAC encoder and decoder hardware have been provided to standards bodies, as well as business partners. Software implementation of real-time decoder algorithm is available on PCs and workstations, as well as low cost general-purpose DSPs, making it suitable for mass-market applications. The decoder typically consumes only a fraction of the CPU processing time (even on a 486-PC). Sophisticated encoders run on current workstations and RISC-PCs; simpler real-time encoders that provide moderate compression or quality are realizable on correspondingly less inexpensive hardware.

In the remainder of this chapter we present a detailed overview of the various elements of PACs, its applications, audio quality, and complexity issues. The organization of the chapter is as follows. In Section 4.2, some of applications of PAC and its performance on formalized audio quality evaluation tests is discussed. In Section 4.3, we begin with a look at the defining blocks of a perceptual coding scheme followed by the description of the PAC structure and its key components (i.e., filterbank, perceptual model, stereo threshold, noise allocation, etc.). In this context we also describe the switched MDCT/wavelet filterbank (WFB) scheme employed in the EPAC codec. Section 4.4 focuses on the multichannel version of PAC. Discussions on bitstream formation and decoder complexity are presented in Sections 4.5 and 4.6, respectively, followed by concluding remarks in Section 4.7.

4.2 Applications and Test Results

In the most recent test of audio quality, [4] PAC was shown to be the best available audio quality choice [4] for audio compression applications concerning 5-channel audio. This test evaluated both backward compatible audio coders (MPEG Layer II, MPEG Layer III) and nonbackward compatible coders, including PAC. The results of these tests showed that PAC's performance far exceeded that of the next best coder in the test.

Among the emerging applications of PAC audio compression technology, the Internet offers one of the best opportunities. High-quality audio on demand is increasingly popular and promises both to make existing Internet services more compelling as well as open avenues for new services. Since most Internet users connect to the network using as low bandwidth modem (14.4–28.8 kbps) or at best an ISDN link, high quality low bit rate compression is essential to make audio streaming (i.e., real-time playback) applications feasible. PAC is particularly suitable for such applications as it offers near CD quality stereo sound at the ISDN rates and the audio quality continues to be reasonably good for bit rates as low as 12–16 kbps. PAC is therefore finding increasing acceptance in the Internet world.

Another application currently in the process of standardization is DAR. In the United States this may have one of several realizations: a terrestrial broadcast in the existing FM band, with the digital audio available as an adjunct to the FM signal and transmitted either coincident with the analog FM, or in an adjacent transmission slot; alternatively, it can be a direct broadcast via satellite (DBS), providing a commercial music service in an entirely new transmission band. In each of the above potential services, AT&T and Lucent Technologies have entered or partnered with other companies or agencies, providing PAC audio compression at a stereo coding rate of 128–160 kbps as the audio compression algorithm proposed for that service.

Some other applications where PAC has been shown to be the best audio compression quality choice is compression of the audio portion of television services, such as high-definition television (HDTV) or advanced television (ATV).

Still other potential applications of PAC that require compression but are broadcast over wired channels or dedicated networks are DAR, HDTV or ATV delivered via cable TV networks, public switched ISDN, or local area networks. In the last case, one might even envision an "entertainment bus" for the home that broadcasts audio, video, and control information to all rooms in a home.

Another application that entails transmitting information from databases of compressed audio are network-based music servers using LAN or ISDN. This would permit anyone with a networked decoder to have a "virtual music catalog" equal to the size of the music server. Considering only compression, one could envision a "CD on a chip," in which an artist's CD is compressed and stored in a semiconductor ROM and the music is played back by inserting it into a robust, low-power palm-sized music player. Audio compression is also important for read-only applications such as multimedia (audio plus video/stills/text) on CD-ROM or on a PC's hard drive. In each case, video or image data compete with audio for the limited storage available and all signals must be compressed.

Finally, there are applications in which point-to-point transmission requires compression. One is radio station studio to transmitter links, in which the studio and the final transmitter amplifier and antenna may be some distance apart. The on-air audio signal might be compressed and carried to the transmitter via a small number of ISDN B-channels. Another application is the creation of a "virtual studio" for music production. In this case, collaborating artists and studio engineers may each be in different studio, perhaps very far apart, but seamlessly connected via audio compression links running over ISDN.

4.3 Perceptual Coding

PAC, as already mentioned, is a "Perceptual Coder" [6], as opposed to a source modeling coder. For typical examples of source, perceptual, and combined source and perceptual coding, see Figures 4.1 through 4.3. Figure 4.1 shows typical block diagrams of source coders, here exemplified by DPCM, ADPCM, LPC, and transform coding [5]. Figure 4.2 illustrates a basic perceptual coder. Figure 4.3 shows a combined source and perceptual coder.

"Source model" coding describes a method that eliminates redundancies in the source material in the process of reducing the bit rate of the coded signal. A source coder can be either lossless, providing perfect reconstruction of the input signal or lossy. Lossless source coders remove no information from the signal; they remove redundancy in the encoder and restore it in the decoder. Lossy coders remove information from (add noise to) the signal; however, they can maintain a constant compression ratio

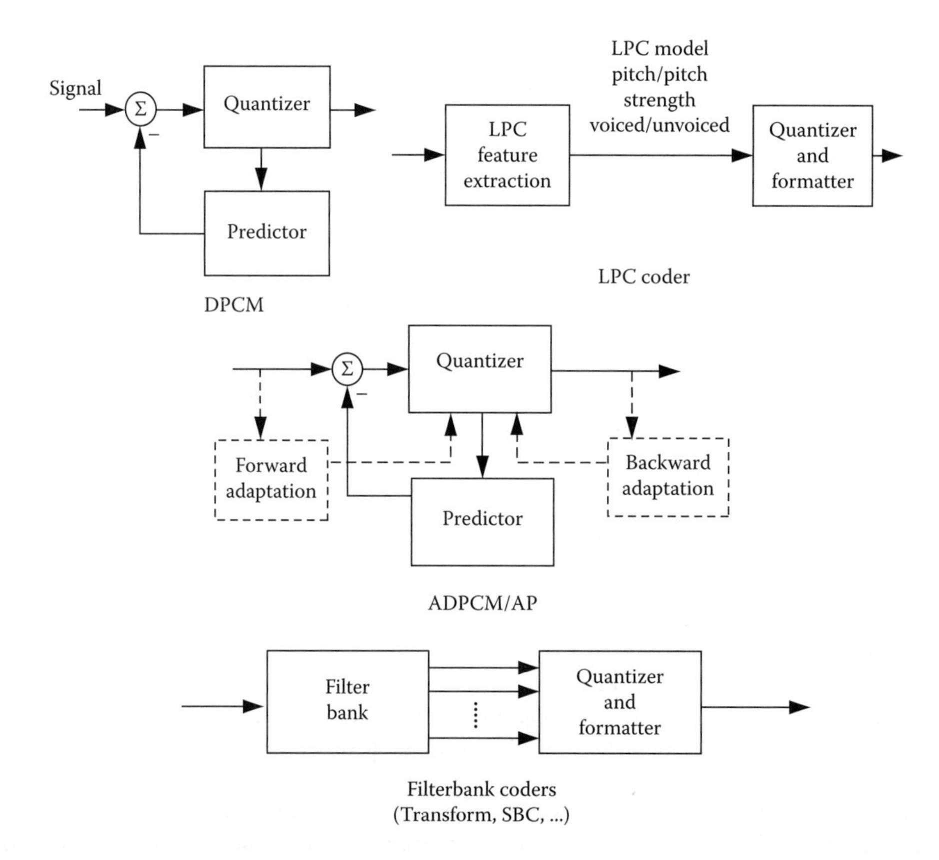

FIGURE 4.1 Block diagrams of selected source-coders.

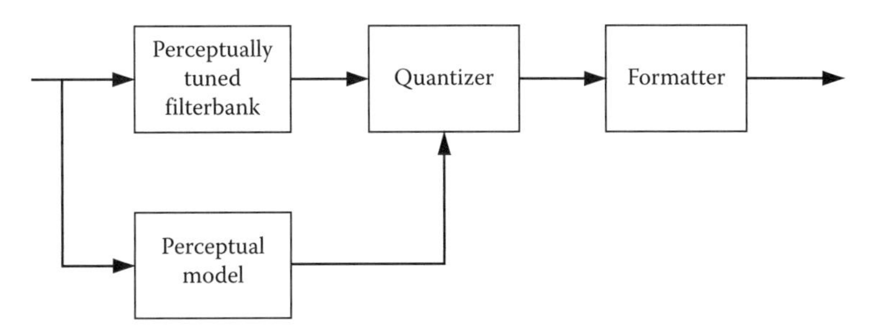

FIGURE 4.2 Block diagrams of a simple perceptual coder.

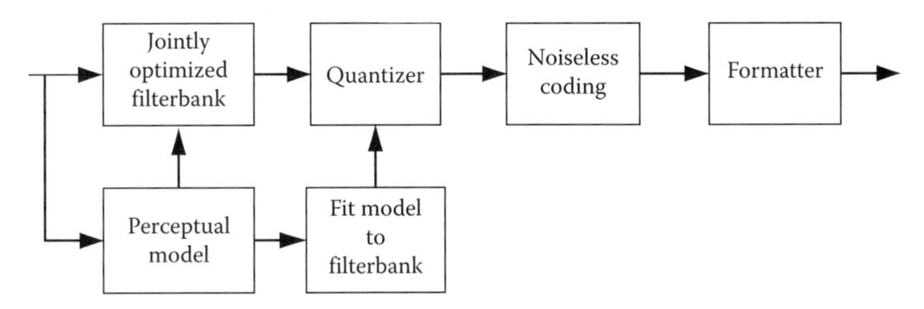

FIGURE 4.3 Block diagrams of an integrated source-perceptual coder.

regardless of the information present in a signal. In practice, most source coders used for audio signals are quite lossy [3].

The particular blocks in source coders, e.g., Figure 4.1, may vary substantially, as shown in [5], but generally include one or more of the following.

- Explicit source model, for example an LPC model
- Implicit source model, for example DCPM with a fixed predictor
- Filterbank, in other words a method of isolating the energy in the signal
- Transform, which also isolates (or "diagonalizes") the energy in the signal

All of these methods serve to identify and potentially remove redundancies in the source signal. In addition, some coders may use sophisticated quantizers and information-theoretic compression techniques to efficiently encode the data, and most if not all coders use a bitstream formatter in order to provide data organization. Typical compression methods do not rely on information-theoretic coding alone; explicit source models and filterbanks provide superior source modeling for audio signals.

All perceptual coders are lossy. Rather than exploit mathematical properties of the signal or attempt to understand the producer, perceptual coders model the listener, and attempt to remove irrelevant (undetectable) parts of the signal. In some sense, one could refer to it as a "destination" rather than "source" coder. Typically, a perceptual coder will have a lower SNR than an equivalent rate source coder, but will provide superior perceived quality to the listener.

The perceptual coder shown in Figure 4.2 has the following functional blocks.

- Filterbank—Converts the input signal into a form suitable for perceptual processing
- Perceptual model—Determines the irrelevancies in the signal, generating a perceptual threshold
- Quantization—Applies the perceptual threshold to the output of the filterbank, thereby removing the irrelevancies discovered by the perceptual model
- Bitstream former—Converts the quantized output and any necessary side information into a form suitable for transmission or storage

The combined source and perceptual coder shown in Figure 4.3 has the following functional blocks.

- Filterbank—Converts the input signal into a form that extracts redundancies and is suitable for perceptual processing
- Perceptual model—Determines the irrelevancies in the signal, generates a perceptual threshold, and relates the perceptual threshold to the filterbank structure
- Fitting of perceptual model to filtering domain—Converts the outputs of the perceptual model into a form relevant to the filterbank
- Quantization—Applies the perceptual threshold to the output of the filterbank, thereby removing the irrelevancies discovered by the perceptual model
- Information-theoretic compression—Removes redundancy from the output of the quantizer
- Bitstream former—Converts the compressed output and any necessary side information into a form suitable for transmission or storage

Most coders referred to as perceptual coders are combined source and perceptual coders. Combining a filterbank with a perceptual model provides not only a means of removing perceptual irrelevancy, but also, by means of the filterbank, provides signal diagonalization, ergo source coding gain. A combined coder may have the same block diagram as a purely perceptual coder; however, the choice of filterbank and quantizer will be different. PAC is a combined coder, removing both irrelevancy and redundancy from audio signals to provide efficient compression.

4.3.1 PAC Structure

Figure 4.4 shows a more detailed block diagram of the monophonic PAC algorithm, and illustrates the flow of data between the algorithmic blocks. There are five basic parts.

1. Analysis filterbank—The filterbank converts the time domain audio signal to the short-term frequency domain. Each block is selectably coded by 1024 or 128 uniformly spaced frequency bands, depending on the characteristics of the input signal. PAC's filterbank is used for source coding and cochlear modeling (i.e., perceptual coding).
2. Perceptual model—The perceptual model takes the time domain signal and the output of the filterbank and calculates a frequency domain threshold of masking. A threshold of masking is a frequency-dependent calculation of the maximum noise that can be added to the audio material without perceptibly altering it. Threshold values are of the same time and frequency resolution as the filterbank.
3. Noise allocation—Noise is added to the signal in the process of quantizing the filterbank outputs. As mentioned above, the perceptual threshold is expressed as a noise level for each filterbank frequency; quantizers are adjusted such that the perceptual thresholds are met or exceeded in a perceptually gentle fashion. While it is always possible to meet the perceptual threshold in a unlimited rate coder, coding at high compression ratios requires both overcoding (adding less

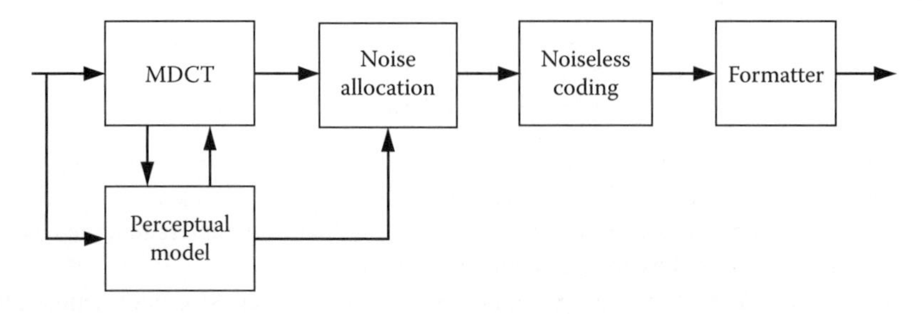

FIGURE 4.4 Block diagram of monophonic PAC encoder.

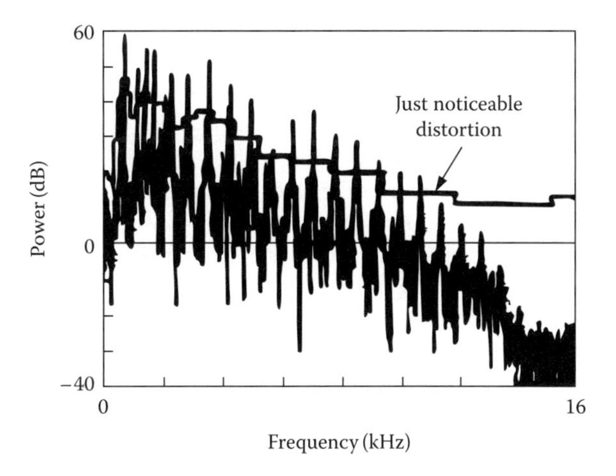

FIGURE 4.5 Example of masking threshold and signal spectrum.

noise to the signal than the perceptual threshold requires) and undercoding (adding more noise to the signal than the perceptual threshold requires). PAC's noise allocation allows for some time buffering, smoothing local peaks and troughs in the bit rate demand.

4. Noiseless compression—Many of the quantized frequency coefficients produced by the noise allocator are zero; the rest have a nonuniform distribution. Information-theoretic methods are employed to provide an efficient representation of the quantized coefficients.
5. Bitstream former—Forms the bitstream, adds any transport layer, and encodes the entire set of information for transmission or storage.

As an example, Figure 4.5 shows the perceptual threshold and spectrum for a typical (trumpet) signal. The staircase curve is the calculated perceptual threshold, and the varying curve is the short-term spectrum of the trumpet signal. Note that a great deal of the signal is below the perceptual threshold, and therefore redundant. This part of the signal is what we discard in the perceptual coder.

4.3.2 The PAC Filterbank

The filterbank normally used in PAC is referred to as the MDCT [15]. It may be viewed as a modulated, maximally decimated perfect reconstruction filterbank. The subband filters in a MDCT filterbank are linear phase FIR filters with impulse responses twice as long as the number of subbands in the filterbank. Equivalently, MDCT is a lapped orthogonal transform with a 50% overlap between two consecutive transform blocks; i.e., the number of transform coefficients is equal to one half the block length. Various efficient forms of this algorithm are detailed in [11]. Previously, Ferreira [10] has created an alternate form of this filterbank where the decimation is done by dropping the imaginary part of an odd-frequency FFT, yielding and odd-frequency FFT and an MDCT from the same calculations.

In an audio coder it is quite important to appropriately choose the frequency resolution of the filterbank. During the development of the PAC algorithm, a detailed study of the effect of filterbank resolution for a variety of signals was examined. Two important considerations in perceptual coding, i.e., coding gain and nonstationarity within a block, were examined as a function of block length. In general the coding gain increases with the block length indicating a better signal representation for redundancy removal. However, increasing nonstationarity within a block forces the use of more conservative perceptual masking thresholds to ensure the masking of quantization noise at all times. This reduces the realizable or net coding gain. It was found that for a vast majority of music samples the realizable coding gain peaks at the frequency resolution of about 1024 lines or subbands, i.e., a window of 2048 points (this is true for sampling rates in the range of 32–48 kHz). PAC therefore employs a 1024 line MDCT as the normal "long" block representation for the audio signal.

In general, some variation in the time–frequency resolution of the filterbank is necessary to adapt to the changes in the statistics of the signal. Using a high-frequency resolution filterbank to encode a signal segment with a sharp attack leads to significant coding inefficiencies or pre-echo conditions. Pre-echos occur when quantization errors are spread over the block by the reconstruction filter. Since premasking by an attack in the audio signal lasts for only about 1 ms (or even less for stereo signals), these reconstruction errors are potentially audible as pre-echos unless significant readjustments in the perceptual thresholds are made resulting in coding inefficiencies.

PAC offers two strategies for matching the filterbank resolution to the signal appropriately. A lower computational complexity version is offered in the form of "window switching" approach whereby the MDCT filterbank is switched to a lower 128 line spectral resolution in the presence of attacks. This approach is quite adequate for the encoding of attacks at moderate to higher bit rates (96 kbps or higher for a stereo pair). Another strategy offered as an enhancement in the EPAC codec is the switched MDCT/WFB scheme mentioned earlier. The advantages of using such a scheme as well as its functional details are presented below.

4.3.3 The EPAC Filterbank and Structure

The disadvantage of the window switching approach is that the resulting time resolution is uniformly higher for all frequencies. In other words, one is forced to increase the time resolution at the lower frequencies to increase it to the necessary extent at higher frequencies. The inefficient coding of lower frequencies becomes increasingly burdensome at lower bit rates, i.e., 64 kbps and lower. An ideal filterbank for sharp attacks is a nonuniform structure whose subband matches the critical band scale. Moreover, it is desirable that the high-frequency filters in the bank be proportionately shorter. This is achieved in EPAC by employing a high spectral resolution MDCT for stationary portions of the signal and switching to a nonuniform (tree structured) WFB during nonstationarities.

WFBs are quite attractive for the encoding of attacks [17]. Besides the fact that wavelet representation of such signals is more compact than the representation derived from a high-resolution MDCT, wavelet filters have desirable temporal characteristics. In a WFB, the high-frequency filters (with a suitable moment condition as discussed below) typically have a compact impulse response. This prevents excessive time spreading of quantization errors during synthesis.

The overview of an encoder based on the switched filterbank idea is illustrated in Figure 4.6. This structure entails the design of a suitable WFB which is discussed next.

The WFB in EPAC consists of a tree-structured WFB which approximates the critical band scale. The tree structure has the natural advantage that the effective support (in time) of the subband filters is progressively smaller with increasing center frequency. This is because the critical bands are wider at higher frequency so fewer cascading stages are required in the tree to achieve the desired frequency resolution. Additionally, proper design of the prototype filters used in the tree decomposition ensures (see below) that the high-frequency filters in particular are compactly localized in time.

The decomposition tree is based on sets of prototype filterbanks. These provide two or more bands of split and are chosen to provide enough flexibility to design a tree structure that approximates the critical band partition closely. The three filterbanks were designed by optimizing parametrized para-unitary filterbanks using standard optimization tools and an optimization criterion based on weighted stopband energy [20]. In this design, the moment condition plays an important role in achieving desirable temporal characteristics for the high frequency filters. An M band para-unitary filterbank with subband filters $\{H_i\}_{i=1}^{i=M}$ is said to satisfy a Pth order moment condition if $H_i(e^{jw})$ for $i = 2, 3, \ldots, M$ has a Pth order zero at $\omega = 0$ [20]. For a given support for the filters, K, requiring $P > 1$ in the design yields filters for which the "effective" support decreases with increasing P. In the other words, most of the energy is concentrated in an interval $K' < K$ and K' is smaller for higher P (for a similar stopband error criterion).

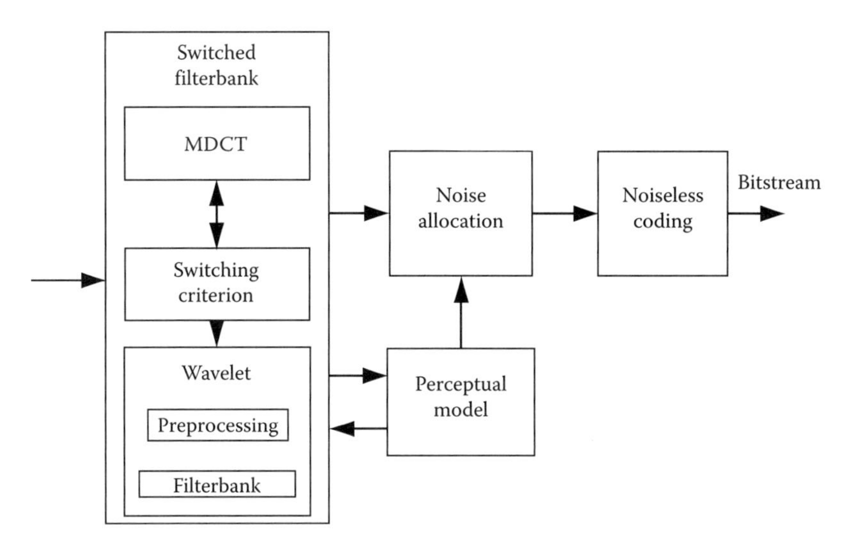

FIGURE 4.6 Block diagram of the switched filterbank audio encoder.

The improvement in the temporal response of the filters occurs at the cost of an increased transition band in the magnitude response. However, requiring at least a few vanishing moments yields filters with attractive characteristics.

The impulse response of a high-frequency wavelet filter (in a 4-band split) is illustrated in Figure 4.7. For comparison, the impulse response of a filter from a modulated filterbank with similar frequency characteristics is also shown. It is obvious that the wavelet filter offers superior localization in time.

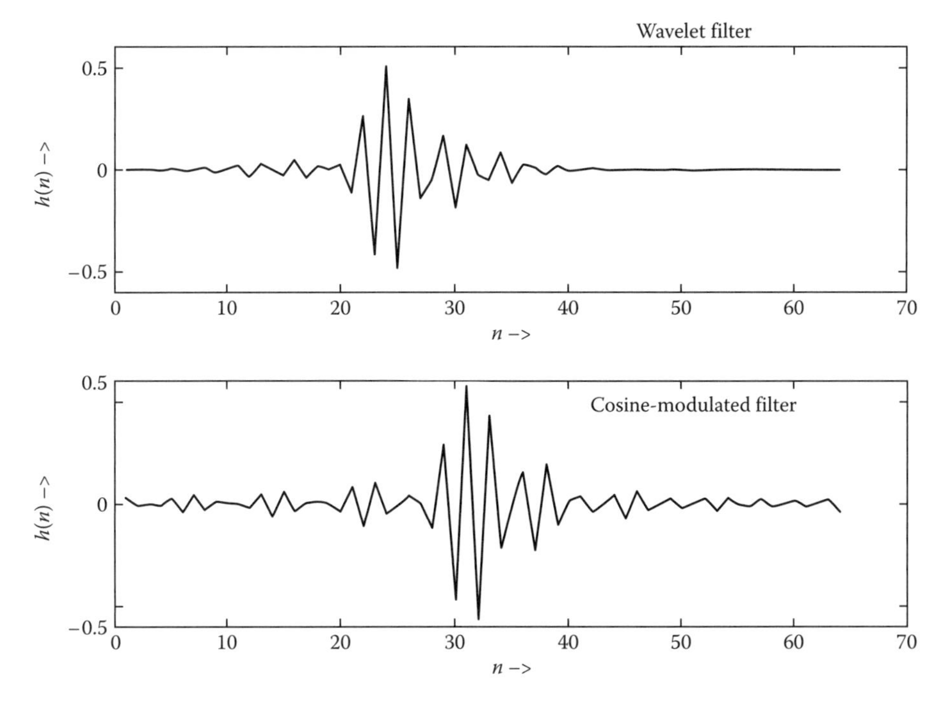

FIGURE 4.7 High-frequency wavelet and cosine-modulated filters.

4.3.3.1 Switching Mechanism

The MDCT is a lapped orthogonal transform. Therefore, switching to a WFB requires orthogonalization in the overlap region. While it is straightforward to set up a general orthogonalization problem, the resulting transform matrix is inefficient computationally. The orthogonalization algorithm can be simplified by noting that a MDCT operation over a block of $2 * N$ samples is equivalent to a symmetry operation on the windowed data (i.e., outer $N/2$ samples from either end of the window are folded into the inner $N/2$ samples) followed by an N point orthogonal block transform Q over these N samples. Perfect reconstruction is ensured irrespective of the choice of a particular block orthogonal transform Q. Therefore, Q may be chosen to be a DCT for one block and a wavelet transform matrix for the subsequent or any other block. The problem with this approach is that the symmetry operation extends the wavelet filter (or its translates) in time and also introduces discontinuities in these filters. Thus, it impairs the temporal as well as frequency characteristics of the wavelet filters. In the present encoder, this impairment is mitigated by the following two steps: (1) start and stop windows are employed to switch between MDCT and WFB (this is similar to the window switching scheme in PAC), and (2) the effective overlap between the transition and wavelet windows is reduced by the application of a new family of smooth windows [19]. The resulting switching sequence is illustrated in Figure 4.8.

The next design issue in the switched filterbank scheme is the design of a $N \times N$ orthogonal matrix Q^{WFB} based on the prototype filters and the chosen tree structure. To avoid circular convolutions, we employ transition filters at the edge of the blocks. Given a subband filter, c_k, of length K a total of $K_1 = (K/M) - 1$ transition filters are needed at the two ends of the block. The number at a particular end is determined by the rank of a $K \times (K_1 + 1)$ matrix formed by the translations of c_k. The transition filters are designed through optimization in a subspace constrained by the predetermined rows of Q^{WFB}.

4.3.4 Perceptual Modeling

Current versions of PAC utilize several perceptual models. Simplest is the monophonic model which calculates an estimated JND in frequency for a single channel. Others add MS (i.e., sum and difference) thresholds and noise-imaging protected thresholds for pairs of channels as well, as "global thresholds" for multiple channels. In this section we discuss the calculation of monophonic thresholds, MS thresholds, and noise-imaging protected thresholds.

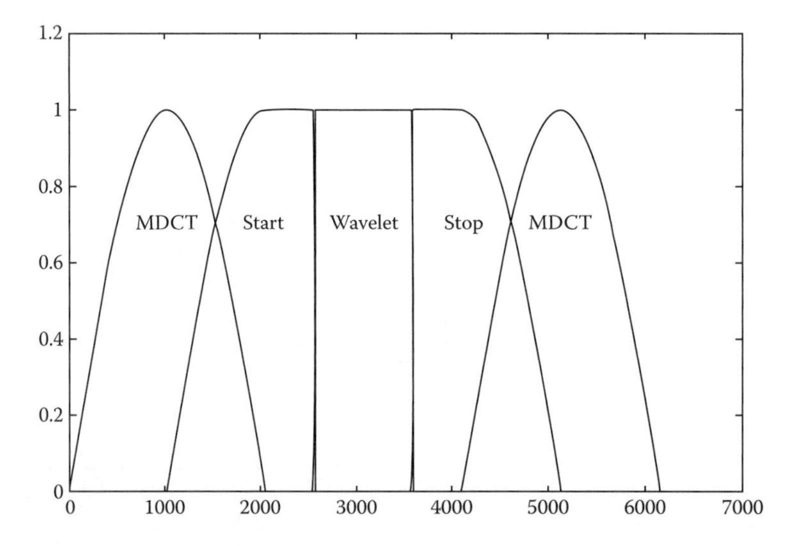

FIGURE 4.8 A filterbank switching sequence.

4.3.4.1 Monophonic Perceptual Model

The perceptual model in PAC is similar in method to the model shown as "Psychoacoustic Model II" in the MPEG-1 audio standard annexes [14]. The following steps are used to calculate the masking threshold of a signal.

- Calculate the power spectrum of the signal in 1/3 critical band partitions
- Calculate the tonal or noise-like nature of the signal in the same partitions, called the tonality measure
- Calculate the spread of masking energy, based on the tonality measure and the power spectrum
- Calculate the time domain effects on the masking energy in each partition
- Relate the masking energy to the filterbank outputs

4.3.4.2 Application of Masking to the Filterbank

Since PAC uses the same filterbank for perceptual modeling and source coding, converting masking energy into terms meaningful to the filterbank is straightforward. However, the noise allocator quantizes filterbank coefficients in fixed blocks, called coder bands, which differ from the 1/3 critical band partitions used in perceptual modeling. Specifically, 49 coder bands are used for the 1024-line filterbank, and 14 for the 128-line filterbank. Perceptual thresholds are mapped to coder bands by using the minimum threshold that overlaps the band.

In EPAC additional processing is necessary to apply the threshold to the WFB. The thresholds for the quantization of wavelet coefficients are based on an estimate of time-varying spread energy in each of the subbands and a tonality measure as estimated above. The spread energy is computed by considering the spread of masking across frequency as well as time. In other words, an interfrequency as well as a temporal spreading function is employed. The shape of these spreading functions may be derived from the cochlear filters [1]. The temporal spread of masking is frequency dependent and is roughly determined by the (inverse of) bandwidth of the cochlear filter at that frequency. A fixed temporal spreading function for a range of frequencies (wavelet subbands) is employed. The coefficients in a subband are grouped in a coder band as above and one threshold value per coderband is used in quantization. The coderband span ranges from 10 ms in the lowest frequency subband to about 2.5 ms in the highest frequency subband.

4.3.4.3 Stereo Threshold Calculation

Experiments have demonstrated that the monaural perceptual model does not extend trivially to the binaural case. Specifically, even if one signal is masked by both the L (left) and R (right) signals individually, it may not be masked when the L and R signals are presented binaurally. For further details, see the discussion of binary masking level difference (BLMD) in [12].

In stereo PAC, Figure 4.9, we used a model of BLMD in several ways, all based on the calculation of the M (mono, $L+R$) and S (stereo, $L-R$) thresholds in addition to the independent L and R thresholds. To compute the M and S thresholds, the following steps are added after the computation of the masking energy:

- Calculate the spread of masking energy for the other channel, assuming a tonal signal and adding BMLD protection
- Choose the more restrictive, or smaller, masking energy

For the L and R thresholds, the following step is added after the computation of the masking energy:

- Calculation of the spread of masking energy for the other channel. If the two masking energies are similar, add BMLD protection to both.

These four thresholds are used for the calculation of quantization, rate, and so on. An example set of spectra and thresholds for a vocal signal are shown in Figure 4.10. In Figure 4.10, compare the threshold values and energy values in the S (or "Difference") signal. As is clear, even with the BMLD protection, most of the S signal can be coded as zero, resulting in substantial coding gain. Because the signal is more

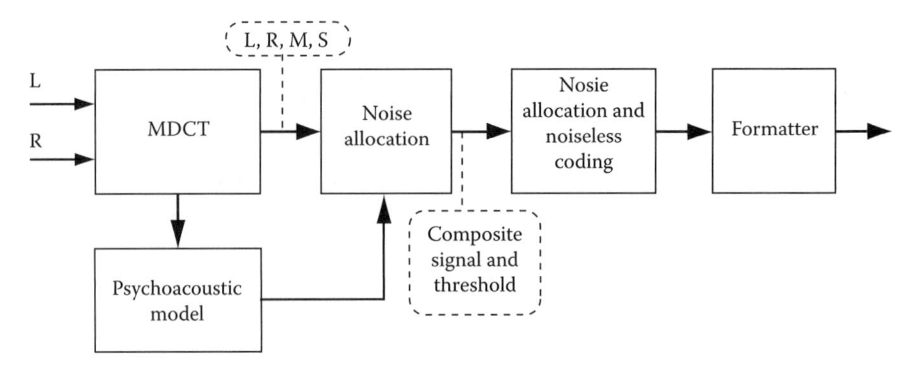

FIGURE 4.9 Stereo PAC block diagram.

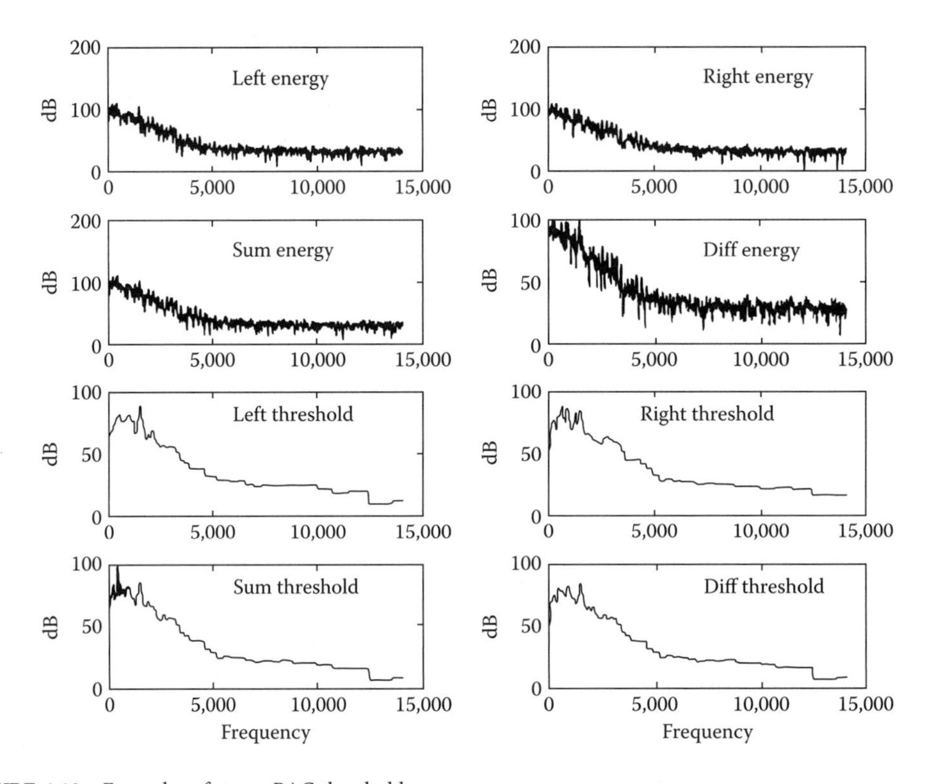

FIGURE 4.10 Examples of stereo PAC thresholds.

efficiently coded as MS even at low frequencies where the BLMD protection is in effect, that protection can be greatly reduced for the more energetic M channel because the noise will image in the same location as the signal, and not create an unmasking condition for the M signal, even at low frequencies. This provides increases in both audio quality and compression rate.

4.3.5 MS versus LR Switching

In PAC, unlike the MPEG Layer III codec [13] MS decisions are made independently for each group of frequencies. For instance, the coder may alternate coding each group as MS or LR, if that proves most efficient. Each of the L, R, M, and S filterbank coefficients are quantized using the appropriate thresholds, and the number of bits required to transmit coefficients is computed. For each group of frequencies, the more efficient of LR or MS is chosen; this information is encoded with a Huffman codebook and transmitted as part of the bitstream.

4.3.6 Noise Allocation

Compression is achieved by quantizing the filterbank outputs into small integers. Each coder band's threshold is mapped onto 1 of 128 exponentially distributed quantizer step sizes, which is used to quantize the filterbank outputs for that coder band.

PAC controls the instantaneous rate of transmission by adjusting the thresholds according to an equal-loudness calculation. Thresholds are adjusted so that the compression ratio is met, plus or minus a small amount to allow for short term irregularities in demand. This noise allocation system is iterative, using a single estimator that represents the absolute loudness of the noise relative to the perceptual threshold. Noise allocation is made across all frequencies for all channels, regardless of stereo coding decision: ergo the bits are allocated in a perceptually effective sense between L, R, M, and S, without regard to any measure of how many bits are assigned to L, R, M, and S.

4.3.7 Noiseless Compression

After the quantizers and quantized coefficients for a block are determined, information-theoretic methods are employed to yield an efficient representation.

Coefficients for each coder band are encoded using one of eight Huffman codebooks. One of the tables encodes only zeros; the rest encode coefficients with increasing absolute value. Each codebook encodes groups of two or four coefficients, with the exception of the zero codebook which encodes all of the coefficients in the band. See Table 4.1 for details. In Table 4.1, LAV refers to the largest absolute value in a given codebook, and dimension refers to the number of quantized outputs that are coded together in one codeword. Two codebooks are special, and require further mention. The zero codebook is of indeterminate size, it indicates that all quantized values that the zero codebook applies to are in fact zero, and no further information is transmitted about those values. Codebook seven is also a special codebook. It is of size $-16{:}16$ by $-16{:}16$, but the entry of absolute value 16 is not a data value, it is, rather, an escape indicator. For each escape indicator sent in codebook seven (there can be zero, one, or two per codeword), there is an additional escape word sent immediately after the Huffman codeword. This additional codeword, which is generated by rule, transmits the value of the escaped codeword. This generation by rule is a process that has no bounds; therefore, any quantized value can be transmitted by the use of an escape sequence.

Communicating the codebook used for each band constitutes a significant overhead; therefore, similar codebooks are grouped together in sections, with only one codebook transmitted and used for encoding each section.

TABLE 4.1 PAC Huffman Codebook

Codebook	LAV	Dimension
0	0	*
1	1	4
2	1	4
3	2	4
4	4	2
5	7	2
6	12	2
7	ESC	2

Since the possible quantizers are precomputed, the indices of the quantizers are encoded rather than the quantizer values. Quantizer indices for coder bands which have only zero coefficients are discarded; the rest are differentially encoded, and the differences are Huffman encoded.

4.4 Multichannel PAC

The multichannel perceptual audio coder (MPAC) extends the stereo PAC algorithm to the coding of multiple audio channels. In general, the MPAC algorithm is software configurable to operate in 2, 4, 5, and 5.1 channel mode. In this chapter we will describe the MPAC algorithm as it is applied to a 5-channel system consisting of the five full bandwidth channels: Left (L), Right (R), Center (C), Left surround (Ls), and Right surround (Rs).

The MPAC 5-channel audio coding algorithm is illustrated in Figure 4.11. Below we describe the various modules, concentrating in particular on the ones that are different from the stereo algorithm.

4.4.1 Filterbank and Psychoacoustic Model

Like the stereo coder, MPAC employs a MDCT filterbank with two possible resolutions, i.e., the usual long block which has 1024 uniformly spaced frequency outputs and a short bank which has 128 uniformly spaced frequency bins. A window switching algorithm, as described above, is used to switch to a short block in the presence of strong nonstationarities in the signal. In the 5-channel setup it desirable to be able to switch the resolution independently for various subsets of channels. For example, one possible, scenario is to apply the window switching algorithm to the front channels (L, R, and C) independently of the surround channels (Ls and Rs). However, this somewhat inhibits the possibilities for composite coding (see below) among the channels. Therefore, one needs to examine the relative gain of independent window switching vs. the gain from a higher level of composite coding. In the present implementation different filterbank resolutions for the front and surround channels are allowed.

The individual masking threshold for the five channels are computed using the PAC psychoacoustic model described above. In addition, the front pair LR and the surround pair Ls/Rs are used to generate two pairs of MS thresholds (c.f., Section 4.3.4.3). The five channels are coded with their individual thresholds excepting in the case where joint stereo coding is being used (either for the front or the surround pair), in which case the appropriate MS thresholds are used. In addition to the five individual

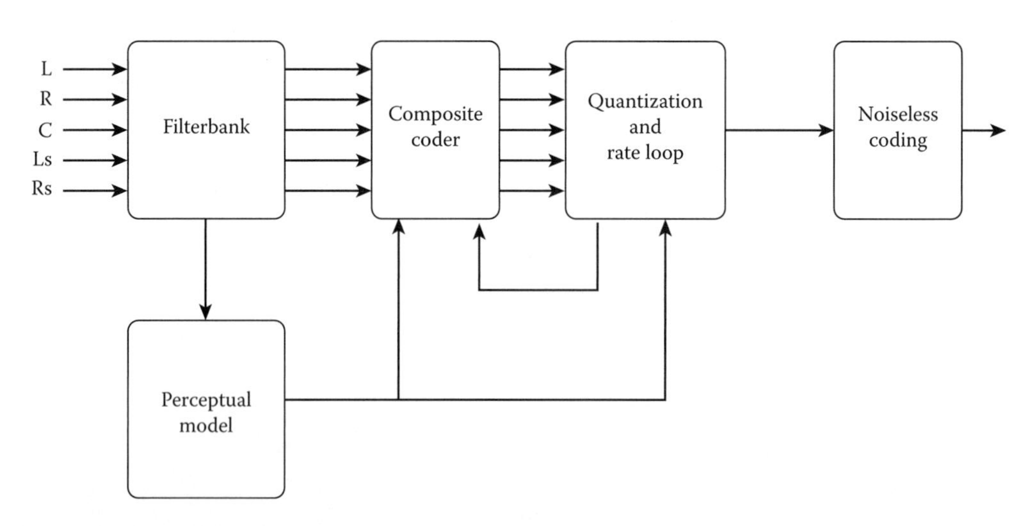

FIGURE 4.11 Block diagram of MPAC.

and four stereo thresholds, a joint (or "global") threshold based on all channels is also computed. The computation and role of the global threshold will be discussed later in this section.

4.4.2 Composite Coding Methods

The MPAC algorithm extends the MS coding of the stereo algorithm to a more elaborate composite coding scheme. Like the MS coding algorithm, the MPAC algorithm uses adaptive composite coding in both time and frequency: the composite coding mode is chosen separately for each of the coder bands at every analysis instance. This selection is based on a PE criterion and attempts to minimize the bit rate requirement as well as exercise some control over noise localization. The coding scheme uses two complementary sets of interchannel combinations as described below:

- MS coding for the front and surround pair
- Interchannel prediction

MS coding is a basis transformation operation and is therefore performed with the uncoded samples of the corresponding pair of channels. The resulting M or S channel is then coded using its own threshold (which is computed separately from the individual channel threshold). Interchannel prediction, on the other hand, is performed using the quantized samples of the predicting channel. This is done to prevent the propagation of quantization errors (or "cross talk"). The predicted value for each channel is subtracted from the channel samples and the resulting difference is encoded using the original channel threshold. It may be noted that the two sets of channel combinations are nested so that either, both, or none may be employed for a particular coder band. The coder currently employs the following possibilities for interchannel prediction.

For the front channels (L, R, and C): Front L and R channels are coded as LR or MS. In addition, one of the following two possibilities for interchannel prediction may be used.

1. Center predicts LR (or M if MS coding mode is on)
2. Front M channel predicts the center

For the surround channels (Ls and Rs): Ls and Rs channels are coded as Ls/Rs or Ms/Ss (where Ms and Ss are, respectively, the surround M and surround S). In addition, one or both of following two modes of interchannel prediction may be employed:

1. Front L, R, M channels predict Ls/Rs or Ms
2. Center channel predicts Ls/Rs or Ms

In the present implementation, the predictor coefficients in all of the above interchannel prediction equations are all fixed to either zero or one.

Note that the possibility of completely independent coding is implicit in the above description, i.e., the possibility of turning off any possible prediction is always included. Furthermore, any of these conditions may be independently used in any of the 49 coder bands (long filter band length) or in the 14 coder bands (short filter band length), for each block of filterbank output. Also note that for the short filterbank where the outputs are grouped into 8 groups of 128 (each group of 128 has 14 bands), each of these 8 groups has independently calculated composite coding.

The decisions for composite coding are based primarily on the PE criterion; i.e., the composite coding mode is chosen to minimize the bit requirement for the perceptual coding of the filterbank outputs from the five channels. The decision for MS coding (for the front and surround pair) is also governed in part by noise localization considerations. As a consequence, the MPAC coding algorithm ensures that signal and noise images are localized at the same place in the front and rear planes. The advantage of this coding scheme is that the quantization noise usually remains masked not only in a listening room environment but also during headphone reproduction of a stereo downmix of the five coded channels (i.e., when two downmixed channels of the form $Lc = L + \alpha C + \beta Ls$, and $Rc = R + \alpha C + \beta Rs$ are produced and fed to a headphone).

The method used for composite coding is still in the experimental phase and subject to refinements/modifications in future.

4.4.3 Use of a Global Masking Threshold

In addition to the five individual thresholds and the four MS thresholds, the MPAC coder also makes use of a global threshold to take advantage of masking across the various channels. This is done when the bit demand is consistently high so that the bit reservoir is close to depletion. The global threshold is taken to be the maximum of five individual thresholds minus a "safety margin." This global threshold is phased in gradually when the bit reservoir is really low (e.g., less than 20%) and in that case it is used as a lower limit for the individual thresholds.

The reason that global threshold is useful is because results in [12] indicate that if the listener is more than a "critical distance" away from the speakers, then the spectrum at either of listener's ear may be well approximated by the sum of power spectrums due to individual speakers.

The computation of a global threshold also involves a safety margin. This safety margin is frequency-dependent and is larger for the lower frequencies and smaller for higher frequencies. The safety margin changes with the bit reservoir state.

4.5 Bitstream Formatter

PAC is a block processing algorithm; each block corresponds to 1024 input samples from each channel, regardless of the number of channels. The encoded filterbank outputs, codebook sections, quantizers, and channel combination information for one 1024-sample chunk or eight 128-sample chunks are packed into one frame.

Depending on the application, various extra information is added to first frame or to every frame. When storing information on a reliable media, such as a hard disk, one header indicating version, sample rate, number of channels, and encoded rate is placed at the beginning of the compressed music. For extremely unreliable transmission channels, like DAR, a header is added to each frame. This header contains synchronization, error recovery, sample rate, number of channels, and the transmission bit rate.

4.6 Decoder Complexity

The PAC decoder is of approximately equal complexity to other decoders currently known in the art. Its memory requirements are approximately

- 1100 words each for MDCT and WFB workspace
- 512 words per channel for MDCT memory
- (optional) 1024 words per channel for error mitigation
- 1024 samples per channel for output buffer
- 12 kB ROM for codebooks

The calculation requirements for the PAC decoder are slightly more than doing a 512-point complex FFT per 1024 samples per channel. On an Intel 486 based platform, the decoder executes in real time using up approximately 30–40.

4.7 Conclusions

PAC has been tested both internally and externally by various organizations. In the 1993 ISO-MPEG-2 5-channel test, PAC demonstrated the best decoded audio signal quality available from any algorithm at 320 kbps, far outperforming all algorithms, including the backward compatible algorithms. PAC is the audio coder in three of the submissions to the U.S. DAR project, at bit rates of 160 or 128 kbps for two-channel audio compression.

PAC presents innovations in the stereo switching algorithm, the psychoacoustic model, filterbank, the noise-allocation method, and the noiseless compression technique. The combination provides either better quality or lower bit rates than techniques currently on the market.

In summary, PAC offers a single encoding solution that efficiently codes signals from AM bandwidth (5–10 kHz) to full CD bandwidth, over dynamic ranges that match the best available analog to digital converters, from one monophonic channel to a maximum of 16 front, 7 back, 7 auxiliary, and at least 1 effects channel. It operates from 16 kbps up to a maximum of more than 1000 kbps for the multiple-channel case. It is currently implemented in 2-channel hardware encoder and decoder, and 5-channel software encoder and hardware decoder. Versions of the bitstream that include an explicit transport layer provide very good robustness in the face of burst-error channels, and methods of mitigating the effects of lost audio data.

In the future, we will continue to improve PAC. Some specific improvements that are already in motion are the improvement of the psychoacoustic threshold for unusual signals, reduction of the overhead in the bitstream at low bit rates, improvements of the filterbanks for higher coding efficiency, and the application of vector quantization techniques.

References

1. Allen, J.B., Ed., *The ASA Edition of Speech Hearing in Communication*, Acoustical Society of America, Woodbury, NY, 1995.
2. Brandenburg, K. and Johnston, J.D., ASPEC: Adaptive spectral entropy coding of high quality music signals, *AES 90th Convention*, Paris, France, 1991.
3. G722. The G722 CCITT *Standard for Audio Transmission.*
4. ISO-II, Report on the MPEG/audio multichannel formal subjective listening tests, ISO/MPEG document MPEG94/063. ISO/MPEG-II Audio Committee, 1994.
5. Jayant, N.S. and Noll, P., *Digital Coding of Waveforms, Principles and Applications to Speech and Video*, Prentice-Hall, Englewood Cliffs, NJ, 1984.
6. Jayant, N.S., Johnston, J., and Safranek, R.J., Signal compression based on models of human perception, *Proc. IEEE*, 81(10), 1385–1422, 1993.
7. Johnston, J.D., Estimation of perceptual entropy using noise masking criteria, *ICASSP-88 Conference Record*, New York, pp. 2524–2527, 1988.
8. Johnston, J.D., Transform coding of audio signals using perceptual noise criteria, *IEEE J. Select. Areas Commun.*, 6(2), 314–323, Feb. 1988.
9. Johnston, J.D., Perceptual coding of wideband stereo signals, *ICASSP-89 Conference Record*, Glasgow, U.K., pp. 1993–1996, 1989.
10. Johnston, J.D. and Ferreira, A.J., Sum-difference stereo transform coding, *ICASSP-92 Conference Record*, San Francisco, CA, pp. II-569–II-572, 1992.
11. Malvar, H.S., *Signal Processing with Lapped Transforms*, Artech House, Norwood, MA, 1992.
12. Moore, B.C.J., *An Introduction to the Psychology of Hearing*, Academic Press, New York, 1989.
13. MPEG, *ISO-MPEG-1/Audio Standard.*
14. Mussmann, H.G., The ISO audio coding standard, *Proceedings of the IEEE-Globecom*, San Diego, CA, pp. 511–517, 1990.
15. Princen, J.P. and Bradlen, A.B., Analysis/synthesis filter bank design based on time domain aliasing cancellation, *IEEE Trans. ASSP*, 34(5), 1153–1161, 1986.
16. Quackenbush, S.R., Ordentlich, E., and Snyder, J.H., Hardware implementation of a 128-kbps monophonic audio coder, *IEEE ASSP Workshop on Applications of Signal Processing to Audio and Acoustics*, Mohonk, NY, 1989.

17. Sinha, D. and Tewfik, A.H., Low bit rate transparent audio compression using adapted wavelets, *IEEE Trans. Signal Process.*, 41(12), 3463–3479, Dec. 1993.
18. Sinha, D. and Johnston, J.D., Audio compression at low bit rates using a signal adaptive switched filterbank, *Proceedings of the IEEE International Conference on Acoustics Speech and Signal Processing*, Atlanta, GA, pp. II-1053–II-1056, May 1996.
19. Ferreira, A.J.S. and Sinha, D., A new class of smooth power complementary Windows and their application to audio signal processing, *AES Convention*, 119, 6604, Oct. 2005.
20. Vaidyanathan, P.P., Multirate digital filters, filter banks, polyphase networks, and applications: A tutorial, *Proc. IEEE*, 78(1), 56–92, Jan. 1990.

5 Sony Systems

Kenzo Akagiri,
Masayuki Katakura,
H. Yamauchi,
E. Saito,
M. Kohut,
Masayuki Nishiguchi,
Kyoya Tsutsui,
and Keisuke Toyama
Sony Corporation

5.1 Introduction

Kenzo Akagiri

In digital signal processing, manipulation of the signal is defined as an essentially mathematical procedure, while the AD and DA converters, the front end and the final stage devices of the processing, include analog factor/limitation. Therefore, the performance of the devices determines the degradation from the theoretical performance defined by the format of the system.

Until the 1970s, AD and DA converters with around 16-bit resolution, which were fabricated by module or hybrid technology, were very expensive devices for industry applications. At the beginning of the 1980s, the compact disc (CD) player, the first mass-production digital audio product, was introduced, and it required low cost and monolithic type DA converters with 16 bit resolution. The two-step dual slope method [1] and the dynamic element matching (DEM) [2] method were used in the first generation DA converters for CD players. These were methods which relieved the accuracy and matching requirements of the elements to guarantee conversion accuracy by circuit technology. Introducing new ideas on circuit and trimming, such as segment decode and laser trimming of the thin film fabricated on monolithic silicon die, classical circuit topologies using binary weighted current source were also used. For AD conversion at same generation, successive approximation topology and two-step dual slope method were also used.

In the mid-1980s, introduction of the oversampling and the noise shaping technology to the AD and DA converters for audio applications were investigated [3]. The converters using these technologies are the most popular devices for recent audio applications, especially as DA converters.

5.2 Oversampling AD and DA Conversion Principle

Masayuki Katakura

5.2.1 Concept

The concept of the oversampling AD and DA conversion, $\Delta\Sigma$ or $\Sigma\Delta$ modulation, was known in the 1950s; however, the technology to fabricate actual devices was impracticable until the 1980s [4].

The oversampling AD and DA conversion is characterized by the following three technologies.

1. Oversampling
2. Noise shaping
3. Fewer bit quantizer (converters used one bit quantizer called the DS or SD type)

It is well-known that the quantization noise shown in the next equation is determined by only quantization step Δ and distributed in bandwidth limited by Nyquist frequency (fs/2), and the spectrum is almost similar to white noise when the step size is smaller than the signal level.

$$V_{\mathrm{n}} = \Delta/\sqrt{12} \tag{5.1}$$

As shown in Figure 5.1, oversampling expands the capacity of the quantization noise cavity on the frequency axis and reduces the noise density in the audio band, and the noise shaping moves it to out of the band. Figure 5.2 is first-order noise shaping to show the principle of noise shaping, in which the quantizer is represented by the adder fed input $U(n)$ and a quantization noise $Q(n)$. $Y(n)$ and $U(n)$, the output and input signals of the quantizer, respectively, are given as follows:

$$Y(n) = U(n) + Q(n) \tag{5.2}$$

$$U(n) = X(n) + Q(n-1) \tag{5.3}$$

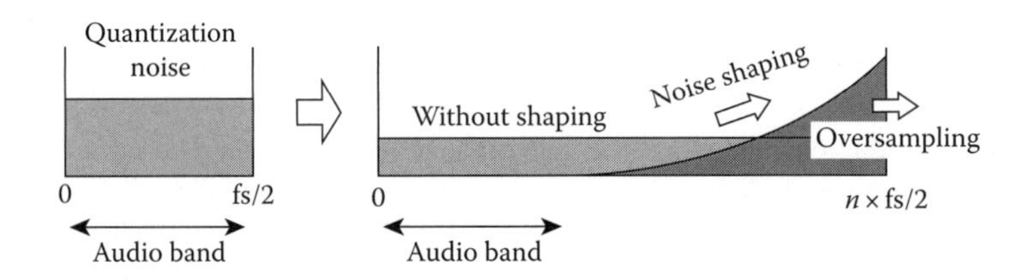

FIGURE 5.1 Quantization noise of the oversampling conversion.

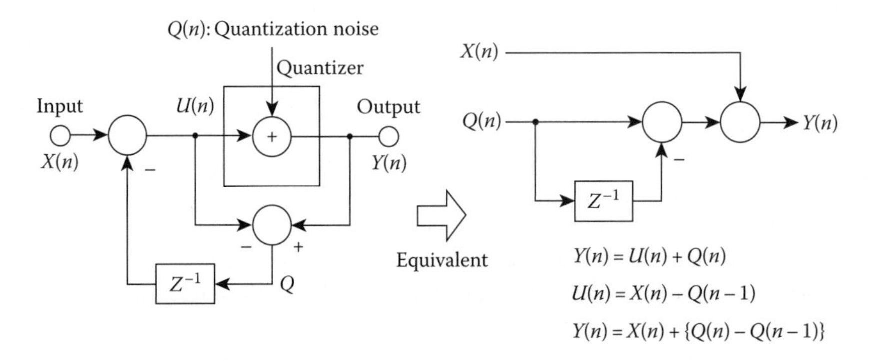

FIGURE 5.2 First-order noise shaping.

As a result, the output $Y(n)$ is

$$Y(n) = X(n) + \{Q(n) - Q(n-1)\} \tag{5.4}$$

The quantization noise in output $Y(n)$, which is a differentiation of the original quantization noise $Q(n)$ and $Q(n-1)$ shifted a time step, has high frequency boosted spectrum. Equation 5.4 is written as follows using z

$$Y(z) = X(z) + Q(z)(1 - Z^{-1}) \tag{5.5}$$

The oversampling conversion using one bit quantizer is called $\Delta\Sigma$ or $\Sigma\Delta$ AD/DA converters. Regarding one bit quantizer, a mismatch of the elements does not affect differential error; in other words, it has no nonlinear error. Assume output swing of the quantizer is $\pm\Delta$, quantization noise $Q(z)$ is white noise, and the magnitude $|Q(\omega T)|$ is $\Delta/\sqrt{3}$, which corresponds to four times in power of Equation 5.1 since the step size is twice that. If θ is $2\pi \cdot f_{max}/f_s$, where f_{max} and f_s are the audio bandwidth and the sampling frequency, respectively, then the in-band noise in Equation 5.5 becomes

$$\bar{N}^2 = |Q(\omega T)|^2 \frac{1}{2\pi} \int_{-\theta}^{\theta} |H(\omega T)|^2 d(\omega T) = \frac{\Delta^2}{3} \frac{1}{2\pi} \int_{-\theta}^{\theta} \left|1 - e^{-j\omega T}\right|^2 d(\omega T) = \frac{\Delta^2}{3} \frac{2}{\pi}(\theta - \sin\theta) \doteqdot \frac{\Delta^2}{9\pi}\theta^3 \tag{5.6}$$

The oversampling conversion has the following remarkable advantages compared with traditional methods.

1. It is easy to realize "good" one bit converters without superior device accuracy and matching
2. Analog antialiasing filters with sharp cutoff characteristics are unnecessary due to oversampling

Using the oversampling converting technology, requirements for analog parts are relaxed; however, they require large scale digital circuits because interpolation filters in front of the DA conversion, which increase sampling frequency of the input digital signal, and decimation filters after the AD conversion, which reject quantization noise in high frequency and reduce sampling frequency, are required.

Figure 5.3 shows the block diagram of the DA converter including an interpolation filter. Though the scheme of the noise shaper is different from that of Figure 5.2, the function is equivalent. Figure 5.4 shows the block diagram of the AD converter including a decimation filter. Note that the AD converter is almost the same as with the DA converters regarding the noise shapers; however, the details of the hardware are different depending on whether the block handles analog or digital signal. For example, to handle digital signals the delay units and the adders should use latches and digital adders; on the other hand, to handle analog signals delay units and adders using switched capacitor topology should be used. In the $\Delta\Sigma$ type, the quantizer is just reduction data length to one bit for the DA converter, and is a comparator for the AD converter by the same rule.

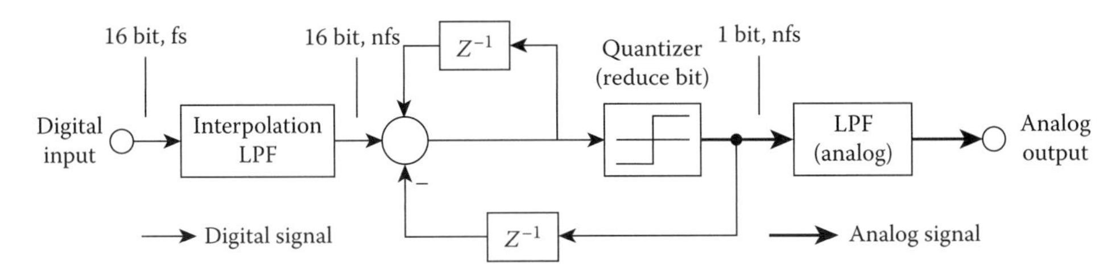

FIGURE 5.3 Oversampling DA converter.

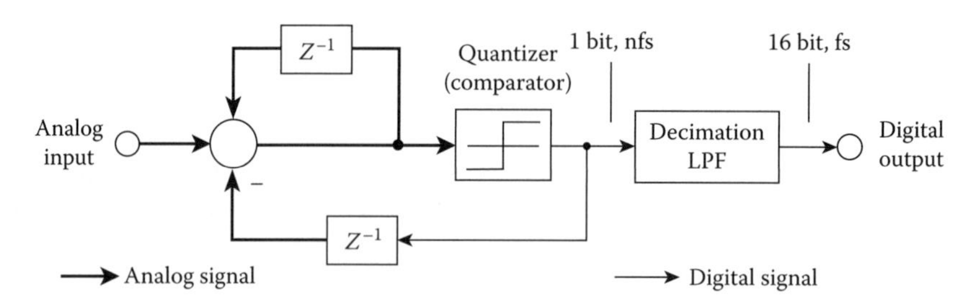

FIGURE 5.4 Oversampling AD converter.

5.2.2 Actual Converters

To achieve resolution of 16 bits or more for digital audio applications, first-order noise shaping is not acceptable because it requires an extra high oversampling ratio, and the following technologies are actually adopted.

- High-order noise shaping
- Multistage (feedforward) noise shaping
- Interpolative conversion

1. High-order noise shaping
   Figure 5.5 shows quantization noise spectrum for order of the noise shaping. The third-order noise shaping achieves 16 bit dynamic range using less than an oversampling ratio of 100. Figure 5.6 shows a third-order noise shaping for example of the high order. Order of the noise shaping used is 2–5 for audio applications.
   In Figure 5.6 output $Y(z)$ is given

$$Y(z) = X(z) + Q(z)(1 - Z^{-1})^3 \tag{5.7}$$

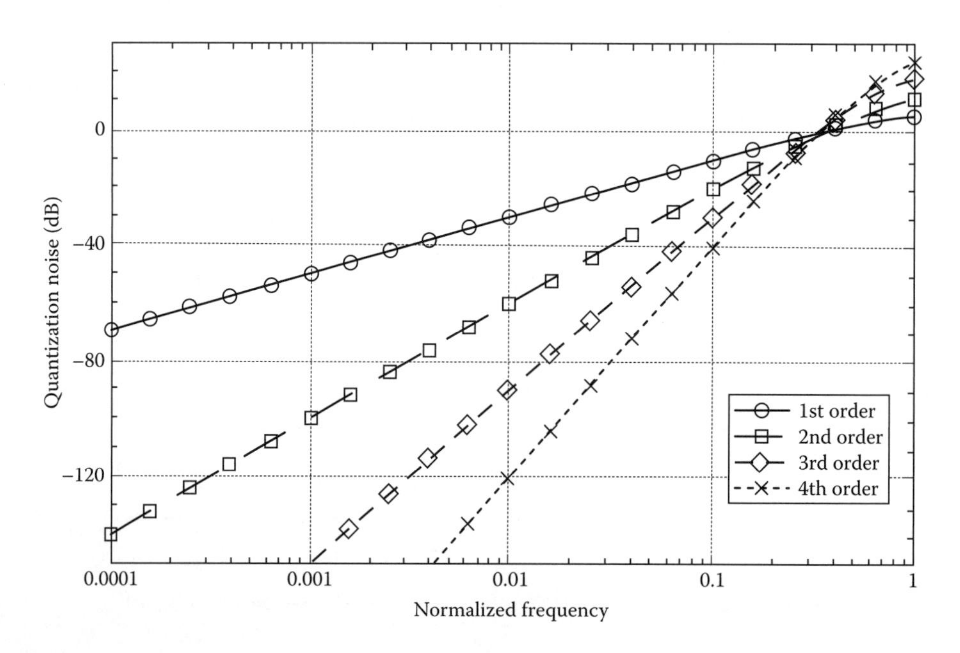

FIGURE 5.5 Quantization noise vs. order of noise shaping.

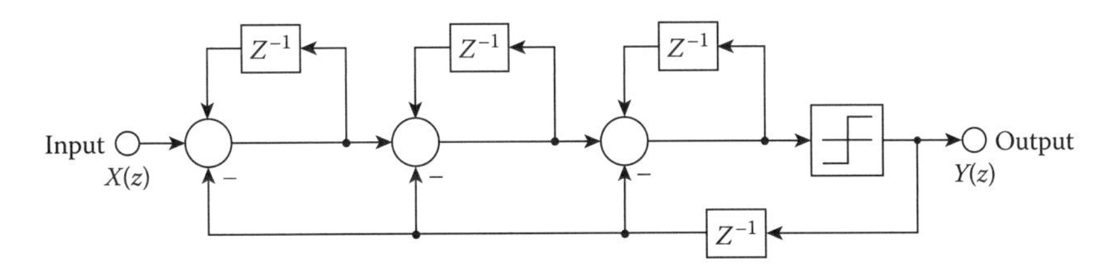

FIGURE 5.6 Third-order noise shaping.

The high-order noise shaping has a stability problem because the phase shift of the open loop in more than a third-order noise shaping exceeds 180°. In order to guarantee the stability, an amplitude limiter at the integrator outputs is used, and modification of the loop transfer function is done, although it degrades the noise shaping performance slightly.

2. Multistage (feedforward) noise shaping [5]
   Multistage (feedforward) noise shaping (called MASH) achieves high-order noise shaping transfer functions using high-order feedforward, and is shown in Figure 5.7. Though two-stage (two-order) is shown in Figure 5.7, three-stage (three-order) is usually used for audio applications.
3. Interpolative converters [6]
   This is a method which uses a few bit resolution converters instead of one bit. The method reduces the oversampling ratio and order of the noise shaping to guarantee specified dynamic range and improve the loop stability. Since absolute value of the quantization noise becomes small, it is relatively easy to guarantee noise level; however, linearity of large signal conditions affects the linearity error of the AD/DA converters used in the noise shaping loop.

Oversampling conversion has become a major technique in digital audio application, and one of the distinctions is that it does not inherently zero cross distort. For recent device technology, it is not so difficult to guarantee 18 bit accuracy. Thus far, the available maximum dynamic range is slightly less than 20 bit (120 dB) without noise weighting (wide band) due to analog limitation. On the other hand, converters with 20 bit or more resolution have been reported [7] and are expected to improve sound quality in very small signal levels from the standpoint of hearing.

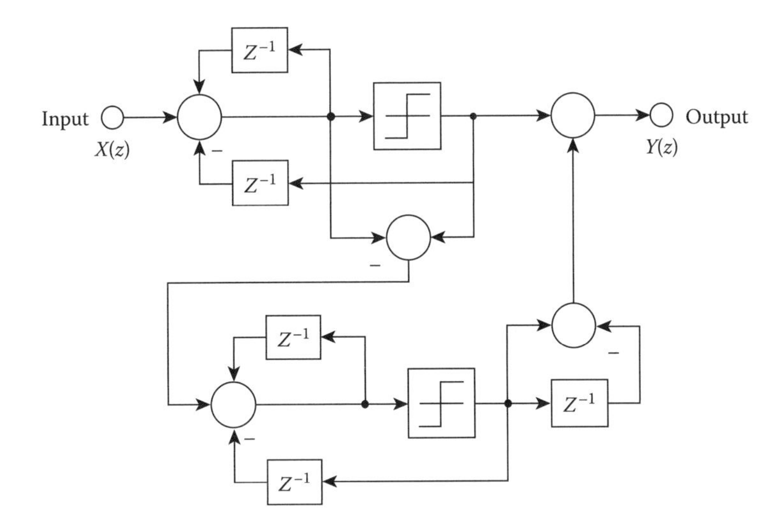

FIGURE 5.7 Multistage noise shaping.

5.3 The SDDS System for Digitizing Film Sound

H. Yamauchi, E. Saito, and M. Kohut

5.3.1 Film Format

There are three basic concepts for developing the SDDS format. They can

1. Provide sound quality similar to that of a CD. We adapt ATRAC (Adaptive TRansform Acoustic Coding) to obtain good sound quality equivalent to that of CDs. ATRAC is the compression method used in the mini disc (MD) which has been in sale since 1992. ATRAC enables one record digital sound data by compressing about 1/5 of the original sound.
2. Provide enough numbers of sound channels with good surround effects. We have eight discrete channel systems and six channels to the screen in the front and two channels in the rear as surround speakers shown in Figure 5.8. We have discrete channel systems, making a good channel separation which provides superior surround effects even in a large theater with no sound defects.
3. Be compatible with the current widespread analogue sound system. There are limited spaces between the sprockets, picture frame, and in the external portion of the sprocket hole where the digital sound could be recorded because the analogue sound track is left as usual. As in the cinema scope format, it may be difficult to obtain enough space between picture frames. Because the signal for recording and playback would become intermittent between sprockets, special techniques would be required to process such signals. As shown in Figure 5.9, we therefore establish track P and track S on a film external portion where continuous recordings are possible and where space can be obtained in the digital sound recording region on the SDDS format.

Data bits are recorded on the film with black and white dot patterns. The size of a bit is decided to overcome the effects caused by film scratch and is able to correct errors. In order to obtain the certainty of reading data, we set a guard band area to the horizontal and track direction.

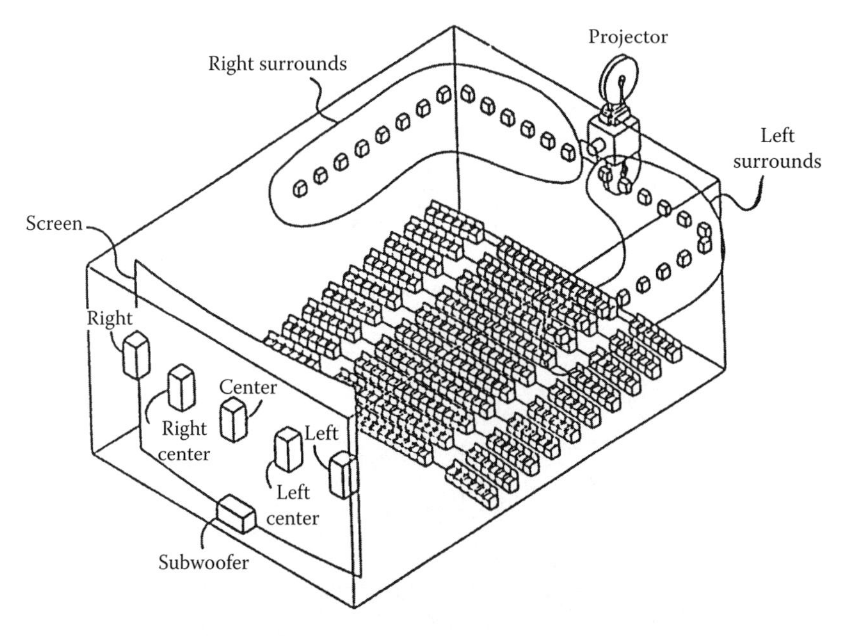

FIGURE 5.8 Speaker arrangement in theater.

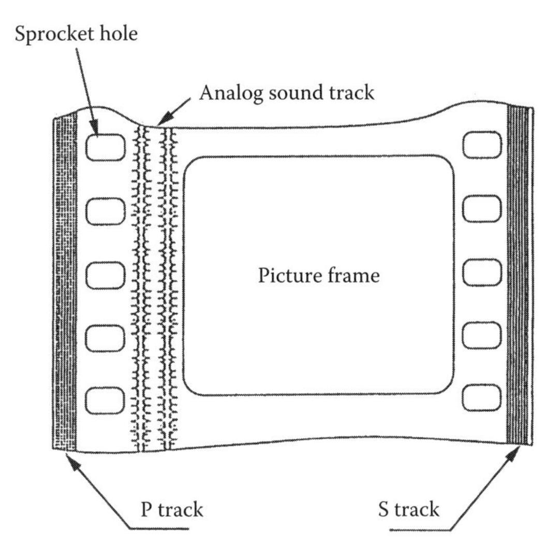

FIGURE 5.9 SDDS track designation.

Now, the method to record digital sound data on these two tracks is to separate eight channels and record four channels each in track P and in track S. A redundant data is also recorded about 18 frames later on the opposite track. By this method, it is possible to obtain equivalent data from track S if any error occurs on track P and the correction is unable to be made, or vice versa. This is called the "Digital Backup System."

Figure 5.10 shows the block structure for the SDDS format. A data compression block of the ATRAC system has 512 bit sound data per film block. A vertical sync region is set at the head of the film block. A film block ID is recorded in this region to reproduce the sound data and picture frame with the right timing and to prevent the "lip sync" offset from discordance; for example, the time accordance between an actor's/actress' lip movement and his/her voice. Also, a horizontal sync is set on the left-hand side of the film block and is referred to correctly detect the head of the data in reading with the line sensor.

5.3.2 Playback System for Digital Sound

The digital playback sound system for the SDDS system consists of a reader unit, DFP-R2000, and a decoder unit, DFP-D2000 as shown in Figure 5.11. The reader unit is set between the supply reel and the projector.

The principle of digital sound reading for the reader unit DFP-R2000 is described in Figure 5.12. The LED light source is derived from the optical fiber and it scans the data portion recorded on track P and track S of the film. Transparent lights through the film form an image on the line sensor through the lens. These optical systems are designed to have appropriate structures which can hardly be affected by scratches on the film. The output of a sensor signal is transmitted to the decoder after signal processing such as the waveform equalization is made.

The block diagram of the decoder unit DFP-D2000 is shown in Figure 5.13. The unit consists of EQ, DEC, DSP, and APR blocks.

In the EQ, signals become digital signals after being equalized. Then the digital signals are transmitted to the DEC together with the regenerated clock signal.

In the DEC, jitters elimination and lip sync control are done by the time base collector circuit, and errors caused by scratches and dust on the film are corrected by the strong error correction algorithm. Also in the DEC, signals for track P and track S which have been compressed by the ATRAC system are decoded. This data is transmitted to the DSP as a linear PCM signal.

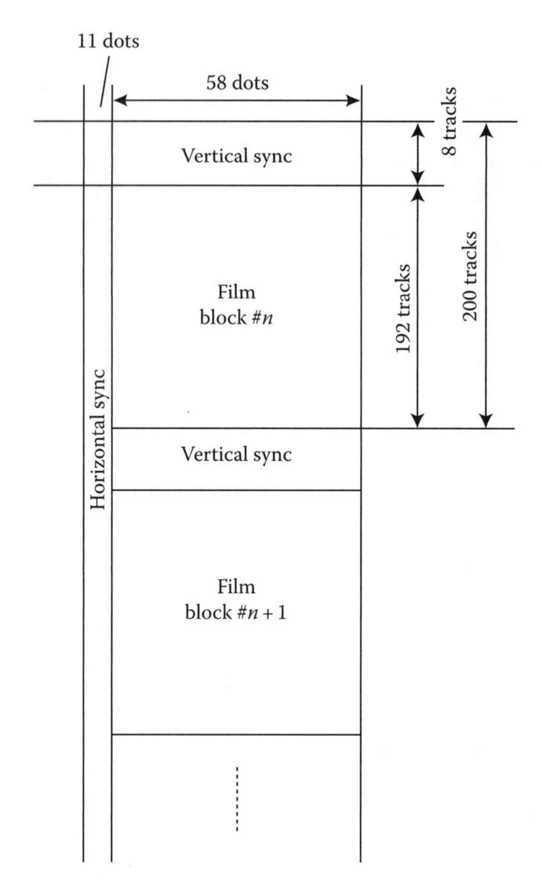

FIGURE 5.10 Data block configuration.

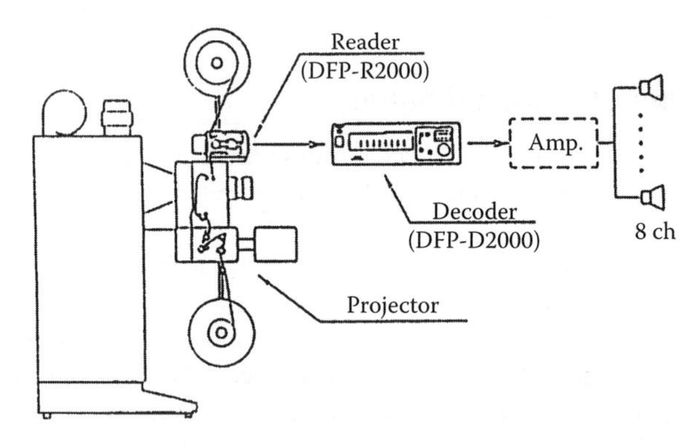

FIGURE 5.11 Playback system.

In the DSP, the sound field of the theater is adjusted and concealment modes are controlled. A CPU is installed in the DSP to control the entire decoder, and control the front panel display and reception and transmission of external control data.

Finally in the APR, 10 channels of digital filter including monitors, D/A converter, and line amplifier are installed. Also, it is possible to directly bypass an analogue input signal by relay as necessary. This bypass is prepared to cope with analogue sound if digital sound would not play back.

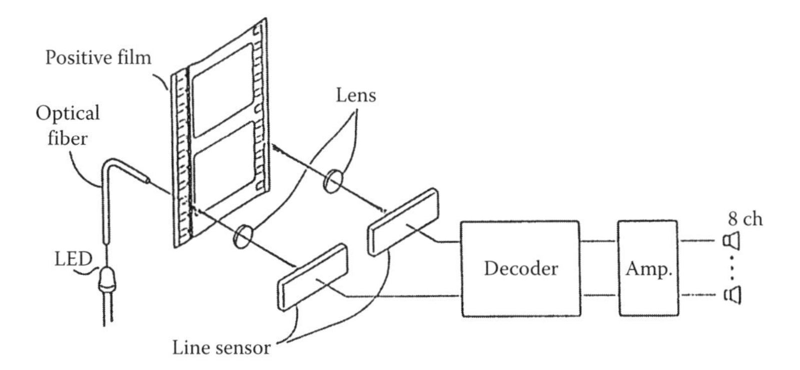

FIGURE 5.12 Optical reader concept.

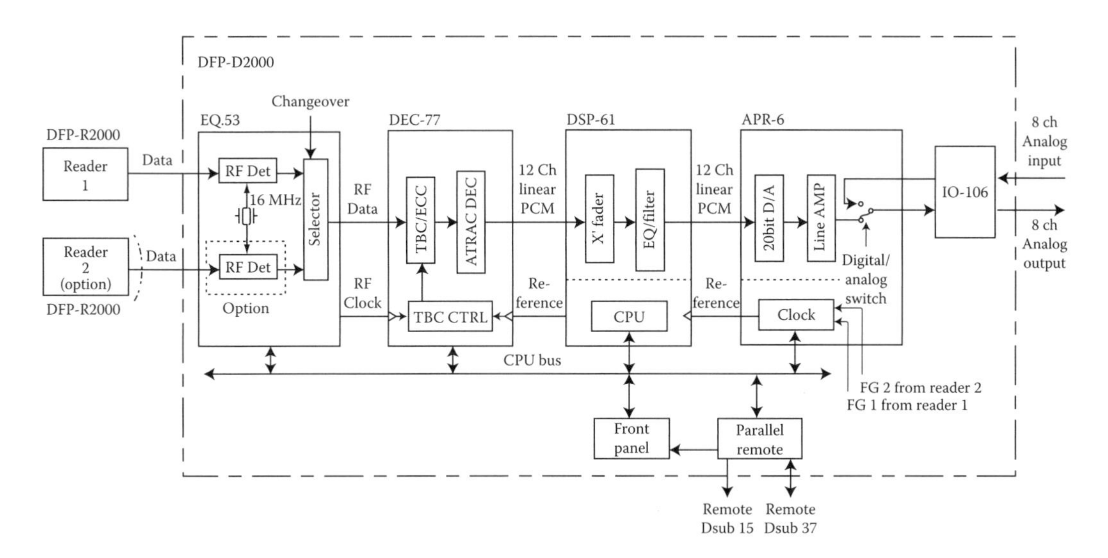

FIGURE 5.13 Overall block diagram.

5.3.3 The SDDS Error Correction Technique

The SDDS system adapts the "Reed Solomon" code for error correction. An error correction technique is essential for maintaining high sound quality and high picture quality for digital recording and playback systems, such as CD, MD, digital VTR, etc. Such C1 parity +C2 parity data necessary for error correction are added and recorded in advance to cope with cases when correct data cannot be obtained. It enables recovery of the correct data by using this additional data even if a reading error occurs.

If the error rate is 10^{-4} (1 bit for every 10,000 bits), the error rate for C1 parity after correction would normally be 10^{-11}. In other words, an error would occur only once every 1.3 years if a film were showed 24 h a day. Errors will be extremely close to "zero" by using C2 parity erasure correction. A strong error correction capability is installed in the SDDS digital sound playback system against random errors.

Other errors besides random errors are

- Errors caused by a scratch in the film running direction
- Errors caused by dust on the film
- Errors caused by splice points of films
- Errors caused by defocusing during printing or playback

TABLE 5.1 SDDS Player System Electrical Specifications

Item	Specification
Sampling frequency	44.1 kHz
Dynamic range	Over 90 dB
Channel	Max 8 ch
Frequency band	20 Hz–20 kHz ± 1.0 dB
K.F	$< 0.7\%$
Crosstalk	< -80 dB
Reference output level	-10 dB/unbalanced, $+4$ dB/balanced
Head room	>20 dB/balanced

These are considered burst errors which occur consistently. Scratch errors in particular will increase more and more every time the film is shown. SDDS has the capability of dealing with such burst errors. Therefore, in spite of the scratch on the film width direction, error correction towards the film length would be possible up to 1.27 mm and in spite of the scratch on the film running direction, error correction towards the film width would be possible up to 336 μm.

5.3.4 Features of the SDDS System

The specification characteristics for the SDDS player are shown in Table 5.1. It is not easy to obtain high fidelity in audio data compression compared to the linear recording system of CDs with regard to a sound quality. By adapting a system with high compression efficiency and making use of the human hearing characteristics, we were able to maintain a sound quality equivalent to CDs by adapting the ATRAC system which restrains deterioration to the minimum.

One of the biggest features of the SDDS is the adaption of a digital backup system. This is a countermeasure system to make up for the damage to the splicing parts of the digital data or the parts of data missing by using the opposite side of the track with a digital data recorded on the backup channel. By this system, it would be possible to obtain an equivalent quality. Next, when finally the film is worn out, the system switches over to an analogue playback signal.

This system also has a digital room EQ function. This supplies 28 bands of graphic EQ with 1/3 octave characteristics and a high- and low-pass filter. Moreover, a simple operation to control the sound field in the theater will become possible by using a graphic user interface panel of an external personal computer.

Such control usually took hours, but it can be completed in about 30 min with this SDDS player. The stability of its features, reproducibility, and reliability of digitizing is well appreciated.

Furthermore, the SDDS player carries a backup function and a reset function for setting parameters by using memories.

5.4 Switched Predictive Coding of Audio Signals for the CD-I and CD-ROM XA Format

Masayuki Nishiguchi

5.4.1 Introduction

An audio bit rate reduction system for the CD-I and CD-ROM XA format based on switched predictive coding algorithm is described. The principal feature of the system is that the coder provides multiple prediction error filters, each of which has fixed coefficients. The prediction error filter that best matches the input signal is selected every 28 samples (1 block). A first-order and two kinds of second-order prediction error filters are used for signals in the low and middle frequencies, and the straight PCM is used for high-frequency signals. The system also uses near-instantaneous companding to expand the dynamic range. A noise-shaping filter is incorporated in the quantization stage, and its frequency response is varied to minimize the energy of the output noise. With a complexity of less than 8MIPS/channel, audio quality almost the same as CD audio can be achieved at 310 kbps (8.2 bits/sample), near transparent audio can be achieved at 159 kbps (4.2 bits/sample), and mid-fidelity audio can be achieved at 80 kbps (4.2 bits/sample).

5.4.2 Coder Scheme

Figure 5.14 is a block diagram of the encoder and decoder system. The input signal, prediction error, quantization error, encoder output, decoder input, and decoder output are respectively expressed as $x(n)$, $d(n), e(n), \hat{d}(n), \hat{d}'(n)$, and $\hat{x}'(n)$. The z-transforms of the signals are expressed as $X(z), D(z), E(z), \hat{D}(z)$, $\hat{D}'(z)$, and $\hat{X}'(z)$. The encoder response can then be expressed as

$$\hat{D}(z) = G \cdot X(z) \cdot \{1 - P(z)\} + E(z) \cdot \{1 - R(z)\}, \tag{5.8}$$

and the decoder response as

$$\hat{X}'(z) = \frac{G^{-1} \cdot \hat{D}'(z)}{1 - P(z)}. \tag{5.9}$$

Assuming that there is no channel error, we can write $\hat{D}'(z) = \hat{D}(z)$. Using Equations 5.8 and 5.9, we can write the decoder output in terms of the encoder input as

$$X'(z) = X(z) + G^{-1} \cdot E(z) \cdot \frac{1 - R(z)}{1 - P(z)}, \tag{5.10}$$

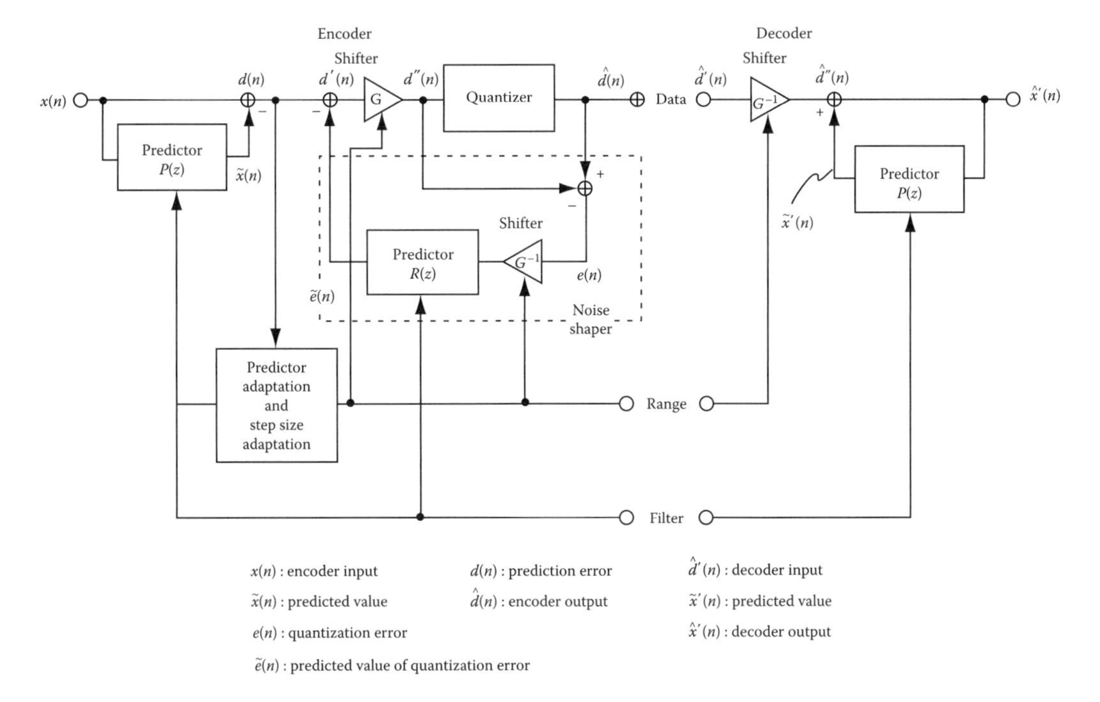

FIGURE 5.14 Block diagram of the bit rate reduction system.

where

$$P(z) = \sum_{k=1}^{P} \alpha_k \cdot z^{-k} \quad \text{and} \quad R(z) = \sum_{k=1}^{R} \beta_k \cdot z^{-k}. \tag{5.11}$$

Here α_k and β_k are, respectively, the coefficients of predictor $P(z)$ and $R(z)$. Equation 5.10 shows the encoder–decoder performance characteristics of the system. It shows that the quantization error $E(z)$ is reduced by the extent of the noise-reduction effect G^{-1}. The distribution of the noise spectrum that appears at the decoder output is

$$N(z) = E(z) \cdot \frac{1 - R(z)}{1 - P(z)}. \tag{5.12}$$

$R(z)$ can be varied according to the spectral shape of the input signal in order to have a maximum masking effect, but we have set $R(z) = P(z)$ to keep from coloring the quantization noise.

G can be regarded as the normalization factor for the peak prediction error (over 28 residual samples) from the chosen prediction error filter. The value of G changes according to the frequency response of the prediction gain:

$$G \propto \frac{|X(z)|}{|D(z)|}. \tag{5.13}$$

This is also proportional to the inverse of the prediction error filter, $1/|1 - P(z)|$. So, in order to maximize G, it is necessary to change the frequency response of the prediction error filter $1 - P(z)$ according to the frequency distribution of the input signals.

5.4.2.1 Selection of the Optimum Filter

Several different strategies of selecting filters are possible in the CD-I/CD-ROM XA format, but the simplest way for the encoder to choose which predictor is most suitable is the following:

- The predictor adaptation section compares the peak value of the prediction errors (over 28 samples) from each prediction error filter $1 - P(z)$ and selects the filter that generates the minimum peak
- The group of prediction errors chosen is then gain controlled (normalized by its maximum value) and noise shaping is executed at the same time

As a result, a high signal-to-noise ratio (SNR) is obtained by using a first-order and two kinds of second-order prediction error filters for signals with the low and middle frequencies and by using the straight pulse code modulation (PCM) for high-frequency signals.

5.4.2.2 Coder Parameters

This system provides three bit rates for the CD-I/CD-ROM XA format, and data encoded at any bit rate can be decoded by a single decoder. The following sections explain how the parameters used in the decoder and the encoder change according to the level of sound quality. Table 5.2 lists the parameters for each level.

5.4.2.2.1 Level A

We can obtain the highest quality audio sound with Level A, which uses only two prediction error filters. Either the straight PCM or the first-order differential PCM is selected. The transfer functions of the prediction error filters are as follows:

TABLE 5.2 The Parameters for Each Level

	Level A	Level B	Level C
Sampling frequency (kHz)	37.8	37.8	18.9
Residual word length (bits per sample)	8	4	4
Block length (number of samples)	28	28	28
Range data (bits per block)	4	4	4
Range values	0–8	0–12	0–12
Filter data (bits per block)	1	2	2
Number of prediction error filters used	2	3	4
Average of bits used per	8.18	4.21	4.21
sample (bits per sample)	$=(8\times 28+4+1)/28$	$=(4\times 28+4+2)/28$	$=(4\times 28+4+2)/28$
Bit rate (kbps)	309	159	80

$$H(z) = 1 \tag{5.14}$$

and

$$H(z) = 1 - 0.975z^{-1}, \tag{5.15}$$

where $H(z) = 1 - P(z)$.

5.4.2.2.2 Level B

The bit rate at Level B is half as high as that at Level A. By using this level, we can obtain high-fidelity audio sound from most high-quality sources. This level uses three filters: the straight PCM, the first-order differential PCM, or the second-order differential PCM-1 is selected. The transfer functions of the first two filters are the same as in Level A, and that for the second-order differential PCM-1 mode is

$$H(z) = 1 - 1.796875z^{-1} + 0.8125z^{-2}. \tag{5.16}$$

5.4.2.2.3 Level C

We can obtain mid-fidelity audio sound at Level C, and a monoaural audio program 16 h long can be recorded on a single CD. Four filters are used for this level. The transfer function of the first three filters are the same as in Level B. The transfer function of the second-order differential PCM-2 mode, used only at this level, is

$$H(z) = 1 - 1.53125z^{-1} + 0.859375z^{-2}. \tag{5.17}$$

At all levels, the noise-shaping filter and the inverse-prediction-error filter in the decoder have the same coefficients as the prediction error filter in the encoder.

5.4.3 Applications

The simple structure and low complexity of this CD-I/CD-ROM XA audio compression algorithm make it suitable for applications with PCs, workstations, and video games.

5.5 ATRAC Family

Keisuke Toyama and Kyoya Tsutsui

5.5.1 ATRAC

ATRAC is a coding system designed to meet the following criteria for the MiniDisc system:

- Compression of 16 bit 44.1 kHz audio (705.6 kbps) into 146 kbps with minimal reduction in sound quality
- Simple hardware implementation suitable for portable players and recorders

Block diagrams of the encoder and decoder structures are shown in Figures 5.15 and 5.16, respectively. The time-frequency analysis block of the encoder decomposes the input signal into spectral coefficients grouped into 52 block floating units (BFUs). The bit allocation block divides the available bits among the BFUs adaptively based on their psychoacoustic characteristics. The spectrum quantization block normalizes spectral coefficients by the scale factor assigned to each BFU, and then quantizes each of them to the specified word length. These processes are performed in every sound unit, which is a block consisting of 512 samples per channel.

In order to generate the BFUs, the time-frequency analysis block first divides the input signal into three subbands (Figure 5.17). Then, each of the subbands is converted to the frequency domain by a modified discrete cosine transform (MDCT), producing a set of spectral coefficients. Finally, the spectral coefficients are nonuniformly grouped into BFUs. Cascaded 48-tap quadrature mirror filters (QMFs) perform the subband decomposition: The first QMF divides the input signal into upper and lower frequency bands, and the second QMF further divides the lower frequency band. Although the output samples of each filter are decimated by two, the use of QMFs for reconstruction cancels the aliasing caused by subband decomposition. The MDCT block length is adaptively determined based on the characteristics of the signal in each band. There are two block-length modes: long mode (11.6 ms for $f_s = 44.1$ kHz) and short mode (1.45 ms in the high frequency band, 2.9 ms in the others). Normally, the long mode is chosen, as this provides good frequency resolution. However, problems can occur during attack portions of the signal because the quantization noise is spread over the entire block and the

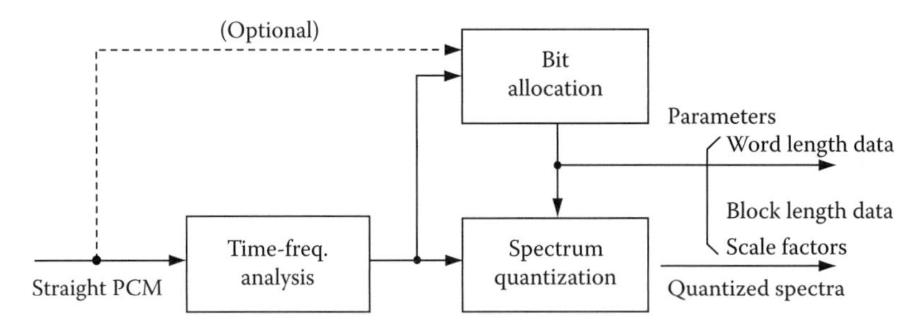

FIGURE 5.15 ATRAC encoder.

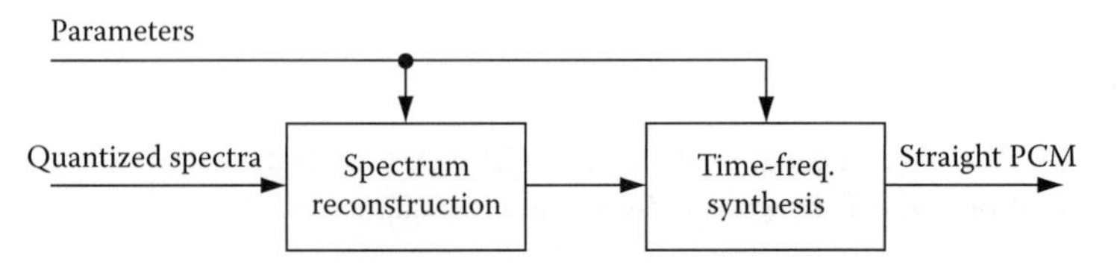

FIGURE 5.16 ATRAC decoder.

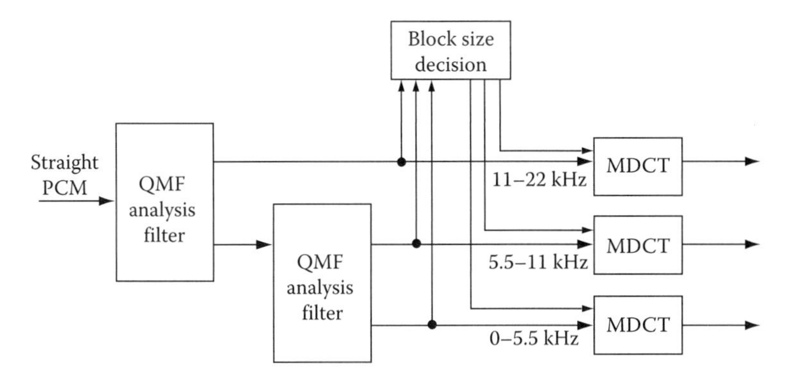

FIGURE 5.17 ATRAC time-frequency analysis.

initial quantization noise is not masked by simultaneous masking. To prevent this degradation, which is called pre-echo, ATRAC switches to short mode when it detects an attack signal. In this case, since the noise before the attack exists for only a very short period of time, it is masked by backward masking. The window form is symmetric for both long and short modes, and the window form in the non-zero-nor-one region of the long mode is the same as that for the short mode. Although this window form has a somewhat negative impact on the separability of the spectrum, it provides the following advantages:

- The transform mode can be determined based solely on the existence of an attack signal in the current sound unit, thus eliminating the need for an extra buffer in the encoder
- A smaller amount of buffer memory in the encoder and decoder is required to store the overlapped samples for the next sound unit

The mapping structure of ATRAC is shown in Figure 5.18.

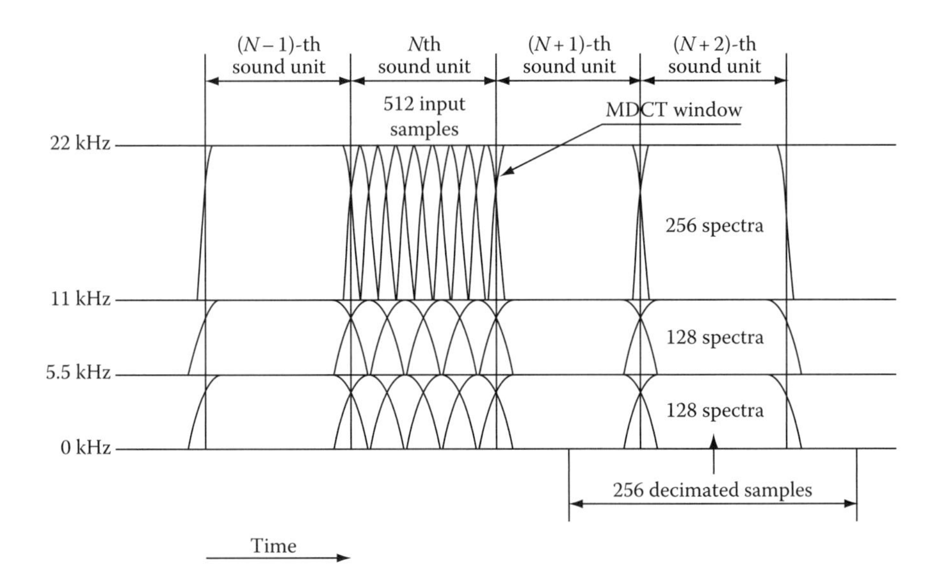

FIGURE 5.18 ATRAC mapping structure.

5.5.2 ATRAC2

The ATRAC2 system takes advantage of the progress in LSI technologies and allows audio signals with 16 bits per sample and a sampling frequency of 44.1 kHz (705.6 kbps) to be compressed into 64 kbps with almost no degradation in audio quality. The design focused on the efficient coding of tonal signals because the human ear is particularly sensitive to distortions in them.

Block diagrams of the encoder and decoder structures are shown in Figures 5.19 and 5.20, respectively. The encoder extracts psychoacoustically important tone components from the input signal spectrum so that they may be efficiently encoded separately from less important spectral data. A tone component is a group of consecutive spectral coefficients and is defined by several parameters, including its location and the width of the data. The remaining spectral coefficients are grouped into 32 nonuniform BFUs. Both the tone components and the remaining spectral coefficients can be encoded with Huffman coding, as shown in Table 5.3, for which simple decoding with a look-up table is practical due to their small size.

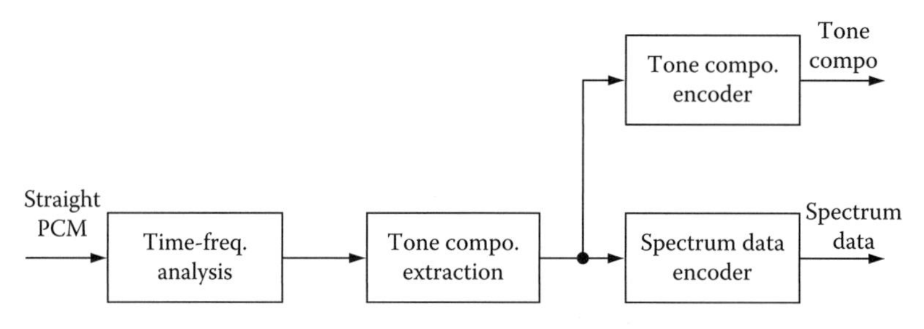

FIGURE 5.19 ATRAC2 encoder.

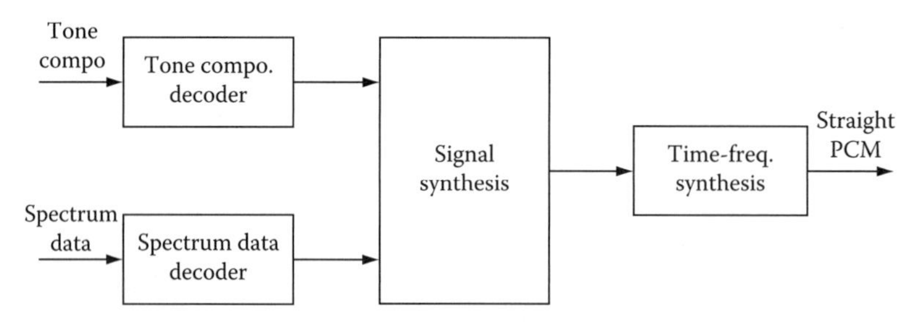

FIGURE 5.20 ATRAC2 decoder.

TABLE 5.3 Huffman Code Table of ATRAC2 and ATRAC3

ID	Quantization Step Number	Dimension (Spectr. Num.)	Maximum Code Length	Look-Up Table Size
0	1	—	—	—
1	3	2	5	32
2	5	1	3	8
3	7	1	4	16
4	9	1	5	32
5	15	1	6	64
6	31	1	7	128
7	63	1	8	256

Note: Total = 536.

Although the number of quantization steps is limited to 63, a high SNR can be obtained by repeatedly extracting tone components from the same frequency range.

The mapping structure of ATRAC2 is shown in Figure 5.21. The frequency resolution is twice that of ATRAC; and to ensure frequency separability, ATRAC2 performs a signal analysis using the combination of a 96-tap polyphase quadrature filter (PQF) and a fixed-length 50%-overlap MDCT with different forward and backward window forms. ATRAC2 prevents pre-echo by adaptively amplifying the signal preceding an attack before the encoder transforms it into spectral coefficients and the decoder restores it to the original level after the inverse transformation. This technique is called gain modification and it simplifies the spectral structure of the system (Figure 5.22).

Subband decomposition provides frequency scalability; and simpler decoders can be constructed simply by limiting decoding to lower-band data. The use of a PQF reduces the computational complexity.

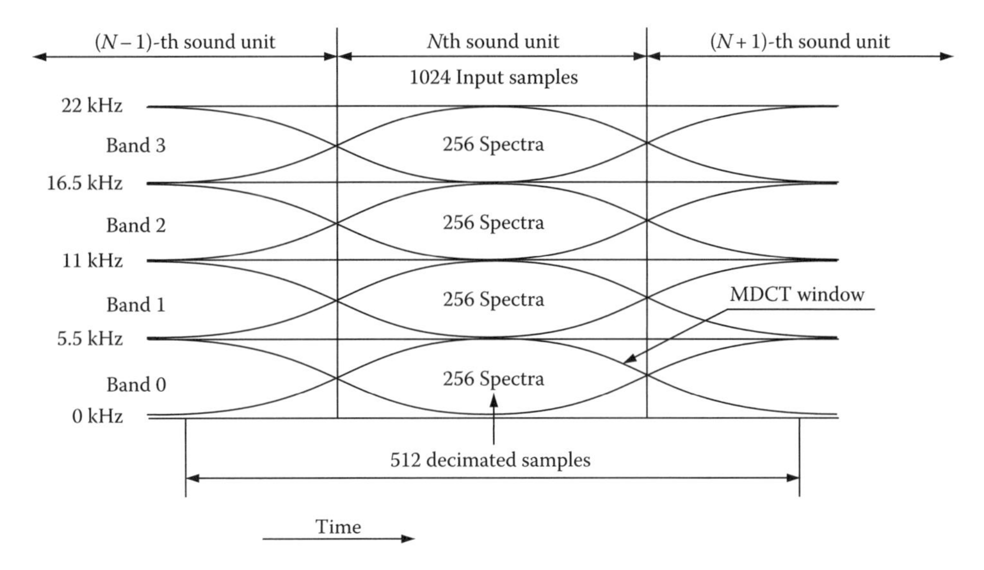

FIGURE 5.21 ATRAC2 mapping structure.

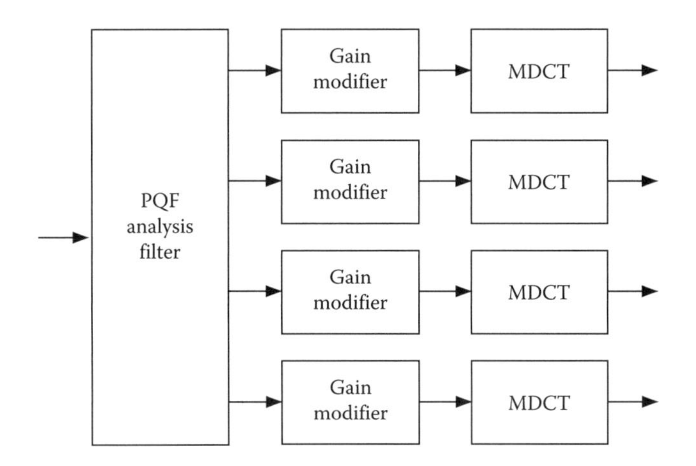

FIGURE 5.22 ATRAC2 time-frequency analysis.

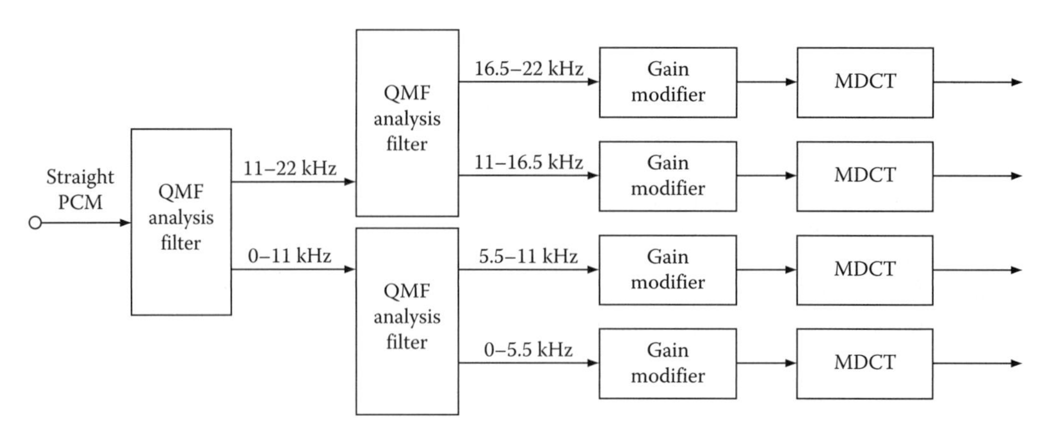

FIGURE 5.23 ATRAC3 time-frequency analysis.

5.5.3 ATRAC3

ATRAC3 was developed not only for specific products, such as MiniDisc, but also for general use. In fact, ATRAC3 is now being used in a large number of products, applications, standards, and services around the world.

The structure of ATRAC3 is very similar to that of ATRAC2. ATRAC3 uses cascaded 48-tap QMFs to split signals into four subbands. The first QMF divides the input signals into upper and lower frequency bands, and the second QMF further divides both bands (Figure 5.23). The main reason for employing QMFs instead of a PQF is to facilitate direct transformation between the bit-stream of ATRAC and that of ATRAC3 because, as mentioned in Section 5.5.1, ATRAC uses QMFs of a similar type.

5.5.4 ATRAC3plus

ATRAC3plus provides twice the coding efficiency of ATRAC3 with almost the same audio quality. Through the use of the technologies listed below, ATRAC3plus achieves such a good compression ratio and such excellent sound quality that it is also being used in many products, applications, standards, and services around the world.

Number of band divisions

To ensure high accuracy in the analysis, a 384-tap PQF divides input signals into 16 subbands (bandwidth: approx. 1.38 kHz; sampling frequency: 44.1 kHz), which is four times the number of divisions in ATRAC3 (Figures 5.24 and 5.25).

Twice the transform block length

The transform block length of ATRAC3plus is twice that of ATRAC3, which enhances the frequency resolution so as to enable efficient bit allocation. The mapping structure of ATRAC3plus is shown in Figure 5.26.

Time domain analysis

Gain modification reduces the quantization noise. That is, adjusting the gain within a block during encoding and decoding reduces unpleasant pre-echo and post-echo noise. In addition, stable signals and transient signals are separated beforehand to make gain modification more effective. These processes are carried out in each PQF subband.

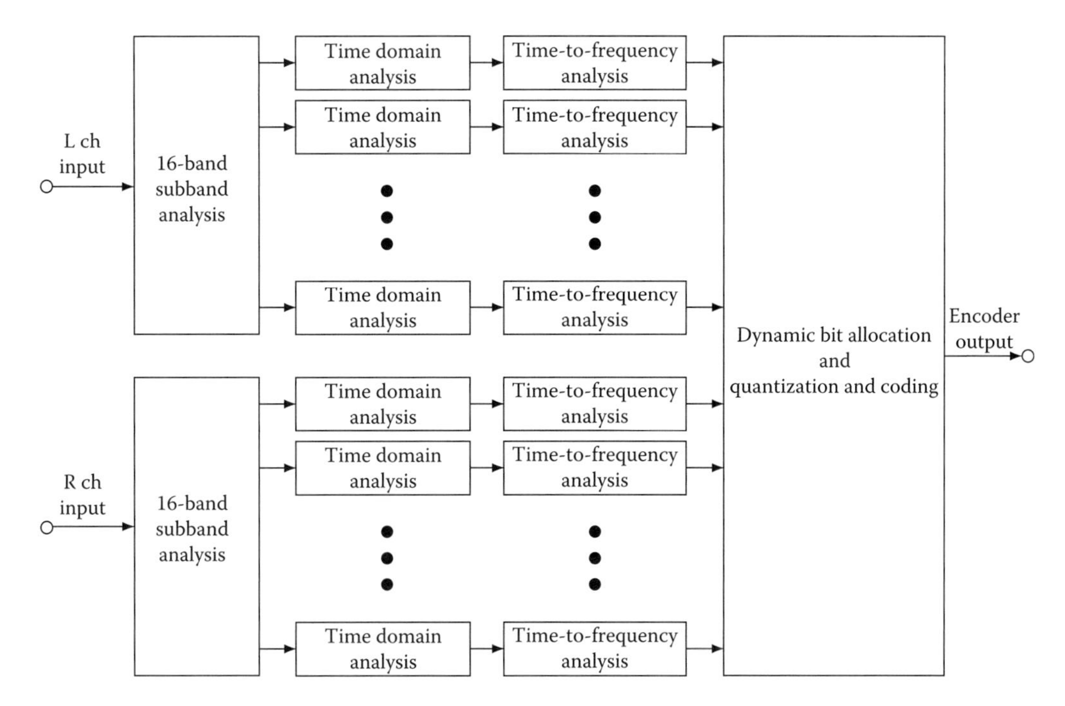

FIGURE 5.24 ATRAC3plus encoder.

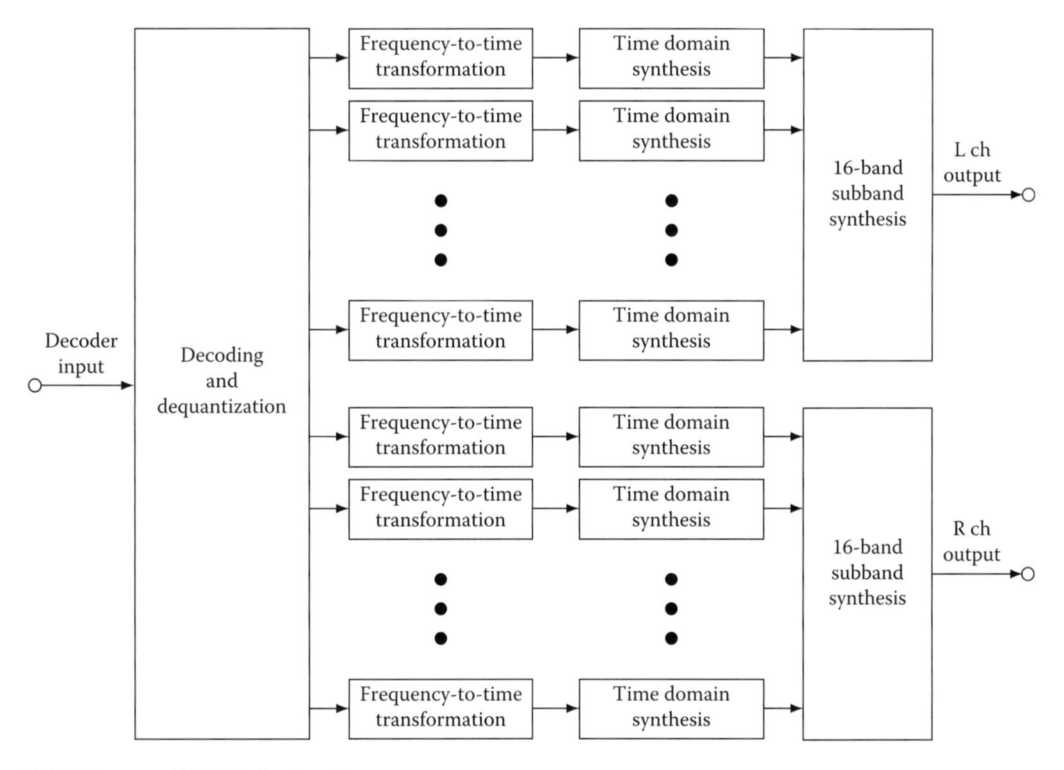

FIGURE 5.25 ATRAC3plus decoder.

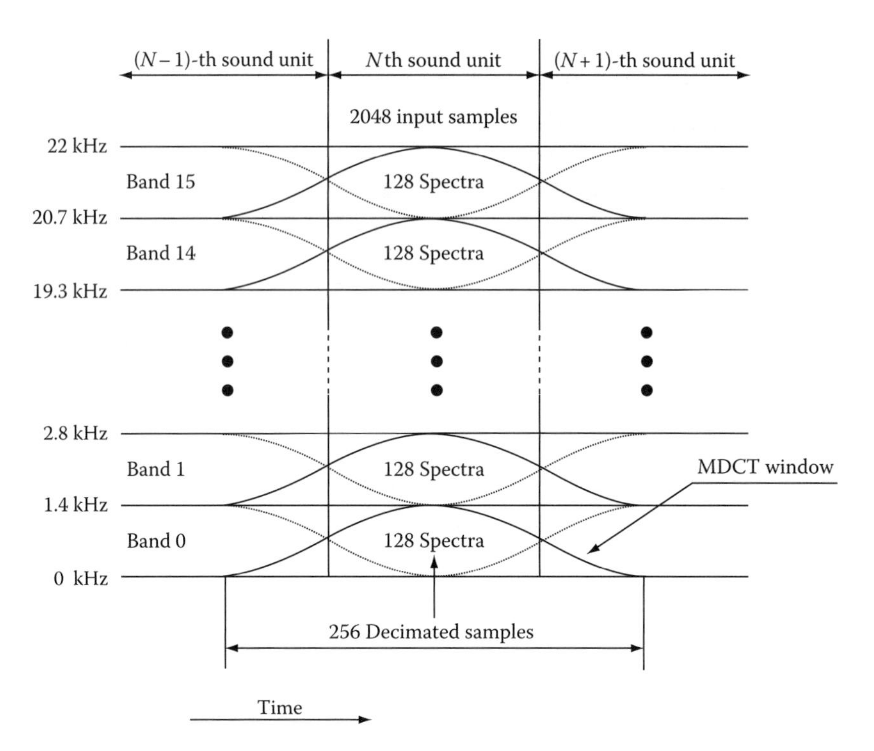

FIGURE 5.26 ATRAC3plus mapping structure.

Multidimensional Huffman coding

Multiple spectra are Huffman-coded at the same time, rather than separately, enabling more efficient compression. ATRAC3 uses only one set of Huffman tables, which have a maximum dimension of two. In contrast, ATRAC3plus employs 16 types of Huffman tables with a maximum dimension of four.

Efficient encoding of side information

Spectrum normalization information, quantization accuracy information, and other side information are also compressed efficiently by using channel and frequency band correlations.

Multichannel support

ATRAC3plus supports multichannel coding (max. 64).

5.5.5 ATRAC Advanced Lossless

ATRAC Advanced Lossless was developed for people who demand true CD sound quality. It can compress CD data without any loss of information and completely restore the data for playback, delivering the same sound quality as CDs with data that is only 30%–80% the size of the original data.

ATRAC Advanced Lossless records the information of ATRAC3 or ATRAC3plus, as well as the residual information that they eliminate. This means that either the ATRAC3 or ATRAC3plus data can be extracted as is, or the residual information can be added to perfectly reproduce the information on the original CD. This type of data compression technique is called scalable compression, which means that it allows a playback device to select only those data layers that suit its capabilities.

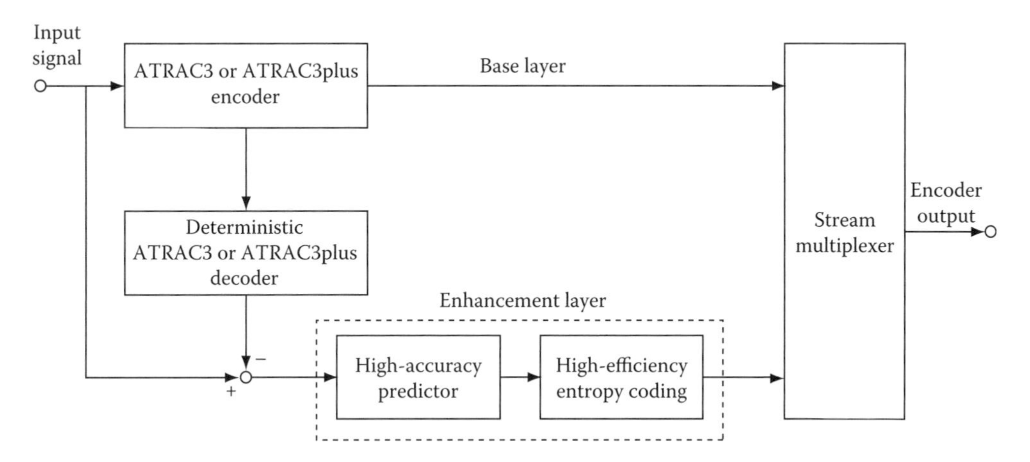

FIGURE 5.27 ATRAC advanced lossless encoder.

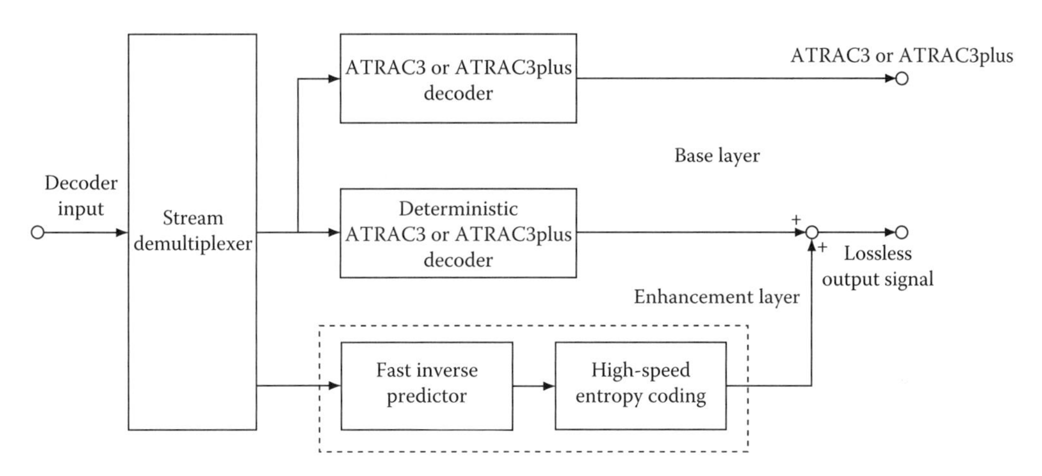

FIGURE 5.28 ATRAC advanced lossless decoder.

The structures of the scalable encoder and decoder are shown in Figures 5.27 and 5.28, respectively. The ATRAC3 or ATRAC3plus encoder quantizes and encodes input audio signals, and outputs the bit stream of ATRAC3 or ATRAC3plus as a base-layer bit stream. The base-layer bit stream is decoded by the deterministic ATRAC3 or ATRAC3plus decoder. The decoder restores a large portion of the original audio signals. Since the deterministic decoder is platform independent, its output signals are identical on any device. The residual signals, which are the difference signals between the original input and the output of the deterministic decoder, are encoded by the enhancement layer encoder, a high-accuracy predictor, and high-efficiency entropy coding. Finally, the bit stream of the base layer and that of the enhancement layer are multiplexed and output.

References

1. Kayanuma, A. et al., An integrated 16 bit A/D converter for PCM audio systems, *ISSCC Dig. Tech. Papers,* pp. 56–57, Feb. 1981.
2. Plassche, R. J. et al., A monolithic 14 bit D/A converter., *IEEE J. Solid State Circuits,* SC-14:552–556, 1979.

3. Naus, P. J. A. et al., A CMOS stereo 16 bit D/A converter for digital audio, *IEEE J. Solid State Circuits,* SC-22:390–395, June 1987.
4. Hauser, M. W., Overview of oversampling A/D converters, Audio Engineering Society, San Francisco, CA, Preprint #2973, 1990.
5. Matsuya, Y. et al., A 16-bit oversampling A to D conversion technology using triple-integration noise shaping, *IEEE J. Solid State Circuits*, SC-22:921–929, Dec. 1987.
6. Schouwenaars, H. J. et al., An oversampling multibit CMOS D/A converter for digital audio with 115dB dynamic range, *IEEE J. Solid State Circuits*, SC-26:1775–1780, Dec. 1991.
7. Maruyama, Y. et al., A 20-bit stereo oversampling D to A converter, *IEEE Trans. Consumer Electron.*, 39:274–276, Aug. 1993.
8. Nishiguchi, M., Akagiri, K., and Suzuki, T., A new audio bit-rate reduction system for the CD-I format, Preprint 81st AES Convention, San Francisco, CA, November 1986.
9. Rabiner, L. R. and Schafer, R. W., *Digital Processing of Speech Signals*, Prentice-Hall, Englewood Cliffs, NJ, 1978.
10. Oppenheim, A. V. and Schafer, R. W., *Digital Signal Processing*, Prentice-Hall, Englewood Cliffs, NJ, 1975.
11. Tsutsui, K. et al., ATRAC: Adaptive transform acoustic coding for MiniDisc, Audio Engineering Society, San Francisco, CA, Preprint #3456, 1992.
12. DVD CA zone option codec, DVD specifications for read-only disc Part 4: Audio Specifications Ver.1.2, Optional specifications: Compressed audio files for DVD-audio Rev.1.0, DVD specifications for DVD-RAM/DVD-RW/DVD-R for general discs Part 4: Audio Recording Ver.1.0, Reference Book name: *ATRAC3plus specifications for DVD.*
13. DLNA option codec, DLNA home networked device interoperability guidelines v1.0, Addendum optional media format guidelines v1.0.
14. IEC 61937-7, Digital audio—Interface for non-linear PCM encoded audio bitstreams applying IEC 60958—Part 7: Non-linear PCM bitstreams according to the ATRAC, ATRAC2/3 and ATRAC-X formats.
15. IEC 61909, Audio recording—Minidisc system.
16. Bluetooth Optional Codec, Audio codec interoperability requirements—4.2 Support of codecs, in *Advanced Audio Distribution Profile*, Advanced Audio Distribution Profile Specification Adopted version 1.04.
17. http://www.sony.net/Products/ATRAC3/index.html

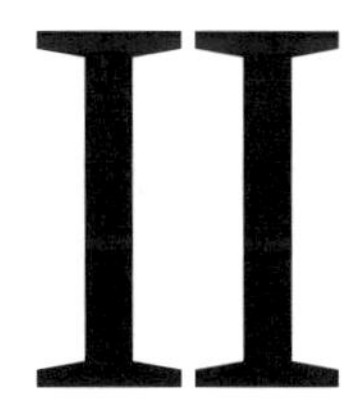

Speech Processing

Richard V. Cox
AT&T Research Labs

Lawrence R. Rabiner
Rutgers University and AT&T Research Labs

WITH THE ADVENT OF CHEAP, HIGH-SPEED PROCESSORS, and with the ever- decreasing cost of memory, the cost of speech processing has been driven down to the point where it can be (and has been) embedded in almost any system, from a low-cost consumer product (e.g., solid-state digital answering machines, voice-controlled telephones, etc.), to a desktop application (e.g., voice dictation of a first draft quality manuscript), to an application embedded in a voice or data network (e.g., voice dialing, packet telephony, voice browser for the Internet, etc.). It is the purpose of this part of the handbook to provide discussions of several of the key technologies in speech processing and to illustrate how the technologies are implemented using special-purpose digital signal processor (DSP) chips or via standard software packages running on more conventional processors.

The broad area of speech processing can be broken down into several individual areas according to both applications and technology. These include the following:

Chapter 6: In order to understand how the characteristics of a speech signal can be exploited in the different application areas, it is necessary to understand the properties and constraints of the human vocal apparatus (to understand how speech is generated by humans). It is also necessary to understand the way in which models can be built that simulate speech production as well as the ways in which they can be implemented as digital systems, since such models form the basis for almost all practical speech-processing systems.

Chapter 7: Speech coding is the process of compressing the information in a speech signal so as to either transit it or store it economically over a channel whose bandwidth is significantly smaller than that of the uncompressed signal. Speech coding is used as the basis for most modern voice-messaging and voice-mail systems, for voice-response systems, for digital cellular and for satellite transmission of speech, for packet telephony, for ISDN teleconferencing, and for digital answering machines and digital voice-encryption machines.

Chapter 8: Speech synthesis is the process of creating a synthetic replica of a speech signal so as to transmit a message from a machine to a person, with the purpose of conveying the information in the message. Speech synthesis is often called "text-to-speech" (TTS), to convey the idea that, in general, the input to the system is ordinary ASCII text, and the output of the system is ordinary speech. The goal of most speech synthesis systems is to provide a broad range of capability for having a machine speak information (stored in the machine) to a user. Key aspects of synthesis systems are the intelligibility and the naturalness of the resulting speech. The major applications of speech synthesis include acting as a voice server for text-based information services (e.g., stock prices, sports scores, flight information); providing a means for reading e-mail, or the text portions of FAX messages over ordinary phone lines; providing a means for previewing text stored in documents (e.g., document drafts, Internet files); and finally as a voice readout for handheld devices, (e.g., phrase book translators, dictionaries, etc.)

Chapter 9: Speech recognition is the process of extracting the message information in a speech signal so as to control the action of a machine in response to spoken commands. In a sense, speech recognition is the complementary process to speech synthesis, and together they constitute the building blocks of a voice-dialogue system with a machine. There are many factors that influence the type of speech-recognition system that is used for different applications, including the mode of speaking to the machine (e.g., single commands, digit sequences, fluent sentences), the size and complexity of the vocabulary that the machine understands, the task that the machine is asked to accomplish, the environment in which the recognition system must run, and finally the cost of the system. Although there is a wide range of applications of speech-recognition systems, the most generic systems are simple "command-and-control" systems (with menu-like interfaces), and the most advanced systems support full voice dialogues for dictation, forms entry, catalog ordering, reservation services, etc.

Chapter 10: Speaker verification is the process of verifying the claimed identity of a speaker for the purpose of restricting access to information (e.g., personal or private records), networks (computer, private branch exchange [PBX]), or physical premises. The basic problem of speaker verification is to decide whether or not an unknown speech sample was spoken by the individual whose identity was claimed. A key aspect of any speaker-verification system is to accept the true speaker as often as possible while rejecting the impostor as often as possible. Since these are inherently conflicting goals, all practical systems arrive at some compromise between levels of these two types of system errors. The major area of application for speaker verification is in access control to information, credit, banking, machines, computer networks, PBXs, and even premises. The concept of a "voice lock" that prevents access until the appropriate speech by the authorized individual(s) (e.g., "Open Sesame") is "heard" by the system is made a reality using speaker-verification technology.

Chapter 11: Until a few years ago, almost all speech-processing systems were implemented on low-cost DSP fixed-point processors because of their high efficiency in realizing the computational aspects of the various signal-processing algorithms. A key problem in the realization of any digital system in integer DSP code is how to map an algorithm efficiently (in both time and space), which is typically running in floating point C code on a workstation to integer C code that takes advantage of the unique characteristics of different DSP chips. Furthermore, because of the rate of change of technology, it is essential that the conversion to DSP code occur rapidly (e.g., on the order of 3 person months) or else by the time a given algorithm is mapped to a specific DSP processor, a new (faster, cheaper) generation of DSP chips will have evolved, obsoleting the entire process.

Chapter 12: The field of speech processing has become a complex one, where an investigator needs a broad range of tools to record, digitize, display, manipulate, process, store, format, analyze, and listen to speech in its different file forms and manifestations. Although it is conceivable that an individual could create a suite of software tools for an individual application, that process would be highly inefficient and would undoubtedly result in tools that were significantly less powerful than those developed in the commercial sector, such as the Entropic Signal Processing System, MATLAB®, Waves, Interactive Laboratory System (ILS), or the commercial packages for TTS and speech recognition such as the Hidden Markov Model Toolkit (HTK).

The material presented in this part should provide the reader with a framework for understanding the signal-processing aspects of speech processing and some pointers into the literature for further investigation of this fascinating and rapidly evolving field.

6

Speech Production Models and Their Digital Implementations

M. Mohan Sondhi
Lucent Technologies

Juergen Schroeter
AT&T Research Labs

6.1 Introduction

The characteristics of a speech signal that are exploited for various applications of speech signal processing to be discussed later in this section on speech processing (e.g., coding, recognition, etc.) arise from the properties and constraints of the human vocal apparatus. It is, therefore, useful in the design of such applications to have some familiarity with the process of speech generation by humans. In this chapter we will introduce the reader to (1) the basic physical phenomena involved in speech production, (2) the simplified models used to quantify these phenomena, and (3) the digital implementations of these models.

6.1.1 Speech Sounds

Speech is produced by acoustically exciting a time-varying cavity—the vocal tract, which is the region of the mouth cavity bounded by the vocal cords and the lips. The various speech sounds are produced by adjusting both the type of excitation as well as the shape of the vocal tract.

There are several ways of classifying speech sounds [1]. One way is to classify them on the basis of the type of excitation used in producing them:

- Voiced sounds are produced by exciting the tract by quasiperiodic puffs of air produced by the vibration of the vocal cords in the larynx. The vibrating cords modulate the air stream from the lungs at a rate which may be as low as 60 times per second for some males to as high as 400 or 500 times per second for children. All vowels are produced in this manner. So are laterals, of which *l* is the only exemplar in English.

- Nasal sounds such as *m*, *n*, *ng*, and nasalized vowels (as in the French word *bon*) are also voiced. However, part or all of the airflow is diverted into the nasal tract by opening the velum.
- Plosive sounds are produced by exciting the tract by a sudden release of pressure. The plosives *p*, *t*, *k* are voiceless, while *b*, *d*, *g* are voiced. The vocal cords start vibrating before the release for the voiced plosives.
- Fricatives are produced by exciting the tract by turbulent flow created by air flow through a narrow constriction. The sounds *f*, *s*, *sh* belong to this category.
- Voiced fricatives are produced by exciting the tract simultaneously by turbulence and by vocal cord vibration. Examples are *v*, *z*, and *zh* (as in *pleasure*).
- Affricates are sounds that begin as a stop and are released as a fricative. In English, *ch* as in *check* is a voiceless affricate and *j* as in *John* is a voiced affricate.

In addition to controlling the type of excitation, the shape of the vocal tract is also adjusted by manipulating the tongue, lips, and lower jaw. The shape determines the frequency response of the vocal tract. The frequency response at any given frequency is defined to be the amplitude and phase at the lips in response to a sinusoidal excitation of unit amplitude and zero phase at the source. The frequency response, in general, shows concentration of energy in the neighborhood of certain frequencies, called formant frequencies.

For vowel sounds, three or four resonances can usually be distinguished clearly in the frequency range 0–4 kHz. (On average, over 99% of the energy in a speech signal is in this frequency range.) The configuration of these resonance frequencies is what distinguishes different vowels from each other.

For fricatives and plosives, the resonances are not as prominent. However, there are characteristic broad frequency regions where the energy is concentrated.

For nasal sounds, besides formants there are antiresonances, or zeros in the frequency response. These zeros are the result of the coupling of the wave motion in the vocal and nasal tracts. We will discuss how they arise in a later section.

6.1.2 Speech Displays

We close this section with a description of the various ways of displaying properties of a speech signal. The three common displays are (1) the pressure waveform, (2) the spectrogram, and (3) the power spectrum. These are illustrated for a typical speech signal in Figure 6.1a through c.

Figure 6.1a shows about half a second of a speech signal produced by a male speaker. What is shown is the pressure waveform (i.e., pressure as a function of time) as picked up by a microphone placed a few centimeters from the lips. The sharp click produced at a plosive, the noise-like character of a fricative, and the quasiperiodic waveform of a vowel are all clearly discernible.

Figure 6.1b shows another useful display of the same speech signal. Such a display is known as a spectrogram [2]. Here the x-axis is time. But the y-axis is frequency and the darkness indicates the intensity at a given frequency at a given time. [The intensity at a time t and frequency f is just the power in the signal averaged over a small region of the time–frequency plane centered at the point (t,f)]. The dark bands seen in the vowel region are the "formants." Note how the energy is much more diffusely spread out in frequency during a plosive or fricative.

Finally, Figure 6.1c shows a third representation of the same signal. It is called the power spectrum. Here the power is plotted as a function of frequency, for a short segment of speech surrounding a specified time instant. A logarithmic scale is used for power and a linear scale for frequency. In this particular plot, the power is computed as the average over a window of duration 20 ms. As indicated in the figure, this spectrum was computed in a voiced portion of the speech signal. The regularly spaced peaks—the fine structure—in the spectrum are the harmonics of the fundamental frequency. The spacing is seen to be about 100 Hz, which checks with the time period of the wave seen in the pressure waveform

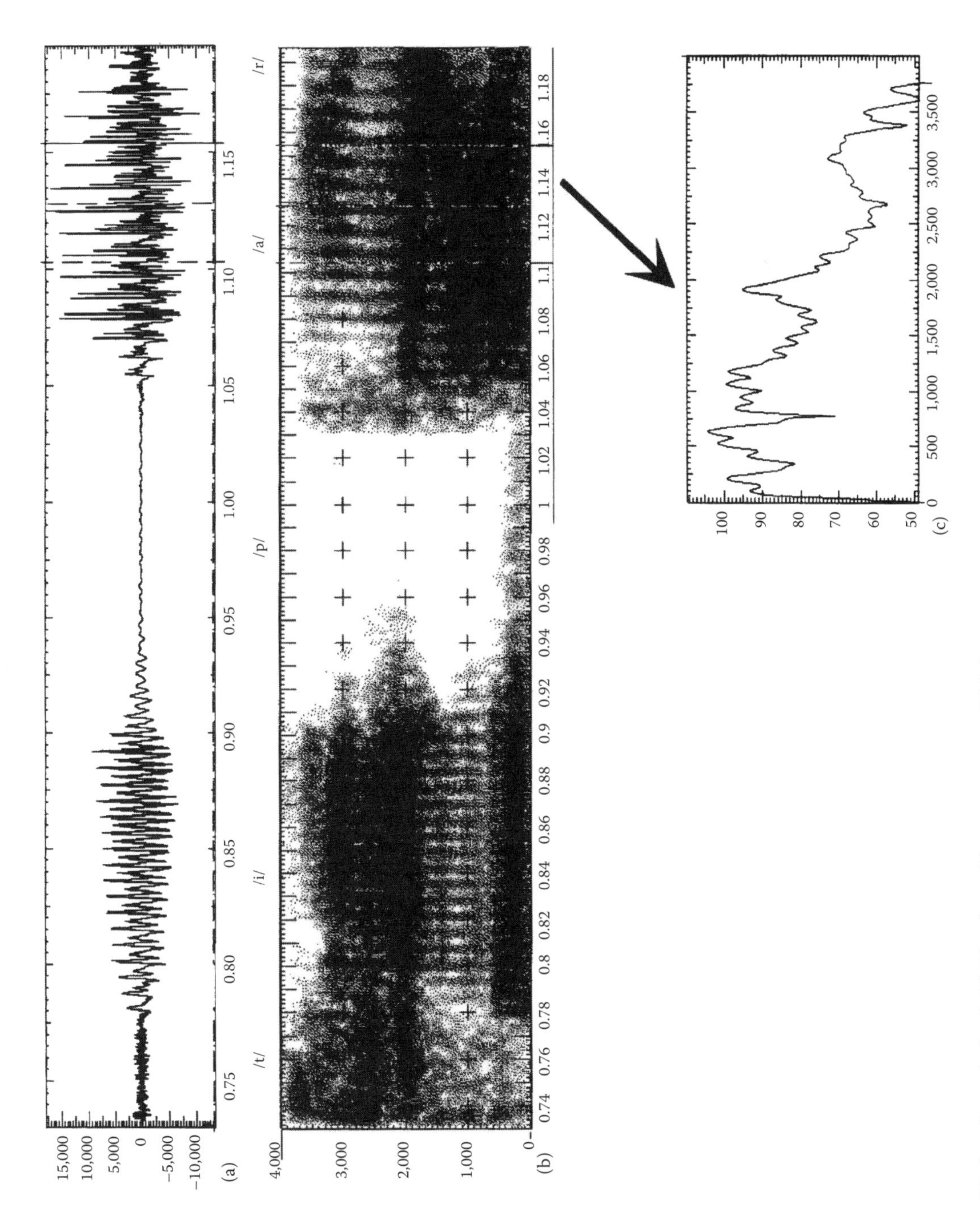

FIGURE 6.1 Display of speech signal: (a) Waveform, (b) spectrogram, and (c) frequency response.

in Figure 6.1a. The peaks in the envelope of the harmonic peaks are the formants. These occur at about 650, 1100, 1900, and 3200 Hz, which checks with the positions of the formants seen in the spectrogram of the same signal displayed in Figure 6.1b.

6.2 Geometry of the Vocal and Nasal Tracts

Much of our knowledge of the dimensions and shapes of the vocal tract is derived from a study of x-ray photographs and x-ray movies of the vocal tract taken while subjects utter various specific speech sounds or connected speech [3]. In order to keep x-ray dosage to a minimum, only one view is photographed, and this is invariably the side view (a view of the midsagittal plane). Information about the cross-dimensions is inferred from static vocal tracts using frontal x-rays, dental molds, etc.

More recently, magnetic resonance imaging (MRI) [4] has also been used to image the vocal and nasal tracts. The images obtained by this technique are excellent and provide three-dimensional reconstructions of the vocal tract. However, at present MRI is not capable of providing images at a rate fast enough for studying vocal tracts in motion.

Other techniques have also been used to study vocal tract shapes. These include

1. Ultrasound imaging [5]. This provides information concerning the shape of the tongue but not about the shape of the vocal cavity.
2. Acoustical probing of the vocal tract [6]. In this technique, a known acoustic wave is applied at the lips. The shape of the time-varying vocal cavity can be inferred from the shape of the time-varying reflected wave. However, this technique has thus far not achieved sufficient accuracy. Also, it requires the vocal tract to be somewhat constrained while the measurements are made.
3. Electropalatography [7]. In this technique, an artificial palate with an array of electrodes is placed against the hard palate of a subject. As the tongue makes contact with this palate during speech production, it closes an electrical connection to some of the electrodes. The pattern of closures gives an estimate of the shape of the contact between tongue and palate. This technique cannot provide details of the shape of the vocal cavity, although it yields important information on the production of consonants.
4. Finally, the movement of the tongue and lips has also been studied by tracking the positions of tiny coils attached to them [8]. The motion of the coils is tracked by the currents induced in them as they move in externally applied electromagnetic fields. Again, this technique cannot provide a detailed shape of the vocal tract.

Figure 6.2 shows an x-ray photograph of a female vocal tract uttering the vowel sound /u/. It is seen that the vocal tract has a very complicated shape, and without some simplifications it would be very difficult to just specify the shape, let alone compute its acoustical properties. Several models have been proposed to specify the main features of the vocal tract shape. These models are based on studies of x-ray photographs of the type shown in Figure 6.2, as well as on x-ray movies taken of subjects uttering various speech materials. Such models are called articulatory models because they specify the shape in terms of the positions of the articulators (i.e., the tongue, lips, jaw, and velum).

Figure 6.3 shows such an idealization, similar to one proposed by Coker [9], of the shape of the vocal tract in the midsagittal plane. In this model, a fixed shape is used for the palate, and the shape of the vocal cavity is adjusted by specifying the positions of the articulators. The coordinates used to describe the shape are labeled in the figure. They are the position of the tongue center, the radius of the tongue body, the position of the tongue tip, the jaw opening, the lip opening and protrusion, the position of the hyoid, and the opening of the velum. The cross-dimensions (i.e., perpendicular to the sagittal plane) are estimated from static vocal tracts. These dimensions are assumed fixed during speech production. In this manner, the three-dimensional shape of the vocal tract is modeled.

Whenever the velum is open, the nasal cavity is coupled to the vocal tract, and its dimensions must also be specified. The nasal cavity is assumed to have a fixed shape which is estimated from static measurements.

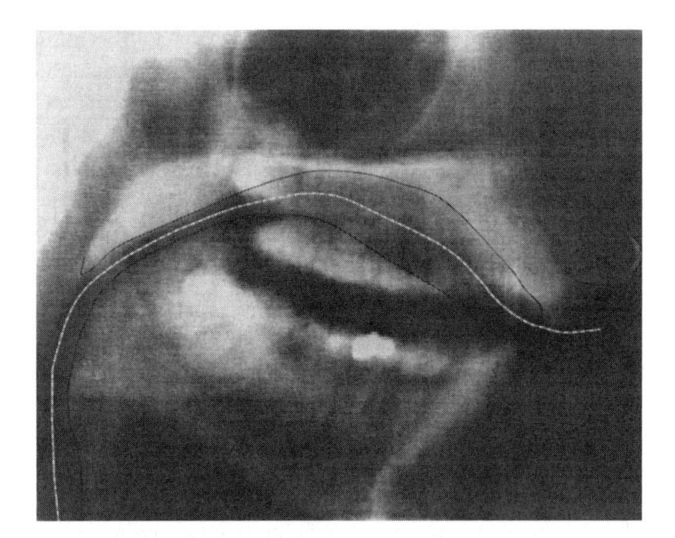

FIGURE 6.2 X-ray side view of a female vocal tract. The tongue, lips, and palate have been outlined to improve visibility. (Modified from a single frame from Laval Film 55, Side 2 of Munhall, K.G. et al., X-ray film data-base for speech research, ATR Technical Report Tr-H-116, December 28, 1994, ATR Human Information Processing Research Laboratories, Kyoto, Japan; From Dr. Claude Rochette, Departement de Radiologie de l'Hotel-Dieu de Quebec, Quebec, Canada. With permission.)

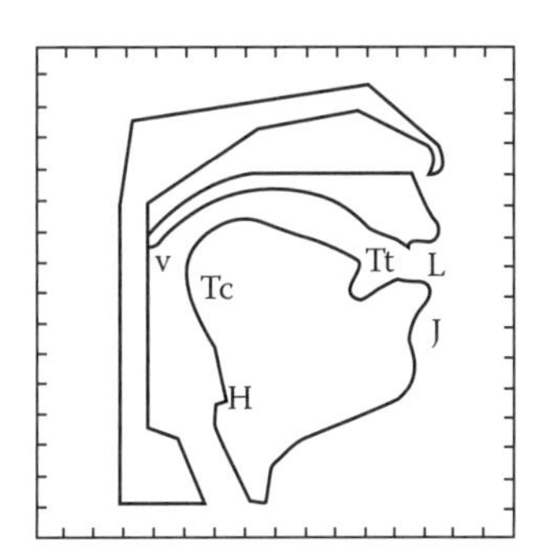

H = Hytoid position
J = Angle of jaw opening
L = Lip protrusion and elevation
Tc = Tongue center
Tt = Position of tongue tip
v = Velum opening

FIGURE 6.3 An idealized articulatory model similar to that of Coker. (From Coker, C.H., *Proc. IEEE*, 64(4), 452, April 1976.)

6.3 Acoustical Properties of the Vocal and Nasal Tracts

Exact computation of the acoustical properties of the vocal (and nasal) tract is difficult even for the idealized models described in the previous section. Fortunately, considerable further simplification can be made without affecting most of the salient properties of speech signals generated by such a model. Almost without exception, three assumptions are made to keep the problem tractable. These assumptions are justifiable for frequencies below about 4 kHz [10,11].

6.3.1 Simplifying Assumptions

1. It is assumed that the vocal tract can be "straightened out" in such a way that a center line drawn through the tract (shown dotted in Figure 6.3) becomes a straight line. In this way, the tract is converted to a straight tube with a variable cross section.

2. Wave propagation in the straightened tract is assumed to be planar. This means that if we consider any plane perpendicular to the axis of the tract, then every quantity associated with the acoustic wave (e.g., pressure, density, etc.) is independent of position in the plane.
3. The third assumption that is invariably made is that wave propagation in the vocal tract is linear. Nonlinear effects appear when the ratio of particle velocity to sound velocity (the Mach number) becomes large. For wave propagation in the vocal tract the Mach number is usually less than 0.02, so that nonlinearity of the wave is negligible. There are, however, two exceptions to this. The flow in the glottis (i.e., the space between the vocal folds), and that in the narrow constrictions used to produce fricative sounds, is nonlinear. We will show later how these special cases are handled in current speech production models.

We ought to point out that some computations have been made without the first two assumptions, and wave phenomena studied in two or three dimensions [12]. Recently there has been some interest in removing the third assumption as well [13]. This involves the solution of the so-called Navier–Stokes equation in the complicated three-dimensional geometry of the vocal tract. Such analyses require very large amounts of high speed computations making it difficult to use them in speech production models. Computational cost and speed, however, are not the only limiting factors. An even more basic barrier is that it is difficult to specify accurately the complicated time-varying shape of the vocal tract. It is, therefore, unlikely that such computations can be used directly in a speech production model. These computations should, however, provide accurate data on the basis of which simpler, more tractable, approximations may be abstracted.

6.3.2 Wave Propagation in the Vocal Tract

In view of the assumptions discussed above, the propagation of waves in the vocal tract can be considered in the simplified setting depicted in Figure 6.4 As shown there, the vocal tract is represented as a variable area tube of length L with its axis taken to be the x-axis. The glottis is located at $x = 0$ and the lips at $x = L$, and the tube has a cross-sectional area $A(x)$ which is a function of the distance x from the glottis. Strictly speaking, of course, the area is time-varying. However, in normal speech the temporal variation in the area is very slow in comparison with the propagation phenomena that we are considering. So, the cross-sectional area may be represented by a succession of stationary shapes.

We are interested in the spatial and temporal variation of two interrelated quantities in the acoustic wave: the pressure $p(x, t)$ and the volume velocity. The latter is $A(x)v(x, t)$, where v is the particle velocity. For the assumption of linearity to be valid, the pressure p in the acoustic wave is assumed to be small compared to the equilibrium pressure P_0, and the particle velocity v is assumed to be small compared to the velocity of sound, c. Two equations can be written down that relate $p(x, t)$ and $u(x, t)$: the equation of motion and the equation of continuity [14]. A combination of these equations will give us the basic equation of wave propagation in the variable area tube. Let us derive these equations first for the case when the walls of the tube are rigid and there are no losses due to viscous friction, thermal conduction, etc.

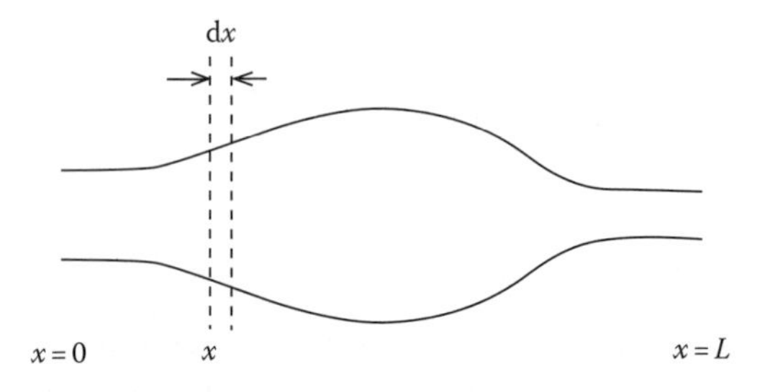

FIGURE 6.4 The vocal tract as a variable area tube.

6.3.3 The Lossless Case

The equation of motion is just a statement of Newton's second law. Consider the thin slice of air between the planes at x and $x + \mathrm{d}x$ shown in Figure 6.4. By equating the net force acting on it due to the pressure gradient to the rate of change of momentum one gets

$$\frac{\partial p}{\partial x} = -\frac{\rho}{A}\frac{\partial u}{\partial t} \tag{6.1}$$

(To simplify notation, we will not always explicitly show the dependence of quantities on x and t.)

The equation of continuity expresses conservation of mass. Consider the slice of tube between x and $x + \mathrm{d}x$ shown in Figure 6.4. By balancing the net flow of air out of this region with a corresponding decrease in the density of air we get

$$\frac{\partial u}{\partial x} = -\frac{A}{\rho}\frac{\partial \delta}{\partial t} \tag{6.2}$$

where $\delta(x, t)$ is the fluctuation in density superposed on the equilibrium density ρ. The density is related to pressure by the gas law. It can be shown that pressure fluctuations in an acoustic wave follow the adiabatic law, so that $p = (\gamma P/\rho)\delta$, where γ is the ratio of specific heats at constant pressure and constant volume. Also, $(\gamma P/\rho) = c^2$, where c is the velocity of sound. Substituting this into Equation 6.2 gives

$$\frac{\partial u}{\partial x} = -\frac{A}{\rho c^2}\frac{\partial p}{\partial t} \tag{6.3}$$

Equations 6.1 and 6.3 are the two relations between p and u that we set out to derive. From these equations it is possible to eliminate u by subtracting $\partial/\partial t$ of Equation 6.3 from $\partial/\partial x$ of Equation 6.1. This gives

$$\frac{\partial}{\partial x} A \frac{\partial p}{\partial x} = \frac{A}{c^2}\frac{\partial^2 p}{\partial t^2} \tag{6.4}$$

Equation 6.4 is known in the literature as "Webster's horn equation" [15]. It was first derived for computations of wave propagation in horns, hence the name. By eliminating p from Equations 6.1 and 6.3, one can also derive a single equation in u.

It is useful to write Equations 6.1, 6.3, and 6.4 in the frequency domain by taking Laplace transforms. Defining $P(x, s)$ and $U(x, s)$ as the Laplace transforms of $p(x, t)$ and $u(x, t)$, respectively, and remembering that $\partial/\partial t \rightarrow s$, we get

$$\frac{\mathrm{d}p}{\mathrm{d}x} = \frac{\rho s}{A} U \tag{6.1a}$$

$$\frac{\mathrm{d}U}{\mathrm{d}x} = -\frac{sA}{\rho c^2} P \tag{6.3a}$$

and

$$\frac{\mathrm{d}}{\mathrm{d}x} A \frac{\mathrm{d}p}{\mathrm{d}x} = \frac{s^2}{c^2} AP \tag{6.4a}$$

It is important to note that in deriving these equations we have retained only first order terms in the fluctuating quantities p and u. Inclusion of higher order terms gives rise to nonlinear equations of propagation. By and large these terms are quite negligible for wave propagation in the vocal tract. However, there is one second order term, neglected in Equation 6.1, which becomes important in the description of flow through the narrow constriction of the glottis. In deriving Equation 6.1 we neglected the fact that the slice of air to which the force is applied is moving away with the velocity v. When this effect is correctly taken into account, it turns out that there is an additional term $\rho v(\partial v/\partial x)$ appearing on the left-hand side of that equation, and the corrected form of Equation 6.1 is

$$\frac{\partial}{\partial x}\left[p+\frac{\rho}{2}\left({}^{u}/_{A^2}\right)\right]=-\rho\frac{\mathrm{d}}{\mathrm{d}t}\left[\frac{u}{A}\right] \tag{6.5}$$

The quantity $\rho/2(u/A)^2$ has the dimensions of pressure, and is known as the Bernoulli pressure. We will have occasion to use Equation 6.3 when we discuss the motion of the vocal cords in the section on sources of excitation.

6.3.4 Inclusion of Losses

The equations derived in the previous section can be used to approximately derive the acoustical properties of the vocal tract. However, their accuracy can be considerably increased by including terms that approximately take account of the effect of viscous friction, thermal conduction, and yielding walls [16]. It is most convenient to introduce these effects in the frequency domain.

The effect of viscous friction can be approximated by modifying the equation of motion, Equation 6.1a as follows:

$$\frac{\mathrm{d}p}{\mathrm{d}x}=-\frac{\rho s}{A}U-R(x,s)U \tag{6.6}$$

Recall that Equation 6.1a states that the force applied per unit area equals the rate of change of momentum per unit area. The added term in Equation 6.6 represents the viscous drag which reduces the force available to accelerate the air. The assumption that the drag is proportional to velocity can be approximately validated. The dependence of R on x and s can be modeled in various ways [16].

The effect of thermal conduction and yielding walls can be approximated by modifying the equation of continuity as follows:

$$\rho\frac{\mathrm{d}u}{\mathrm{d}x}=\frac{A}{c^2}sP-Y(x,s)P \tag{6.7}$$

Recall that the left hand side of Equation 6.3a represents net outflow of air in the longitudinal direction, which is balanced by an appropriate decrease in the density of air. The term added in Equation 6.7 represents net outward volume velocity into the walls of the vocal tract. This velocity arises from (1) a temperature gradient perpendicular to the walls that is due to the thermal conduction by the walls, and (2) due to the yielding of the walls. Both these effects can be accounted for by appropriate choice of the function $Y(x, s)$, provided the walls can be assumed to be locally reacting. By that we mean that the motion of the wall at any point depends on the pressure at that point alone. Models for the function $Y(x, s)$ may be found in [16].

Finally, the lossy equivalent of Equation 6.4a is

$$\frac{\mathrm{d}}{\mathrm{d}x}\frac{A}{\rho s+AR}\frac{\mathrm{d}P}{\mathrm{d}x}=\left(\frac{As}{\rho c^2}\right)P \tag{6.8}$$

6.3.5 Chain Matrices

All properties of linear wave propagation in the vocal tract can be derived from Equations 6.1a, 6.3a, 6.4a or the corresponding Equations 6.6 through 6.8 for the lossy tract. The most convenient way to derive these properties is in terms of chain matrices, which we now introduce.

Since Equation 6.8 is a second order linear ordinary differential equation, its general solution can be written as a linear combination of two independent solutions, say $\phi(x, s)$ and $\Psi(x, s)$. Thus

$$P(x, s) = a\phi(x, s) + b\psi(x, s) \tag{6.9}$$

where a and b are, in general, functions of s. Hence, the pressure at the input of the tube ($x = 0$) and at the output ($x = L$) are linear combinations of a and b. The volume velocity corresponding to the pressure given in Equation 6.9 is obtained from Equation 6.6 to be

$$U(x, s) = -\frac{A}{\rho s + AR}\left[\frac{a\mathrm{d}\phi}{\mathrm{d}x} + \frac{b\mathrm{d}\psi}{\mathrm{d}x}\right] \tag{6.10}$$

Thus, the input and output volume velocities are seen to be linear combinations of a and b. Eliminating the parameters a and b from these relationships shows that the input pressure and volume velocity are linear combinations of the corresponding output quantities. Thus, the relationship between the input and output quantities may be represented in terms of a 2×2 matrix as follows:

$$\begin{aligned}\begin{bmatrix} P_{\text{in}} \\ U_{\text{in}} \end{bmatrix} &= \begin{bmatrix} \mathbf{k}_{11} & \mathbf{k}_{12} \\ \mathbf{k}_{21} & \mathbf{k}_{22} \end{bmatrix}\begin{bmatrix} P_{\text{out}} \\ U_{\text{out}} \end{bmatrix} \\ &= \mathbf{K}\begin{bmatrix} P_{\text{out}} \\ U_{\text{out}} \end{bmatrix}\end{aligned} \tag{6.11}$$

The matrix $\mathbf{K}$ is called a chain matrix or ABCD matrix [17]. Its entries depend on the values of ϕ and Ψ at $x = 0$ and $x = L$. For an arbitrarily specified area function $A(x)$ the functions ϕ and ψ are hard to find. However, for a uniform tube, i.e., a tube for which the area and the losses are independent of x, the solutions are very easy. For a uniform tube, Equation 6.8 becomes

$$\frac{\mathrm{d}^2 P}{\mathrm{d}x^2} = \sigma^2 P \tag{6.12}$$

where σ is a function of s given by

$$\sigma^2 = (\rho s + AR)\left(\frac{s}{\rho c^2} + \frac{Y}{A}\right)$$

Two independent solutions of Equation 6.12 are well known to be $\cosh(\sigma x)$ and $\sinh(\sigma x)$, and a bit of algebra shows that the chain matrix for this case is

$$\mathbf{K} = \begin{bmatrix} \cosh(\sigma L) & (1/\beta)\sinh(\sigma L) \\ \beta\sinh(\sigma L) & \cosh(\sigma L) \end{bmatrix} \tag{6.13}$$

where

$$\beta = \sqrt{Y + \frac{As}{\rho c^2} \bigg/ \left[R + \frac{\rho s}{A}\right]}$$

For an arbitrary tract, one can utilize the simplicity of the chain matrix of a uniform tube by approximating the tract as a concatenation of N uniform sections of length $\Delta = L/N$. Now the output quantities of the ith section become the input quantities for the $i + 1$ th section. Therefore, if $\mathbf{K}_i$ is the chain matrix for the ith section, then the chain matrix for the variable-area tract is approximated by

$$\mathbf{K} = \mathbf{K}_1\mathbf{K}_2 \ldots \mathbf{K}_N \tag{6.14}$$

This method can, of course, be used to relate the input–output quantities for any portion of the tract, not just the entire vocal tract. Later we shall need to find the input–output relations for various sections of the tract, for example, the tract from the glottis to the velum for nasal sounds, from the narrowest constriction to the lips for fricative sounds, etc.

As stated above, all linear properties of the vocal tract can be derived in terms of the entries of the chain matrix. Let us give several examples.

Let us associate the input with the glottal end, and the output with the lip end of the tract. Suppose the tract is terminated by the radiation impedance Z_{R} at the lips. Then, by definition, $P_{\mathrm{out}} = Z_{\mathrm{R}}U_{\mathrm{out}}$. Substituting this in Equation 6.11 gives

$$\begin{bmatrix} P_{\mathrm{in}}/U_{\mathrm{out}} \\ U_{\mathrm{in}}/U_{\mathrm{out}} \end{bmatrix} = \begin{bmatrix} \mathbf{k}_{11} & \mathbf{k}_{12} \\ \mathbf{k}_{21} & \mathbf{k}_{22} \end{bmatrix} \begin{bmatrix} Z_{\mathrm{R}} \\ 1 \end{bmatrix} \tag{6.15}$$

From Equation 6.15 it follows that

$$\frac{U_{\mathrm{out}}}{U_{\mathrm{in}}} = \frac{1}{k_{21}Z_{\mathrm{R}}K_{22}} \tag{6.16a}$$

Equation 6.16a gives the transfer function relating the output volume velocity to the input volume velocity. Multiplying this by Z_R gives the transfer function relating output pressure to the input volume velocity. Other transfer functions relating output pressure or volume velocity to input pressure may be similarly derived.

Relationships between pressure and volume velocity at a single point may also be derived. For example,

$$\frac{P_{\mathrm{in}}}{U_{\mathrm{in}}} = \frac{K_{11}Z_R + K_{12}}{K_{21}Z_R + K_{22}} \tag{6.16b}$$

Also, formant frequencies, which we mentioned in the Introduction, can be computed from the transfer function of Equation 6.16a. They are just the values of s at which the denominator on the right-hand side becomes zero. For a lossy vocal tract, the zeros are complex and have the form $s_n = -\alpha_n + j\omega_n, n = 1, 2, \ldots$. Then ω_n is the frequency (in rad/s) of the nth formant, and α_n is its half bandwidth.

Finally, the chain matrix formulation also leads to linear prediction coefficients (LPC), which are the most commonly used representation of speech signals today. Strictly speaking, the representation is valid for speech signals for which the excitation source is at the glottis (i.e., voiced or aspirated speech sounds). Modifications are required when the source of excitation is at an interior point.

To derive the LPC formulation, we will assume the vocal tract to be lossless, and the radiation impedance at the lips to be zero. From Equation 6.16a we see that to compute the output volume velocity from the input volume velocity, we need only the $\mathbf{k}_{22}$ element of the chain matrix for the entire vocal tract. This chain matrix is obtained by a concatenation of matrices as shown in Equation 6.14. The individual matrices $\mathbf{K}_i$ are derived from Equation 6.13, with $N = L/\Delta$. In the lossless case, R and Y are zero, so $\sigma = s/c$ and $\beta = A/\rho c$. Also, if we define $z = \mathrm{e}^{2s\Delta/c}$, then the matrix $\mathbf{K}_i$ becomes

$$\mathbf{K}_i = z^{N/2}\begin{pmatrix} \frac{1}{2}(1+z^{-1}) & \frac{A_i}{2\rho c}(1-z^{-1}) \\ \frac{\rho c}{2A_i}(1-z^{-1}) & \frac{1}{2}(1+z^{-1}) \end{pmatrix} \tag{6.17}$$

Clearly, therefore, $\mathbf{k}_{22}$ is $z^{N/2}$ times an Nth degree polynomial in z^{-1}. Hence, Equation 6.16a can be written as

$$\sum_{k=0}^{N} a_k z^{-k} U_{\text{out}} = z^{-\left(\frac{N}{2}\right)} U_{\text{in}} \tag{6.18}$$

where a_k are the coefficients of the polynomial. The frequency domain factor $z = \mathrm{e}^{-2s\Delta/c}$ represents a delay of $2\Delta/cs$. Thus, the time domain equivalent of Equation 6.18 is

$$\sum_{k=0}^{N} a_k u_{\text{out}}(t - 2k\Delta/c) = u_{\text{in}}(t - N\Delta/c) \tag{6.19}$$

Now $u_{\text{out}}(t)$ is the volume velocity in the speech signal, so we will call it $s(t)$ for brevity. Similarly, since $u_{\text{in}}(t)$ is the input signal at the glottis, we will call it $g(t)$. To get the time-sampled version of Equation 6.19 we set $t = 2n\Delta/c$ and define $s(2n\Delta/c) = s_n$ and $g((2n - N)\Delta/c) = g_n$. Then Equation 6.19 becomes

$$\sum_{k=0}^{N} a_k s_{n-k} = \varepsilon_n \tag{6.20}$$

Equation 6.20 is the LPC representation of a speech signal.

6.3.6 Nasal Coupling

Nasal sounds are produced by opening the velum and thereby coupling the nasal cavity to the vocal tract. In nasal consonants, the vocal tract itself is closed at some point between the velum and the lips, and all the airflow is diverted into the nostrils. In nasal vowels the vocal tract remains open. (Nasal vowels are common in French and several other languages. They are not nominally phonemes of English. However, some nasalization of vowels commonly occurs in English speech.)

In terms of chain matrices, the nasal coupling can be handled without too much additional effort. As far as its acoustical properties are concerned, the nasal cavity can be treated exactly like the vocal tract, with the added simplification that its shape may be regarded as fixed. The common assumption is that the nostrils are symmetric, in which case the cross-sectional areas of the two nostrils can be added and the nose replaced by a single, fixed, variable-area tube.

The description of the computations is easier to follow with the aid of the block diagram shown in Figure 6.5 From a knowledge of the area functions and losses for the vocal and nasal tracts three chain matrices $\mathbf{K}_{\text{gv}}, \mathbf{K}_{\text{vt}}$, and $\mathbf{K}_{\text{vn}}$ are first computed. These represent, respectively, the matrices from glottis to velum, velum to tract closure (or velum to lips, in case of a nasal vowel), and velum to nostrils.

From $\mathbf{K}_{\text{vn}}$ with some assumed impedance termination at the nostrils, the input impedance of the nostrils at the velum may be computed as indicated in Equation 6.16b. Similarly, $\mathbf{K}_{\text{vt}}$ gives the input impedance at the velum, of the vocal tract looking toward the lips. At the velum, these two impedances are combined in parallel to give a total impedance, say Z_{v}. With this as termination, the velocity to velocity transfer function, T_{gv}, from glottis to velum can be computed from $\mathbf{K}_{\text{gv}}$ as shown

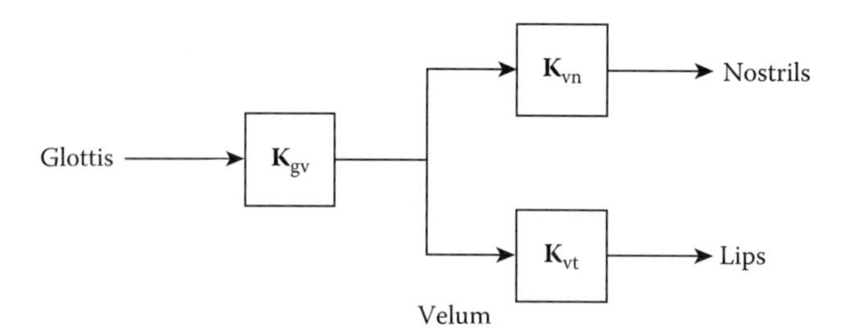

FIGURE 6.5 Chain matrices for synthesizing nasal sounds.

in Equation 6.17a. For a given volume velocity at the glottis, U_g, the volume velocity at the velumis $U_v = T_{gv}U_g$, and the pressure at the velum is $P_v = Z_vU_v$. Once P_v and U_v are known, the volume velocity and/or pressure at the nostrils and lips can be computed by inverting the matrices $\mathbf{K}_{vn}$ and $\mathbf{K}_{vt}$.

6.4 Sources of Excitation

As mentioned earlier, speech sounds may be classified by type of excitation: periodic, turbulent, or transient. All of these types of excitation are created by converting the potential energy stored in the lungs due to excess pressure into sound energy in the audible frequency range of 20 Hz to 20 kHz.

The lungs of a young adult male may have a maximum usable volume ("vital capacity") of about 5 l. While reading aloud the pressure in the lungs is typically in the range of 6–15 cm of water (6,000–15,000 Pa). Vocal cord vibrations can be sustained with a pressure as low as 2 cm of water. At the other extreme, a pressure as high as 195 cm of water has been recorded for a trumpet player. Typical average airflow for normal speech is about 0.1 L/s. It may peak as high as 5 L/s during rapid inhales in singing.

Periodic excitation originates mainly at the vibrating vocal folds, turbulent excitation originates primarily downstream of the narrowest constriction in the vocal tract, and transient excitations occur whenever a complete closure of the vocal pathway is suddenly released. In the following, we will explore these three types of excitation in some detail. The interested reader is referred to [18] for more information.

6.4.1 Periodic Excitation

Many of the acoustic and perceptual features of an individual's voice are believed to be due to specific characteristics of the quasiperiodic excitation signal provided by the vocal folds. These, in turn, depend on the morphology of the voice organ, the "larynx." The anatomy of the larynx is quite complicated, and descriptions of it may be found in the literature [19]. From an engineering point of view, however, it suffices to note that the larynx is the structure that houses the vocal folds whose vibration provides the periodic excitation. The space between the vocal folds, called the glottis, varies with the motion of the vocal folds, and thus modulates the flow of air through them. As late as 1950 Husson postulated that each movement of the folds is in fact induced by individual nerve signals sent from the brain (the Neurochronaxis hypothesis) [20]. We now know that the larynx is a self-oscillating acousto-mechanical oscillator. This oscillator is controlled by several groups of tiny muscles also housed in the larynx. Some of these muscles control the rest position of the folds, others control their tension, and still others control their shape. During breathing and production of fricatives, for example, the folds are pulled apart (abducted) to allow free flow of air. To produce voiced speech, the vocal folds are brought close together (adducted). When brought close enough together, they go into a spontaneous periodic oscillation. These oscillations are driven by Bernoulli pressure (the same mechanism that keeps

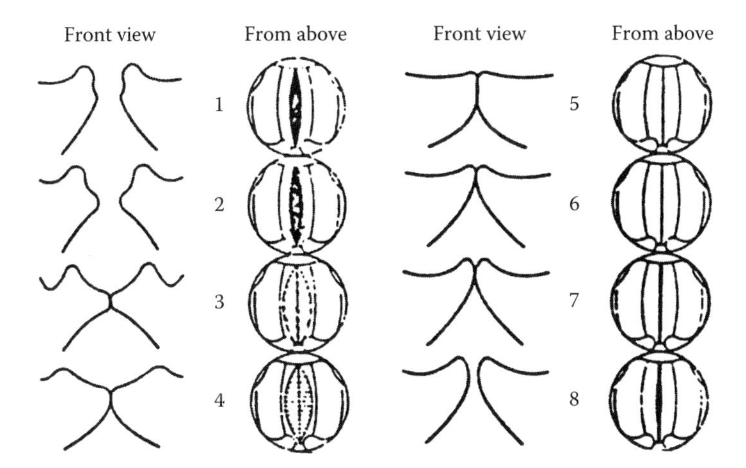

FIGURE 6.6 One cycle of vocal fold oscillation seen from the front and from above. (After Schönhärl, E., *Die Stroboskopie in der praktischen Laryngologie*, Georg Thieme Verlag, Stuttgart, Germany, 1960. With permission.)

airplanes aloft) created by the airflow through the glottis. If the opening of the glottis is small enough, the Bernoulli pressure due to the rapid flow of air is large enough to pull the folds toward each other, eventually closing the glottis. This, of course, stops the flow and the laryngeal muscles pull the folds apart. This sequence repeats itself until the folds are pulled far enough away, or if the lung pressure becomes too low. We will discuss this oscillation in greater detail later in this section.

Besides the laryngeal muscles, the lung pressure and the acoustic load of the vocal tract also affect the oscillation of the vocal folds.

The larynx also houses many mechanoreceptors that signal to the brain the vibrational state of the vocal folds. These signals help control pitch, loudness, and voice timbre.

Figure 6.6 shows stylized snapshots taken from the side and above the vibrating folds. The view from above can be obtained on live subjects with high speed (or stroboscopic) photography, using a laryngeal mirror or a fiber optic bundle for illumination and viewing.

The view from the side is the result of studies on excised (mostly animal) larynges. From studies such as these, we know that, during glottal vibration, the folds carry a mechanical wave that starts at the tracheal (lower) end of the folds and moves upwards to the pharyngeal (upper) end. Consequently, the edge of the folds that faces the vocal tract usually lags behind the edge of the folds that faces the lungs. This phenomenon is called vertical phasing. Higher eigenmodes of these mechanical waves have been observed and have been modeled.

Figure 6.7 shows typical acoustic flow waveforms, called flow glottograms, and their first time derivatives. In a normal glottogram, the closed phase of the glottal cycle is characterized by zero flow. Often, however, the closure is not complete. Also, in some cases, although the folds close completely, there is a parallel path—a chink—which stays open all the time.

In the open phase the flow gradually builds up, reaches a peak, and then falls sharply. The asymmetry is due to the inertia of the airflow in the vocal tract and the subglottal cavities. The amplitude of the fundamental frequency is governed mainly by the peak of the flow while the amplitudes of the higher harmonics is governed mainly by the (negative) peak rate of change of flow, which occurs just before closure.

6.4.1.1 Voice Qualities

Depending on the adjustment of the various parameters mentioned above, the glottis can produce a variety of phonations (i.e., excitations for voiced speech), resulting in different perceptual voice qualities.

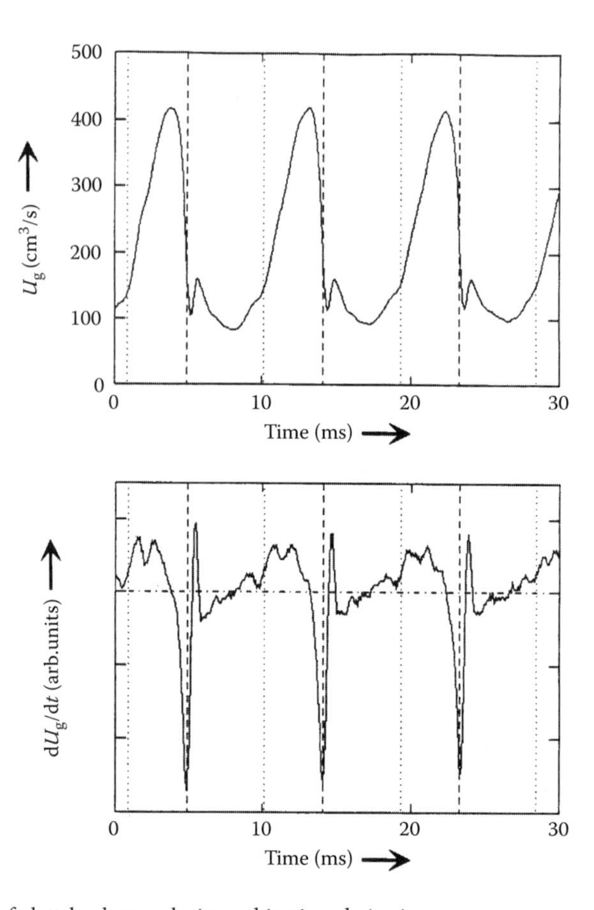

FIGURE 6.7 Example of glottal volume velocity and its time derivative.

Some perceptual qualities vary continuously whereas others are essentially categorical (i.e., they change abruptly when some parameters cross a threshold).

Voice timbre is an important continuously variable quality which may be given various labels ranging from "mellow" to "pressed." The spectral slope of the glottal waveform is the main physical correlate of this perceptual quality. On the other hand, nasality and aspiration may be regarded as categorical qualities.

The physical properties that distinguish a "male" voice from a "female" voice are still not well understood, although many distinguishing features are known. Besides the obvious cue of fundamental frequency, the perceptual quality of "breathiness" seems to be important for producing a female-sounding voice. It occurs when the glottis does not close completely during the glottal cycle. This results in a more sinusoidal movement of the folds which makes the amplitude of the fundamental frequency much larger compared to those of the higher harmonics. The presence of leakage in the abducted glottis also increases the damping of the lower formants, thus increasing their bandwidths. Also, the continuous airflow through the leaking glottis gives rise to increased levels of glottal noise (aspiration noise) that masks the higher harmonics of the glottal spectrum. Finally, in glottograms of female voices, the open phase is a larger proportion of the glottal cycle (about 80%) than in glottograms of male voices (about 60%). The points of closure are also smoother for female voices, which results in lower high frequency energy relative to the fundamental.

Finally, the individuality of a voice (which allows us to recognize the speaker) appears to be dependent largely on the exact relationships between the amplitudes of the first few harmonics.

6.4.1.2 Models of the Glottis

A study of the mechanical and acoustical properties of the larynx is still an area of active interdisciplinary research. Modeling in the mechanical and acoustical domains requires making simplifying assumptions about the tissue movements and the fluid mechanics of the airflow. Depending on the degree to which the models incorporate physiological knowledge, one can distinguish three categories of glottal models:

Parametrization of glottal flow is the "black-box" approach to glottal modeling. The glottal flow wave or its first time derivative is parametrized in segments by analytical functions. It seems doubtful that any simple model of this kind can match all kinds of speakers and speaking styles. Examples of speech sounds that are difficult to parametrize in this way are nasal and mixed-excitation sounds (i.e., sounds with an added fricative component) and "simple" high-pitch female vowels.

Parametrization of glottal area is more realistic. In this model, the area of the glottal opening is parametrized in segments, but the airflow is computed from the propagation equations, and includes its interaction with the acoustic loads of the vocal tract and the subglottal structures. Such a model is capable of reproducing much more of the detail and individuality of the glottal wave than the black box approach. Problems are still to be expected for mixed glottal/fricative sounds unless the tract model includes an accurate mechanism for frication (see the section on turbulent excitation below).

In a complete, self-oscillating model of the glottis described below, the amplitude of the glottal opening as well as the instants of glottal closure are automatically derived, and depend in a complicated manner on the laryngeal parameters, lung pressure, and the past history of the flow. The area-driven model has the disadvantage that amplitude and instants of closure must be specified as side information. However, the ability to specify the points of glottal closure can, in fact, be an advantage in some applications; for example, when the model is used to mimic a given speech signal.

Self-oscillating physiological models of the glottis attempt to model the complete interaction of the airflow and the vocal folds which results in periodic excitation. The input to a model of this type is slowly varying physical parameters such as lung pressure, tension of the folds, prephonatory glottal shape, etc. Of the many models of this type that have been proposed, the one most often used is the two-mass model of Ishizaka and Flanagan (I&F). In the following we will briefly review this model.

The I&F two-mass model is depicted in Figure 6.8 As shown there, the thickness of the vocal folds that separates the trachea from the vocal tract is divided into two parts of length d_1 and d_2, respectively, where the subscript 1 refers to the part closest to the trachea and 2 refers to the part closest to the vocal tract. These portions of the vocal folds are represented by damped spring–mass systems coupled to each other. The division into two portions is a refinement of an earlier version that represented the folds by a single

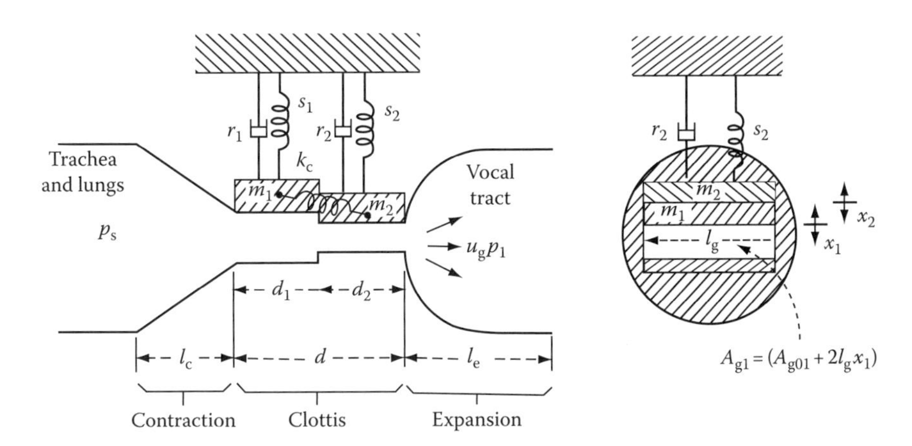

FIGURE 6.8 The two-mass model of Ishizaka and Flanagan. (From Ishizaka, K. and Flanagan, J.L., *Bell Syst. Tech. J.*, 51(6), 1233, July–August 1972.)

spring–mass system. By using two sections the model comes closer to reality and exhibits the phenomenon of vertical phasing mentioned earlier.

In order to simulate tissue, all the springs and dampers are chosen to be nonlinear. Before discussing the choice of these nonlinear elements, let us first consider the relationship between the airflow and the pressure variations from the lungs to the vocal tract.

6.4.1.3 Airflow in the Glottis

The dimensions d_1 and d_2 are very small—about 1.5 mm each. This is a very small fraction of the wavelength even at the highest frequencies of interest. (The wavelength of a sound wave in air at 100 kHz is about 3 mm!). Therefore we may assume the flow through the glottis to be incompressible. With this assumption the equation of continuity, Equation 6.2, merely states that the volume velocity is the same everywhere in the glottis. We will call this volume velocity u_g. The relationship of this velocity to the pressure is governed by the equation of motion. Since the particle velocity in the glottis can be very large, we need to consider the nonlinear version given in Equation 6.5. Also, since the cross section of the glottis is very small, viscous drag cannot be neglected. So we will include a term representing viscous drag proportional to the velocity. With this addition, Equation 6.5 becomes

$$\frac{\partial}{\partial x}\left[p + \frac{\rho}{2}(u_g/A)^2\right] = -\rho\frac{\partial}{\partial t}\left(\frac{u_g}{A}\right) - R_v\left(\frac{u_g}{A}\right) \tag{6.21}$$

The drag coefficient R_v can be estimated for simple geometries. In the present application a rectangular aperture is appropriate. If the length of the aperture is l, its width (corresponding to the opening between the folds) is w and its depth in the direction of flow is d, then $R_v = (12\mu d/lw^3)$, where μ is the coefficient of shear viscosity. The pressure distribution is obtained by repeated use of Equation 6.22, using the appropriate value of A (and hence of R_v) in the different parts of the glottis. In this manner, the pressure at any point in the glottis may be determined in terms of the volume velocity, u_g, the lung pressure, P_s, and the pressure at the input to the vocal tract, p_1.

The detailed derivation of the pressure distribution is given in [21]. The derivation shows that the total pressure drop across the glottis, $P_s - p_1$, is related to the glottal volume velocity, u_g, by an equation of the form

$$p_s - p_1 = Ru_g + \frac{d}{dt}(Lu_g) + \frac{\rho}{2}\left(\frac{u_g}{\alpha^2}\right) \tag{6.22}$$

With the analogy of pressure to voltage and volume velocity to current, the quantity R is analogous to resistance and L to inductance. The term in u_g^2 may be regarded as u_g times a current-dependent resistance. The quantity α has the dimensions of an area.

6.4.1.4 Models of Vocal Fold Tissue

When the pressure distribution derived above is coupled to the mechanical properties of the vocal folds, we get a self-oscillating system with properties quite similar to those of a real larynx. The mechanical properties of the vocal folds have been modeled in many ways with varying degrees of complexity ranging from a single spring–mass system to a distributed parameter flexible tube. In the following, by way of example, we will summarize only the original 1972 I&F model.

Returning to Figure 6.8, we observe that the mechanical properties of the folds are represented by the masses m_1 and m_2, the (nonlinear) springs s_1 and s_2, the coupling spring k_c, and the nonlinear dampers r_1 and r_2. The opening in each section of the glottis is assumed to have a rectangular shape with length l_g. The widths of the two sections are $2x_j, j = 1, 2$. Assuming a symmetrical glottis, the cross-sectional areas of the two sections are

$$A_{gj} = A_{g0j} + 2l_g x_j, \quad j = 1, 2 \tag{6.23}$$

where A_{g01} and A_{g02} are the areas at rest. From this equation, we compute the lateral displacements $x_{j\,\min}, j = 1, 2$ at which the two folds touch each other in each section to be $x_{j\,\min} = -A_{g0j}/(2l_g)$. Displacements more negative than these indicate a collision of the folds. The springs s_1 and s_2 are assumed to have restoring forces of the form $ax + bx^3$, where the constants a and b take on different values for the two sections and for the colliding and noncolliding conditions.

The dampers r_1 and r_2 are assumed to be linear, but with different values in the colliding and non-colliding cases. The coupling spring k_c is assumed to be linear. With these choices, the coupled equations of motion for the two masses are:

$$m_1 \frac{d^2 x_1}{dt^2} + r_1 + f_{s1}(x_1) + k_c(x_1 - x_2) = F_1 \tag{6.24a}$$

and

$$m_2 \frac{d^2 x_2}{dt^2} + r_2 \frac{dx^2}{dt} + f_{s2}(x_2) + k_c(x_2 - x_1) = F_2 \tag{6.24b}$$

Here f_{s1} and f_{s2} are the cubic nonlinear springs. The parameters of these springs as well as the damping constants r_1 and r_2 change when the folds go from a colliding state to a noncolliding state and vice versa. The driving forces F_1 and F_2 are proportional to the average acoustic pressures in the two sections of the glottis. Whenever a section is closed (due to the collision of its sides) the corresponding driving force is zero. Note that it is these forces that provide the feedback of the acoustic pressures to the mechanical system. This feedback is ignored in the area-driven models of the glottis.

We close this section with an example of ongoing research in glottal modeling. In the introduction to this section we had stated that breathiness of a voice is considered important for producing a natural-sounding synthetic female voice. Breathiness results from incomplete closures of the folds. We had also stated that incomplete glottal closures due to abducted folds lead to a steep spectral roll-off of the glottal excitation and a strong fundamental. However, practical experience shows that many voices show clear evidence for breathiness but do not show a steep spectral roll-off, and have relatively weak fundamentals instead. How can this mystery be solved? It has been suggested that the glottal "chink" mentioned in the discussion of Figure 6.7 might be the answer. Many high-speed videos of the vocal folds show evidence of a separate leakage path in the "posterior commissure" (where the folds join) which stays open all the time. Analysis of such a permanently open path produces the stated effect [22].

6.4.2 Turbulent Excitation

Turbulent airflow shows highly irregular fluctuations of particle velocity and pressure. These fluctuations are audible as broadband noise. Turbulent excitation occurs mainly at two locations in the vocal tract: near the glottis and at constriction(s) between the glottis and the lips. Turbulent excitation at a constriction downstream of the glottis produces fricative sounds or voiced fricatives depending on whether or not voicing is simultaneously present. Also, stressed versions of the vowel *i*, and liquids *l* and *r* are usually accompanied by turbulent flow. Measurements and models for turbulent excitation are even more difficult to establish than for the periodic excitation produced by the glottis because, usually, no vibrating surfaces are involved. Because of the lack of a comprehensive model, much confusion exists over the proper subclassification of fricatives. The simplest model for turbulent excitation is a "nozzle" (narrow orifice) releasing air into free space. Experimental work has shown that half (or more) of the noise power generated by a jet of air originates within the so-called mixing region that starts at the nozzle outlet and extends as far as a distance four times the diameter of the orifice. The noise source is therefore distributed. Several scaling relations hold between the acoustic output and the nozzle geometry. One of these scaling properties is the so-called Reynolds number, *Re*, that characterizes the amount of turbulence generated as the air from the jet mixes with the ambient air downstream from the orifice:

$$Re = \frac{u}{A}\frac{x}{\nu} \tag{6.25}$$

where
- u is the volume velocity
- A is the area of the orifice (hence, u/A is the particle velocity)
- x is a characteristic dimension of the orifice (the width for a rectangular orifice)
- $\nu = \mu/\rho$ is the kinematic viscosity of air

Beyond a critical value of the Reynolds number, Re_{crit} (which is about 1200 for the case of a free jet), the flow becomes fully turbulent; below this value, the flow is partly turbulent and becomes fully laminar at very low velocities. Another scaling equation defines the so-called Strouhal number, *S*, that relates the frequency F_{max} of the (usually broad) peak in the power spectrum of the generated noise to the width of the orifice and the velocity:

$$S = F_{max}\frac{x}{u/A} \tag{6.26}$$

For the case of a free jet, the Strouhal number *S* is 0.15. Within the jet, higher frequencies are generated closer to the orifice and lower frequencies further away.

Distributed sources of turbulence can be modeled by expanding them in terms of monopoles (i.e., pulsating spheres), dipoles (two pulsating spheres in opposite phase), quadrupoles (two dipoles in opposite phase), and higher-order representations. The total power generated by a monopole source in free space is proportional to the fourth power of the particle velocity of the flow, that of a dipole source obeys a $(u/A)^6$ power law, and that of a quadrupole source obeys a $(u/A)^8$ power law. Thus, the low order sources are more important at low flow rates, while the reverse is the case at high flow rates. In a duct, however, the exponents of the power laws decrease by 2, that is, a dipole source's noise power is proportional to $(u/A)^4$, etc.

Thus far, we have summarized noise generation in a free jet or air. A much stronger noise source is created when a jet of air hits an obstacle. Depending on the angle between the surface of the obstacle and the direction of flow, the surface roughness, and the obstacle geometry, the noise generated can be up to 20 dB higher than that generated by the same jet in free space. Because of the spatially concentrated source, modeling obstacle noise is easier than modeling the noise in a free jet. Experiments reveal that obstacle noise can be approximated by a dipole source located at the obstacle.

The above theoretical findings qualitatively explain the observed phenomenon that the fricatives *th* and *f* (and the corresponding voiced *dh* and *v*) are weak compared to the fricatives *s* and *sh*. The teeth (upper for *s* and lower for *sh*) provide the obstacle on which the jet impinges to produce the higher noise levels. A fricative of intermediate strength results from a distributed obstacle (the "wall" case) when the jet is forced along the roof of the mouth as for the sound *y*.

In a synthesizer, dipole noise sources can be implemented as series pressure sources. One possible implementation is to make the source pressure proportional to $Re^2 - Re^2_{crit}$ for $Re > Re_{crit}$ and zero otherwise [11]. Another option [23] is to relate the noise source power to the Bernoulli pressure $B = .5\rho(u/A)^2$. Since the power of a dipole source located at the teeth (and radiating into free space) is $(u/A)^6$, it is also proportional to B^3, and the noise source pressure $p_n \propto B^{3/2}$. On the other hand, for wall sources located further away from the lips, we need multiple (distributed) dipole sources with source pressures proportional either to $Re^2 - Re^2_{crit}$ or to B. In either case, the source should have a broadband spectrum with a peak at a frequency given by Equation 6.26.

When a noise source is located at some point inside the tract, its effect on the acoustic output at the lips is computed in terms of two chain matrices—the matrix $\mathbf{K}_F$ from the glottis to the noise source, and the matrix $\mathbf{K}_L$ from the noise source to the lips. For fricative sounds, the glottis is wide open, so the termination impedance at the glottis end may be assumed to be zero. With this termination, the impedance at the noise source looking toward the glottis is computed from $\mathbf{K}_F$ as explained in the section on chain matrices. Call this impedance Z_1. Similarly, knowledge of the radiation impedance at the lips and the matrix $\mathbf{K}_L$ allows us to compute the input impedance Z_2 looking toward the lips. The volume velocity at the source is then just $P_n/(Z_1 + Z_2)$, where P_n is the pressure generated by the noise source. The transfer function obtained from Equation 6.16a for the matrix $\mathbf{K}_L$ then gives the volume velocity at the lips.

It can be shown that the series noise source P_n excites all formants of the entire tract (i.e., the ones we would see if the source were at the glottis). However, the spectrum of fricative noise usually has a high pass character. This can be understood qualitatively by the following considerations.

When the tract has a very narrow constriction, the front and back cavities are essentially decoupled, and the formants of the tract are the formants of the back cavity plus those of the front cavity. If now the noise source is just downstream of the constriction, the formants of the back cavity are only slightly excited because the impedance Z_1 also has poles at those frequencies. Since the back cavity is usually much longer than the front cavity for fricatives, the lower formants are missing in the velocity at the lips. This gives it a high pass character.

6.4.3 Transient Excitation

Transient excitation of the vocal tract occurs whenever pressure is built up behind a total closure of the tract and suddenly released. This sudden release produces a step-function of input pressure at the point of release. The output velocity is therefore proportional to the integral of the impulse response of the tract from the point of release to the lips. In the frequency domain, this is just P_r/s times the transfer function, where P_r is the step change in pressure. Hence, the velocity at the lips may be computed in the same way as in the case of turbulent excitation, with P_n replaced by P_r/s. In practice, this step excitation is usually followed by the generation of fricative noise for a short period after release when the constriction is still narrow enough. Sometimes, if the glottis is also being constricted (e.g., to start voicing) some aspiration might also result.

6.5 Digital Implementations

The models of the various parts of the human speech production apparatus which we have described above can be assembled to produce fluent speech. Here we will consider how a digital implementation of this process may be carried out. Basically, the standard theory of sampling in the time and frequency domains is used to convert the continuous signals considered above to sampled signals, and the samples are represented digitally to the desired number of bits per sample.

6.5.1 Specification of Parameters

The parameters that drive the synthesizer need to be specified about every 20 ms. (The assumed quasistationarity is valid over durations of this size.)

Two sets of parameters are needed—the parameters that specify the shape of the vocal tract and those that control the glottis. The vocal tract parameters implicitly control nasality (by specifying the opening area of the velum) and also frication (by specifying the size of the narrowest constriction).

6.6 Synthesis

The vocal tract is approximated by a concatenation of about 20 uniform sections. The cross-sectional areas of these sections is either specified directly, or computed from a specification of articulatory parameters as shown in Figure 6.3 The chain matrix for each section is computed at an adequate sampling rate in the frequency domain to avoid time-aliasing of the corresponding time functions. (Computation of the chain matrices requires a specification of the losses also. Several models exist which assign the losses in terms of the cross-sectional area [11,16]).

The chain matrices for the individual sections are combined to derive the matrices for various portions of the tract, as appropriate for the particular speech sound being synthesized. For voiced sounds, the matrices for the sections from the glottis to the lips are sequentially multiplied to give the matrix from the glottis to the lips. From the $\mathbf{k}_{11}, \mathbf{k}_{12}, \mathbf{k}_{21}, \mathbf{k}_{22}$ components of this matrix, the transfer function U_{out}/U_{in} and the input impedance are obtained as in Equations 6.16a and b. Knowing the radiation impedance Z_R at the lips we can compute the transfer function for output pressure, $H = (U_{out}/U_{in})Z_R$. The inverse FFT of the transfer function H and the input impedance Z_{in} give the corresponding time functions $h(n)$ and $z_{in}(n)$, respectively. These functions are computed every 20 ms, and the intermediate values are obtained by linear interpolation.

For the current time sampling instant n, the current pressure $p_1(n)$ at the input to the vocal tract is then computed by convolving z_{in} with the past values of the glottal volume velocity u_g. With p_1 known, the pressure difference $P_s - p_1$ on the left hand side of Equation 6.22 is known. Equation 6.18 is discretized by using a backward difference for the time derivative. Thus, a new value of the glottal volume velocity is derived. This, together with the current values of the displacements of the vocal folds, gives us new values for the driving forces F_1 and F_2 for the coupled oscillator Equations 6.24a and b. The coupled oscillator equations are also discretized by backward differences for time derivatives. Thus, the new values of the driving forces give new values for the displacements of the vocal folds. The new value of volume velocity also gives a new value for p_1, and the computational cycle repeats, to give successive samples of p_1, u_g, and the vocal fold displacements.

The glottal volume velocity obtained in this way, is convolved with the impulse response $h(n)$ to produce voiced speech.

If the speech sound calls for frication, the chain matrix of the tract is derived as the product of two matrices—from the glottis to the narrowest constriction and from the constriction to the lips, as discussed in the section on turbulent excitation. This enables us to compute the volume velocity at the constriction, and thus introduce a noise source on the basis of the Reynolds number.

Finally, to produce nasal sounds, the chain matrix for the nasal tract is also computed, and the output at the nostrils computed as discussed in the section on chain matrices. If the lips are open, the output from the lips is also computed and added to the output from the nostrils to give the total speech signal. Details of the synthesis procedure may be found in [24].

References

1. Edwards, H.T., *Applied Phonetics: The Sounds of American English*, Singular Publishing Group, San Diego, CA, 1992, Chap. 3.
2. Olive, J.P., Greenwood, A., and Coleman, J., *Acoustics of American English Speech*, Springer Verlag, New York, 1993.
3. Fant, G., *Acoustic Theory of Speech Production*, Mouton Book Co., Gravenhage, the Netherlands, 1960, Chap. 2.1, pp. 93–95.
4. Baer, T., Gore, J.C., Gracco, L.C., and Nye, P.W., Analysis of vocal tract shape and dimensions using magnetic resonance imaging: Vowels, *J. Acoust. Soc. Am.*, 90(2), 799–828, Aug. 1991.
5. Stone, M., A three-dimensional model of tongue movement based on ultrasound and microbeam data, *J. Acoust. Soc. Am.*, 87(5), 2207–2217, May 1990.

6. Sondhi, M.M. and Resnick, J.R., The inverse problem for the vocal tract: Numerical methods, acoustical experiments, and speech synthesis, *J. Acoust. Soc. Am.*, 73(3), 985–1002, Mar. 1983.
7. Hardcastle, W.J., Jones, W., Knight, C., Trudgeon, A., and Calder, G., New developments in electropalatography: A state of the art report, *Clin. Linguist. Phonet.*, 3, 1–38, 1989.
8. Perkell, J.S., Cohen, M.H., Svirsky, M.A., Mathies, M.L., Garabieta, I., and Jackson, M.T.T., Electromagnetic midsagittal articulometer systems for transducing speech articulatory movements, *J. Acoust. Soc. Am.*, 92(6), 3078–3096, Dec. 1992.
9. Coker, C.H., A model of articulatory dynamics and control, *Proc. IEEE*, 64(4), 452–460, April 1976.
10. Sondhi, M.M., Resonances of a bent vocal tract, *J. Acoust. Soc. Am.*, 79(4), 1113–1116, April 1986.
11. Flanagan, J.L., *Speech Analysis, Synthesis and Perception*, 2nd ed., Springer Verlag, New York, 1972, Chap. 3.
12. Lu, C., Nakai, T., and Suzuki, H., Three-dimensional FEM simulation of the effects of the vocal tract shape on the transfer function, *International Conference on Spoken Language Processing*, Banff, Alberta, vol. 1, pp. 771–774, 1992.
13. Richard, G., Liu, M., Sinder, D., Duncan, H., Lin, O., Flanagan, J.L., Levinson, S.E., Davis, D.W. and Slimon, S., Numerical simulations of fluid flow in the vocal tract, *Proceedings of the Eurospeech'95*, European Speech Communication Assocation, Madrid, Spain, Sept. 18–21, 1995.
14. Morse, P.M., *Vibration and Sound*, McGraw-Hill, New York, 1948, Chap. 6.
15. Pierce, A.D., *Acoustics*, 2nd ed., McGraw-Hill, New York, p. 360, 1981.
16. Sondhi, M.M., Model for wave propagation in a lossy vocal tract, *J. Acoust. Soc. Am.*, 55(5), 1070–1075, May 1974.
17. Siebert, W. McC., *Circuits, Signals and Systems*, MIT Press/McGraw-Hill, Cambridge, MA/New York, p. 97, 1986.
18. Sundberg, J., *The Science of the Singing Voice*, Northern Illinois University Press, DeKalb, IL, 1987.
19. Zemlin, W.R., *Speech and Hearing Science, Anatomy, and Physiology*, Prentice-Hall, Englewood Cliffs, NJ, 1968.
20. Husson, R., Etude des phénomenes physiologiques et acoustiques fondamentaux de la voix cantée, Disp edit Rev Scientifique, pp. 1–91, 1950. For a discussion see Diehl, C.F., *Introduction to the Anatomy and Physiology of the Speech Mechanisms*, Charles C Thomas, Springfield, IL, pp. 110–111, 1968.
21. Ishizaka, K. and Flanagan, J.L., Synthesis of voiced sounds from a two-mass model of the vocal cords, *Bell Syst. Tech. J.*, 51(6), 1233–1268, July–Aug. 1972.
22. Cranen, B. and Schroeter, J., Modeling a leaky glottis, *J. Phonetics*, 23, 165–177, 1995.
23. Stevens, K.N., Airflow and turbulence noise for fricative and stop consonants: Static considerations, *J. Acoust. Soc. Am.*, 50(4), 1180–1192, 1971.
24. Sondhi, M.M. and Schroeter, J., A hybrid time-frequency domain articulatory speech synthesizer, *IEEE Trans. Acoust. Speech Signal Process.*, ASSP-35(7), 955–967, July 1987.
25. Schönhärl, E., *Die Stroboskopie in der praktischen Laryngologie*, Georg Thieme Verlag, Stuttgart, Germany, 1960.

7

Speech Coding

Richard V. Cox
AT&T Research Labs

7.1 Introduction

Digital speech coding is used in a wide variety of everyday applications that the ordinary person takes for granted, such as network telephony or telephone answering machines. By speech coding we mean a method for reducing the amount of information needed to represent a speech signal for transmission or storage applications. For most applications this means using a lossy compression algorithm because a small amount of perceptible degradation is acceptable. This section reviews some of the applications, the basic attributes of speech coders, methods currently used for coding, and some of the most important speech coding standards.

7.1.1 Examples of Applications

Digital speech transmission is used in network telephony. The speech coding used is just sample-by-sample quantization. The transmission rate for most calls is fixed at 64 kilobits per second (kbps). The speech is sampled at 8000 Hz (8 kHz) and a logarithmic 8-bit quantizer is used to represent each sample as one of 256 possible output values. International calls over transoceanic cables or satellites are often reduced in bit rate to 32 kbps in order to boost the capacity of this relatively expensive equipment. Digital wireless transmission has already begun. In North America, Europe, and Japan there are digital cellular phone systems already in operation with bit rates ranging from 6.7 to 13 kbps for the speech coders. Secure telephony has existed since World War II, based on the first vocoder. (Vocoder is a contraction of the words voice coder.) Secure telephony involves first converting the speech to a digital form, then digitally encrypting it and then transmitting it. At the receiver, it is decrypted, decoded, and reconverted back to analog. Current videotelephony is accomplished through digital transmission of both the speech and the video signals. An emerging use of speech coders is for simultaneous voice and data. In these applications, users exchange data (text, images, FAX, or any other form of digital information) while carrying on a conversation.

All of the above examples involve real-time conversations. Today, we use speech coders for many storage applications that make our lives easier. For example, voice mail systems and telephone answering machines allow us to leave messages for others. The called party can retrieve the message when they wish, even from halfway around the world. The same storage technology can be used to broadcast announcements to many different individuals. Another emerging use of speech coding is multimedia. Most forms of multimedia involve only one-way communications, so we include them with storage applications. Multimedia documents on computers can have snippets of speech as an integral part. Capabilities currently exist to allow users to make voice annotations onto documents stored on a personal computer (PC) or workstation.

7.1.2 Speech Coder Attributes

Speech coders have attributes that can be placed in four groups: "bit rate, quality, complexity, and delay." For a given application, some of these attributes are predetermined while trade-offs can be made among the others. For example, the communications channel may set a limit on bit rate, or cost considerations may limit complexity. Quality can usually be improved by increasing bit rate or complexity, and sometimes by increasing delay. In the following sections, we discuss these attributes.

Primarily we will be discussing "telephone bandwidth speech." This is a slightly nebulous term. In the telephone network, speech is first band-pass filtered from roughly 200–3200 Hz. This is often referred to as 3 kHz speech. Speech is sampled at 8 kHz in the telephone network. The usual telephone bandwidth filter rolls off to about 35 dB by 4 kHz in order to eliminate the aliasing artifacts caused by sampling.

There is a second bandwidth of interest. It is referred to as "wideband speech." The sampling rate is doubled to 16 kHz. The low-pass filter is assumed to begin rolling off at 7 kHz. At the low end, the speech is assumed to be uncontamined by line noise and only the DC component needs to be filtered out. Thus, the high-pass filter cutoff frequency is 50 Hz. When we refer to wideband speech, we mean speech with a bandwidth of 50 to 7000 Hz and a sampling rate of 16 kHz. This is also referred to as 7 kHz speech.

7.1.2.1 Bit Rate

Bit rate tells us the degree of compression that the coder achieves. Telephone bandwidth speech is sampled at 8 kHz and digitized with an 8 bit logarithmic quantizer, resulting in a bit rate of 64 kbps. For telephone bandwidth speech coders, we measure the degree of compression by how much the bit rate is lowered from 64 kbps. International telephone network standards currently exist for coders operating from 64 kbps down to 5.3 kbps. The speech coders for regional cellular standards span the range from 13 to 3.45 kbps, and those for secure telephony span the range from 16 kbps to 800 bps. Finally, there are proprietary speech coders that are in common use which span the entire range.

Speech coders need not have a constant bit rate. Considerable compression can be gained by not transmitting speech during the silence intervals of a conversation. Nor is it necessary to keep the bit rate fixed during the talk spurts of a conversation.

7.1.2.2 Delay

The communication delay of the coder is more important for transmission than for storage applications. In real-time conversations, a large communication delay can impose an awkward protocol on talkers. Large communication delays of 300 ms or greater are particularly objectionable to users even if there are no echoes.

Most low bit rate speech coders are block coders. They encode a block of speech, also known as a frame, at a time. Speech coding delay can be allocated as follows. First, there is algorithmic delay. Some coders have an amount of look-ahead or other inherent delays in addition to their frame size. The sum of frame size and other inherent delays constitutes algorithmic delay. The coder requires computation.

The amount of time required for this is called processing delay. It is dependent on the speed of the processor used. Other delays in a complete system are the multiplexing delay and the transmission delay.

7.1.2.3 Complexity

The degree of complexity is a determining factor in both the cost and power consumption of a speech coder. Cost is almost or always a factor in the selection of a speech coder for a given application. With the advent of wireless and portable communications, power consumption has also become an important factor. Simple scalar quantizers, such as linear or logarithmic pulse code modulation (PCM), are necessary in any coding system and have the lowest possible complexity.

More complex speech coders are first simulated on host processors, then implemented on digital signal processor (DSP) chips and may later be implemented on special purpose VLSI devices. Speed and random access memory (RAM) are the two most important contributing factors of complexity. The faster the chip or the greater the chip size, the greater the cost. In fact, complexity is a determining factor for both cost and power consumption. Generally 1 word of RAM takes up as much on-chip area as 4 to 6 words of read-only memory (ROM). Most speech coders are implemented on fixed-point DSP chips, so one way to compare the complexity of coders is to measure their speed and memory requirements when efficiently implemented on commercially available fixed-point DSP chips.

DSP chips are available in both 16 bit fixed point and 32 bit floating point. 16 bit DSP chips are generally preferred for dedicated speech coder implementations because the chips are usually less expensive and consume less power than implementations based on floating-point DSPs. A disadvantage of fixed-point DSP chips is that the speech coding algorithm must be implemented using 16 bit arithmetic. As part of the implementation process, a representation must be selected for each and every variable. Some can be represented in a fixed format, some in block floating point, and still others may require double precision. As VLSI technology has advanced, fixed-point DSP chips contain a richer set of instructions to handle the data manipulations required to implement representations such as block floating point. The advantage of floating-point DSP chips is that implementing speech coders is much quicker. Their arithmetic precision is about the same as that of a high-level language simulation, so the steps of determining the representation of each and every variable, and how these representations affect performance can be omitted.

7.1.2.4 Quality

The attribute of quality has many dimensions. Ultimately, quality is determined by how the speech sounds to a listener. Some of the factors that affect the performance of a coder are whether the input speech is clean or noisy, whether the bit stream has been corrupted by errors, and whether multiple encodings have taken place.

Speech coder quality ratings are determined by means of subjective listening tests. The listening is done in a quiet booth and may use specified telephone handsets, headphones, or loudspeakers. The speech material is presented to the listeners at specified levels and is originally prepared to have particular frequency characteristics. The most often used test is the absolute category rating (ACR) test. Subjects hear pairs of sentences and are asked to give one of the following ratings: "excellent, good, fair, poor, or bad." A typical test contains a variety of different talkers and a number of different coders or reference conditions. The data resulting from this test can be analyzed in many ways. The simplest way is to assign a numerical ranking to each response, giving a 5 to the best possible rating, 4 to the next best, down to a 1 for the worst rating, then computing the mean rating for each of the conditions under test. This is a referred to as a mean opinion score (MOS) and the ACR test is often referred to as a MOS test.

There are many other dimensions to quality besides those pertaining to noiseless channels. Bit error sensitivity is another aspect of quality. For some low bit rate applications such as secure telephones over 2.4 or 4.8 kbps modems, it might be reasonable to expect the distribution of bit errors to be random and coders should be made robust for low random bit error rates up to 1%–2%. For radio channels, such as in digital cellular telephony, provision is made for additional bits to be used for channel coding to protect the information bearing bits. Errors are more likely to occur in bursts and the speech coder requires a mechanism to recover from an entire lost frame. This is referred to as frame erasure concealment, another aspect of quality for cellular speech coders.

For the purposes of conserving bandwidth, voice activity detectors are sometimes used with speech coders. During nonspeech intervals, the speech coder bit stream is discontinued. At the receiver "comfort noise" is injected to simulate the background acoustic noise at the encoder. This method is used for some cellular systems and also in digital speech interpolation (DSI) systems to increase the effective number of channels or circuits. Most international phone calls carried on undersea cables or satellites use DSI systems. There is some impact on quality when these techniques are used. Subjective testing can determine the degree of degradation.

7.2 Useful Models for Speech and Hearing

7.2.1 The LPC Speech Production Model

Human speech is produced in the vocal tract by a combination of the vocal cords in the glottis interacting with the articulators of the vocal tract. The vocal tract can be approximated as a tube of varying diameter. The shape of the tube gives rise to resonant frequencies called formants. Over the years, the most successful speech coding techniques have been based on linear prediction coding (LPC). The LPC model is derived from a mathematical approximation to the vocal tract representation as a variable diameter tube. The essential element of LPC is the linear prediction filter. This is an all pole filter which predicts the value of the next sample based on a linear combination of previous samples.

Let x_n be the speech sample value at sampling instant n. The object is to find a set of prediction coefficients $\{a_i\}$ such that the prediction error for a frame of size M is minimized:

$$\varepsilon = \sum_{m=0}^{M-1} \left(\sum_{i=1}^{I} a_i x_{n+m-i} + x_{n+m} \right)^2 \tag{7.1}$$

where I is the order of the linear prediction model. The prediction value for x_n is given by

$$\tilde{x}_n = -\sum_{i=1}^{I} a_i x_{n-i} \tag{7.2}$$

The prediction error signal $\{e_n\}$ is also referred to as the residual signal. In z-transform notation we can write

$$A(z) = 1 + \sum_{i=1}^{I} a_i z^{-i} \tag{7.3}$$

$1/A(z)$ is referred to as the LPC synthesis filter and (ironically) $A(z)$ is referred to as the LPC inverse filter.

LPC analysis is carried out as a block process on a frame of speech. The most often used techniques are referred to as the autocorrelation and the autocovariance methods [1–3]. Both methods involve inverting matrices containing correlation statistics of the speech signal. If the poles of the LPC filter are close to the unit circle, then these matrices become more ill-conditioned, which means that the techniques used for inversion are more sensitive to errors caused by finite numerical precision. Various techniques for dealing with this aspect of LPC analysis include windows for the data [1,2], windows for the correlation statistics [4], and bandwidth expansion of the LPC coefficients.

For forward adaptive coders, the LPC information must also be quantized and transmitted or stored. Direct quantization of LPC coefficients is not efficient. A small quantization error in a single coefficient can render the entire LPC filter unstable. Even if the filter is stable, sufficient precision is required and too many bits will be needed. Instead, it is better to transform the LPC coefficients to another domain in which stability is more easily determined and fewer bits are required for representing the quantization levels.

The first such domain to be considered is the reflection coefficient [5]. Reflection coefficients are computed as a byproduct of LPC analysis. One of their properties is that all reflection coefficients must have magnitudes less than 1, making stability easily verified. Direct quantization of reflection coefficients is still not efficient because the sensitivity of the LPC filter to errors is much greater when reflection coefficients are nearly 1 or -1. More efficient quantizers have been designed by transforming the individual reflection coefficients with a nonlinearity that makes the error sensitivity more uniform. Two such nonlinear functions are the inverse sine function, $\arcsin(k_i)$, and the logarithm of the area ratio, $\log((1 + k_i)/(1 - k_i))$.

A second domain that has attracted even greater interest recently, is the line spectral frequency (LSF) domain [6]. The transformation is given as follows. We first use $A(z)$ to define two polynomials:

$$P(z) = A(z) + z^{-(I+1)}A(z^{-1}) \tag{7.4a}$$

$$Q(z) = A(z) - z^{-(I+1)}A(z^{-1}) \tag{7.4b}$$

These polynomials can be shown to have two useful properties: all zeros of $P(z)$ and $Q(z)$ lie on the unit circle and they are interlaced with each other. Thus, stability is easily checked by assuring both the interlaced property and that no two zeros are too close together. A second property is that the frequencies tend to be clustered near the formant frequencies; the closer together two LSFs are, the sharper the formant. LSFs have attracted more interest recently because they typically result in quantizers having either better representations or using fewer bits than reflection coefficient quantizers.

The simplest quantizers are scalar quantizers [8]. Each of the values (in whatever domain is being used to represent the LPC coefficients) is represented by one of the possible quantizer levels. The individual values are quantized independently of each other. There may also be additional redundancy between successive frames, especially during stationary speech. In such cases, values may be quantized differentially between frames.

A more efficient, but also more complex method of quantization is called vector quantization [9]. In this technique, the complete set of values is quantized jointly. The actual set of values is compared against all sets in the codebook using a distance metric. The set that is nearest is selected. In practice, an exhaustive codebook search is too complex. For example, a 10-bit codebook has 1024 entries. This seems like a practical limit for most codebooks, but does not give sufficient performance for typical 10th order LPC. A 20 bit codebook would give increased performance, but would contain over 1 million vectors. This is both too much storage and too much computational complexity to be practical. Instead of using large codebooks, product codes are used. In one technique, an initial codebook is used, and then the remaining error vector is quantized by a second stage codebook. In the second technique, the vector is subdivided and each subvector is quantized using its own codebook. Both of these techniques lose efficiency compared to a full-search vector quantizer, but represent a good means for reducing computational complexity and codebook size for bit rate or quality.

7.2.2 Models of Human Perception for Speech Coding

Our ears have a limited dynamic range that depends on both the level and the frequency content of the input signal. The typical band-pass telephone filter has a stopband of only about 35 dB. Also, the logarithmic quantizer characteristics specified by CCITT Rec. G.711 result in a signal-to-quantization noise ratio of about 35 dB. Is this a coincidence? Of course not! If a signal maintains an signal-to-noise ratio (SNR) of about 35 dB or greater for telephone bandwidth, then most humans will perceive little or no noise.

Conceptually, the masking property tells us that we can permit greater amounts of noise in and near the formant regions and that noise will be most audible in the spectral valleys. If we use a coder that produces a white noise characteristic, then the noise spectrum is flat. The white noise would probably be audible in all but the formant regions.

In modern speech coders, an additional linear filter is added to weight the difference between the original speech signal and the synthesized signal. The object is to minimize the error in a space whose metric is like that of the human auditory system. If the LPC filter information is available, it constitutes the best available estimate of the speech spectrum. It can be used to form the basis for this "perceptual weighting filter" [10]. The perceptual weighting filter is given by

$$W(z) = \frac{1 - A(z/\gamma_1)}{1 - A(z/\gamma_2)} \quad 0 < \gamma_2 < \gamma_1 < 1 \tag{7.5}$$

The perceptual weighting filter de-emphasizes the importance of noise in the formant region and emphasizes its importance in spectral valleys. The quantization noise will have a spectral shape that is similar to that of the LPC spectral estimate, making it easier to mask.

The adaptive postfilter is an additional linear filter that is combined with the synthesis filter to reduce noise in the spectral valleys [11]. Once again the LPC synthesis filter is available as the estimate of the speech spectrum. As in the perceptual weighting filter, the synthesis filter is modified. This idea was later further extended to include a long-term (pitch) filter. A tilt-compensation filter was added to correct for the low-pass characteristic that causes a muffled sound. A gain control strategy helped prevent any segments from being either too loud or too soft. Adaptive postfilters are now included as a part of many standards.

7.3 Types of Speech Coders

This part of the section describes a variety of speech coders that are widely used. They are divided into two categories: waveform-following coders and model-based coders. Waveform-following coders have the property that if there were no quantization error, the original speech signal would be exactly reproduced. Model-based coders are based on parametric models of speech production. Only the values of the parameters are quantized. If there were no quantization error, the reproduced signal would not be the original speech.

7.3.1 Model-Based Speech Coders

7.3.1.1 LPC Vocoders

A block diagram of the LPC vocoder is shown in Figure 7.1 LPC analysis is performed on a frame of speech and the LPC information is quantized and transmitted. A voiced/unvoiced determination is made.

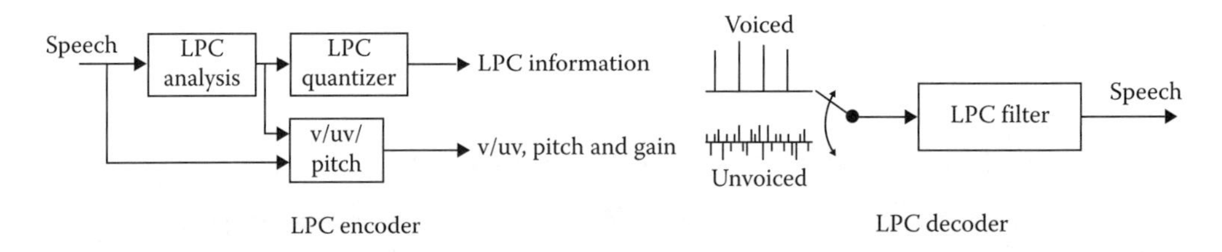

FIGURE 7.1 Block diagram of LPC vocoder.

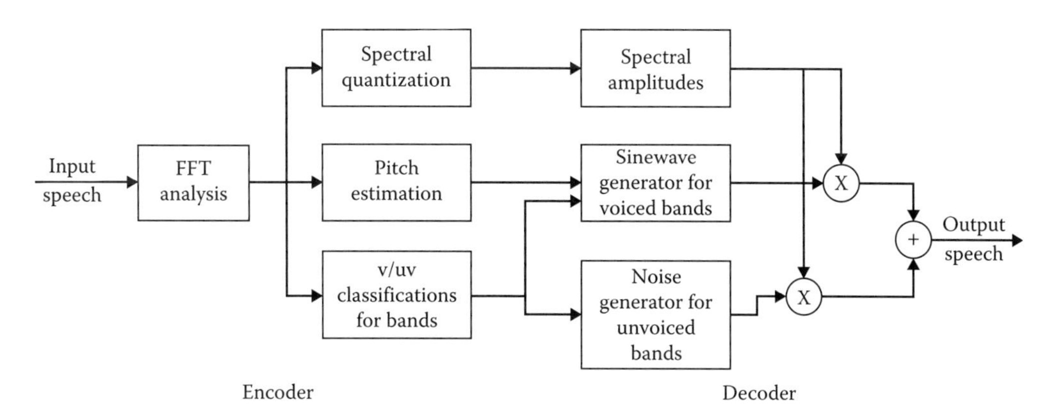

FIGURE 7.2 Block diagram of multiband excitation coder.

The decision may be based on either the original speech or the LPC residual signal, but it will always be based on the degree of periodicity of the signal. If the frame is classified as unvoiced, the excitation signal is white noise. If the frame is voiced, the pitch period is transmitted and the excitation signal is a periodic pulse train. In either case, the amplitude of the output signal is selected such that its power matches that of the original speech. For more information on the LPC vocoder, the reader is referred to [12].

7.3.1.2 Multiband Excitation Coders

Figure 7.2 is a block diagram of a multiband sinusoidal excitation coder. The basic premise of these coders is that the speech waveform can be modeled as a combination of harmonically related sinusoidal waveforms and narrowband noise. Within a given bandwidth, the speech is classified as periodic or aperiodic. Harmonically related sinusoids are used to generate the periodic components and white noise is used to generate the aperiodic components. Rather than transmitting a single voiced/unvoiced decision, a frame consists of a number of voiced/unvoiced decisions corresponding to the different bands. In addition, the spectral shape and gain must be transmitted to the receiver. LPC may or may not be used to quantize the spectral shape. Most often the analysis of the encoder is performed via fast Fourier transform (FFT). Synthesis at the decoder is usually performed by a number of parallel sinusoid and white noise generators. Multiband excitation (MBE) coders are model-based because they do not transmit the phase of the sinusoids, nor do they attempt to capture anything more than the energy of the aperiodic components. For more information the reader is referred to [13–16].

7.3.1.3 Waveform Interpolation Coders

Figure 7.3 is a block diagram of a waveform interpolation coder. In this coder, the speech is assumed to be composed of a slowly evolving periodic waveform (SEW) and a rapidly evolving noise-like waveform (REW). A frame is analyzed first to extract a "characteristic waveform." The evolution of these waveforms is filtered to separate the REW from the SEW. REW updates are made several times more often than SEW updates. The LPC, the pitch, the spectra of the SEW and REW, and the overall energy are all transmitted independently. At the receiver, a parametric representation of the SEW and REW information is constructed, summed, and passed through the LPC synthesis filter to produce output speech. For more information the reader is referred to [17,18].

7.3.2 Time Domain Waveform-Following Speech Coders

All of the time domain waveform coders described in this section include a prediction filter. We begin with the simplest.

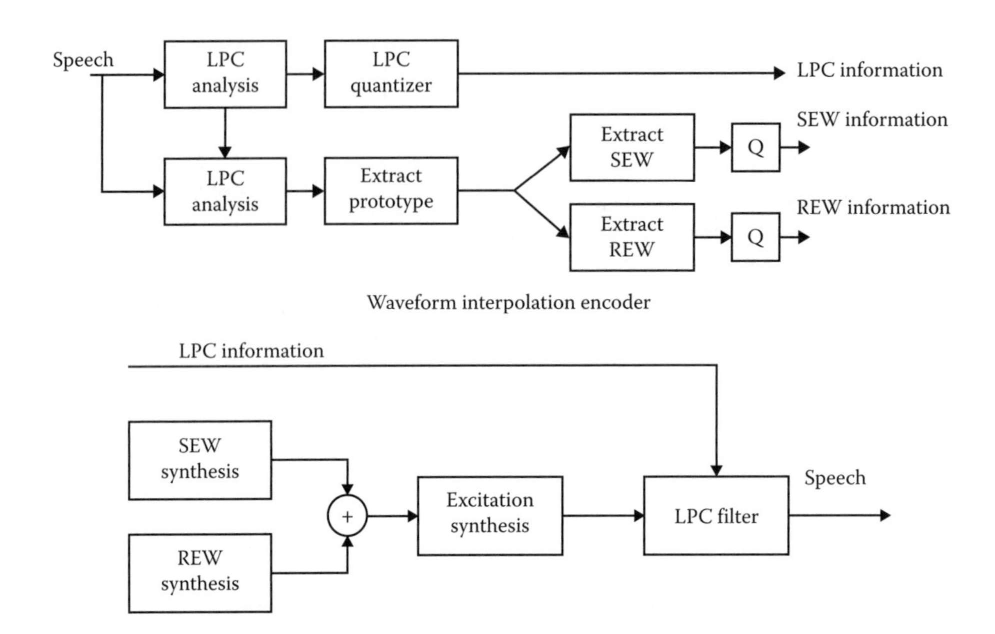

FIGURE 7.3 Block diagram of waveform interpolation coder.

7.3.2.1 Adaptive Differential Pulse Code Modulation

Adaptive differential pulse code modulation (ADPCM) [19] is based on sample-by-sample quantization of the prediction error. A simple block diagram is shown in Figure 7.4 Two parts of the coder may be adaptive: the quantizer step size and/or the prediction filter. ITU Recommendations G.726 and G.727 adapt both. The adaptation may be either forward or backward adaptive. In a backward adaptive system, the adaptation is based only on the previously quantized sample values and the quantizer codewords. At the receiver, the backward adaptive parameter values must be recomputed. An important feature of such adaptation schemes is that they must use predictors that include a "leakage factor" that allows the effects of erroneous values caused by channel errors to die out over time. In a forward adaptive system, the adapted values are quantized and transmitted. This additional "side information" uses bit rate, but can improve quality. Additionally, it does not require recomputation at the decoder.

7.3.2.2 Delta Modulation Coders

In delta modulation coders [20], the quantizer is just the sign bit. The quantization step size is adaptive. Not all the adaptation schemes used for ADPCM will work for delta modulation because the quantization

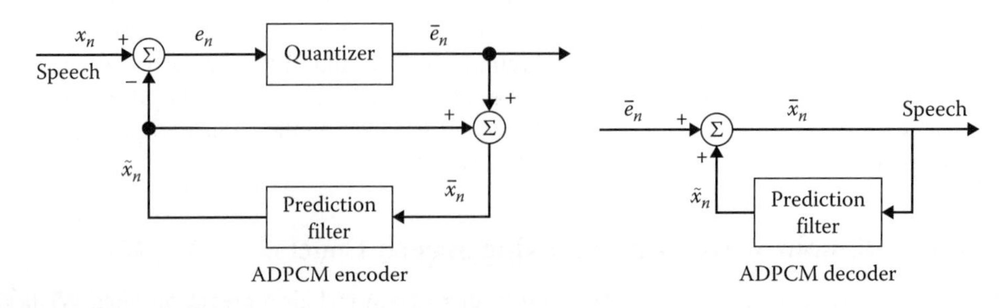

FIGURE 7.4 ADPCM encoder and decoder block diagrams.

is so coarse. The quality of delta modulation coders tends to be proportional to their sampling clock: the greater the sampling clock, the greater the correlation between successive samples, and the finer the quantization step size that can be used. The block diagram for delta modulation is the same as that of ADPCM.

7.3.2.3 Adaptive Predictive Coding

The better the performance of the prediction filter, the lower the bit rate needed to encode a speech signal. This is the basis of the adaptive predictive coder [21] shown in Figure 7.5 A forward adaptive higher order linear prediction filter is used. The speech is quantized on a frame-by-frame basis. In this way the bit rate for the excitation can be reduced compared to an equivalent quality ADPCM coder.

7.3.2.4 Linear Prediction Analysis-by-Synthesis Speech Coders

Figure 7.6 shows a typical linear prediction analysis-by-synthesis speech coder [22]. Like APC, these are frame-by-frame coders. They begin with an LPC analysis. Typically the LPC information is forward

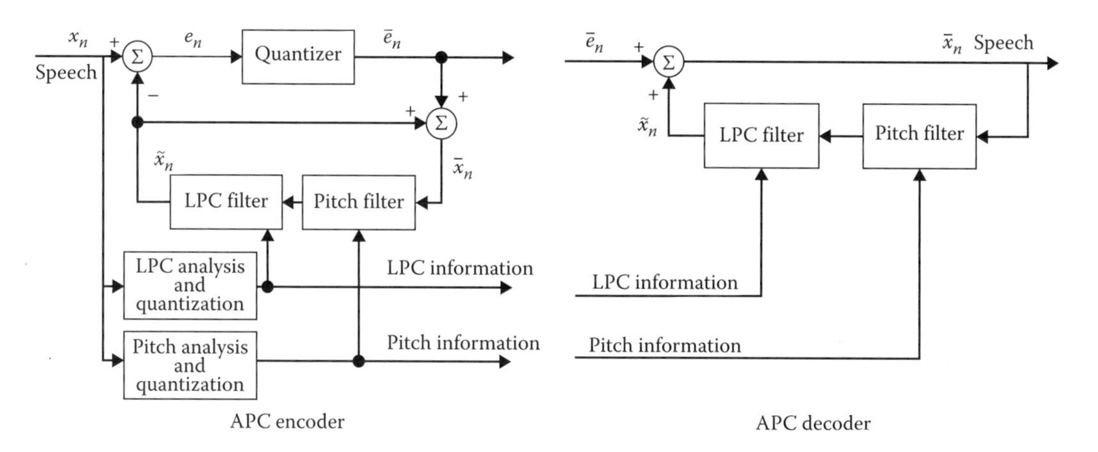

FIGURE 7.5 Adaptive predictive coding encoder and decoder.

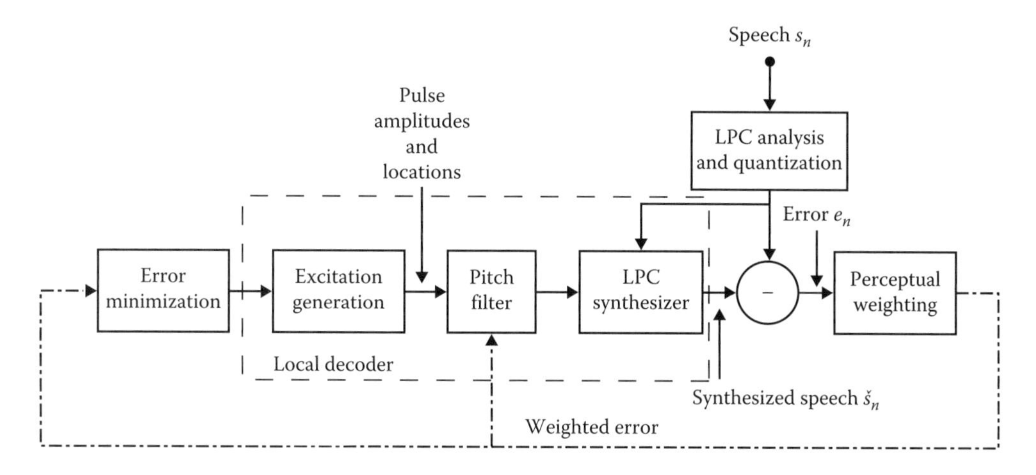

FIGURE 7.6 Linear prediction analysis-by-synthesis coder.

adaptive, but there are exceptions. LPAS coders borrow the concept from ADPCM of having a locally available decoder. The difference between the quantized output signal and the original signal is passed through a perceptual weighting filter. Possible excitation signals are considered and the best (minimum mean square error in the perceptual domain) is selected. The long-term prediction filter removes long-term correlation (the pitch structure) in the signal. If pitch structure is present in the coder, the parameters for the long-term predictor are determined first. The most commonly used system is the adaptive codebook, where samples from previous excitation sequences are stored. The pitch period and gain that result in the greatest reduction of perceptual error are selected, quantized, and transmitted. The fixed codebook excitation is next considered and, again, the excitation vector that most reduces the perceptual error energy is selected and its index and gain are transmitted. A variety of different possible fixed excitation codebooks and their corresponding names have been created for coders that fall into this class. Our enumeration touches only the highlights.

Multipulse linear predictive coding (MPLPC) assumes that the speech frame is subdivided into smaller subframes. After determining the adaptive codebook contribution, the fixed codebook consists of a number of pulses. Typically the number of pulses is about one-tenth the number of samples in a subframe. The pulse that makes the greatest contribution to reducing the error is selected first, then the pulse making the next largest contribution, etc. Once the requisite number of pulses have been selected, determination of the pulses is complete. For each pulse, its location and amplitude must be transmitted.

Codebook excited linear predictive coding (CELP) assumes that the fixed codebook is composed of vectors. This is similar in nature to the vector excitation Coder (VXC). In the first CELP coder, the codebooks were composed of Gaussian random numbers. It was subsequently discovered that center-clipping of these random number codebooks resulted in better quality speech. This had the effect of making the codebook look more like a collection of multipulse LPC excitation vectors. One means for reducing the fixed codebook search is if the codebook consists of overlapping vectors.

Vector sum excitation linear predictive coding (VSELP) assumes that the fixed codebook is composed of a weighted sum of a set of basis vectors. The basis vectors are orthogonal to each other. The weights on any basis vector are always either -1 or $+1$. A fast search technique is possible based on using a pseudo-Gray code method of exploration. VSELP was used for several first or second generation digital cellular phone standards [23].

7.3.3 Frequency Domain Waveform-Following Speech Coders

7.3.3.1 Subband Coders

Figure 7.7 shows the structures of a typical subband encoder and decoder [19,24]. The concept behind subband coding is quite simple: divide the speech signal into a number of frequency bands and quantize

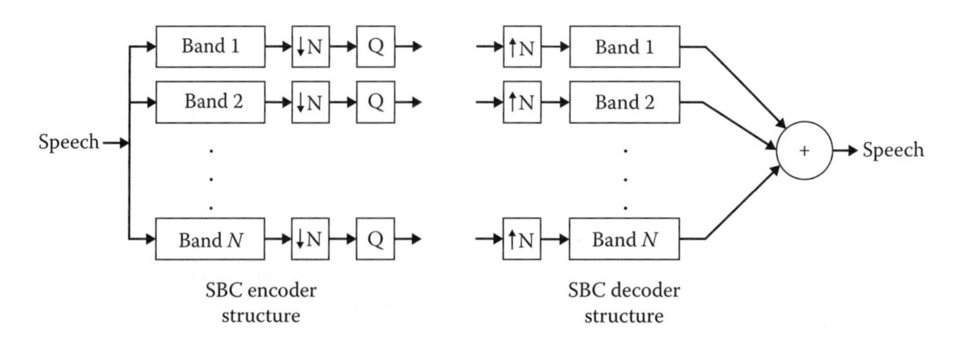

FIGURE 7.7 Subband coder.

each band separately. In this way, the quantization noise is kept within the band. Typically quadrature mirror or wavelet filterbanks are used. These have the properties that (1) in the absence of quantization error all aliasing caused by decimation in the analysis filterbank is canceled in the synthesis filterbank and (2) the bands can be critically sampled, i.e., the number of frequency domain samples is the same as the number of time domain samples. The effectiveness of these coders depends largely on the sophistication of the quantization algorithm. Generally, algorithms that dynamically allocate the bits according to the current spectral characteristics of the speech give the best performance.

7.3.3.2 Adaptive Transform Coders

Adaptive transform coding (ATC) can be viewed as a further extension to subband coding [19,24]. The filterbank structure of SBC is replaced with a transform such as the FFT, the discrete cosine transform (DCT), wavelet transform or other transform-filterbank. They provide a higher resolution analysis than the subband filterbanks. This allows the coder to exploit the pitch harmonic structure of the spectrum. As in the case of SBC, the ATC coders that use sophisticated quantization techniques that dynamically allocate the bits usually give the best performance. Most recently, work has combined transform coding with LPC and time-domain pitch analysis [25]. The residual signal is coded using ATC.

7.4 Current Standards

This part of the section is divided into descriptions of current speech coder standards and activities. The subsections contain information on speech coders that have been or will soon be standardized. We begin first by briefly describing the standards organizations who formulate speech coding standards and the processes they follow in making these standards.

The International Telecommunications Union (ITU) is an agency of the United Nations Economic, Scientific and Cultural Organization (UNESCO) charged with all aspects of standardization in telecommunications and radio networks. Its headquarters are in Geneva, Switzerland. The ITU Telecommunications Standardization Sector (ITU-T) formulates standards related to both wireline and wireless telecommunications. The ITU Radio Standardization Sector (ITU-R) handles standardization related to radio issues. There is also a third branch, the ITU–Telecommunications Standards Bureau (ITU-B) is the bureaucracy handling all of the paperwork. Speech coding standards are handled jointly by Study Groups 16 and 12 within the ITU-T. Other Study Groups may originate requests for speech coders for specific applications. The speech coding experts are found in SG16. The experts on speech performance are found in SG12. When a new standard is being formulated, SG16 draws up a list of requirements based on the intended applications. SG12 and other interested bodies may review the requirements before they are finalized. SG12 then creates a test plan and enlists the help of subjective testing laboratories to measure the quality of the speech coders under the various test conditions. The process of standardization can be time consuming and take between 2 to 6 years.

Three different standards bodies make regional cellular standards, including those for the speech coders. In Europe, the parent body is the European Telecommunications Standards Institute (ETSI). ETSI is an organization that is composed mainly of telecommunications equipment manufacturers. In North America, the parent body is the American National Standards Institute (ANSI). The body charged with making digital cellular standards is the Telecommunications Industry Association (TIA). In Japan, the body charged with making digital cellular standards is the Research and Development Center for Radio Systems (RCR).

There are also speech coding standards for satellite, emergencies, and secure telephony. Some of these standards were promulgated by government bodies, while others were promulgated by private organizations.

Each of these standards organizations works according to its own rules and regulations. However, there is a set of common threads among all of the organizations. These are the standards making process.

Creating a standard is a long process, not to be undertaken lightly. First, a consensus must be reached that a standard is needed. In most cases this is obvious. Second, the terms of reference need to be created. This becomes the governing document for the entire effort. If defines the intended applications. Based on these applications, requirements can be set on the attributes of the speech coder: "quality, complexity, bit rate, and delay." The requirements will later determine the test program that is needed to ascertain whether any candidates are suitable.

Finally, the members of the group need to define a schedule for doing the work. There needs to be an initial period to allow proponents to design coders that are likely to meet the requirements. A deadline is set for submissions. The services of one or more subjective test labs need to be secured and a test plan needs to be defined. A host lab is also needed to process all of the data that will be used in the selection test. Some criteria are needed for determining how to make the selection. Based on the selection, a draft standard needs to be written. Only after the standard is fully specified can manufacturers begin to produce implementations of the standard.

7.4.1 Current ITU Waveform Signal Coders

Table 7.1 describes current ITU speech coding recommendations that are based on sample-by-sample scalar quantization. Three of these coders operate in the time domain on the original sampled signal while the fourth is based on a two-band subband coder for wideband speech.

The CCITT standardized two 64 kbps companded PCM coders in 1972. North America and Japan use μ-law PCM. The rest of the world uses A-law PCM. Both coders use 8 bits to represent the signal. Their effective SNR is about 35 dB. The tables for both of the G.711 quantizer characteristics are contained in [19]. Both coders are considered equivalent in overall quality. A tandem encoding with either coder is considered equivalent to dropping the least significant bit (which is equivalent to reducing the bit rate to 56 kbps). Both coders are extremely sensitive to bit errors in the most significant bits. Their complexity is very low.

32 kbps ADPCM was first standardized by the ITU in 1984 [26–28]. Its primary application was intended to be digital circuit multiplication equipment (DCME). In combination with DSI, a 5:1 increase in the capacity of undersea cables and satellite links was realized for voice conversations. An additional reason for its creation was that such links often encountered the problem of having μ-law PCM at one end and A-law at the other. G.726 can accept either μ-law or A-law PCM as inputs or outputs. Perhaps, its most unique feature is a property called synchronous tandeming. If a circuit

TABLE 7.1 ITU Waveform Speech Coders

Standard Body	ITU	ITU	ITU	ITU
Number	G.711	G.726	G.727	G.722
Year	1972	1990	1990	1988
Type of coder	Companded PCM	ADPCM	ADPCM	SBC/ADPCM
Bit rate	64 kb/s	16–40 kb/s	16–40 kb/s	48, 56 64 kb/s
Quality	Toll	$\leq$ Toll	$\leq$ Toll	Commentary
Complexity				
MIPS	$\ll 1$	1	1	10
RAM	1 byte	$<$ 50 bytes	$<$ bytes	1 K words
Delay				
Frame size	0.125 ms	0.125 ms	0.125 ms	1.5 ms
Specification type				
Fixed point	Bit exact	Bit exact	Bit exact	Bit exact

involves two ADPCM codings with a μ-law or A-law encoding in-between, no additional degradation occurs because of the second encoding. The second bit stream will be identical to the first! In 1986 the Recommendation was revised to eliminate the all-zeroes codeword and so that certain low rate modem signals would be passed satisfactorily. In 1988 extensions for 24 and 40 kbps were added and in 1990 the 16 kbps rate was added. All of these additional rates were added for use in DCME applications.

G.727 includes the same rates as G.726, but all of the quantizers have an even number of levels. The 2 bit quantizer is embedded in the 3 bit quantizer, which is embedded in the 4 bit quantizer, which is embedded in the 5 bit quantizer. This is needed for packet circuit multiplex equipment (PCME) where the least significant bits in the packet can be discarded when there is an overload condition.

Recommendation G.722 is a wideband speech coding standard. Its principal applications are teleconferences and video teleconferences [29]. The wider bandwidth (50–7000 Hz) is more natural sounding and less fatiguing than telephone bandwidth (200–3200 Hz). The wider bandwidth increases the intelligibility of the speech, especially for fricative sounds such as /f/ and /s/, which are difficult to distinguish for telephone bandwidth. The G.722 coder is a two-band subband coder with ADPCM coding in both bands. The ADPCM is similar in structure to that of the G.727 recommendation. The upper band uses an ADPCM coder with a 2 bit adaptive quantizer. The lower band uses an ADPCM coder with an embedded 4–5–6 bit adaptive quantizer. This makes the rates of 48, 56, and 64 kbps all possible. A 24-tap quadrature mirror filter is used to efficiently split the signal.

7.4.2 ITU Linear Prediction Analysis-by-Synthesis Speech Coders

Table 7.2 describes three current analysis-by-synthesis speech coder recommendations of the ITU. All three are block coders based on extensions of the original multipulse LPC speech coder.

G.728 low-delay CELP (LD-CELP) [30] is a backward adaptive CELP coder whose quality is equivalent to that of 32 kbps ADPCM. It was initially specified as a floating-point CELP coder that required implementers to follow exactly the algorithm specified in the recommendation. A set of test vectors for verifying correct implementation was created. Subsequently, a bit exact fixed-point specification was requested and completed in 1994. The performance of G.728 has been extensively tested by

TABLE 7.2 ITU Linear Prediction Analysis-by-Synthesis Speech Coders

Standard Body	ITU	ITU	ITU
Number	G.728	G.729	G.723.1
Year	1992 and 1994	1995	1995
Type of coder	LD-CELP	CS-ACELP	MPC-MLQ and ACELP
Bit rate	16 kb/s	8 kb/s	6.3 & 5.3 kb/s
Quality	Toll	Toll	≤Toll
Complexity			
MIPS	30	≤22	≤16
RAM	2 K	<2.5 K	2.2 K
Delay			
Frame size	0.625 ms	10 ms	30 ms
Look ahead	0	5 ms	7.5 ms
Specification type			
Floating point	Algorithm exact	None	None
Fixed point	Bit exact	Bit exact C	Bit exact C

SG12. It gives robust performance for signals with background noise or music. It is very robust to random bit errors, more so than previous ITU standards G.711, G.726, G.727, and the newer standards described below. In addition to passing low bit rate modem signals as high as 2400 bps, it passes all network signaling tones.

In response to a request from CCIR Task Group 8/1 for a speech coder for wireless networks as envisioned in the future public land mobile telecommunication service (FPLMTS), the ITU initiated a work program for a toll quality 8 kbps speech coder which resulted in G.729. It is a forward adaptive CELP coder with a 10 ms frame size that uses algebraic CELP (ACELP) excitation.

The work program for G.723.1 was initiated in 1993 by the ITU as part of a group of standards to specify a low bit rate videophone for use on the public switched toll networks (PSTN) carried over a high speed modem. Other standards in this group include the video coder, modem, and data multiplexing scheme. A dual rate coder was selected. The two rates differ primarily by their excitation scheme. The higher rate used multipulse LPC with maximum likelihood quantization (MPC-MLQ) while the lower rate used ACELP. G.723.1 and G.729 are the first ITU coders to be specified by a bit exact fixed-point ANSI C code simulation of the encoder and decoder.

7.4.3 Digital Cellular Speech Coding Standards

Table 7.3 describes the first and second generation of speech coders to be standardized for digital cellulartelephony. The first generation coders provided adequate quality. Two of the second generation coders are so-called half-rate coders that have been introduced in order to double the capacity of the rapidly growing digital cellular industry. Another generation of coders will soon follow them in order to bring the voice quality of digital cellular service up to that of current wireline network telephony.

The RPE-LTP coder [33] was standardized by the Group Special Mobile (GSM) of CEPT in 1987 for pan-European digital cellular telephony. RPE-LTP stands for regular pulse excitation with long-term predictor. The GSM full-rate channel supports 22.8 kbps. The additional 9.8 kbps is used for channel coding to protect the coder from bit errors in the radio channel. Voice activity detection and discontinuous transmission are included as part of this standard. In addition to digital cellular telephony, this coder has since been used for other applications, such as messaging, because of its low complexity.

TABLE 7.3 Digital Cellular Telephony Speech Coders

Standard Body	CEPT	ETSI	TIA	TIA	RCR	RCR
Standard name	GSM	GSM 1/2 Rate	IS-54	IS-96	PDC	PDC 1/2 Rate
Type of coder	RPE-LTP	VSELP	VSELP	CELP	VSELP	PSI-CELP
Date	1987	1994	1989	1993	1990	1993
Bit rate	13 kb/s	5.6 kb/s	7.95 kb/s	0.8 to 8.5	6.7 kb/s	3.45 kb/s
Quality	<toll	= GSM	= GSM	<GSM	<GSM	= PDC
Est. complexity						
MIPS	4.5	30	20	20	20	50
RAM	1K	4K	2K	2K	2K	4K
Delay						
Frame size	20 ms	20 ms	20 ms	20 ms	20 ms	40 ms
Look ahead	0	5 ms	5 ms	5 ms	5 ms	10 ms
Specification type fixed point	Bit exact	Bit exact C	Bit stream	Bit stream	Bit stream	Bit stream

The GSM half-rate coder was standardized by ETSI (an off-shoot of CEPT) in order to double the capacity of the GSM cellular system. The coder is a 5.6 kbps VSELP coder [23]. A greater percentage of the channel bits are used for error protection because the half-rate channel has less frequency diversity than the full-rate system. The overall performance was measured to be similar to that of RPE-LTP, except for certain signals with background noise.

Vector sum excitation liner prediction coding (VSELP) was standardized by the Telecommunications Industry Association (TIA) for time division multiple access (TDMA) digital cellular telephony in North America as a part of Interim Standard 54 (IS-54). It was selected on the basis of subjective listening tests in 1989. The quality of this coder and RPE-LTP are somewhat different in the character of their distortion, but they usually receive about the same MOS in subjective listening tests. IS-54 does not have a bit exact specification. Implementations need only conform to the bit stream specification. The TIA does have a qualification procedure IS-85 to verify whether the performance of an implementation is good enough to be used for digital cellular [34]. In addition, Motorola provided a floating-point C program for their version of the coder, which implementers may use as a guideline.

The IS-96 coder [35] was standardized by the TIA for code division multiple access (CDMA) digital cellular telephony in North America. It is a part of IS-96 and is used in the system specified by IS-95. CDMA system capacity is its most attractive feature. When there is no speech, the rate of the channels is reduced. IS-96 is a variable rate CELP coder which uses DSI to achieve this rate reduction. It runs at 8.5 kbps during most of a talk spurt. When there is no speech on the channel, it drops down to just 0.8 kbps. At this rate, it is just supplying statistics about the background noise. These two rates are the ones most often used during operation of IS-96, although the coder does transition through the two intermediate rates of 2 and 4 kbps. The validation procedure for this coder is similar to that of IS-85.

The personal digital cellular (PDC) full-rate speech coder was standardized by the Research and Development Center for Radio Systems (RCR) for TDMA digital cellular telephone service in Japan as RCR STD-27B. The coder is very similar to IS-54 VSELP. The principal difference is that instead of two vector sum excitation codebooks, there is only one.

The PDC half-rate coder [37] was standardized by RCR to double the capacity of the Japanese TDMA PDC system. Pitch synchronous innovation CELP (PSI-CELP) uses fixed codebooks that are modified as a function of the pitch in order to improve the speech quality for such a low rate coder. If the pitch period is less than the frame size, then all vectors in the fixed codebook for that frame are made periodic. It has a background noise preprocessor as part of the standard. When it senses that the background noise exceeds a certain threshold, the preprocessor attempts to improve the quality of the speech. To date, this coder appears to be the most complex yet standardized.

7.4.4 Secure Voice Standards

Table 7.4 presents information about three secure voice standards. Two are existing standards, while the third describes a standard that the U.S. government hopes to promulgate in 1996.

FS1015 [12] is a U.S. Federal Standard 2.4 kbps LPC vocoder that was created over a long period of time beginning in the late 1970s. It was standardized by the U.S. Department of Defense (DoD) and later the North Atlantic Treaty Organization (NATO) before becoming a U.S. Federal Standard in 1984. It was always intended for secure voice terminals. It does not produce natural sounding speech, but over the years its intelligibility has been greatly improved through a series of changes to both its encoder and decoder. Remarkably, these changes never required changes to the bit stream. Presently, the intelligibility of FS1015 for clean input speech having telephone bandwidth is almost equivalent to that of the source material as measured by the diagnostic rhyme test (DRT). Most recently, an 800 bps vector quantized version of FS1015 has been standardized by NATO [39].

TABLE 7.4 Secure Telephony Speech Coding Standards

Standard Body	U.S. Dept. of Defense	U.S. Dept. of Defense	U.S. Dept. of Defense
Standard number	FS-1015	FS-1016	?
Type of coder	LPC vocoder	CELP	Model-based
Year	1984	1991	1996
Bit rate	2.4 kb/s	4.8 kb/s	2.4 kb/s
Quality	high DRT	<IS-54	=FS-1016
Complexity			
MIPS	20	19	41[a]
RAM	2K	1.5K	Unknown
Delay			
Frame size	22.5 ms	30 ms	22.5
Look ahead	90 ms	7.5 ms	23
Specification type	Bit stream	Bit stream	Bit stream

[a] Actual goal is 40 MIPS floating point or 80 MIPS fixed point.

FS1016 [40] is the result of a project undertaken by DoD to increase the naturalness of the secure telephone unit III (STU-3) by the introduction of 4.8 kbps modem technology. DoD surveyed available 4.8 kbps speech coder technology in 1988 and 1989. It selected a CELP-based coder having a so-called ternary codebook, meaning that all excitation amplitudes are $+1$, -1, or 0 before scaling by the gain for that subframe. This allows an easier codebook search. FS1016 definitely preserves far more of the naturalness of the original speech than FS1015, but the speech still contains many artifacts and the quality is substantially below that of the cellular coders such as GSM of IS-54. Both FS1015 and FS1016 have bit stream specifications, but there are C code simulations of them available from the government.

The next coder to be standardized by DoD is a new 2.4 kbps coder to replace both FS1015 and FS1016. A 3-year project was initiated in 1993 which should culminate in a new standard in 1997. Subjective testing was done in 1993 and 1994 on software versions of potential coders and a real-time hardware evaluation took place in 1995 and 1996 to select a best candidate. The mixed excitation linear prediction (MELP) coder was selected [41–43]. The need for this coder is due to the lack of a sufficient number of satellite channels at 4.8 kbps. The quality target for this coder is to match or exceed the quality and intelligibility of FS1016 for most scenarios. Many of the scenarios include severe background noise and noisy channel conditions. At 2.4 kbps, there is not enough bit rate available for explicit channel coding, so the speech coder itself must be designed to be robust for the channel conditions. The noisy background conditions have proven to be difficult for vocoders making voiced/unvoiced classification decisions, whether the decisions are made for all bands or for individual bands.

7.4.5 Performance

Figure 7.8 is included to give an impression of the relative performance for clean speech of most of the standard coders that were included above. There has never been a single subjective test which included all of the above coders. Figure 7.8 is based on the relative performances of these coders across a number of tests that have been reported. In the case of coders that are not yet standards, their performance is projected and shown as a circle. The vertical axis of Figure 7.8 gives the approximate single encoding quality for clean input speech. The horizontal axis is a logarithmic scale of bit rate. Figure 7.8 only includes telephone bandwidth speech coders. The 7 kHz speech coders have been omitted. Figure 7.9 compares the complexity as measured in MIPS and RAM for a fixed-point DSP implementation for most of the same standard coders. The horizontal axis is in RAM and the vertical axis is in MIPS.

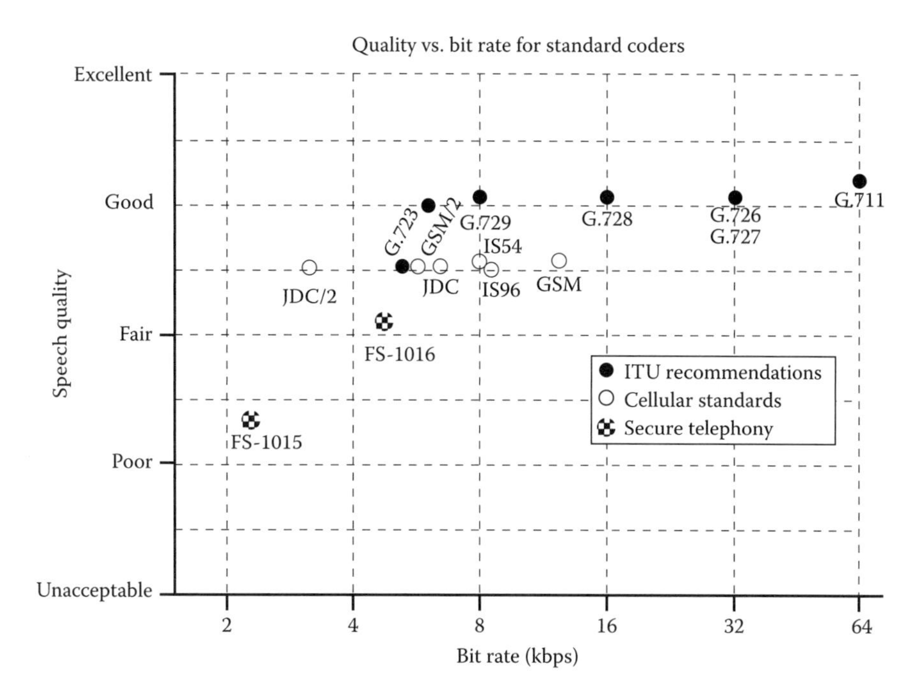

FIGURE 7.8 Approximate speech quality of speech coding standards.

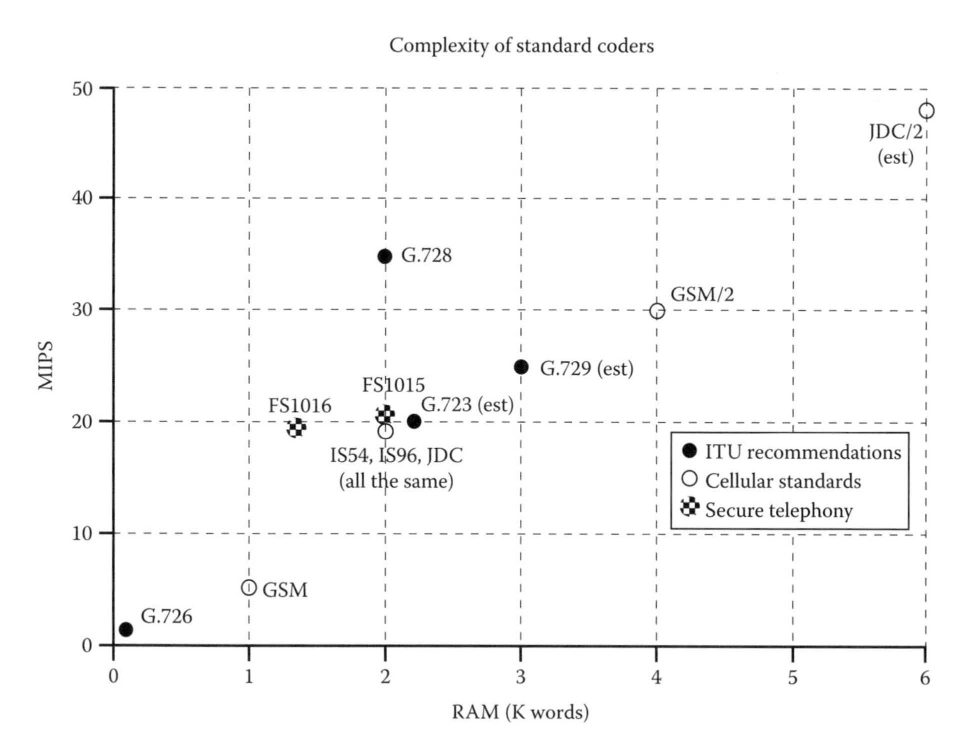

FIGURE 7.9 Approximate complexity of speech coding standards.

References

1. Markel, J.D. and Gray, Jr., A.H., *Linear Prediction of Speech*, Springer-Verlag, Berlin, 1976.
2. Rabiner, L.R. and Schafer, R.W., *Digital Processing of Speech Signals*, Prentice-Hall, Englewood Cliffs, NJ, 1978.
3. LeRoux, J. and Gueguen, C., A fixed point computation of partial correlation coefficients, *IEEE Trans. ASSP*, ASSP-27, 257–259, 1979.
4. Tohkura, Y., Itakura, F., and Hashimoto, S., Spectral smoothing technique in PARCOR speech analysis/synthesis, *IEEE Trans. ASSP*, 27, 257–259, 1978.
5. Viswanathan, R. and Makhoul, J., Quantization properties of transmission parameters in linear predictive systems, *IEEE Trans. ASSP*, 23, 309–321, 1975.
6. Sugamura, N. and Itakura, F., Speech analysis and synthesis methods developed at ECL in NTT—From LPC to LSP, *Speech Commun.*, 5, 199–215, 1986.
7. Soong, F. and Juang, B.-H., Optimal quantization of LSP parameters, *IEEE Trans. Speech Audio Process.*, 1, 15–24, 1993.
8. Lloyd, S.P., Least squares quantization in PCM, *IEEE Trans. Inf. Theory*, 28, 129–137, 1982.
9. Gersho, A. and Gray, R.M., *Vector Quantization and Signal Compression*, Kluwer-Academic Publishers, Dordrecht, Holland, 1991.
10. Schroeder, M.R., Atal, B.S., and Hall, J.L., Optimizing digital speech coders by exploiting masking properties of the human ear, *J. Acoust. Soc. Am.*, 66, 1647–1652, Dec. 1979.
11. Chen, J.-H. and Gersho, A., Adaptive postfiltering for quality enhancement of coded speech, *IEEE Trans. Speech Audio Process.*, 3, 59–71, 1995.
12. Tremain, T., The government standard linear predictive coding algorithm: LPC-10, *Speech Technol.*, 1(2), 40–49, Apr. 1982. Federal Standard 1015 is available from the U.S. government, as is C source code.
13. McAulay, R.J. and Quatieri, T.F., Speech analysis/synthesis based on a sinusoidal representation, *IEEE Trans. ASSP*, 34, 744–754, 1986.
14. McAulay, R.J. and Quatieri, T.F., Low-rate speech coding based on the sinusoidal model, in *Advances in Acoustics and Speech Processing*, Sondhi, M. and Furui, S., Eds., Marcel-Dekker, New York, 1992, pp. 165–207.
15. Griffin, D.W. and Lim, J.S., Multiband excitation vocoder, *IEEE Trans. ASSP*, 36, 1223–1235, 1988.
16. Hardwick, J.C. and Lim, J.S., The application of the IMBE speech coder to mobile communications, *Proc. ICASSP '91*, 249–252, 1991.
17. Kleijn, W.B. and Haagen, J., Transformation and decomposition of the speech signal for coding, *IEEE Signal Process. Lett.*, 136–138, 1994.
18. Kleijn, W.B. and Haagen, J., A general waveform interpolation structure for speech coding, in *Signal Processing VII*, Holt, M.J.J., Grant, P.M. and Sandham, W.A., Eds., Kluwer Academic Publishers, Dordrecht, Holland, 1994.
19. Jayant, N.S. and Noll, P., *Digital Coding of Waveforms*, Prentice-Hall, Englewood Cliffs, NJ, 1984, pp. 232–233.
20. Steele, R., *Delta Modulation Systems*, Halsted Press, New York, 1975.
21. Atal, B.S., Predictive coding of speech at low bit rates, *IEEE Trans. Comm.*, 30, 600–614, 1982.
22. Gersho, A., Advances in speech and audio compression, *Proc. IEEE*, 82, 900–918, 1994.
23. Gerson, I.A. and Jasiuk, M.A., Techniques for improving the performance of CELP-type speech coders, *IEEE JSAC*, 10, 858–865, 1992.
24. Crochiere, R.E. and Tribolet, J., Frequency domain coding of speech, *Proc. IEEE Trans. ASSP*, ASSP-27, 512–530, 1979.
25. Lefebvre, R., Salami, R., Laflamme, C., and Adoul, J.-P., High quality coding of wideband audio signals using transform coded excitation (TCX), *Proceedings of the ICASSP '94*, Adelaide, Australia, II-193–pp. 196, Apr. 1994.

26. Petr, D.W., 32 kb/s ADPCM-DLQ coding for network applications, *Proceedings of the IEEE GLOBECOM '82*, pp. A8.3-1–A8.3-5, 1982.
27. Daumer, W.R., Maitre, X., Mermelstein, P., and Tokizawa, I., Overview of the 32 kb/s ADPCM algorithm, *Proceedings of the IEEE GLOBECOM '84*, Atlanta, GA, 774–777, 1984.
28. Taka, M., Maruta, R., and LeGuyader, A., Synchronous tandem algorithm for 32 kb/s ADPCM, *Proceedings of the IEEE GLOBECOM '84*, 791–795, 1984.
29. Taka, M. and Maitre, X., CCITT standardizing activities in speech coding, *Proceedings of the ICASSP '86*, 817–820, 1986.
30. Chen, J.-H., Cox, R.V., Lin, Y.-C., Jayant, N., and Melchner, M.J., A low-delay CELP coder for the CCITT 16 kb/s speech coding standard, *IEEE JSAC*, 10, 830–849, 1992.
31. Johansen, F.T., A non bit-exact approach for implementation verification of the CCITT LD-CELP speech coder, *Speech Commun.*, 12, 103–112, 1993.
32. South, C.R., Rugelbak, J., Usai, P., Kitawaki, N., Irii, H., Rosenberger, J., Cavanaugh, J.R., Adesanya, C.A., Pascal, D., Gleiss, N., and Barnes, G.J., Subjective performance assessment of CCITT's 16 kbit/s speech coding algorithm, *Speech Commun.*, 12, 113–134, 1993.
33. Vary, P., Hellwig, K., Hofmann, R., Sluyter, R.J., Galand, C., and Russo, M., Speech codec for the European mobile radio system, *Proceedings of the ICASSP '88*, New York, pp. 227–230, 1988.
34. TIA/EIA Interim Standard 85, Recommended minimum performance standards for full rate speech codes, May 1992.
35. DeJaco, A., Gardner, W., Jacobs, P., and Lee, C., QCELP: the North American CDMA digital cellular variable rate speech coding standard, *Proceedings of the IEEE Workshop on Speech Coding for Telecommunications*, Sainte-Adele, Quebec, Canada, pp. 5–6, 1993.
36. TIA/EIA Interim Standard 125, Recommended minimum performance for digital cellular wide-band spread spectrum speech service option 1, Aug. 1994.
37. Miki, T., 5.6 kb/s PSI-CELP for digital cellular mobile radio, *Proceedings of the First International Workshop on Mobile Multimedia Communications*, Tokyo, Japan, Dec. 7–10, 1993.
38. Ohya, T., Suda, H., and Miki, T., 5.6 kb/s PSI-CELP of the half-rate PDC speech coding standard, *Proceedings of the IEEE Vehicular Technology Conference*, Stockholm, Sweden, pp. 1680–1684, June 1994.
39. Nouy, B., de la Noue, P., and Goudezeune, G., NATO stanag 4479, a standard for an 800 bps vocoder and redundancy protection in HF-ECCM system, *Proceedings of the ICASSP '95*, Detroit, MI, pp. 480–483, May 1995.
40. Campbell, J.P., Welch, V.C., and Tremain, T.E., The new 4800 bps voice coding standard, *Proc. Military Speech Tech. 89*, 64–70, Nov. 1989. Copies of Federal Standard 1016 are available from the U.S. Government, as is C source code.
41. McCree, A., Truong, K., George, E., Barnwell, T., and Viswanathan, V., A 2.4 kbit/s MELP coder candidate for the new U.S. federal standard, *Proceedings of the ICASSP '96*, Arlington, VA, May 1996.
42. Kohler, M., A comparison of the new 2400 bps MELP federal standard with other standard coders, *Proceedings of the ICASSP'97*, Atlanta, GA, pp. 1587–1590, April 1997.
43. Supplee, L., Cohn, R., Collura, J., and McCree, A., MELP: The new federal standard at 2400 bps, *Proceedings of the ICASSP'97*, Munich, Germany, pp. 1591–1594, April 1997.

8

Text-to-Speech Synthesis

Richard Sproat
Lucent Technologies

Joseph Olive
Lucent Technologies

8.1 Introduction

Text-to-speech (TTS) synthesis has had a long history, one that can be traced back at least to Dudley's "Voder," developed at Bell Laboratories and demonstrated at the 1939 World's Fair [1]. Practical systems for automatically generating speech parameters from a linguistic representation (such as a phoneme string) were not available until the 1960s, and systems for converting from ordinary text into speech were first completed in the 1970s, with MITalk being the best-known such system [2]. Many projects in TTS conversion have been initiated in the intervening years, and papers on many of these systems have been published.*

It is tempting to think of the problem of converting written text into speech as "speech recognition in reverse": current speech recognition systems are generally deemed successful if they can convert speech input into the sequence of words that was uttered by the speaker, so one might imagine that a TTS synthesizer would start with the words in the text, convert each word one-by-one into speech (being careful to pronounce each word correctly), and concatenate the result together. However, when one considers what literate native speakers of a language must do when they read a text aloud, it quickly becomes clear that things are much more complicated than this simplistic view suggests. Pronouncing words correctly is only part of the problem faced by human readers: in order to sound natural and to sound as if they understand what they are reading, they must also appropriately emphasize (accent) some words, and de-emphasize others; they must "chunk" the sentence into meaningful (intonational) phrases; they must pick an appropriate F0 (fundamental frequency) contour; they must control certain aspects of their voice quality; they must know that a word should be pronounced longer if it appears in some positions in the sentence than if it appears in others because "segmental durations" are affected by various factors, including phrasal position.

What makes reading such a difficult task is that, all writing systems systematically fail to specify many kinds of information that are important in speech. While the written form of a sentence (usually) completely specifies the words that are present, it will only partly specify the intonational phrases (typically with some form of punctuation), will usually not indicate which words to accent or deaccent,

* For example, [3] gives an overview of recent Dutch efforts in this area. Audio examples of several current projects on TTS can be found at the WWW URL http:/www.cs.bham.ac.uk/~jpi/synth/museum.html.

and hardly ever give information on segmental duration, voice quality, or intonation. (One might think that a question mark "?" indicates that a sentence should be pronounced with a rising intonation: generally, though, a question mark merely indicates that a sentence is a question, leaving it up to the reader to judge whether this question should be rendered with a rising intonation.) The orthographies of some languages—e.g., Chinese, Japanese, and Thai—fail to give information on where word boundaries are, so that even this needs to be figured out by the reader.* Humans are able to perform these tasks because, in addition to being knowledgeable about the grammar of their language, they also (usually) understand the content of the text that they are reading, and can thus appropriately manipulate various extragrammatical "affective" factors, such as appropriate use of intonation and voice quality.

The task of a TTS system is thus a complex one that involves mimicking what human readers do. But a machine is hobbled by the fact that it generally "knows" the grammatical facts of the language only imperfectly, and generally can be said to "understand" nothing of what it is reading. TTS algorithms thus have to do the best they can making use, where possible, of purely grammatical information to decide on such things as accentuation, phrasing, and intonation—and coming up with a reasonable "middle ground" analysis for aspects of the output that are more dependent on actual understanding.

It is natural to divide the TTS problem into two broad subproblems. The first of these is the conversion of text—an imperfect representation of language, as we have seen—into some form of linguistic representation that includes information on the phonemes (sounds) to be produced, their duration, the locations of any pauses, and the F0 contour to be used. The second—the actual synthesis of speech—takes this information and converts it into a speech waveform. Each of these main tasks naturally breaks down into further subtasks, some of which have been alluded to. The first part, text and linguistic analysis, may be broken down as follows:

- *Text preprocessing*: Including end-of-sentence detection, "text normalization" (expansion of numerals and abbreviations), and limited grammatical analysis, such as grammatical part-of-speech assignment
- *Accent assignment*: The assignment of levels of prominence to various words in the sentence
- *Word pronunciation*: Including the pronunciation of names and the disambiguation of homographs†
- *Intonational phrasing*: The breaking of (usually long) stretches of text into one or more intonational units
- *Segmental durations*: The determination, on the basis of linguistic information computed thus far, of appropriate durations for phonemes in the input
- *F0 contour computation*

Speech synthesis breaks down into two parts:

- The selection and concatenation of appropriate concatenative units given the phoneme string
- The synthesis of a speech waveform given the units, plus a model of the glottal source

8.2 Text Analysis and Linguistic Analysis

8.2.1 Text Preprocessing

The input to TTS systems is text encoded using an electronic coding scheme appropriate for the language, such as ASCII, JIS (Japanese), or Big-5 (Chinese). One of the first tasks facing a TTS system is that of dividing the input into reasonable chunks, the most obvious chunk being the sentence. In some

* Even in English, single orthographic words e.g., AT&T, can actually represent multiple words–AT and T.

† A *homograph* is a single written word that represents two or more different lexical entries, often having different pronunciation: an example would be *bass* which could be the word for a musical range—with pronunciation /beʲs/–or a fish–with pronunciation/baes/. We transcribe pronunciations using the International Phonetic Association's (IPA) symbol set. Symbols used in this chapter are defined in Table 8.1.

TABLE 8.1 IPA Symbols Used in This Chapter

IPA Symbol	Phonetic Value
æ	**a** as in *cat*
b	**b** as in *bass*
e^{j}	**a** as in *bake*
ɛ	**e** as in *bet*
ə	**a** as in *data*
i	**i** as in *linguini*
ɪ	**i** as in *in*
s	**s** as in *soggy*
ʃ	**sh** as in *ship*
t	**t** as in *test*
tʃ	**ch** as in *chase*
ð	**th** as in *the*
r	**r** as in *are*
v	**v** as in *voodoo*
w	**w** as in *we*
ˈ	Primary stress

writing systems there is a designated symbol used for marking the end of a declarative sentence and for nothing else—in Chinese, for example, a small circle is used—and in such languages end-of-sentence detection is generally not a problem. For English and other languages we are not so fortunate because a period, in addition to its use as a sentence delimiter, is also used, for example, to mark abbreviations: if one sees the period in *Mr.*, one would not (normally) want to analyze this as an end-of-sentence marker. Thus, before one concludes that a period does in fact mark the end of a sentence, one needs to eliminate some other possible analyses. In a typical TTS system, text analysis would include an abbreviation-expansion module; this module is invoked to check for common abbreviations which might allow one to eliminate one or more possible periods from further consideration. For example, if a preprocessor for English encounters the string *Mr.* in an appropriate context (e.g., followed by a capitalized word), it will expand it as *mister* and remove the period.

Of course, abbreviation expansion itself is not trivial, since many abbreviations are ambiguous. For example, is *St.* to be expanded as *Street* or *Saint*? Is *Dr.*, *Doctor* or *Drive*? Such cases can be disambiguated via a series of heuristics. For *St.*, for example, the system might first check to see if the abbreviation is followed by a capitalized word (i.e., a potential name), in which case it would be expanded as *Saint*; otherwise, if it is preceded by a capitalized word, a number, or an alphanumeric (*th*), it would be expanded as *Street*. Another problem that must be dealt with is the conversion of numbers into words: should usually be expanded as *two hundred thirty two*, whereas if the same sequence occurs as part of -3142—a likely telephone number—it would normally be read *two three two*.

In languages like English, tokenization into words can to a large extent be done on the basis of white space. In contrast, in many Asian languages, including Chinese, the situation is not so simple because spaces are never used to delimit words. For the purposes of text analysis it is therefore generally necessary to "reconstruct" word boundary information. A minimal requirement for word segmentation is an online dictionary that enumerates the word forms of the language. This is not enough on its own, however, since there are many words that will not be found in the dictionary; among these are personal names, foreign names in transliteration, and morphological derivatives of words that do not occur in the dictionary. It is therefore necessary to build models of these nondictionary words; see [4] for further discussion.

In addition to lexical analysis, the text-analysis portion of a TTS system will typically perform syntactic analysis of various kinds. One commonly performed analysis is grammatical part-of-speech assignment, as information on the part of speech of words can be useful for accentuation and phrasing, among other things. Thus, in a sentence like *they can can cans*, it is useful for accentuation purposes to know that the first *can* is a *function word*—an auxiliary verb, whereas the second and third are *content words*—respectively a verb and a noun. There are a number of part-of-speech algorithms available, perhaps the best known being the stochastic method of [5], which computes the most likely analysis of a sequence of words, maximizing the product of the *lexical probabilities* of the parts-of-speech in the sentence (i.e., the possible parts of speech of each word and their probabilities), and the *n-gram probabilities* (probabilities of *n*-grams of parts of speech), which provide a model of the context.

8.2.2 Accentuation

In languages like English, various words in a sentence are associated with *accents*, which are usually manifested as upward or downward movements of fundamental frequency. Usually, not every word in the sentence bears an accent, however, and the decision on which words should be accented and which should be unaccented is one of the problems that must be addressed as part of text analysis. It is common in prosodic analysis to distinguish three levels of *prominence*. Two are accented and unaccented, as just described, and the third is *cliticized*. Cliticized words are unaccented but in addition have lost their word stress, so that they tend to be durationally short: in effect, they behave like unstressed affixes, even though they are written as separate words.

A good first step in assigning accents is to make the accentual determination on the basis of broad lexical categories or parts of speech. Content words—nouns, verbs, adjectives, and perhaps adverbs, tend in general to be accented; function words, including auxiliary verbs and prepositions tend to be deaccented; short function words tend to be cliticized. But accenting has a wider function than merely communicating lexical category distinctions between words. In English, one important set of constructions where accenting is more complicated than what might be inferred from the above discussion are complex noun phrases—basically, a noun preceded by one or more adjectival or nominal modifiers. In a "discourse-neutral" context, some constructions are accented on the final word (*Madison **Avenue***), some on the penultimate (***Wall** Street*, *kitchen **towel** rack*), and some on an even earlier word (***sump** pump factory*). The assignment of accent to complex noun phrases depends on complex lexical and semantic factors; see [6].

Accenting is not only sensitive to syntactic structure and semantics, but also to properties of the discourse. One straightforward effect is *contrast*, as in the example *I didn't ask for **cherry** pie, I asked for **apple** pie*. For most speakers, the "discourse neutral" accent would be on *pie*, but in this example there is a clear intention to contrast the ingredients in the pies, and *pie* is thus deaccented to affect the contrast between *cherry* and *apple*. See [7] for a discussion of how these kind of effects are handled in a TTS system for English. Note, while humanlike accenting capabilities are possible in many cases, there are still some intractable problems. For example, just as one would often deaccent a word that had been previously mentioned, so would one often deaccent a word if a supercategory of that word had been mentioned: *My son wants a Labrador, but I'm **allergic** to dogs*. Handling such cases in any general way is beyond the capabilities of current TTS systems.

8.2.3 Word Pronunciation

The next stage of analysis involves computing pronunciations for the words in the input, given the orthographic representation of those words. The simplest approach is to have a set of "letter-to-sound" rules that simply map sequences of graphemes into sequences of phonemes, along with possible diacritic information, such as stress placement. This approach is naturally best suited to languages where there is a relatively simple relation between orthography and phonology: languages such as Spanish or Finnish fall into this category. However, languages like English manifestly do not, so it has generally been recognized that a highly accurate word pronunciation module must contain a pronouncing dictionary that, at the very least, records words whose pronunciation could not be predicted on the basis of general rules. However, having a dictionary that is merely a list of words presents us with familiar problems of coverage: many text words occur that are not to be found in the dictionary, including morphological derivatives from known words, or previously unseen personal names.

For morphological derivatives, standard techniques for morphological analysis [2,8] can be applied to achieve a morphological decomposition for a word. The pronunciation of the whole can then, in general, be computed from the (presumably known) pronunciation of the morphological parts, applying appropriate phonological rules of the language. For novel personal names, additional mechanisms may be necessary since novel names cannot always be related *morphologically* to previously seen ones. One such additional method involves computing the pronunciation of a new name by analogy with the pronunciation of a similar name [9,10]. For example, imagine that we have the name *Califano* in our dictionary and that we know its pronunciation: then we could compute the pronunciation of a hypothetical name *Balifano* by noting that both names share the "suffix" *alifano*. The pronunciation of *Balifano* can then be computed by removing the phoneme /k/, corresponding to the letter *C* in *Califano*, and replacing it with the phoneme /b/.

There are some word forms that are inherently ambiguous in pronunciation, and for which a word pronunciation module as just described can only return a set of possible pronunciations, from which one must then be chosen. A straightforward example is the word *Chevy*, which is most commonly pronounced

/ʃ'ɛvi/, but is /tʃ'ɛvi/ in the name *Chevy Chase*, so in this case one could succeed by simply storing the bigram *Chevy Chase*. But *n*-gram models do not solve all cases of homograph disambiguation. So, the word *bass*, is most likely to be pronounced /b æ s/ in a "fishy" context like *he was fishing for bass*, but /beʲ s/ in a musical context like *he plays bass*. What defines the context as being musical or "fishy" is not characterizable in terms of *n*-grams, but rather relates to the occurrence of certain words (e.g., *fish, lake, boat* vs. *play, sing, orchestra*) in a wider context. A method proposed by Yarowsky [11,12] allows for both local (*n*-gram) context and wide context to be used in homograph disambiguation, and excellent results have been achieved using this approach.*

8.2.4 Intonational Phrasing

In reading a long sentence, speakers will typically break the sentence up into several phrases, each of which can be said to "stand alone" as an intonational unit. If punctuation is used liberally so that there are relatively few words between the commas, semicolons, or periods, then a reasonable guess at an appropriate phrasing would be simply to break the sentence at the punctuation marks (though this is not always appropriate [13]). The real problem comes when long stretches occur without punctuation; in such cases, human readers would normally break the string of words into phrases, and the problem then arises of where to place these breaks.

The simplest approach is to have a list of words, typically function words, that are likely indicators of good places to break [1]. One has to use some caution, however, because while a particular function word such as *and* may coincide with a plausible phrase break in some cases (*He got out of the car and walked towards the house*), in other examples it might coincide with a particularly *poor* place to break as in *I was forced to sit through a dog and pony show that lasted most of Wednesday afternoon*. Other approaches to intonational phrasing have been proposed in the literature, including methods that depend on syntactic parsers of various degrees of sophistication [13,14]. An alternative approach, described in [15], uses a decision tree model [16,17] that is trained on a corpus of text annotated with prosodic phrase-boundary information.

* Clearly the above-described method for homograph disambiguation can also be applied to other formally similar problems in TTS, such as whether St. to be expanded as Saint or Street, or 747 is to be read as a number seven hundred and forty seven or the name of an aircraft seven forty seven.

8.2.5 Segmental Durations

Having computed which phonemes are to be produced by the synthesizer, it is necessary to decide how long to make each one. In this section we briefly describe the methods used for computing segmental durations: the reader is referred to [18] for an extended discussion of this topic.

What duration to assign to a phonemic segment depends on many factors, including

- The identity of the segment in question. For example, in many dialects of English, the vowel /æ/ has a longer *intrinsic duration* than the vowel /ɪ/.
- The stress of the syllable of which the segment is a member. For example, vowels in stressed syllables tend to be longer than vowels in unstressed syllables.
- Whether the syllable of which the segment is a member bears an accent. Accented syllables tend to be longer than otherwise identical unaccented syllables.
- The quality of the surrounding segments. For example, a vowel preceding a voiced consonant in the same syllable tends to be longer than the same vowel preceding a voiceless consonant.
- The position of the segment in the phrase: elements close to the ends of phrases tend to be longer than elements more internal to the phrase.

Various approaches have been taken to modeling segmental durations in TTS systems. One method involves *duration rules*, which are rules of the form "if the segment is *X* and it is in phrase-final position, then lengthen *X* by *n* milliseconds" [19,20]. In rule-based systems of this kind, it is not unusual for the duration of a given segment to be rewritten several times as the conditions for the application of the various rules are considered. The rule-based approach can be formalized explicitly in terms of the second approach—*duration models*—which are mathematical expressions that prescribe how the various conditioning factors are to be used in computing the duration of a segment [19]; the successive application of the rules can, in effect, be "compiled" into a single mathematical expression that implements the combined effect of the rules. As argued in [18], all extant duration models can be viewed as instances of a more general *sum-of-products* model, where the duration of a segment is predicted by a formula of the general form:

$$DUR(\mathbf{f}) = \sum_{i \in T} \prod_{j \in I_i} S_{i,j}(f_j) \tag{8.1}$$

Here the duration assigned to a feature vector—$DUR(\mathbf{f})$—is computed by scaling each factor f_j in the ith product term by a factor scale $S_{i,j}$; computing the product of all scaled factors within each product term; and then summing over all i product terms. Rather than deciding *a priori* on a particular sums-of-products model (or set of such models) within the space of all possible models, one approach taken to segmental duration is to use exploratory data analysis to arrive at models whose predictions show a good fit to durations from a corpus of labeled speech [18].

More specifically, we start with a text corpus that (ideally) has a good coverage both of various phonemes and of the factors (and their combinations) that are deemed likely to be relevant for duration. A native speaker of the language reads this text and the speech is segmented and labeled. Using the text-analysis modules of TTS, with some possible hand correction, we automatically compute the sequence of phonemes, and the feature vectors (including features on stress, accent, phrasal position, etc.) associated with each phoneme. Given the feature vectors, various sums-of-products models are compared and their predictions of the values of the observed segmental durations are evaluated. In general, different specific duration models may be better suited to different sets of conditions than others: for example, in the English duration system, intervocalic consonants are associated with a different sums-of-products model than consonants that occur in clusters. In the actual implementation of segmental duration predictions, a decision tree is used to determine, on the basis of contextual factors appropriate to the segment at hand, what particular sums-of-products model to use; this model is then used to compute the duration of the segment.

Designing a corpus with good coverage of relevant factors is a nontrivial task in itself: the basic problem is to provide a set that has maximal coverage with the minimal amount of text to be read by a speaker, and analyzed. The method that we use involves starting with a large corpus of text in a language and automatically predicting the phonemic segments along with their features (again, using text analysis components for the language). A *greedy* algorithm is then applied to arrive at a minimal set of sentences that have good (ideally total) coverage of the desired feature vectors.

8.2.6 Intonation

Having computed linguistic information such as the sequence of segments to be produced, their duration, the prominence of the various words, and the locations of prosodic boundaries, the next thing that a TTS system needs to compute is an intonation contour. There are almost as many models of intonation implemented in TTS systems as there are TTS systems, and we do not have the space to review these different approaches here. Suffice it to say that most intonation models that have actually been incorporated into working TTS systems can be classified into one of three "schools":

- The *Fujisaki* school [21,22]. An intonation contour for a phrase is computed from a phrase impulse and some number of accent impulses. These impulses are convolved with a smoothing function to produce phrase and accent curves, which are then summed to produce the final contour.
- The *Dutch* school [23]. Intonation contours are represented as sequences of connected line segments which are chosen so as to perceptually closely approximate real (smooth) intonation contours.
- The *autosegmental/metrical* school [24,25]. Intonation contours are represented abstractly as sequences of high and low targets.

The computation of an intonation contour from a phonological representation can be illustrated by considering the Bell Labs English TTS system, which currently uses a version of the Pierrehumbert autosegmental model [26–28]. As the first stage in the computation of an intonation contour, a tone-timing function sets up nominal times for each accent in the sentence. Separate routines are called for initial boundary tones, final boundary tones, pitch accents and phrase accents. Roughly, initial boundary tones are aligned with the silence that is placed at the beginning of each minor phrase, whereas final boundary tones are aligned with the final vowel of each minor phrase. Phrase accents are aligned after the final word accent of the minor phrase, if there is one; otherwise at the end of the first vowel of the first word, or else at the end of the first phoneme. Finally, accents on words are aligned with their associated syllables using a complex set of contextual factors. These nominal accent times are then converted into actual F0/time pairs, by another function. F0 values are computed dependent on the prominence of the accent (either determined automatically, or else definable by the user), and various phrasal parameters from the intonation model, as well as the particular type of accent involved. Finally, an *F0 contour* is produced by interpolating the computed pitch/time pairs, and smoothing via convolution with a rectangular window.

8.3 Speech Synthesis

Once the text has been transformed into phonemes, and their associated durations and a fundamental frequency contour have been computed, the system is ready to compute the speech parameters for synthesis.

There are two independent variables in the choice of parametric computation in a TTS system. One variable is the choice between a rule-based scheme for the computation of the parameters on the one hand, and a concatenative scheme involving concatenation of short segments of previously uttered speech on the other. The second variable is the actual parametric representation chosen: possible choices include articulatory parameters, formants, linear predictive coding (LPC), spectral parameters, or time domain parameters. In a concatenative scheme, any parametric representation that permits independent control of loudness, F0, voicing, timing, and possibly spectral manipulations is appropriate. Rule-based systems are more restrictive of the choice of parameters since such schemes rely both on our understanding of the relation between the parameters and the acoustic signals they represent, and on our ability to compute the dynamics of the parameters as they move from one sound to another. Thus far only articulatory parameters and formants have been used in rule-based systems. The best-known examples of a formant-based synthesizer are the Klatt synthesizer and its commercial offshoot DECtalk.

Rule-based systems are space-efficient because they eliminate the need to store speech segments. Rule-based approaches also make it easier, in principle, to implement new speaker characteristics for different voices, as well as different phone inventories for new dialects and languages. However, since the dynamics of the parameters are very difficult to model it requires a great deal more effort to produce a rule-based system than it does to produce a concatenative system of comparable quality. Given the right choice of units, a concatenative scheme is able to store the dynamics of the speech signal and thus produce high quality synthetic sound. The choice of the exact parameters depends on what the designer values in such a system. Waveform representations—such as PSOLA [29]—have a high sound quality, but they are limiting in terms of the ability to alter the sound, and thus far, no one has been able to change the spectral parameters in a time domain system. Articulatory parameters or formants, on the other hand, can be successfully manipulated. However, the speech quality produced by using these parameters is somewhat degraded because there are no reliable methods to extract these parameters and even in a plain coding application (analysis and resynthesis without manipulations) these methods produce degradation of the speech signal.

Other systems use a concatenative approach. In this approach, parametrized short speech segments of natural speech are connected to form a representation of the synthetic speech. The majority of the natural speech segments are merely transitions between pairs of phonemes. However, due to the large contextual variation of some phonemes, some segments consisting of three or more phoneme elements are often necessary; such elements consist of the transition from the first phoneme to the second, and a transition from the penultimate phoneme to the last, but the intermediate phonemes are stored completely. For the Bell Labs English system, there are approximately 2900 different speech elements—also called *dyads*—in the acoustic inventory, and these elements are sufficient to make up all the legal phoneme combinations for English.

The concatenative approach to speech synthesis requires that speech samples be stored in some parametric representation that will be suitable for connecting the segments and changing the signal's characteristics of loudness, F0, and spectrum. One method for changing the characteristics of natural speech is to analyze the speech in terms of a source/filter model, as diagramed in Figure 8.1

This model of speech synthesis has a variety of independent input controls. Starting at the left side of the figure, we show two possible source generators: a noise generator and a simple pulse generator. The noise generator has no controlling input whereas the pulse generator is controlled by the F0

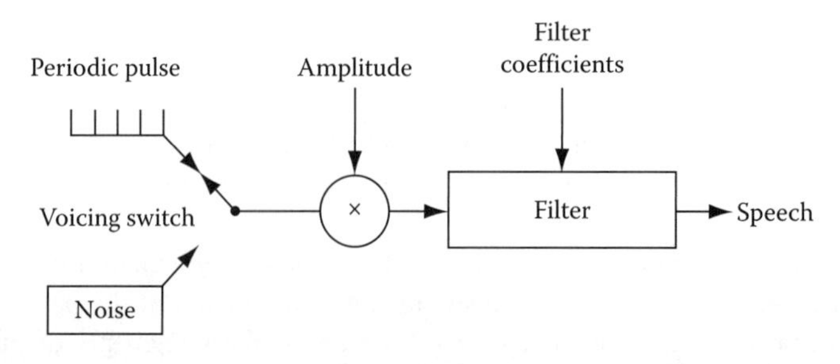

FIGURE 8.1 Source/filter model for speech synthesis.

parameter; the F0 parameter specifies the distance between any two pulses thus controlling the F0 of the periodic source. These inputs are selected by a switch which is controlled by a voicing flag. It is also possible to have a mixer to control the relative contribution of the noise and pulse source, and to insert a glottal pulse with additional controls for the shape of the glottal source in place of the simple pulse generator. To the right of the switch, we have a multiplier that multiplies the source by an amplitude parameter. This serves as the loudness control for the system. The signal from the multiplier is fed into a filter controlled by the filter coefficients which are varied slowly to shape the speech spectrum.

The source/filter model can be used to replicate naturally spoken speech when the parameters are obtained by analysis of natural speech. Speech can be parametrized by an amplitude control, voiced/voiceless flag, F0, and filter coefficients at a small interval (on the order of 5–15 ms) The loudness control is determined from the power of the speech at the time frame of the analysis. F0 extraction algorithms determine the voicing of the speech as well as the fundamental frequency. The filter parameters can be determined by various analysis techniques. The parameters obtained from the analysis can be used to drive the source/filter model to reproduce the analyzed speech. However, these parameters can also be varied independently to change the speech. The ability to alter the analysis parameters is crucial to a concatenative approach, where the spectral parameters have to be smoothed and interpolated whenever two elements from different utterances are connected, or when the duration of the speech has to be altered. Of course, the fundamental frequency of the original speech is completely discarded and replaced during synthesis by a rule-generated F0, as described earlier.

8.4 The Future of TTS

Using current methods such as those outlined in this chapter, it is possible to produce speech output that is of high intelligibility and reasonable naturalness, given unrestricted input text. (See [30] for a discussion of methods for evaluating TTS systems.) However, there is still much work to be done in all areas of the problem, including: improving voice quality, and allowing for greater user control over aspects of voice quality; producing better models of intonation to allow for more natural-sounding F0 contours; and improving linguistic analysis so that more accurate information on contextually appropriate word pronunciation, accenting, and phrasing can be computed automatically. This latter area—linguistic analysis—is particularly crucial: most high-quality TTS systems allow for user control of the output speech by means of various "escape sequences," which can be inserted into the input text. By use of such escape sequences, it is possible to produce highly appropriate and natural-sounding output. What is still lacking in many cases are natural-language analysis techniques that can mimic what a human annotator is able to do.

References

1. Klatt, D., Review of text-to-speech conversion for English, *J. Acoust. Soc. Am.*, 82, 737–793, 1987.
2. Allen, J., Hunnicutt, M.S., and Klatt, D., *From Text to Speech*, Cambridge University Press, Cambridge, MA, 1987.
3. van Heuven, V. and Pols, L., *Analysis and Synthesis of Speech: Strategic Research towards High-Quality Text-to-Speech Generation*, Mouton de Gruyter, Berlin, 1993.
4. Sproat, R., Shih, C., Gale, W., and Chang, N., A stochastic finite-state word-segmentation algorithm for Chinese, in *Association for Computational Linguistics, Proceedings of 32nd Annual Meeting*, Las Cruces, NM, pp. 66–73, 1994.
5. Church, K., A stochastic parts program and noun phrase parser for unrestricted text, in *Proceedings of the Second Conference on Applied Natural Language Processing*, Association for Computational Linguistics, Morristown, NJ, pp. 136–143, 1988.
6. Sproat, R., English noun-phrase accent prediction for text-to-speech, *Comput. Speech Lang.*, 8, 79–94, 1994.

7. Hirschberg, J., Pitch accent in context: Predicting intonational prominence from text, *Artif. Intell.*, 63, 305–340, 1993.
8. Koskenniemi, K., Two-level morphology: A general computational model for word-form recognition and production, PhD thesis, University of Helsinki, Helsinki, Finland, 1983.
9. Coker, C., Church, K., and Liberman, M., Morphology and rhyming: Two powerful alternatives to letter-to-sound rules for speech synthesis, in *Proceedings of the ESCA Workshop on Speech Synthesis*, Bailly, G. and Benoit, C., Eds., Autrans, France, pp. 83–86, 1990.
10. Golding, A., Pronouncing names by a combination of case-based and rule-based reasoning, PhD thesis, Stanford University, Stanford, CA, 1991.
11. Yarowsky, D., Homograph disambiguation in speech synthesis, in *Proceedings of the Second ESCA/IEEE Workshop on Speech Synthesis*, New Paltz, NY, 244–247, 1994.
12. Sproat, R., Hirschberg, J., and Yarowsky, D., A corpus-based synthesizer, in *Proceedings of the International Conference on Spoken Language Processing*, Banff, ICSLP, pp. 563–566, Oct. 1992.
13. O'Shaughnessy, D., Parsing with a small dictionary for applications such as text to speech, *Comput. Linguist.*, 15, 97–108, 1989.
14. Bachenko, J. and Fitzpatrick, E., A computational grammar of discourse-neutral prosodic phrasing in English, *Comput. Linguist.*, 16, 155–170, 1990.
15. Wang, M. and Hirschberg, J., Automatic classification of intonational phrase boundaries, *Comput. Speech Lang.*, 6, 175–196, 1992.
16. Breiman, L., Friedman, J.H., Olshen, R.A., and Stone, C.J., *Classification and Regression Trees*, Wadsworth & Brooks, Pacific Grove, CA, 1984.
17. Riley, M., Some applications of tree-based modelling to speech and language, in *Proceedings of the Speech and Natural Language Workshop*, DARPA, Morgan Kaufmann, Cape Cod, MA, pp. 339–352, Oct. 1989.
18. van Santen, J., Assignment of segmental duration in text-to-speech synthesis, *Comput. Speech Lang.*, 8, 95–128, 1994.
19. Klatt, D., Linguistic uses of segmental duration in English: Acoustic and perceptual evidence, *J. Acoust. Soc. Am.*, 59, 1209–1221, 1976.
20. Syrdal, A.K., Improved duration rules for text-to-speech synthesis, *J. Acoust. Soc. Am.*, 85, S1(Q4), 1989.
21. Fujisaki, H., Dynamic characteristics of voice fundamental frequency in speech and singing, in *The Production of Speech*, MacNeilage, P., Ed., Springer, New York, pp. 39–55, 1983.
22. Möbius, B., *Ein quantitatives Modell der deutschen Intonation*, Niemeyer, Tübingen, 1993.
23. 't Hart, J., Collier, R., and Cohen, A., *A Perceptual Study of Intonation: An Experimental-Phonetic Approach to Speech Melody*, Cambridge University Press, Cambridge, MA, 1990.
24. Pierrehumbert, J.B., The phonology and phonetics of English intonation, PhD thesis, Massachusetts Institute of Technology, Cambridge, MA, Sept. 1980.
25. Ladd, D.R., *The Structure of Intonational Meaning*, Indiana University Press, Bloomington, IN, 1980.
26. Liberman, M. and Pierrehumbert, J., Intonational invariants under changes in pitch range and length, in *Language Sound Structure*, Aronoff, M. and Oehrle, R., Eds., MIT Press, Cambridge, MA, 1984.
27. Anderson, M., Pierrehumbert, J., and Liberman, M., Synthesis by rule of English intonation patterns, in *Proceedings of the International Conference on Acoustics, Speech, and Signal Processing*, vol.1, ICASSP, San Diego, CA, pp. 2.8.1–2.8.4, 1984.
28. Silverman, K., Utterance-internal prosodic boundaries, in *Proceedings of the Second Australian International Conference on Speech Science and Technology*, Sydney, Australia, pp. 86–91, 1988.
29. Charpentier, F. and Moulines, E., Pitch-synchronous waveform processing techniques for text-to-speech synthesis using diphones, *Speech Commun.*, 9(5/6), 453–467, 1990.
30. van Santen, J., Perceptual experiments for diagnostic testing of text-to-speech systems, *Comput. Speech Lang.*, 7, 49–100, 1993.

9

Speech Recognition by Machine

Lawrence R. Rabiner
Rutgers University

and

AT&T Research Labs

Biing-Hwang Juang
Georgia Institute of Technology

9.1 Introduction

Over the past several decades, the need has arisen to enable humans to communicate with machines in order to control their actions or obtain information. Initial attempts at providing human–machine communications led to the development of the keyboard, the mouse, the trackball, the touch screen, and the joy stick. However, none of these communication devices provides the richness or the ease of use of speech which has been the most natural form of communication between humans for tens of centuries. This need of a natural voice interface between humans and machines has been met, to a limited extent, by speech processing systems which enable a machine to speak (speech synthesis systems) and which enable a machine to understand (speech recognition systems) human speech. We concentrate on speech recognition systems in this section.

Speech recognition by machine refers to the capability of a machine to convert human speech to a textual form, providing a transcription or interpretation of everything that the human speaks while the machine is listening. This capability is required for tasks in which the human is controlling the actions of the machine using only limited speaking capability, e.g., while speaking simple commands or sequences of words from a limited vocabulary (e.g., digit sequences for a telephone number). In the more general case, usually referred to as speech understanding, the machine needs to only recognize a limited subset of the user input speech, namely, the speech that specifies enough about the action requested, so that the machine can either respond appropriately, or initiate some action in response to what was understood.

Speech recognition systems have been deployed in applications ranging from control of desktop computers to telecommunication services and business services, and have achieved varying degrees of success and commercialization.

In this section, we discuss the principle and a range of issues involved in the design and implementation of speech recognition systems and provide a brief summary of where the technology stands in terms of its applications and deployment.

9.2 Characterization of Speech Recognition Systems

An automatic speech recognition system can be defined along a number of technological dimensions. These include

1. The manner in which a user speaks to the machine. There are generally three modes of speaking, including
   a. Isolated word (or phrase) mode in which the user speaks individual words (or phrases) drawn from a specified (and usually limited) vocabulary
   b. Connected word mode in which the user speaks fluent speech consisting entirely of words from a specified vocabulary (e.g., telephone or account numbers)
   c. Continuous speech mode in which the user fluently can speak from a large (often unlimited) vocabulary
2. The size of the recognition vocabulary, including
   a. Small vocabulary systems which provide recognition capability for up to 100 words
   b. Medium vocabulary systems which provide recognition capability from 100 to 1000 words
   c. Large vocabulary systems which provide recognition capability for over 1000 words
3. The intended coverage of the user population, including
   a. Speaker-dependent (SD) systems which have been custom-tailored to each individual talker
   b. Speaker-independent (SI) systems which work on broad populations of talkers most of which the system has never encountered or adapted to
4. The ability to adapt to the changing operating conditions, including
   a. Nonadaptive systems which do not automatically change the system parameters nor the operating configurations during deployment
   b. Adaptive systems which customize their knowledge (over time) to each individual user while the system is being utilized by the individual user, or to the specific acoustic environment in which the systems are deployed
5. The amount of explicit linguistic knowledge used in the system, including
   a. Simple acoustic systems which use only sound patterns for recognition without explicit linguistic knowledge about the structure of the sound
   b. Systems which integrate acoustic and linguistic knowledge, where the linguistic knowledge is generally represented via syntactical and semantic constraints on the hypothesized output of the recognition system
6. The degree of dialogue between the human and the machine including
   a. One-way (passive) communication in which each user's spoken input is acted upon by the machine, also called a user-initiative mode
   b. A system-initiative mode in which the system is the sole initiator of the dialog, requesting information from the user via verbal input
   c. A natural dialog in which the machine conducts a conversation with the speaker, solicits inputs, acts in response to user inputs, or even tries to clarify ambiguity in the conversation, also called a natural or mixed-initiative mode

9.3 Sources of Variability of Speech

Speech recognition by machine is inherently difficult because of the variability in the signal. Sources of this variability include

1. Within-speaker variability in maintaining consistent pronunciation and use of words and phrases
2. Across-speaker variability due to physiological differences (e.g., different vocal tract lengths), regional accents, influence of foreign languages, etc.
3. Transducer variability while speaking over different microphone/telephone handsets
4. Variability introduced by the transmission system (the media through which speech is transmitted or acquired such as landline telecommunication networks, wireless channels, etc.)
5. Variability in the speaking environment, including extraneous conversations and acoustic background events (e.g., noise, door slams)

Failure to properly take the above variability into account in a speech recognition system design will generally result in unsatisfactory performance.

9.4 Approaches to ASR by Machine

9.4.1 The Acoustic-Phonetic Approach [1]

The earliest approaches to speech recognition were based on finding speech sounds and providing appropriate labels to these sounds according to the theory of acoustic-phonetics which postulates that there exist finite, distinctive phonetic units (phonemes) in a spoken language, and that these units can be broadly characterized by a set of acoustic properties that are manifest in the speech signal over time. Even though the acoustic properties of phonetic units are highly variable, both with speakers and with neighboring sounds (the so-called coarticulation effect), it is assumed in the acoustic-phonetic approach that the rules governing the variability are straightforward and can be readily learned by a machine. The first step in the acoustic-phonetic approach is a segmentation and labeling phase in which the speech signal is segmented into stable acoustic regions, followed by attaching one or more phonetic labels to each segmented region, resulting in a phoneme lattice characterization of the speech (see Figure 9.1). The second step attempts to determine a valid word (or string of words) from the phonetic label sequences produced in the first step. In the validation process, linguistic constraints of the task (i.e., the vocabulary, the syntax, and other semantic rules) are invoked in order to access the lexicon for word decoding based on the phoneme lattice. The acoustic-phonetic approach has not been widely used in most commercial applications due to the lack of reliable and robust segmentation and labeling techniques that are consistent with the theory of acoustic-phonetics. Recent efforts by Lee [2] try to integrate results from the acoustic-phonetic approach with those from more modern statistical systems based on so-called automatic speech attribute transcription (ASAT) methods.

9.4.2 The Pattern-Matching Approach [3]

The "pattern-matching approach" involves two essential steps, namely, pattern training and pattern comparison. The essential feature of this approach is that it uses a well-formulated mathematical framework and establishes consistent speech pattern representations for reliable pattern comparison from a set of labeled training samples via a formal training algorithm. A speech pattern representation can be in the form of a speech template or a statistical model, and can be applied to a sound (smaller than a word), a word, or a phrase. In the pattern-comparison stage of the approach, a direct comparison is made between the unknown speech pattern (the speech to be recognized) with each possible pattern learned in the training stage, in order to determine the identity of the unknown according to the goodness of match of the patterns. The pattern-matching approach was the predominant method of speech recognition in the 1970s and 1980s and we shall elaborate on it in subsequent sections.

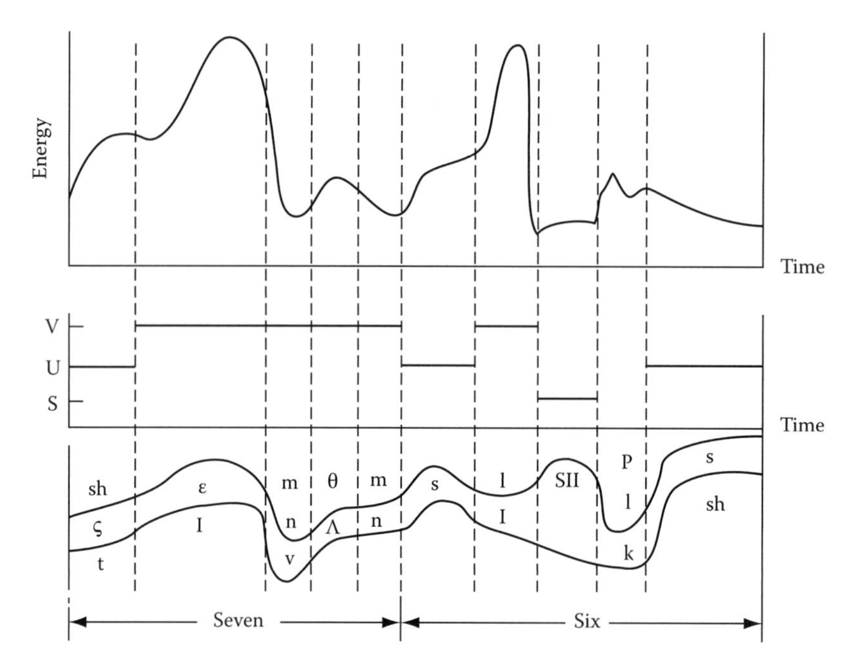

FIGURE 9.1 Segmentation and labeling for word sequence "seven-six."

9.4.3 Artificial Intelligence Approach [4,5]

The "artificial intelligence approach" attempts to mechanize the recognition procedure as inspired by the way a person applies knowledge in visualizing, analyzing, and characterizing speech based on a set of measured acoustic features. Among the techniques used within this class of methods is the use of a biologically inspired system (e.g., a neural network as a computational model of the human brain) which attempts to integrate phonemic, lexical, syntactic, semantic, and even pragmatic knowledge for segmentation and labeling, and uses tools such as artificial neural networks for learning the relationships among phonetic events. The focus in this approach has been mostly in the representation of knowledge and integration of knowledge sources. This method has not been used widely in commercial systems.

9.5 Speech Recognition by Pattern Matching

Figure 9.2 is a block diagram that depicts the pattern-matching framework. The speech signal is first analyzed and a feature representation is obtained for comparison with either stored reference templates or statistical models in the pattern-matching block. A decision scheme determines the word or phonetic class of the unknown speech based on the matching scores with respect to the stored reference patterns.

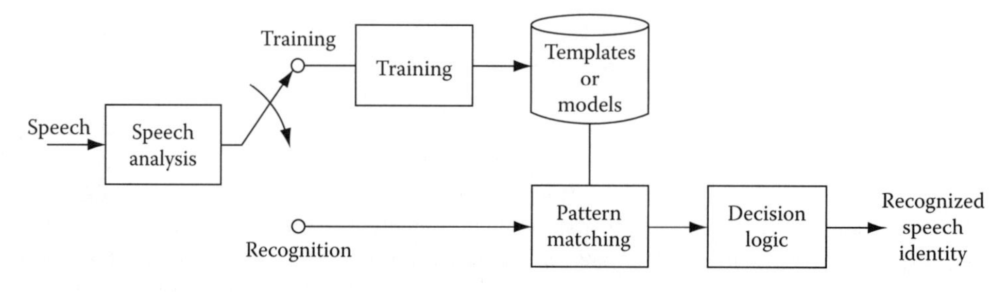

FIGURE 9.2 Block diagram of a pattern-matching speech recognizer.

There are two types of reference patterns that can be used with the model of Figure 9.2. The first type, called a nonparametric reference pattern [6] (or often a template), is a pattern created from one or more spoken tokens (exemplars) of the sound associated with the pattern. The second type, called a statistical reference model, is created as a statistical characterization (via a fixed type of model) of the statistical behavior of a collection of tokens of the sound associated with the pattern. The hidden Markov model (HMM) [7] is an example of the statistical model.

The model of Figure 9.2 has been used (either explicitly or implicitly) for almost all commercial and industrial speech recognition systems for the following reasons:

1. It is invariant to different speech vocabularies, user sets, feature sets, pattern-matching algorithms, and decision rules
2. It is easy to implement
3. It works well in practice

We now discuss the elements of the pattern recognition model and show how it has been used in isolated word, connected word, and continuous speech recognition systems.

9.5.1 Speech Analysis

The purpose of the speech analysis block is to transform the speech waveform into a parsimonious representation which characterizes the time-varying properties of the speech. The transformation is normally done on successive and possibly overlapped short intervals 10–40 ms in duration (i.e., short-time analysis) due to the time-varying nature of speech. The representation [8] could be spectral parameters, such as the output from a filter bank, a discrete Fourier transform (DFT), or a linear predictive coding (LPC) analysis, or they could be temporal parameters, such as the locations of various zeros or level crossing times in the speech signal.

Empirical knowledge gained over decades of psychoacoustic studies suggests that the power spectrum has the necessary acoustic information for highly accurate determination of sound identity. Studies in psychoacoustics also suggest that our auditory perception of sound power and loudness involves both compression and frequency scale warping, leading to the use of the logarithmic power spectrum and the mel-scale cepstrum [9], which is the Fourier transform of the mel-scale log-spectrum. The low-order mel-scale cepstral coefficients (up to 20) provide a parsimonious representation of the short-time speech segment which is usually sufficient for phonetic identification.

The mel-scale cepstral parameters are often augmented by the so-called delta (and often delta-delta) mel-scale cepstrum [10] which characterizes dynamic aspects of the time-varying speech process.

9.5.2 Pattern Training

Pattern training is the method by which representative sound patterns (for the unit being trained) are converted into reference patterns for use by the pattern-matching algorithm. There are several ways in which pattern training can be performed, including

1. Casual training in which a single sound pattern is used directly to create either a template or a crude statistical model (due to the paucity of data)
2. Robust training in which several versions of the sound pattern (usually extracted from the speech of a single talker) are used to create a single merged template or statistical model
3. Clustering training in which a large number of versions of the sound pattern (extracted from a wide range of talkers) are used to create one or more templates or a reliable statistical model of the sound pattern

In order to better understand how and why statistical models are so broadly used in speech recognition, we now formally define an important class of statistical models, namely the HMM [7].

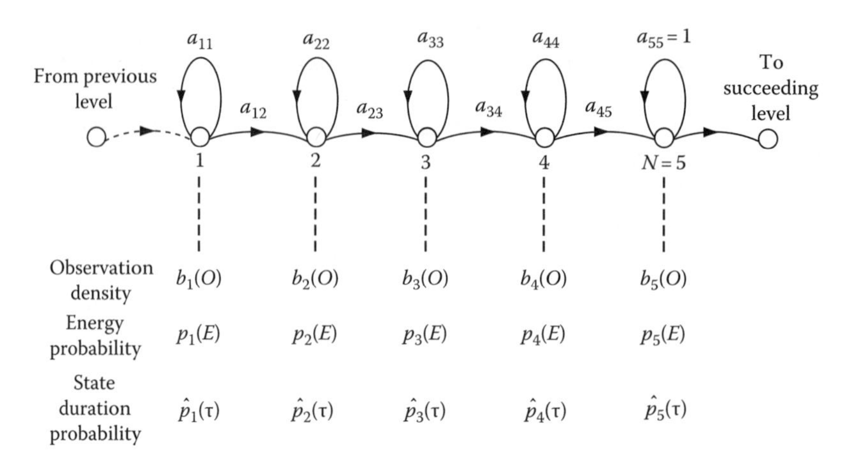

FIGURE 9.3 Characterization of a word (or phrase, or subword) using a $N(5)$-state, left-to-right, HMM, with continuous observation densities in each state of the model.

9.5.2.1 The HMM

The HMM is a statistical characterization of both the dynamics (time-varying nature) and statics (the spectral characterization of sounds) of speech during speaking of a subword unit, a word, or even a phrase. The basic premise of the HMM is that a Markov chain can be used to describe the probabilistic nature of the temporal sequence of sounds in speech, i.e., the phonemes in the speech, via a probabilistic state sequence. The states in the sequence are not observed with certainty because the correspondence between linguistic codes and the speech waveform is probabilistic in nature; hence, the concept of a hidden model. Instead, the states manifest themselves through the second component of the HMM which is a set of output distributions governing the production of the speech features in each state (the spectral characterization of the sounds). In other words, the output distribution represents the local statistical knowledge of the speech pattern (which is observed) within the state, and the Markov chain characterizes, through a set of state transition probabilities, how these sound processes evolve from one sound to another. Integrated together, the HMM is particularly well suited for modeling speech processes.

An example of an HMM of a speech pattern is shown in Figure 9.3. The model has five states (corresponding to five distinct "sounds or phonemes" within the speech), and the state (corresponding to the sound being spoken) proceeds from left-to-right (as time progresses). Within each state (assumed to represent a stable acoustical distribution), the spectral features of the speech signal are characterized by a mixture Gaussian density of spectral features (called the observation density), along with (in this specific case) an energy distribution, and a state duration probability. The states represent the changing temporal nature of the speech signal; hence, indirectly they represent the speech sounds within the pattern.

The training problem for HMMs consists of estimating the parameters of the statistical distributions within each state (e.g., means, variances, mixture gains, etc.), along with the state transition probabilities. Well-established techniques (e.g., the Baum–Welch method [11] or the segmental K-means method [12]) have been developed and tested for doing this pattern training efficiently and robustly. These techniques are developed based on mathematically justified objectives which lead to consistency in the trained models and thus their recognition performance. These optimization objectives include maximum likelihood [11], maximum mutual information (MMI) [13], and minimum classification error (MCE) [14].

9.5.3 Pattern Matching

Pattern matching refers to the process of assessing the similarity between two speech patterns, one of which represents the unknown speech and one of which represents the reference pattern (derived from

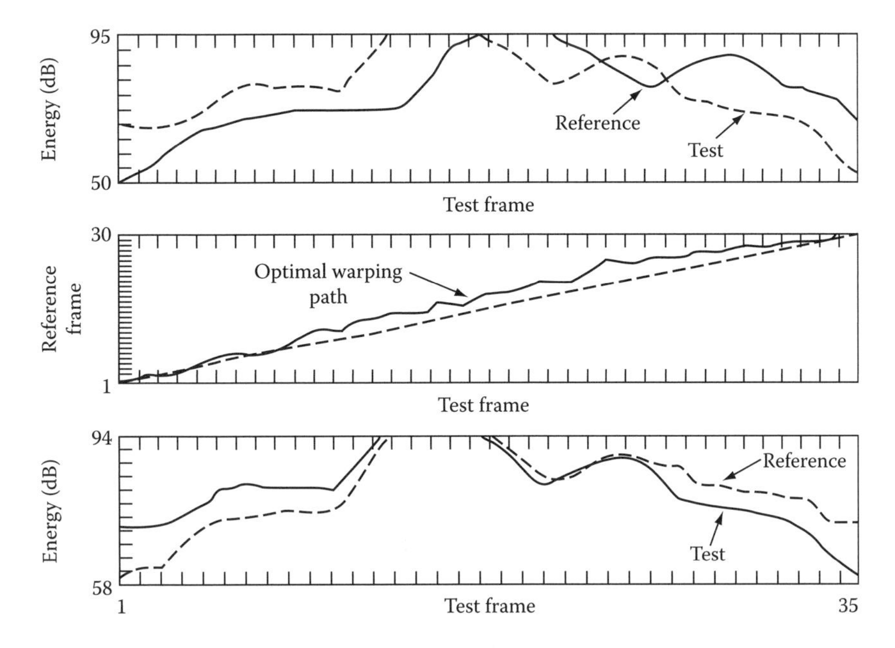

FIGURE 9.4 Result of time-aligning two versions of the word "seven," showing linear alignment of the two utterances (top panel); optimal time-alignment path (middle panel); and nonlinearly aligned patterns (lower panel).

the training process) of each element in the reference set that can be recognized. When the reference pattern is a typical utterance template, pattern matching produces a gross similarity (or dissimilarity) score. When the reference pattern consists of a probabilistic model, such as an HMM, the process of pattern matching is equivalent to using the statistical knowledge contained in the probabilistic model to assess the likelihood of the speech (which led to the model) being realized as the unknown pattern.

A major problem in comparing speech patterns is due to speaking rate variations. HMMs provide an implicit time normalization as part of the process for measuring likelihood. However, for template-based approaches, explicit time normalization is required. Figure 9.4 demonstrates the effect of explicit time normalization between two patterns representing isolated word utterances. The top panel of the figure shows the log energy contour of the two patterns (for the spoken word seven)–one called the reference (known) pattern and the other called the test (or unknown input) pattern. It can be seen that the inherent duration of the two patterns, 30 and 35 frames (where each frame is a 15 ms segment of speech), is different and that linear alignment is grossly inadequate for internally aligning events within the two patterns (compare the locations of the vowel peaks in the two patterns) and may lead to unreliable pattern-matching scores. A basic principle of time alignment is to nonuniformly warp the time scale so as to achieve the best possible matching score between the two patterns (regardless of whether the two patterns are of the same word identity or not). This can be accomplished by a dynamic programming procedure, often called dynamic time warping (DTW) [15] when applied to speech template matching. The "optimal" nonlinear alignment result of DTW is shown at the bottom of Figure 9.4 in contrast to the linear alignment of the patterns at the top. It is clear that the nonlinear alignment provides a more realistic measure of similarity between the patterns.

9.5.4 Decision Strategy

The decision strategy takes all the matching scores (from the unknown pattern to each of the stored reference patterns) into account, finds the "closest" match, and decides if the quality of the match is good enough to make a recognition decision. If not, the user is asked to provide another token of the

TABLE 9.1 Performance of Isolated Word Recognizers

Vocabulary		Mode	Word Error Rate (%)
10	Digits	SI	0.1
		SD	0.0
39	Alphadigits	SI	7.0
		SD	4.5
129	Airline terms	SI	2.9
		SD	1.0
1109	Basic English	SD	4.3

speech (e.g., the word or phrase) for another recognition attempt. This is necessary because often the user may speak words that are incorrect in some sense (e.g., hesitation, incorrectly spoken word, etc.) or simply outside of the vocabulary of the recognition system.

9.5.5 Results of Isolated Word Recognition

Using the pattern recognition model of Figure 9.2 and using either the nonparametric template approach or the statistical HMM method to derive reference patterns, a wide variety of tests of the recognizer have been performed on telephone speech with isolated word inputs in both SD and SI modes. Vocabulary sizes have ranged from as few as 10 words (i.e., the digits zero–nine) to as many as 1109 words. Table 9.1 gives a summary of recognizer performance under the conditions described above.

9.6 Connected Word Recognition

The systems we have been describing in the previous sections have all been isolated word recognition systems. In this section, we consider extensions of the basic processing methods described in the previous sections in order to handle recognition of sequences of words, the so-called connected word recognition system.

The basic approach to connected word recognition is shown in Figure 9.5. Assume we are given a fluently spoken sequence of words, represented by the (unknown) test pattern T and we are also given a set of V reference patterns, $\{R_1, R_2, \ldots, R_V\}$ each representing one of the words in the vocabulary. The connected word recognition problem consists of finding the concatenated reference pattern, R^S,

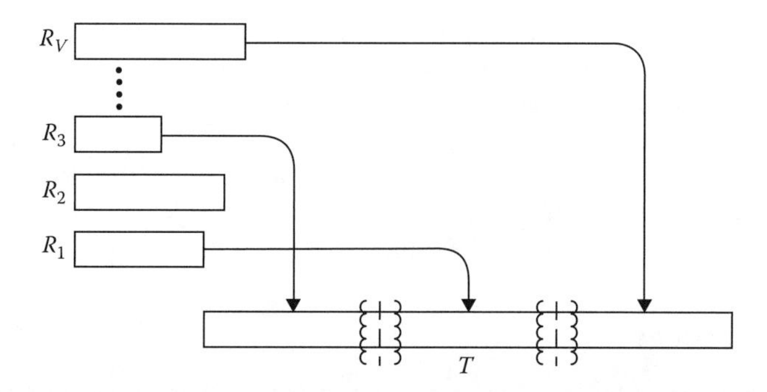

FIGURE 9.5 Illustration of the problem of matching a connected word string, spoken fluently, using whole word patterns concatenated together to provide the best match.

TABLE 9.2 Performance of Connected Digit Recognizers

Vocabulary	Mode	Word Error Rate (%)	Task	String Error Rate (%)
10 digits	SD	0.1	Variable length digit	0.4
	SI	0.2	strings (1–7 digits)	0.8
26 letters of the	SD	10.0	Name retrieval from	4.0
alphabets	SI	10.0	directory of 1700 names	10.0
129 airline terms	SD	0.1	Sentences in a grammar	1.0
	SI	3.0		10.0

which best matches the test pattern, in the sense that the overall similarity between T and R^S is maximum over all sequence lengths and over all combinations of vocabulary words.

There are several problems associated with solving the connected word recognition problem, as formulated above. First of all, we do not know how many words were spoken; hence, we have to consider solutions with a range on the number of words in the utterance. Second, we do not know nor can we reliably find word boundaries within the test pattern. Hence, we cannot use word boundary information to segment the problem into simple word-matching recognition problems. Finally, since the combinatorics of trying to solve the problem exhaustively (by trying to match every possible string) are exponential in nature, we need to devise efficient algorithms to solve this problem. Such efficient algorithms have been developed and they solve the connected word recognition problem by iteratively building up time-aligned matches between sequences of reference patterns and the unknown test pattern, one frame at a time [16–18].

9.6.1 Performance of Connected Word Recognizers

Typical recognition performance for connected word recognizers is given in Table 9.2 for a range of vocabularies, and for a range of associated tasks. In the next section, we will see how we exploit linguistic constraints of the task to improve recognition accuracy for word strings beyond the level one would expect on the basis of word error rates of the system.

9.7 Continuous Speech Recognition

The techniques used in connected word recognition systems cannot be extended to the problem of continuous speech recognition for several reasons. First of all, as the size of the vocabulary of the recognizer grows, it becomes impractical to train patterns for each individual word in the vocabulary. Hence, continuous speech recognizers generally use subword speech units as the basic patterns to be trained, and use a lexicon to define the structure of word patterns in terms of the subword unit sequences. Second, the words spoken during continuous speech generally have a syntax associated with the word order, i.e., they are spoken according to a grammar. In order to achieve good recognition performance, account must be taken of the word grammar so as to constrain the set of possible recognized sentences. Finally, the spoken sentence often must make sense according to a semantic model of the task which the recognizer is asked to perform. Again, by explicitly including these semantic constraints on the spoken sentence, as part of the recognition process, performance of the system improves.

Based on the discussion above, there are three distinct new problems associated with continuous speech recognition [19] namely:

1. Choice of subword unit used to represent the sounds of speech, and methods of creating appropriate acoustic models for these subword units
2. Choice of a representation of words in the recognition vocabulary, in terms of the subword units
3. Choice of a method for integrating syntactic (and possibly semantic) information into the recognition process so as to properly constrain the sentences that are allowed by the system

9.7.1 Subword Speech Units and Acoustic Modeling

For the basic subword speech recognition unit, one could consider a range of linguistic units, including syllables, half syllables, dyads, dyphones, or phonemes. The most common choice is a simple phoneme set, which for English comprises about 40–50 units, depending on fine choices as to what constitutes a unique phoneme. Since the number of phonemes is limited, it is usually straightforward to collect sufficient speech training data for reliable estimation of statistical models of the phonemes. The resulting set of subword speech models are usually referred to as "context-independent" phone-like units (CI-PLU) since each unit is trained independent of the context of neighboring units. The problem with using such CI-PLU models is that phonemes are highly variable according to different contexts, and therefore using models which cannot represent this variability properly leads to inferior speech recognition performance.

A straightforward way to improve the modeling of phonemes is to augment the CI-PLU set with phoneme models that are context-dependent. In this manner, a target phoneme is modeled differently depending on the phonemes that precede and follow it. A phoneme is thus represented by a number of unit models, each of which is differentiated by the context in which the target phoneme appears. By using such context-dependent PLUs (in addition to the CI-PLUs), the "resolution" of the acoustic models is increased, and the performance of the recognition system improves.

9.7.2 Word Modeling from Subword Units

Once the base set of subword units is chosen, one can use standard lexical modeling techniques to represent words in terms of these units. The key problem here is variability of word pronunciation across talkers with different regional accents. Hence, for each word in the recognition vocabulary the lexicon contains a baseform (or standard) pronunciation of the word, as well as alternative pronunciations, as appropriate.

The lexicon used in most recognition systems is extracted from a standard pronouncing dictionary, and each word pronunciation is represented as a linear sequence of phonemes. This lexical definition is basically speech data-independent because no speech or text data are used to derive the pronunciation. Hence, the lexical variability of a word in speech is not modeled explicitly but characterized only indirectly through the statistical distributions contained in subword unit models. To improve lexical modeling capability, the use of (multiple) pronunciation networks has been shown to work well [20].

9.7.3 Language Modeling within the Recognizer

In order to determine the best match to a spoken sentence, a continuous speech recognition system has to evaluate both an acoustic-model match score (corresponding to the "local" acoustic matches of the words in the sentence) and a language-model match score (corresponding to the match of the words to the grammar and syntax of the task). The acoustic-matching score is readily determined using dynamic programming methods much like those used in connected word recognition systems. The language match scores are computed according to a production model of the syntax and the semantics. The language match score indicates the degree at which a tentatively decoded sequence of speech codes (e.g., words) would conform to a given set of constraints that the generating language implies. Most often the language model is represented as a finite state network (FSN) for which the language score is computed according to arc scores along the best decoded path (according to an integrated model where acoustic and language modeling are combined) in the network. Other models of language include word pair models as well as N-gram word probabilities.

The language model, when defined over word sequences, can also be used to compute a quantity, called the "perplexity," which measures the average number of words that may follow any given word. The perplexity is thus an indicator of the complexity of the recognition task, before the acoustic variability is taken into account.

TABLE 9.3 Performance of Continuous Speech Recognition Systems

Task	Syntax	Mode	Vocabulary	Word Error Rate (%)
Resource management (RM)	Finite-state grammar (perplexity = 60)	SI, fluent input	~1,000 words	4.4
Air travel information system (ATIS)	Backoff trigram (perplexity = 18)	SI, natural language	~2,500 words	3.6
North America business broadcast (NAB)	Backoff 5-gram (perplexity = 173)	SI, fluent input	~60,000 words	10.8
Telephone speech (Switchboard)	4-gram interpolated with trigram (perplexity ~60)	SI, conversational	~45,000 words	~25
Casual conversation in telephone calls (CallHome)	Backoff 4-gram (perplexity ~190)	SI, casual, and spontaneous	~28,000 words	~40

9.7.4 Performance of Continuous Speech Recognizers

Table 9.3 illustrates current capabilities in continuous speech recognition, for five distinct tasks, namely database access (Resource Management, RM), natural spoken language queries (ATIS) for air travel reservations, read text from a set of business publications (NAB), telephone conversational speech (Switchboard), and casual colloquial phone calls between family members (CallHome).

9.8 Speech Recognition System Issues

This section discusses some key issues in building "real world" speech recognition systems.

9.8.1 Robust Speech Recognition

Robust speech recognition refers to the problem of designing an ASR system that works equally well in various unknown or adverse operating environments [21]. Robustness is important because the performance of existing ASR systems, whose designs are predicated on known or clean environments, often degrades rapidly under field conditions.

There are basically four types of sound degradation, namely, noise, distortion, articulation effects, and pronunciation variations. Noise is an inevitable component of the acoustic environment and is normally considered additive with the speech. Distortion refers to the modification of the spectral characteristics of the signal by the room, the transducer (microphone), the channel (e.g., transmission), etc. Articulation effects result from the factors that affect a talker's speaking manner when responding to a machine rather than a human. One well-known phenomenon is the Lombard effect which is related to the changes in articulation when the talker speaks in a noisy environment. Finally, different speakers will pronounce a word differently depending on the regional accent. These conditions are often not known a priori when the recognizer is trained in the laboratory and are often detrimental to the recognizer performance.

There are essentially two broad categories of techniques that have been proposed for dealing with adverse conditions. These are invariant methods and adaptive methods, respectively.

Invariant methods use speech features (or the associated similarity measures) that are invariant under a wide range of conditions, e.g., liftering and RelAtive SpecTrAl (RASTA) [22] (which suppress speech features that are more susceptible to signal variability), the short-time modified coherence (SMC) [23] (which has a built-in noise averaging advantage), and the ensemble interval histogram (EIH) [24] (which mimics the human auditory mechanism). Robust distortion measures include the group-delay measure [25] and a family of distortion measures based on the projection operator [26] which were shown to be effective in conditions involving additive noise.

Adaptive methods differ from invariant methods in the way the characteristics of the operating environment are taken into account. Invariant methods assume no explicit knowledge of the signal environment, while adaptive methods attempt to estimate the adverse condition and adjust the signal or the reference models accordingly in order to achieve reliable matching results.

When channel or transducer distortions are the major factor, it is convenient to assume that the linear distortion effect appears as an additive signal bias in the cepstral domain. This distortion model leads to the method of cepstral mean subtraction and, more generally, signal bias removal [27], which makes a maximum likelihood estimate of the bias due to distortion and subtracts the estimated bias from the cepstral features before pattern matching is performed.

9.8.2 Speaker Adaptation

Given sufficient training data, an SD recognition system usually performs better than an SI system for the same task [28]. Many systems are designed for SI applications, however, due to the fact that it is often difficult to collect speaker-specific training data that would be adequate for reliable performance. One way to bridge the performance gap is to apply the method of speaker adaptation which uses a very limited amount of speaker-specific data to modify the model parameters of an SI recognition system in order to achieve a recognition accuracy approaching that of a well-trained SD system. Adaptation techniques can also be applied for the purpose of automatically optimizing the system parameters to match the environment in which the system is deployed.

9.8.3 Keyword Spotting and Utterance Verification

An automatic speech recognition system needs to have both high accuracy and a user-friendly interface in order to be acceptable to the users. One major component in a friendly user interface is to allow the user to speak naturally and spontaneously without imposing a rigid speaking format. In a typical spontaneously spoken utterance, however, we usually observe various kinds of disfluency, such as hesitation and extraneous sounds like um and ah and false starts, and unanticipated ambient noise, such as mouth clicks and lip smacks, etc. In the conventional paradigm, which formulates speech recognition as decoding of an unknown utterance into a contiguous sequence of phonetic units, the task is equivalent to designing an unlimited vocabulary continuous speech recognition and understanding system which is, unfortunately, beyond reach with today's technology.

One alternative to the above approach, particularly when implementing domain-specific services, is to focus on a finite set of vocabulary words most relevant to the intended task and design the system using the technology of keyword spotting [29] and, more generally, utterance verification (UV) [30]. With UV incorporated into the speech recognition system, the user is allowed to speak spontaneously so long as the requisite keywords appear somewhere in the spoken utterance. The system then detects and identifies the in-vocabulary words (i.e., keywords), while rejecting all other superfluous acoustic events in the utterance (which include out-of-vocabulary words, invalid inputs—any form of disfluency as well as lack of keywords—and ambient sounds). In such cases, no critical constraints are imposed on the users' speaking format, making the user interface natural and effective.

9.8.4 Barge-In

In natural human–human conversation, talkers often interrupt each other during speaking. This is called "barge-in." For human–machine interactions, in which machine prompts are often routine messages or instructions, the capability of allowing talkers to "barge-in" becomes an important enabling technology for a natural human–machine interface.

Two key technologies are integrated in the implementation of "barge-in," namely, an echo canceller (to remove the spoken message from the machine to the recognizer) and a partial rejection mechanism.

With "barge-in," the recognizer needs to be activated and to listen starting from the beginning of the system prompt. An echo canceller, with a proper double talk detector, is used to cancel the system prompt while attempting to detect if the near-end signal from the talker (i.e., speech to be recognized) is present. The tentatively detected signal is then passed through the recognizer with rejection thresholds to produce the partial recognition results. The rejection technique is critical because extraneous input is very likely to be present, both from the ambient background and from the talker (breathing, lip smacks, etc.) during the long period when the recognizer is activated.

9.9 Practical Issues in Speech Recognition

As progress is made in fundamental recognition technologies, we need to examine carefully the key attributes that a recognition system must possess in order for it to be useful. These include: high recognition performance in terms of speed and accuracy, ease of use, and low cost. A recognizer must be able to deliver high recognition accuracy without excessive delay. A system that does not provide high performance often adds to users' frustration and may even be considered counterproductive. A recognition system must also be easy to use. The more naturally a system interacts with the user (e.g., does not require words in a sentence to be spoken in isolation), the higher the perceived effectiveness. Finally, the recognition system must be low cost to be competitive with alternative technologies such as keyboard or mouse devices in computer interface applications.

9.10 ASR Applications

Speech recognition has been successfully applied in a range of systems. We categorize these applications into five broad classes.

1. *Office or business system*: Typical applications include data entry onto forms, database management and control, keyboard enhancement, and dictation. Examples of voice-activated dictation machines include the IBM Tangora system [31] and the Dragon Dictate system [32] (currently part of Nuance).
2. *Manufacturing*: ASR is used to provide "eyes-free, hands-free" monitoring of manufacturing processes (e.g., parts inspection) for quality control.
3. *Telephone or telecommunications*: Applications include automation of operator-assisted services (the voice recognition call processing system by AT&T to automate operator service routing according to call types), inbound and outbound telemarketing, information services (the Automatic Answer Network System for Electrical Request (ANSER) system by NTT for limited home banking services, the stock price quotation system by Bell Northern Research, Universal Card services by Conversant/AT&T for account information retrieval), voice dialing by name/number (AT&T VoiceLine, 800 Voice Calling services, Conversant FlexWord, etc.), directory assistance call completion, catalog ordering, and telephone calling feature enhancements (AT&T VIP—voice interactive phone for easy activation of advanced calling features such as call waiting, call forwarding, etc. by voice rather than by keying in the code sequences).
4. *Medical*: The application is primarily in voice creation and editing of specialized medical reports (e.g., Kurtweil's system).
5. *Other*: This category includes voice-controlled and voice-operated toys and games, aids for the handicapped and voice control of nonessential functions in moving vehicles (such as climate or navigation control and the audio system).

References

1. Hemdal, J. F. and Hughes, G. W., A feature based computer recognition program for the modeling of vowel perception, in *W Wath-enDunned Models for the Perception of Speech and Visual Form*, pp. 440–453, MIT Press, Cambridge, MA, 1964.
2. Lee, C.-H., An overview on automatic speech attribute transcription (ASAT), *Proceedings of the Interspeech*, Antwerp, Belgium, August 2007.
3. Itakura, F., Minimum prediction residual principle applied to speech recognition. *IEEE Trans. Acoust. Speech Signal Process.*, 23(1), 67–72, Feb. 1975.
4. Lesser, V. R., Fennell, R. D., Erman, L. D., and Reddy, D. R., Organization of the Hearsay-II speech understanding system. *IEEE Trans. Acoust. Speech Signal Process.*, 23(1), 11–23, 1975.
5. Lippmann, R., An introduction to computing with neural networks, *IEEE ASSP Mag.*, 4(2), 4–22, Apr. 1987.
6. Rabiner, L. R. and Levinson, S. L., Isolated and connected word recognition–theory and selected applications. *IEEE Trans. Commun.*, COM-29(5), 621–659, May 1981.
7. Rabiner, L. R., A tutorial on hidden Markov models and selected applications in speech recognition, *Proc. IEEE*, 77(2), 257–286, Feb. 1989.
8. Rabiner, L. R. and Juang, B. H, *Fundamentals of Speech Recognition*, Prentice-Hall, Englewood Cliffs, NJ, 1993.
9. Davis, S. B. and Mermelstein, P., Comparison of parametric representations for monosyllabic word recognition in continuously spoken sentences, *IEEE Trans. Acoust. Speech Signal Process.*, 28(4), 357–366, Aug. 1980.
10. Furui, S., Speaker independent isolated word recognition using dynamic features of speech spectrum, *IEEE Trans. Acoust. Speech Signal Process.*, 34(1), 52–59, Feb. 1986.
11. Baum, L. E., Petrie, T., Soules, G., and Weiss, N., A maximization technique occurring in the statistical analysis of probabilistic functions of Markov chains, *Ann. Math. Statist.*, 41(1), 164–171, 1970.
12. Juang, B. H. and Rabiner. L. R., The segmental k-means algorithm for estimating parameters of hidden Markov models, *IEEE Trans. Acoust. Speech Signal Process.*, 38(9), 1639–1641, Sept. 1990.
13. Bahl, L. R., Brown, P. F., de Souza, P. V., and Mercer, L. R., Maximum mutual information estimation of hidden Markov model parameters for speech recognition, *Proceedings of the ICASSP-86*, pp. 49–52, Tokyo, Japan, April 1986.
14. Juang, B. H., Chou, W., and Lee, C.-H., Minimum classification error rate methods for speech recognition, *IEEE Trans. Speech Audio Process.*, 5(3), 257–265, May 1997.
15. Sakoe, H. and Chiba, S., Dynamic programming optimization for spoken word recognition, *IEEE Trans. Acoust. Speech Signal Process.*, 26(1), 43–49, Feb. 1978.
16. Sakoe, H., Two-level DP matching—a dynamic programming-based pattern matching algorithm for connected word recognition, *IEEE Trans. Acoust. Speech Signal Process.*, 27(6), 588–595, Dec. 1979.
17. Myers, C. S. and Rabiner, L. R., A level building dynamic time warping algorithm for connected word recognition. *IEEE Trans. Acoust. Speech Signal Process.*, 29(3), 351–363, June 1981.
18. Bridle, J. S., Brown, M. D., and Chamberlain, R. M., An algorithm for connected word recognition, *Proceedings of the ICASSP-82*, Paris, France, pp. 899–902, May 1982.
19. Lee, C.-H., Rabiner, L. R., and Pieraccini, R., Speaker independent continuous speech recognition using continuous density hidden Markov models, in *Proceedings of the NAW-ASI, Speech Recognition and Understanding: Recent Advances Trends and Applications*. Laface, P. and DeMori, R., Eds., Springer-Verlag, pp. 135–163, Cetraro, Italy, 1992.
20. Riley, M. D., A statistical model for generating pronunciation networks, *Proceedings of the ICASSP-91*, Toronto, Ontario, Canada, vol. 2, pp. 737–740, 1991.
21. Juang, B. H., Speech recognition in adverse environments, *Comput. Speech Lang.*, 5, 275–294, 1991.

22. Hermansky, H. et al., RASTA-PLP speech analysis technique, *Proceedings of the ICASSP-92*, San Francisco, CA, pp. 121–124, 1992.
23. Mansour, D. and Juang, B. H., The short-tune modified coherence representation and noisy speech recognition, *IEEE Trans. Acoust. Speech Signal Process.*, 37(6), 795–804, June 1989.
24. Ghitza, O., Auditory nerve representation as a front-end for speech recognition in a noisy environment, *Comput. Speech Lang.*, 1(2), 109–130, Dec. 1986.
25. Itakura, F. and Umezaki, T., Distance measure for speech recognition based on the smoothed group delay spectrum, *Proceedings of the 1CASSP-87*, Dallas, TX, vol. 12, pp. 1257–1260, Apr. 1987.
26. Mansour, D. and Juang, B. H., A family of distortion measures based upon projection operation for robust speech recognition, *Proceedings of the ICASSP-88*, New York, pp. 36–39, Apr. 1988. Also in *IEEE Trans. Acoust. Speech Signal Process.*, 37(11), 1659–1671, Nov. 1989.
27. Rahim, M. G. and Juang, B. H., Signal bias removal for robust telephone speech recognition in adverse environments, *Proceedings of the ICASSP-94*, Apr. 1994.
28. Lee, C.-H., Lin, C.-H., and Juang, B. H., A study on speaker adaptation of the parameters of continuous density hidden Markov models, *IEEE Trans. Acoust. Speech Signal Process.*, 39(4), 806–814, Apr. 1991.
29. Wilpon, J. G., Rabiner, L. R., Lee, C.-H., and Goldman, E., Automatic recognition of keywords in unconstrained speech using hidden Markov models, *IEEE Trans. Acoust. Speech Signal Process.*, 38(11), 1870–1878, Nov. 1990.
30. Rahim, M., Lee, C.-H., and Juang, B. H., Robust utterance verification for connected digit recognition, *Proceedings of the ICASSP-95*, WAO2.02, Detroit, MI, May 1995.
31. Jelinek, F., The development of an experimental discrete dictation recognizer, *IEEE Proc.*, 73(11), 1616–1624, Nov. 1985.
32. Baker, J. M., Large vocabulary speech recognition prototype, *Proceedings of the DARPA Speech and Natural Language Workshop*, Hidden Valley, PA, pp. 414–415, June 1990.

10

Speaker Verification

Sadaoki Furui
Tokyo Institute of Technology

Aaron E. Rosenberg
Rutgers University

10.1 Introduction

Speaker recognition is the process of automatically extracting personal identity information by analysis of spoken utterances. In this section, speaker recognition is taken to be a general process, whereas speaker identification and speaker verification refer to specific tasks or decision modes associated with this process. Speaker identification refers to the task of determining who is speaking and speaker verification is the task of validating a speaker's claimed identity.

Many applications have been considered for automatic speaker recognition. These include secure access control by voice, customizing services or information to individuals by voice, indexing or labeling speakers in recorded conversations or dialogues, surveillance, and criminal and forensic investigations involving recorded voice samples. Currently, the most frequently mentioned application is access control. Access control applications include voice dialing, banking transactions over a telephone network, telephone shopping, database access services, information and reservation services, voice mail, and remote access to computers. Speaker recognition technology, as such, is expected to create new services and make our daily lives more convenient. Another potentially important application of speaker recognition technology is its use for forensic purposes [31].

For access control and other important applications, speaker recognition operates in a speaker verification task decision mode. For this reason, the section is entitled speaker verification. However, the term speaker recognition is used frequently in this section when referring to general processes.

This section is not intended to be a comprehensive review of speaker recognition technology. Rather, it is intended to give an overview of the recent advances and the problems that must be solved in the future. The reader is referred to papers by Doddington [6], Furui [13,14,16,17], O' Shaughnessy [49], Rosenberg and Soong [59], and Bimbot et al. [3] for more general reviews.

10.2 Personal Identity Characteristics

A universal human faculty is the ability to distinguish one person from another by personal identity characteristics. The most prominent of these characteristics are facial and vocal features. Organized, scientific efforts to make use of personal identifying characteristics for security and forensic purposes began about 100 years ago. The most successful of such efforts was fingerprint classification which has gained widespread use in forensic investigations.

Today, there is a rapidly growing technology based on biometrics, the measurement of human physiological or behavioral characteristics, for the purpose of identifying individuals or verifying the claimed or asserted identity of an individual [44]. The goal of these technological efforts is to produce completely automated systems for personal identity identification or verification that are convenient to use and offer high performance and reliability. Some of the personal identity characteristics which have received serious attention are blood typing, DNA analysis, hand shape, retinal and iris patterns, and signatures, in addition to fingerprints, facial features, and voice characteristics. In general, characteristics that are subject to the least amount of contamination or distortion and variability provide the greatest accuracy and reliability. Difficulties arise, for example, with smudged fingerprints, inconsistent signature handwriting, recording and channel distortions, and inconsistent speaking behavior for voice characteristics. Indeed, behavioral characteristics, intrinsic to signature and voice features, although potentially an important source of identifying information, are also subject to large amounts of variability from one sample to another.

The demand for effective biometric techniques for personal identity verification comes from forensic and security applications. For security applications, especially, there is a great need for techniques that are not intrusive, are convenient and efficient, and are fully automated. For these reasons, techniques such as signature verification or speaker verification are attractive even if they are subject to more sources of variability than other techniques. Speaker verification, in addition, is particularly useful for remote access, since voice characteristics are easily recorded and transmitted over telephone lines.

10.3 Voice-Related Personal Identity Characteristics

Both physiology and behavior underlie personal identity characteristics of the voice. Physiological correlates are associated with the size and configuration of the components of the vocal tract (see Figure 10.1).

For example, variations in the size of vocal tract cavities are associated with characteristic variations in the spectral distributions in the speech signal for different speech sounds. The most prominent of these spectral features is the characteristic resonances associated with voiced speech sounds known as formants [9]. Vocal cord variations are associated with the average pitch or fundamental frequency of voiced speech sounds. Variations in the velum and nasal cavities are associated with characteristic variations in the spectrum of nasalized speech sounds. Atypical anatomical variations, in the

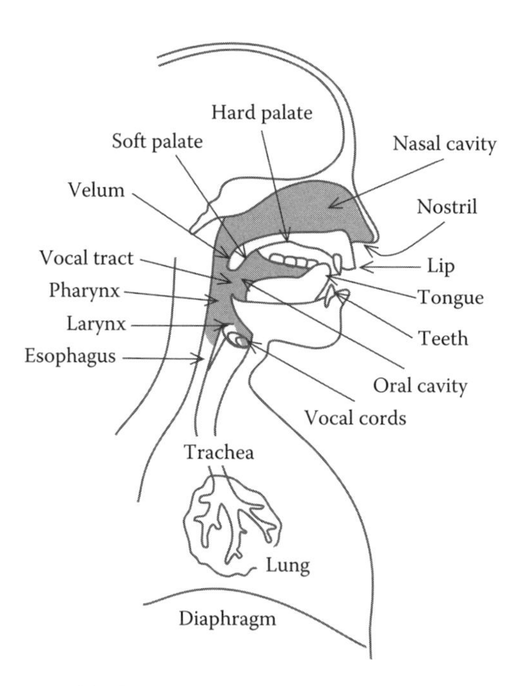

FIGURE 10.1 Schematic diagram of the human vocal mechanism. The size and shape of the articulators differ from person to person.

configuration of the teeth or the structure of the palate, are associated with atypical speech sounds such as lisps or abnormal nasality.

Behavioral correlates of speaker identity in the speech signal are more difficult to specify. "Low-level" behavioral characteristics are associated with individuality in articulating speech sounds, characteristic pitch contours, rhythm, timing, etc. Characteristics of speech that have to do with individual speech sounds, or phones, are referred to as "segmental," while those that pertain to speech phenomena over a sequence of phones are referred to as "suprasegmental." Phonetic or articulatory suprasegmental "settings" distinguishing speakers have been identified which are associated with characteristic "breathy," nasal, and other voice qualities [48]. "High-level" speaker behavioral characteristics refer to individual choice of words and phrases and other aspects of speaking styles.

10.4 Basic Elements of a Speaker Recognition System

The basic elements of a speaker recognition system are shown in Figure 10.2. An input utterance from an unknown speaker is analyzed to extract speaker characteristic features. The measured features are compared with prototype features obtained from known speaker models.

Speaker recognition systems can operate in either an identification decision mode (Figure 10.2a) or a verification decision mode (Figure 10.2b). The fundamental difference between these two modes is the number of decision alternatives.

In the identification mode, a speech sample from an unknown speaker is analyzed and compared with models of known speakers. The unknown speaker is identified as the speaker whose model best matches the input speech sample. In the "closed set" identification mode, the number of decision alternatives is equal to the size of the population. In the "open set" identification mode, a reference model for the unknown speaker may not exist. In this case, an additional alternative, "the unknown does not match any of the models," is required.

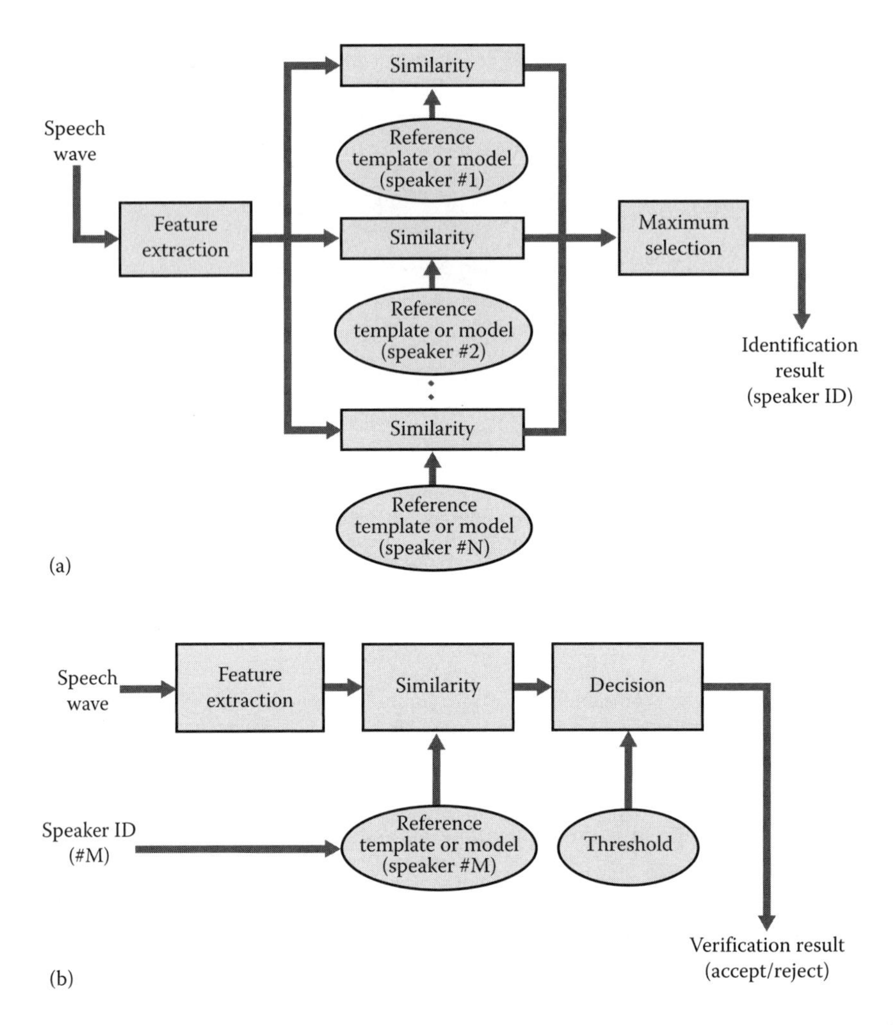

FIGURE 10.2 Basic structures of speaker recognition systems. (a) Speaker identification. (b) Speaker verification.

In the verification decision mode, an identity claim is made by, or asserted for, the unknown speaker. The unknown speaker's speech sample is compared with the model for the speaker whose identity is claimed. If the match is good enough, as indicated by passing a threshold test, the identity claim is verified. In the verification mode, there are two decision alternatives, accept or reject the identity claim, regardless of the size of the population. Verification can be considered as a special case of the "open set" identification mode in which the known population size is one.

Crucial to the operation of a speaker recognition system is the establishment and maintenance of speaker models. One or more enrollment sessions are required in which training utterances are obtained from known speakers. Features are extracted from the training utterances and compiled into models. In addition, if the system operates in the "open set" or verification decision mode, decision thresholds must also be set. Many speaker recognition systems include an updating facility in which test utterances are used to adapt speaker models and decision thresholds.

A list of terms commonly found in the speaker recognition literature can be found at the end of this chapter. In the remaining sections of the chapter, the following subjects are treated: how speaker characteristic features are extracted from speech signals, how these features are used to represent speakers, how speaker models are constructed and maintained, how speech utterances from unknown speakers are compared with speaker models and scored to make speaker recognition decisions, and how speaker verification performance is measured. The chapter concludes with a discussion of outstanding issues in speaker recognition.

10.5 Extracting Speaker Information from Speech

10.5.1 Signal-Level Features

Explicit measurements of speaker characteristics in the speech signal are often difficult to carry out. Segmenting, labeling, and measuring specific segmental speech events that characterize speakers, such as nasalized speech sounds, are difficult because of variable speech behavior and variable and distorted recording and transmission conditions. Overall qualities, such as breathiness, are difficult to correlate with specific speech signal measurements and are subject to variability in the same way as segmental speech events.

Even though voice characteristics are difficult to specify and measure explicitly, most characteristics are captured implicitly in the kinds of speech measurements that can be performed relatively easily. Such measurements as short-time and long-time spectral energy, overall energy, and fundamental frequency are relatively easy to obtain. They can often resolve differences in speaker characteristics surpassing human discriminability. Although subject to distortion and variability, features based on these analysis tools form the basis for most automatic speaker recognition systems.

The most important analysis tool is short-time spectral analysis. It is no coincidence that short-time spectral analysis also forms the basis for most speech recognition systems [52]. Short-time spectral analysis not only resolves the characteristics that differentiate one speech sound from another, but also many of the characteristics already mentioned that differentiate one speaker from another. There are two principal modes of short-time spectral analysis: filter bank analysis and linear predictive coding (LPC) analysis.

In filter bank analysis, the speech signal is passed through a bank of band-pass filters covering the available range of frequencies associated with the signal. Typically, this range is 200–3000 Hz for telephone band speech and 50–8000 Hz for wide band speech. A typical filter bank for wide band speech contains 16 band-pass filters spaced uniformly 500 Hz apart. The output of each filter is usually implemented as a windowed, short-time Fourier transform (using fast Fourier transform [FFT] techniques) at the center frequency of the filter. The speech is typically windowed using a 10–30 ms Hamming window. Instead of uniformly spacing the band-pass filters, a nonuniform spacing is often carried out reflecting perceptual criteria that allot approximately equal perceptual contributions for each such filter. Such mel-scale or bark-scale filters [52] provide a spacing linear in frequency below 1000 Hz and logarithmic above.

LPC-based spectral analysis is widely used for speech and speaker recognition. The LPC model of the speech signal specifies that a speech sample at time t, $s(t)$, can be represented as a linear sum of the p previous samples plus an excitation term, as follows:

$$s(t) = a_1 s(t-1) + a_2 s(t-2) + \cdots + a_p s(t-p) + Gu(t) \tag{10.1}$$

The LPC coefficients, a_i, are computed by solving a set of linear equations resulting from the minimization of the mean-squared error between the signal at time t and the linearly predicted estimate of the signal. Two generally used methods for solving the equations, the autocorrelation method and the covariance method, are described in Rabiner and Juang [52]. The LPC representation is computationally efficient and easily convertible to other types of spectral representations.

An important spectral representation for speech and speaker recognition is the cepstrum. The cepstrum is the (inverse) Fourier transform of the log of the signal spectrum. Thus, the log spectrum can be represented as a Fourier series expansion in terms of a set of cepstral coefficients c_n

$$\log S(\omega) = \sum_{n=-\infty}^{\infty} c_n e^{-nj\omega} \tag{10.2}$$

The cepstrum can be calculated from the filter bank spectrum or from LPC coefficients by a recursion formula [52]. In the latter case, it is known as the LPC cepstrum indicating that it is based on an all-pole representation of the speech signal. The cepstrum has many interesting properties. Since the cepstrum represents the log of the signal spectrum, signals that can be represented as the cascade of two effects which are products in the spectral domain are additive in the cepstral domain. Also, pitch harmonics, which produce prominent ripples in the spectral envelope, are associated with high-order cepstral coefficients. Thus, the set of cepstral coefficients truncated, for example, at order 12–24 can be used to reconstruct a relatively smooth version of the speech spectrum. The spectral envelope obtained is associated with vocal tract resonances and does not have the variable, oscillatory effects of the pitch excitation. It is considered that one of the reasons that cepstral representation has been found to be more effective than other representations for speech and speaker recognition is this property of separability of source and tract. Since the excitation function is considered to have speaker-dependent characteristics, it may seem contradictory that a representation which largely removes these effects works well for speaker recognition. However, in short-time spectral analysis the effects of the source spectrum are highly variable so that they are not especially effective in providing consistent representations of the source spectrum.

Other spectral features such as partial autocorrelation (PARCOR) coefficients, log area ratio coefficients, and line spectral pair (LSP) coefficients have been used for both speech and speaker recognition [52]. Generally speaking, however, the cepstral representation is most widely used and usually associated with better speaker recognition performance than other representations.

Cruder measures of spectral energy, such as waveform zero-crossing or level-crossing measurements, have also been used for speech and speaker recognition in the interest of saving computation with some success.

Additional features have been proposed for speaker recognition which are not used often or considered to be marginally useful for speech recognition. For example, pitch and energy features, particularly when measured as a function of time over a sufficiently long utterance, have been shown to be useful for speaker recognition [34]. Such time sequences or "contours" are thought to represent characteristic speaking inflections and rhythms associated with individual speaking behavior. Pitch and energy measurements have an advantage over short-time spectral measurements in that they are more robust to many different kinds of transmission and recording variations and distortions, since they are not sensitive to spectral amplitude variability. However, since speaking behavior can be highly variable due to both voluntary and involuntary activity, pitch and energy can acquire more variability than short-time spectral features and are more susceptible to imitation.

The time course of feature measurements, as represented by the so-called feature contours, provides valuable speaker characterizing information. This is because such contours provide overall, suprasegmental information characterizing speaking behavior and also because they contain information on a more local, segmental time scale describing transitions from one speech sound to another. This latter kind of information can be obtained explicitly by measuring the local trajectory in time of a measured feature at each analysis frame. Such measurements can be obtained by averaging successive differences of the feature in a window around each analysis frame, or by fitting a polynomial in time to the successive feature measurements in the window. The window size is typically five to nine analysis frames. The polynomial fit provides a less noisy estimate of the trajectory than averaging successive differences. The order of the polynomial is typically 1 or 2, and the polynomial coefficients are called delta- and delta–delta-feature coefficients. It has been shown in experiments that such dynamic feature measurements are fairly uncorrelated with the original static feature measurements and provide improved speech and speaker recognition performance [12].

10.5.2 High-Level Features

Recently, high-level features such as word idiolect, pronunciation, phone usage, etc. have been successfully used in text-independent speaker verification. Typically, high-level-feature recognition systems produce a sequence of symbols from the acoustic signal and then perform recognition using the frequency and co-occurrence of symbols. In Doddington's idiolect work [7], word unigrams and bigrams from manually transcribed conversations were used to characterize a particular speaker in a traditional target/background likelihood ratio framework. Campbell et al. [4] proposed the use of support vector machines for performing the speaker verification task based on phone and word sequences obtained using phone recognizers. The benefit of these features was demonstrated in the "NIST extended data" task for speaker verification; with enough conversational data, a recognition system can become "familiar" with a speaker and achieve excellent accuracy. The corpus was a combination of phases 2 and 3 of the switchboard-2 corpora. Each training utterance in the corpus consisted of a conversation side that was nominally of length 5 min (approximately 2.5 min of speech) recorded over a landline telephone. Speaker models were trained using 1–16 conversation sides. In order for this approach to work, utterances must be at least several minutes long, much longer than those used in conventional speaker recognition methods.

10.6 Feature Similarity Measurements

Much of the originality and distinctiveness in the design of a speaker recognition system is found in how features are combined and compared with reference models. Underlying this design is the basic representation of features in some space and the formation of a distance or distortion measurement to use when one set of features is compared with another. The distortion measure can be used to partition the feature vectors representing a speaker's utterances into regions representative of the most prominent speech sounds for that speaker, as in the vector quantization (VQ) codebook representation (Section 10.9.2). It can be used to segment utterances into speech sound units. And it can be used to score an unknown speaker's utterances against a known speaker's utterance models.

A general approach for calculating a distance between two feature vectors is to make use of a distance metric from the family of L_p norm distances d_p, such as the absolute value of the difference between the feature vectors

$$d_1 = \sum_{i=1}^{D} |f_i - f_i'| \tag{10.3}$$

or the Euclidean distance

$$d_2 = \sum_{i=1}^{D} (f_i - f_i')^2 \tag{10.4}$$

where $f_i, f_i', i = 1, 2, \ldots, D$ are the coefficients of two feature vectors f and f'. The feature vectors, for example, could comprise filter bank outputs or cepstral coefficients described in the previous section. (It is not common, however, to use filter bank outputs directly, as previously mentioned, because of the variability associated with these features due to harmonics from the pitch excitation.)

For example, a weighted Euclidean distance distortion measure for cepstral features of the form

$$d_{cw}^2 = \sum_{i=1}^{D} \left(w_i(c_i - c_i')\right)^2 \tag{10.5}$$

where

$$w_i = 1/\sigma_i \tag{10.6}$$

σ_i^2 is an estimate of the variance of the ith coefficient has been shown to provide good performance for both speech and speaker recognition

A still more general formulation is the Mahalanobis distance formulation which accounts for the interactions between the coefficients with a full covariance matrix.

An alternate approach to comparing vectors in a feature space with a distortion measurement is to establish a probabilistic formulation of the feature space. It is assumed that the feature vectors in a subspace associated with, for example, a particular speech sound for a particular speaker can be specified by some probability distribution. A common assumption is that the feature vector is a random variable x whose probability distribution is Gaussian

$$p(x|\lambda) = \frac{1}{(2\pi)^{D/2} |\sum|^{1/2}} \exp\left[-\frac{1}{2}(x-\mu)^{T} {\sum}^{-1} (x-\mu)\right] \tag{10.7}$$

where λ represents the parameters of the distribution, which are the mean vector μ and covariance matrix Σ.

When x is a feature vector sample, $p(x|\lambda)$ is referred to as the likelihood of x with respect to λ. Suppose there is a population of n speakers each modeled by a Gaussian distribution of feature vectors, λ_i, $i = 1, 2, \ldots, n$. In the maximum likelihood (ML) formulation, a sample x is associated with speaker I if

$$p(x|\lambda_I) > p(x|\lambda_i), \quad \text{for all } i \neq I \tag{10.8}$$

where $p(x|\lambda_i)$ is the likelihood of the test vector x for speaker model λ_i. It is common to use log likelihoods to evaluate Gaussian models.

From Equation 10.7

$$L(x|\lambda_i) = \log p(x|\lambda_i) = -\frac{D}{2}\log 2\pi - \frac{1}{2}\log\left|{\sum}_i\right| - \frac{1}{2}(x-\mu_i)^{T} {\sum}_i^{-1} (x-\mu_i) \tag{10.9}$$

It can be seen from Equation 10.9 that, using log likelihoods, the ML classifier is equivalent to the minimum distance classifier using a Mahalanobis distance formulation.

A more general probabilistic formulation is the Gaussian mixture distribution of a feature vector x

$$p(x|\lambda) = \sum_{i=1}^{M} w_i b_i(x) \tag{10.10}$$

where

$b_i(x)$ is the Gaussian probability density function with mean μ_i and covariance Σ_i
w_i is the weight associated with the ith component
M is the number of Gaussian components in the mixture

The weights w_i are constrained so that $\sum_{i=1}^{n} w_i = 1$. The model parameters λ are

$$\lambda = \{\mu_i, \Sigma_i, w_i, \ i = 1, 2, \ldots, M\} \tag{10.11}$$

The Gaussian mixture probability function is capable of approximating a wide variety of smooth, continuous, probability functions.

10.7 Units of Speech for Representing Speakers

An important consideration in the design of a speaker recognition system is the choice of a speech unit to model a speaker's utterances. The choice of units includes phonetic or linguistic units such as whole sentences or phrases, words, syllables, and phone-like units. It also includes acoustic units such as subword segments, segmented from utterances and labeled on the basis of acoustic rather than phonetic criteria. Some speaker recognition systems model speakers directly from single feature vectors rather than through an intermediate speech unit representation. Such systems usually operate in a text-independent mode (see Sections 10.8 and 10.9) and seek to obtain a general model of a speaker's utterances from a usually large number of training feature vectors. Direct models might include long-time averages, VQ codebooks, segment and matrix quantization codebooks, or Gaussian mixture models of the feature vectors.

Most speech recognizers of moderate to large vocabulary are based on subword units such as phones so that large numbers of utterances transcribed as sequences of phones can be represented as concatenations of phone models. For speaker recognition, there is no absolute need to represent utterances in terms of phones or other phonetically based units because there is no absolute need to account for the linguistic or phonetic content of utterances in order to build speaker recognition models. Generally speaking, systems in which phonetic representations are used are more complex than other representations because they require phonetic transcriptions for both training and testing utterances and also because they require accurate and reliable segmentations of utterances in terms of these units. The case in which phonetic representations are required for speaker recognition is the same as for speech recognition; where there is a need to represent utterances as concatenations of smaller units. Speaker recognition systems based on subword units have been described by Rosenberg et al. [57] and Matsui and Furui [41].

10.8 Input Modes

Speaker recognition systems typically operate in one of the two input modes: text-dependent or text-independent. In the text-dependent mode, speakers must provide utterances of the same text for both training and recognition trials. In the text-independent mode, speakers are not constrained to provide specific texts in recognition trials. Since the text-dependent mode can directly exploit the voice individuality associated with each phoneme or syllable, it generally achieves higher recognition performance than the text-independent mode.

10.8.1 Text-Dependent (Fixed Passwords)

The structure of a system using fixed passwords is rather simple; input speech is time aligned with reference templates or models created by using training utterances for the passwords. If the fixed passwords are different from speaker to speaker, the difference can also be used as additional individual information. This helps to increase performance.

10.8.2 Text-Independent (No Specified Passwords)

There are several applications in which predetermined passwords cannot be used. In addition, human beings can recognize speakers irrespective of the content of the utterance. Therefore, text-independent methods have recently been actively investigated. Another advantage of text-independent recognition is that it can be done sequentially, until a desired significance level is reached, without the annoyance of having to repeat passwords again and again.

10.8.3 Text-Dependent (Randomly Prompted Passwords)

Both text-dependent and text-independent methods have a potentially serious problem. Namely, these systems can be defeated because someone who plays back the recorded voice of a registered speaker uttering key words or sentences into the microphone could be accepted as the registered speaker. To cope with this problem, there are methods in which a small set of words, such as digits, are used as keywords and each user is prompted to utter a given sequence of keywords that is randomly chosen every time the system is used [27,58].

A text-prompted speaker recognition method was also proposed in which password sentences are completely changed every time [41,43]. The system accepts the input utterance only when it judges that the registered speaker has uttered the prompted sentence. Because the vocabulary is unlimited, prospective impostors cannot know in advance the sentence they will be prompted to say. This method cannot only accurately recognize speakers, but can also reject utterances whose text differs from the prompted text, even if it is uttered by a registered speaker. Thus, a recorded and played-back voice can be correctly rejected.

10.9 Representations

10.9.1 Representations That Preserve Temporal Characteristics

The most common approach to automatic speaker recognition in the text-dependent mode uses representations that preserve temporal characteristics. Each speaker is represented by a sequence of feature vectors (generally, short-term spectral feature vectors), analyzed for each test word or phase. This approach is usually based on template matching techniques in which the time axes of an input speech sample and each reference template of registered speakers are aligned, and the similarity between them accumulated from the beginning to the end of the utterance is calculated.

Trial-to-trial timing variations of utterances of the same talker, both local and overall, can be normalized by aligning the analyzed feature vector sequence of a test utterance to the template feature vector sequence using a dynamic programming (DP) time warping algorithm or DTW [12,14,52]. Since the sequence of phonetic events is the same for training and testing, there is an overall similarity among these sequences of feature vectors. Ideally the intraspeaker differences are significantly smaller than the interspeaker differences.

Another approach using representations that preserve temporal characteristics is based on the hidden Markov model (HMM) technique [52]. In this approach, a reference model for each speaker is represented by an HMM instead of directly using a time series of feature vectors. An HMM can efficiently model statistical variation in spectral features. Therefore, HMM-based methods have achieved significantly better recognition accuracies than the DTW-based methods [46,58,66].

10.9.2 Representations That Do Not Preserve Temporal Characteristics

In a text-independent system, the words or phrases used in recognition trials generally cannot be predicted. Therefore, it is impossible to model or match speech events at the level of phrases. Classical text-independent speaker recognition techniques are based on measurements for which the time dimension is collapsed. Text-independent speaker verification techniques based on short-duration speech events have also been studied. The new approaches extract and measure salient acoustic and phonetic events. The bases for these approaches lie in statistical techniques for extracting and modeling reduced sets of optimally representative feature vectors or feature vector sequences or segment. These techniques fall under the related categories of VQ, matrix and segment quantization, probabilistic mixture models, and HMM.

A set of short-term training feature vectors of a speaker can be used directly to represent the essential characteristics of that speaker. However, such a direct representation is impractical when the number of training vectors is large, since the memory and amount of computation required become prohibitively large. Therefore, efficient ways of compressing the training data have been tried using VQ techniques.

In this method, VQ codebooks consisting of a small number of representative feature vectors are used as an efficient means of characterizing speaker-specific features [32,38,56,64]. A speaker-specific codebook is generated by clustering the training feature vectors of each speaker. In the recognition stage, an input utterance is vector-quantized using the codebook of each reference speaker, and the VQ distortion accumulated over the entire input utterance is used in making the recognition decision.

In contrast with the memoryless VQ-based method, source coding algorithms with memory have also been studied using a segment (matrix) quantization technique [29]. The advantage of a segment quantization codebook over a VQ codebook representation is its characterization of the sequential nature of speech events. Higgins and Wohlford [26] proposed a segment modeling procedure for constructing a set of representative time normalized segments, which they called "filler templates." The procedure, a combination of *K*-means clustering and DP time alignment, provides a way to handle temporal variation.

On a longer time scale, temporal variation in speech signal parameters can be represented by stochastic Markovian transitions between states. Poritz [51] proposed using a five-state ergodic HMM (i.e., all the possible transitions between the states are allowed) to classify speech segments into one of the broad phonetic categories corresponding to the HMM states. A linear predictive HMM was used to characterize the output probability function. Poritz characterized the automatically obtained categories as strong voicing, silence, nasal/liquid, stop burst/post silence, and frication.

Savic and Gupta [61] also used a five-state ergodic linear predictive HMM for broad phonetic categorization. After identifying frames belonging to particular phonetic categories, feature selection was performed. In the training phase, reference templates are generated and verification thresholds are computed for each phonetic category. In the verification phase, after the phonetic categorization, a comparison with the reference template for each particular category provides a verification score for that category. The final verification score is a weighted linear combination of the scores for each category. The weights are chosen to reflect the effectiveness of particular categories of phonemes in discriminating between speakers and are adjusted to maximize the verification performance.

Gauvain et al. [19] investigated a statistical modeling approach, where each speaker was viewed as a source of phonemes, modeled by a fully connected Markov chain. Maximum *a posteriori* (MAP) estimation was used to generate speaker-specific models from a set of speaker-independent seed models. The lexical and syntactic structures of the language were approximated by local phonotactic constraints. The unknown speech is recognized by all of the speakers' models in parallel, and the hypothesized identity is that associated with the model set having the highest likelihood. Since phonemes and speakers are simultaneously recognized by using speaker-specific Markov chains, this method can be considered as an extension of the ergodic-HMM-based method.

The performances of speaker recognition based on a VQ-based method and that using discrete/continuous ergodic HMM-based methods have been compared, in particular from the viewpoint of robustness against utterance variations [40]. It was shown that a continuous ergodic HMM method is far superior to a discrete ergodic HMM method, and that a continuous ergodic HMM method is as robust as a VQ-based method when enough training data is available. However, when little data is available, the VQ-based method is more robust than a continuous HMM method. It was also shown that the information on transitions between different states is ineffective for text-independent speaker recognition; so the speaker recognition rates using a continuous ergodic HMM are strongly correlated with the total number of mixtures, irrespective of the number of states.

Rose and Reynolds [55] investigated a technique based on ML estimation of a Gaussian mixture model representation of speaker identity. This method corresponds to the single-state continuous ergodic HMM. Gaussian mixtures are noted for their robustness as a parametric model and for their ability to form smooth estimates of rather arbitrary underlying densities.

Traditionally, long-term sample statistics of various spectral features, for example, the mean and variance of spectral components averaged over a series of utterances, have been used for speaker recognition [10,35]. However, long-term spectral averages are extreme condensations of the spectral characteristics of a speaker's utterances and, as such, lack the discriminating power obtained in the sequence of short-term spectral features used as models in text-dependent systems. Moreover, recognition based on long-term spectral averages tends to be less tolerant of recording and transmitting variations since many of these variations are themselves associated with long-term spectral averages.

Studies on the use of statistical dynamic features have also been reported. Montacie et al. [45] used a multivariate autoregression (MAR) model to characterize speakers, and reported good speaker recognition results. Griffin et al. [25] studied distance measures for the MAR-based method, and reported that the identification and verification rates were almost the same as those obtained by an HMM-based method. In these experiments, the MAR model was applied to the time series of cepstral vectors. It was also reported that the optimum order of the MAR model was 2 or 3, and that distance normalization was essential to obtain good results in speaker verification.

Speaker recognition based on feed-forward neural net models has been investigated [50]. Each registered speaker has a personalized neural net that is trained to be activated only by that speaker's utterances. It is assumed that including speech from many people in the training data of each net enables direct modeling of the differences between the registered person's speech and an impostor's speech. It has been found that while the net architecture and the amount of training utterances strongly affect the recognition performance, it is comparable to the performance of the VQ approach based on personalized codebooks.

As an expansion of the VQ-based method, a connectionist approach has also been developed based on the learning vector quantization (LVQ) algorithm [2].

10.10 Optimizing Criteria for Model Construction

The establishment of effective speaker models is fundamental for good performing speaker recognition. In the previous section, we described different kinds of representations for speaker models. In this section, we describe some of the techniques for optimizing model representations.

Statistical and discriminative training techniques are based on optimizing criteria for constructing models. Typical criteria for optimizing the model parameters include likelihood maximization, *a posteriori* probability maximization, linear discriminant analysis (LDA), and discriminative error minimization.

The ML approach is widely used in statistical model parameter estimation, such as for HMM parameter training [52]. Although ML estimation has good asymptotic properties, it often requires a large amount of training data to achieve reliable results.

LDA techniques have been used in a speaker verification system reported by Netsch and Doddington [47]. A set of LDA weights applied to word-level feature vectors is found by maximizing the ratio of between-speaker to within-speaker covariances obtained from pooled customer and impostor training data.

In contrast to conventional ML training, which estimates a model based only on training utterances from the same speaker, discriminative training takes into account the models of other competing speakers and formulates the optimization criterion so that speaker separation is enhanced. In the minimum classification error/generalized probabilistic descent (MCE/GPD) method [30], the optimum solution is obtained with a steepest descent algorithm minimizing recognition error rate for the training data. Unlike the statistical framework, this method does not require estimating the probability distributions, which usually cannot be reliably obtained. However, discriminative training methods require a sufficient amount of representative reference speaker training data, which is often difficult to obtain, to be effective. This method has been applied to speaker recognition with good results [33].

Neural nets are capable of discriminative training. Various investigations have been conducted to cope with training problems, such as overtuning to training data. A typical implementation is the neural tree network (NTN) classifier [8]. In this system each speaker is represented by a VQ codebook and an NTN classifier. The NTN classifier is trained on both customer and impostor training data.

10.11 Model Training and Updating

Trial-to-trial variations have a major impact on the performance of speaker recognition systems. Variations arise from the speaker himself/herself, from differences in recording and transmission conditions, and from noise. Speakers cannot repeat an utterance precisely the same way from trial to trial. It has been found that tokens of the same utterance recorded in one session are much more highly correlated than tokens recorded in separate sessions. There are also long-term trends in voices [10,11].

There are two approaches for dealing with variability. One, discussed in this section, is to construct and update models to accommodate variability. Another, discussed in the next section, is to condition or normalize the acoustic features or the recognition scores to manage some sources of variability.

Training difficulties are closely related to training conditions. The key training conditions include the number of training sessions, the number of tokens, and the transmission channel and recording conditions. Since tokens from the same utterance recorded in one session are much more highly correlated than tokens recorded in separate sessions, wherever it is practicable, it is desirable to collect training utterances for each speaker in multiple sessions to accommodate trial-to-trial variability. For example, Gish and Schmidt [24] reported a text-independent speaker identification system in which multiple models of a speaker were constructed from multiple session training utterances.

It is inconvenient to request speakers to utter training tokens at many sessions before being allowed to use a speaker recognition system. It is possible, however, to compensate for small amounts of training data collected in a small number of enrollment sessions, often only one, by updating models with utterances collected in recognition sessions. Updating is especially important for speaker verification systems used for access control, where it can be expected that user trials will take place periodically over long periods of time in which trial-to-trial variations are likely. Updating models in this way incorporates into the models the effects of trial-to-trial variations we have mentioned. Rosenberg and Soong [56] reported significant improvements in performance in a text-independent speaker verification system based on VQ speaker models in which the VQ codebooks were updated with the test utterance data. A hazard associated with updating models using test session data is the possibility of adapting a customer model with impostor data.

10.12 Signal Feature and Score Normalization Techniques

Some sources of variability can be managed by normalization techniques applied to signal features or the scores. For example, as noted in Section 10.9.1, it is possible to adjust for trial-to-trial timing variations by aligning test utterances with model parameters using DTW or Viterbi alignment techniques.

10.12.1 Signal Feature Normalization

A typical normalization technique in the parameter domain, spectral equalization, also called "blind equalization" or "blind deconvolution," has been shown to be effective in reducing linear channel effects and long-term spectral variation [1,12]. This method is especially effective for text-dependent speaker recognition applications using sufficiently long utterances. In this method, cepstral coefficients are averaged over the duration of an entire utterance, and the averaged values are subtracted from the cepstral coefficients of each frame. This method can compensate fairly well for additive variation in the log spectral domain. However, it unavoidably removes some text-dependent and speaker-specific features, and is therefore inappropriate for short utterances in speaker recognition applications.

Gish [22] demonstrated that by simply prefiltering the speech transmitted over different telephone lines with a fixed filter; text-independent speaker recognition performance can be significantly improved. Gish et al. [20,23] have also proposed using multivariate Gaussian probability density functions to model channels statistically. This can be achieved if enough training samples of channels to be modeled are available. It was shown that time derivatives (short-time spectral dynamic features) of cepstral coefficients (delta-cepstral coefficients) are resistant to linear channel mismatch between training and testing [63].

10.12.2 HMM Adaptation for Noisy Conditions

Increasing the robustness of speaker recognition techniques against noisy speech or speech distorted by a telephone is a crucial issue in real applications. Rose et al. [54] applied the HMM composition (PMC) method [18] to speaker identification under noisy conditions. The HMM composition is a technique to combine a clean speech HMM and a background noise HMM to create a noise-added speech HMM. In order to cope with the problem of the variation of the signal-to-noise ratio (SNR), Matsui and Furui [39] proposed a method in which several noise-added HMMs with various SNRs were created and the HMM that had the highest likelihood value for the input speech was selected. A speaker decision was made using the likelihood value corresponding to the selected model. Experimental application of this method to text-independent speaker identification and verification in various kinds of noisy environments demonstrated considerable improvement in speaker recognition.

10.12.3 Likelihood and Normalized Scores

Likelihood measures (see Section 10.6) are commonly used in speaker recognition systems based on statistical models, such as HMMs, to compare test utterances with models. Since likelihood values are highly subject to intersession variability, it is essential to normalize these variations.

Higgins et al. [27] proposed a normalization method that uses a likelihood ratio. The likelihood ratio is defined as the ratio of the conditional probability of the observed measurements of the utterance given the claimed identity to the conditional probability of the observed measurements given the speaker is an impostor. A mathematical expression in terms of log likelihoods is given as

$$\log l(x) = \log p(x|S = S_c) - \log p(x|S \neq S_c) \tag{10.12}$$

Generally, a positive value of indicates a valid claim, whereas a negative value indicates an impostor. The second term of the right-hand side of Equation 10.12 is called the normalization term. Some proposals for calculating the normalization term are described.

The density at point x for all speakers other than the true speaker S can be dominated by the density for the nearest reference speaker, if we assume that the set of reference speakers is representative of all speakers. We can, therefore, arrive at the decision criterion

$$\log l(x) = \log p(x|S = S_c) - \max_{S \in \text{Ref}, S \neq S_c} \log p(x|S) \tag{10.13}$$

This shows that likelihood ratio normalization is approximately equal to optimal scoring in Bayes' sense. However, this decision criterion is unrealistic for two reasons. First, in order to choose the nearest reference speaker, conditional probabilities must be calculated for all the reference speakers, which involve a high computational cost. Second, the maximum conditional probability value is rather variable from speaker to speaker, depending on how close the nearest speaker is in the reference set.

10.12.4 Cohort or Speaker Background Models

A set of speakers, "cohort speakers," has been chosen for calculating the normalization term of Equation 10.12. Higgins et al. [27] proposed the use of speakers that are representative of the population near the claimed speaker:

$$\log l(x) = \log p(x|S = S_c) - \log \sum_{S \in \text{Cohort}, S \neq S_c} p(x|S) \tag{10.14}$$

Experimental results show that this normalization method improves speaker separability and reduces the need for speaker-dependent or text-dependent thresholding, compared with scoring using only the model of the claimed speaker. Another experiment in which the size of the cohort speaker set was varied from 1 to 5 showed that speaker verification performance increases as a function of the cohort size, and that the use of normalization significantly compensates for the degradation obtained by comparing verification utterances recorded using an electret microphone with models constructed from training utterances recorded with a carbon button microphone [60].

This method using speakers that are representative of the population near the claimed speaker is expected to increase the selectivity of the algorithm against voices similar to the claimed speaker. However, this method has a serious problem in that it is vulnerable to attack by impostors of the opposite gender. Since the cohorts generally model only same-gender speakers, the probability of opposite-gender impostor speech is not well modeled, and the likelihood ratio is based on the tails of the distributions giving rise to unreliable values. Another way of choosing the cohort speaker set is to use speakers who are typical of the general population. Reynolds [53] reported that a randomly selected, gender-balanced background speaker population outperformed a population near the claimed speaker.

Matsui and Furui [41] proposed a normalization method based on *a posteriori* probability:

$$\log l(x) = \log p(x|S = S_c) - \log \sum_{S \in \text{Ref}} p(x|S) \tag{10.15}$$

The difference between the normalization method based on the likelihood ratio and that based on *a posteriori* probability is in whether or not the claimed speaker is included in the speaker set for normalization; the cohort speaker set in the likelihood-ratio-based method does not include the claimed speaker, whereas the normalization term for the *a posteriori*-probability-based method is calculated using all the reference speakers, including the claimed speaker. Matsui and Furui approximated the summation in Equation 10.15 by the summation over a small set of speakers having relatively high likelihood values. Experimental results indicate that the two normalization methods are almost equally effective.

Carey and Parris [5] proposed a method in which the normalization term is approximated by the likelihood for a world model representing the population in general. This method has the advantage that the computational cost for calculating the normalization term is much smaller than in the original method as it does not need to sum the likelihood values for cohort speakers. Matsui and Furui [42] proposed a method based on tied-mixture HMMs in which the world model is made as a pooled mixture model representing the parameter distribution for all the registered speakers. This model is created by averaging the mixture-weighting factors of each registered speaker calculated using speaker-independent mixture distributions. Therefore, the pooled model can be easily updated when a new speaker is added as a registered speaker. In addition, this method has been shown to give much better results than either of the original normalization methods. The use of a single background model for calculating the normalization term has become the predominate approach used in speaker verification systems.

Since these normalization methods neglect the absolute deviation between the claimed speaker's model and the input speech, they cannot differentiate highly dissimilar speakers. Higgins et al. [27] reported that a multilayer network decision algorithm makes effective use of the relative and absolute scores obtained from the matching algorithm.

10.12.5 Score Normalization Techniques

A family of normalization techniques has recently been proposed, in which the scores are normalized by subtracting the mean and then dividing by standard deviation, both terms having been estimated from the (pseudo) imposter score distribution. Various methods are available for computing the imposter score distribution: Znorm, Hnorm, Tnorm, Htnorm, Cnorm and Dnorm [3]. State-of-the-art text-independent speaker verification techniques associate one or several parameterization-level normalization schemes (cepstral mean subtraction [CMS]), feature variance normalization, feature warping, etc.) with a world model normalization and one or more score normalizations.

10.13 Decision Process

10.13.1 Specifying Decision Thresholds and Measuring Performance

A "tight" decision threshold makes it difficult for impostors to be falsely accepted by the system. However, it increases the possibility of rejecting legitimate users (customers). Conversely, a "loose" threshold enables customers to be consistently accepted, while also falsely accepting impostors. To set the threshold at a desired level of customer acceptance and impostor rejection, the distribution of customer and impostor scores must be known. In practice, samples of impostor and customer scores of a reasonable size that will provide adequate estimates of distributions are not readily available. A satisfactory empirical procedure for setting the threshold is to assign a relatively loose initial threshold and then allow it to adapt by setting it to the average, or some other statistic, of recent trial scores, plus some margin that allows a reasonable rate of customer acceptance. For the first few verification trials, the threshold may be so loose that it does not adequately protect against impostor attempts. To prevent impostor acceptance during initial trials, they may be carried out as part of an extended enrollment.

10.13.2 ROC Curves

Measuring the false rejection (FR) and false acceptance (FA) rates for a given threshold condition is an incomplete description of system performance. A general description can be obtained by varying the threshold over a sufficiently large range and tabulating the resulting FR and FA rates. A tabulation of this kind can be summarized in a receiver-operating characteristic (ROC) curve, first used in psychophysics. An ROC curve, shown as the probability of correct acceptance versus the probability of incorrect (false) acceptance is shown in Figure 10.3 [14].

The figure exemplifies the curves for three systems: A, B, and C. Clearly, the performance of curve B is consistently superior to that of curve A, and C corresponds to the limiting case of purely chance

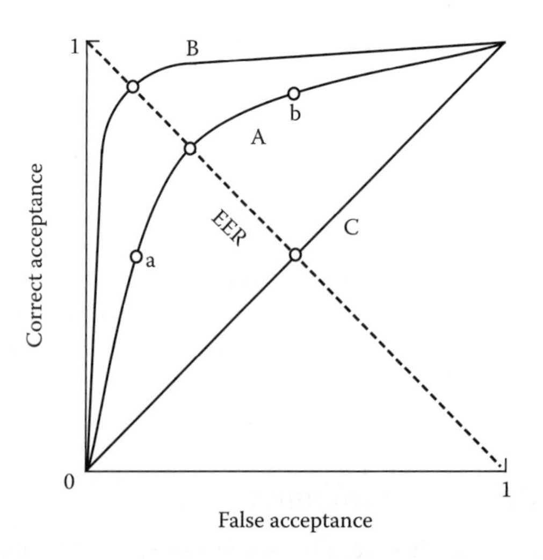

FIGURE 10.3 ROC curves; performance examples of three speaker recognition systems: A, B, and C.

performance. Position a in the figure corresponds to the case in which a strict decision criterion is employed, and position b corresponds to a case involving a lax criterion.

The point-by-point knowledge of the ROC curve provides a threshold-independent description of all possible functioning conditions of the system. For example, if a FR rate is specified, the corresponding FA rate is obtained as the intersection of the ROC curve with the horizontal straight line indicating the FR.

Equal-error rate is a commonly accepted summary of system performance. It corresponds to a threshold at which the rate of FA is equal to the rate of FR. The equal-error rate point corresponds to the intersection of the ROC curve with the straight line of 45°, indicated in the figure.

10.13.3 DET Curves

It has recently become a standard to plot the error curve on a normal deviate scale [36], in which case the curve is known as the detection error trade-offs (DETs) curve. With the normal deviate scale, a speaker verification system whose customer and impostor scores are normally distributed, regardless of variance, will result in a linear scale with a slope equal to −1. The better the system is, the closer to the origin the curve will be. In practice, the score distributions are not exactly Gaussians but are quite close to it. The DET curve representation is therefore more easily readable and allows for a comparison of the system's performances over a large range of operating conditions. Figure 10.4 shows a typical example of DET curves. EER corresponds to the intersection of the DET curve with the first bisector curve.

In NIST speaker recognition evaluations, a cost function defined as a weighted sum of the two types of errors has been chosen for use as the basic performance measure [37]. This cost, referred to as the C_{DET} cost, is defined as

$$C_{\text{DET}} = (C_{\text{FR}} \times P_{\text{FR}} \times P_{\text{C}}) + (C_{\text{FA}} \times P_{\text{FA}} \times (1 - P_{\text{C}})) \tag{10.16}$$

where P_{FR} and P_{FA} are FR and FA rates, respectively. The required parameters in this function are the cost of FR (C_{FR}), the cost of FA (C_{FA}), and the a priori probability of a customer (P_{C}).

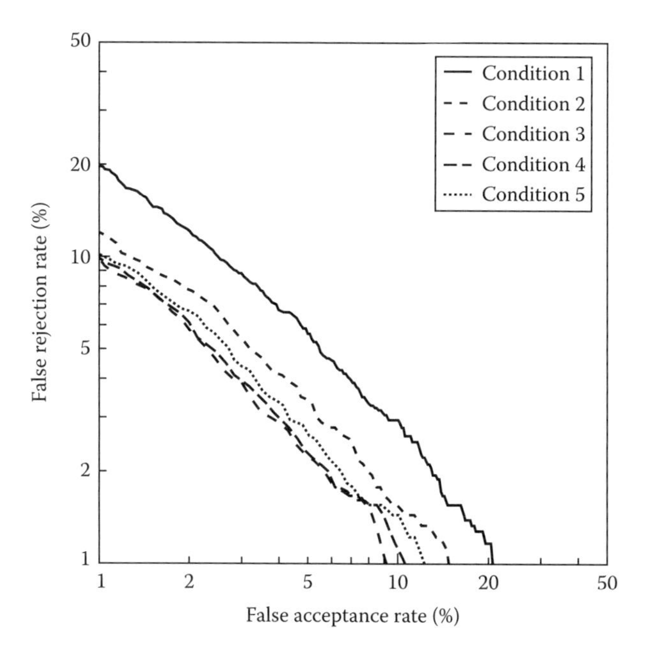

FIGURE 10.4 Examples of the DET curve.

10.13.4 Adaptive Thresholds

An issue related to model updating is the selection of a strategy for updating thresholds. A threshold updating strategy must be specified that tolerates trial-to-trial variations while, at the same time, ensures the desired level of performance.

10.13.5 Sequential Decisions (Multiattempt Trials)

In either the verification or identification mode, an additional threshold test can be applied to determine whether the match is good enough to accept the decision or whether the decision should be deferred to a new trial.

10.14 Outstanding Issues

There are many outstanding issues and problems in the area of speaker recognition. The most pressing issues, providing challenges for implementing practical and uniformly reliable systems for speaker verification, are rooted in problems associated with variability and insufficient data. As described earlier, variability is associated with trial-to-trial variations in recording and transmission conditions and speaking behavior. The most serious variations occur between enrollment sessions and subsequent test sessions resulting in models that are mismatched to test conditions. Most applications require reliable system operation under a variety of environmental and channel conditions and require that variations in speaking behavior will be tolerated. Insufficient data refers to the unavailability of sufficient amounts of data to provide representative models and accurate decision thresholds. Insufficient data is a serious and common problem because most applications require systems that operate with the smallest practicable amounts of training data recorded in the fewest number of enrollment sessions, preferably one. The challenge is to find techniques that compensate for these deficiencies. A number of techniques have been mentioned which provide partial solutions, such as cepstral subtraction techniques for channel normalization and spectral subtraction for noise removal. An especially effective technique for combating both variability and insufficient data is updating models with data extracted from test utterances. Studies have shown that model adaptation, properly implemented, can improve verification performance significantly with a small number of updates. It is difficult, however, for model adaptation to respond to large, precipitous changes. Moreover, adaptation provides for the possibility that customer models might be updated and possibly captured by impostors. Another effective tool for making speaker verification more robust is the use of likelihood-ratio scoring. An utterance recorded in conditions mismatched to the conditions of enrollment will experience degraded scores for both the customer reference model and the cohort or background model so that the ratio of these two scores remains relatively stable. Ongoing research is directed towards constructing efficient and effective background models for which likelihood-ratio scores that behave in this manner can be reliably obtained.

A desirable feature for a practical speaker verification system is reasonably uniform performance across a population of speakers. Unfortunately, it is typical to observe in a speaker verification experiment a substantial discrepancy between the best performing individuals, the "sheep," and the worst, the "goats." This additional problem in variability has been widely observed, but there are virtually no studies focusing on its origin. Speakers with no observable speech pathologies, and for whom apparently good reference models have been obtained, are often observed to be "goats." It is possible that such speakers exhibit large amounts of trial-to-trial variability, beyond the ability of the system to provide adequate compensation.

Finally, there are fundamental research issues which require additional study to promote further advances in speaker recognition technology. First, and most important, is the selection of effective features for speaker discrimination and the specification of robust, efficient acoustic measurements for representing these features. Currently, as we have described, the most effective speaker recognition features are short-time spectral features, the same features used for speech recognition. These features are mainly correlated with segmental speech phenomena and have been shown to be capable of resolving very fine spectral differences, possibly exceeding human perceptual resolving ability. Suprasegmental features, such as pitch and energy, are generally acknowledged to be less effective for speaker recognition. However, it may be that suprasegmental features are not being measured or used effectively since human listeners make effective use of such features in their speaker recognition judgments.

Perhaps the single most fundamental speaker recognition research issue is the intrinsic discriminability of speakers. A related issue is whether intrinsic discriminability should be calibrated by the ability of listeners to discriminate speakers. It is not at all clear that the intrinsic discriminability of speakers is the same order as the discriminability that can be obtained using other personal identification characteristics, such as fingerprints and facial features. Speakers' voices differ on the basis of physiological and behavioral characteristics. But it is not clear precisely which characteristics are significant, what acoustic measurements are correlated with specific features, and how close features of different speakers must be, to be acoustically and perceptually indistinguishable. Fundamental research on these questions will provide answers for developing better speaker recognition technology.

Speaker recognition techniques are related to research on improving speech recognition accuracy by speaker adaptation [15], improving synthesized speech quality by adding the natural characteristics of voice individuality, and converting synthesized voice individuality from one speaker to another. Studies on automatically extracting the individual utterances of each speaker from conversations/dialogues/meetings involving more than two people have appeared as an extension of speaker recognition technology [21,62,65]. Increasingly, speaker segmentation and clustering techniques are being used to aid adapting speech recognizers and for supplying metadata for audio indexing and searching.

Defining Terms

Acceptance: A decision outcome which involves a positive response to a speaker (or speaker class) verification task.

***A posteriori* equal error threshold:** A decision threshold which is set *a posteriori* on the test data so that the FR rate and FA rate become equal. Although this method cannot be put into actual practice, it is the most common constraint because it is a simple way to summarize the overall performance of the system into a single figure.

***A priori* threshold:** A decision threshold which is set beforehand usually based on estimates from a set of training data.

False acceptance: Erroneous acceptance of an impostor in open-set identification or speaker verification.

False rejection: Erroneous rejection of a genuine speaker in open-set speaker identification or speaker verification.

Genuine speaker: A speaker whose real identity is in accordance with the claimed identity. Alternative terms: true speaker, correct speaker.

Impostor: In the context of speaker identification, a speaker who does not belong to the set of registered speakers. In the context of speaker verification, a speaker whose real identity is different from his/her claimed identity.

Misclassification: Erroneous identity assignment to a registered speaker in speaker identification.

Registered speaker: A speaker who belongs to the list of known (registered) users for a given speaker recognition system. Alternative terms: reference speaker, customer.

Rejection: A decision outcome which involves refusal to assign a registered identity (or class) in the context of open-set speaker identification or speaker verification.

References

1. Atal, B. S., Effectiveness of linear prediction characteristics of the speech wave for automatic speaker identification and verification, *J. Acoust. Soc. Am.*, 55(6), 1304–1312, 1974.
2. Bennani, Y., Fogelman Soulie, F., and Gallinari, P., A connectionist approach for automatic speaker identification, *Proceedings of the IEEE International Conference on Acoustics, Speech, Signal Processing*, Albuquerque, NM, pp. 265–268, 1990.
3. Bimbot, F. J., Bonastre, F., Fredouille, C., Gravier, G., Magrin-Chagnolleau, I., Meignier, S., Merlin, T., Ortega-Garcia, J., Petrovska-Delacretaz D., and Reynolds, D. A., A tutorial on text-independent speaker verification, *EURASIP J. Appl. Signal Process.*, 430–451, 2004.
4. Campbell, W. M., Campbell, J. P., Reynolds, D. A., Jones, D. A., and Leek, T. R., High-level speaker verification with support vector machines, *Proceedings of the IEEE International Conference on Acoustics, Speech, Signal Processing*, Montreal, Quebec, Canada, pp. I-73–I-76, 2004.
5. Carey, M. J. and Parris, E. S., Speaker verification using connected words, *Proc. Inst. Acoust.*, 14(6), 95–100, 1992.
6. Doddington, G. R., Speaker recognition-identifying people by their voices, *Proc. IEEE*, 73(11), 1651–1664, 1985.
7. Doddington, G. R., Speaker recognition based on idiolectal differences between speakers, *Proceedings of the Eurospeech*, Aalborg, Denmark, pp. 2521–2524, 2001.
8. Farrel, K. R., Mammone, R. J., and Assaleh, K. T., Speaker recognition using neural networks and conventional classifiers, *IEEE Trans. Speech Audio Process.*, 2(1), 194–205, 1993.
9. Flanagan, J. L. *Speech Analysis, Synthesis and Perception*, Springer-Verlag, New York, 1972.
10. Furui, S., Itakura, F., and Saito, S., Talker recognition by longtime averaged speech spectrum, *Trans. IECE*, 55-A, 1(10), 549–556, 1972.
11. Furui, S., An analysis of long-term variation of feature parameters of speech and its application to talker recognition, *Trans. IECE*, 57-A, 12, 880–887, 1974.
12. Furui, S., Cepstral analysis technique for automatic speaker verification, *IEEE Trans. Acoust. Speech Signal Process.*, 29(2), 254–272, 1981.
13. Furui, S., Research on individuality features in speech waves and automatic speaker recognition techniques, *Speech Commun.*, 5(2), 183–197, 1986.
14. Furui, S., *Digital Speech Processing, Synthesis, and Recognition*, Marcel Dekker, New York, 1989.
15. Furui, S., Speaker-independent and speaker-adaptive recognition techniques, in Furui, S. and Sondhi, M.M. (Eds.), *Advances in Speech Signal Processing*, Marcel Dekker, New York, pp. 597–622, 1991.
16. Furui, S., Speaker-dependent-feature extraction, recognition and processing techniques, *Speech Commun.*, 10(5–6), 505–520, 1991.
17. Furui, S., An overview of speaker recognition technology, *ESCA Workshop on Automatic Speaker Recognition, Identification and Verification*, Martigny, Switzerland, pp. 1–9, 1994.
18. Gales, M. J. F. and Young, S. J., HMM recognition in noise using parallel model combination, *Proceedings of the Eurospeech*, Berlin, pp. II-837–II-840, 1993.
19. Gauvain, J. L., Lamel, L. F., and Prouts, B., Experiments with speaker verification over the telephone, *Proceedings of the Eurospeech*, Madrid, Spain, pp. 651–654, 1995.
20. Gish, H., Krasner, M., Russell, W., and Wolf, J., Methods and experiments for text-independent speaker recognition over telephone channels, *Proceedings of the IEEE International Conference on Acoustics, Speech, Signal Processing*, Tokyo, Japan, pp. 865–868, 1986.
21. Gish, H., Siu, M., and Rohlicek, R., Segregation of speakers for speech recognition and speaker identification, *Proceedings of the IEEE International Conference on Acoustics, Speech, Signal Processing*, Toronto, Canada, S13.11, pp. 873–876, 1991.
22. Gish, H., Robust discrimination in automatic speaker identification, *Proceedings of the IEEE International Conference on Acoustics, Speech, Signal Processing*, Albuquerque, NM, pp. 289–292, 1990.

23. Gish, H., Karnofsky, K., Krasner, K., Roucos, S., Schwartz, R., and Wolf, J., Investigation of text-independent speaker identification over telephone channels, *Proceedings of the IEEE International Conference on Acoustics, Speech, Signal Processing*, Tampa, FL, pp. 379–382, 1985.
24. Gish, H. and Schmidt, M., Text-independent speaker identification, *IEEE Signal Process. Mag.*, 11(4), 18–32, 1994.
25. Griffin, C., Matsui, T., and Furui, S., Distance measures for text-independent speaker recognition based on MAR model, *Proceedings of the IEEE International Conference on Acoustics, Speech, Signal Processing*, Adelaide, Australia, pp. I-309–I-312, 1994.
26. Higgins, A. L. and Wohlford, R. E., A new method of text-independent speaker recognition, *Proceedings of the IEEE International Conference on Acoustics, Speech, Signal Processing*, Tokyo, Japan, pp. 869–872, 1986.
27. Higgins, A. L., Bahler, L., and Porter, J., Speaker verification using randomized phrase prompting, *Dig. Signal Process.*, 1, 89–106, 1991.
28. Juang, B.-H., Rabiner, L. R., and Wilpon, J. G., On the use of bandpass filtering in speech recognition, *IEEE Trans. Acoust., Speech Signal Process.*, Dallas, TX, ASSP-35, 947–954, 1987.
29. Juang, B.-H. and Soong, F. K., Speaker recognition based on source coding approaches, *Proceedings of the IEEE International Conference on Acoustics, Speech, Signal Processing*, Albuquerque, NM, pp. 613–616, 1990.
30. Juang, B.-H. and Katagiri, S., Discriminative learning for minimum error classification, *IEEE Trans. Signal Process.*, 40, 3043–3054, 1992.
31. Kunzel, H. J., Current approaches to forensic speaker recognition, *ESCA Workshop on Automatic Speaker Recognition, Identification and Verification*, Martigny, Switzerland, pp. 135–141, 1994.
32. Li, K.-P. and Wrench Jr., E. H., An approach to text-independent speaker recognition with short utterances, *Proceedings of the IEEE International Conference on Acoustics, Speech, Signal Processing*, Paris, France, pp. 555–558, 1983.
33. Liu, C.-S., Lee, C.-H., Chou, W., Juang, B.-H., and Rosenberg, A. E., A study on minimum error discriminative training for speaker recognition, *J. Acoust. Soc. Am.*, 97(1), 637–648, 1995.
34. Lummis, R. C., Speaker verification by computer using speech intensity for temporal registration, *IEEE Trans. Audio Electroacoust.*, AU-21, 80–89, 1973.
35. Markel, J. D., Oshika, B. T., and Gray, A. H., Long-term feature averaging for speaker recognition, *IEEE Trans. Acoust. Speech Signal Process.*, ASSP-25(4), 330–337, 1977.
36. Martin, A., Doddington, G., Kamm, T., Ordowski, M., and Przybocki, M., The DET curve in assessment of detection task performance, *Proceedings of the Eurospeech*, Rhodes, Greece, 4, pp. 1895–1898, 1997.
37. Martin, A. and Przybocki, M., The NIST speaker recognition evaluations: 1996–2001, *Proceedings of the Odyssey Workshop*, Crete, Greece, 2001.
38. Matsui, T. and Furui, S., Text-independent speaker recognition using vocal tract and pitch information, *Proceedings of the International Conference on Spoken Language Processing*, Kobe, Japan, 1, pp. 137–140, 1990.
39. Matsui T. and Furui, S., Speaker recognition using HMM composition in noisy environments, *Comput. Speech Lang.*, 10, 107–116, 1996.
40. Matsui, T. and Furui, S., Comparison of text-independent speaker recognition methods using VQ-distortion and discrete/continuous HMMs, *Proceedings of the IEEE International Conference on Acoustics, Speech, Signal Processing*, San Francisco, CA, II, pp. 157–160, 1992.
41. Matsui, T. and Furui, S., Concatenated phoneme models for text-variable speaker recognition, *Proceedings of the IEEE International Conference on Acoustics, Speech, Signal Processing*, Minneapolis, MN, II, pp. 391–394, 1993.
42. Matsui, T. and Furui, S., Similarity normalization method for speaker verification based on *a posteriori* probability, *ESCA Workshop on Automatic Speaker Recognition, Identification and Verification*, Martigny, Switzerland, pp. 59–62, 1994.

43. Matsui, T. and Furui, S., Speaker adaptation of tied-mixture-based phoneme models for text-prompted speaker recognition, *Proceedings of the IEEE International Conference on Acousties, Speech, Signal Processing*, Adelaide, Australia, I, pp. 125–128, 1994.
44. Miller, B., Vital signs of identity, *IEEE Spectrum*, 31(2), 22–30, Feb. 1994.
45. Montacie, C. et al., Cinematic techniques for speech processing: temporal decomposition and multivariate linear prediction, *Proceedings of the IEEE International Conference on Acoustics, Speech, Signal Processing*, San Francisco, CA, I, pp. 153–156, 1992.
46. Naik, J. M., Netsch, L. P., and Doddington, G. R., Speaker verification over long distance telephone lines, *Proceedings of the IEEE International Conference on Acoustics, Speech, Signal Processing*, Glasgow, Scotland, pp. 524–527, 1989.
47. Netsch, L. P. and Doddington, G. R., Speaker verification using temporal decorrelation post-processing, *Proceedings of the IEEE International Conference on Acoustics, Speech, Signal Processing*, San Francisco, CA, II, pp. 181–184, 1992.
48. Nolan, F., *The Phonetic Bases of Speaker Recognition*, Cambridge University Press, Cambridge, U.K., 1983.
49. O' Shaughnessy, D., Speaker recognition, *IEEE ASSP Mag.*, 3(4), 4–17, 1986.
50. Oglesby, J. and Mason, J. S., Optimization of neural models for speaker identification, *Proceedings of the IEEE International Conference on Acoustics, Speech, Signal Processing*, Albuquerque, NM, pp. 261–264, 1990.
51. Poritz, A. B., Linear predictive hidden Markov models and the speech signal, *Proceedings of the IEEE International Conference on Acousties, Speech, Signal Processing*, Paris, France, pp. 1291–1294, 1982.
52. Rabiner, L. R. and Juang, B.-H., *Fundamentals of Speech Recognition*, Prentice-Hall, Englewood Cliffs, NJ, 1993.
53. Reynolds, D., Speaker identification and verification using Gaussian mixture speaker models, *ESCA Workshop on Automatic Speaker Recognition, Identification and Verification*, Martigny, Switzerland, pp. 27–30, 1994.
54. Rose, R., Hofstetter, E. M., and Reynolds, D. A., Integrated models of signal and background with application to speaker identification in Noise, *IEEE Trans. Speech Audio Process.*, 2(2), 245–257, 1994.
55. Rose, R. and Reynolds, R. A., Text independent speaker identification using automatic acoustic segmentation, *Proceedings of the IEEE International Conference on Acoustics, Speech, Signal Processing*, Albuquerque, NM, pp. 293–296, 1990.
56. Rosenberg, A. E. and Soong, F. K., Evaluation of a vector quantization talker recognition system in text independent and text dependent modes, *Comput. Speech Lang.*, 2, 143–157, 1987.
57. Rosenberg, A. E., Lee, C.-H., Soong, F. K., and McGee, M. A., Experiments in automatic talker verification using sub-word unit hidden Markov models, *Proceedings of the International Conference on Spoken Language Processing*, Kobe, Japan, vol. 1, pp. 141–144, 1990.
58. Rosenberg, A. E., Lee, C.-H., and Gokcen, S., Connected word talker verification using whole word hidden Markov models, *Proceedings of the IEEE International Conference on Acoustics, Speech, Signal Processing*, Toronto, Ontario, Canada, pp. 381–384, 1991.
59. Rosenberg, A. E. and Soong, F. K., Recent research in automatic speaker recognition, in *Advances in Speech Signal Processing*, Furui, S. and Sondhi, M. M., Eds., Marcel Dekker, New York, pp. 701–737, 1991.
60. Rosenberg, A. E., Delong, J., Lee, C.-H., Juang, B.-H., and Soong, F. K., The use of cohort normalized scores for speaker verification, *Proceedings of the International Conference on Spoken Language Processing*, Banff, Scotland, pp. 599–602, 1992.
61. Savic, M. and Gupta, S. K., Variable parameter speaker verification system based on hidden Markov modeling, *Proceedings of the IEEE International Conference on Acoustics, Speech, Signal Processing*, Albuquerque, NM, pp. 281–284, 1990.

62. Siu, M., Yu, G., and Gish, H., An unsupervised, sequential learning algorithm for the segmentation of speech waveforms with multiple speakers, *Proceedings of the IEEE International Conference on Acoustics, Speech, Signal Processing*, San Francisco, CA, pp. I-189–I-192, 1992.
63. Soong, F. K. and Rosenberg, A. E., On the use of instantaneous and transitional spectral information in speaker recognition, *IEEE Trans. Acoust. Speech Signal Process.*, ASSP-36(6), 871–879, 1988.
64. Soong, F. K., Rosenberg, A. E., Juang, B.-H., and Rabiner, L. R., A vector quantization approach to speaker recognition, *AT&T Tech. J.*, 66, 14–26, 1987.
65. Wilcox, L., Chen, F., Kimber, D., and Balasubramanian, V., Segmentation of speech using speaker identification, *Proceedings of the IEEE International Conference on Acoustics, Speech, Signal Processing*, I-Adelaide, Australia, pp. 161–164, 1994.
66. Zheng, Y.-C. and Yuan, B.-Z., Text-dependent speaker identification using circular hidden Markov models, *Proceedings of the IEEE International Conference on Acoustics, Speech, Signal Processing*, New York, pp. 580–582, 1988.

11

DSP Implementations of Speech Processing

Kurt Baudendistel
Momentum Data Systems

Implementations of digital speech processing algorithms in software can be distinguished from those resulting from general-purpose algorithms basically in the "type of arithmetic" and the "algorithmic constructs" used in their realization. In addition, many speech processing algorithms are realized with programmable digital signal processors (PDSPs) as the software development target—this leads to important considerations in the languages and paradigms used to realize the algorithms.

Although they are important topics in their own right, this section does not discuss the historical development of PDSPs to explain why these devices provide the architectural features that they do, and it does not provide a primer on PDSP architectures, either in general or in specific. Brief synopses of these topics are presented in the text, however, where they are appropriate.

11.1 Software Development Targets

PDSPs were developed as specialized microprocessors in the late 1970s in response to the needs of speech processing algorithms, and the vast majority of these devices have remained to this day basically audio-rate and, hence, speech processing devices [1–5]. These processors present a unique venue in which to both examine and implement speech processing algorithms since even a cursory examination of the device architectures quickly reveals the strong synergy between PDSP features and speech processing algorithms. As a result, specialized but restricted software development skills are necessary to realize speech processing algorithms on these devices.

Within the context of speech processing application realization, PDSPs which provide fixed-point data processing capabilities in hardware rather than floating-point capabilities are significantly more important. The simple reason for this is that fixed-point PDSPs are significantly less expensive than floating-point PDSPs but still provide the required computational capabilities for this class of applications. The fixed-point hardware capabilities are used to realize various types of arithmetic or data abstractions for the infinite-precision mathematical constructs used in algorithms. These abstractions include the well-known *integer arithmetic,* as well as various forms of "fixed-point arithmetic" that use fixed shifts to control the scale of values within a computation and "block floating-point arithmetic" which performs run-time scale manipulations under programmer control.

General-purpose microprocessors also present an implementation medium that is well suited to many speech processing operations, although not one so well tailored to the task as that provided by PDSPs. All of the algorithmic structures presented here can be realized via microprocessors, and in fact many software libraries have been specifically designed to allow such realization [6,7].

11.2 Software Development Paradigms

As with general-purpose algorithms, a single software development paradigm cannot be described under which speech processing algorithms are always implemented. A small set of such paradigms do exist, however, and they are distinguished by just a few salient features.

11.2.1 Imperative vs. Applicative Language

Imperative programming languages specify a program as a sequence of commands to be performed in the order given. All of the familiar high-level programming languages, such as C, C++, or FORTRAN, as well as the assembly languages of most PDSPs, are imperative.

Applicative programming languages, on the other hand, describe a program via a collection of relationships that must be maintained between variables. Applicative languages intended to be programmed directly by the user such as Silage, SIGNAL, LUCID/Lustre, and Esterel, as well as the assembly language of data-flow PDSPs, can all be used to specify speech processing algorithms in a nonimperative manner, but their use to date in real applications is quite limited. And, although usually described as a hardware-description language and used as an intermediate language generated by other tools rather than directly by programmers, from the point of view of this discussion VHDL is an applicative language that can be used to describe speech processing algorithms directly.

Graphical programming environments such as Ptolemy, GOSPL, COSSAP, and SPW also provide an applicative "language" in which speech processing algorithms can be described. However, these environments universally rely on atomic elements that are programmed with a separate paradigm, usually an imperative one.

Most speech processing applications are implemented using imperative languages, and this programming model will be used here. Note, however, that the important distinguishing features of speech processing algorithms, arithmetic and algorithmic constructs, are applicable within any programming paradigm.

11.2.2 High Level Language vs. Assembly Language

Given that an imperative programming paradigm is to be used, the choice of a high level language (HLL) or assembly language as an implementation vehicle seems very straightforward [8,9]. The common wisdom holds that (1) assembly language should be chosen where execution speed is of the essence, in realizing "signal-processing kernels," since HLL compilers cannot produce object code of the same efficiency as can be obtained with hand-coded assembly language. However, (2) a HLL should be used otherwise, in the realization of "control code," since this allows effective software development and the use of a top-down code development strategy.

In PDSP implementations, however, this sensible arrangement is often not possible. The reason for this is that use of a HLL compiler and run-time system makes untoward demands, relatively speaking, on an embedded system where resources such as registers, memory, and instruction cycles are quite scarce. In particular:

1. The settings in the control registers of the processor are often different between signal-processing and control code, and the device is more often than not "in the wrong mode"
2. The run-time memory organization demanded by a HLL, typically including a stack on which automatic variables are to be allocated but lacking memory bank control, is one that most system designers are not willing to provide
3. The standard function-call mechanism of a HLL does not fit well with the customized register usage demanded in embedded systems programming

Thus, more often than not, HLL programming is not currently utilized in PDSP systems. This will change, however, as PDSP HLL compilers become more sophisticated and as PDSP architectures become more "microprocessor-like."

11.2.3 Specialized vs. Standard High Level Languages

Specialized languages are often developed as dialects of standard high level programming languages by the authors of compilers. DSP/C, for example, is an extension of the C language that contains special vector and signal-processing operations [10]. While they appear to be quite useful for target code development for speech processing applications, the lack of general support means that these languages are not often used for either algorithm or target code development.

Extensible languages, on the other hand, allow "dialects" of standard programming languages to be created by the end-user. C++ and Ada allow the construction of specialized arithmetic support via + class + and + generic + constructs, respectively. While these languages are quite useful for algorithm development, they generally cannot produce efficient realizations of the kind desired in target code for speech processing applications.

More often than not, when standard HLLs are used, they are simply augmented by libraries of operations. The signal-processing toolbox for MATLAB® and the basic operators for the C language used in standard speech codecs are good examples of these [6,7].

11.2.4 Block vs. Single-Sample Processing

Speech coding applications lend themselves quite well to block processing, where individual time-domain signal samples are buffered into vectors or frames [11]. This is often done for algorithmic reasons, as in linear prediction coefficients (LPC) analysis, but significant performance gains can be realized by choosing this processing structure as well, when this is possible.*

Buffered data can be processed much more efficiently than single samples with typical PDSP architectures because the overhead associated with data transfer and instruction pipelining in these devices can be amortized over the entire vector rather than occurring for each sample. For example, the ubiquitous multiply-accumulate operation can be performed in a single instruction cycle by most PDSPs, but only within the instruction execution pipeline, meaning that overhead of several instruction cycles are required to set up for this level of performance. In single-sample processing, this instruction execution rate cannot be achieved.

Frames can be processed in "toto" or divided into "subframes" that are to be processed individually. This technique provides algorithmic flexibility without sacrificing the significant performance enhancement to be achieved with block processing.

* Not all algorithms can use block processing—modems and other signaling system with very low delay requirements cannot. This technique is generally useful, however, for speech processing applications.

TABLE 11.1 Static vs. Dynamic Operation

Resource	Static Operation	Dynamic Operation
Memory allocation	Global	Stack/heap
Address computation	Fixed	Stack-relative dynamic
Vector size	Fixed	Data dependent
Execution time	Fixed	Data dependent
Branch paths	Time-equivalent	Time-disparate
Wait state insertion[a]	Must be computed	Can be ignored
Data transfer	Polling possible	DMA required
FIFO buffers	Not necessary	Required
FIFO overflow	Impossible	Possible
Operating system	Not typical	Typical

[a] Wait states may be inserted by an interlocked pipeline.

11.2.5 Static vs. Dynamic Run-Time Operation

Two disparate philosophies on the operation of any real-time software system are particularly evident in speech processing implementations. Static and dynamic here indicate that run-time resource requirements, outlined in Table 11.1, can be computed and known at compile-time or only at run-time, respectively. Of course, some mix of these two philosophies can be found in any system, but the emphasis will usually be placed on one or the other.

11.2.6 Exact vs. Approximate Arithmetic

The terms exact and approximate here refer to the concern on the part of the programmer as to whether the results produced by a given arithmetic operation are fully specified by the programmer in a "bit-exact" manner, or whether the best, approximate numerical performance that can be produced by a particular processor is acceptable [12]. For example, IEEE floating-point arithmetic is exact, while machine-dependent floating-point formats can be considered approximate from the point of view of a programmer porting code to that architecture from another. The most important form of exact arithmetic for speech processing applications is that provided by the basic operators, which are used in the C language specification provided as part of modern speech coding standards [6,7]. As a general rule, the integral and fractional fixed-point arithmetic forms, discussed in Section 11.4, can be considered exact and approximate, respectively.

Approximate arithmetic is much simpler to specify than exact arithmetic, but it is harder to evaluate. In the former case, implementation details are left up to the target architecture, but if the numerical performance of a particular realization does not meet some criteria, gross changes are required in the source code. The problem here is that the criteria are not defined as part of the source code and must be supplied elsewhere. Exact arithmetic, on the other hand, requires excruciating detail in the specification of the algorithm from the outset, but no evaluation of the realization is required since this realization must adhere to the specification.

Approximate arithmetic is the form promoted by the C language where, for example, the data type + int + does not define the precision of the integer or the results of operations that overflow.* It is

* This is not to say that the C language cannot be used to realize exact arithmetic, which it often is through the machine-dependent declaration of data types such as 16 and 32, but rather that the language was not designed for use with exact arithmetic.

also the form preferred by software developers working in a native code development environment.*
Exact arithmetic, on the other hand, is preferred by developers who produce standards and who work in cross-code development environments because it eases the task of porting the algorithm from one environment to the other. Care must be exercised in this case, however, as any cross-development introduces inherent biases into an implementation that may be difficult or impractical to realize on a particular target processor [7].

It is well-known that a trade-off always exists between numerical and execution performance, as discussed in Section 11.4. It is not so well-known, however, that approximate arithmetic will always allow an equivalent or better balance to be struck in this trade-off than exact arithmetic. This is because the excruciating detail provided as part of an exact specification supplies not a minimum numerical requirement, but an exact one. In the case where a particular architecture can provide more precision than is specified, extra code must be inserted to remove that precision, resulting in less efficient execution performance. And, precisely because of this, an exact specification is in fact always targeted to a particular PDSP or microprocessor architecture—no algorithm can be specified in an exact manner and be truly portable or architecturally neutral.

11.3 Assembly Language Basics

Assembly languages for PDSPs are closely matched with the PDSP architecture for which they are designed, but they all share common elements [1–5]. In particular, multiple processing units must be programmed at the same time:

- Adder
- Multiplier
- Fixed-point logic, such as shifter(s), rounding logic, saturation logic, etc.
- Address generation unit
- Program memory, for instruction fetch or data fetch
- Data memories, perhaps multiple

In some cases, these units operate by default. For example, instruction fetches occur each machine cycle unless program memory is otherwise used. And in other cases, these units are utilized in combination. For example, (1) the DSP56000 multiply-accumulate instructions and (2) all address generation and memory fetch operations are indivisible and not pipelined. In all other cases, however, these processing units must be programmed within the "instruction execution pipeline" in which the outputs of one processing unit are connected directly to the inputs of another.

11.3.1 Coding Paradigms

Distinct coding paradigms are required by the architectures of various PDSPs, basically determined by the pipeline of that device, in order to perform this programming [13,14]. Several assembly language forms are presented by PDSPs to realize these coding paradigms:

Data stationary coding specifies ultimately the "data" that is operated on by an instruction, but not the "time" at which the operation takes place—the latter is implicit in the form of the instruction. For example, the AT&T DSP32 instruction

$$*r0++ = a0 = *r1 ++ + *r2++ \tag{11.1}$$

* *Native* indicates that code for a particular processor is developed on that processor, while *cross* indicates that the host and target processors are different.

specifies the locations in memory from which the addends should be read and to which the sum should be written, but it is implicit that the sum will be written to memory in the third instruction cycle following this one.

Because of such delays, illegal and erroneous instruction combinations can be written that cause conflicts in the use of data from both memory and registers—the former can be detected by the assembler, but the latter will simply produce data manipulations different from those intended by programmer.

Time stationary coding specifies the operations that should occur at the "time" that this instruction is executed, while the "data" to be used is whatever is present in the "pipeline registers" at this time. For example, the AT&T DSP16 instruction

$$a1 = a0 + yy = *r0 + + \tag{11.2}$$

specifies that a sum should occur at this time between the named registers and that a memory read should occur to the **y** register in parallel. No illegal or erroneous instruction combinations are possible in this case.

Interlocked coding solves the instruction combination problems of data stationary coding by automatically introducing extra machine cycles or wait states to ensure that conflicts do not occur. While this is convenient for the programmer, it does not produce more efficient execution than pure data stationary coding—on the contrary, it encourages programmers to be less savvy about their product.

Data flow coding is appropriate for machines that realize an applicative paradigm directly, such as the Hughes DFSP or the NEC μ PD7281.

It must be pointed out that a mixture of these coding paradigms is often used in real PDSPs for control of different processing units. For example, the AT&T DSP16, while ostensibly a time-stationary device, utilizes a form of interlocking to allow multiple accesses to the same memory bank in a single instruction cycle [1].

11.3.2 Assembly Languages Forms

Within the four coding paradigms presented above, several assembly language forms can be utilized. First, either an infix form as given in Equation 11.1 or the traditional assembly language prefix form using instruction mnemonics, as shown in Equation 11.3 for the Motorola DSP56000, can be used:

$$\text{clr} \quad \text{a} \tag{11.3}$$

Second, the instruction may consist of a single field, as in Equation 11.1 or Equation 11.3, or it may contain multiple fields to be executed in parallel, as in Equation 11.2 or Equation 11.4:

$$\text{mac} \quad \text{x0, y0, } a \quad \text{x}:(\text{r0})+\text{, x0} \quad \text{y}:(\text{r4})+\text{, y0} \tag{11.4}$$

Note, however, that even within the multiple fields more than one operation is specified—in both Equations 11.2 and 11.4 address register updates are specified along with the memory move. Pure "horizontal microcode," in which a dedicated field in each instruction word controls a particular processing unit, is used in only a few modern PDSP architectures, but the multiple-field instructions are similar.

Additionally, all PDSPs contain "mode registers" which control operation of particular elements of the device. For example, the + auc + register of the AT&T DSP16 controls the multiplier-shift, and thus the type of arithmetic realized by this processor's $p = x * y$ instruction. Such mode registers, while prevalent and powerful in extending the effective instruction encoding space of a PDSP, are quite difficult to manage in large programming systems, especially in the design of function libraries.

11.4 Arithmetic

The most fundamental problem encountered during the implementation of speech processing algorithms is that the algorithm must be realized (1) using the finite-precision arithmetic capabilities of real processors rather than the infinite-precision available in mathematic formulae (2) under typically severe cost constraints in terms of the processing capabilities of the target system [15–17]. Any arbitrary level of arithmetic performance can be achieved by any processor, but the cost of this performance in terms of machine cycles can be prohibitive, and so an engineering trade-off is required.

Finite-precision arithmetic effects can be broadly classified as "representational" and "operational errors":

- The bit pattern used to represent a finite-precision value can be of many forms, but all restrict the range of values over which a representation can be provided as well as the precision or number of bits used for the representation of a given value. No forms of arithmetic allow values outside the range to be represented, but some invoke an exception handler when such is requested. This is not appropriate in most speech processing systems, however, and in this case a finite-precision representation must be provided to approximate this value.

The difference between an infinite-precision value and its finite-precision representation is the representational error, and there are two sources of such error: "truncation error" results from finite precision and "overflow error" results from range violations.

- Finite-precision operators used to transform values can also introduce error. In the case of simple arithmetic operators, this is equivalent to representational error, but it is often useful to conceptualize more complicated operators, such as an finite impulse response (FIR) or infinite impulse response (IIR) filter, and to characterize the error introduced by that entity.

The engineering trade-off thus becomes an exercise in balancing the numerical performance of a realization of an algorithm in terms of truncation error and overflow error under the considerations introduced by possibly wide variance in input signal strengths or dynamic range, against implementation cost constraints in terms of target processor choice and available machine cycles on that processor. Because of the importance of this trade-off, it is important to examine different types of arithmetic and to evaluate the numerical performance and implementation cost of each type.

For example, floating-point arithmetic produces adequate numerical performance for most speech processing applications. However, the cost of floating-point processors is often prohibitive in dollar terms, and the cost of realizing floating-point arithmetic on a less expensive, fixed-point processor is prohibitive in terms of machine cycles. For this reason, some other type of arithmetic is often a better choice even though it may be numerically inferior and much harder to implement.

Regardless of the type of arithmetic chosen, however, it will be used in speech processing applications as a proxy or abstraction for the real-valued, infinite-precision arithmetic of mathematics. An important aspect that must be considered in evaluating finite-precision arithmetic types, then, is the effectiveness of the abstraction they provide for real-valued arithmetic. For example, all arithmetic needed for speech processing applications can be provided by integers, but determining what bit pattern to use to represent π or how to add two values of different scales can be quite difficult with this data abstraction.

11.4.1 Arithmetic Errors as Noise

Considering that most numerical values used in a speech processing algorithm are "signals," in that they take on distinct values at distinct sample points, the difference between a finite-precision realization and the infinite-precision mathematical model on which it is based can be considered an error or noise signal that is injected into an algorithm at the point at which that arithmetic is used, as illustrated in Figure 11.1.

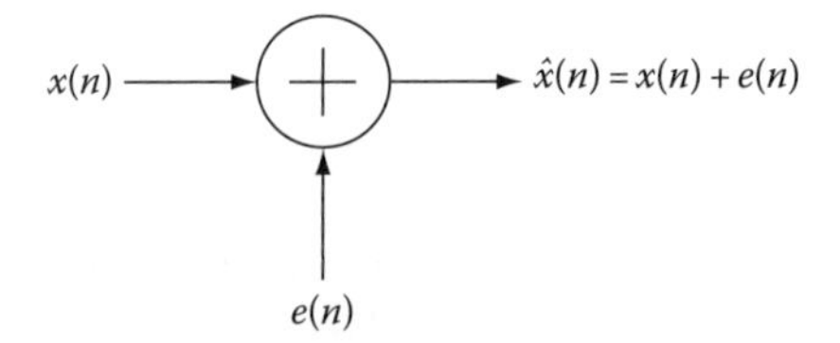

$x(n)$: infinite-precision signal
$e(n)$: error signal
$\hat{x}(n)$: finite-precision signal

FIGURE 11.1 Noise model of arithmetic error.

Given this model for the error as simply a noise source, finite-precision arithmetic effects can be analyzed in a manner similar to that used for other noise sources in a signal-processing system.

An important corollary to this fact is that speech processing algorithms should be, and typically are, designed to be robust in the presence of arithmetic noise, just as they are designed to be robust in the presence of other noise sources.

The model for the noise that is injected at each point is a function of the type of arithmetic used in that operation, however. This noise model is an important element in understanding the motivation for using various types of arithmetic, and is presented where appropriate in the sections that follow.

11.4.2 Floating Point

A floating-point number consists of a sign bit, a mantissa, and an exponent, and it presents a well-known model for realizing an approximation to real-valued arithmetic, where the value of the number V is given by

$$V = M \cdot \beta^{E} \tag{11.5}$$

with β the radix of the representation, usually 2, and M and E the effective values of the signed mantissa and the exponent, respectively. A wide variety of floating-point formats exist, especially for PDSPs, of which the IEEE 754 Floating-Point Standard is the most widely utilized for general-purpose processors. These different formats are distinguished chiefly by the precision of the exponent and the mantissa, and the behavior of the arithmetic at the limits of the representable range.

Floating-point arithmetic is usually used only in applications in which such arithmetic capabilities are provided in hardware by the processor—it is not often simulated via software by a processor that provides only fixed-point arithmetic capabilities, but rather another, similar data abstraction is used. While quite powerful and easy-to-use, floating-point arithmetic is actually of little practical value in the realization of speech processing algorithms.

11.4.3 Block Floating Point

A block floating-point representation of a vector of length N of numbers $\bar{v}$ consists of a single signed, 2's complement integer of precision B_e representing the exponent e for the block computed as*

$$e = \max_{i \in N} \lceil \log_2 (|v_i|) \rceil \tag{11.6}$$

* This is a simplified exponent definition used for purposes of illustration. The actual value used in any particular implementation will be machine dependent, but this is of no consequence expect as regards the point at which exponent overflow or underflow occurs, rare occurrences in most systems.

along with an array of N signed, 2s complement fractions of precision $B_m = b_m + 1$ representing the mantissas m_i to which the exponent can be applied to yield the represented values $\hat{v}_i$ as

$$\hat{v}_i = m_i \cdot 2^{b_m - e} \tag{11.7}$$

The precision of the exponent B_e and of the mantissas B_m are almost always chosen as the word length of the target machine, yielding a "single-precision block floating-point vector."

Arithmetic on the exponent and mantissas in a block floating-point representation are controlled separately, since significant savings in computation can often be supplied directly by the programmer. For example, if a block floating-point vector is to be computed as the result of a correlation, it is known that the zeroth lag will produce the value with the largest magnitude, and so the exponent for the vector can be immediately determined. In the absence of such direct support from the programmer, block floating-point computations require either (1) that high precision results be saved in a temporary buffer to be scaled after all values have been computed and the maximum exponent found or (2) that all results be computed twice—once to determine the exponent and a second time to compute the mantissas.

An array of length L of block floating-point vectors of length N can be constructed, yielding a construct consisting of L exponents and $L \cdot N$ mantissa values. This "segmented block floating-point representation" allows better representation of values over a wide dynamic range than is available with a single exponent. It is also quite suited to applications in which a segment of values is known to be of one scale that can be quite different from that of neighboring segments.

In the limit with $N = 1$, (segmented) block floating-point yields the "scalar (segmented) block floating-point representation" which is quite like the well-known (vector) floating-point representation, except that normalization occurs only on demand. This is an appropriate representation to use for quantities of large dynamic range in speech processing applications realized on fixed-point processors where true floating-point would be prohibitively expensive.

11.4.4 Fixed Point

A fixed-point number consists of a field of $B = b + 1$ data bits that is interpreted as a binary, 2's complement number relative to a scale factor or size that is multiplied by the field to yield a value. The two basic forms of fixed-point numbers are the "integral" and "fractional" forms, in which the "justification" of the data bits within the field determines how the value of a bit pattern is interpreted:

Justification	Field	Size	Value	Range
Right	Integer i	Stepsize Δ	$\Delta\, i$	$[-\Delta\, 2^b, \Delta\, 2^b)$
Left	Fraction f	Fieldsize ϕ	$\phi \cdot f$	$[-\phi, \phi)$

Regardless of the representation, note that the stepsize Δ and fieldsize ϕ are always related as $\phi = \Delta \cdot 2^b$ for quantities of precision $B = b + 1$.

Among other possible fixed-point representations, center-justified or mixed numbers are quite rare in speech processing applications, and all other common representations are easily derived from the integral and fractional forms.

Given the basic machine word length or precision, usually 16 or 24 bits, fixed-point PDSPs universally provide signed, single-precision multiplication producing a double-precision product, along with double-precision addition, which allows numerically efficient computation of a sum-of-products. Multiple-precision operations of greater precision, discussed below, must be simulated in software.

The additive operators (addition, subtraction, negation, and absolute value) are equivalent for any fixed-point representation, with the caveat that only numbers of the same type can be combined with the binary additive operators. That is, only numbers of the same precision, form, and size can be added

TABLE 11.2 Multiplier-Shift Determines

Processor Type	Shift[a]
Integer	0
Fractional	1
Biquadratic[b]	2
Summation[b]	$-N$

[a] This value is a relative one—the value zero could just as easily have been assigned to the fractional machine.
[b] These names derive from the use of this type of arithmetic in second-order IIR filter sections and long summations, respectively.

together directly—other combinations require conversion of one or both quantities to another, possibly a third, type before the operation can take place. Given this equivalency, it can be seen that it is the kind of multiplication, controlled by the shift that occurs at the output of the multiplier and the input to the ALU in all processors, which determines the type of arithmetic realized by a device, as shown in Table 11.2.

Fixed-point PDSPs abound with shifters—at the ALU inputs, the multiplier output, accumulator outputs, and perhaps within an independent barrel shifter. Because of a dearth of instruction encoding space, however, these are often fixed or controlled from mode registers rather than instructions or general registers, as discussed in Section 11.3.

The kind of multiplication realized by a processor also defines the kinds of data abstractions that are most useful given that machine architecture:

Q-notation is a natural extension of integer notation that is useful for right-justified arithmetic. A B-bit Q n fixed-point number is defined to have a binary point to the right of bit n, where bit 0 is the least significant bit (LSB), yielding a stepsize $\Delta = 2^B$ a range $[-2^{B-n-1}, 2^{B-n-1})$. Multiplication is defined as producing a product with a precision and Q-value that are the sums of those of the multiplicands, respectively:

$$B_{x*y} = B_x + B_y \tag{11.8}$$

$$n_{x*y} = n_x + n_y \tag{11.9}$$

When precision is increased or reduced, it is naturally done on the left of a right-justified quantity, as with an integer. This seemingly simple operation is catastrophic when Q-notation is used to model real-valued arithmetic, however, since it produces overflow. Thus, precision must not be omitted at any point when Q-notation is in use—the term "a Q n number" should always be qualified as "a B-bit, Q n number."

Scaled fractions are a natural extension of fractions that are useful for left-justified arithmetic. A $b+1-$bit fractional number of fieldsize ϕ has a range $[-\phi, \phi)$, and multiplication is defined as producing a product with a precision and fieldsize that are the sum and product of those of the multiplicands, respectively:

$$b_{x*y} = b_x + b_y \tag{11.10}$$

$$\phi_{x*y} = \phi_x \cdot \phi_y \tag{11.11}$$

With this notation, biquadratic quantities can be seen to be simply scaled-fractions of fieldsize 2.0.

Precision is much less important for scaled-fractions than for Q-values. This is because increasing or reducing the precision of a left-justified quantity naturally occurs on the right, which simply raises or lowers the accuracy of the representation. Thus, while important as regards numerical performance, precision is not required in describing a quantity as "a scaled-fractional of fieldsize ϕ."

It should be pointed out that use of a right-justified data abstraction on a left-justified machine, or vice versa, is quite difficult.

Reduction describes the common response to overflow in fixed-point additive operations, where a sum is simply allowed to "wrap around" in the 2's complement representation:

$$x + y \equiv \text{sgn}(x + y) \cdot [(|x + y| + \phi) \text{ mod } 2\phi - \phi] \tag{11.12}$$

Saturation describes an alternate response to overflow where the result is set to the maximum representable value of the appropriate sign:

$$x + y \equiv \begin{cases} \phi - \Delta & x + y \geq \phi \\ x + y & -\phi \geq x + y < \phi \\ -\phi & x + y < -\phi \end{cases} \tag{11.13}$$

The bit patterns that result from saturation are + 0x7f + $\cdots$ + f + and + 0x80 + $\cdots$ + 0 + in the cases of positive and negative overflow, respectively. Fixed-point PDSPs typically provide hardware to realize saturation because it gives a significant boost to the numerical performance of many speech processing algorithms in the presence of overflow. In most cases, when reduction arithmetic is in use, no overflow can be tolerated, even in extremely unlikely situations, while some overflow can be tolerated with saturation arithmetic in most algorithms.

General-purpose microprocessors traditionally provide only a single overflow-detection bit. Fixed-point PDSPs, on the other hand, typically provide $N > 1$ overflow bits for each register that can be the destination of an additive operation in the ALU, usually termed "accumulators." This feature allows summations of up to 2^N terms to be performed while the result can be saturated correctly if overflow does occur. The overflow bits are alternately called secondary overflow bits, guard bits, or extension words by different manufacturers.

For summations involving more than 2^N terms, it is often useful to determine if overflow occurred during the summation, even though enough information to saturate the result is not available—this capability is also required for the support of block floating-point operations. Sticky or permanent overflow bits are set when overflow occurs, but they are only cleared under programmer control, allowing such overflow detection. And, it is sometimes useful to provide such permanent overflow detection at a saturation value other than the range, as noted in Section 11.5.

Another option in the case of summations involving more than 2^N terms is to scale the inputs to the summation and then perform saturation at the end of the summation during a rescaling operation. As with all scaling operations, however, this one trades off overflow error for truncation error, and it may introduce unacceptable noise levels. For example, in the case of a summation of K i.i.d. Gaussian random variables, prescaling introduces a $3\lceil \log_2 K \rceil$ dB SNR degradation relative to an unscaled summation.

The nature of fixed-point PDSPs as single-precision multiply/double-precision add machines means that conversions between single- and double-precision quantities is quite common. Extension from single- to double-precision always takes place on the right for fixed-point quantities, except in the rare cases where integers are involved, and the extension is always with zeroes. Conversion from double- to single-precision, however, can be performed by "truncation" where the extra bits are simply removed,

$$(x \,\&\, ((-1) \ll B) \tag{11.14}$$

where B is the basic machine precision, or by rounding:

$$((x + (1 \ll B - 1)) \,\&\, ((-1) \ll B) \tag{11.15}$$

TABLE 11.3 Double-Precision Formats

Format	High Word	Low Word
Native	0_x89AB	0_xCDEF
DPF	0_x89AB	0_x66F7
DRF	0_x89AC	0_xCDEF

Fixed-point PDSPs typically provide hardware to realize rounding because it gives a significant boost to the numerical performance of many speech processing algorithms. In most applications, it can be safely assumed that the low bits of the 2s complement value that are removed as part of a conversion operation are neither deterministic nor correlated and that they represent values that are uniformly distributed over the range $[0, \Delta)$. In this case, rounding produces errors that statistically are approximately zero-mean, while truncation produces errors with mean $\mu \approx \frac{1}{2}\Delta$, and this *bias error* can be significant in many situations.

Multiple-precision operations can be simulated in software in many ways, but usually one or more of the following formats are used to represent them:

Native format represents double- and higher-precision numbers as simply the appropriate bit pattern broken into multiple machine words. High-precision additive operations can be directly realized in this format using a carry flag, but multiplication of such quantities requires unsigned multiplication capabilities, which are lacking in most PDSPs and many general-purpose processors.

Double precision format (DPF) allows double-precision fractional multiplication, with double-precision inputs and double-precision output, to be realized using signed multiplier capabilities by representing a double-precision value as the concatenation of the high-order word with the low-order word logically right-shifted by one bit.

Double round format (DRF) allows double-precision multiplication, both integral and fractional, to be realized using signed multiplier capabilities by representing a double-precision value as the concatenation of the high-order word that would result from rounding the double-precision quantity to single-precision, with the original low-order word.

These representations are illustrated in Table 11.3.

11.5 Algorithmic Constructs

The second major distinction between implementations of digital speech processing algorithms and general-purpose algorithms concerns the algorithmic constructs used in their realization, and the most important of these are discussed below.

11.5.1 Delay Lines

Delay lines, which allow the storage of sample values from one operational cycle to the next, are an important component of speech processing systems, and they can be realized in a variety of ways with PDSPs:

Registers including implicit pipeline registers, can be used to effectively realize short delays, including the one- and two-tap delays required in IIR filters.

Modulo addressing causes an address register to "wrap around" within a defined range to the start of a buffer when an attempt is made to increment that register past the end of the defined range. A delay line can be realized using modulo addressing by utilizing the location containing the expired data at a given step for the new data and by bumping the address register accordingly.

Most PDSPs do provide modulo-addressing capabilities, but often in only a limited manner. For example, strides greater than one or negative strides may not be supported, and the buffer may require a certain alignment in memory.

Writeback causes a delay line element to be written back to memory at a new location after it is read and used in a computation. While this technique is quite powerful as regards the rearrangement of data in memory, it is quite expensive in terms of memory bandwidth requirements.

These techniques are most useful for fixed delays, but equivalent methods can be used to realize variable delays.

11.5.2 Transforms

Modern PDSPs provide specialized support for transforms, and inverse transforms as well, especially the radix-2 FFT. This can include the ability to

- Compute both a sum and a difference on the same data in parallel.
- Compute addresses using "reverse-carry addition," where the carry propagates to the right rather than to the left as in ordinary addition. This allows straightforward computation of the bit-reversed addresses needed to unscramble the results of many transform calculations.
- Detect overflow in fixed-point computations at a point other than the saturation point. This can be used to predict that overflow is likely to occur at the current transform stage based on the output of the previous stage before computation of the current stage begins. With this capability, the data can be scaled as part of the current processing stage if and only if it is necessary, efficiently producing an optimally scaled transform output.

11.5.3 Vector Structure Organization

Vectors of atomic components are always laid out simply as an array of the elements. When the components are not atomic, however, as with segmented block floating-point or complex quantities, an alternative is to organize the vector as two arrays: an exponent array and a mantissa array for segmented block floating-point quantities, or a real array and an imaginary array for complex quantities.

The choice of "interleaved" or "separate" arrays, as these two techniques are known, is a trade-off between resource demands, in terms of the number of address registers needed to access a single element, vs. flexibility, in terms of the order of access and stride control that is possible.

11.5.4 Zipping

Zipping is a generic term that is used to refer to the process of performing a sequence of multiply-accumulate operations on input arrays to realize the signal-processing tasks of scaling, windowing, convolution, auto- and cross-correlation, and FIR filtering. The only real difference between these conceptually distinct tasks is (1) the choice of data or constant input arrays and (2) the order of access within these arrays.

PDSPs are designed to implement this operation, above all others, efficiently—their performance here is what distinguishes them most from general-purpose and RISC microprocessors. Regardless of the coding paradigm used,* all PDSPs allow in a single instruction cycle the following:

- Two memory accesses, either data-constant or data-data
- Two address register updates
- A single-precision multiply
- A double-precision accumulate

Programming contortions are often required to achieve this throughput in the face of processor limitations and memory access penalties, but the Holy Grail single-cycle operation is always attainable.

---

* Coding paradigms are discussed in Section 11.3.

11.5.5 Mathematical Functions

As in general-purpose programming, higher level mathematical functions can be realized within speech processing applications in one of three ways:

Bitwise computation can be used to build an exact representation one bit at a time. This technique is often used to implement single-precision division and square root functions, and some PDSPs even include special iterative instructions to accomplish these operations in a single cycle per output bit. For example, unsigned division can be realized for the Motorola 56000 as follows:

```
and  #$fe,ccr ;  Clear quotient sign bit
rep  #24      ;  Form 24 bit quotient,          (11.16)
div  x0,a     ;  ...one bit at a time.
```

Approximate computation such as Newton's method, is often used to produce a double-precision result from a single-precision estimate.

Table lookup is often used, along with linear interpolation between sample points, especially for trigonometric, logarithmic, and inverse functions. Several PDSPs even include the necessary tables in ROM.

11.5.6 Looping Constructs

Loop counting can be done with general registers, and this is required in deeply nested loops, but hardware support is often provided by PDSPs for low- and zero-overhead loops. "Low-overhead loops" utilize a special counter register to realize a branching construct similar to the well-known decrement-and-branch instruction of the Motorola 68000 microprocessor. They are "low overhead" in that the cost of the loop is typically only that of the branch instruction per iteration—separate increment (or decrement) and test instructions are not needed. "Zero-overhead loops" go one step further and eliminate even the cost of the branch instruction per iteration. They do this via special-purpose hardware to perform the program counter manipulations normally handled in the branch instruction. There is an overhead cost at the start of the loop, but the cost per iteration is truly zero.

Loop reversal is an important concept that often allows more efficient coding of speech processing constructs. In its simplest form, a loop counter is run backward to allow more efficient counting of iterations, or an address register is run backward to allow it to be reused without having to reinitialize it. In both of these cases, the reversal of the counter or address register is only possible when there are no dependencies from one loop iteration to the next. More powerful, however, is to perform memory access via a temporary register to allow loops that need to run in one direction for algorithmic reasons to be coded in the opposite direction. This technique can be used to exploit the pipelined nature of PDSPs to great effect.

References

1. AT&T Microelectronics, *DSP1610 Digital Signal Processor Information Manual*, 1992.
2. Motorola, Inc., *DSP65000 Digital Signal Processor User's Manual*, 1990.
3. Texas Instruments, Inc., *TMS320C25 User's Guide*, 1986.
4. Analog Devices, Inc., *ADSP-2100 User's Guide*, 1988.
5. NEC, Corp., *NEC μ PD7720 User's Manual*, 1984.
6. *Draft Recommendation G.723—Dual Rate Speech Coder for Multimedia Telecommunication Transmitting at 5.3 & 6.3 kbit/s*, International Telecommunication Union Telecommunications Standardization Sector (ITU) Study Group 15, 1995.
7. *Draft Recommendation G.729—Coding of Speech at 8 kbit/s using Conjugate-Structure Algebraic-Code-Excited Linear Prediction (CS-ACELP)*, ITU Study Group 15, 1995.

8. Chassaing, C., *Digital Signal Processing with C and the TMS 320C30*, John Wiley & Sons, New York, 1992.
9. Baudendistel, K., Code generation for the AT&T DSP32, in *Proceedings of the ICASSP-90*, Albuquerque, NM, pp. 1073–76, Apr. 1990.
10. Leary, K. and Waddington, W., DSP/C: A standard high level language for DSP and numeric processing, in *Proceedings of the ICASSP-90*, Albuquerque, NM, pp. 1065–1068, Apr. 1990.
11. Sridharan, S. and Dickman, G., Block floating-point implementation of digital filters using the DSP56000, *Microprocessors and Microsystems*, 12, 299–308, July/Aug. 1988.
12. Baudendistel, K., Compiler development for fixed-point processors, PhD thesis, Georgia Institute of Technology, Atlanta, GA, 1992.
13. Madisetti, V. K., *VLSI Digital Signal Processors, An Introduction to Rapid Prototyping and Design Synthesis*, Butterworth-Heinemann, Newton, MA, 1995.
14. Lee, E.A., Programmable DSP architectures: Parts I & II, *IEEE ASSP Magazine*, 5 & 6, 4–19 & 4–14, Oct. 1988 & Jan. 1989.
15. Oppenheim, A.V. and Schafer, R.W., *Digital Signal Processing*, Prentice-Hall, Englewood Cliffs, NJ, 1975.
16. Jackson, L., Roundoff-noise analysis for fixed-point digital filters realized in cascade or parallel form, *IEEE Transactions on Audio and Electroacoustics*, AU-18, 102–122, June 1970.
17. Parks, T.W. and Burrus, C.S., *Digital Filter Design*, John Wiley & Sons, New York, 1987.

12

Software Tools for Speech Research and Development

John Shore
Entropic Research Laboratory, Inc.

12.1 Introduction

Experts in every field of study depend on specialized tools. In the case of speech research and development, the dominant tools today are computer programs. In this chapter, we present an overview of key technical approaches and features that are prevalent today.

We restrict the discussion to software intended to support R&D, as opposed to software for commercial applications of speech processing. For example, we ignore DSP programming (which is discussed in chapter 11 of this book). Also, we concentrate on software intended to support the specialities of speech analysis, coding, synthesis, and recognition, since these are the main subjects of this chapter. However, much of what we have to say applies as well to the needs of those in such closely related areas as psychoacoustics, clinical voice analysis, sound and vibration, etc.

We do not attempt to survey available software packages, as the result would likely be obsolete by the time this book is printed. The examples mentioned are illustrative, and not intended to provide a thorough or balanced review. Our aim is to provide sufficient background so that readers can assess their needs and understand the differences among available tools. Up-to-date surveys are readily available online (see Section 12.13).

In general, there are three common uses of speech R&D software:

- Teaching, e.g., homework assignments for a basic course in speech processing
- Interactive, free-form exploration, e.g., designing a filter and evaluating its effects on a speech processing system
- Batch experiments, e.g., training and testing speech coders or speech recognizers using a large database

The relative importance of various features differs among these uses. For example, in conducting batch experiments, it is important that large signals can be handled, and that complicated algorithms execute efficiently. For teaching, on the other hand, these features are less important than simplicity, quick experimentation, and ease-of-use. Because of practical limitations, such differences in priority mean that no one software package today can meet all needs.

To explain the variation among current approaches, we identify a number of distinguishing characteristics. These characteristics are not independent (i.e., there is considerable overlap), but they do help to present the overall view.

For simplicity, we will refer to any particular speech R&D software as "the speech software."

12.2 Historical Highlights

Early or significant examples of speech R&D software include "Visible Speech" [5], "MITSYN" [1], and Lloyd Rice's "WAVE" program of the mid 1970s (not to be confused with David Talkin's "waves" [8]).

The first general, commercial system that achieved widespread acceptance was the Interactive Laboratory System (ILS) from Signal Technology Incorporated, which was popular in the late 1970s and early 1980s. Using the terminology defined below, ILS is compute-oriented software with an operating-system-based environment. The first popular, display-oriented, workspace-based speech software was David Shipman's LISP-machine application called "Spire" [6].

12.3 The User's Environment (OS-Based vs. Workspace-Based)

In some cases, the user sees the speech software as an extension of the computer's operating system. We call this "operating-system-based" (or OS-based); an example is the entropic signal processing system (ESPS) [7].

In other cases, the software provides its own operating environment. We call this "workspace-based" (from the term used in implementations of the programming language "APL"); an example is MATLAB® (from The Mathworks).

12.3.1 Operating-System-Based Environment

In this approach, signals are represented as files under the native operating system (e.g., Unix, DOS), and the software consists of a set of programs that can be invoked separately to process or display signals in various ways. Thus, the user sees the software as an extension of an already familiar operating system. Because signals are represented as files, the speech software inherits file manipulation capabilities from the operating system. Under Unix, for example, signals can be copied and moved respectively using the *cp* and *mv* programs, and they can be organized as directory trees in the Unix hierarchical file system (including NFS).

Similarly, the speech software inherits extension capabilities inherent in the operating system. Under Unix, for example, extensions can be created using shell scripts in various languages (*sh*, *csh*, *Tcl*, *perl*, etc.), as well as such facilities as pipes and remote execution. OS-based speech software packages are often called command-line packages because usage typically involves providing a sequence of commands to some type of shell.

12.3.2 Workspace-Based Environment

In this approach, the user interacts with a single application program that takes over from the operating system. Signals, which may or may not correspond to files, are typically represented as variables in some kind of virtual space. Various commands are available to process or display the signals. Such a workspace is often analogous to a personal blackboard.

Workspace-based systems usually offer means for saving the current workspace contents and for loading previously saved workspaces.

An extension mechanism is typically provided by a command interpreter for a simple language that includes the available operations and a means for encapsulating and invoking command sequences (e.g., in a function or procedure definition). In effect, the speech software provides its own shell to the user.

12.4 Compute-Oriented vs. Display-Oriented

This distinction concerns whether the speech software emphasizes computation or visualization or both.

12.4.1 Compute-Oriented Software

If there is a large number of signal processing operations relative to the number of signal display operations, we say that the software is compute-oriented. Such software typically can be operated without a display device and the user thinks of it primarily as a computation package that supports such functions as spectral analysis, filtering, linear prediction, quantization, analysis/synthesis, pattern classification, hidden Markov model (HMM) training, speech recognition, etc.

Compute-oriented software can be either OS-based or workspace based. Examples include "ESPS," MATLAB, and the hidden Markov model toolkit (HTK) (from Cambridge University and Entropic).

12.4.2 Display-Oriented Software

In contrast, display-oriented speech software is not intended to and often cannot operate without a display device. The primary purpose is to support visual inspection of waveforms, spectrograms, and other parametric representations. The user typically interacts with the software using a mouse or other pointing device to initiate display operations such as scrolling, zooming, enlarging, etc.

While the software may also provide computations that can be performed on displayed signals (or marked segments of displayed signals), the user thinks of the software as supporting visualization more than computation. An example is the "waves" program [8].

12.4.3 Hybrid Compute/Display-Oriented Software

Hybrid compute/display software combines the best of both. Interactions are typically by means of a display device, but computational capabilities are rich. The computational capabilities may be built-in to works pace-based speech software, or may be OS-based but accessible from the display program. Examples include the Computerized Speech Lab (CSL) from Kay Elemetrics Corp., and the combination of "ESPS and waves."

12.5 Compiled vs. Interpreted

Here we distinguish according to whether the bulk of the signal processing or display code (whether written by developers or users) is interpreted or compiled.

12.5.1 Interpreted Software

The interpreter language may be specially designed for the software (e.g., S-PLUS from Statistical Sciences, Inc., and MATLAB), or may be an existing, general purpose language (e.g., LISP is used in "N!Power" from Signal Technology, Inc.).

Compared to compiler languages, interpreter languages tend to be simpler and easier to learn. Furthermore, it is usually easier and faster to write and test programs under an interpreter. The disadvantage, relative to compiled languages, is that the resulting programs can be quite slow to run. As a result, interpreted speech software is usually better suited for teaching and interactive exploration than for batch experiments.

12.5.2 Compiled Software

Compared to interpreted languages, compiled languages (e.g., FORTRAN, C, C++) tend to be more complicated and harder to learn. Compared to interpreted programs, compiled programs are slower to write and test, but considerably faster to run. As a result, compiled speech software is usually better suited for batch experiments than for teaching.

12.5.3 Hybrid Interpreted/Compiled Software

Some interpreters make it possible to create new language commands with an underlying implementation that is compiled. This allows a hybrid approach that can combine the best of both.

Some languages provide a hybrid approach in which the source code is precompiled quickly into intermediate code that is then (usually!) interpreted. "Java" is a good example.

If compiled speech software is OS-based, signal processing scripts can typically be written in an interpretive language (e.g., a *sh* script containing a sequence of calls to "ESPS" programs). Thus, hybrid systems can also be based on compiled software.

12.5.4 Computation vs. Display

The distinction between compiled and interpreted languages is relevant mostly to the computational aspects of the speech software. However, the distinction can apply as well to display software, since some display programs are compiled (e.g., using Motif) while others exploit interpreters (e.g., *Tcl/Tk*, *Java*).

12.6 Specifying Operations among Signals

Here we are concerned with the means by which users specify what operations are to be done and on what signals. This consideration is relevant to how speech software can be extended with user-defined operations (see Section 12.7), but is an issue even in software that is not extensible.

The main distinction is between a text-based interface and a visual ("point-and-click") interface. Visual interfaces tend to be less general but easier to use.

12.6.1 Text-Based Interfaces

Traditional interfaces for specifying computations are based on a textual-representation in the form of scripts and programs. For OS-based speech software, operations are typically specified by typing the name of a command (with possible options) directly to a shell. One can also enter a sequence of such commands into a text editor when preparing a script.

This style of specifying operations also is available for workspace-based speech software that is based on a command interpreter. In this case, the text comprises legal commands and programs in the interpreter language.

Both OS-based and workspace-based speech software may also permit the specification of operations using source code in a high-level language (e.g., C) that gets compiled.

12.6.2 Visual ("Point-and-Click") Interfaces

The point-and-click approach has become the ubiquitous user-interface of the 1990s. Operations and operands (signals) are specified by using a mouse or other pointing device to interact with on-screen graphical user-interface (GUI) controls such as buttons and menus. The interface may also have a text-based component to allow the direct entry of parameter values or formulas relating signals.

12.6.2.1 Visual Interfaces for Display-Oriented Software

In display-oriented software, the signals on which operations are to be performed are visible as waveforms or other directly representative graphics.

A typical user-interaction proceeds as follows: A relevant signal is specified by a mouse-click operation (if a signal segment is involved, it is selected by a click-and-drag operation or by a pair of mouse-click operations). The operation to be performed is then specified by mouse click operations on screen buttons, pull-down menus, or pop-up menus.

This style works very well for unary operations (e.g., compute and display the spectrogram of a given signal segment), and moderately well for binary operations (e.g., add two signals). But it is awkward for operations that have more than two inputs. It is also awkward for specifying chained calculations, especially if you want to repeat the calculations for a new set of signals.

One solution to these problems is provided by a "calculator-style" interface that looks and acts like a familiar arithmetic calculator (except the operands are signal names and the operations are signal processing operations).

Another solution is the "spreadsheet-style" interface. The analogy with spreadsheets is tight. Imagine a spreadsheet in which the cells are replaced by images (waveforms, spectrograms, etc.) connected logically by formulas. For example, one cell might show a test signal, a second might show the results of filtering it, and a third might show a spectrogram of a portion of the filtered signal. This exemplifies a spreadsheet-style interface for speech software.

A spreadsheet-style interface provides some means for specifying the "formulas" that relate the various "cells." This formula interface might itself be implemented in a point-and-click fashion, or it might permit direct entry of formulas in some interpretive language. Speech software with a spreadsheet-style interface will maintain consistency among the visible signals. Thus, if one of the signals is edited or replaced, the other signal graphics change correspondingly, according to the underlying formulas.

DADisp (from DSP Development Corporation) is an example of a spreadsheet-style interface.

12.6.2.2 Visual Interfaces for Compute-Oriented Software

In a visual interface for display-oriented software, the focus is on the signals themselves. In a visual interface for compute-oriented software, on the other hand, the focus is on the operations. Operations among signals typically are represented as icons with one or more input and output lines that interconnect the operations. In effect, the representation of a signal is reduced to a straight line indicating its relationship (input or output) with respect to operations. Such visual interfaces are often called block-diagram interfaces. In effect, a block-diagram interface provides a visual representation of the computation chain. Various point-and-click means are provided to support the user in creating, examining, and modifying block diagrams.

"Ptolemy" [4] and "N!Power" are examples of systems that provide a block-diagram interface.

12.6.2.3 Limitations of Visual Interfaces

Although much in vogue, visual interfaces are inherently limited as a means for specifying signal computations.

For example, the analogy between spreadsheets and spreadsheet-style speech software continues. For simple signal computations, the spreadsheet-style interface can be very useful; computations are simple to set up and informative when operating. For complicated computations, however, the spreadsheet-style interface inherits all of the worst features of spreadsheet programming. It is difficult to encapsulate common subcalculations, and it is difficult to organize the "program" so that the computational structure is self-evident. The result is that spreadsheet-style programs are hard to write, hard to read, and error-prone.

In this respect, block-diagram interfaces do a better job since their main focus is on the underlying computation rather than on the signals themselves. Thus, screen "real-estate" is devoted to the computation rather than to the signal graphics. However, as the complexity of computations grows, the geometric and visual approach eventually becomes unwieldy. When was the last time you used a flowchart to design or document a program?

It follows that visual interfaces for specifying computations tend to be best suited for teaching and interactive exploration.

12.6.3 Parametric Control of Operations

Speech processing operations often are based on complicated algorithms with numerous parameters. Consequently, the means for specifying parameters is an important issue for speech software.

The simplest form of parametric control is provided by command-line options on command-line programs. This is convenient, but can be cumbersome if there are many parameters. A common alternative is to read parameter values from parameter files that are prepared in advance. Typically, command-line values can be used to override values in the parameter file. A third input source for parameter values is directly from the user in response to prompts issued by the program.

Some systems offer the flexibility of a hierarchy of inputs for parameter values, for example:

- Default values
- Values from a global parameter file read by all programs
- Values from a program-specific parameter file
- Values from the command line
- Values from the user in response to run-time prompts

In some situations, it is helpful if a current default value is replaced by the most recent input from a given parameter source. We refer to this property as "parameter persistence."

12.7 Extensibility (Closed vs. Open Systems)

Speech software is "closed" if there is no provision for the user to extend it. There is a fixed set of operations available to process and display signals. What you get is all you get.

OS-based systems are always extensible to a degree because they inherit scripting capabilities from the OS, which permits the creation of new commands. They may also provide programming libraries so that the user can write and compile new programs and use them as commands.

Workspace-based systems may be extensible if they are based on an interpreter whose programming language includes the concept of an encapsulated procedure. If so, then users can write scripts that define new commands. Some systems also allow the interpreter to be extended with commands that are implemented by underlying code in C or some other compiled language.

In general, for speech software to be extensible, it must be possible to specify operations (see Section 12.6) and also to reuse the resulting specifications in other contexts. A block-diagram interface is extensible, for example, if a given diagram can be reduced to an icon that is available for use as a single block in another diagram.

For speech software with visual interfaces, extensibility considerations also include the ability to specify new GUI controls (visible menus and buttons), the ability to tie arbitrary internal and external computations to GUI controls, and the ability to define new display methods for new signal types.

In general, extended commands may behave differently from the built-in commands provided with the speech software. For example, built-in commands may share a common user interface that is difficult to implement in an independent script or program (such a common interface might provide standard parameters for debug control, standard processing of parameter files, etc.).

If user-defined scripts, programs, and GUI components are indistinguishable from built-in facilities, we say that the speech software provides seamless extensibility.

12.8 Consistency Maintenance

A speech processing chain involves signals, operations, and parameter sets. An important consideration for speech software is whether or not consistency is maintained among all of these. Thus, for example, if one input signal is replaced with another, are all intermediate and output signals recalculated automatically? Consistency maintenance is primarily an issue for speech software with visual interfaces, namely whether or not the software guarantees that all aspects of the visible displays are consistent with each other.

Spreadsheet-style interfaces (for display-oriented software) and block-diagram interfaces (for compute-oriented software) usually provide consistency maintenance.

12.9 Other Characteristics of Common Approaches

12.9.1 Memory-Based vs. File-Based

"Memory-based" speech software carries out all of its processing and display operations on signals that are stored entirely within memory, regardless of whether or not the signals also have an external representation as a disk file. This approach has obvious limitations with respect to signal size, but it simplifies programming and yields fast operation. Thus, memory-based software is well-suited for teaching and the interactive exploration of small samples.

In "file-based" speech software, on the other hand, signals are represented and manipulated as disk files. The software partially buffers portions of the signal in memory as required for processing and display operations. Although programming can be more complicated, the advantage is that there are no inherent limitations on signal size. The file-based approach is, therefore, well-suited for large-scale experiments.

12.9.2 Documentation of Processing History

Modern speech processing involves complicated algorithms with many processing steps and operating parameters. As a result, it is often important to be able to reconstruct exactly how a given signal was produced. Speech software can help here by creating appropriate records as signal and parameter files are processed.

The most common method for recording this information about a given signal is to put it in the same file as the signal. Most modern speech software uses a file format that includes a "file header" that is used for this purpose. Most systems store at least some information in the header, e.g., the sampling rate of the signal. Others, such as "ESPS," attempt to store all relevant information. In this approach, the header of a signal file produced by any program includes the program name, values of processing parameters, and the names and headers of all source files. The header is a recursive structure, so that the headers of the source files themselves contain the names and headers of files that were prior sources. Thus, a signal file header contains the headers of all source files in the processing chain. It follows that files contain a complete history of the origin of the data in the file and all the intermediate processing steps. The importance of record keeping grows with the complexity of computation chains and the extent of available parametric control.

12.9.3 Personalization

There is considerable variation in the extent to which speech software can be customized to suit personal requirements and tastes. Some systems cannot be personalized at all; they start out the same way, every time. But most systems store personal preferences and use them again next time. Savable preferences may include color selections, button layout, button semantics, menu contents, currently loaded signals, visible windows, window arrangement, and default parameter sets for speech processing operations.

At the extreme, some systems can save a complete "snapshot" that permits exact resumption. This is particularly important for the interactive study of complicated signal configurations across repeated software sessions.

12.9.4 Real-Time Performance

Software is generally described as "real-time" if it is able to keep up with relevant, changing inputs. In the case of speech software, this usually means that the software can keep up with input speech.

Even this definition is not particularly meaningful unless the input speech is itself coming from a human speaker and digitized in real-time. Otherwise, the real-issue is whether or not the software is fast enough to keep up with interactive use.

For example, if one is testing speech recognition software by directly speaking into the computer, real-time performance is important. It is less important, on the other hand, if the test procedure involves running batch scripts on a database of speech files.

If the speech software is designed to take input directly from devices (or pipes, in the case of Unix), then the issue becomes one of CPU speed.

12.9.5 Source Availability

It is unfortunate but true that the best documentation for a given speech processing command is often the source code. Thus, the availability of source code may be an important factor for this reason alone. Typically, this is more important when the software is used in advanced R&D applications. Sources also are needed if users have requirements to port the speech software to additional platforms. Source availability may also be important for extensibility, since it may not be possible to extend the speech software without the sources.

If the speech software is interpreter-based, sources of interest will include the sources for any built-in operations that are implemented as interpreter scripts.

12.9.6 Hardware Requirements

Speech software may require the installation of special-purpose hardware. There are two main reasons for such requirements: to accelerate particular computations (e.g., spectrograms), and to provide speech I/O with A/D and D/A converters.

Such hardware has several disadvantages. It adds to the system cost, and it decreases the overall reliability of the system. It may also constrain system software upgrades; for example, the extra hardware may use special device drivers that do not survive OS upgrades. Special-purpose hardware used to be common, but is less so now owing to the continuing increase in CPU speeds and the prevalence of built-in audio I/O. It is still important, however, when maximum speed and high-quality audio I/O are important. "CSL" is a good example of an integrated hardware/software approach.

12.9.7 Cross-Platform Compatibility

If your hardware platform may change or your site has a variety of platforms, then it is important to consider whether the speech software is available across a variety of platforms. Source availability (Section 12.9.5) is relevant here.

If you intend to run the speech software on several platforms that have different underlying numeric representations (a byte order difference being most likely), then it is important to know whether the file formats and signal I/O software support transparent data exchange.

12.9.8 Degree of Specialization

Some speech software is intended for general purpose work in speech (e.g., "ESPS/waves," MATLAB). Other software is intended for more specialized usage. Some of the areas where specialized software tools may be relevant include linguistics, recognition, synthesis, coding, psychoacoustics, clinical-voice, music, multimedia, sound and vibration, etc. Two examples are HTK for recognition, and Delta (from Eloquent Technology) for synthesis.

12.9.9 Support for Speech Input and Output

In the past, built-in speech I/O hardware was uncommon in workstations and PCs, so speech software typically supported speech I/O by means of add-on hardware supplied with the software or available from other third parties. This provided the desired capability, albeit with the disadvantages mentioned earlier (see Section 12.9.6).

Today most workstations and PCs have built-in audio support that can be used directly by the speech software. This avoids the disadvantages of add-on hardware, but the resulting A/D-D/A quality can be too noisy or otherwise inadequate for use in speech R&D (the built-in audio is typically designed for more mundane requirements). There are various reasons why special-purpose hardware may still be needed, including

- Need for more than two channels
- Need for very high sampling rates
- Compatibility with special hardware (e.g., DAT tape)

12.10 File Formats (Data Import/Export)

Signal file formats are fundamentally important because they determine how easy it is for independent programs to read and write the files (interoperability). Furthermore, the format determines whether files can contain all of the information that a program might need to operate on the file's primary data (e.g., can the file contain the sampling frequency in addition to a waveform itself?).

The best way to design speech file formats is hotly debated, but the clear trend has been towards "self-describing" file formats that include information about the names, data types, and layout of all data in the file. (For example, this permits programs to retrieve data by name.)

There are many popular file formats, and various programs are available for converting among them (e.g., SOX). For speech sampled data, the most important file format is Sphere (from NIST), which is used in the speech databases available from the Linguistic Data Consortium (LDC). Sphere supports several data compression formats in a variety of standard and specialized formats.

Sphere works well for sampled data files, but is limited for more general speech data files. A general purpose, public-domain format ("Esignal") has recently been made available by Entropic.

12.11 Speech Databases

Numerous databases (or corpora) of speech are available from various sources. For a current list, see the comp. speech frequently asked questions (FAQ) (see Section 12.13). The largest supplier of speech data is the LDC, which publishes a large number of CDs containing speech and linguistic data.

12.12 Summary of Characteristics and Uses

In Section 12.1, we mentioned that the three most common uses for speech software are teaching, interactive exploration, and batch experiments. And at various points during the discussion of speech software characteristics, we mentioned their relative importance for the different classes of software uses. We attempt to summarize this in Table 12.1, where the symbol "•" indicates that a characteristic is particularly useful or important.

It is important not to take Table 12.1 too seriously. As we mentioned at the outset, the various distinguishing characteristics discussed in this section are not independent (i.e., there is considerable overlap). Furthermore, the three classes of software use are broad and not always easily distinguishable; i.e., the importance of particular software characteristics depends a lot on the details of intended use. Nevertheless, Table 12.1 is a reasonable starting point for evaluating particular software in the context of intended use.

TABLE 12.1 Relative Importancce of Software Charcteristics

	Teaching	Interactive Exploration	Batch Experiments
OS-based (12.3.1)			•
Workspace-based (12.3.2)	•		
Compute-oriented (12.4.1)			•
Display-oriented (12.4.2)	•	•	
Compiled (12.5.2)			•
Interpreted (12.5.1)		•	
Text-base interface (12.6.1)		•	•
Visual interface (12.6.2)	•	•	
Memory-based (12.9.1)	•	•	
File-based (12.9.1)			•
Parametric control (12.6.3)	•	•	•
Consistency maintenance (12.8.0)	•	•	
Historic documentation (12.9.2)		•	•
Extensibility (12.7.0)		•	•
Personalization (12.9.3)		•	•
Real-time performance (12.9.4)	•	•	
Source availability (12.9.5)		•	•
Cross-platform compatibility (12.9.7)	•	•	•
Support for speech I/O (12.9.9)	•	•	

12.13 Sources for Finding Out What Is Currently Available

The best single online source of general information is the Internet news group comp. speech, and in particular its FAQ (see +http://svr-www.eng.cam.ac.uk/comp.speech/+). Use this as a starting point.

Here are some other WWW sites that (at this writing) contain speech software information or pointers to other sites:

- +http://svr-www.eng.cam.ac.uk+
- +http://mambo.ucsc.edu/psl/speech.html+
- +http://www.bdti.com/faq/dspfaq.html+
- +http://www.ldc.upenn.edu+
- +http://www.entropic.com+

12.14 Future Trends

From the user's viewpoint, speech software will continue to become easier to use, with a heavier reliance on visual interfaces with consistency maintenance.

Calculator, spreadsheet, and block-diagram interfaces will become more common, but will not eliminate text-based (programming, scripting) interfaces for specifying computations and system extensions.

Software will become more open. Seamless extensibility will be more common, and extensions will be easier. GUI extensions as well as computation extensions will be supported.

There will be less of a distinction between compute-oriented and display-oriented speech software. Hybrid compute/display-oriented software will dominate.

Visualization will become more important and more sophisticated, particularly for multidimensional data. "Movies" will be used to show arbitrary 3D data. Sound will be used to represent an arbitrary dimension. Various methods will be available to project N-dimensional data into 2- or 3-space. (This will be used, for example, to show aspects of vector quantization or HMM clustering.)

Public-domain file formats will dominate proprietary formats. Networked computers will be used for parallel computation if available. *Tcl/Tk and Java* will grow in popularity as a base for graphical data displays and user interfaces.

References

1. Henke, W.L., Speech and audio computer-aided examination and analysis facility, Quarterly Progress Rep. No. 95, MIT Research Laboratory for Electronics, 1969, pp. 69–73.
2. Henke, W.L., MITSYN—An interactive dialogue language for time signal processing, MIT Research Laboratory for Electronics, report RLE TM-1, 1975.
3. Kopec, G., The integrated signal processing system ISP, *IEEE Transactions on Acoustics, Speech, and Signal Processing*, ASSP-32(4), 842–851, Aug. 1984.
4. Pino, J.L., Ha, S., Lee, E.A., and Buck, J.T., Software synthesis for DSP using ptolemy, *Journal on VLSI Signal Processing*, 9(1), 7–21, Jan. 1995.
5. Potter, R.K., Kopp, G.A., and Green, H.C., *Visible Speech*, D. Van Nostrand Company, New York, 1946.
6. Shipman, D., SpireX: Statistical analysis in the SPIRE acoustic-phonetic workstation, *Proceedings of the ICASSP*, Boston, MA, 1983.
7. Shore, J., Interactive signal processing with UNIX, *Speech Technology*, 4(2), 70–79, Mar./Apr. 1988.
8. Talkin, D., Looking at speech, *Speech Technology*, 4, 4, Apr./May 1989.

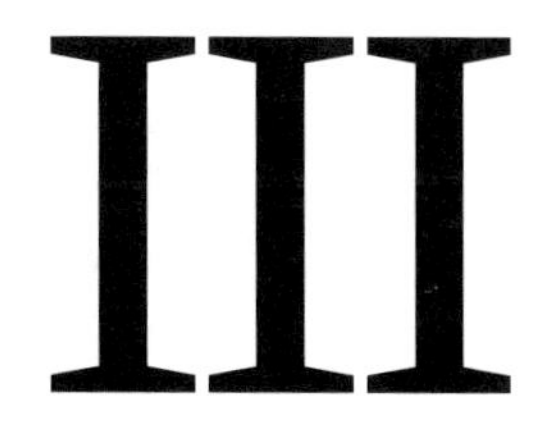

Image and Video Processing

Jan Biemond
Delft University of Technology

Russell M. Mersereau
Georgia Institute of Technology

IMAGE AND VIDEO-SIGNAL PROCESSING are quite different from other forms of signal processing for a variety of reasons. The most obvious difference lies in the fact that these signals are two or three dimensional. This means that some familiar techniques used for processing one-dimensional signals, for example, those that require factorization of polynomials, have to be abandoned. Other techniques for filtering, sampling, and transform computation have to be modified. Even more compromises have to be made, however, because of the signals' size. Images and sequences of images can be huge. For example, processing sequences of color images each of which contains 780 rows and 1024 columns at a frame rate of 30 fps requires a data rate of 72 MBps. Successful image processing techniques reward careful attention to problem requirements, algorithmic complexity, and machine architecture. The past decade has been particularly exciting as each new wave of faster computing hardware has opened the door to new applications. This is a trend that will likely continue for some time.

The following chapters, written by experts in their fields, highlight the state of the art in several aspects of image and video processing. The range of topics is quite broad. While it includes some discussions of techniques that go back more than a decade, the emphasis is on current practice. There is some danger in this, because the field is changing very rapidly, but, on the other hand, many of the concepts on which these current techniques are based should be around for some time.

Chapter 13 is a very long and thorough discussion of image processing fundamentals. For a novice to the field, this material is important for a complete understanding. It discusses the basics of how images differ from other types of signals and how the limitations of cameras, displays, and the human visual system affect the kinds of processing that can be done. It also defines the basic theory of multidimensional digital signal processing, particularly with respect to how linear and nonlinear filtering, transform computation, and sampling are generalized from the one-dimensional case. Other topics treated include statistical models for images, models for recording distortions, histogram-based methods for image processing, and image segmentation.

Probably the most visible image processing occurs in the development of standards for image and video compression. JPEG, MPEG, and digital television are all highly visible success stories. Chapter 14 looks at methods for still image compression including JPEG, wavelet, and fractal coders. Image compression is successful because image samples are spatially correlated with their neighbors. Operators such as the discrete cosine transform (DCT) largely remove this correlation and capture the essence of an image block in a few parameters that can be quantized and transmitted. The transform domain also enables these coders to exploit limitations in the human visual system. Chapters 20 and 21 extend these approaches to video and television compression, respectively. Video compression achieves significant additional compression gains by exploiting the temporal redundancy that is present in video sequences. This is done by using simple models for modeling object motion within a scene, using these models to predict the current frame, and then encoding only the model parameters and the quantized prediction errors.

Images are often distorted when they are recorded. This might be caused by out-of-focus optics, motion blur, camera noise, or coding errors. Chapter 15 looks at methods for image and video restoration. This is the most mathematically based area of image processing, and it is also one of the areas with the longest history. It has applications in the analysis of astronomical images, in forensic imaging, and in the production of high-quality stills from video sequences.

Chapter 16 looks at methods for motion estimation and video scan conversion. Motion estimation is a key technique for removing temporal redundancy in image sequences and, as a result, it is a key component in all of video compression standards. It is also, however, a highly time-consuming numerically ill-posed operation. As a result, it continues to be highly studied, particularly with respect to more sophisticated motion models. A related problem is the problem of scanning format conversion. This is a major issue in television systems where both interlaced and progressively scanned images are encountered.

Chapter 22 explores stereoscopic and multiview image processing. Traditional image processing assumes that only one camera is present. As a result, depth information in a three-dimensional scene is lost. When explicit depth information is needed, multiple cameras can be used. Differences in the displacement of objects in the left and right images can be converted to depth measurements. Mammals do this naturally with their two eyes. Stereoscopic image-processing techniques are becoming increasingly used in problems of computer vision and computer graphics. This chapter discusses the state of the art in this emerging area.

The final two chapters in this part, Chapters 23 and 24, look at software and hardware systems for doing image processing. Chapter 23 provides an overview of a representative set of image software packages that embody the core capabilities required by many image-processing applications. It also provides a list of Internet addresses for a number of image databases. Chapter 24 provides an overview of VLSI architectures for implementing many of the video compression standards.

13

Fundamentals of Image Processing

Ian T. Young
Delft University of Technology

Jan J. Gerbrands
Delft University of Technology

Lucas J. van Vliet
Delft University of Technology

13.1 Introduction

Modern digital technology has made it possible to manipulate multidimensional signals with systems that range from simple digital circuits to advanced parallel computers. The goal of this manipulation can be divided into three categories:

- Image processing—*image in* → *image out*
- Image analysis—*image in* → *measurements out*
- Image understanding—*image in* → *high-level description out*

We will focus on the fundamental concepts of "image processing." Space does not permit us to make more than a few introductory remarks about "image analysis." "Image understanding" requires an approach that differs fundamentally from the theme of this book. Further, we will restrict ourselves to two-dimensional (2D) image processing although most of the concepts and techniques that are to be described can be extended easily to three or more dimensions. Readers interested in either greater detail than presented here or in other aspects of image processing are referred to [1–10].

We begin with certain basic definitions. An image defined in the "real world" is considered to be a function of two real variables, for example, $a(x,y)$ with a as the amplitude (e.g., brightness) of the image at the real coordinate position (x,y). An image may be considered to contain sub-images sometimes referred to as regions-of-interest (ROIs), or simply "regions." This concept reflects the fact that images frequently contain collections of objects each of which can be the basis for a region. In a sophisticated image processing system it should be possible to apply specific image processing operations to selected regions. Thus one part of an image (region) might be processed to suppress motion blur while another part might be processed to improve color rendition.

The amplitudes of a given image will almost always be either real numbers or integers. The latter is usually a result of a quantization process that converts a continuous range (say, between 0% and 100%) to a discrete number of levels. In certain image-forming processes, however, the signal may involve photon counting which implies that the amplitude would be inherently quantized. In other image forming procedures, such as magnetic resonance imaging, the direct physical measurement yields a complex number in the form of a real magnitude and a real phase. For the remainder of this book we will consider amplitudes as real numbers or integers unless otherwise indicated.

13.2 Digital Image Definitions

A digital image $a[m,n]$ described in a 2D discrete space is derived from an analog image $a(x,y)$ in a 2D continuous space through a sampling process that is frequently referred to as digitization. The mathematics of that sampling process will be described in Section 13.5. For now we will look at some basic definitions associated with the digital image. The effect of digitization is shown in Figure 13.1.

The 2D continuous image $a(x,y)$ is divided into N rows and M columns. The intersection of a row and a column is termed a "pixel." The value assigned to the integer coordinates $[m,n]$ with $\{m=0,1,2,\ldots,M-1\}$ and $\{n=0,1,2,\ldots,N-1\}$ is $a[m, n]$. In fact, in most cases $a(x,y)$—which we might consider to be the physical signal that impinges on the face of a 2D sensor—is actually a function

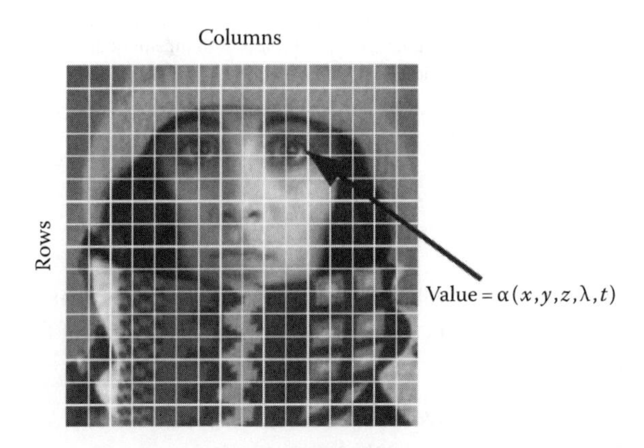

FIGURE 13.1 Digitization of a continuous image. The pixel at coordinates $[m=10, n=3]$ has the integer brightness value of 110.

of many variables including depth (z), color (λ), and time (t). Unless otherwise stated, we will consider the case of 2D, monochromatic, static images in this chapter.

The image shown in Figure 13.1 has been divided into $N = 16$ rows and $M = 16$ columns. The value assigned to every pixel is the average brightness in the pixel rounded to the nearest integer value. The process of representing the amplitude of the 2D signal at a given coordinate as an integer value with L different gray levels is usually referred to as amplitude quantization or simply quantization.

13.2.1 Common Values

There are standard values for the various parameters encountered in digital image processing. These values can be caused by video standards, by algorithmic requirements, or by the desire to keep digital circuitry simple. Table 13.1 gives some commonly encountered values.

Quite frequently we see cases of $M = N = 2^K$ where $\{K = 8, 9, 10, 11, 12\}$. This can be motivated by digital circuitry or by the use of certain algorithms such as the (fast) Fourier transform (see Section 13.3.3).

The number of distinct gray levels is usually a power of 2, that is, $L = 2^B$ where B is the number of bits in the binary representation of the brightness levels. When $B > 1$ we speak of a "gray-level image"; when $B = 1$ we speak of a "binary image." In a binary image there are just two gray levels which can be referred to, for example, as "black" and "white" or "0" and "1."

13.2.2 Characteristics of Image Operations

Image operations can be classified and characterized in a variety of ways. The reason for doing so is to understand what type of results we might expect to achieve with a given type of operation or what might be the computational burden associated with a given operation.

13.2.2.1 Types of Operations

The types of operations that can be applied to digital images to transform an input image $a[m, n]$ into an output image $b[m, n]$ (or another representation) can be classified into three categories as shown in Table 13.2.

This is shown graphically in Figure 13.2.

TABLE 13.1 Common Values of Digital Image Parameters

Parameter	Symbol	Typical Values
Rows	N	256, 512, 525, 625, 1,024, 1,080
Columns	M	256, 512, 768, 1,024, 1,920
Gray levels	L	2,64, 256, 1,024, 4,096, 16,384

TABLE 13.2 Types of Image Operations. Image Size $= N \times N$; Neighborhood Size $= P \times P$

Operation	Characterization	Generic Complexity/Pixel
Point	The output value at a specific coordinate is dependent only on the input value at that same coordinate.	Constant
Local	The output value at a specific coordinate is dependent on the input values in the neighborhood of that same coordinate.	P^2
Global	The output value at a specific coordinate is dependent on all the values in the input image.	N^2

Note: Note that the complexity is specified in operations per pixel.

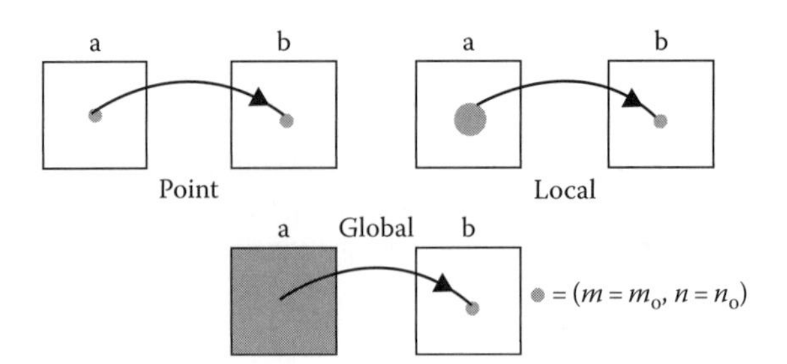

FIGURE 13.2 Illustration of various types of image operations.

13.2.2.2 Types of Neighborhoods

Neighborhood operations play a key role in modern digital image processing. It is therefore important to understand how images can be sampled and how that relates to the various neighborhoods that can be used to process an image.

- Rectangular sampling—In most cases, images are sampled by laying a rectangular grid over an image as illustrated in Figure 13.1. This results in the type of sampling shown in Figure 13.3a and b.
- Hexagonal sampling—An alternative sampling scheme is shown in Figure 13.3c and is termed hexagonal sampling.

Both sampling schemes have been studied extensively [1] and both represent a possible periodic tiling of the continuous image space. We will restrict our attention, however, to only rectangular sampling as it remains, due to hardware and software considerations, the method of choice.

Local operations produce an output pixel value $b[m = m_o, n = n_o]$ based upon the pixel values in the neighborhood of $a[m = m_o, n = n_o]$. Some of the most common neighborhoods are the 4-connected neighborhood and the 8-connected neighborhood in the case of rectangular sampling and the 6-connected neighborhood in the case of hexagonal sampling illustrated in Figure 13.3.

13.2.3 Video Parameters

We do not propose to describe the processing of dynamically changing images in this introduction. It is appropriate—given that many static images are derived from video cameras and frame grabbers—to mention the standards that are associated with the three standard video schemes that are currently in worldwide use—NTSC, PAL, and SECAM. This information is summarized in Table 13.3.

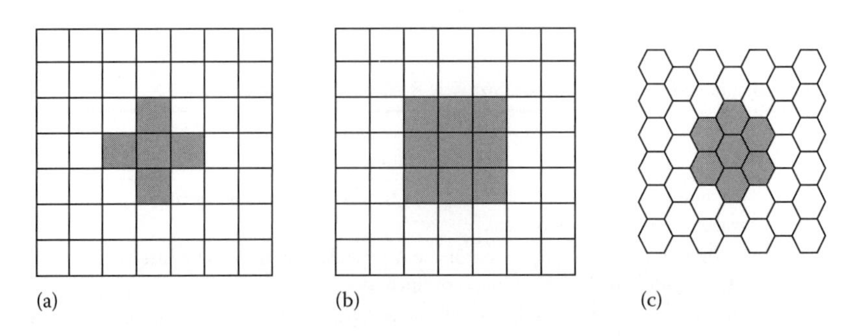

FIGURE 13.3 (a) Rectangular sampling 4-connected. (b) Rectangular sampling 8-connected. (c) Hexagonal sampling 6-connected.

TABLE 13.3 Standard Video Parameters

Standard Property	NTSC	PAL	SECAM
Images/s	29.97	25	25
ms/image	33.37	40.0	40.0
Lines/image	525	625	625
(horiz./vert.) = aspect ratio	4:3	4:3	4:3
Interlace	2:1	2:1	2:1
μs/line	63.56	64.00	64.00

In an interlaced image the odd numbered lines $(1, 3, 5, \ldots)$ are scanned in half of the allotted time (e.g., 20 ms in PAL) and the even numbered lines $(2, 4, 6, \ldots)$ are scanned in the remaining half. The image display must be coordinated with this scanning format (see Section 13.8.2). The reason for interlacing the scan lines of a video image is to reduce the perception of flicker in a displayed image. If one is planning to use images that have been scanned from an interlaced video source, it is important to know if the two half-images have been appropriately "shuffled" by the digitization hardware or if that should be implemented in software. Further, the analysis of moving objects requires special care with interlaced video to avoid "zigzag" edges.

The number of rows (N) from a video source generally corresponds one-to-one with lines in the video image. The number of columns, however, depends on the nature of the electronics that is used to digitize the image. Different frame grabbers for the same video camera might produce $M = 384$, 512, or 768 columns (pixels) per line.

13.3 Tools

Certain tools are central to the processing of digital images. These include mathematical tools such as convolution, Fourier analysis, and statistical descriptions, and manipulative tools such as chain codes and run codes. We will present these tools without any specific motivation. The motivation will follow in later sections.

13.3.1 Convolution

Several possible notations can be used to indicate the convolution of two (multidimensional) signals to produce an output signal. The most common are

$$c = a \otimes b = a * b \tag{13.1}$$

We shall use the first form, $c = a \otimes b$, with the following formal definitions.

In 2D continuous space:

$$c(x, y) = a(x, y) \otimes b(x, y) = \int_{-\infty}^{+\infty}\int_{-\infty}^{+\infty} a(\chi, \zeta) b(x - \chi, y - \zeta) d\chi d\zeta \tag{13.2}$$

In 2D discrete space:

$$c[m, n] = a[m, n] \otimes b[m, n] = \sum_{j=-\infty}^{+\infty} \sum_{k=-\infty}^{+\infty} a[j, k] b[m - j, n - k] \tag{13.3}$$

13.3.2 Properties of Convolution

A number of important mathematical properties are associated with convolution.

- Convolution is commutative.

$$c = a \otimes b = b \otimes a \tag{13.4}$$

- Convolution is associative.

$$c = a \otimes (b \otimes d) = (a \otimes b) \otimes d = a \otimes b \otimes d \tag{13.5}$$

- Convolution is distributive.

$$c = a \otimes (b + d) = (a \otimes b) + (a \otimes d) \tag{13.6}$$

where a, b, c, and d are all images, either continuous or discrete.

13.3.3 Fourier Transforms

The Fourier transform produces another representation of a signal, specifically a representation as a weighted sum of complex exponentials. Because of Euler's formula:

$$e^{jq} = \cos(q) + j\sin(q) \tag{13.7}$$

where $j^2 = -1$, we can say that the Fourier transform produces a representation of a (2D) signal as a weighted sum of sines and cosines. The defining formulas for the forward Fourier and the inverse Fourier transforms are as follows. Given an image a and its Fourier transform A, then the forward transform goes from the spatial domain (either continuous or discrete) to the frequency domain which is always continuous.

$$\text{Forward}-A = \mathcal{F}\{a\} \tag{13.8}$$

The inverse Fourier transform goes from the frequency domain back to the spatial domain.

$$\text{Inverse}-a = \mathcal{F}^{-1}\{A\} \tag{13.9}$$

The Fourier transform is a unique and invertible operation so that

$$a = \mathcal{F}^{-1}\{\mathcal{F}\{a\}\} \quad \text{and} \quad A = \mathcal{F}\{\mathcal{F}^{-1}\{A\}\} \tag{13.10}$$

The specific formulas for transforming back and forth between the spatial domain and the frequency domain are given below.

In 2D continuous space:

$$\text{Forward}-A(u, v) = \int_{-\infty}^{+\infty}\int_{-\infty}^{+\infty} a(x, y)e^{-j(ux+vy)}dxdy \tag{13.11}$$

$$\text{Inverse}-a(x, y) = \frac{1}{4\pi^2}\int_{-\infty}^{+\infty}\int_{-\infty}^{+\infty} A(u, v)e^{+j(ux+vy)}dudv \tag{13.12}$$

In 2D discrete space:

$$\text{Forward}-A(\Omega,\Psi)=\sum_{m=-\infty}^{+\infty}\sum_{n=-\infty}^{+\infty}a[m,n]e^{-j(\Omega m+\Psi n)} \tag{13.13}$$

$$\text{Inverse}-a[m,n]=\frac{1}{4\pi^2}\int_{-\pi}^{+\pi}\int_{-\pi}^{+\pi}A(\Omega,\Psi)e^{+j(\Omega m+\Psi n)}\mathrm{d}\Omega\mathrm{d}\Psi \tag{13.14}$$

13.3.4 Properties of Fourier Transforms

There are a variety of properties associated with the Fourier transform and the inverse Fourier transform. The following are some of the most relevant for digital image processing.

- The Fourier transform is, in general, a complex function of the real frequency variables. As such the transform can be written in terms of its magnitude and phase.

$$A(u,v)=|A(u,v)|e^{j\varphi(u,v)} \quad A(\Omega,\Psi)=|A(\Omega,\Psi)|e^{j\varphi(\Omega,\Psi)} \tag{13.15}$$

- A 2D signal can also be complex and thus written in terms of its magnitude and phase.

$$a(x,y)=|a(x,y)|e^{j\vartheta(x,y)} \quad a[m,n]=|a[m,n]|e^{j\vartheta[m,n]} \tag{13.16}$$

- If a 2D signal is real, then the Fourier transform has certain symmetries.

$$A(u,v)=A^*(-u,-v) \quad A(\Omega,\Psi)=A^*(-\Omega,-\Psi) \tag{13.17}$$

The symbol (*) indicates complex conjugation. For real signals Equation 13.17 leads directly to

$$\begin{aligned}|A(u,v)|=|A(-u,-v)| \qquad \varphi(u,v)=-\varphi(-u,-v)\\ |A(\Omega,\Psi)|=|A(-\Omega,-\Psi)| \quad \varphi(\Omega,\Psi)=-\varphi(-\Omega,-\Psi)\end{aligned} \tag{13.18}$$

- If a 2D signal is real and even, then the Fourier transform is real and even.

$$A(u,v)=A(-u,-v) \quad A(\Omega,\Psi)=A(-\Omega,-\Psi) \tag{13.19}$$

- The Fourier and the inverse Fourier transforms are linear operations.

$$\begin{aligned}\mathcal{F}\{w_1a+w_2b\}=\mathcal{F}\{w_1a\}+\mathcal{F}\{w_2b\}=w_1A+w_2B\\ \mathcal{F}^{-1}\{w_1A+w_2B\}=\mathcal{F}^{-1}\{w_1A\}+\mathcal{F}^{-1}\{w_2B\}=w_1a+w_2b\end{aligned} \tag{13.20}$$

where
a and b are 2D signals (images)
w_1 and w_2 are arbitrary, complex constants

- The Fourier transform in discrete space, $A(\Omega,\Psi)$, is periodic in both Ω and Ψ. Both periods are 2π.

$$A(\Omega+2\pi j,\Psi+2\pi k)=A(\Omega,\Psi) \quad j,k \text{ integers} \tag{13.21}$$

- The energy, E, in a signal can be measured either in the spatial domain or the frequency domain. For a signal with finite energy,

Parseval's theorem (2D continuous space):

$$E = \int_{-\infty}^{+\infty}\int_{-\infty}^{+\infty} |a(x,y)|^2 \mathrm{d}x\mathrm{d}y = \frac{1}{4\pi^2}\int_{-\infty}^{+\infty}\int_{-\infty}^{+\infty} |A(u,v)|^2 \,\mathrm{d}u\mathrm{d}v \tag{13.22}$$

Parseval's theorem (2D discrete space):

$$E = \sum_{m=-\infty}^{+\infty}\sum_{n=-\infty}^{+\infty} |a[m,n]|^2 = \frac{1}{4\pi^2}\int_{-\pi}^{+\pi}\int_{-\pi}^{+\pi} |A(\Omega,\Psi)|^2 \,\mathrm{d}\Omega\mathrm{d}\Psi \tag{13.23}$$

This "signal energy" is not to be confused with the physical energy in the phenomenon that produced the signal. If, for example, the value $a[m,n]$ represents a photon count, then the physical energy is proportional to the amplitude, a, and not the square of the amplitude. This is generally the case in video imaging.

- Given three, multidimensional signals a, b, and c and their Fourier transforms A, B, and C:

$$\begin{aligned} c = a \otimes b \quad &\overset{\mathcal{F}}{\leftrightarrow} \quad C = A \cdot B \\ \text{and} \qquad c = a \cdot b \quad &\overset{\mathcal{F}}{\leftrightarrow} \quad C = \frac{1}{4\pi^2} A \otimes B \end{aligned} \tag{13.24}$$

In words, convolution in the spatial domain is equivalent to multiplication in the Fourier (frequency) domain and vice versa. This is a central result which provides not only a methodology for the implementation of a convolution but also insight into how two signals interact with each other—under convolution—to produce a third signal. We shall make extensive use of this result later.

- If a 2D signal $a(x,y)$ is scaled in its spatial coordinates, then

$$\begin{aligned} &\text{If} \quad a(x,y) \rightarrow a(M_x \cdot x, M_y \cdot y) \\ &\text{Then} \quad A(u,v) \rightarrow A(u/M_x, v/M_y)/|M_x \cdot M_y| \end{aligned} \tag{13.25}$$

- If a 2D signal $a(x,y)$ has Fourier spectrum $A(u,v)$, then

$$\begin{aligned} A(u=0,v=0) &= \int_{-\infty}^{+\infty}\int_{-\infty}^{+\infty} a(x,y)\,\mathrm{d}x\mathrm{d}y \\ a(x=0,y=0) &= \frac{1}{4\pi^2}\int_{-\infty}^{+\infty}\int_{-\infty}^{+\infty} A(u,v)\,\mathrm{d}x\mathrm{d}y \end{aligned} \tag{13.26}$$

- If a 2D signal $a(x,y)$ has Fourier spectrum $A(u,v)$, then

$$\begin{aligned} &\frac{\partial a(x,y)}{\partial x} \overset{\mathcal{F}}{\leftrightarrow} juA(u,v) \qquad \frac{\partial a(x,y)}{\partial y} \overset{\mathcal{F}}{\leftrightarrow} jvA(u,v) \\ &\frac{\partial^2 a(x,y)}{\partial x^2} \overset{\mathcal{F}}{\leftrightarrow} -u^2A(u,v) \qquad \frac{\partial^2 a(x,y)}{\partial y^2} \overset{\mathcal{F}}{\leftrightarrow} -v^2A(u,v) \end{aligned} \tag{13.27}$$

13.3.4.1 Importance of Phase and Magnitude

Equation 13.15 indicates that the Fourier transform of an image can be complex. This is illustrated below in Figure 13.4a through c. Figure 13.4a shows the original image $a[m, n]$, Figure 13.4b the magnitude in a scaled form as $\log(|A(\Omega, \Psi)|)$, and Figure 13.4c the phase $\varphi(\Omega, \Psi)$.

Both the magnitude and the phase functions are necessary for the complete reconstruction of an image from its Fourier transform. Figure 13.5a shows what happens when Figure 13.4a is restored solely on the basis of the magnitude information and Figure 13.5b shows what happens when Figure 13.4a is restored solely on the basis of the phase information.

Neither the magnitude information nor the phase information is sufficient to restore the image. The magnitude-only image (Figure 13.5a) is unrecognizable and has severe dynamic range problems. The phase-only image (Figure 13.5b) is barely recognizable, that is, severely degraded in quality.

13.3.4.2 Circularly Symmetric Signals

An arbitrary 2D signal $a(x, y)$ can always be written in a polar coordinate system as $a(r, \theta)$. When the 2D signal exhibits a circular symmetry this means that

$$a(x, y) = a(r, \theta) = a(r) \tag{13.28}$$

where $r^2 = x^2 + y^2$ and $\tan\theta = y/x$. As a number of physical systems such as lenses exhibit circular symmetry, it is useful to be able to compute an appropriate Fourier representation.

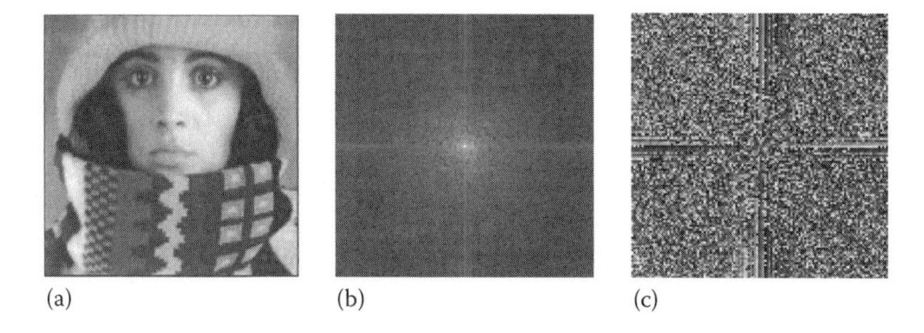

FIGURE 13.4 (a) Original. (b) $\log(|A(\Omega, \Psi)|)$. (c) $\varphi(\Omega, \Psi)$.

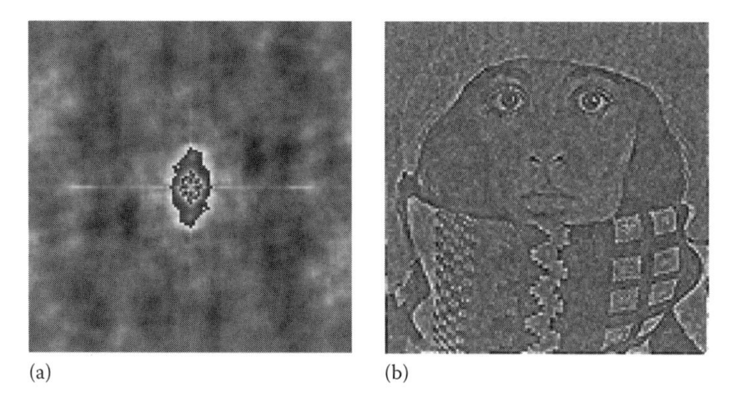

FIGURE 13.5 (a) $\varphi(\Omega, \Psi) = 0$. (b) $|A(\Omega, \Psi)| = \text{Constant}$.

The Fourier transform $A(u, v)$ can be written in polar coordinates $A(q, \zeta)$ and then, for a circularly symmetric signal, rewritten as a Hankel transform:

$$A(u, v) = \mathcal{F}\{a(x, y)\} = 2\pi \int_0^{\infty} a(r) J_o(rq) r dr = A(q) \tag{13.29}$$

where $q^2 = u^2 + v^2$ and $\tan \xi = v/u$ and $J_o(\cdot)$ is a Bessel function of the first kind of order zero.

The inverse Hankel transform is given by

$$a(r) = \frac{1}{2\pi} \int_0^{\infty} A(q) J_o(rq) q dq \tag{13.30}$$

The Fourier transform of a circularly symmetric 2D signal is a function of only the radial frequency, q. The dependence on the angular frequency, ξ, has vanished. Further, if $a(x, y) = a(r)$ is real, then it is automatically even due to the circular symmetry. According to Equation 13.19, $A(q)$ will then be real and even.

13.3.4.3 Examples of 2D Signals and Transforms

Table 13.4 shows some basic and useful signals and their 2D Fourier transforms. In using the table entries in the remainder of this chapter we will refer to a spatial domain term as the "point spread function"

TABLE 13.4 2D Images and Their Fourier Transforms

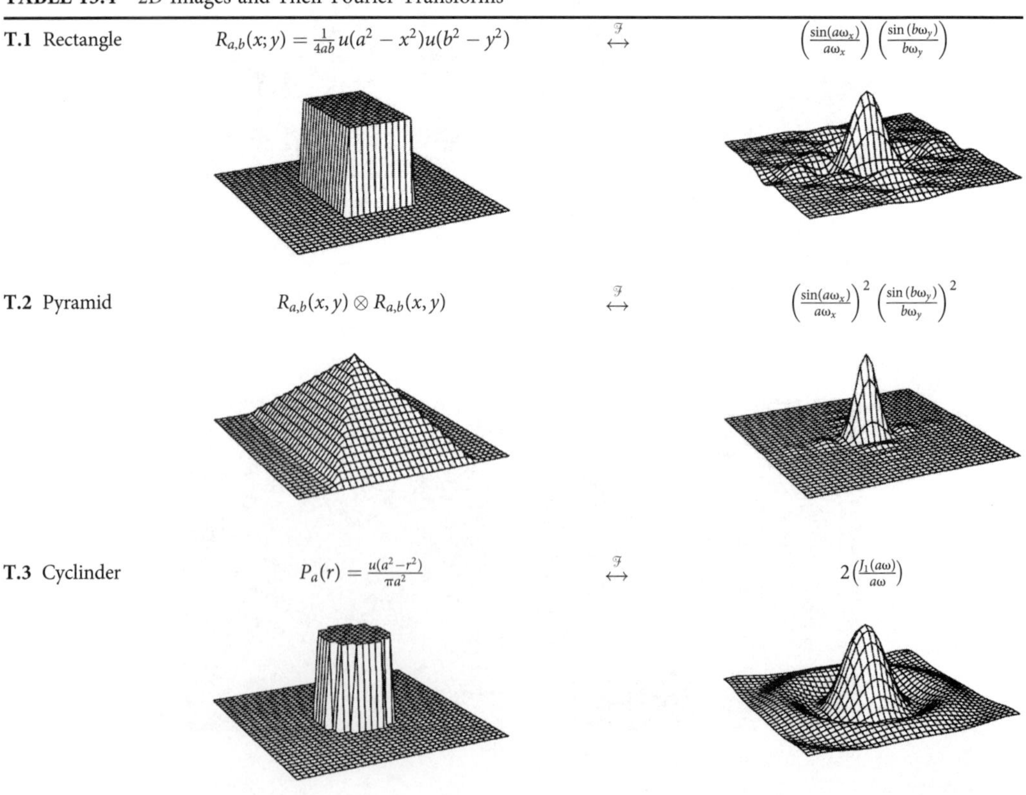

T.1 Rectangle	$R_{a,b}(x; y) = \frac{1}{4ab} u(a^2 - x^2) u(b^2 - y^2)$	$\overset{\mathcal{F}}{\leftrightarrow}$	$\left(\frac{\sin(a\omega_x)}{a\omega_x}\right) \left(\frac{\sin(b\omega_y)}{b\omega_y}\right)$
T.2 Pyramid	$R_{a,b}(x, y) \otimes R_{a,b}(x, y)$	$\overset{\mathcal{F}}{\leftrightarrow}$	$\left(\frac{\sin(a\omega_x)}{a\omega_x}\right)^2 \left(\frac{\sin(b\omega_y)}{b\omega_y}\right)^2$
T.3 Cyclinder	$P_a(r) = \frac{u(a^2 - r^2)}{\pi a^2}$	$\overset{\mathcal{F}}{\leftrightarrow}$	$2\left(\frac{J_1(a\omega)}{a\omega}\right)$

TABLE 13.4 (continued) 2D Images and Their Fourier Transforms

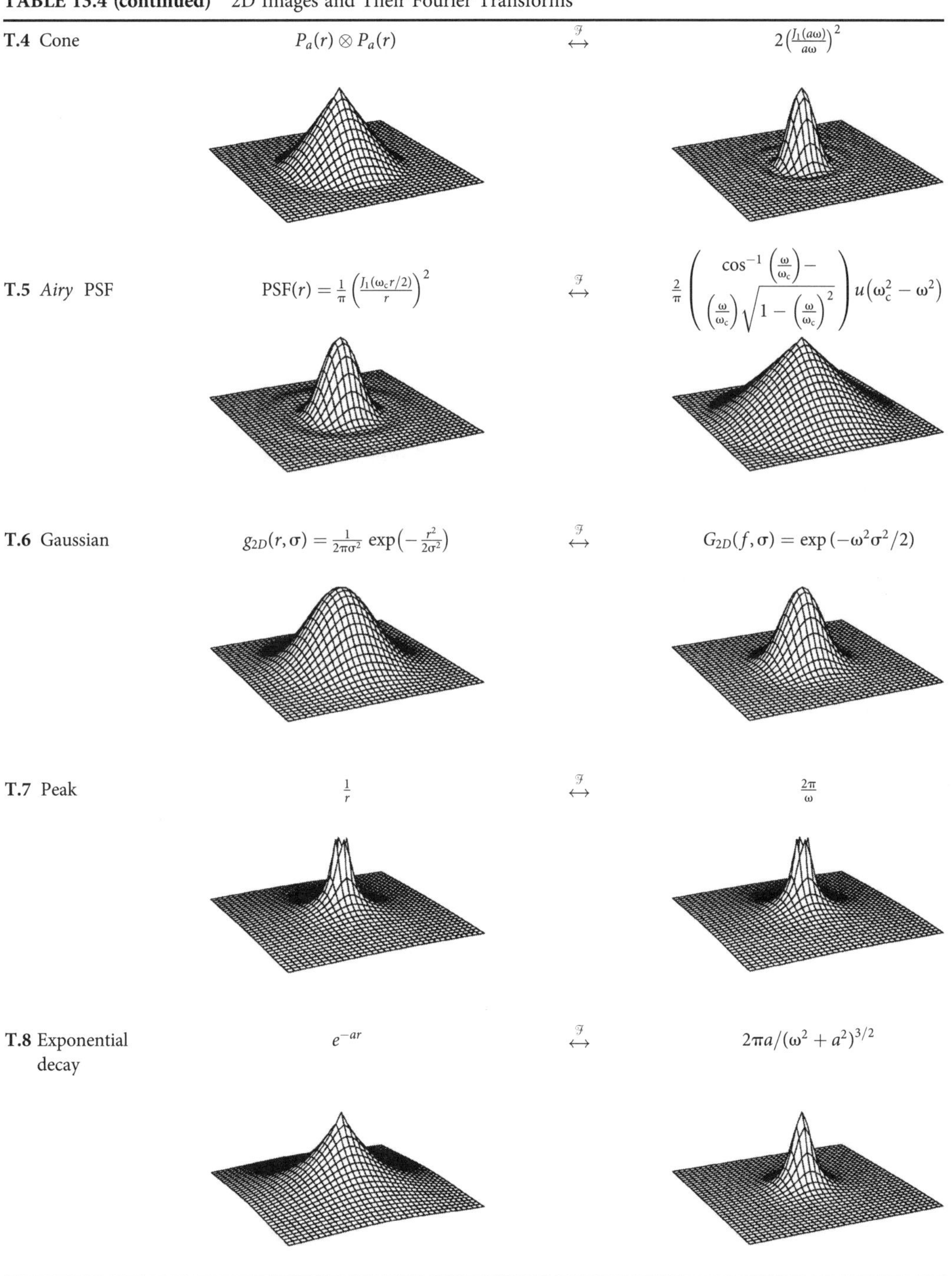

T.4 Cone	$P_a(r) \otimes P_a(r)$	$\overset{\mathcal{F}}{\leftrightarrow}$	$2\left(\frac{J_1(a\omega)}{a\omega}\right)^2$
T.5 *Airy* PSF	$\mathrm{PSF}(r) = \frac{1}{\pi}\left(\frac{J_1(\omega_c r/2)}{r}\right)^2$	$\overset{\mathcal{F}}{\leftrightarrow}$	$\frac{2}{\pi}\left(\cos^{-1}\left(\frac{\omega}{\omega_c}\right) - \left(\frac{\omega}{\omega_c}\right)\sqrt{1-\left(\frac{\omega}{\omega_c}\right)^2}\right)u\left(\omega_c^2 - \omega^2\right)$
T.6 Gaussian	$g_{2D}(r,\sigma) = \frac{1}{2\pi\sigma^2}\exp\left(-\frac{r^2}{2\sigma^2}\right)$	$\overset{\mathcal{F}}{\leftrightarrow}$	$G_{2D}(f,\sigma) = \exp\left(-\omega^2\sigma^2/2\right)$
T.7 Peak	$\frac{1}{r}$	$\overset{\mathcal{F}}{\leftrightarrow}$	$\frac{2\pi}{\omega}$
T.8 Exponential decay	e^{-ar}	$\overset{\mathcal{F}}{\leftrightarrow}$	$2\pi a/(\omega^2 + a^2)^{3/2}$

(PSF) or the "2D impulse response" and its Fourier transforms as the "optical transfer function" (OTF) or simply "transfer function." Two standard signals used in this table are $u(\cdot)$, the unit step function, and $J_1(\cdot)$, the Bessel function of the first kind. Circularly symmetric signals are treated as functions of r as in Equation 13.28.

13.3.5 Statistics

In image processing it is quite common to use simple statistical descriptions of images and subimages. The notion of a statistic is intimately connected to the concept of a probability distribution, generally the distribution of signal amplitudes. For a given region, which could conceivably be an entire image, we can define the probability distribution function of the brightnesses in that region and the probability density function of the brightnesses in that region. We will assume in the discussion that follows that we are dealing with a digitized image $a[m, n]$.

13.3.5.1 Probability Distribution Function of the Brightnesses

The probability distribution function, $P(a)$, is the probability that a brightness chosen from the region is less than or equal to a given brightness value a. As a increases from $-\infty$ to $+\infty$, $P(a)$ increases from 0 to 1. $P(a)$ is monotonic, nondecreasing in a and thus $\mathrm{d}P/\mathrm{d}a \geq 0$.

13.3.5.2 Probability Density Function of the Brightnesses

The probability that a brightness in a region falls between a and $a+\Delta a$, given the probability distribution function $P(a)$, can be expressed as $p(a)\Delta a$ where $p(a)$ is the probability density function:

$$p(a)\Delta a = \left(\frac{\mathrm{d}P(a)}{\mathrm{d}a}\right)\Delta a \quad (13.31)$$

Because of the monotonic, nondecreasing character of $P(a)$ we have that

$$p(a) \geq 0 \quad \text{and} \quad \int_{-\infty}^{+\infty} p(a)\mathrm{d}a = 1 \quad (13.32)$$

For an image with quantized (integer) brightness amplitudes, the interpretation of Δa is the width of a brightness interval. We assume constant width intervals. The brightness probability density function is frequently estimated by counting the number of times that each brightness occurs in the region to generate a histogram, $h[a]$. The histogram can then be normalized so that the total area under the histogram is 1 (Equation 13.32). Said another way, the $p[a]$ for a region is the normalized count of the number of pixels, Λ, in a region that have quantized brightness a:

$$p[a] = \frac{1}{\Lambda}h[a] \quad \text{with} \quad \Lambda = \sum_a h[a] \quad (13.33)$$

The brightness probability distribution function for the image shown in Figure 13.4a is shown in Figure 13.6a. The (unnormalized) brightness histogram of Figure 13.4a which is proportional to the estimated brightness probability density function is shown in Figure 13.6b. The height in this histogram corresponds to the number of pixels with a given brightness.

Both the distribution function and the histogram as measured from a region are a statistical description of that region. It must be emphasized that both $P[a]$ and $p[a]$ should be viewed as estimates of true distributions when they are computed from a specific region. That is, we view an image and a specific region as one realization of the various random processes involved in the formation of that image and that region. In the same context, the statistics defined below must be viewed as estimates of the underlying parameters.

13.3.5.3 Average

The average brightness of a region is defined as the sample mean of the pixel brightnesses within that region. The average, m_a, of the brightnesses over the Λ pixels within a region ($\Re$) is given by

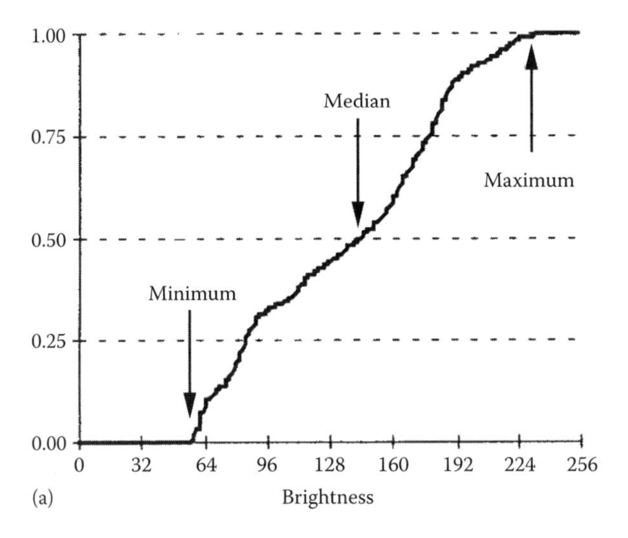

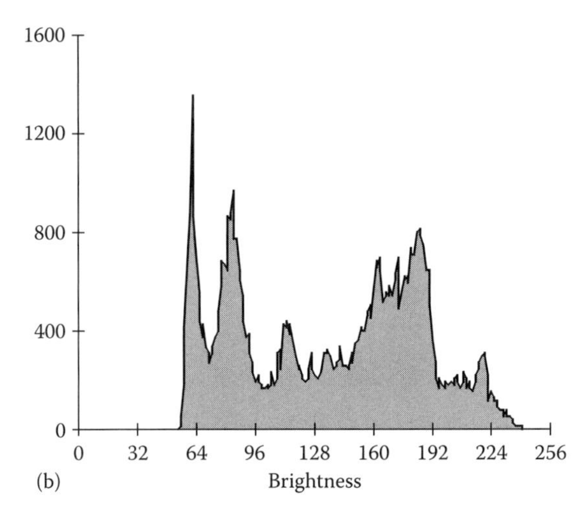

FIGURE 13.6 (a) Brightness distribution function of Figure 13.4a with minimum, median, and maximum indicated. See text for explanation. (b) Brightness histogram of Figure 13.4a.

$$m_{\mathrm{a}} = \frac{1}{\Lambda} \sum_{(m,n)\in\Re} a[m,n] \tag{13.34}$$

Alternatively, we can use a formulation based upon the (unnormalized) brightness histogram, $h(a) = \Lambda \cdot p(a)$, with discrete brightness values a. This gives

$$m_{\mathrm{a}} = \frac{1}{\Lambda} \sum_{a} a \cdot h[a] \tag{13.35}$$

The average brightness, m_{a}, is an estimate of the mean brightness, μ_{a}, of the underlying brightness probability distribution.

13.3.5.4 Standard Deviation

The unbiased estimate of the standard deviation, s_a, of the brightnesses within a region ($\Re$) with Λ pixels is called the sample standard deviation and is given by

$$s_a = \sqrt{\frac{1}{\Lambda - 1} \sum_{m,n \in \Re} (a[m,n] - m_a)^2}$$
$$= \sqrt{\frac{\sum_{m,n \in \Re} a^2[m,n] - \Lambda m_a^2}{\Lambda - 1}} \quad (13.36)$$

Using the histogram formulation gives

$$s_a = \sqrt{\frac{\left(\sum_a a^2 \cdot h[a]\right) - \Lambda \cdot m_a^2}{\Lambda - 1}} \quad (13.37)$$

The standard deviation, s_a, is an estimate of σ_a of the underlying brightness probability distribution.

13.3.5.5 Coefficient-of-Variation

The dimensionless coefficient-of-variation, CV, is defined as

$$CV = \frac{s_a}{m_a} \times 100\% \quad (13.38)$$

13.3.5.6 Percentiles

The percentile, $p\%$, of an unquantized brightness distribution is defined as that value of the brightness a such that

$$P(a) = p\%$$

or equivalently

$$\int_{-\infty}^{a} p(\alpha)d\alpha = p\% \quad (13.39)$$

Three special cases are frequently used in digital image processing.

- 0% the minimum value in the region
- 50% the median value in the region
- 100% the maximum value in the region

All three of these values can be determined from Figure 13.6a.

13.3.5.7 Mode

The mode of the distribution is the most frequent brightness value. There is no guarantee that a mode exists or that it is unique.

13.3.5.8 Signal-to-Noise Ratio

The signal-to-noise ratio (SNR) can have several definitions. The noise is characterized by its standard deviation, s_n. The characterization of the signal can differ. If the signal is known to lie between two boundaries, $a_{min} \leq a \leq a_{max}$, then the SNR is defined as

$$\text{Bounded signal—SNR} = 20 \log_{10}\left(\frac{a_{max} - a_{min}}{s_n}\right) \text{dB} \tag{13.40}$$

If the signal is not bounded but has a statistical distribution then two other definitions are known
Stochastic signal

$$\text{S and N interdependent SNR} = 20 \log_{10}\left(\frac{m_a}{s_n}\right) \text{dB} \tag{13.41}$$

$$\text{S and N independent SNR} = 20 \log_{10}\left(\frac{s_a}{s_n}\right) \text{dB} \tag{13.42}$$

where m_a and s_a are defined above.

The various statistics are given in Table 13.5 for the image and the region shown in Figure 13.7.

An SNR calculation for the entire image based on Equation 13.40 is not directly available. The variations in the image brightnesses that lead to the large value of s (=49.5) are not, in general, due to noise but to the variation in local information. With the help of the region there is a way to estimate the SNR. We can use the $s_{\Re}$ (=4.0) and the dynamic range, $a_{max} - a_{min}$, for the image (=241–56) to calculate a global SNR (=33.3 dB). The underlying assumptions are that (1) the signal is approximately constant in that region and the variation in the region is therefore due to noise, and, (2) that the noise is the same over the entire image with a standard deviation given by $s_n = s_{\Re}$.

13.3.6 Contour Representations

When dealing with a region or object, several compact representations are available that can facilitate manipulation of and measurements on the object. In each case we assume that we begin with an image representation of the object as shown in Figure 13.8a and b. Several techniques exist to represent the region or object by describing its contour.

13.3.6.1 Chain Code

This representation is based upon the work of Freeman [11]. We follow the contour in a clockwise manner and keep track of the directions as we go from one contour pixel to the next. For the

TABLE 13.5 Statistics from Figure 13.7

Statistic	Image	ROI
Average	137.7	219.3
Standard deviation	49.5	4.0
Minimum	56	202
Median	141	220
Maximum	241	226
Mode	62	220
SNR (dB)	NA	33.3

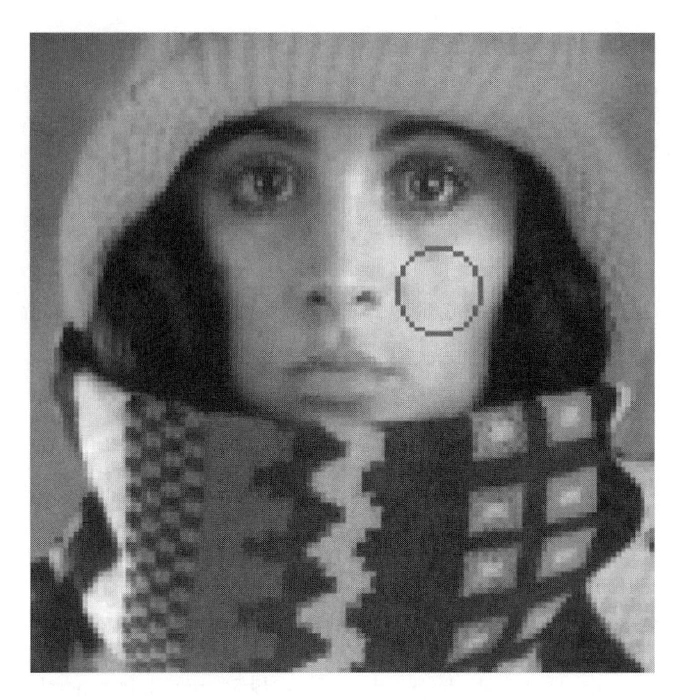

FIGURE 13.7 Region is the interior of the circle.

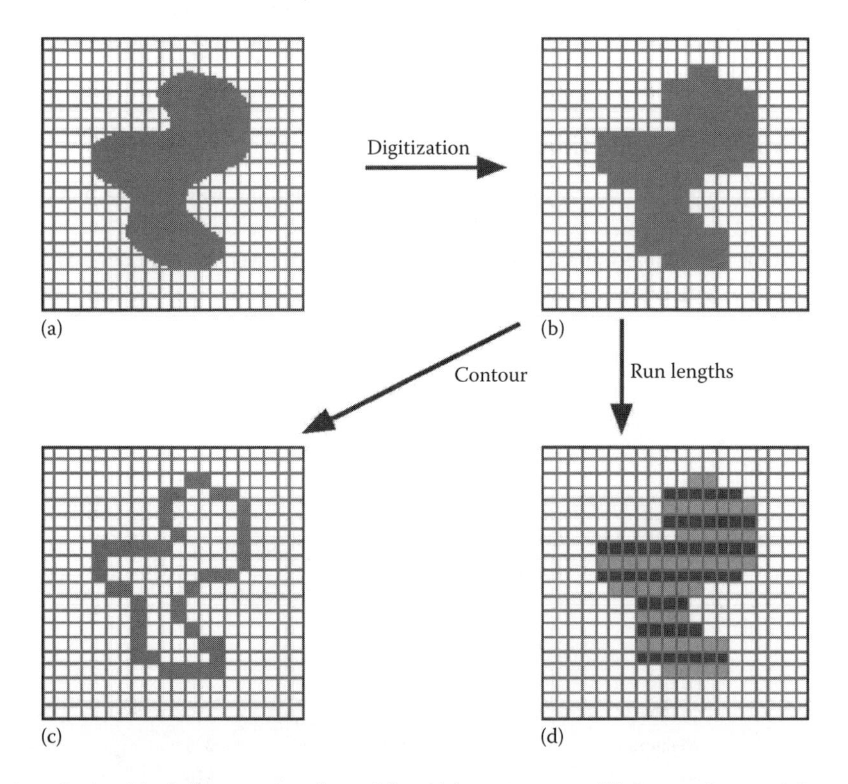

FIGURE 13.8 Region (shaded) as it is transformed from (a) continuous to (b) discrete form and then considered as a (c) contour or (d) run lengths illustrated in alternating colors.

standard implementation of the chain code we consider a contour pixel to be an object pixel that has a background (nonobject) pixel as one or more of its 4-connected neighbors. See Figures 13.3a and 13.8c.

The codes associated with eight possible directions are the chain codes and, with x as the current contour pixel position, the codes are generally defined as

$$\text{Chain codes} = \begin{matrix} 3 & 2 & 1 \\ 4 & x & 0 \\ 5 & 6 & 7 \end{matrix} \tag{13.43}$$

13.3.6.2 Chain Code Properties

- Even codes $\{0, 2, 4, 6\}$ correspond to horizontal and vertical directions; odd codes $\{1, 3, 5, 7\}$ correspond to the diagonal directions.
- Each code can be considered as the angular direction, in multiples of 45° that we must move to go from one contour pixel to the next.
- The absolute coordinates $[m, n]$ of the first contour pixel (e.g., top, leftmost) together with the chain code of the contour represent a complete description of the discrete region contour.
- When there is a change between two consecutive chain codes, then the contour has changed direction. This point is defined as a corner.

13.3.6.3 "Crack" Code

An alternative to the chain code for contour encoding is to use neither the contour pixels associated with the object nor the contour pixels associated with background but rather the line, the "crack," in between. This is illustrated with an enlargement of a portion of Figure 13.8 in Figure 13.9.

The "crack" code can be viewed as a chain code with four possible directions instead of eight.

$$\text{Crack codes} = \begin{matrix} & 1 & \\ 2 & x & 0 \\ & 3 & \end{matrix} \tag{13.44}$$

The chain code for the enlarged section of Figure 13.9b, from top to bottom, is $\{5, 6, 7, 7, 0\}$. The crack code is $\{3, 2, 3, 3, 0, 3, 0, 0\}$.

13.3.6.4 Run Codes

A third representation is based on coding the consecutive pixels along a row—a run—that belong to an object by giving the starting position of the run and the ending position of the run. Such runs are illustrated in Figure 13.8d. There are a number of alternatives for the precise definition of the positions. Which alternative should be used depends upon the application and thus will not be discussed here.

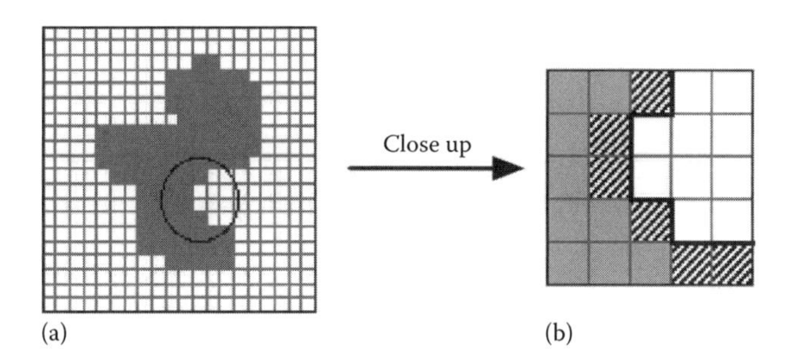

FIGURE 13.9 (a) Object including part to be studied. (b) Contour pixels as used in the chain code are diagonally shaded. The "crack" is shown with the thick black line.

13.4 Perception

Many image processing applications are intended to produce images that are to be viewed by human observers (as opposed to, say, automated industrial inspection.) It is therefore important to understand the characteristics and limitations of the human visual system—to understand the "receiver" of the 2D signals. At the outset it is important to realize that (1) the human visual system is not well understood, (2) no objective measure exists for judging the quality of an image that corresponds to human assessment of image quality, and, (3) the "typical" human observer does not exist. Nevertheless, research in perceptual psychology has provided some important insights into the visual system. See, for example, Stockham [12].

13.4.1 Brightness Sensitivity

There are several ways to describe the sensitivity of the human visual system. To begin, let us assume that a homogeneous region in an image has an intensity as a function of wavelength (color) given by $I(\lambda)$. Further let us assume that $I(\lambda) = I_o$, a constant.

13.4.1.1 Wavelength Sensitivity

The perceived intensity as a function of λ, the spectral sensitivity, for the "typical observer" is shown in Figure 13.10 [13].

13.4.1.2 Stimulus Sensitivity

If the constant intensity (brightness) I_o is allowed to vary then, to a good approximation, the visual response, R, is proportional to the logarithm of the intensity. This is known as the Weber–Fechner law:

$$R = \log(I_o) \tag{13.45}$$

The implications of this are easy to illustrate. Equal perceived steps in brightness, $\Delta R = k$, require that the physical brightness (the stimulus) increases exponentially. This is illustrated in Figure 13.11a and b.

A horizontal line through the top portion of Figure 13.11a shows a linear increase in objective brightness (Figure 13.11b) but a logarithmic increase in subjective brightness. A horizontal line through the bottom portion of Figure 13.11a shows an exponential increase in objective brightness (Figure 13.11b) but a linear increase in subjective brightness.

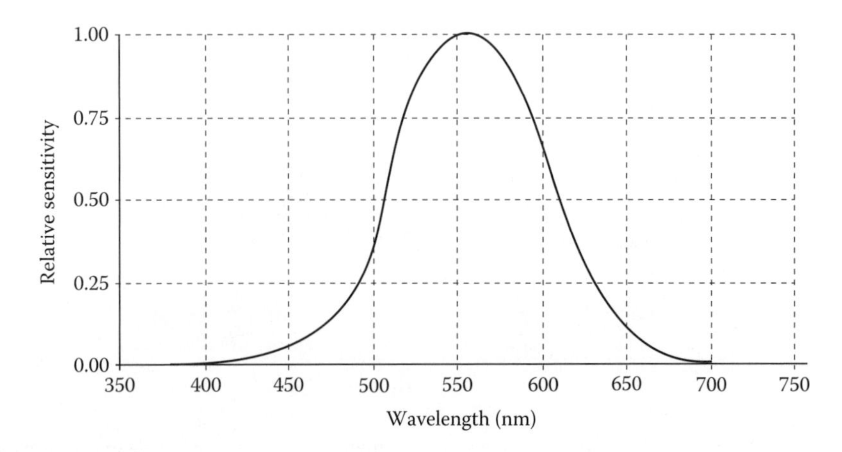

FIGURE 13.10 Spectral Sensitivity of the "typical" human observer.

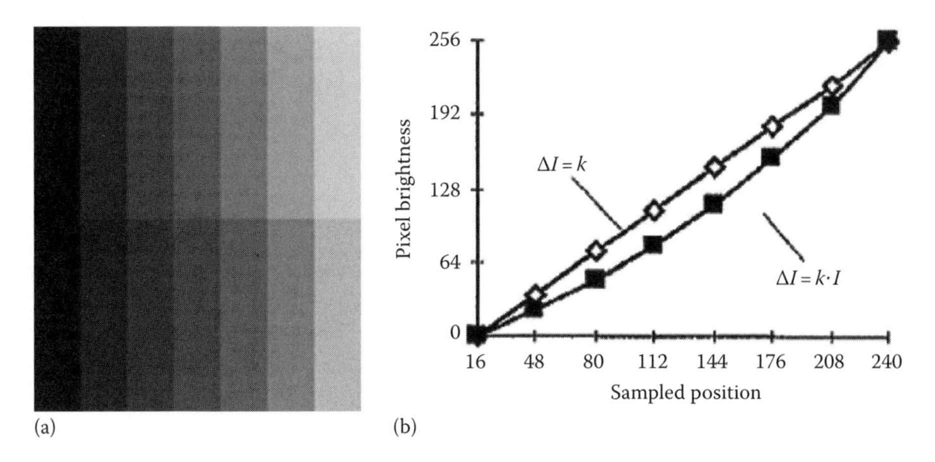

FIGURE 13.11 (a) (top) Brightness step $\Delta I = k$. (bottom) Brightness step $\Delta I = k \cdot I$. (b) Actual brightnesses plus interpolated values.

The Mach band effect is visible in Figure 13.11a. Although the physical brightness is constant across each vertical stripe, the human observer perceives an "undershoot" and "overshoot" in brightness at what is physically a step edge. Thus, just before the step, we see a slight decrease in brightness compared to the true physical value. After the step we see a slight overshoot in brightness compared to the true physical value. The total effect is one of increased, local, perceived contrast at a step edge in brightness.

13.4.2 Spatial Frequency Sensitivity

If the constant intensity (brightness) I_o is replaced by a sinusoidal grating with increasing spatial frequency (Figure 13.12a), it is possible to determine the spatial frequency sensitivity. The result is shown in Figure 13.12b [14,15].

To translate these data into common terms, consider an "ideal" computer monitor at a viewing distance of 50 cm. The spatial frequency that will give maximum response is at 10 cycles per degree (see Figure 13.12b). The one degree at 50 cm translates to 50 tan(1°) = 0.87 cm on the computer screen. Thus the spatial frequency of maximum response f_{max} = 10 cycles/0.87 cm = 11.46 cycles/cm at this viewing distance. Translating this into a general formula gives

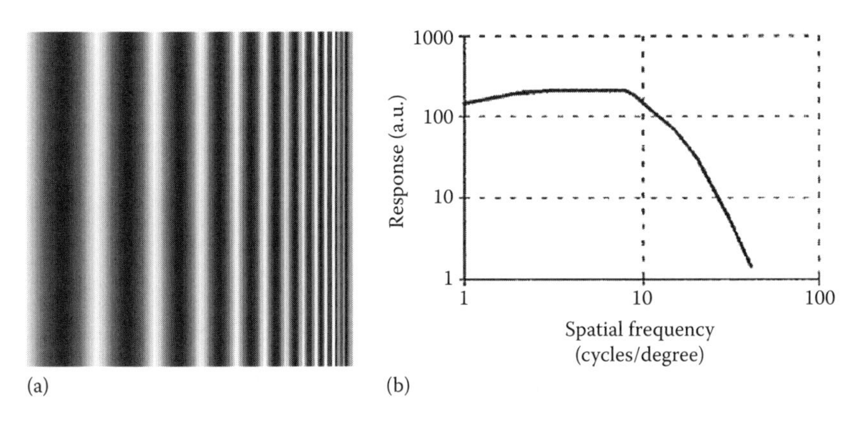

FIGURE 13.12 (a) Sinusoidal test grating. (b) Spatial frequency sensitivity.

$$f_{\max} = \frac{10}{d \cdot \tan(1^\circ)} = \frac{572.9}{d} \text{ cycles/cm} \tag{13.46}$$

where d is the viewing distance measured in cm.

13.4.3 Color Sensitivity

Human color perception is an exceedingly complex topic. As such we can only present a brief introduction here. The physical perception of color is based upon three color pigments in the retina.

13.4.3.1 Standard Observer

Based upon psychophysical measurements, standard curves have been adopted by the Commission Internationale de l'Eclairage (CIE) as the sensitivity curves for the "typical" observer for the three "pigments" $\bar{x}(\lambda)$, $\bar{y}(\lambda)$, and $\bar{z}(\lambda)$. These are shown in Figure 13.13. These are not the actual pigment absorption characteristics found in the "standard" human retina but rather sensitivity curves derived from actual data [10].

For an arbitrary homogeneous region in an image that has an intensity as a function of wavelength (color) given by $I(\lambda)$, the three responses are called the tristimulus values:

$$X = \int_0^\infty I(\lambda)\bar{x}(\lambda)d\lambda \quad Y = \int_0^\infty I(\lambda)\bar{y}(\lambda)d\lambda \quad Z = \int_0^\infty I(\lambda)\bar{z}(\lambda)d\lambda \tag{13.47}$$

13.4.3.2 CIE Chromaticity Coordinates

The chromaticity coordinates which describe the perceived color information are defined as

$$x = \frac{X}{X+Y+Z} \quad y = \frac{Y}{X+Y+Z} \quad z = 1-(x+y) \tag{13.48}$$

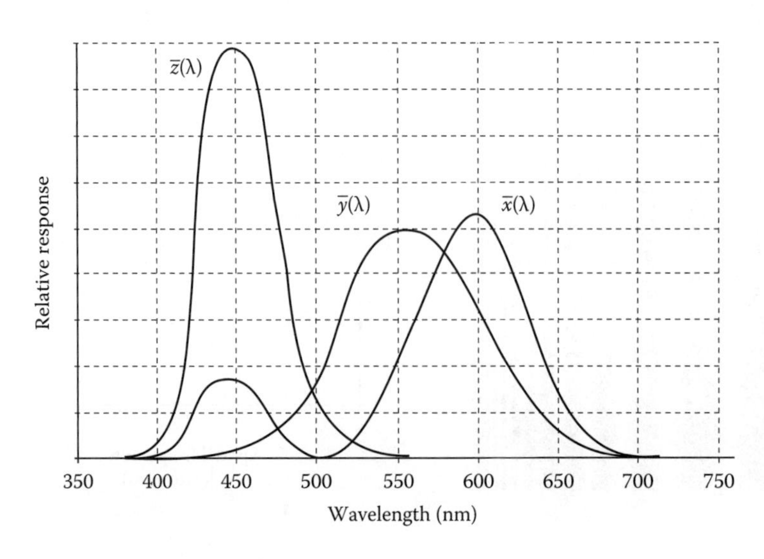

FIGURE 13.13 Standard observer spectral sensitivity curves.

The red chromaticity coordinate is given by x and the green chromaticity coordinate by y. The tristimulus values are linear in $I(\lambda)$ and thus the absolute intensity information has been lost in the calculation of the chromaticity coordinates $\{x, y\}$. All color distributions, $I(\lambda)$, that appear to an observer as having the same color will have the same chromaticity coordinates.

If we use a tunable source of pure color (such as a dye laser), then the intensity can be modeled as $I(\lambda) = \delta(\lambda - \lambda_o)$ with $\delta(\cdot)$ as the impulse function. The collection of chromaticity coordinates $\{x, y\}$ that will be generated by varying λ_o gives the CIE chromaticity triangle as shown in Figure 13.14.

Pure spectral colors are along the boundary of the chromaticity triangle. All other colors are inside the triangle. The chromaticity coordinates for some standard sources are given in Table 13.6.

The description of color on the basis of chromaticity coordinates not only permits an analysis of color but provides a synthesis technique as well. Using a mixture of two color sources, it is possible to generate any of the colors along the line connecting their respective chromaticity coordinates. Since we cannot have a negative number of photons, this means the mixing coefficients must be positive. Using three color sources such as the red, green, and blue phosphors on cathode-ray tube (CRT) monitors leads to the set of colors defined by the interior of the "phosphor triangle" shown in Figure 13.14.

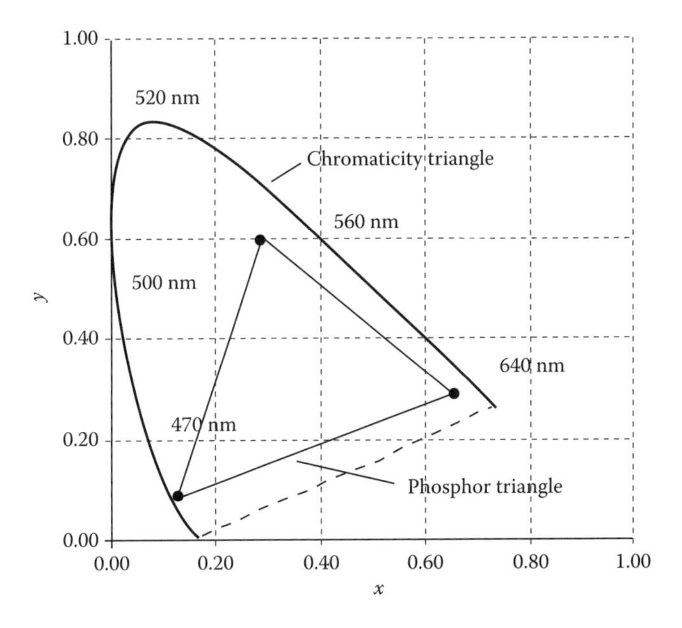

FIGURE 13.14 Chromaticity diagram containing the CIE chromaticity triangle associated with pure spectral colors and the triangle associated with CRT phosphors.

TABLE 13.6 Chromaticity Coordinates for Standard Sources

Source	x	y
Fluorescent lamp at 4800°K	0.35	0.37
Sun at 6000°K	0.32	0.33
Red phosphor (europium yttrium vanadate)	0.68	0.32
Green phosphor (zinc cadmium sulfide)	0.28	0.60
Blue phosphor (zinc sulfide)	0.15	0.07

The formulas for converting from the tristimulus values (X, Y, Z) to the well-known CRT colors (R, G, B) and back are given by

$$\begin{bmatrix} R \\ G \\ B \end{bmatrix} = \begin{bmatrix} 1.9107 & -0.5326 & -0.2883 \\ -0.9843 & 1.9984 & -0.0283 \\ 0.0583 & -0.1185 & 0.8986 \end{bmatrix} \cdot \begin{bmatrix} X \\ Y \\ Z \end{bmatrix} \tag{13.49}$$

and

$$\begin{bmatrix} X \\ Y \\ Z \end{bmatrix} = \begin{bmatrix} 0.6067 & 0.1736 & 0.2001 \\ 0.2988 & 0.5868 & 0.1143 \\ 0.0000 & 0.0661 & 1.1149 \end{bmatrix} \cdot \begin{bmatrix} R \\ G \\ B \end{bmatrix} \tag{13.50}$$

As long as the position of a desired color (X, Y, Z) is inside the phosphor triangle in Figure 13.14, the values of R, G, and B as computed by Equation 13.49 will be positive and can therefore be used to drive a CRT monitor.

It is incorrect to assume that a small displacement anywhere in the chromaticity diagram (Figure 13.14) will produce a proportionally small change in the perceived color. An empirically derived chromaticity space where this property is approximated is the (u', v') space

$$u' = \frac{4x}{-2x + 12y + 3} \qquad v' = \frac{9y}{-2x + 12y + 3}$$

and

$$x = \frac{9u'}{6u' - 16v' + 12} \qquad y = \frac{4v'}{6u' - 16v' + 12} \tag{13.51}$$

Small changes almost anywhere in the (u', v') chromaticity space produce equally small changes in the perceived colors.

13.4.4 Optical Illusions

The description of the human visual system presented above is couched in standard engineering terms. This could lead one to conclude that there is sufficient knowledge of the human visual system to permit modeling the visual system with standard system analysis techniques. Two simple examples of optical illusions, shown in Figure 13.15, illustrate that this system approach would be a gross oversimplification. Such models should only be used with extreme care.

The left illusion induces the illusion of gray values in the eye that the brain "knows" does not exist. Further, there is a sense of dynamic change in the image due, in part, to the saccadic movements of the

FIGURE 13.15 Optical Illusions.

eye. The right illusion, Kanizsa's triangle, shows enhanced contrast and false contours [14] neither of which can be explained by the system-oriented aspects of visual perception described above.

13.5 Image Sampling

Converting from a continuous image $a(x, y)$ to its digital representation $b[m, n]$ requires the process of sampling. In the ideal sampling system $a(x, y)$ is multiplied by an ideal 2D impulse train:

$$\begin{aligned} b_{\text{ideal}}[m.n] &= a(x,y) \cdot \sum_{m=-\infty}^{+\infty} \sum_{n=-\infty}^{+\infty} \delta(x - mX_o, y - nY_o) \\ &= \sum_{m=-\infty}^{+\infty} \sum_{n=-\infty}^{+\infty} a(mX_o, nY_o)\delta(x - mX_o, y - nY_o) \end{aligned} \quad (13.52)$$

where

X_o and Y_o are the sampling distances or intervals
$\delta(\cdot,\cdot)$ is the ideal impulse function

(At some point, of course, the impulse function $\delta(x, y)$ is converted to the discrete impulse function $\delta[m, n]$.) Square sampling implies that $X_o = Y_o$. Sampling with an impulse function corresponds to sampling with an infinitesimally small point. This, however, does not correspond to the usual situation as illustrated in Figure 13.1. To take the effects of a finite sampling aperture $p(x, y)$ into account, we can modify the sampling model as follows.

$$b[m.n] = (a(x,y) \otimes p(x,y)) \cdot \sum_{m=-\infty}^{+\infty} \sum_{n=-\infty}^{+\infty} \delta(x - mX_o, y - nY_o) \quad (13.53)$$

The combined effect of the aperture and sampling are best understood by examining the Fourier domain representation

$$B(\Omega, \Psi) = \frac{1}{4\pi^2} \sum_{m=-\infty}^{+\infty} \sum_{n=-\infty}^{+\infty} A(\Omega - m\Omega_s, \Psi - n\Psi_s) \cdot P(\Omega - m\Omega_s, \Psi - n\Psi_s) \quad (13.54)$$

where

$\Omega_s = 2\pi/X_o$ is the sampling frequency in the x-direction
$\Psi_s = 2\pi/Y_o$ is the sampling frequency in the y-direction

The aperture $p(x, y)$ is frequently square, circular, or Gaussian with the associated $P(\Omega, \Psi)$ (see Table 13.4). The periodic nature of the spectrum, described in Equation 13.21 is clear from Equation 13.54.

13.5.1 Sampling Density for Image Processing

To prevent the possible aliasing (overlapping) of spectral terms that is inherent in Equation 13.54, two conditions must hold:

- Bandlimited $A(u, v)$

$$|A(u, v)| \equiv 0 \quad \text{for} \quad |u| > u_c \quad \text{and} \quad |v| > v_c \quad (13.55)$$

- Nyquist sampling frequency

$$\Omega_s > 2 \cdot u_c \quad \text{and} \quad \Psi_s > 2 \cdot v_c \quad (13.56)$$

where u_c and v_c are the cutoff frequencies in the *x*- and *y*-directions, respectively. Images that are acquired through lenses that are circularly-symmetric, aberration-free, and diffraction-limited will, in general, be bandlimited. The lens acts as a low-pass filter with a cutoff frequency in the frequency domain (Equation 13.11) given by

$$u_c = v_c = \frac{2\text{NA}}{\lambda} \tag{13.57}$$

where

NA is the numerical aperture of the lens
λ is the shortest wavelength of light used with the lens [16]

If the lens does not meet one or more of these assumptions then it will still be bandlimited but at lower cutoff frequencies than those given in Equation 13.57. When working with the *F*-number (*F*) of the optics instead of the NA and in air (with index of refraction = 1.0), Equation 13.57 becomes:

$$u_c = v_c = \frac{2}{\lambda}\left(\frac{1}{\sqrt{4F^2+1}}\right) \tag{13.58}$$

13.5.1.1 Sampling Aperture

The aperture $p(x, y)$ described above will have only a marginal effect on the final signal if the two conditions Equations 13.56 and 13.57 are satisfied. Given, for example, the distance between samples X_o equals Y_o and a sampling aperture that is not wider than X_o, the effect on the overall spectrum—due to the $A(u, v)P(u, v)$ behavior implied by Equation 13.53—is illustrated in Figure 13.16 for square and Gaussian apertures.

The spectra are evaluated along one axis of the 2D Fourier transform. The Gaussian aperture in Figure 13.16 has a width such that the sampling interval X_o contains $\pm 3\sigma$ (99.7%) of the Gaussian. The rectangular apertures have a width such that one occupies 95% of the sampling interval and the other occupies 50% of the sampling interval. The 95% width translates to a fill factor of 90% and the 50% width to a fill factor of 25%. The fill factor is discussed in Section 13.7.5.2.

13.5.2 Sampling Density for Image Analysis

The "rules" for choosing the sampling density when the goal is image analysis—as opposed to image processing—are different. The fundamental difference is that the digitization of objects in an image into a collection of pixels introduces a form of spatial quantization noise that is not bandlimited. This leads to

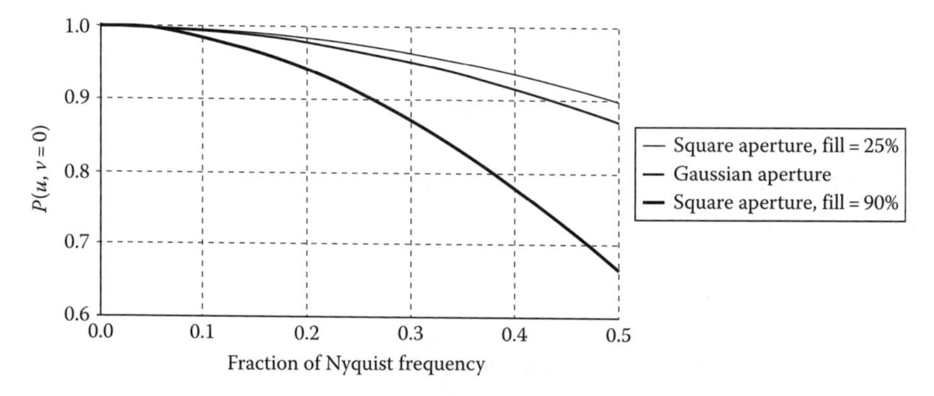

FIGURE 13.16 Aperture spectra $P(u, v=0)$ for frequencies up to half the Nyquist frequency. For explanation of "fill" see text.

the following results for the choice of sampling density when one is interested in the measurement of area and (perimeter) length.

13.5.2.1 Sampling for Area Measurements

Assuming square sampling, $X_o = Y_o$ and the unbiased algorithm for estimating area which involves simple pixel counting, the CV (see Equation 13.38) of the area measurement is related to the sampling density by [17]

$$\text{2D:} \quad \lim_{S\to\infty} \mathrm{CV}(S) = k_2 S^{-3/2} \quad \text{3D:} \quad \lim_{S\to\infty} \mathrm{CV}(S) = k_3 S^{-2} \tag{13.59}$$

and in D dimensions:

$$\lim_{S\to\infty} \mathrm{CV}(S) = k_D S^{-(D+1)/2} \tag{13.60}$$

where S is the number of samples per object diameter. In 2D the measurement is area, in 3D volume, and in D-dimensions hypervolume.

13.5.2.2 Sampling for Length Measurements

Again assuming square sampling and algorithms for estimating length based upon the Freeman chain-code representation (see Section 13.3.6.1), the CV of the length measurement is related to the sampling density per unit length as shown in Figure 13.17 (see [18,19].)

The curves in Figure 13.17 were developed in the context of straight lines but similar results have been found for curves and closed contours. The specific formulas for length estimation use a chain code representation of a line and are based upon a linear combination of three numbers:

$$L = \alpha \cdot N_e + \beta \cdot N_o + \gamma \cdot N_c \tag{13.61}$$

where

N_e is the number of even chain codes
N_o the number of odd chain codes
N_c the number of corners

The specific formulas are given in Table 13.7.

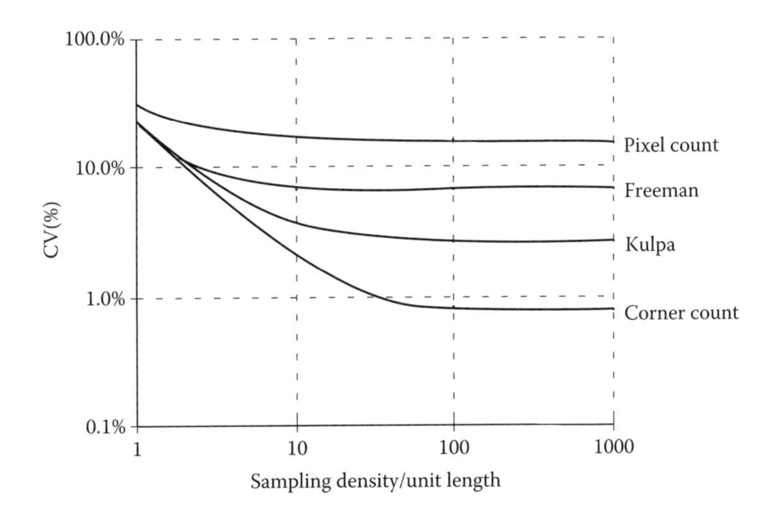

FIGURE 13.17 CV of length measurement for various algorithms.

TABLE 13.7 Length Estimation Formulas Based on Chain Code Counts (N_e, N_o, N_c)

Coefficients Formula	α	β	γ	Reference
Pixel count	1	1	0	[18]
Freeman	1	$\sqrt{2}$	0	[11]
Kulpa	0.9481	$0.9481 \bullet \sqrt{2}$	0	[20]
Corner count	0.980	1.406	−0.091	[21]

13.5.2.3 Conclusions on Sampling

If one is interested in image processing, one should choose a sampling density based upon classical signal theory, that is, the Nyquist sampling theory. If one is interested in image analysis, one should choose a sampling density based upon the desired measurement accuracy (bias) and precision (CV). In a case of uncertainty, one should choose the higher of the two sampling densities (frequencies).

13.6 Noise

Images acquired through modern sensors may be contaminated by a variety of noise sources. By noise we refer to stochastic variations as opposed to deterministic distortions such as shading or lack of focus. We will assume for this section that we are dealing with images formed from light using modern electro-optics. In particular we will assume the use of modern, charge-coupled device (CCD) cameras where photons produce electrons that are commonly referred to as photoelectrons. Nevertheless, most of the observations we shall make about noise and its various sources hold equally well for other imaging modalities.

While modern technology has made it possible to reduce the noise levels associated with various electro-optical devices to almost negligible levels, one noise source can never be eliminated and thus forms the limiting case when all other noise sources are "eliminated."

13.6.1 Photon Noise

When the physical signal that we observe is based upon light, then the quantum nature of light plays a significant role. A single photon at $\lambda = 500$ nm carries an energy of $E = h\nu = hc/\lambda = 3.97 \times 10^{-19}$ Joules. Modern CCD cameras are sensitive enough to be able to count individual photons. (Camera sensitivity will be discussed in Section 13.7.2.). The noise problem arises from the fundamentally statistical nature of photon production. We cannot assume that, in a given pixel for two consecutive but independent observation intervals of length T, the same number of photons will be counted. Photon production is governed by the laws of quantum physics which restrict us to talking about an average number of photons within a given observation window. The probability distribution for p photons in an observation window of length T seconds is known to be Poisson:

$$P(p|\rho, T) = \frac{(\rho T)^p e^{-\rho T}}{p!} \tag{13.62}$$

where ρ is the rate or intensity parameter measured in photons per second. It is critical to understand that even if there were no other noise sources in the imaging chain, the statistical fluctuations associated with photon counting over a finite time interval T would still lead to a finite SNR. If we use the appropriate formula for the SNR (Equation 13.41), then due to the fact that the average value and the standard deviation are given by

$$\text{Poisson process—} \quad \begin{aligned} \text{average} &= \rho T \\ \sigma &= \sqrt{\rho T} \end{aligned} \tag{13.63}$$

we have for the SNR:

Photon noise— $$\text{SNR} = 10 \log_{10}(\rho T) \text{ dB} \tag{13.64}$$

The three traditional assumptions about the relationship between signal and noise do not hold for photon noise:

- Photon noise is not independent of the signal
- Photon noise is not Gaussian
- Photon noise is not additive

For very bright signals, where ρT exceeds 10^5, the noise fluctuations due to photon statistics can be ignored if the sensor has a sufficiently high saturation level. This will be discussed further in Section 13.7.3 and, in particular, Equation 13.73.

13.6.2 Thermal Noise

An additional, stochastic source of electrons in a CCD well is thermal energy. Electrons can be freed from the CCD material itself through thermal vibration and then, trapped in the CCD well, be indistinguishable from "true" photoelectrons. By cooling the CCD chip it is possible to reduce significantly the number of "thermal electrons" that give rise to thermal noise or dark current. As the integration time T increases, the number of thermal electrons increases. The probability distribution of thermal electrons is also a Poisson process where the rate parameter is an increasing function of temperature. There are alternative techniques (to cooling) for suppressing dark current and these usually involve estimating the average dark current for the given integration time and then subtracting this value from the CCD pixel values before the A/D converter. While this does reduce the dark current average, it does not reduce the dark current standard deviation and it also reduces the possible dynamic range of the signal.

13.6.3 On-Chip Electronic Noise

This noise originates in the process of reading the signal from the sensor, in this case through the field-effect transistor (FET) of a CCD chip. The general form of the power spectral density of readout noise is

Readout noise— $$S_{nn}(\omega) \propto \begin{cases} \omega^{-\beta} & \omega < \omega_{\min} & \beta > 0 \\ k & \omega_{\min} < \omega < \omega_{\max} & \\ \omega^{\alpha} & \omega > \omega_{\max} & \alpha > 0 \end{cases} \tag{13.65}$$

where
α and β are constants
ω is the (radial) frequency at which the signal is transferred from the CCD chip to the "outside world"

At very low readout rates ($\omega < \omega_{\min}$) the noise has a $1/f$ character. Readout noise can be reduced to manageable levels by appropriate readout rates and proper electronics. At very low signal levels (see Equation 13.64), however, readout noise can still become a significant component in the overall SNR [22].

13.6.4 *kTC* Noise

Noise associated with the gate capacitor of an FET is termed *kTC* noise and can be nonnegligible. The output rms value of this noise voltage is given by

kTC noise (voltage)— $$\sigma_{kTC} = \sqrt{\frac{kT}{C}} \tag{13.66}$$

where

C is the FET gate switch capacitance
k is Boltzmann's constant
T is the absolute temperature of the CCD chip measured in K

Using the relationships $Q = C \cdot V = N_{e^-} \cdot e^-$, the output rms value of the *KTC* noise expressed in terms of the number of photoelectrons (N_{e^-}) is given by

$$kTC \text{ noise (electrons)—} \qquad \sigma_{N_e} = \sqrt{\frac{kTC}{e^-}} \tag{13.67}$$

where e^- is the electron charge. For $C = 0.5$ pF and $T = 233$ K this gives $N_{e^-} = 252$ electrons. This value is a "one time" noise per pixel that occurs during signal readout and is thus independent of the integration time (see Sections 13.6.1 and 13.7.7). Proper electronic design that makes use, for example, of correlated double sampling and dual-slope integration can almost completely eliminate *kTC* noise [22].

13.6.5 Amplifier Noise

The standard model for this type of noise is additive, Gaussian, and independent of the signal. In modern well-designed electronics, amplifier noise is generally negligible. The most common exception to this is in color cameras where more amplification is used in the blue color channel than in the green channel or red channel leading to more noise in the blue channel (see also Section 13.7.6).

13.6.6 Quantization Noise

Quantization noise is inherent in the amplitude quantization process and occurs in the analog-to-digital converter (ADC). The noise is additive and independent of the signal when the number of levels $L \geq 16$. This is equivalent to $B \geq 4$ bits (see Section 13.2.1). For a signal that has been converted to electrical form and thus has a minimum and maximum electrical value, Equation 13.40 is the appropriate formula for determining the SNR. If the ADC is adjusted so that 0 corresponds to the minimum electrical value and $2^B - 1$ corresponds to the maximum electrical value then

$$\text{Quantization noise—} \qquad \text{SNR} = 6B + 11 \text{ dB} \tag{13.68}$$

For $B \geq 8$ bits, this means an SNR ≥ 59 dB. Quantization noise can usually be ignored as the total SNR of a complete system is typically dominated by the smallest SNR. In CCD cameras this is photon noise.

13.7 Cameras

The cameras and recording media available for modern digital image processing applications are changing at a significant pace. To dwell too long in this section on one major type of camera, such as the CCD camera, and to ignore developments in areas such as charge injection device (CID) cameras and CMOS cameras is to run the risk of obsolescence. Nevertheless, the techniques that are used to characterize the CCD camera remain "universal" and the presentation that follows is given in the context of modern CCD technology for purposes of illustration.

13.7.1 Linearity

It is generally desirable that the relationship between the input physical signal (e.g., photons) and the output signal (e.g., voltage) be linear. Formally this means (as in Equation 13.20) that if we have two images, a and b, and two arbitrary complex constants, w_1 and w_1 and a linear camera response, then

TABLE 13.8 Comparison of γ of Various Sensors

Sensor	Surface	γ	Possible Advantages
CCD chip	Silicon	1.0	Linear
Vidicon Tube	Sb_2S_3	0.6	Compresses dynamic range $\rightarrow$ high contrast scenes
Film	Silver halide	<1.0	Compresses dynamic range $\rightarrow$ high contrast scenes
Film	Silver halide	>1.0	Expands dynamic range $\rightarrow$ low contrast scenes

$$c = \Re\{w_1 a + w_2 b\} = w_1 \Re\{a\} + w_2 \Re\{b\} \tag{13.69}$$

where

$\Re\{\bullet\}$ is the camera response
c is the camera output

In practice, the relationship between input a and output c is frequently given by

$$c = \text{gain} \cdot a^{\gamma} + \text{offset} \tag{13.70}$$

where γ is the gamma of the recording medium. For a truly linear recording system we must have $\gamma = 1$ and offset $= 0$. Unfortunately, the offset is almost never zero and thus we must compensate for this if the intention is to extract intensity measurements. Compensation techniques are discussed in Section 13.10.1.

Typical values of γ that may be encountered are listed in Table 13.8. Modern cameras often have the ability to switch electronically between various values of γ.

13.7.2 Sensitivity

There are two ways to describe the sensitivity of a camera. First, we can determine the minimum number of detectable photoelectrons. This can be termed the absolute sensitivity. Second, we can describe the number of photoelectrons necessary to change from one digital brightness level to the next, that is, to change one analog-to-digital unit (ADU). This can be termed the relative sensitivity.

13.7.2.1 Absolute Sensitivity

To determine absolute sensitivity, we need a characterization of the camera in terms of its noise. If the total noise has a σ of, say, 100 photoelectrons, then to ensure detectability of a signal we could then say that, at the 3σ level, the minimum detectable signal (or absolute sensitivity) would be 300 photoelectrons. If all the noise sources listed in Section 13.6, with the exception of photon noise, can be reduced to negligible levels, this means that an absolute sensitivity of less than 10 photoelectrons is achievable with modern technology.

13.7.2.2 Relative Sensitivity

The definition of relative sensitivity, S, given above when coupled to the linear case, Equation 13.70 with $\gamma = 1$, leads immediately to the result:

$$S = 1/\text{gain} = \text{gain}^{-1} \tag{13.71}$$

The measurement of the sensitivity or gain can be performed in two distinct ways.

- If, following Equation 13.70, the input signal a can be precisely controlled by either "shutter" time or intensity (through neutral density filters), then the gain can be estimated by estimating the slope of the resulting straight-line curve. To translate this into the desired units, however, a standard

TABLE 13.9 Sensitivity Measurements

Camera (Label)	Pixels	Pixel Size (μm $\times$ μm)	Temp. (K)	S (e^-/ADU)	Bits
C–1	1320 × 1035	6.8 × 6.8	231	7.9	12
C–2	578 × 385	22.0 × 22.0	227	9.7	16
C–3	1320 × 1035	6.8 × 6.8	293	48.1	10
C–4	576 × 384	23.0 × 23.0	238	90.9	12
C–5	756 × 581	11.0 × 5.5	300	109.2	8

Note: A more sensitive camera has a lower value of S.

source must be used that emits a known number of photons onto the camera sensor and the quantum efficiency (η) of the sensor must be known. The quantum efficiency refers to how many photoelectrons are produced—on the average—per photon at a given wavelength. In general $0 \leq \eta(\lambda) \leq 1$.

- If, however, the limiting effect of the camera is only the photon (Poisson) noise (see Section 13.6.1), then an easy-to-implement, alternative technique is available to determine the sensitivity. Using Equations 13.63, 13.70, and 13.71 and after compensating for the offset (see Section 13.10.1), the sensitivity measured from an image c is given by

$$S = \frac{E\{c\}}{\mathrm{Var}\{c\}} = \frac{m_c}{s_c^2} \tag{13.72}$$

where m_c and s_c are defined in Equations 13.34 and 13.36.

Measured data for five modern (1995) CCD camera configurations are given in Table 13.9.

The extraordinary sensitivity of modern CCD cameras is clear from these data. In a scientific-grade CCD camera (C–1), only 8 photoelectrons (approximately 16 photons) separate two gray levels in the digital representation of the image. For a considerably less expensive video camera (C–5), only about 110 photoelectrons (approximately 220 photons) separate two gray levels.

13.7.3 SNR

As described in Section 13.6, in modern camera systems the noise is frequently limited by

- Amplifier noise in the case of color cameras
- Thermal noise which, itself, is limited by the chip temperature K and the exposure time T
- Photon noise which is limited by the photon production rate ρ and the exposure time T

13.7.3.1 Thermal Noise (Dark Current)

Using cooling techniques based upon Peltier cooling elements it is straightforward to achieve chip temperatures of 230–250 K. This leads to low thermal electron production rates. As a measure of the thermal noise, we can look at the number of seconds necessary to produce a sufficient number of thermal electrons to go from one brightness level to the next, an ADU, in the absence of photoelectrons. This last condition—the absence of photoelectrons—is the reason for the name dark current. Measured data for the five cameras described above are given in Table 13.10.

The video camera (C–5) has on-chip dark current suppression (see Section 13.6.2). Operating at room temperature this camera requires more than 20 s to produce one ADU change due to thermal noise. This means at the conventional video frame and integration rates of 25–30 images per second (see Table 13.3), the thermal noise is negligible.

TABLE 13.10 Thermal Noise Characteristics

Camera (Label)	Temp. (K)	Dark Current (s/ADU)
C-1	231	526.3
C-2	227	0.2
C-3	293	8.3
C-4	238	2.4
C-5	300	23.3

13.7.3.2 Photon Noise

From Equation 13.64 we see that it should be possible to increase the SNR by increasing the integration time of our image and thus "capturing" more photons. The pixels in CCD cameras have, however, a finite well capacity. This finite capacity, *C*, means that the maximum SNR for a CCD camera per pixel is given by

$$\text{Capacity-limited photon noise}-\text{SNR} = 10\log_{10}(C)\ \text{dB} \tag{13.73}$$

Theoretical as well as measured data for the five cameras described above are given in Table 13.11.

Note that for certain cameras, the measured SNR achieves the theoretical maximum, indicating that the SNR is, indeed, photon and well capacity limited. Further, the curves of SNR versus T (integration time) are consistent with Equations 13.64 and 13.73 (data not shown). It can also be seen that, as a consequence of CCD technology, the "depth" of a CCD pixel well is constant at about $0.7ke^{-}/\mu m^2$.

13.7.4 Shading

Virtually all imaging systems produce shading. By this we mean that if the physical input image $a(x, y) =$ constant, then the digital version of the image will not be constant. The source of the shading might be outside the camera such as in the scene illumination or the result of the camera itself where a gain and offset might vary from pixel to pixel. The model for shading is given by

$$c[m, n] = \text{gain}[m, n] \cdot a[m, n] + \text{offset}[m, n] \tag{13.74}$$

where $a[m, n]$ is the digital image that would have been recorded if there were no shading in the image, that is, $a[m, n] =$ constant. Techniques for reducing or removing the effects of shading are discussed in Section 13.10.1.

TABLE 13.11 Photon Noise Characteristics

Camera (Label)	$C(\#e^{-})$	Theor. SNR (dB)	Meas. SNR (dB)	Pixel size ($\mu m \times \mu m$)	Well Depth ($\#e^{-}/\mu m^2$)
C-1	32,000	45	45	6.8 × 6.8	692
C-2	340,000	55	55	22.0 × 22.0	702
C-3	32,000	45	43	6.8 × 6.8	692
C-4	400,000	56	52	23.0 × 23.0	756
C-5	40,000	46	43	11.0 × 5.5	661

13.7.5 Pixel Form

While the pixels shown in Figure 13.1 appear to be square and to "cover" the continuous image, it is important to know the geometry for a given camera/digitizer system. In Figure 13.18 we define possible parameters associated with a camera and digitizer and the effect they have upon the pixel.

The parameters X_o and Y_o are the spacing between the pixel centers and represent the sampling distances from Equation 13.52. The parameters X_a and Y_a are the dimensions of that portion of the camera's surface that is sensitive to light. As mentioned in Section 13.2.3, different video digitizers (frame grabbers) can have different values for X_o while they have a common value for Y_o.

13.7.5.1 Square Pixels

As mentioned in Section 13.5, square sampling implies that $X_o = Y_o$ or alternatively $X_o/Y_o = 1$. It is not uncommon, however, to find frame grabbers where $X_o/Y_o = 1.1$ or $X_o/Y_o = 4/3$. (This latter format matches the format of commercial television. (see Table 13.3). The risk associated with nonsquare pixels is that isotropic objects scanned with nonsquare pixels might appear isotropic on a camera-compatible monitor but analysis of the objects (such as length-to-width ratio) will yield nonisotropic results. This is illustrated in Figure 13.19.

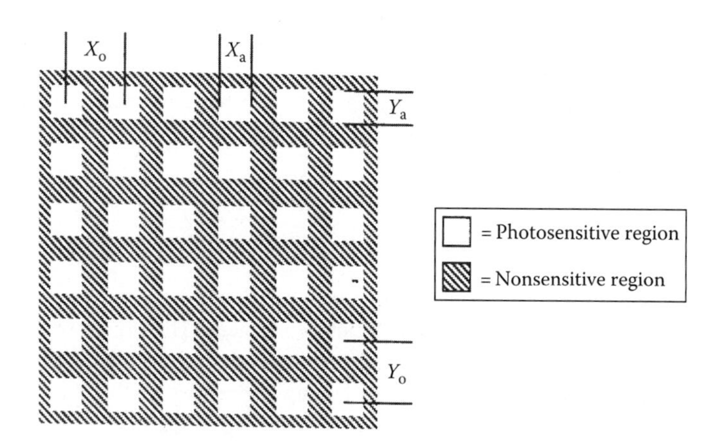

FIGURE 13.18 Pixel form parameters.

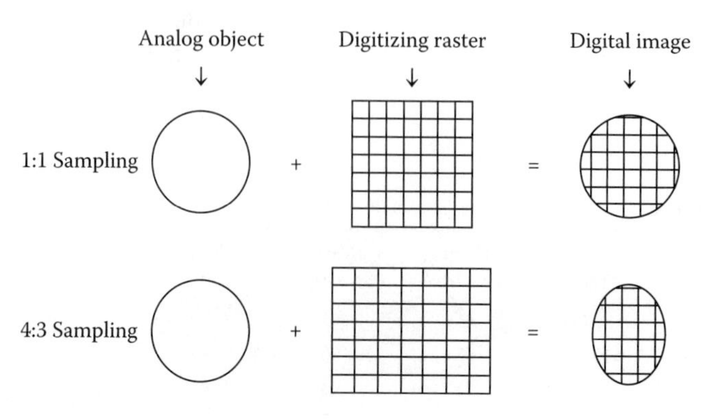

FIGURE 13.19 Effect of nonsquare pixels.

The ratio X_o/Y_o can be determined for any specific camera/digitizer system by using a calibration test chart with known distances in the horizontal and vertical direction. These are straightforward to make with modern laser printers. The test chart can then be scanned and the sampling distances X_o and Y_o determined.

13.7.5.2 Fill Factor

In modern CCD cameras it is possible that a portion of the camera surface is not sensitive to light and is instead used for the CCD electronics or to prevent blooming. Blooming occurs when a CCD well is filled (see Table 13.11) and additional photoelectrons spill over into adjacent CCD wells. Antiblooming regions between the active CCD sites can be used to prevent this. This means, of course, that a fraction of the incoming photons are lost as they strike the nonsensitive portion of the CCD chip. The fraction of the surface that is sensitive to light is termed the fill factor and is given by

$$\text{fill factor} = \frac{X_a \cdot Y_a}{X_o \cdot Y_o} \times 100\% \tag{13.75}$$

The larger the fill factor, the more light will be captured by the chip up to the maximum of 100%. This helps improve the SNR. As a trade-off, however, larger values of the fill factor mean more spatial smoothing due to the aperture effect described in Section 13.5.1.1. This is illustrated in Figure 13.16.

13.7.6 Spectral Sensitivity

Sensors, such as those found in cameras and film, are not equally sensitive to all wavelengths of light. The spectral sensitivity for the CCD sensor is given in Figure 13.20.

The high sensitivity of silicon in the infrared means that, for applications where a CCD (or other silicon-based) camera is to be used as a source of images for digital image processing and analysis, consideration should be given to using an IR blocking filter. This filter blocks wavelengths above 750 nm and thus prevents "fogging" of the image from the longer wavelengths found in sunlight. Alternatively, a CCD-based camera can make an excellent sensor for the near infrared wavelength range of 750–1000 nm.

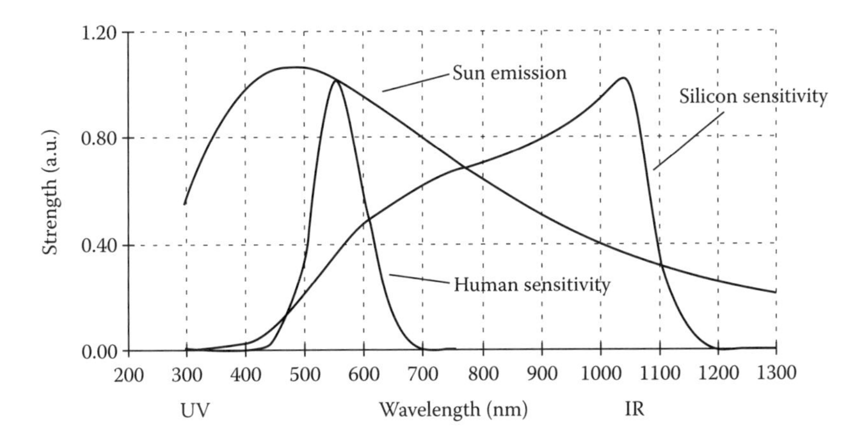

FIGURE 13.20 Spectral characteristics of silicon, the sun, and the human visual system. UV = ultraviolet and IR = infra-red.

TABLE 13.12 Video Camera Readout Rates

Format	Lines/s	Pixels/Line	R (MHz)
NTSC	15,750	(4/3)*525	$\approx$11.0
PAL/SECAM	15,625	(4/3)*625	$\approx$13.0

13.7.7 Shutter Speeds (Integration Time)

The length of time that an image is exposed—that photons are collected—may be varied in some cameras or may vary on the basis of video formats (see Table 13.3). For reasons that have to do with the parameters of photography, this exposure time is usually termed shutter speed although integration time would be a more appropriate description.

13.7.7.1 Video Cameras

Values of the shutter speed as low as 500 ns are available with commercially available CCD video cameras although the more conventional speeds for video are 33.37 ms (NTSC) and 40.0 ms (PAL, SECAM). Values as high as 30 s may also be achieved with certain video cameras although this means sacrificing a continuous stream of video images that contain signal in favor of a single integrated image amongst a stream of otherwise empty images. Subsequent digitizing hardware must be capable of handling this situation.

13.7.7.2 Scientific Cameras

Again values as low as 500 ns are possible and, with cooling techniques based on Peltier cooling or liquid nitrogen cooling, integration times in excess of one hour are readily achieved.

13.7.8 Readout Rate

The rate at which data is read from the sensor chip is termed the readout rate. The readout rate for standard video cameras depends on the parameters of the frame grabber as well as the camera. For standard video (see Section 13.2.3), the readout rate is given by

$$R = \left(\frac{\text{images}}{\text{s}}\right) \cdot \left(\frac{\text{lines}}{\text{images}}\right) \cdot \left(\frac{\text{pixels}}{\text{line}}\right) \tag{13.76}$$

While the appropriate unit for describing the readout rate should be pixels/second, the term Hz is frequently found in the literature and in camera specifications; we shall therefore use the latter unit. For a video camera with square pixels (see Section 13.7.5), this means:

Note that the values in Table 13.12 are approximate. Exact values for square-pixel systems require exact knowledge of the way the video digitizer (frame grabber) samples each video line.

The readout rates used in video cameras frequently means that the electronic noise described in Section 13.6.3 occurs in the region of the noise spectrum (Equation 13.65) described by $\omega > \omega_{max}$ where the noise power increases with increasing frequency. Readout noise can thus be significant in video cameras.

Scientific cameras frequently use a slower readout rate in order to reduce the readout noise. Typical values of readout rate for scientific cameras, such as those described in Tables 13.9 through 13.11, are 20 kHz, 500 kHz, and 1–8 MHz.

13.8 Displays

The displays used for image processing—particularly the display systems used with computers—have a number of characteristics that help determine the quality of the final image.

13.8.1 Refresh Rate

The refresh rate is defined as the number of complete images that are written to the screen per second. For standard video the refresh rate is fixed at the values given in Table 13.3, either 29.97 or 25 images/s. For computer displays the refresh rate can vary with common values being 67 images/s and 75 images/s. At values above 60 images/s visual flicker is negligible at virtually all illumination levels.

13.8.2 Interlacing

To prevent the appearance of visual flicker at refresh rates below 60 images/s, the display can be interlaced as described in Section 13.2.3. Standard interlace for video systems is 2:1. Since interlacing is not necessary at refresh rates above 60 images/s, an interlace of 1:1 is used with such systems. In other words, lines are drawn in an ordinary sequential fashion: $1, 2, 3, 4, \ldots, N$.

13.8.3 Resolution

The pixels stored in computer memory, although they are derived from regions of finite area in the original scene (see Sections 13.5.1 and 13.7.5), may be thought of as mathematical points having no physical extent. When displayed, the space between the points must be filled in. This generally happens as a result of the finite spot size of a CRT. The brightness profile of a CRT spot is approximately Gaussian and the number of spots that can be resolved on the display depends on the quality of the system. It is relatively straightforward to obtain display systems with a resolution of 72 spots per in. (28.3 spots per cm) This number corresponds to standard printing conventions. If printing is not a consideration, then higher resolutions, in excess of 30 spots per cm, are attainable.

13.9 Algorithms

In this section we will describe operations that are fundamental to digital image processing. These operations can be divided into four categories: operations based on the image histogram, on simple mathematics, on convolution, and on mathematical morphology. Further, these operations can also be described in terms of their implementation as a point operation, a local operation, or a global operation as described in Section 13.2.2.1.

13.9.1 Histogram-Based Operations

An important class of point operations is based upon the manipulation of an image histogram or a region histogram. The most important examples are described below.

13.9.1.1 Contrast Stretching

Frequently, an image is scanned in such a way that the resulting brightness values do not make full use of the available dynamic range. This can be easily observed in the histogram of the brightness values shown in Figure 13.6. By stretching the histogram over the available dynamic range we attempt to correct this situation. If the image is intended to go from brightness 0 to brightness $2^B - 1$ (see Section 13.2.1), then one generally maps the 0% value (or minimum as defined in Section 13.3.5.2) to the value 0% and the 100% value (or maximum) to the value $2^B - 1$. The appropriate transformation is given by

$$b[m, n] = (2^B - 1) \cdot \frac{a[m, n] - \text{minimum}}{\text{maximum} - \text{minimum}} \tag{13.77}$$

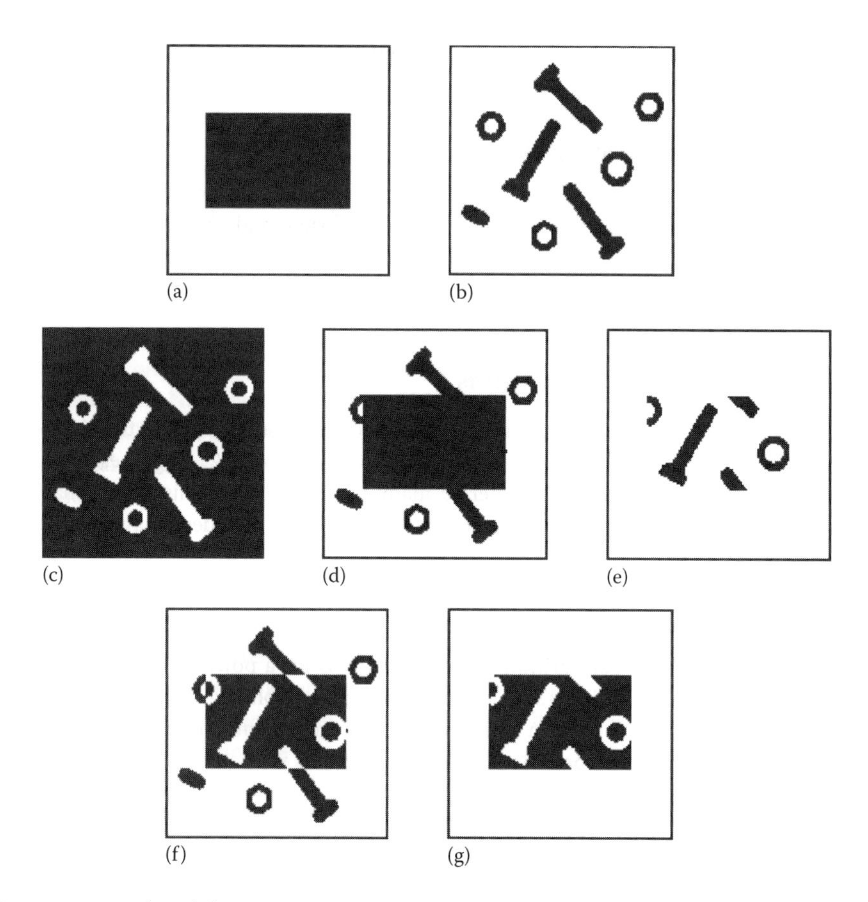

FIGURE 13.22 Examples of the various binary point operations. (a) Image a. (b) Image b. (c) NOT (b) $= \bar{b}$. (d) OR$(a, b) = a + b$. (e) AND$(a, b) = a \cdot b$. (f) XOR$(a, b) = a \oplus b$. (g) SUB$(a, b) = a/b$.

13.9.2.2 Arithmetic-Based Operations

The gray-value point operations that form the basis for image processing are based on ordinary mathematics and include

Operation	Definition	Preferred Data Type
ADD	$c = a + b$	Integer
SUB	$c = a - b$	Integer
MUL	$c = a \cdot b$	Integer or floating point
DIV	$c = a/b$	Floating point
LOG	$c = \log(a)$	Floating point
EXP	$c = \exp(a)$	Floating point
SQRT	$c = \mathrm{sqrt}(a)$	Floating point
TRIG.	$c = \sin/\cos/\tan(a)$	Floating point
INVERT	$c = (2^B - 1) - a$	Integer

(13.83)

13.9.3 Convolution-Based Operations

Convolution, the mathematical, local operation defined in Section 13.3.1 is central to modern image processing. The basic idea is that a window of some finite size and shape—the support—is scanned across the image. The output pixel value is the weighted sum of the input pixels within the window where the weights are the values of the filter assigned to every pixel of the window itself. The window with its weights is called the convolution kernel. This leads directly to the following variation on Equation 13.3. If the filter $h[j, k]$ is zero outside the (odd sized rectangular) window of size $J \times K$ centered around the origin $\{j = {}^{-}J_0, {}^{-}J_0 + 1, \ldots, -1, 0, 1, \ldots, J_0 - 1, J_0;\ k = -K_0, -K_0 + 1, \ldots, -1, 0, 1, \ldots, K_0 - 1, K_0\}$, then, using Equation 13.4, the convolution can be written as the following finite sum:

$$c[m,n] = a[m,n] \otimes h[m,n] = \sum_{j=-J_0}^{J_0} \sum_{k=-K_0}^{K_0} h[j,k]a[m-j, n-k] \tag{13.84}$$

This equation can be viewed as more than just a pragmatic mechanism for smoothing or sharpening an image. Further, while Equation 13.84 illustrates the local character of this operation, Equations 13.10 and 13.24 suggest that the operation can be implemented through the use of the Fourier domain which requires a global operation, the Fourier transform. Both of these aspects will be discussed below.

13.9.3.1 Background

In a variety of image-forming systems an appropriate model for the transformation of the physical signal $a(x, y)$ into an electronic signal $c(x, y)$ is the convolution of the input signal with the impulse response of the sensor system. This system might consist of both an optical as well as an electrical subsystem. If each of these systems can be treated as a linear shift-invariant (LSI) system then the convolution model is appropriate. The definitions of these two, possible, system properties are given below:

Linearity—

$$\begin{array}{ll} \text{If} & a_1 \to c_1 \quad \text{and} \quad a_2 \to c_2 \\ \text{Then} & w_1 \cdot a_1 + w_2 \cdot a_2 \to w_1 \cdot c_1 + w_2 \cdot c_2 \end{array} \tag{13.85}$$

Shift-invariance—

$$\begin{array}{ll} \text{If} & a(x,y) \to c(x,y) \\ \text{Then} & a(x - x_o, y - y_o) \to c(x - x_o, y - y_o) \end{array} \tag{13.86}$$

where

w_1 and w_2 are arbitrary complex constants
x_o and y_o are coordinates corresponding to arbitrary spatial translations

Two remarks are appropriate at this point. First, linearity implies (by choosing $w_1 = w_2 = 0$) that "zero in" gives "zero out." The offset described in Equation 13.70 means that such camera signals are not the output of a linear system and thus (strictly speaking) the convolution result is not applicable. Fortunately, it is straightforward to correct for this nonlinear effect (see Section 13.10.1).

Second, optical lenses with a magnification, M, other than $1\times$ are not shift invariant; a translation of 1 unit in the input image $a(x, y)$ produces a translation of M units in the output image $c(x, y)$. Due to the Fourier property described in Equation 13.25 this case can still be handled by linear system theory.

If an impulse point of light $\delta(x, y)$ is imaged through an LSI system then the impulse response of that system is called the PSF. The output image then becomes the convolution of the input image with the PSF. The Fourier transform of the PSF is called the OTF. For optical systems that are circularly-symmetric, aberration-free, and diffraction-limited the PSF is given by the Airy disk shown in Table 13.4–T.5. The OTF of the Airy disk is also presented in Table 13.4–T.5.

If the convolution window is not the diffraction-limited PSF of the lens but rather the effect of defocusing a lens then an appropriate model for $h(x, y)$ is a pill box of radius a as described in Table 13.4–T.3. The effect on a test pattern is illustrated in Figure 13.23.

The effect of the defocusing is more than just simple blurring or smoothing. The almost periodic negative lobes in the OTF in Table 13.4–T.3 produce a 180° phase shift in which black turns to white and vice versa. The phase shift is clearly visible in Figure 13.23b.

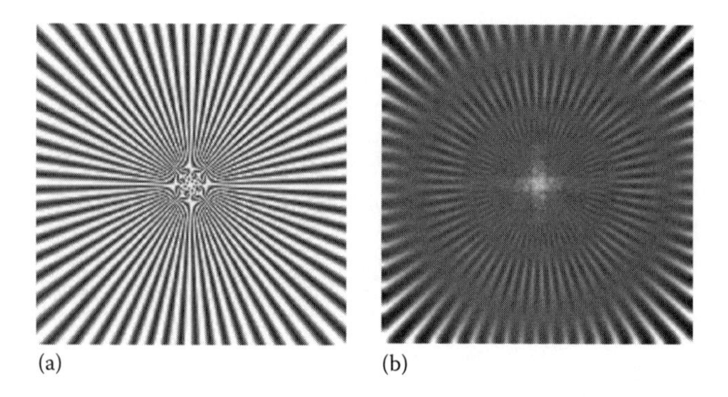

FIGURE 13.23 Convolution of test pattern with a pill box of radius $a=4.5$ pixels. (a) Test pattern. (b) Defocused image.

13.9.3.2 Convolution in the Spatial Domain

In describing filters based on convolution we will use the following convention. Given a filter $h[j,k]$ of dimensions $J \times K = 2J_0 + 1 \times 2K_0 + 1$, we will consider the coordinate $[j=0, k=0]$ to be in the center of the filter matrix, **h**. This is illustrated in Figure 13.24. The "center" is well-defined when the filter sizes are odd.

$$\mathbf{h} = \begin{bmatrix} h[-J_0, -K_0] & \cdots & \cdots & h[0, -K_0] & \cdots & \cdots & h[J, -K_0] \\ \vdots & \ddots & \vdots & \vdots & \vdots & \iddots & \vdots \\ \vdots & \cdots & h[-1,-1] & h[0,-1] & h[1,-1] & \cdots & \vdots \\ h[-J_0, 0] & \cdots & h[-1,0] & h[0,0] & h[1,0] & \cdots & h[J_0, 0] \\ \vdots & \cdots & h[-1,1] & h[0,1] & h[1,+1] & \cdots & \vdots \\ \vdots & \iddots & \vdots & \vdots & \vdots & \ddots & \vdots \\ h[-J_0, K_0] & \cdots & \cdots & h[0, K_0] & \cdots & \cdots & h[J_0, K_0] \end{bmatrix}$$

When we examine the convolution sum (Equation 13.84) closely, several issues become evident.

- Evaluation of formula (13.84) for $m=n=0$ while rewriting the limits of the convolution sum based on the "centering" of $h[j,k]$ shows that values of $a[j,k]$ can be required that are outside the image boundaries:

$$c[0,0] = \sum_{j=-J_0}^{+J_0} \sum_{k=-K_0}^{+K_0} h[j,k]a[-j,-k] \tag{13.87}$$

$$\mathbf{h} = \begin{bmatrix} h\left[-\left(J-\tfrac{1}{2}\right), -\left(K-\tfrac{1}{2}\right)\right] & \cdots & \cdots & h\left[0, -\left(K-\tfrac{1}{2}\right)\right] & \cdots & \cdots & h\left[\left(J-\tfrac{1}{2}\right), -\left(K-\tfrac{1}{2}\right)\right] \\ \vdots & \ddots & \vdots & \vdots & \vdots & \iddots & \vdots \\ \vdots & \cdots & h[-1,-1] & h[0,-1] & h[1,-1] & \cdots & \vdots \\ h\left[-\left(J-\tfrac{1}{2}\right), 0\right] & \cdots & h[-1,0] & h[0,0] & h[1,0] & \cdots & h\left[\left(J-\tfrac{1}{2}\right), 0\right] \\ \vdots & \cdots & h[-1,1] & h[0,1] & h[1,+1] & \cdots & \vdots \\ \vdots & \iddots & \vdots & \vdots & \vdots & \ddots & \vdots \\ h\left[-\left(J-\tfrac{1}{2}\right), \left(K-\tfrac{1}{2}\right)\right] & \cdots & \cdots & h\left[0, \left(K-\tfrac{1}{2}\right)\right] & \cdots & \cdots & h\left[\left(J-\tfrac{1}{2}\right), \left(K-\tfrac{1}{2}\right)\right] \end{bmatrix}$$

FIGURE 13.24 Coordinate system for describing $h[j,k]$.

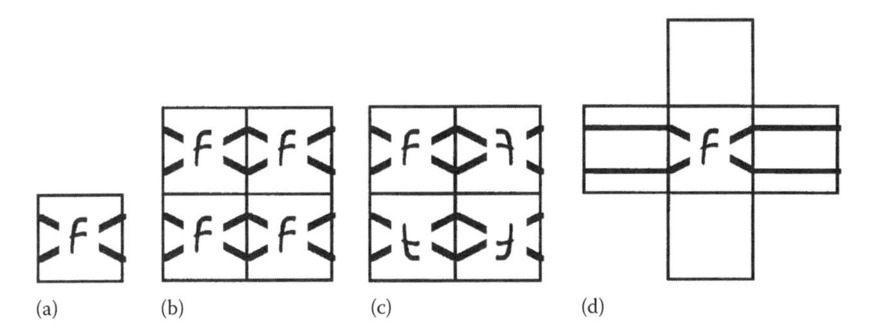

FIGURE 13.25 Examples of various alternatives to extend an image outside its formal boundaries. See text for explanation.

The question arises—what values should we assign to the image $a[m, n]$ for $m < 0$, $m \geq M$, $n < 0$, and $n \geq N$? There is no "answer" to this question. There are only alternatives among which we are free to choose assuming we understand the possible consequences of our choice. The standard alternatives are (1) extend the images with a constant (possibly zero) brightness value, (2) extend the image periodically, (3) extend the image by mirroring it at its boundaries, or (4) extend the values at the boundaries indefinitely. These alternatives are illustrated in Figure 13.25.

- When the convolution sum is written in the standard form (Equation 13.3) for an image $a[m, n]$ of size $M \times N$:

$$c[m, n] = \sum_{j=0}^{M-1} \sum_{k=0}^{N-1} a[j, k]h[m - j, n - k] \tag{13.88}$$

we see that the convolution kernel $h[j, k]$ is mirrored around $j = k = 0$ to produce $h[-j, -k]$ before it is translated by $[m, n]$ as indicated in Equation 13.88. While some convolution kernels in common use are symmetric in this respect, $h[j, k] = h[-j, -k]$, many are not (see Section 13.9.5). Care must therefore be taken in the implementation of filters with respect to the mirroring requirements.

- The computational complexity for a $J \times K$ convolution kernel implemented in the spatial domain on an image of $N \times N$ is $O(J \cdot K)$ where the complexity is measured per pixel on the basis of the number of multiplies-and-adds (MADDs).
- The value computed by a convolution that begins with integer brightnesses for $a[m, n]$ may produce a rational number or a floating point number in the result $c[m, n]$. Working exclusively with integer brightness values will, therefore, cause round off errors.
- Inspection of Equation 13.84 reveals another possibility for efficient implementation of convolution. If the convolution kernel $h[j, k]$ is separable, that is, if the kernel can be written as

$$h[j, k] = h_{\text{row}}[k] \cdot h_{\text{col}}[j] \tag{13.89}$$

then the filtering can be performed as follows:

$$c[m, n] = \sum_{j=-J_0}^{J_0} \left\{ \sum_{k=-K_0}^{K_0} h_{\text{row}}[k]a[m - j, n - k] \right\} h_{\text{col}}[j] \tag{13.90}$$

This means that instead of applying one 2D filter it is possible to apply two one-dimensional filters, the first one in the k direction and the second one in the j direction. For an $N \times N$ image this, in general, reduces the computational complexity per pixel from $O(J \cdot K)$ to $O(J + K)$.

An alternative way of writing separability is to note that the convolution kernel (Figure 13.24) is a matrix **h** and, if separable, **h** can be written as

$$\begin{aligned}[\mathbf{h}] &= [\mathbf{h}_{\mathbf{col}}] \cdot [\mathbf{h}_{\mathbf{row}}]^{t} \\ (J \times K) &= (J \times 1) \cdot (1 \times K)\end{aligned} \tag{13.91}$$

where "t" denotes the matrix transpose operation. In other words, **h** can be expressed as the outer product of a column vector $[\mathbf{h}_{\mathbf{col}}]$ and a row vector $[\mathbf{h}_{\mathbf{row}}]$.

- For certain filters it is possible to find an incremental implementation for a convolution. As the convolution window moves over the image (see Equation 13.88), the leftmost column of image data under the window is shifted out as a new column of image data is shifted in from the right. Efficient algorithms can take advantage of this [24] and, when combined with separable filters as described above, this can lead to algorithms where the computational complexity per pixel is *O* (constant).

13.9.3.3 Convolution in the Frequency Domain

In Section 13.3.4 we indicated that there was an alternative method to implement the filtering of images through convolution. Based on Equation 13.24 it appears possible to achieve the same result as in Equation 13.84 by the following sequence of operations:

$$\begin{aligned}&1.\ \text{Compute } A(\Omega, \Psi) = \mathcal{F}\{a[m, n]\} \\ &2.\ \text{Multiply } A(\Omega, \Psi) \text{ by the precomputed } H(\Omega, \Psi) = \mathcal{F}\{h[m, n]\} \\ &3.\ \text{Compute the result } c[m, n] = \mathcal{F}^{-1}\{A(\Omega, \Psi) \cdot H(\Omega, \Psi)\}\end{aligned} \tag{13.92}$$

- While it might seem that the "recipe" given above in Equation 13.92 circumvents the problems associated with direct convolution in the spatial domain—specifically, determining values for the image outside the boundaries of the image—the Fourier domain approach, in fact, simply "assumes" that the image is repeated periodically outside its boundaries as illustrated in Figure 13.25b. This phenomenon is referred to as circular convolution.

  If circular convolution is not acceptable then the other possibilities illustrated in Figure 13.25 can be realized by embedding the image $a[m, n]$ and the filter $H(\Omega, \Psi)$ in larger matrices with the desired image extension mechanism for $a[m, n]$ being explicitly implemented.
- The computational complexity per pixel of the Fourier approach for an image of $N \times N$ and for a convolution kernel of $K \times K$ is $O({}^{2}\log N)$ complex MADDs independent *of* K. Here we assume that $N > K$ and that N is a highly composite number such as a power of two (see also Section 13.2.1). This latter assumption permits use of the computationally efficient Fast Fourier Transform (FFT) algorithm. Surprisingly then, the indirect route described by Equation 13.92 can be faster than the direct route given in Equation 13.84. This requires, in general, that $K^2 \gg {}^{2}\log N$. The range of K and N for which this holds depends on the specifics of the implementation. For the machine on which this manuscript is being written and the specific image processing package that is being used, for an image of $N = 256$ the Fourier approach is faster than the convolution approach when $K \geq 15$. (It should be noted that in this comparison the direct convolution involves only integer arithmetic while the Fourier domain approach requires complex floating point arithmetic.)

13.9.4 Smoothing Operations

These algorithms are applied in order to reduce noise and/or to prepare images for further processing such as segmentation. We distinguish between linear and non linear algorithms where the former are amenable to analysis in the Fourier domain and the latter are not. We also distinguish between implementations based on a rectangular support for the filter and implementations based on a circular support for the filter.

13.9.4.1 Linear Filters

Several filtering algorithms will be presented together with the most useful supports.

13.9.4.1.1 Uniform Filter

The output image is based on a local averaging of the input filter where all of the values within the filter support have the same weight. In the continuous spatial domain (x, y) the PSF and OTF are given in Table 13.4–T.1 for the rectangular case and in Table 13.4–T.3 for the circular (pill box) case. For the discrete spatial domain $[m, n]$ the filter values are the samples of the continuous domain case. Examples for the rectangular case $(J = K = 5)$ and the circular case $(R = 2.5)$ are shown in Figure 13.26.

Note that in both cases the filter is normalized so that $\Sigma h[j, k] = 1$. This is done so that if the input $a[m, n]$ is a constant then the output image $c[m, n]$ is the same constant. The justification can be found in the Fourier transform property described in Equation 13.26. As can be seen from Table 13.4, both of these filters have OTFs that have negative lobes and can, therefore, lead to phase reversal as seen in Figure 13.23. The square implementation of the filter is separable and incremental; the circular implementation is incremental [24,25].

13.9.4.1.2 Triangular Filter

The output image is based on a local averaging of the input filter where the values within the filter support have differing weights. In general, the filter can be seen as the convolution of two (identical) uniform filters either rectangular or circular and this has direct consequences for the computational complexity [24,25] (see Table 13.13). In the continuous spatial domain the PSF and OTF are given in Table 13.4–T.2 for the rectangular support case and in Table 13.4–T.4 for the circular (pill box) support

$$h_{\text{rect}}[j,k] = \frac{1}{25}\begin{bmatrix} 1 & 1 & 1 & 1 & 1 \\ 1 & 1 & 1 & 1 & 1 \\ 1 & 1 & 1 & 1 & 1 \\ 1 & 1 & 1 & 1 & 1 \\ 1 & 1 & 1 & 1 & 1 \end{bmatrix} \qquad h_{\text{circ}}[j,k] = \frac{1}{21}\begin{bmatrix} 0 & 1 & 1 & 1 & 0 \\ 1 & 1 & 1 & 1 & 1 \\ 1 & 1 & 1 & 1 & 1 \\ 1 & 1 & 1 & 1 & 1 \\ 0 & 1 & 1 & 1 & 0 \end{bmatrix}$$

(a) (b)

FIGURE 13.26 Uniform filters for image smoothing. (a) Rectangular filter $(J = K = 5)$. (b) Circular filter $(R = 2.5)$.

TABLE 13.13 Characteristics of Smoothing Filters

Algorithm	Domain	Type	Support	Separable/Incremental	Complexity/Pixel
Uniform	Space	Linear	Square	Y/Y	$O(\text{Constant})$
Uniform	Space	Linear	Circular	N/Y	$O(K)$
Triangle	Space	Linear	Square	Y/N	$O(\text{Constant})$[a]
Triangle	Space	Linear	Circular	N/N	$O(K)$[a]
Gaussian	Space	Linear	∞[a]	Y/N	$O(\text{Constant})$[a]
Median	Space	Nonlinear	Square	N/Y	$O(K)$[a]
Kuwahara	Space	Nonlinear	Square[a]	N/N	$O(J \cdot K)$
Other	Frequency	Linear	—	—/—	$O(\log N)$

[a] See text for additional explanation.

$$h_{\text{rect}}[j,k] = \frac{1}{81}\begin{bmatrix} 1 & 2 & 3 & 2 & 1 \\ 2 & 4 & 6 & 4 & 2 \\ 3 & 6 & 9 & 6 & 3 \\ 2 & 4 & 6 & 4 & 2 \\ 1 & 2 & 3 & 2 & 1 \end{bmatrix} \qquad h_{\text{circ}}[j,k] = \frac{1}{25}\begin{bmatrix} 0 & 0 & 1 & 0 & 0 \\ 0 & 2 & 2 & 2 & 0 \\ 1 & 2 & 5 & 2 & 1 \\ 0 & 2 & 2 & 2 & 0 \\ 0 & 0 & 1 & 0 & 0 \end{bmatrix}$$

(a) (b)

FIGURE 13.27 Triangular filters for image smoothing. (a) Pyramidal filter ($J = K = 5$). (b) Cone filter ($R = 2.5$).

case. As seen in Table 13.4 the OTFs of these filters do not have negative lobes and thus do not exhibit phase reversal.

Examples for the rectangular support case ($J = K = 5$) and the circular support case ($R = 2.5$) are shown in Figure 13.27. The filter is again normalized so that $\Sigma h[j,k] = 1$.

13.9.4.1.3 Gaussian Filter

The use of the Gaussian kernel for smoothing has become extremely popular. This has to do with certain properties of the Gaussian (e.g., the central limit theorem, minimum space–bandwidth product) as well as several application areas such as edge finding and scale space analysis. The PSF and OTF for the continuous space Gaussian are given in Table 13.4–T6. The Gaussian filter is separable:

$$\begin{aligned} h(x,y) = g_{2D}(x,y) &= \left(\frac{1}{\sqrt{2\pi}\sigma}e^{-(x^2/2\sigma^2)}\right)\cdot\left(\frac{1}{\sqrt{2\pi}\sigma}e^{-(y^2/2\sigma^2)}\right) \\ &= g_{1D}(x)\cdot g_{1D}(y) \end{aligned} \tag{13.93}$$

There are four distinct ways to implement the Gaussian:

1. Convolution using a finite number of samples (N_o) of the Gaussian as the convolution kernel. It is common to choose $N_o = \lceil 3\sigma \rceil$ or $\lceil 5\sigma \rceil$.

$$g_{1D}[n] = \begin{cases} \dfrac{1}{\sqrt{2\pi}\sigma}e^{-(n^2/2\sigma^2)} & |n| \le N_o \\ 0 & |n| > N_o \end{cases} \tag{13.94}$$

2. Repetitive convolution using a uniform filter as the convolution kernel.

$$\begin{aligned} g_{1D}[n] &\approx u[n]\otimes u[n]\otimes u[n] \\ u[n] &= \begin{cases} 1/(2N_o+1) & |n| \le N_o \\ 0 & |n| > N_o \end{cases} \end{aligned} \tag{13.95}$$

The actual implementation (in each dimension) is usually of the form:

$$c[n] = ((a[n]\otimes u[n])\otimes u[n])\otimes u[n] \tag{13.96}$$

This implementation makes use of the approximation afforded by the central limit theorem. For a desired σ with Equation 13.96, we use $N_o = \lceil \sigma \rceil$ although this severely restricts our choice of σ's to integer values.

3. Multiplication in the frequency domain. As the Fourier transform of a Gaussian *is* a Gaussian (see Table 13.4–T.6), this means that it is straightforward to prepare a filter $H(\Omega,\Psi) = G_{2D}(\Omega,\Psi)$

for use with Equation 13.92. To avoid truncation effects in the frequency domain due to the infinite extent of the Gaussian it is important to choose a σ that is sufficiently large. Choosing $\sigma > k/\pi$, where $k = 3$ or 4 will usually be sufficient.

4. Use of a recursive filter implementation. A recursive filter has an infinite impulse response and thus an infinite support. The separable Gaussian filter can be implemented [26] by applying the following recipe in each dimension when $\sigma \geq 0.5$.

$$\begin{array}{l} \text{a. Choose the } \sigma \text{ based on the desired goal of the filtering} \\ \text{b. Determine the parameter } q \text{ based on Equation 13.98} \\ \text{c. Use Equation 13.99 to determine the filter coefficients } \{b_0, b_1, b_2, b_3, B\} \\ \text{d. Apply the forward difference equation, Equation 13.100} \\ \text{e. Apply the backward difference equation, Equation 13.101} \end{array} \tag{13.97}$$

The relation between the desired σ and q is given by

$$q = \begin{cases} .98711\sigma - 0.96330 & \sigma \geq 2.5 \\ 3.97156 - 4.14554\sqrt{1 - .26891\sigma} & 0.5 \leq \sigma \leq 2.5 \end{cases} \tag{13.98}$$

The *filter coefficients* $\{b_0, b_1, b_2, b_3, B\}$ are defined by

$$\begin{aligned} b_0 &= 1.57825 + (2.44413q) + (1.4281q^2) + (0.422205q^3) \\ b_1 &= (2.44413q) + (2.85619q^2) + (1.26661q^3) \\ b_2 &= -(1.4281q^2) - (1.26661q^3) \\ b_3 &= 0.422205q^3 \\ B &= 1 - (b_1 + b_2 + b_3)/b_0 \end{aligned} \tag{13.99}$$

The one-dimensional forward difference equation takes an input row (or column) $a[n]$ and produces an intermediate output result $w[n]$ given by

$$w[n] = Ba[n] + (b_1 w[n-1] + b_2 w[n-2] + b_3 w[n-3])/b_0 \tag{13.100}$$

The one-dimensional backward difference equation takes the intermediate result $w[n]$ and produces the output $c[n]$ given by

$$c[n] = Bw[n] + (b_1 c[n+1] + b_2 c[n+2] + b_3 c[n+3])/b_0 \tag{13.101}$$

The forward equation is applied from $n = 0$ up to $n = N-1$ while the backward equation is applied from $n = N - 1$ down to $n = 0$.

The relative performance of these various implementations of the Gaussian filter can be described as follows. Using the root-square error $\sqrt{\sum_{n=-\infty}^{+\infty} |g[n|\sigma] - h[n]|^2}$ between a true, infinite-extent Gaussian, $g[n|\sigma]$, and an approximated Gaussian, $h[n]$, as a measure of accuracy, the various algorithms described above give the results shown in Figure 13.28a. The relative speed of the various algorithms in shown in Figure 13.28b.

The root-square error measure is extremely conservative and thus all filters, with the exception of "Uniform 3×" for large σ, are sufficiently accurate. The recursive implementation is the fastest independent of σ; the other implementations can be significantly slower. The FFT implementation, for example, is 3.1 times slower for $N = 256$. Further, the FFT requires that N be a highly composite number.

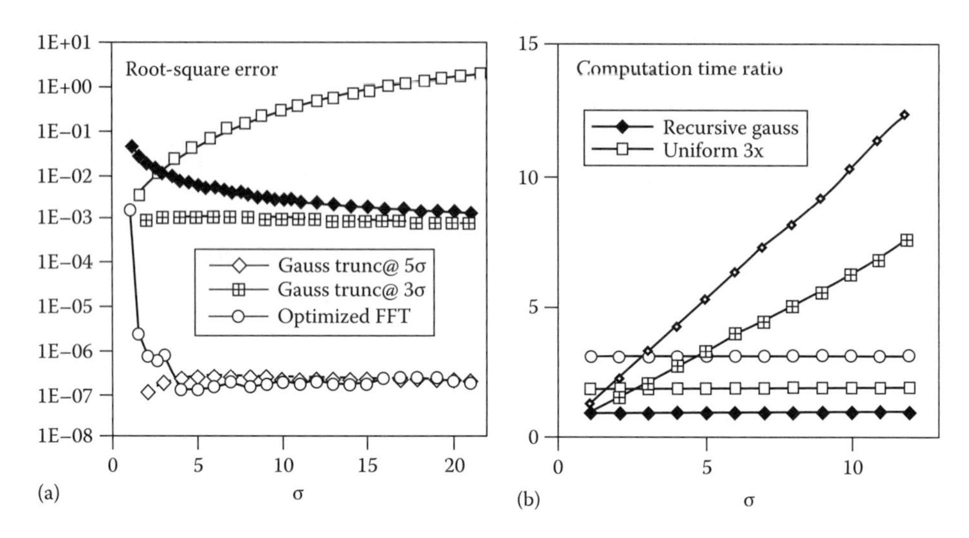

FIGURE 13.28 Comparison of various Gaussian algorithms with $N = 256$. The legend is spread across both graphs. (a) Accuracy comparison. (b) Speed comparison.

13.9.4.1.4 Other

The Fourier domain approach offers the opportunity to implement a variety of smoothing algorithms. The smoothing filters will then be low-pass filters. In general it is desirable to use a low-pass filter that has zero phase so as not to produce phase distortion when filtering the image. The importance of phase was illustrated in Figures 13.5 and 13.23. When the frequency domain characteristics can be represented in an analytic form, then this can lead to relatively straightforward implementations of $H(\Omega, \Psi)$. Possible candidates include the low-pass filters "Airy" and "Exponential Decay" found in Table 13.4–T.5 and Table 13.4–T.8, respectively.

13.9.4.2 Nonlinear Filters

A variety of smoothing filters have been developed that are not linear. While they cannot, in general, be submitted to Fourier analysis, their properties and domains of application have been studied extensively.

13.9.4.2.1 Median Filter

The median statistic was described in Section 13.3.5.2. A median filter is based upon moving a window over an image (as in a convolution) and computing the output pixel as the median value of the brightnesses within the input window. If the window is $J \times K$ in size we can order the $J \cdot K$ pixels in brightness value from smallest to largest. If $J \cdot K$ is odd then the median will be the $(J \cdot K + 1)/2$ entry in the list of ordered brightnesses. Note that the value selected will be exactly equal to one of the existing brightnesses so that no round off error will be involved if we want to work exclusively with integer brightness values. The algorithm as it is described above has a generic complexity per pixel of $O(J \cdot K \cdot \log(J \cdot K))$. Fortunately, a fast algorithm (Huang et al. [23]) exists that reduces the complexity to $O(K)$ assuming $J \geq K$.

A useful variation on the theme of the median filter is the percentile filter. Here the center pixel in the window is replaced not by the 50% (median) brightness value but rather by the p% brightness value where p% ranges from 0% (the minimum filter) to 100% (the maximum filter). Values other than $p = 50\%$ do not, in general, correspond to smoothing filters.

13.9.4.2.2 Kuwahara Filter

Edges play an important role in our perception of images (see Figure 13.15) as well as in the analysis of images. As such it is important to be able to smooth images without disturbing the sharpness and,

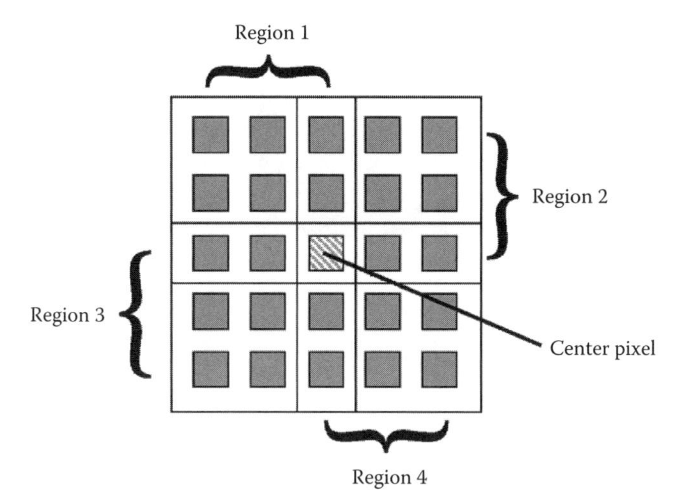

FIGURE 13.29 Four, square regions defined for the Kuwahara filter. In this example $L=1$ and thus $J=K=5$. Each region is $[(J+1)/2]\times[(K+1)/2]$.

if possible, the position of edges. A filter that accomplishes this goal is termed an edge-preserving filter and one particular example is the Kuwahara filter [27]. Although this filter can be implemented for a variety of different window shapes, the algorithm will be described for a square window of size $J=K=4L+1$ where L is an integer. The window is partitioned into four regions as shown in Figure 13.29.

In each of the four regions ($i=1, 2, 3, 4$), the mean brightness, mi in Equation 13.34, and the *variancei*, si^2 in Equation 13.36, are measured. The output value of the center pixel in the window is the mean value of that region that has the smallest variance.

13.9.4.3 Summary of Smoothing Algorithms

The following table summarizes the various properties of the smoothing algorithms presented above. The filter size is assumed to be bounded by a rectangle of $J\times K$ where, without loss of generality, $J\geq K$. The image size is $N\times N$.

Examples of the effect of various smoothing algorithms are shown in Figure 13.30.

13.9.5 Derivative-Based Operations

Just as smoothing is a fundamental operation in image processing, so is the ability to take one or more spatial derivatives of the image. The fundamental problem is that, according to the mathematical definition of a derivative, this cannot be done. A digitized image is not a continuous function $a(x, y)$ of the spatial variables but rather a discrete function $a[m, n]$ of the integer spatial coordinates. As a result the algorithms we will present can only be seen as approximations to the true spatial derivatives of the original spatially-continuous image.

Further, as we can see from the Fourier property in Equation 13.27, taking a derivative multiplies the signal spectrum by either u or v. This means that high frequency noise will be emphasized in the resulting image. The general solution to this problem is to combine the derivative operation with one that suppresses high frequency noise, in short, smoothing in combination with the desired derivative operation.

13.9.5.1 First Derivatives

As an image is a function of two (or more) variables it is necessary to define the direction in which the derivative is taken. For the 2D case we have the horizontal direction, the vertical direction, or an arbitrary direction which can be considered as a combination of the two. If we use $\mathbf{h}_x$ to denote a horizontal

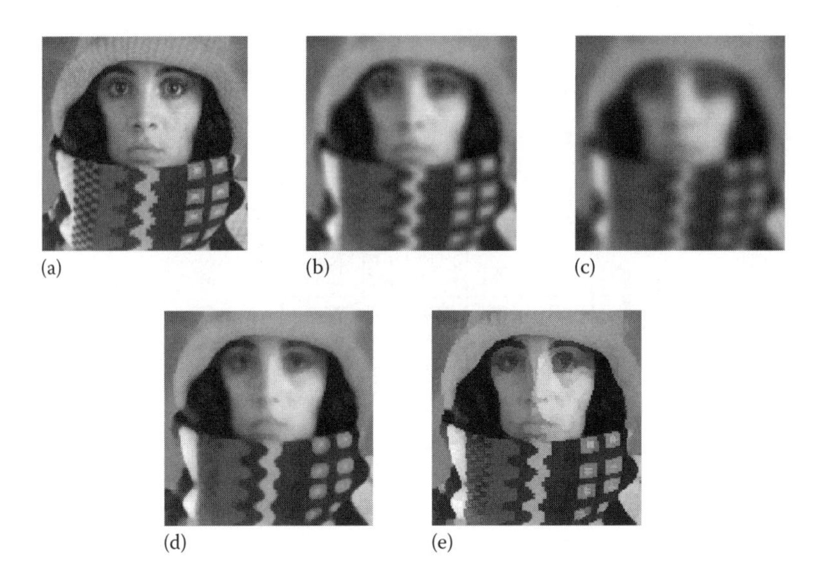

FIGURE 13.30 Illustration of various linear and nonlinear smoothing filters. (a) Original. (b) Uniform 5 × 5. (c) Gaussian ($\sigma = 2.5$). (d) Median 5 × 5 (e) Kuwahara 5 × 5.

derivative filter (matrix), $\mathbf{h}_y$ to denote a vertical derivative filter (matrix), and $\mathbf{h}_\theta$ to denote the arbitrary angle derivative filter (matrix), then

$$[\mathbf{h}_\theta] = \cos\theta \cdot [\mathbf{h}_x] + \sin\theta \cdot [\mathbf{h}_y] \tag{13.102}$$

- *Gradient filters*—It is also possible to generate a vector derivative description as the gradient, $\nabla a[m, n]$, of an image:

$$\nabla a = \frac{\partial a}{\partial x}\vec{i}_x + \frac{\partial a}{\partial y}\vec{i}_y = (h_x \otimes a)\vec{i}_x + (h_y \otimes a)\vec{i}_y \tag{13.103}$$

where $\vec{i}_x$ and $\vec{i}_y$ are unit vectors in the horizontal and vertical direction, respectively. This leads to two descriptions:

Gradient magnitude— $$|\nabla a| = \sqrt{(h_x \otimes a)^2 + (h_y \otimes a)^2} \tag{13.104}$$

and

Gradient direction— $$\psi(\nabla a) = \arctan\{(h_y \otimes a)/(h_x \otimes a)\} \tag{13.105}$$

The gradient magnitude is sometimes approximated by

Approximate gradient magnitude— $$|\nabla a| \cong |h_x \otimes a| + |h_y \otimes a| \tag{13.106}$$

The final results of these calculations depend strongly on the choices of $\mathbf{h}_x$ and $\mathbf{h}_y$. A number of possible choices for $(\mathbf{h}_x, \mathbf{h}_y)$ will now be described.

- *Basic derivative filters*—These filters are specified by

$$\begin{aligned} \text{i. } [\mathbf{h}_x] &= [\mathbf{h}_y]^t = [1 \ \ -1] \\ \text{ii. } [\mathbf{h}_x] &= [\mathbf{h}_y]^t = [1 \ 0 \ \ -1] \end{aligned} \tag{13.107}$$

where "t" denotes matrix transpose. These two filters differ significantly in their Fourier magnitude and Fourier phase characteristics. For the frequency range $0 \leq \Omega \leq \pi$, these are given by

$$\begin{aligned} \text{i. } [\mathbf{h}] = [1 \ \ -1] &\overset{\mathcal{F}}{\leftrightarrow} |H(\Omega)| = 2|\sin(\Omega/2)|; \quad \varphi(\Omega) = (\pi - \Omega)/2 \\ \text{ii. } [\mathbf{h}] = [1 \ 0 \ \ -1] &\overset{\mathcal{F}}{\leftrightarrow} |H(\Omega)| = 2|\sin\Omega|; \quad \varphi(\Omega) = \pi/2 \end{aligned} \tag{13.108}$$

The second form (Equation 13.108ii) gives suppression of high frequency terms ($\Omega \approx \pi$) while the first form (Equation 13.108i) does not. The first form leads to a phase shift; the second form does not.
- *Prewitt gradient filters*—These filters are specified by

$$\begin{aligned} \text{i. } [\mathbf{h}_x] &= \frac{1}{3}\begin{bmatrix} 1 & 0 & -1 \\ 1 & 0 & -1 \\ 1 & 0 & -1 \end{bmatrix} = \frac{1}{3}\begin{bmatrix} 1 \\ 1 \\ 1 \end{bmatrix} \cdot [1 \ 0 \ -1] \\ \text{ii. } [\mathbf{h}_y] &= \frac{1}{3}\begin{bmatrix} 1 & 1 & 1 \\ 0 & 0 & 0 \\ -1 & -1 & -1 \end{bmatrix} = \frac{1}{3}\begin{bmatrix} 1 \\ 0 \\ -1 \end{bmatrix} \cdot [1 \ 1 \ 1] \end{aligned} \tag{13.109}$$

Both $\mathbf{h}_x$ and $\mathbf{h}_y$ are separable. Beyond the computational implications are the implications for the analysis of the filter. Each filter takes the derivative in one direction using Equation 13.107ii and smoothes in the orthogonal direction using a one-dimensional version of a uniform filter as described in Section 13.9.4.1.
- *Sobel gradient filters*—These filters are specified by

$$\begin{aligned} \text{i. } [\mathbf{h}_x] &= \frac{1}{4}\begin{bmatrix} 1 & 0 & -1 \\ 2 & 0 & -2 \\ 1 & 0 & -1 \end{bmatrix} = \frac{1}{4}\begin{bmatrix} 1 \\ 2 \\ 1 \end{bmatrix} \cdot [1 \ 0 \ -1] \\ \text{ii. } [\mathbf{h}_y] &= \frac{1}{4}\begin{bmatrix} 1 & 2 & 1 \\ 0 & 0 & 0 \\ -1 & -2 & -1 \end{bmatrix} = \frac{1}{4}\begin{bmatrix} 1 \\ 0 \\ -1 \end{bmatrix} \cdot [1 \ 2 \ 1] \end{aligned} \tag{13.110}$$

Again, $\mathbf{h}_x$ and $\mathbf{h}_y$ are separable. Each filter takes the derivative in one direction using Equation 13.107ii and smoothes in the orthogonal direction using a one-dimensional version of a *triangular* filter as described in Section 13.9.4.1.
- *Alternative gradient filters*—The variety of techniques available from one-dimensional signal processing for the design of digital filters offer us powerful tools for designing one-dimensional versions of $\mathbf{h}_x$ and $\mathbf{h}_y$. Using the Parks–McClellan filter design algorithm, for example, we can choose the frequency bands where we want the derivative to be taken and the frequency bands where we want the noise to be suppressed. The algorithm will then produce a real, odd filter with a minimum length that meets the specifications.

As an example, if we want a filter that has derivative characteristics in a passband (with weight 1.0) in the frequency range $0.0 \leq \Omega \leq 0.3\pi$ and a stopband (with weight 3.0) in the range $0.32\pi \leq \Omega \leq \pi$, then the algorithm produces the following optimized seven sample filter:

$$[\mathbf{h}_x] = [\mathbf{h}_y] = \frac{1}{16{,}348}[-3{,}571 \quad 8{,}212 \quad -15{,}580 \quad 0 \quad 15{,}580 \quad -8{,}212 \quad 3{,}571] \tag{13.111}$$

The gradient can then be calculated as in Equation 13.103.

- *Gaussian gradient filters*—In modern digital image processing one of the most common techniques is to use a Gaussian filter (see Section 13.9.4.1) to accomplish the required smoothing and one of the derivatives listed in Equation 13.107. Thus, we might first apply the recursive Gaussian in Equation 13.97 followed by Equation 13.110ii to achieve the desired, smoothed derivative filters $\mathbf{h}_x$ and $\mathbf{h}_y$. Further, for computational efficiency, we can combine these two steps as

$$\begin{aligned} w[n] &= \left(\frac{B}{2}\right)(a[n+1] - a[n-1]) + (b_1 w[n-1] + b_2 w[n-2] + b_3 w[n-3])/b_0 \\ c[n] &= Bw[n] + (b_1 c[n+1] + b_2 c[n+2] + b_3 c[n+3])/b_0 \end{aligned} \tag{13.112}$$

  where the various coefficients are defined in Equation 13.99. The first (forward) equation is applied from $n = 0$ up to $n = N - 1$ while the second (backward) equation is applied from $n = N - 1$ down to $n = 0$.
- *Summary*—Examples of the effect of various derivative algorithms on a noisy version of Figure 13.30a (SNR = 29 dB) are shown in Figure 13.31a through c. The effect of various magnitude gradient algorithms on Figure 13.30a are shown in Figure 13.32a through c. After processing, all images are contrast stretched as in Equation 13.77 for display purposes.

The magnitude gradient takes on large values where there are strong edges in the image. Appropriate choice of σ in the Gaussian-based derivative (Figure 13.31c) or gradient (Figure 13.32c) permits

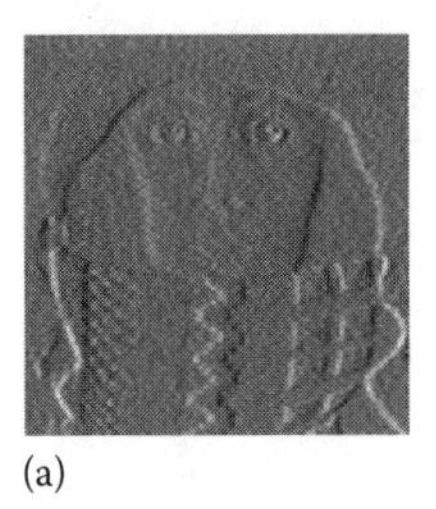
(a)
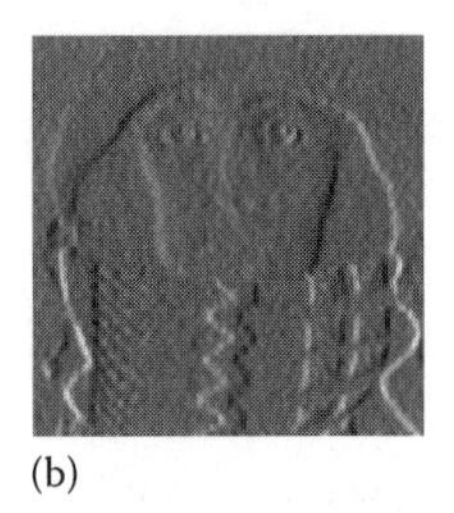
(b)
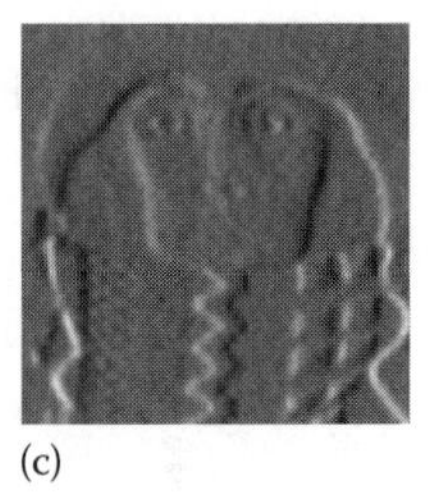
(c)

FIGURE 13.31 Application of various algorithms for $\mathbf{h}_x$—the horizontal derivative. (a) Simple Derivative—Equation 13.107ii. (b) Sobel—Equation 13.110. (c) Gaussian ($\sigma = 1.5$) and Equation 13.107ii.

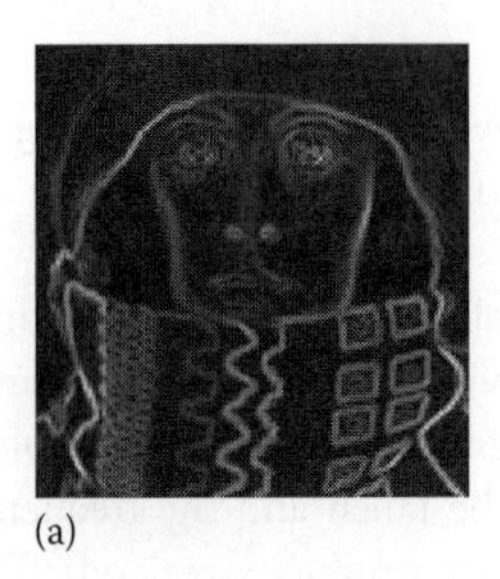
(a)
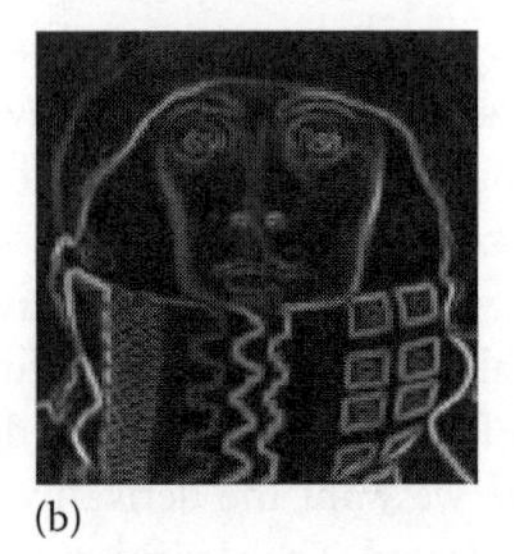
(b)
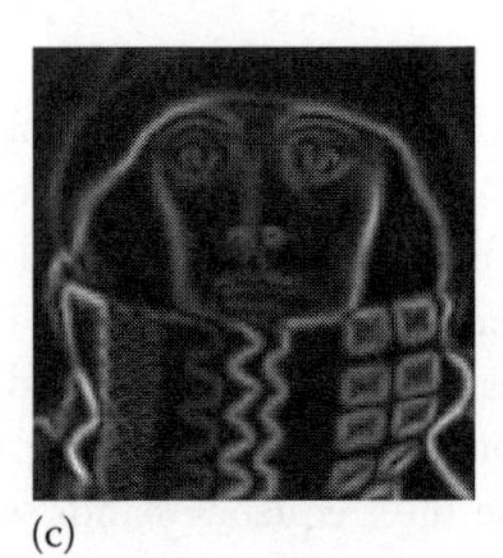
(c)

FIGURE 13.32 Various algorithms for the magnitude gradient, $|\nabla a|$. (a) Simple derivative—Equation 13.107ii. (b) Sobel—Equation 13.110. (c) Gaussian ($\sigma = 1.5$) and Equation 13.107ii.

computation of virtually any of the other forms—simple, Prewitt, Sobel, etc. In that sense, the Gaussian derivative represents a superset of derivative filters.

13.9.5.2 Second Derivatives

It is, of course, possible to compute higher-order derivatives of functions of two variables. In image processing, as we shall see in Sections 13.10.2.1 and 13.10.3.2, the second derivatives or Laplacian play an important role. The Laplacian is defined as

$$\nabla^2 a = \frac{\partial^2 a}{\partial x^2} + \frac{\partial^2 a}{\partial y^2} = (h_{2x} \otimes a) + (h_{2y} \otimes a) \tag{13.113}$$

where $\mathbf{h}_{2x}$ and $\mathbf{h}_{2y}$ are second derivative filters. In the frequency domain we have for the Laplacian filter (from Equation 13.27):

$$\nabla^2 a \quad \overset{\mathcal{F}}{\leftrightarrow} \quad -(u^2 + v^2)A(u, v) \tag{13.114}$$

The OTF of a Laplacian corresponds to a parabola $H(u, v) = -(u^2 + v^2)$.

- *Basic second derivative filter*—This filter is specified by

$$[\mathbf{h}_{2x}] = [\mathbf{h}_{2y}]^{\mathrm{t}} = [1 \quad -2 \quad 1] \tag{13.115}$$

and the frequency spectrum of this filter, in each direction, is given by

$$H(\Omega) = \boldsymbol{F}\{1 \quad -2 \quad 1\} = -2(1 - \cos\Omega) \tag{13.116}$$

over the frequency range $-\pi \le \Omega \le \pi$. The two, one-dimensional filters can be used in the manner suggested by Equation 13.113 or combined into one, 2D filter as

$$[\mathbf{h}] = \begin{bmatrix} 0 & 1 & 0 \\ 1 & -4 & 1 \\ 0 & 1 & 0 \end{bmatrix} \tag{13.117}$$

and used as in Equation 13.84.

- *Frequency domain Laplacian*—This filter is the implementation of the general recipe given in Equation 13.92 and for the Laplacian filter takes the form:

$$c[m, n] = \mathcal{F}^{-1}\{-(\Omega^2 + \Psi^2)A(\Omega, \Psi)\} \tag{13.118}$$

- *Gaussian second derivative filter*—This is the straightforward extension of the Gaussian first derivative filter described above and can be applied independently in each dimension. We first apply Gaussian smoothing with a σ chosen on the basis of the problem specification. We then apply the desired second derivative filter (Equation 13.115 or 13.118). Again there is the choice among the various Gaussian smoothing algorithms.

  For efficiency, we can use the recursive implementation and combine the two steps—smoothing and derivative operation—as follows:

$$\begin{aligned} w[n] &= B(a[n] - a[n-1]) + (b_1 w[n-1] + b_2 w[n-2] + b_3 w[n-3])/b_0 \\ c[n] &= B(w[n+1] - w[n]) + (b_1 c[n+1] + b_2 c[n+2] + b_3 c[n+3])/b_0 \end{aligned} \tag{13.119}$$

where the various coefficients are defined in Equation 13.99. Again, the first (forward) equation is applied from $n = 0$ up to $n = N-1$ while the second (backward) equation is applied from $n = N-1$ down to $n = 0$.

- *Alternative Laplacian filters*—Again, one-dimensional digital filter design techniques offer us powerful methods to create filters that are optimized for a specific problem. Using the Parks–McClellan design algorithm, we can choose the frequency bands where we want the second derivative to be taken and the frequency bands where we want the noise to be suppressed. The algorithm will then produce a real, even filter with a minimum length that meets the specifications. As an example, if we want a filter that has second derivative characteristics in a passband (with weight 1.0) in the frequency range $0.0 \leq \Omega \leq 0.3\pi$ and a stopband (with weight 3.0) in the range $0.32\pi \leq \Omega \leq \pi$, then the algorithm produces the following optimized seven sample filter:

$$[\mathbf{h}_x] = [\mathbf{h}_y]^t = \frac{1}{11{,}043}[-3{,}448 \quad 10{,}145 \quad 1{,}495 \quad -16{,}383 \quad 1{,}495 \quad 10{,}145 \quad -3{,}448] \quad (13.120)$$

The Laplacian can then be calculated as in Equation 13.113.

- *SDGD filter*—A filter that is especially useful in edge finding and object measurement is the second-derivative-in-the-gradient-direction (SDGD) filter. This filter uses five partial derivatives:

$$\begin{aligned} A_{xx} &= \frac{\partial^2 a}{\partial x^2} & A_{xy} &= \frac{\partial^2 a}{\partial x \partial y} & A_x &= \frac{\partial a}{\partial x} \\ A_{yx} &= \frac{\partial^2 a}{\partial x \partial y} & A_{yy} &= \frac{\partial^2 a}{\partial y^2} & A_y &= \frac{\partial a}{\partial y} \end{aligned} \quad (13.121)$$

Note that $A_{xy} = A_{yx}$ which accounts for the five derivatives.
This SDGD combines the different partial derivatives as follows:

$$\text{SDGD}(a) = \frac{A_{xx}A_x^2 + 2A_{xy}A_xA_y + A_{yy}A_y^2}{A_x^2 + A_y^2} \quad (13.122)$$

As one might expect, the large number of derivatives involved in this filter implies that noise suppression is important and that Gaussian derivative filters—both first and second order—are highly recommended if not required [28]. It is also necessary that the first and second derivative filters have essentially the same passbands and stopbands. This means that if the first derivative filter h_{1x} is given by [1 0−1] (Equation 13.107b), then the second derivative filter should be given by $h_{1x} \otimes h_{1x} = h_{2x} = [1\ 0 -2\ 0\ 1]$.

- *Summary*—The effects of the various second derivative filters are illustrated in Figure 13.33a through e. All images were contrast stretched for display purposes using Equation 13.78 and the parameters 1% and 99%.

13.9.5.3 Other Filters

An infinite number of filters, both linear and nonlinear, are possible for image processing. It is therefore impossible to describe more than the basic types in this section. The description of others can be found be in the reference literature (see Section 13.11) as well as in the applications literature. It is important to use a small consistent set of test images that are relevant to the application area to understand the effect of a given filter or class of filters. The effect of filters on images can be frequently understood by the use of images that have pronounced regions of varying sizes to visualize the effect on edges or by the use of test patterns such as sinusoidal sweeps to visualize the effects in the frequency domain. The former have been used above (Figures 13.21, 13.23, and 13.30 through 13.33) and the latter are demonstrated below in Figure 13.34.

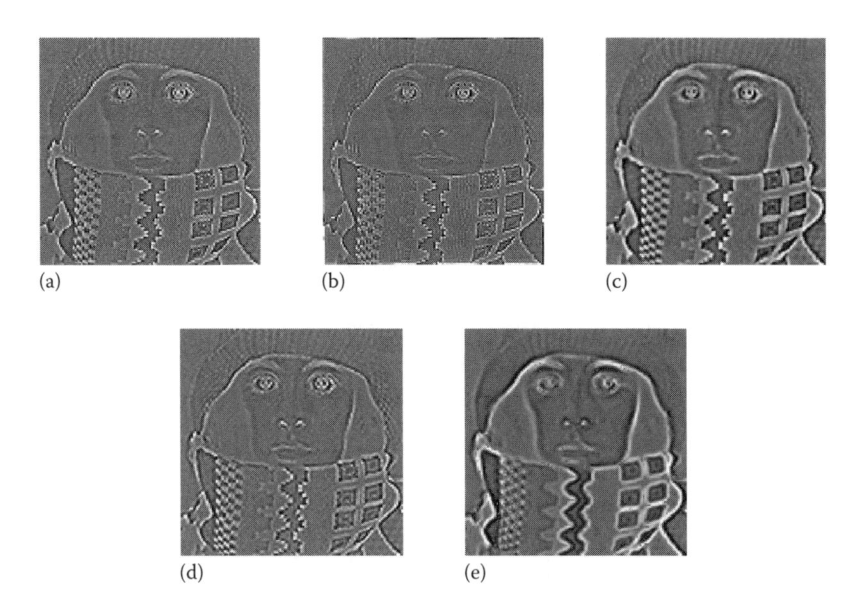

FIGURE 13.33 Various algorithms for the Laplacian and Laplacian-related filters. (a) Laplacian—Equation 13.117. (b) Fourier parabola—Equation 13.118. (c) Gaussian ($\sigma = 1.0$) and Equation 13.117. (d) "Designer"—Equation 13.120. (e) SDGD ($\sigma = 1.0$)—Equation 13.122.

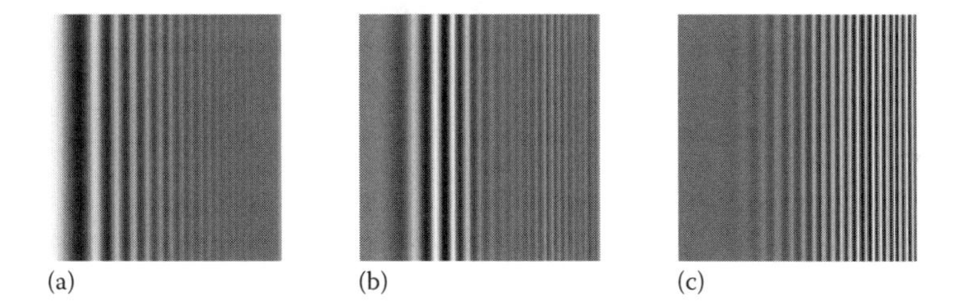

FIGURE 13.34 Various convolution algorithms applied to sinusoidal test image. (a) Low-pass filter. (b) Band-pass filter. (c) High-pass filter.

13.9.6 Morphology-Based Operations

In Section 13.1 we defined an image as an (amplitude) function of two, real (coordinate) variables $a(x, y)$ or two, discrete variables $a[m, n]$. An alternative definition of an image can be based on the notion that an image consists of a set (or collection) of either continuous or discrete coordinates. In a sense the set corresponds to the points or pixels that belong to the objects in the image. This is illustrated in Figure 13.35 which contains two objects or sets $\boldsymbol{A}$ and $\boldsymbol{B}$. Note that the coordinate system is required. For the moment we will consider the pixel values to be binary as discussed in Sections 13.2.1 and 13.9.2.1. Further we shall restrict our discussion to discrete space (Z^2). More general discussions can be found in Refs. [6,7,29].

The object $\boldsymbol{A}$ consists of those pixels α that share some common property:

Object—
$$\boldsymbol{A} = \{\alpha | \text{property}(\alpha) == \text{TRUE}\} \tag{13.123}$$

As an example, object $\boldsymbol{B}$ in Figure 13.35 consists of $\{[0, 0], [1, 0], [0, 1]\}$.

The background of $\boldsymbol{A}$ is given by $\boldsymbol{A}^c$ (the complement of $\boldsymbol{A}$) which is defined as those elements that are not in A:

Background—
$$\boldsymbol{A}^c = \{\alpha | \alpha \notin A\} \tag{13.124}$$

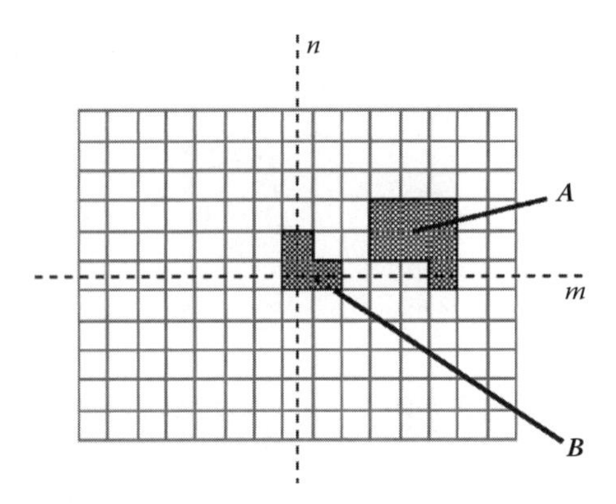

FIGURE 13.35 A binary image containing two object sets $\mathbf{A}$ and $\mathbf{B}$.

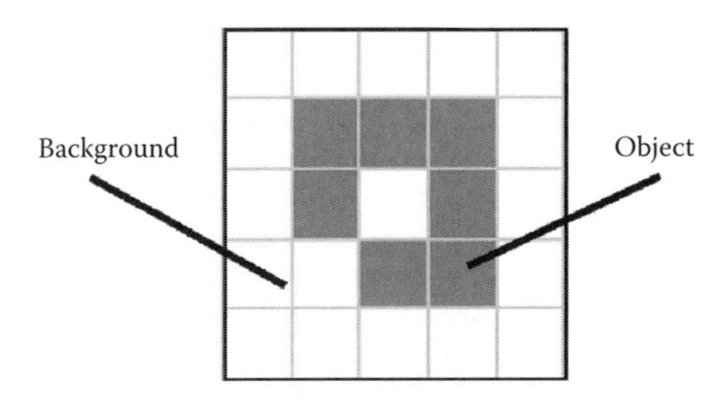

FIGURE 13.36 A binary image requiring careful definition of object and background connectivity.

In Figure 13.3 we introduced the concept of neighborhood connectivity. We now observe that if an object $\mathbf{A}$ is defined on the basis of C-connectivity ($C = 4$, 6, or 8) then the background $\mathbf{A}^c$ has a connectivity given by $12-C$. The necessity for this is illustrated for the Cartesian grid in Figure 13.36.

13.9.6.1 Fundamental Definitions

The fundamental operations associated with an object are the standard set operations union, intersection, and complement $\{\cup, \cap, {}^c\}$ plus translation:

- Translation—Given a vector $\mathbf{x}$ and a set $\mathbf{A}$, the translation, $\mathbf{A} + \mathbf{x}$, is defined as

$$\mathbf{A} + \mathbf{x} = \{\alpha + \mathbf{x} | \alpha \in \mathbf{A}\} \tag{13.125}$$

Note that since we are dealing with a digital image composed of pixels at integer coordinate positions (Z^2), this implies restrictions on the allowable translation vectors $\mathbf{x}$.

The basic Minkowski set operations—addition and subtraction—can now be defined. First we note that the individual elements that comprise $\mathbf{B}$ are not only pixels but also vectors as they have a clear coordinate position with respect to $[0, 0]$. Given two sets $\mathbf{A}$ and $\mathbf{B}$:

Minkowski addition—
$$\mathbf{A} \oplus \mathbf{B} = \bigcup_{\beta \in \mathbf{B}} (\mathbf{A} + \beta) \tag{13.126}$$

Minkowski subtraction—
$$\mathbf{A} \ominus \mathbf{B} = \bigcap_{\beta \in \mathbf{B}} (\mathbf{A} + \beta) \tag{13.127}$$

13.9.6.2 Dilation and Erosion

From these two Minkowski operations we define the fundamental mathematical morphology operations dilation and erosion:

Dilation—
$$D(\boldsymbol{A},\boldsymbol{B}) = \boldsymbol{A} \oplus \boldsymbol{B} = \bigcup_{\beta \in \boldsymbol{B}} (\boldsymbol{A} + \beta) \tag{13.128}$$

Erosion—
$$E(\boldsymbol{A},\boldsymbol{B}) = \boldsymbol{A} \ominus \tilde{\boldsymbol{B}} = \bigcap_{\beta \in \boldsymbol{B}} (\boldsymbol{A} - \beta) \tag{13.129}$$

where $\tilde{\boldsymbol{B}} = \{-\beta | \beta \in \boldsymbol{B}\}$. These two operations are illustrated in Figure 13.37 for the objects defined in Figure 13.35.

While either set $\boldsymbol{A}$ or $\boldsymbol{B}$ can be thought of as an "image," $\boldsymbol{A}$ is usually considered as the image and $\boldsymbol{B}$ is called a structuring element. The structuring element is to mathematical morphology what the convolution kernel is to linear filter theory.

Dilation, in general, causes objects to dilate or grow in size; erosion causes objects to shrink. The amount and the way that they grow or shrink depend upon the choice of the structuring element. Dilating or eroding without specifying the structural element makes no more sense than trying to low-pass filter an image without specifying the filter. The two most common structuring elements (given a Cartesian grid) are the 4-connected and 8-connected sets, $\boldsymbol{N_4}$ and $\boldsymbol{N_8}$. They are illustrated in Figure 13.38.

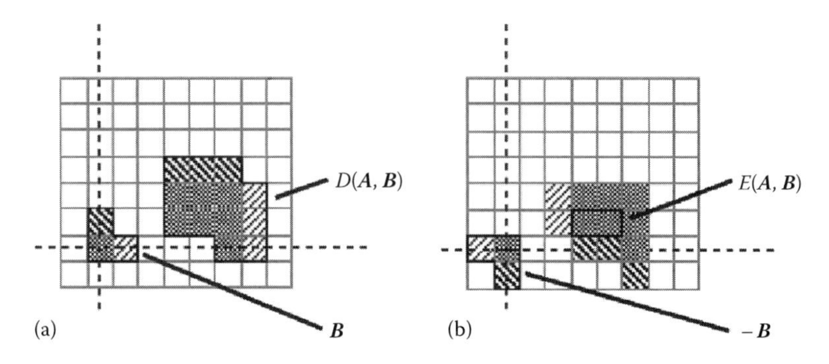

FIGURE 13.37 A binary image containing two object sets $\boldsymbol{A}$ and $\boldsymbol{B}$. The three pixels in B are "color-coded" as is their effect in the result. (a) Dilation $D(\boldsymbol{A}, \boldsymbol{B})$ (b) Erosion $E(\boldsymbol{A}, \boldsymbol{B})$.

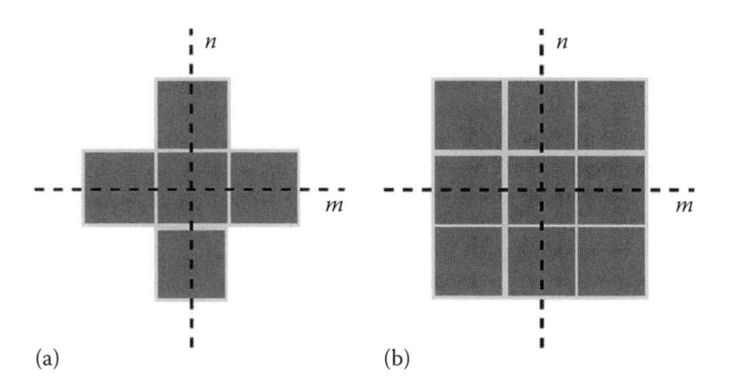

FIGURE 13.38 The standard structuring elements $\boldsymbol{N_4}$ and $\boldsymbol{N_8}$. (a) $\boldsymbol{N_4}$ (b) $\boldsymbol{N_8}$.

Dilation and erosion have the following properties:

Commutative—
$$D(\boldsymbol{A},\boldsymbol{B}) = \boldsymbol{A} \oplus \boldsymbol{B} = \boldsymbol{B} \oplus \boldsymbol{A} = D(\boldsymbol{B},\boldsymbol{A}) \tag{13.130}$$

Noncommutative—
$$E(\boldsymbol{A},\ \boldsymbol{B}) \neq E(\boldsymbol{B},\boldsymbol{A}) \tag{13.131}$$

Associative—
$$\boldsymbol{A} \oplus (\boldsymbol{B} \oplus \boldsymbol{C}) = (\boldsymbol{A} \oplus \boldsymbol{B}) \oplus \boldsymbol{C} \tag{13.132}$$

Translation invariance—
$$\boldsymbol{A} \oplus (\boldsymbol{B} + \mathbf{x}) = (\boldsymbol{A} \oplus \boldsymbol{B}) + \mathbf{x} \tag{13.133}$$

Duality—
$$\begin{aligned} D^c(\boldsymbol{A},\boldsymbol{B}) &= E(\boldsymbol{A}^c,\tilde{\boldsymbol{B}}) \\ E^c(\boldsymbol{A},\boldsymbol{B}) &= D(\boldsymbol{A}^c,\tilde{\boldsymbol{B}}) \end{aligned} \tag{13.134}$$

With $\boldsymbol{A}$ as an object and $\boldsymbol{A}^c$ as the background, Equation 13.134 says that the dilation of an object is equivalent to the erosion of the background. Likewise, the erosion of the object is equivalent to the dilation of the background.

Except for special cases:

Noninverses—
$$D(E(\boldsymbol{A},\boldsymbol{B}),\boldsymbol{B}) \neq \boldsymbol{A} \neq E(D(\boldsymbol{A},\boldsymbol{B}),\boldsymbol{B}) \tag{13.135}$$

Erosion has the following translation property:

Translation invariance—
$$\boldsymbol{A} \ominus (\boldsymbol{B} + \mathbf{x}) = (\boldsymbol{A} + \mathbf{x}) \ominus \boldsymbol{B} = (\boldsymbol{A} \ominus \boldsymbol{B}) + \mathbf{x} \tag{13.136}$$

Dilation and erosion have the following important properties. For any arbitrary structuring element $\boldsymbol{B}$ and two image objects $\boldsymbol{A}_1$ and $\boldsymbol{A}_2$ such that $\boldsymbol{A}_1 \subset \boldsymbol{A}_2$ ($\boldsymbol{A}_1$ is a proper subset of $\boldsymbol{A}_2$):

Increasing in $\boldsymbol{A}$—
$$\begin{aligned} D(\boldsymbol{A}_1,\boldsymbol{B}) &\subset D(\boldsymbol{A}_2,\boldsymbol{B}) \\ E(\boldsymbol{A}_1,\boldsymbol{B}) &\subset E(\boldsymbol{A}_2,\boldsymbol{B}) \end{aligned} \tag{13.137}$$

For two structuring elements $\boldsymbol{B}_1$ and $\boldsymbol{B}_2$ such that $\boldsymbol{B}_1 \subset \boldsymbol{B}_2$:

Decreasing in $\boldsymbol{B}$—
$$E(\boldsymbol{A},\boldsymbol{B}_1) \supset E(\boldsymbol{A},\boldsymbol{B}_2) \tag{13.138}$$

The decomposition theorems below make it possible to find efficient implementations for morphological filters.

Dilation—
$$\boldsymbol{A} \oplus (\boldsymbol{B} \cup \boldsymbol{C}) = (\boldsymbol{A} \oplus \boldsymbol{B}) \cup (\boldsymbol{A} \oplus \boldsymbol{C}) = (\boldsymbol{B} \cup \boldsymbol{C}) \oplus \boldsymbol{A} \tag{13.139}$$

Erosion—
$$\boldsymbol{A} \ominus (\boldsymbol{B} \cup \boldsymbol{C}) = (\boldsymbol{A} \ominus \boldsymbol{B}) \cap (\boldsymbol{A} \ominus \boldsymbol{C}) \tag{13.140}$$

Erosion—
$$(\boldsymbol{A} \ominus \boldsymbol{B}) \ominus \boldsymbol{C} = \boldsymbol{A} \ominus (\boldsymbol{B} \oplus \boldsymbol{C}) \tag{13.141}$$

Multiple dilations—
$$n\boldsymbol{B} = \underbrace{(\boldsymbol{B} \oplus \boldsymbol{B} \oplus \boldsymbol{B} \oplus \cdots \oplus \boldsymbol{B})}_{n\ \text{times}} \tag{13.142}$$

An important decomposition theorem is due to Vincent [30]. First, we require some definitions. A convex set (in R^2) is one for which the straight line joining any two points in the set consists of points that are also in the set. Care must obviously be taken when applying this definition to discrete pixels as the concept of a "straight line" must be interpreted appropriately in Z^2. A set is bounded if each of its elements has a finite magnitude, in this case distance to the origin of the coordinate system. A set is symmetric if $\boldsymbol{B} = \tilde{\mathbf{B}}$. The sets $\boldsymbol{N}_4$ and $\boldsymbol{N}_8$ in Figure 13.38 are examples of convex, bounded, symmetric sets.

Vincent's theorem, when applied to an image consisting of discrete pixels, states that for a bounded, symmetric structuring element $\boldsymbol{B}$ that contains no holes and contains its own center, $[0, 0] \in \boldsymbol{B}$:

$$D(\boldsymbol{A}, \boldsymbol{B}) = \boldsymbol{A} \oplus \boldsymbol{B} = \boldsymbol{A} \cup (\partial\boldsymbol{A} \oplus \boldsymbol{B}) \tag{13.143}$$

where $\partial\boldsymbol{A}$ is the contour of the object. That is, $\partial\boldsymbol{A}$ is the set of pixels that have a background pixel as a neighbor. The implication of this theorem is that it is not necessary to process all the pixels in an object in order to compute a dilation or (using Equation 13.134) an erosion. We only have to process the boundary pixels. This also holds for all operations that can be derived from dilations and erosions. The processing of boundary pixels instead of object pixels means that, except for pathological images, computational complexity can be reduced from $O(N^2)$ to $O(N)$ for an $N \times N$ image. A number of "fast" algorithms can be found in the literature that are based on this result [30–32]. The simplest dilation and erosion algorithms are frequently described as follows.

- *Dilation*—Take each binary object pixel (with value "1") and set all background pixels (with value "0") that are *C*-connected to that object pixel to the value "1"
- *Erosion*—Take each binary object pixel (with value "1") that is *C*-connected to a background pixel and set the object pixel value to "0"

Comparison of these two procedures to Equation 13.143 where $\boldsymbol{B} = \boldsymbol{N}_C = 4$ or $\boldsymbol{N}_C = 8$ shows that they are equivalent to the formal definitions for dilation and erosion. The procedure is illustrated for dilation in Figure 13.39.

13.9.6.3 Boolean Convolution

An arbitrary binary image object (or structuring element) $\boldsymbol{A}$ can be represented as

$$\boldsymbol{A} \leftrightarrow \sum_{k=-\infty}^{+\infty} \sum_{j=-\infty}^{+\infty} a[j, k] \cdot \delta[m - j, n - k] \tag{13.144}$$

where Σ and $\cdot$ are the Boolean operations "OR" and "AND" as defined in Equations 13.81 and 13.82, $a[j, k]$ is a characteristic function that takes on the Boolean values "1" and "0" as follows:

$$a[j, k] = \begin{cases} 1 & a \in \boldsymbol{A} \\ 0 & a \notin \boldsymbol{A} \end{cases} \tag{13.145}$$

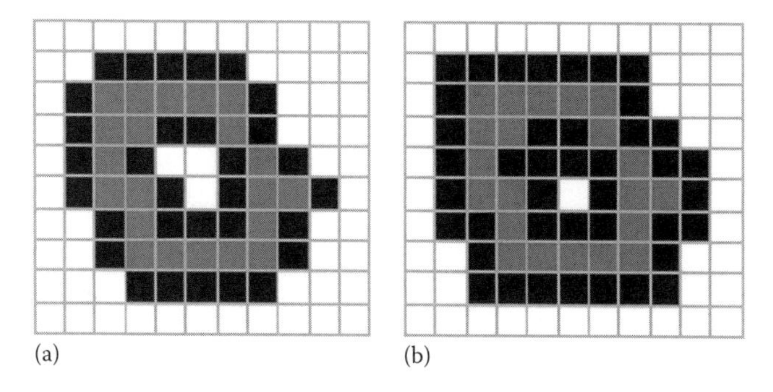

FIGURE 13.39 Illustration of dilation. Original object pixels are in gray; pixels added through dilation are in black. (a) $\boldsymbol{B} = \boldsymbol{N}_4$ (b) $\boldsymbol{B} = \boldsymbol{N}_8$.

and $\delta[m, n]$ is a Boolean version of the Dirac delta function that takes on the Boolean values "1" and "0" as follows:

$$\delta[j, k] = \begin{cases} 1 & j = k = 0 \\ 0 & \text{otherwise} \end{cases} \tag{13.146}$$

Dilation for binary images can therefore be written as

$$D(\boldsymbol{A}, \boldsymbol{B}) = \sum_{k=-\infty}^{+\infty} \sum_{j=-\infty}^{+\infty} a[j, k] \cdot b[m - j, n - k] = \boldsymbol{a} \otimes \boldsymbol{b} \tag{13.147}$$

which, because Boolean "OR" and "AND" are commutative, can also be written as

$$D(\boldsymbol{A}, \boldsymbol{B}) = \sum_{k=-\infty}^{+\infty} \sum_{j=-\infty}^{+\infty} a[m - j, n - k] \cdot b[j, k] = \boldsymbol{b} \otimes \boldsymbol{a} = D(\boldsymbol{B}, \boldsymbol{A}) \tag{13.148}$$

Using De Morgan's theorem

$$\overline{(a + b)} = \bar{a} \cdot \bar{b} \quad \text{and} \quad \overline{(a \cdot b)} = \bar{a} + \bar{b} \tag{13.149}$$

on Equation 13.148 together with Equation 13.134, erosion can be written as

$$E(\boldsymbol{A}, \boldsymbol{B}) = \prod_{k=-\infty}^{+\infty} \prod_{j=-\infty}^{+\infty} \left(a[m - j, n - k] + \bar{b}[-j, -k]\right) \tag{13.150}$$

Thus, dilation and erosion on binary images can be viewed as a form of convolution over a Boolean algebra.

In Section 13.9.3.2 we saw that, when convolution is employed, an appropriate choice of the boundary conditions for an image is essential. Dilation and erosion—being a Boolean convolution—are no exception. The two most common choices are that either everything outside the binary image is "0" or everything outside the binary image is "1."

13.9.6.4 Opening and Closing

We can combine dilation and erosion to build two important higher order operations:

Opening— $$O(\boldsymbol{A}, \boldsymbol{B}) = \boldsymbol{A} \circ \boldsymbol{B} = D(E(\boldsymbol{A}, \boldsymbol{B}), \boldsymbol{B}) \tag{13.151}$$

Closing— $$C(\boldsymbol{A}, \boldsymbol{B}) = \boldsymbol{A} \cdot \boldsymbol{B} = E\big(D(\boldsymbol{A}, \tilde{\boldsymbol{B}}), \tilde{\boldsymbol{B}}\big) \tag{13.152}$$

The opening and closing have the following properties:

Duality— $$\begin{aligned} C^c(\boldsymbol{A}, \boldsymbol{B}) &= O(\boldsymbol{A}^c, \boldsymbol{B}) \\ O^c(\boldsymbol{A}, \boldsymbol{B}) &= C(\boldsymbol{A}^c, \boldsymbol{B}) \end{aligned} \tag{13.153}$$

Translation— $$\begin{aligned} O(\boldsymbol{A} + \boldsymbol{x}, \boldsymbol{B}) &= O(\boldsymbol{A}, \boldsymbol{B}) + \boldsymbol{x} \\ C(\boldsymbol{A} + \boldsymbol{x}, \boldsymbol{B}) &= C(\boldsymbol{A}, \boldsymbol{B}) + \boldsymbol{x} \end{aligned} \tag{13.154}$$

For the opening with structuring element $\boldsymbol{B}$ and images $\boldsymbol{A}$, $\boldsymbol{A}_1$, and $\boldsymbol{A}_2$, where $\boldsymbol{A}_1$ is a subimage of $\boldsymbol{A}_2$ $(\boldsymbol{A}_1 \subseteq \boldsymbol{A}_2)$:

Antiextensivity— $$O(\boldsymbol{A},\boldsymbol{B}) \subseteq \boldsymbol{A} \tag{13.155}$$

Increasing monotonicity— $$O(\boldsymbol{A}_1,\boldsymbol{B}) \subseteq O(\boldsymbol{A}_2,\boldsymbol{B}) \tag{13.156}$$

Idempotence— $$O(O(\boldsymbol{A},\boldsymbol{B}),\boldsymbol{B}) = O(\boldsymbol{A},\boldsymbol{B}) \tag{13.157}$$

For the closing with structuring element $\boldsymbol{B}$ and images $\boldsymbol{A}$, $\boldsymbol{A}_1$, and $\boldsymbol{A}_2$, where $\boldsymbol{A}_1$ is a subimage of $\boldsymbol{A}_2(\boldsymbol{A}_1 \subseteq \boldsymbol{A}_2)$:

Extensivity— $$\boldsymbol{A} \subseteq C(\boldsymbol{A},\boldsymbol{B}) \tag{13.158}$$

Increasing monotonicity— $$C(\boldsymbol{A}_1,\boldsymbol{B}) \subseteq C(\boldsymbol{A}_2,\boldsymbol{B}) \tag{13.159}$$

Idempotence— $$C(C(\boldsymbol{A},\boldsymbol{B}),\boldsymbol{B}) = C(\boldsymbol{A},\boldsymbol{B}) \tag{13.160}$$

The two properties given by Equations 13.155 and 13.84 are so important to mathematical morphology that they can be considered as the reason for defining erosion with $\tilde{\boldsymbol{B}}$ instead of $\boldsymbol{B}$ in Equation 13.129.

13.9.6.5 Hit-and-Miss Operation

The hit-or-miss operator was defined by Serra but we shall refer to it as the hit-and-miss operator and define it as follows. Given an image $\boldsymbol{A}$ and two structuring elements $\boldsymbol{B}_1$ and $\boldsymbol{B}_2$, the set definition and Boolean definition are

Hit-and-mis— $$\text{Hitmiss}(\boldsymbol{A},\boldsymbol{B}_1,\boldsymbol{B}_2) = \begin{cases} E(\boldsymbol{A},\boldsymbol{B}_1) \cap E(\boldsymbol{A}^c,\boldsymbol{B}_2) \\ E(\boldsymbol{A},\boldsymbol{B}_1) \cdot E(\bar{\boldsymbol{A}},\boldsymbol{B}_2) \end{cases} \tag{13.161}$$

where $\boldsymbol{B}_1$ and $\boldsymbol{B}_2$ are bounded, disjoint structuring elements. (Note the use of the notation from Equation 13.81. Two sets are disjoint if $\boldsymbol{B}_1 \cap \boldsymbol{B}_2 = \emptyset$, the empty set. In an important sense the hit-and-miss operator is the morphological equivalent of template matching, a well-known technique for matching patterns based upon cross-correlation. Here, we have a template $\boldsymbol{B}_1$ for the object and a template $\boldsymbol{B}_2$ for the background.

13.9.6.6 Summary of the Basic Operations

The results of the application of these basic operations on a test image are illustrated below. In Figure 13.40 the various structuring elements used in the processing are defined. The value "–" indicates a "don't care." All three structuring elements are symmetric.

The results of processing are shown in Figure 13.41 where the binary value "1" is shown in black and the value "0" in white.

The opening operation can separate objects that are connected in a binary image. The closing operation can fill in small holes. Both operations generate a certain amount of smoothing on an object contour given a "smooth" structuring element. The opening smoothes from the inside of the object

$$B = N_8 = \begin{bmatrix} 1 & 1 & 1 \\ 1 & 1 & 1 \\ 1 & 1 & 1 \end{bmatrix} \qquad B_1 = \begin{bmatrix} - & - & - \\ - & 1 & - \\ - & - & - \end{bmatrix} \qquad B_2 = \begin{bmatrix} - & 1 & - \\ 1 & - & 1 \\ - & 1 & - \end{bmatrix}$$

(a) (b) (c)

FIGURE 13.40 Structuring elements $\boldsymbol{B}$ (a), $\boldsymbol{B}_1$ (b), and $\boldsymbol{B}_2$ (c) that are 3×3 and symmetric.

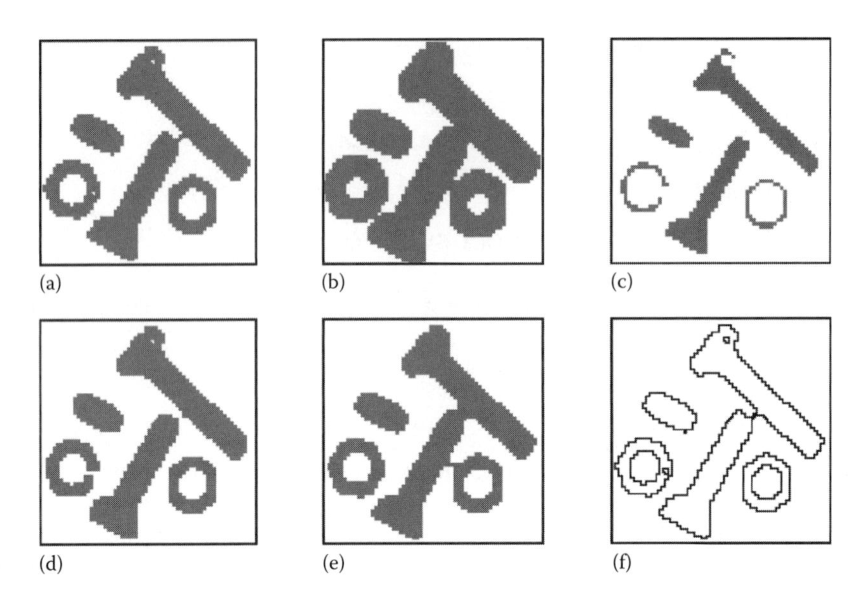

FIGURE 13.41 Examples of various mathematical morphology operations. (a) Image $\boldsymbol{A}$. (b) Dilation with $2\boldsymbol{B}$. (c) Erosion with $2\boldsymbol{B}$. (d) Opening with $2\boldsymbol{B}$. (e) Closing with $2\boldsymbol{B}$. (f) *8-c* contour: $\boldsymbol{A}-\boldsymbol{E}(\boldsymbol{A}, \boldsymbol{N}_8)$.

contour and the closing smoothes from the outside of the object contour. The hit-and-miss example has found the 4-connected contour pixels. An alternative method to find the contour is simply to use the relation:

$$\text{4-connected contour}-\partial \boldsymbol{A} = \boldsymbol{A} - E(\boldsymbol{A}, \boldsymbol{N}_8) \tag{13.162}$$

or

$$\text{8-connected contour}-\partial \boldsymbol{A} = \boldsymbol{A} - E(\boldsymbol{A}, \boldsymbol{N}_4) \tag{13.163}$$

13.9.6.7 Skeleton

The informal definition of a skeleton is a line representation of an object that is

1. One-pixel thick
2. Through the middle of the object (13.164)
3. Preserves the topology of the object

These are not always realizable. Figure 13.42 shows why this is the case.

In the first example, Figure 13.42a, it is not possible to generate a line that is one pixel thick and in the center of an object while generating a path that reflects the simplicity of the object. In Figure 13.42b it is not possible to remove a pixel from the 8-connected object and simultaneously preserve the topology—the notion of connectedness—of the object. Nevertheless, there are a variety of techniques that attempt to achieve this goal and to produce a skeleton.

A basic formulation is based on the work of Lantuéjoul [33]. The skeleton subset $\boldsymbol{S}_k(\boldsymbol{A})$ is defined as

$$\text{Skeleton subsets}-\boldsymbol{S}_k(\boldsymbol{A}) = E(\boldsymbol{A}, k\boldsymbol{B}) - [E(\boldsymbol{A}, k\boldsymbol{B})\boldsymbol{B}] \quad k = 0, 1, \ldots K \tag{13.165}$$

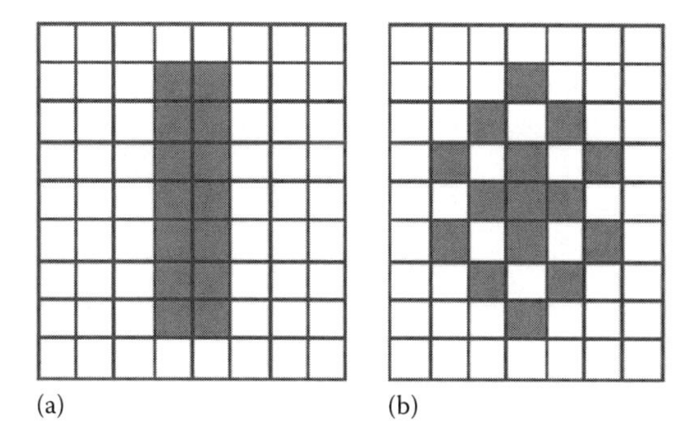

FIGURE 13.42 Counter examples to the three requirements.

where K is the largest value of k before the set $\mathbf{S}_k(\mathbf{A})$ becomes empty. (From Equation 13.156, $E(\mathbf{A}, k\mathbf{B}) \circ \mathbf{B} \subseteq E(\mathbf{A}, k\mathbf{B})$). The structuring element $\mathbf{B}$ is chosen (in Z^2) to approximate a circular disc, that is, convex, bounded and symmetric. The skeleton is then the union of the skeleton subsets:

$$\text{Skeleton}-\mathbf{S}(\mathbf{A}) = \bigcup_{k=0}^{K} \mathbf{S}_k(\mathbf{A}) \tag{13.166}$$

An elegant side effect of this formulation is that the original object can be reconstructed given knowledge of the skeleton subsets $\mathbf{S}_k(\mathbf{A})$, the structuring element $\mathbf{B}$, and K:

$$\text{Reconstruction}-\mathbf{A} = \bigcup_{k=0}^{K} (\mathbf{S}_k(\mathbf{A}) \oplus k\mathbf{B}) \tag{13.167}$$

This formulation for the skeleton, however, does not preserve the topology, a requirement described in Equation 13.164.

An alternative point-of-view is to implement a thinning, an erosion that reduces the thickness of an object without permitting it to vanish. A general thinning algorithm is based on the hit-and-miss operation:

$$\text{Thinning}-\text{Thin}(\mathbf{A}, \mathbf{B}_1, \mathbf{B}_2) = \mathbf{A} - \text{Hitmiss}(\mathbf{A}, \mathbf{B}_1, \mathbf{B}_2) \tag{13.168}$$

Depending on the choice of $\mathbf{B}_1$ and $\mathbf{B}_2$, a large variety of thinning algorithms—and through repeated application skeletonizing algorithms—can be implemented.

A quite practical implementation can be described in another way. If we restrict ourselves to a 3×3 neighborhood, similar to the structuring element $\mathbf{B} = \mathbf{N}_8$ in Figure 13.40a, then we can view the thinning operation as a window that repeatedly scans over the (binary) image and sets the center pixel to "0" under certain conditions. The center pixel is not changed to "0" if and only if

(1) An isolated pixel is found (e.g., Figure 13.43a)
(2) Removing a pixel would change the connectivity (e.g., Figure 13.43b) (13.169)
(3) Removing a pixel would shorten a line (e.g., Figure 13.43c)

As pixels are (potentially) removed in each iteration, the process is called a conditional erosion. Three test cases of Equation 13.169 are illustrated in Figure 13.43. In general all possible rotations and variations

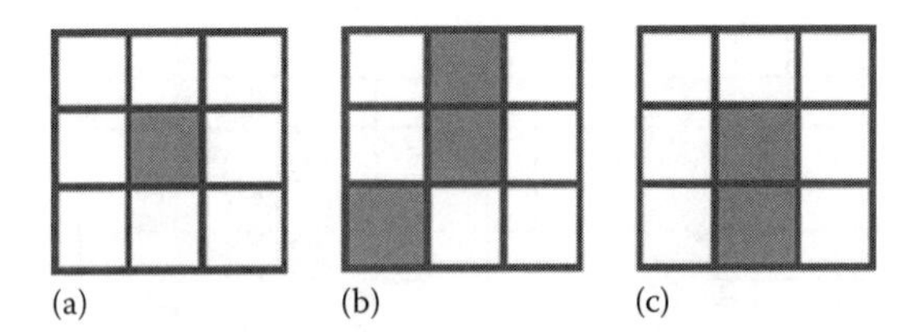

FIGURE 13.43 Test conditions for conditional erosion of the center pixel. (a) Isolated pixel. (b) Connectivity pixel. (c) End pixel.

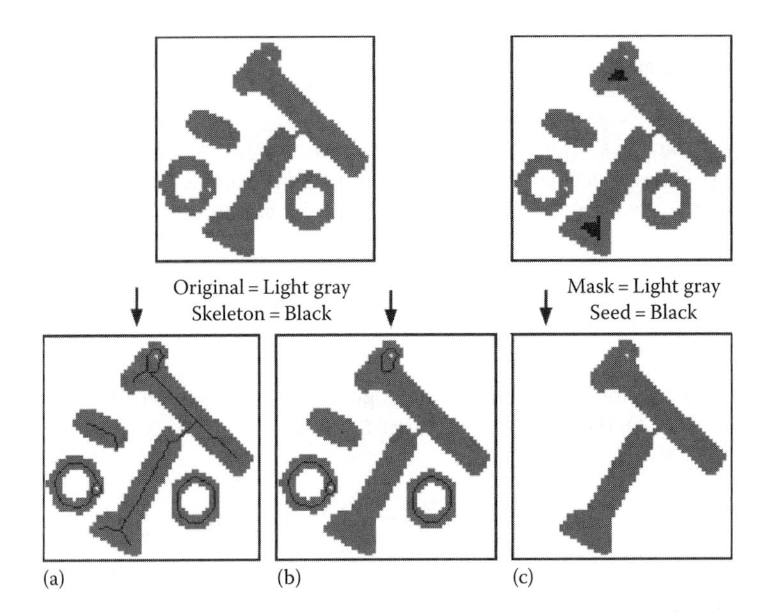

FIGURE 13.44 Examples of skeleton and propagation. (a) Skeleton with end pixels (Equation 13.169 i+ii+iii). (b) Skeleton without end pixels (Equation 13.169 i+ii). (c) Propagation with N_8.

have to be checked. As there are only 512 possible combinations for a 3×3 window on a binary image, this can be done easily with the use of a lookup table.

If only condition 1 is used, then each object will be reduced to a single pixel. This is useful if we wish to count the number of objects in an image. If only condition 2 is used, then holes in the objects will be found. If conditions $1 + 2$ are used each object will be reduced to either a single pixel if it does not contain a hole or to closed rings if it does contain holes. If conditions $1 + 2 + 3$ are used then the "complete skeleton" will be generated as an approximation to Equation 13.56. Illustrations of these various possibilities are given in Figure 13.44a and b.

13.9.6.8 Propagation

It is convenient to be able to reconstruct an image that has "survived" several erosions or to fill an object that is defined, for example, by a boundary. The formal mechanism for this has several names including region-filling, reconstruction, and propagation. The formal definition is given by the following algorithm. We start with a seed image $\boldsymbol{S}(0)$, a mask image $\boldsymbol{A}$, and a structuring element $\boldsymbol{B}$. We then use dilations of $\boldsymbol{S}$ with structuring element $\boldsymbol{B}$ and masked by $\boldsymbol{A}$ in an iterative procedure as follows:

$$\text{Iteration}k\text{–}\boldsymbol{S}^{(k)} = \left[\boldsymbol{S}^{(k-1)} \oplus \boldsymbol{B}\right] \cap \boldsymbol{A} \quad \text{until} \quad \boldsymbol{S}^{(k)} = \boldsymbol{S}^{(k-1)} \tag{13.170}$$

With each iteration the seed image grows (through dilation) but within the set (object) defined by $\boldsymbol{A}$; $\boldsymbol{S}$ propagates to fill $\boldsymbol{A}$. The most common choices for $\boldsymbol{B}$ are $\boldsymbol{N}_4$ or $\boldsymbol{N}_8$. Several remarks are central to the use of propagation. First, in a straightforward implementation, as suggested by Equation 13.170, the computational costs are extremely high. Each iteration requires $O(N^2)$ operations for an $N \times N$ image and with the required number of iterations this can lead to a complexity of $O(N^3)$. Fortunately, a recursive implementation of the algorithm exists in which one or two passes through the image are usually sufficient, meaning a complexity of $O(N^2)$. Second, although we have not paid much attention to the issue of object/background connectivity until now (see Figure 13.36), it is essential that the connectivity implied by $\boldsymbol{B}$ be matched to the connectivity associated with the boundary definition of $\boldsymbol{A}$ (see Equations 13.162 and 13.163). Finally, as mentioned earlier, it is important to make the correct choice ("0" or "1") for the boundary condition of the image. The choice depends upon the application.

13.9.6.9 Summary of Skeleton and Propagation

The application of these two operations on a test image is illustrated in Figure 13.44. In Figure 13.44a,b the skeleton operation is shown with the endpixel condition (Equation 13.169i+ii+iii) and without the end pixel condition (Equation 13.169i+ii). The propagation operation is illustrated in Figure 13.44c. The original image, shown in light gray, was eroded by $E(\boldsymbol{A}, 6\boldsymbol{N}_8)$ to produce the seed image shown in black. The original was then used as the mask image to produce the final result. The border value in both images was "0."

Several techniques based upon the use of skeleton and propagation operations in combination with other mathematical morphology operations will be given in Section 13.10.3.3.

13.9.6.10 Gray-Value Morphological Processing

The techniques of morphological filtering can be extended to gray-level images. To simplify matters we will restrict our presentation to structuring elements, $\boldsymbol{B}$, that comprise a finite number of pixels and are convex and bounded. Now, however, the structuring element has gray values associated with every coordinate position as does the image $\boldsymbol{A}$.

- Gray-level dilation, $D_G(\cdot)$, is given by

$$\text{Dilation}-D_G(\boldsymbol{A}, \boldsymbol{B}) = \max_{[j,k] \in \boldsymbol{B}} \{a[m-j, n-k] + b[j,k]\} \tag{13.171}$$

  For a given output coordinate $[m, n]$, the structuring element is summed with a shifted version of the image and the maximum encountered over all shifts within the $J \times K$ domain of $\boldsymbol{B}$ is used as the result. Should the shifting require values of the image $\boldsymbol{A}$ that are outside the $M \times N$ domain of $\boldsymbol{A}$, then a decision must be made as to which model for image extension, as described in Section 13.9.3.2, should be used.

- Gray-level erosion, $E_G(\cdot)$, is given by

$$\text{Erosion}-E_G(\boldsymbol{A}, \boldsymbol{B}) = \min_{[j,k] \in \boldsymbol{B}} \{a[m+j, n+k] - b[j,k]\} \tag{13.172}$$

  The duality between gray-level erosion and gray-level dilation—the gray-level counterpart of Equation 13.134 is

$$\text{Duality}- \begin{array}{l} E_G(\boldsymbol{A}, \boldsymbol{B}) = -D_G(-\boldsymbol{A}, \tilde{\boldsymbol{B}}) \\ D_G(\boldsymbol{A}, \boldsymbol{B}) = -E_G(-\boldsymbol{A}, \tilde{\boldsymbol{B}}) \end{array} \tag{13.173}$$

  where
  "$\tilde{\boldsymbol{A}}$" means that $a[j,k] \to a[-j,-k]$
  "$-\boldsymbol{A}$" means that $a[j,k] \to -a[j,k]$

The definitions of higher order operations such as gray-level opening and gray-level closing are

$$\text{Opening}-O_G(\boldsymbol{A},\boldsymbol{B}) = D_G(E_G(\boldsymbol{A},\boldsymbol{B}),\boldsymbol{B}) \tag{13.174}$$

$$\text{Closing}-C_G(\boldsymbol{A},\boldsymbol{B}) = E_G\left(D_G(\boldsymbol{A},\tilde{\boldsymbol{B}}),\tilde{\boldsymbol{B}}\right) \tag{13.175}$$

The duality between gray-level opening and gray-level closing is

$$\text{Duality}-\begin{array}{l} O_G(\boldsymbol{A},\boldsymbol{B}) = -C_G(-\boldsymbol{A},\boldsymbol{B}) \\ C_G(\boldsymbol{A},\boldsymbol{B}) = -O_G(-\boldsymbol{A},\boldsymbol{B}) \end{array} \tag{13.176}$$

The important properties that were discussed earlier such as idempotence, translation invariance, increase in A, and so forth are also applicable to gray level morphological processing. The details can be found in Giardina and Dougherty [6].

In many situations the seeming complexity of gray level morphological processing is significantly reduced through the use of symmetric structuring elements where $b[j,k] = b[-j,-k]$. The most common of these is based on the use of $\boldsymbol{B} = \text{constant} = 0$. For this important case and using again the domain $[j,k] \in \boldsymbol{B}$, the definitions above reduce to

$$\text{Dilation}-D_G(\boldsymbol{A},\boldsymbol{B}) = \max_{[j,k]\in\boldsymbol{B}} \{a[m-j,n-k]\} = \max_{\boldsymbol{B}}(\boldsymbol{A}) \tag{13.177}$$

$$\text{Erosion}-E_G(\boldsymbol{A},\boldsymbol{B}) = \min_{[j,k]\in\boldsymbol{B}} \{a[m+j,n+k]\} = \min_{\boldsymbol{B}}(\boldsymbol{A}) \tag{13.178}$$

$$\text{Opening}-O_G(\boldsymbol{A},\boldsymbol{B}) = \max_{\boldsymbol{B}}\left(\min_{\boldsymbol{B}}(\boldsymbol{A})\right) \tag{13.179}$$

$$\text{Closing}-C_G(\boldsymbol{A},\boldsymbol{B}) = \min_{\boldsymbol{B}}\left(\max_{\boldsymbol{B}}(\boldsymbol{A})\right) \tag{13.180}$$

The remarkable conclusion is that the maximum filter and the minimum filter, introduced in Section 13.9.4.2, are gray-level dilation and gray-level erosion for the specific structuring element given by the shape of the filter window with the gray value "0" inside the window. Examples of these operations on a simple one-dimensional signal are shown in Figure 13.45.

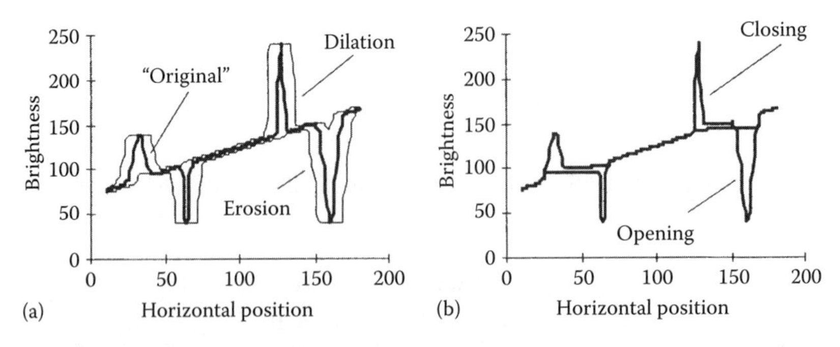

FIGURE 13.45 Morphological filtering of gray-level data. (a) Effect of 15×1 dilation and erosion. (b) Effect of 15×1 opening and closing.

For a rectangular window, $J \times K$, the 2D maximum or minimum filter is separable into two, one-dimensional windows. Further, a one-dimensional maximum or minimum filter can be written in incremental form (see Section 13.9.3.2). This means that gray-level dilations and erosions have a computational complexity per pixel that is O(constant), that is, independent of J and K (see also Table 13.13).

The operations defined above can be used to produce morphological algorithms for smoothing, gradient determination and a version of the Laplacian. All are constructed from the primitives for gray-level dilation and gray-level erosion and in all cases the maximum and minimum filters are taken over the domain $[j, k] \in \boldsymbol{B}$.

13.9.6.11 Morphological Smoothing

This algorithm is based on the observation that a gray-level opening smoothes a gray-value image from above the brightness surface given by the function $a[m, n]$ and the gray-level closing smoothes from below. We use a structuring element $\boldsymbol{B}$ based on Equations 13.84 and 13.178.

$$\begin{aligned} \text{Morphsmooth}(\boldsymbol{A}, \boldsymbol{B}) &= C_{\text{G}}(O_{\text{G}}(\boldsymbol{A}, \boldsymbol{B}), \boldsymbol{B}) \\ &= \min(\max(\max(\min(\boldsymbol{A})))) \end{aligned} \tag{13.181}$$

Note that we have suppressed the notation for the structuring element $\boldsymbol{B}$ under the max and min operations to keep the notation simple. Its use, however, is understood.

13.9.6.12 Morphological Gradient

For linear filters the gradient filter yields a vector representation (Equation 13.103) with a magnitude (Equation 13.104) and direction (Equation 13.105). The version presented here generates a morphological estimate of the gradient magnitude:

$$\begin{aligned} \text{Gradient}(\boldsymbol{A}, \boldsymbol{B}) &= \frac{1}{2}(D_{\text{G}}(\boldsymbol{A}, \boldsymbol{B}) - E_{\text{G}}(\boldsymbol{A}, \boldsymbol{B})) \\ &= \frac{1}{2}(\max(\boldsymbol{A}) - \min(\boldsymbol{A})) \end{aligned} \tag{13.182}$$

13.9.6.13 Morphological Laplacian

The morphologically based Laplacian filter is defined by

$$\begin{aligned} \text{Laplacian}(\boldsymbol{A}, \boldsymbol{B}) &= \frac{1}{2}((D_{\text{G}}(\boldsymbol{A}, \boldsymbol{B}) - \boldsymbol{A}) - (\boldsymbol{A} - E_{\text{G}}(\boldsymbol{A}, \boldsymbol{B}))) \\ &= \frac{1}{2}(D_{\text{G}}(\boldsymbol{A}, \boldsymbol{B}) + E_{\text{G}}(\boldsymbol{A}, \boldsymbol{B}) - 2\boldsymbol{A}) \\ &= \frac{1}{2}(\max(\boldsymbol{A}) + \min(\boldsymbol{A}) - 2\boldsymbol{A}) \end{aligned} \tag{13.183}$$

13.9.6.14 Summary of Morphological Filters

The effect of these filters is illustrated in Figure 13.46. All images were processed with a 3×3 structuring element as described in Equations 13.177 through 13.183. Figure 13.46e was contrast stretched for display purposes using Equation 13.78 and the parameters 1% and 99%. Figure 13.46c through e should be compared to Figures 13.30, 13.32, and 13.33.

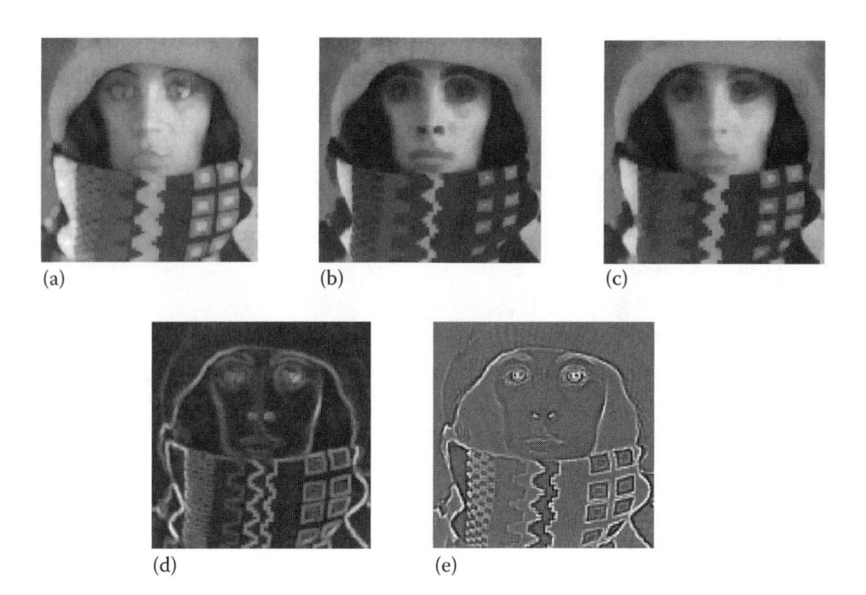

FIGURE 13.46 Examples of gray-level morphological filters. (a) Dilation. (b) Erosion. (c) Smoothing. (d) Gradient. (e) Laplacian.

13.10 Techniques

The algorithms presented in Section 13.9 can be used to build techniques to solve specific image processing problems. Without presuming to present the solution to all processing problems, the following examples are of general interest and can be used as models for solving related problems.

13.10.1 Shading Correction

The method by which images are produced—the interaction between objects in real space, the illumination, and the camera—frequently leads to situations where the image exhibits significant shading across the field-of-view. In some cases the image might be bright in the center and decrease in brightness as one goes to the edge of the field-of-view. In other cases the image might be darker on the left side and lighter on the right side. The shading might be caused by nonuniform illumination, nonuniform camera sensitivity, or even dirt and dust on glass (lens) surfaces. In general this shading effect is undesirable. Eliminating it is frequently necessary for subsequent processing and especially when image analysis or image understanding is the final goal.

13.10.1.1 Model of Shading

In general we begin with a model for the shading effect. The illumination $I_{\text{ill}}(x,y)$ usually interacts in a multiplicative with the object $a(x,y)$ to produce the image $b(x, y)$

$$b(x,y) = I_{\text{ill}}(x,y) \cdot a(x,y) \tag{13.184}$$

with the object representing various imaging modalities such as

$$a(x,y) = \begin{cases} r(x,y) & \text{reflectance model} \\ 10^{-\text{OD}(x,y)} & \text{absorption model} \\ c(x,y) & \text{fluorescence model} \end{cases} \tag{13.185}$$

where at position (x, y), $r(x, y)$ is the reflectance, $\mathrm{OD}(x, y)$ is the optical density, and $c(x, y)$ is the concentration of fluorescent material. Parenthetically, we note that the fluorescence model only holds for low concentrations. The camera may then contribute gain and offset terms, as in Equation 13.74, so that

$$\text{Total shading}- \begin{aligned} c[m,n] &= \mathrm{gain}[m,n] \cdot b[m,n] + \mathrm{offset}[m,n] \\ &= \mathrm{gain}[m,n] \cdot I_{ill}[m,n] \cdot a[m,n] + \mathrm{offset}[m,n] \end{aligned} \quad (13.186)$$

In general we assume that $I_{\mathrm{ill}}[m, n]$ is slowly varying compared to $a[m, n]$.

13.10.1.2 Estimate of Shading

We distinguish between two cases for the determination of $a[m, n]$ starting from $c[m, n]$. In both cases we intend to estimate the shading terms $\{\mathrm{gain}[m, n] \cdot I_{\mathrm{ill}}[m, n]\}$ and $\{\mathrm{offset}[m, n]\}$. While in the first case we assume that we have only the recorded image $c[m, n]$ with which to work, in the second case we assume that we can record two, additional, calibration images.

- *A posteriori estimate*—In this case we attempt to extract the shading estimate from $c[m, n]$. The most common possibilities are the following.

  Low-pass filtering—We compute a smoothed version of $c[m, n]$ where the smoothing is large compared to the size of the objects in the image. This smoothed version is intended to be an estimate of the background of the image. We then subtract the smoothed version from $c[m, n]$ and then restore the desired DC value. In formula

$$\text{Low-pass}-\hat{a}[m,n] = c[m,n] - \mathrm{Lowpass}\{c[m,n]\} + \text{constant} \quad (13.187)$$

  where $\hat{a}[m, n]$ is the estimate of $a[m, n]$. Choosing the appropriate low-pass filter means knowing the appropriate spatial frequencies in the Fourier domain where the shading terms dominate.

  Homomorphic filtering—We note that, if the $\mathrm{offset}[m, n] = 0$, then $c[m, n]$ consists solely of multiplicative terms. Further, the term $\{\mathrm{gain}[m, n] \cdot I_{\mathrm{ill}}[m, n]\}$ is slowly varying while $a[m, n]$ presumably is not. We therefore take the logarithm of $c[m, n]$ to produce two terms one of which is low frequency and one of which is high frequency. We suppress the shading by high-pass filtering the logarithm of $c[m, n]$ and then take the exponent (inverse logarithm) to restore the image. This procedure is based on homomorphic filtering as developed by Oppenheim et al. [34]. In formula

$$\begin{aligned} &\text{i. } c[m,n] = \mathrm{gain}[m,n] \cdot I_{ill}[m,n] \cdot a[m,n] \\ &\text{ii. } \ln\{c[m,n]\} = \ln\left\{ \underbrace{\mathrm{gain}[m,n] \cdot I_{ill}[m,n]}_{\text{slowly varying}} \right\} + \ln\left\{ \underbrace{a[m,n]}_{\text{rapidly varying}} \right\} \\ &\text{iii. } \mathrm{Highpass}\{\ln\{c[m,n]\}\} \approx \ln\{a[m,n]\} \\ &\text{iv. } \hat{a}[m,n] = \exp\{\mathrm{Highpass}\{\ln\{c[m,n]\}\}\} \end{aligned} \quad (13.188)$$

  Morphological filtering—We again compute a smoothed version of $c[m, n]$ where the smoothing is large compared to the size of the objects in the image but this time using morphological smoothing as in Equation 13.181. This smoothed version is the estimate of the background of the image. We then subtract the smoothed version from $c[m, n]$ and then restore the desired DC value. In formula

$$\hat{a}[m,n] = c[m,n] - \mathrm{Morphsmooth}\{c[m,n]\} + \text{constant} \quad (13.189)$$

Choosing the appropriate morphological filter window means knowing (or estimating) the size of the largest objects of interest.

- *A priori estimate*—If it is possible to record test (calibration) images through the camera's system, then the most appropriate technique for the removal of shading effects is to record two images—$\text{BLACK}[m, n]$ and $\text{WHITE}[m, n]$. The BLACK image is generated by covering the lens leading to $b[m, n] = 0$ which in turn leads to $\text{BLACK}[m, n] = \text{offset}[m, n]$. The WHITE image is generated by using $a[m, n] = 1$ which gives $\text{WHITE}[m, n] = \text{gain}[m, n] \cdot I_{\text{ill}}[m, n] + \text{offset}[m, n]$. The correction then becomes

$$\hat{a}[m, n] - \text{constant} \cdot \frac{c[m, n] - \text{BLACK}[m, n]}{\text{WHITE}[m, n] - \text{BLACK}[m, n]} \tag{13.190}$$

The constant term is chosen to produce the desired dynamic range.

The effects of these various techniques on the data from Figure 13.45 are shown in Figure 13.47. The shading is a simple, linear ramp increasing from left to right; the objects consist of Gaussian peaks of varying widths.

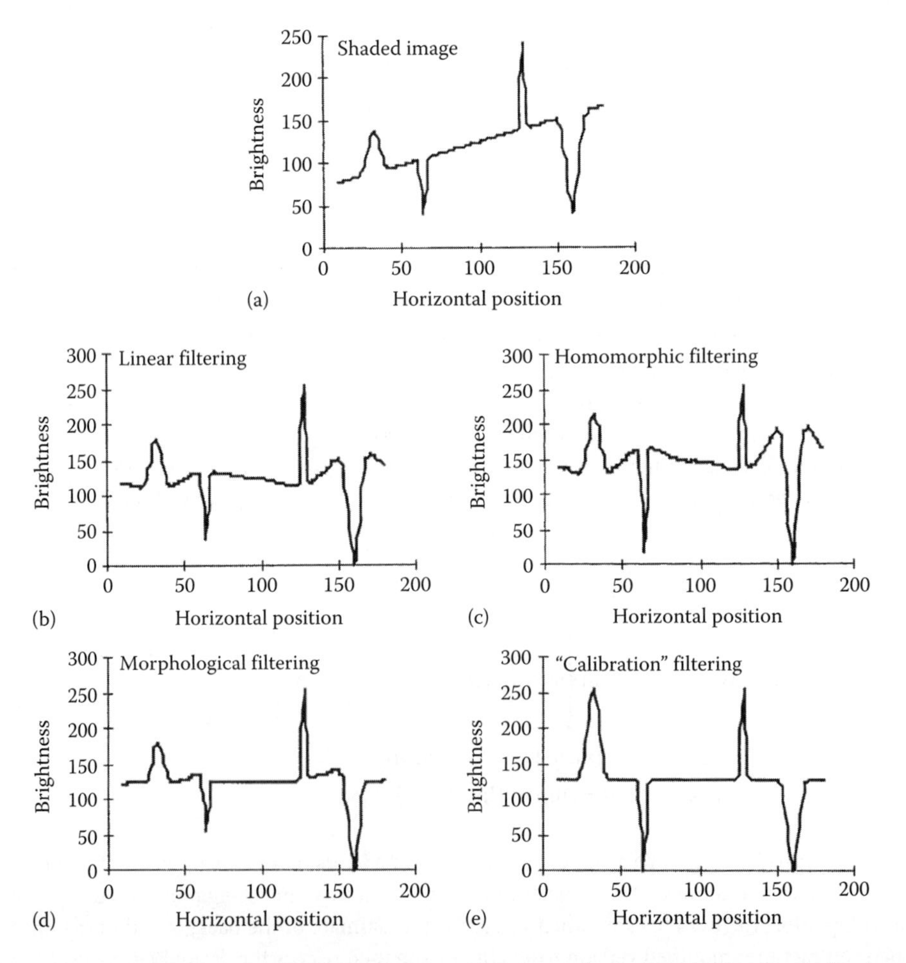

FIGURE 13.47 Comparison of various shading correction algorithms. The final result (e) is identical to the original (not shown). (a) Original. (b) Correction with low-pass filtering. (c) Correction with logarithmic filtering. (d) Correction with max/min filtering. (e) Correction with test images.

In summary, if it is possible to obtain BLACK and WHITE calibration images, then Equation 13.190 is to be preferred. If this is not possible, then one of the other algorithms will be necessary.

13.10.2 Basic Enhancement and Restoration Techniques

The process of image acquisition frequently leads (inadvertently) to image degradation. Due to mechanical problems, out-of-focus blur, motion, inappropriate illumination, and noise, the quality of the digitized image can be inferior to the original. The goal of enhancement is—starting from a recorded image $c[m, n]$—to produce the most visually pleasing image $\hat{a}[m, n]$. The goal of restoration is—starting from a recorded image $c[m, n]$—to produce the best possible estimate $\hat{a}[m, n]$ of the original image $a[m, n]$. The goal of enhancement is beauty; the goal of restoration is truth.

The measure of success in restoration is usually an error measure between the original $a[m, n]$ and the estimate $\hat{a}[m, n]$: $\mathcal{E}\{\hat{a}[m, n], a[m, n]\}$. No mathematical error function is known that corresponds to human perceptual assessment of error. The mean-square error (mse) function is commonly used because:

1. It is easy to compute
2. It is differentiable implying that a minimum can be sought
3. It corresponds to "signal energy" in the total error
4. It has nice properties vis à vis Parseval's theorem, Equations 13.22 and 13.23

The mse is defined by

$$\mathcal{E}\{\hat{a}, a\} = \frac{1}{MN} \sum_{m=0}^{M-1} \sum_{n=0}^{N-1} |\hat{a}[m, n] - a[m, n]|^2 \tag{13.191}$$

In some techniques an error measure will not be necessary; in others it will be essential for evaluation and comparison purposes.

13.10.2.1 Unsharp Masking

A well-known technique from photography to improve the visual quality of an image is to enhance the edges of the image. The technique is called unsharp masking. Edge enhancement means first isolating the edges in an image, amplifying them, and then adding them back into the image. Examination of Figure 13.33 shows that the Laplacian is a mechanism for isolating the gray level edges. This leads immediately to the technique:

$$\hat{a}[m, n] = a[m, n] - \left(k \cdot \nabla^2 a[m, n]\right) \tag{13.192}$$

The term k is the amplifying term and $k > 0$. The effect of this technique is shown in Figure 13.48.

The Laplacian used to produce Figure 13.48 is given by Equation 13.120 and the amplification term $k = 1$.

13.10.2.2 Noise Suppression

The techniques available to suppress noise can be divided into those techniques that are based on temporal information and those that are based on spatial information. By temporal information we mean that a sequence of images $\{a_p[m, n] \mid p = 1, 2, \ldots, P\}$ are available that contain exactly the same objects and that differ only in the sense of independent noise realizations. If this is the case and if the noise is additive, then simple averaging of the sequence:

$$\text{Temporal averaging}-\hat{a}[m, n] = \frac{1}{P} \sum_{p=1}^{P} a_p[m, n] \tag{13.193}$$

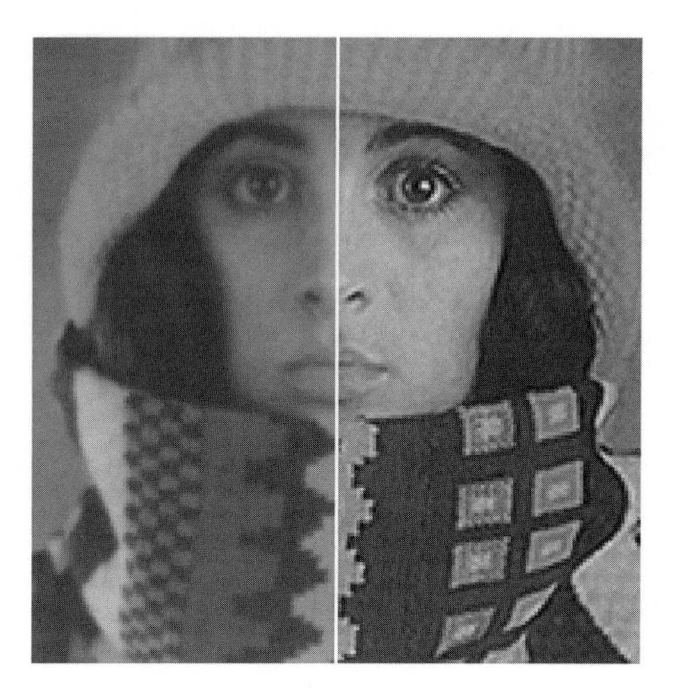

FIGURE 13.48 Edge enhanced compared to original.

will produce a result where the mean value of each pixel will be unchanged. For each pixel, however, the standard deviation will decrease from σ to $\sigma/\sqrt{P}$.

If temporal averaging is not possible, then spatial averaging can be used to decrease the noise. This generally occurs, however, at a cost to image sharpness. Four obvious choices for spatial averaging are the smoothing algorithms that have been described in Section 13.9.4—Gaussian filtering (Equation 13.93), median filtering, Kuwahara filtering, and morphological smoothing (Equation 13.181).

Within the class of linear filters, the optimal filter for restoration in the presence of noise is given by the Wiener filter [2]. The word "optimal" is used here in the sense of minimum mse. Because the square root operation is monotonic increasing, the optimal filter also minimizes the root mean-square (rms) error. The Wiener filter is characterized in the Fourier domain and for additive noise that is independent of the signal it is given by

$$H_W(u, v) = \frac{S_{aa}(u, v)}{S_{aa}(u, v) + S_{nn}(u, v)} \tag{13.194}$$

where $S_{aa}(u, v)$ is the power spectral density of an ensemble of random images $\{a[m, n]\}$ and $S_{nn}(u, v)$ is the power spectral density of the random noise. If we have a single image then $S_{aa}(u, v) = |A(u, v)|^2$. In practice it is unlikely that the power spectral density of the uncontaminated image will be available. Because many images have a similar power spectral density that can be modeled by Table 13.4–T.8, that model can be used as an estimate of $S_{aa}(u, v)$.

A comparison of the five different techniques described above is shown in Figure 13.49. The Wiener filter was constructed directly from Equation 13.113 because the image spectrum and the noise spectrum were known. The parameters for the other filters were determined choosing that value (either σ or window size) that led to the minimum rms.

The rms errors associated with the various filters are shown in Figure 13.49. For this specific comparison, the Wiener filter generates a lower error than any of the other procedures that are examined

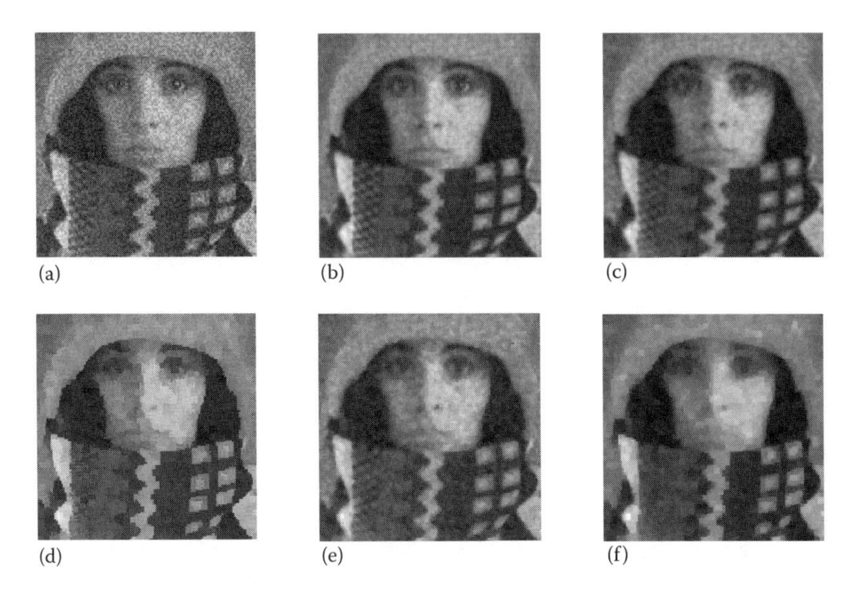

FIGURE 13.49 Noise suppression using various filtering techniques. (a) Noisy image (SNR = 20 dB) rms = 25.7. (b) Wiener filter rms = 20.2. (c) Gauss filter ($\sigma = 1.0$) rms = 21.1. (d) Kuwahara filter (5×5) rms = 22.4. (e) Median filter (3×3) rms = 22.6. (f) Morphology smoothing (3×3) rms = 26.2.

here. The two linear procedures, Wiener filtering and Gaussian filtering, performed slightly better than the three nonlinear alternatives.

13.10.2.3 Distortion Suppression

The model presented above—an image distorted solely by noise—is not, in general, sophisticated enough to describe the true nature of distortion in a digital image. A more realistic model includes not only the noise but also a model for the distortion induced by lenses, finite apertures, possible motion of the camera and/or an object, and so forth. One frequently used model is of an image $a[m, n]$ distorted by a linear, shift-invariant system $h_o[m, n]$ (such as a lens) and then contaminated by noise $\kappa[m, n]$. Various aspects of $h_o[m, n]$ and $\kappa[m, n]$ have been discussed in earlier sections. The most common combination of these is the additive model:

$$c[m, n] = (a[m, n] \otimes h_o[m, n]) + \kappa[m, n] \tag{13.195}$$

The restoration procedure that is based on linear filtering coupled to a minimum mse criterion again produces a Wiener filter [2]:

$$\begin{aligned} H_W(u, v) &= \frac{H_o^*(u, v) S_{aa}(u, v)}{|H_o(u, v)|^2 S_{aa}(u, v) + S_{nn}(u, v)} \\ &= \frac{H_o^*(u, v)}{|H_o(u, v)|^2 + \frac{S_{nn}(u, v)}{S_{aa}(u, v)}} \end{aligned} \tag{13.196}$$

Once again $S_{aa}(u, v)$ is the power spectral density of an image, $S_{nn}(u, v)$ is the power spectral density of the noise, and $H_o(u, v) = \mathrm{F}\{h_o[m, n]\}$. Examination of this formula for some extreme cases can be useful. For those frequencies where $S_{aa}(u, v) \gg S_{nn}(u, v)$, where the signal spectrum dominates the noise spectrum, the Wiener filter is given by $1/H_o(u, v)$, the inverse filter solution. For those frequencies

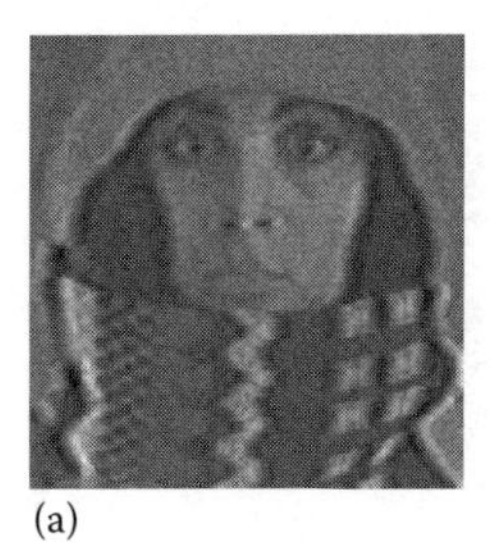
(a)

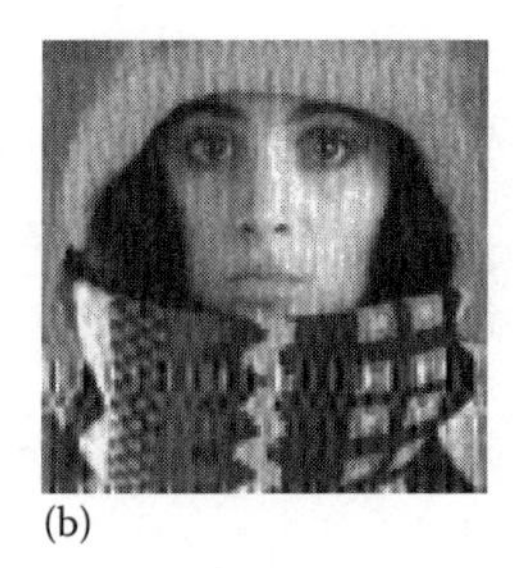
(b)

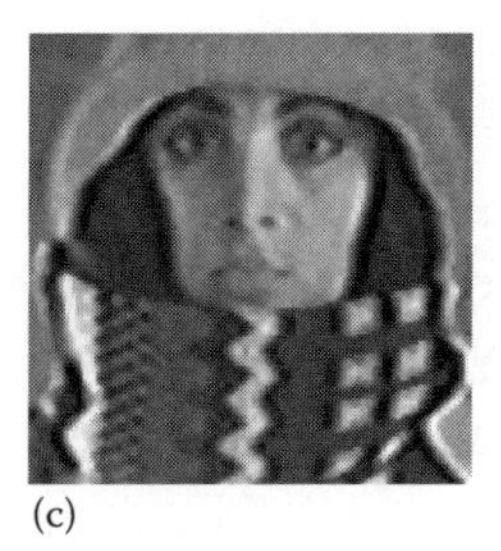
(c)

FIGURE 13.50 Noise and distortion suppression using the Wiener filter, Equation 13.196 and the median filter. (a) Distorted, noisy image. (b) Wiener filter rms = 108.4. (c) Median filter (3 × 3) rms = 40.9.

where $S_{aa}(u, v) \ll S_{nn}(u, v)$, where the noise spectrum dominates the signal spectrum, the Wiener filter is proportional to $H_o^*(u, v)$, the matched filter solution. For those frequencies where $H_o(u, v) = 0$, the Wiener filter $H_W(u, v) = 0$ preventing overflow.

The Wiener filter is a solution to the restoration problem based upon the hypothesized use of a linear filter and the minimum mean-square (or rms) error criterion. In the example below the image $a[m, n]$ was distorted by a band-pass filter and then white noise was added to achieve an SNR = 30 dB. The results are shown in Figure 13.50.

The rms after Wiener filtering but before contrast stretching was 108.4; after contrast stretching with Equation 13.77 the final result as shown in Figure 13.50b has an mse of 27.8. Using a 3 × 3 median filter as shown in Figure 13.50c leads to an rms error of 40.9 before contrast stretching and 35.1 after contrast stretching. Although the Wiener filter gives the minimum rms error over the set of all linear filters, the nonlinear median filter gives a lower rms error. The operation contrast stretching is itself a nonlinear operation. The "visual quality" of the median filtering result is comparable to the Wiener filtering result. This is due in part to periodic artifacts introduced by the linear filter which are visible in Figure 13.50b.

13.10.3 Segmentation

In the analysis of the objects in images it is essential that we can distinguish between the objects of interest and "the rest." This latter group is also referred to as the background. The techniques that are used to find the objects of interest are usually referred to as segmentation techniques—segmenting the foreground from background. In this section we will describe two of the most common techniques—thresholding and edge finding—and we will present techniques for improving the quality of the segmentation result. It is important to understand that

- There is no universally applicable segmentation technique that will work for all images
- No segmentation technique is perfect

13.10.3.1 Thresholding

This technique is based upon a simple concept. A parameter θ called the brightness threshold is chosen and applied to the image $a[m, n]$ as follows:

$$\begin{array}{lll} \textbf{If } a[m,n] \geq \theta & a[m,n] = \text{object} = 1 \\ \textbf{Else} & a[m,n] = \text{background} = 0 \end{array} \tag{13.197}$$

This version of the algorithm assumes that we are interested in light objects on a dark background. For dark objects on a light background we would use

$$\begin{array}{ll}\textbf{If } a[m,n] < \theta & a[m,n] = \text{object} = 1 \\ \textbf{Else} & a[m,n] = \text{background} = 0\end{array} \quad (13.198)$$

The output is the label "object" or "background" which, due to its dichotomous nature, can be represented as a Boolean variable "1" or "0." In principle, the test condition could be based upon some other property than simple brightness (for example, *If* (Redness$\{a[m,n]\} \geq \theta_{red}$), but the concept is clear.

The central question in thresholding then becomes: How do we choose the threshold θ? While there is no universal procedure for threshold selection that is guaranteed to work on all images, there are a variety of alternatives.

- *Fixed threshold*—One alternative is to use a threshold that is chosen independently of the image data. If it is known that one is dealing with very high-contrast images where the objects are very dark and the background is homogeneous (Section 13.10.1) and very light, then a constant threshold of 128 on a scale of 0 to 255 might be sufficiently accurate. By accuracy we mean that the number of falsely classified pixels should be kept to a minimum.
- *Histogram-derived thresholds*—In most cases the threshold is chosen from the brightness histogram of the region or image that we wish to segment (see Sections 13.3.5.2 and 13.9.1). An image and its associated brightness histogram are shown in Figure 13.51.

A variety of techniques have been devised to automatically choose a threshold starting from the gray-value histogram, $\{h[b] \mid b = 0, 1, \ldots, 2^B - 1\}$. Some of the most common ones are presented below. Many of these algorithms can benefit from a smoothing of the raw histogram data to remove small fluctuations but the smoothing algorithm must not shift the peak positions. This translates into a zero-phase smoothing algorithm given below where typical values for W are 3 or 5:

$$h_{smooth}[b] = \frac{1}{W} \sum_{w=-(W-1)/2}^{(W-1)/2} h_{raw}[b-w] \quad W \text{ odd} \quad (13.199)$$

- *Isodata algorithm*—This iterative technique for choosing a threshold was developed by Ridler and Calvard [35]. The histogram is initially segmented into two parts using a starting threshold value such as $\theta_0 = 2^{B-1}$, half the maximum dynamic range. The sample mean ($m_{f,0}$) of the gray values associated with the foreground pixels and the sample mean ($m_{b,0}$) of the gray values associated with the background pixels are computed. A new threshold value θ_1 is now computed as the

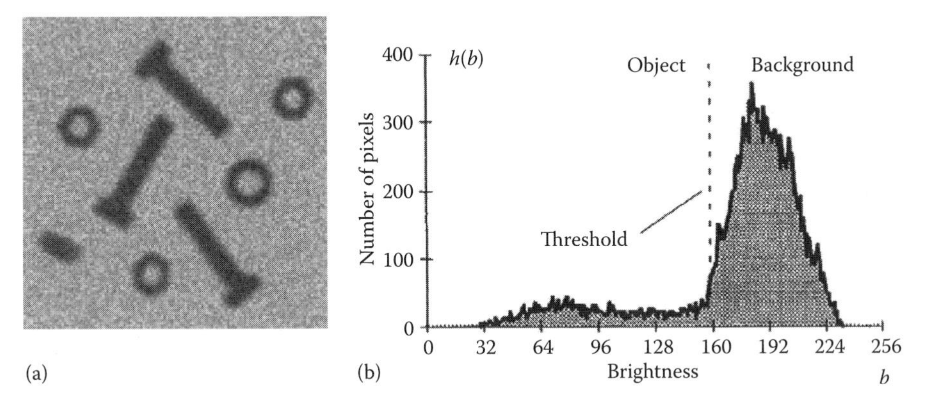

FIGURE 13.51 Pixels below the threshold ($a[m,n] < \theta$) will be labeled as object pixels; those above the threshold will be labeled as background pixels. (a) Image to be thresholded. (b) Brightness histogram of the image.

average of these two sample means. The process is repeated, based upon the new threshold, until the threshold value does not change any more. In formula

$$\theta_k = (m_{f,k-1} + m_{b,k-1})/2 \quad \text{until} \quad \theta_k = \theta_{k-1} \tag{13.200}$$

- *Background-symmetry algorithm*—This technique assumes a distinct and dominant peak for the background that is symmetric about its maximum. The technique can benefit from smoothing as described in Equation 13.199. The maximum peak (maxp) is found by searching for the maximum value in the histogram. The algorithm then searches on the nonobject pixel side of that maximum to find a p% point as in Equation 13.39.

  In Figure 13.51b, where the object pixels are located to the left of the background peak at brightness 183, this means searching to the right of that peak one can locate, as an example, the 95% value. At this brightness value, 5% of the pixels lie to the right (are above) of that value. This occurs at brightness 216 in Figure 13.51b. Because of the assumed symmetry, we use as a threshold a displacement to the left of the maximum that is equal to the displacement to the right where the p% is found. For Figure 13.51b this means a threshold value given by 183 – (216–183) = 150. In formula

$$\theta = \text{maxp} - (p\% - \text{maxp}) \tag{13.201}$$

  This technique can be adapted easily to the case where we have light objects on a dark, dominant background. Further, it can be used if the object peak dominates and we have reason to assume that the brightness distribution around the object peak is symmetric. An additional variation on this symmetry theme is to use an estimate of the sample standard deviation (s in Equation 13.37) based on one side of the dominant peak and then use a threshold based on $\theta = \text{maxp} \pm 1.96s$ (at the 5% level) or $\theta = \text{maxp} \pm 2.57s$ (at the 1% level). The choice of "+" or "−" depends on which direction from maxp is being defined as the object/background threshold. Should the distributions be approximately Gaussian around maxp, then the values 1.96 and 2.57 will, in fact, correspond to the 5% and 1 % level.

- *Triangle algorithm*—This technique due to Zack [36] is illustrated in Figure 13.52. A line is constructed between the maximum of the histogram at brightness b_{max} and the lowest value $b_{min} = (p = 0)\%$ in the image. The distance d between the line and the histogram $h[b]$ is computed for all values of b from $b = b_{min}$ to $b = b_{max}$. The brightness value b_o where the distance between $h[b_o]$ and the line is maximal is the threshold value, that is, $\theta = b_o$. This technique is particularly effective when the object pixels produce a weak peak in the histogram.

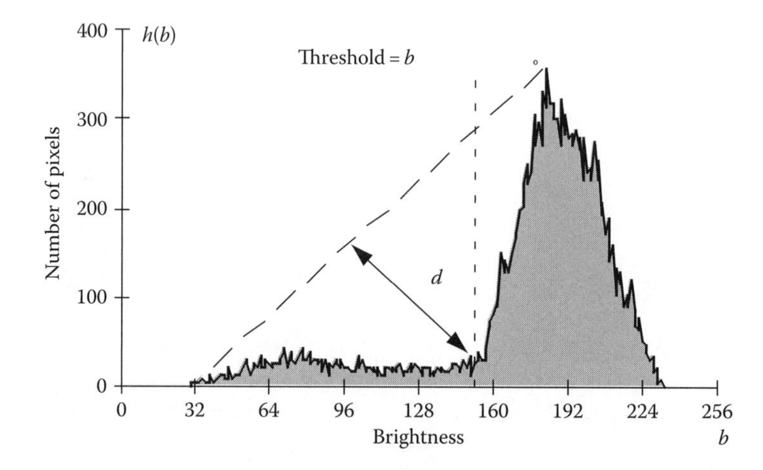

FIGURE 13.52 The triangle algorithm is based on finding the value of b that gives the maximum distance d.

The three procedures described above give the values $\theta = 139$ for the Isodata algorithm, $\theta = 150$ for the background symmetry algorithm at the 5% level, and $\theta = 152$ for the triangle algorithm for the image in Figure 13.51a.

Thresholding does not have to be applied to entire images but can be used on a region by region basis. Chow and Kaneko [37] developed a variation in which the $M \times N$ image is divided into nonoverlapping regions. In each region a threshold is calculated and the resulting threshold values are put together (interpolated) to form a thresholding surface for the entire image. The regions should be of "reasonable" size so that there are a sufficient number of pixels in each region to make an estimate of the histogram and the threshold. The utility of this procedure—like so many others—depends on the application at hand.

13.10.3.2 Edge Finding

Thresholding produces a segmentation that yields all the pixels that, in principle, belong to the object or objects of interest in an image. An alternative to this is to find those pixels that belong to the borders of the objects. Techniques that are directed to this goal are termed edge finding techniques. From our discussion in Section 13.9.6 on mathematical morphology, specifically Equations 13.79, 13.163, and 13.170, we see that there is an intimate relationship between edges and regions.

- *Gradient-based procedure*—The central challenge to edge finding techniques is to find procedures that produce closed contours around the objects of interest. For objects of particularly high SNR, this can be achieved by calculating the gradient and then using a suitable threshold. This is illustrated in Figure 13.53.

  While the technique works well for the 30 dB image in Figure 13.53a, it fails to provide an accurate determination of those pixels associated with the object edges for the 20 dB image in Figure 13.53b. A variety of smoothing techniques as described in Section 13.9.4 and in Equation 13.181 can be used to reduce the noise effects before the gradient operator is applied.

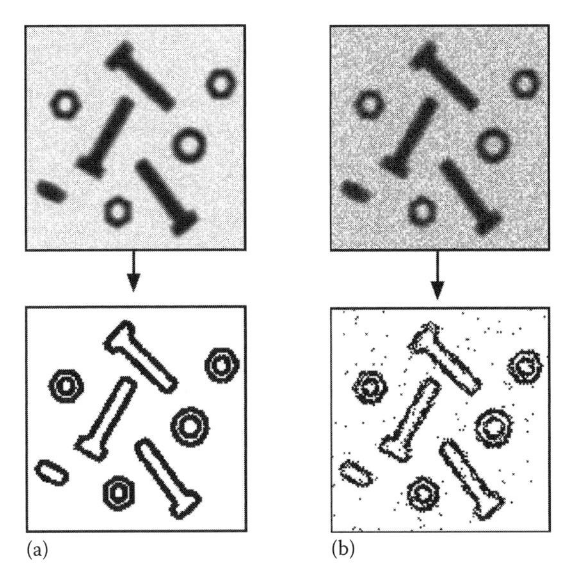

FIGURE 13.53 Edge finding based on the Sobel gradient, Equation 13.110, combined with the isodata thresholding algorithm Equation 13.92. (a) SNR = 30 dB. (b) SNR = 20 dB.

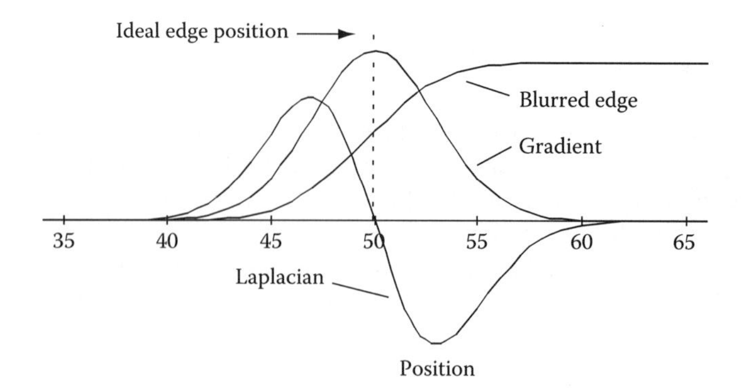

FIGURE 13.54 Edge finding based on the zero crossing as determined by the second derivative, the Laplacian. The curves are not to scale.

- *Zero-crossing based procedure*—A more modern view to handling the problem of edges in noisy images is to use the zero crossings generated in the Laplacian of an image (Section 13.9.5.2). The rationale starts from the model of an ideal edge, a step function, that has been blurred by an OTF such as Table 13.4–T.3 (out-of-focus), T.5 (diffraction-limited), or T.6 (general model) to produce the result shown in Figure 13.54.

  The edge location is, according to the model, at that place in the image where the Laplacian changes sign, the zero crossing. As the Laplacian operation involves a second derivative, this means a potential enhancement of noise in the image at high spatial frequencies (see Equation 13.114). To prevent enhanced noise from dominating the search for zero crossings, a smoothing is necessary.

  The appropriate smoothing filter, from among the many possibilities described in Section 13.9.4, should according to Canny [38] have the following properties:
  - In the frequency domain, (u, v) or (Ω, Ψ), the filter should be as narrow as possible to provide suppression of high frequency noise, and;
  - In the spatial domain, (x, y) or $[m, n]$, the filter should be as narrow as possible to provide good localization of the edge. A too wide filter generates uncertainty as to precisely where, within the filter width, the edge is located.

The smoothing filter that simultaneously satisfies both these properties—minimum bandwidth and minimum spatial width—is the Gaussian filter described in Section 13.9.4. This means that the image should be smoothed with a Gaussian of an appropriate σ followed by application of the Laplacian. In formula

$$\text{Zerocrossing}\{a(x,y)\} = \left\{(x,y)|\nabla^2\{g_{2D}(x,y) \otimes a(x,y)\} = 0\right\} \tag{13.202}$$

where $g_{2D}(x,y)$ is defined in Equation 13.93. The derivative operation is linear and shift-invariant as defined in Equations 13.85 and 13.86. This means that the order of the operators can be exchanged (Equation 13.4) or combined into one single filter (Equation 13.5). This second approach leads to the Marr-Hildreth formulation of the "Laplacian-of-Gaussians" (LoG) filter [39]:

$$\text{Zerocrossing}\{a(x,y)\} = \{(x,y)|\text{LoG}(x,y) \otimes a(x,y) = 0\} \tag{13.203}$$

where

$$\text{LoG}(x,y) = \frac{x^2 + y^2}{\sigma^4} g_{2D}(x,y) - \frac{2}{\sigma^2} g_{2D}(x,y) \tag{13.204}$$

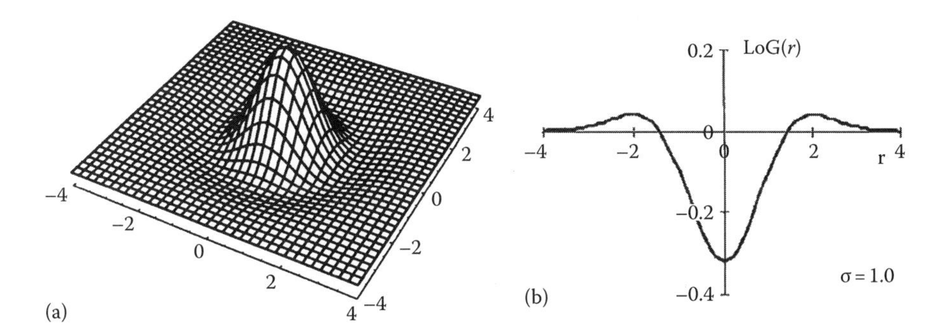

FIGURE 13.55 LoG filter with $\sigma = 1.0$. (a) $-\text{LoG}(x, y)$. (b) $\text{LoG}(r)$.

Given the circular symmetry this can also be written as

$$\text{LoG}(r) = \left(\frac{r^2 - 2\sigma^2}{2\pi\sigma^6}\right) e^{-(r^2/2\sigma^2)} \quad (13.205)$$

This 2D convolution kernel, which is sometimes referred to as a "Mexican hat filter," is illustrated in Figure 13.55.

- *PLUS-based procedure*—Among the zero crossing procedures for edge detection, perhaps the most accurate is the PLUS filter as developed by Verbeek and Van Vliet [40]. The filter is defined, using Equations 13.121 and 13.122, as

$$\begin{aligned} \text{PLUS}(a) &= \text{SDGD}(a) + \text{Laplace}(a) \\ &= \left(\frac{A_{xx}A_x^2 + 2A_{xy}A_xA_y + A_{yy}A_y^2}{A_x^2 + A_y^2}\right) + (A_{xx} + A_{yy}) \end{aligned} \quad (13.206)$$

Neither the derivation of the PLUS's properties nor an evaluation of its accuracy are within the scope of this section. Suffice to say that for positively curved edges in gray value images, the Laplacian-based zero crossing procedure overestimates the position of the edge and the SDGD-based procedure underestimates the position. This is true in both 2D and three-dimensional images with an error on the order of $(\sigma/R)^2$ where R is the radius of curvature of the edge. The PLUS operator has an error on the order of $(\sigma/R)^4$ if the image is sampled at, at least, 3× the usual Nyquist sampling frequency as in Equation 13.56 or if we choose $\sigma \geq 2.7$ and sample at the usual Nyquist frequency.

All of the methods based on zero crossings in the Laplacian must be able to distinguish between zero crossings and zero values. While the former represent edge positions, the latter can be generated by regions that are no more complex than bilinear surfaces, that is, $a(x, y) = a_0 + a_1 \cdot x + a_2 \cdot y + a_3 \cdot x \cdot y$. To distinguish between these two situations, we first find the zero crossing positions and label them as "1" and all other pixels as "0." We then multiply the resulting image by a measure of the edge strength at each pixel. There are various measures for the edge strength that are all based on the gradient as described in Section 13.9.5.1 and Equation 13.182. This last possibility, use of a morphological gradient as an edge strength measure, was first described by Lee et al. [41] and is particularly effective. After multiplication the image is then thresholded (as above) to produce the final result. The procedure is thus as follows [42] (Figure 13.56):

The results of these two edge finding techniques based on zero crossings, LoG filtering and PLUS filtering, are shown in Figure 13.57 for images with a 20 dB SNR.

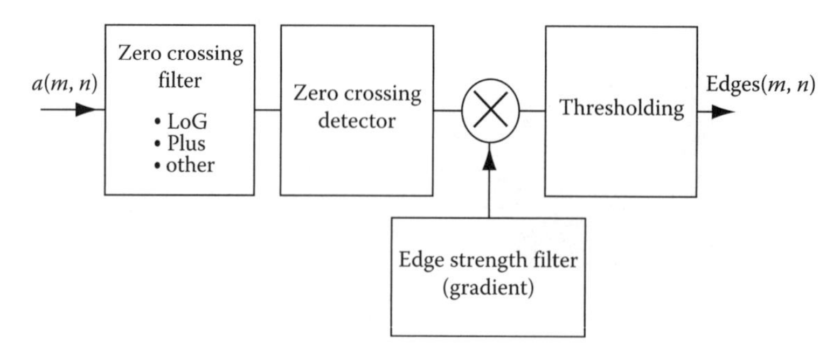

FIGURE 13.56 General strategy for edges based on zero crossings.

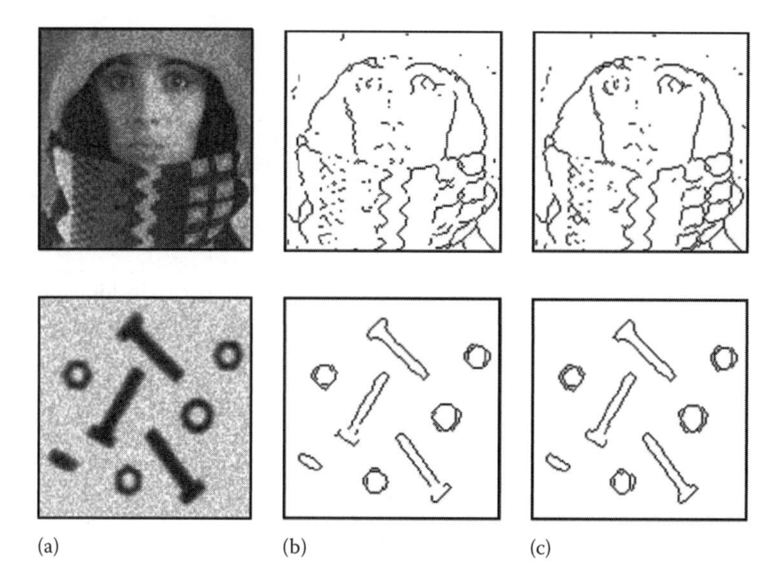

FIGURE 13.57 Edge finding using zero crossing algorithms LoG and PLUS. In both algorithms $\sigma = 1.5$.

Edge finding techniques provide, as the name suggests, an image that contains a collection of edge pixels. Should the edge pixels correspond to objects, as opposed to say simple lines in the image, then a region-filling technique such as Equation 13.170 may be required to provide the complete objects.

13.10.3.3 Binary Mathematical Morphology

The various algorithms that we have described for mathematical morphology in Section 13.9.6 can be put together to form powerful techniques for the processing of binary images and gray level images. As binary images frequently result from segmentation processes on gray level images, the morphological processing of the binary result permits the improvement of the segmentation result.

- *Salt-or-pepper filtering*—Segmentation procedures frequently result in isolated "1" pixels in a "0" neighborhood (salt) or isolated "0" pixels in a "1" neighborhood (pepper). The appropriate neighborhood definition must be chosen as in Figure 13.3. Using the lookup table formulation for Boolean operations in a 3×3 neighborhood that was described in association with Figure 13.43, salt filtering and pepper filtering are straightforward to implement. We describe the weight for different positions in the 3×3 neighborhood as follows:

$$\text{Weights} = \begin{bmatrix} w_4 = 16 & w_3 = 8 & w_2 = 4 \\ w_5 = 32 & w_0 = 1 & w_1 = 2 \\ w_6 = 64 & w_7 = 128 & w_8 = 256 \end{bmatrix} \tag{13.207}$$

For a 3×3 window in $a[m, n]$ with values "0" or "1" we then compute:

$$\begin{aligned} \text{Sum} = {} & w_0 a[m,n] + w_1 a[m+1,n] + w_2 a[m+1,n-1] \\ & + w_3 a[m,n-1] + w_4 a[m-1,n-1] + w_5 a[m-1,n] \\ & + w_6 a[m-1,n+1] + w_7 a[m,n+1] + w_8 a[m+1,n-1] \end{aligned} \tag{13.208}$$

The result, sum, is a number bounded by $0 \leq \text{sum} \leq 511$.

- *Salt filter*—The 4-connected and 8-connected versions of this filter are the same and are given by the following procedure:

$$\begin{aligned} & \text{(i) Compute } sum \\ & \text{(ii) } \textbf{If } ((sum == 1) \quad c[m,n] = 0 \\ & \qquad \textbf{Else} \qquad\qquad c[m,n] = am,n] \end{aligned} \tag{13.209}$$

- *Pepper filter*—The 4-connected and 8-connected versions of this filter are the following procedures:

$$\begin{array}{ll} \textit{4-connected} & \textit{8-connected} \\ \text{(i) Compute } sum & \text{(i) Compute } sum \\ \text{(ii) } \textbf{If}((sum == 170) & \text{(ii) } \textbf{If } ((sum == 510) \\ \quad c[m,n] = 1 & \quad c[m,n] = 1 \\ \textbf{Else} & \textbf{Else} \\ \quad c[m,n] = a[m,n] & \quad c[m,n] = a[m,n] \end{array} \tag{13.210}$$

- *Isolate objects with holes*—To find objects with holes we can use the following procedure which is illustrated in Figure 13.58.

1. Segment image to produce binary mask representation
2. Compute skeleton without end pixels—Equation 13.169
3. Use salt filter to remove single skeleton pixels
4. Propagate remaining skeleton pixels into original binary mask—Equation 13.170

(13.211)

The binary objects are shown in gray and the skeletons, after application of the salt filter, are shown as a black overlay on the binary objects. Note that this procedure uses no parameters other then the

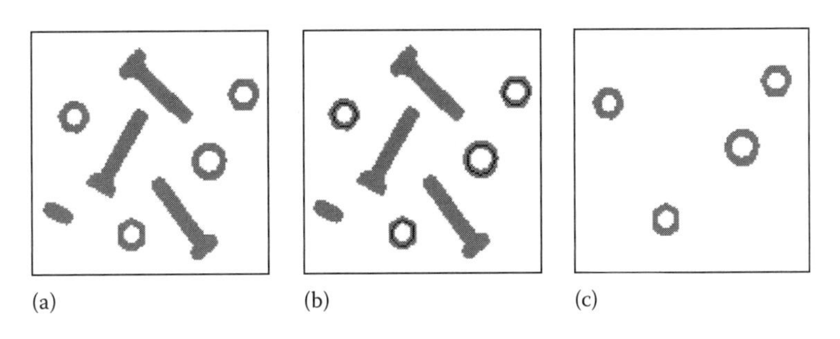

FIGURE 13.58 Isolation of objects with holes using morphological operations. (a) Binary image. (b) Skeleton after salt filter. (c) Objects with holes.

fundamental choice of connectivity; it is free from "magic numbers." In the example shown in Figure 13.58, the 8-connected definition was used as well as the structuring element $\boldsymbol{B} = \boldsymbol{N}_8$.

- *Filling holes in objects*—To fill holes in objects we use the following procedure which is illustrated in Figure 13.59.

$$\begin{aligned}&\text{1. Segment image to produce binary representation of objects}\\&\text{2. Compute complement of binary image as a mask image}\\&\text{3. Generate a seed image as the border of the image}\\&\text{4. Propagate the seed into the mask—Equation 13.97}\end{aligned} \tag{13.212}$$

The mask image is illustrated in *gray* in Figure 13.59a and the seed image is shown in black in that same illustration. When the object pixels are specified with a connectivity of $C = 8$, then the propagation into the mask (background) image should be performed with a connectivity of $C = 4$, that is, dilations with the structuring element $\boldsymbol{B} = \boldsymbol{N}_4$. This procedure is also free of "magic numbers."

- *Removing border-touching objects*—Objects that are connected to the image border are not suitable for analysis. To eliminate them we can use a series of morphological operations that are illustrated in Figure 13.60.

$$\begin{aligned}&\text{1. Segment image to produce binary mask image of objects}\\&\text{2. Generate a seed image as the border of the image}\\&\text{3. Propagate the seed into the mask—Equation 13.97}\\&\text{4. Compute ``XOR'' of the propagation result and the mask image as final result}\end{aligned} \tag{13.213}$$

The mask image is illustrated in gray in Figure 13.60a and the seed image is shown in black in that same illustration. If the structuring element used in the propagation is $\boldsymbol{B} = \boldsymbol{N}_4$, then objects are

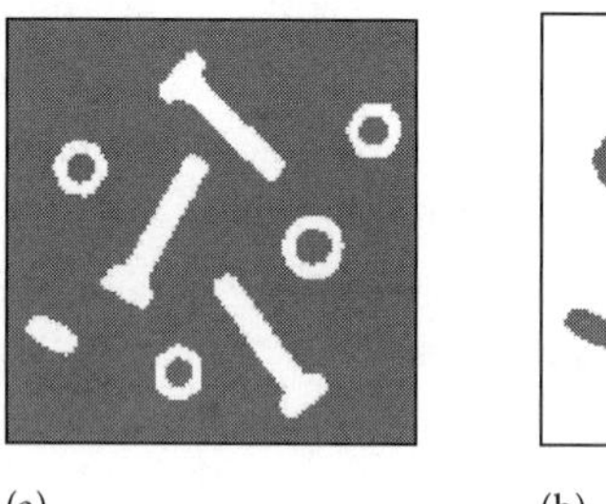
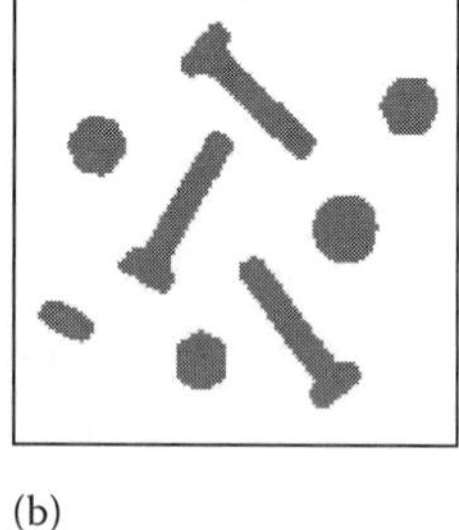

(a) (b)

FIGURE 13.59 Filling holes in objects. (a) Mask and seed images. (b) Objects with holes filled.

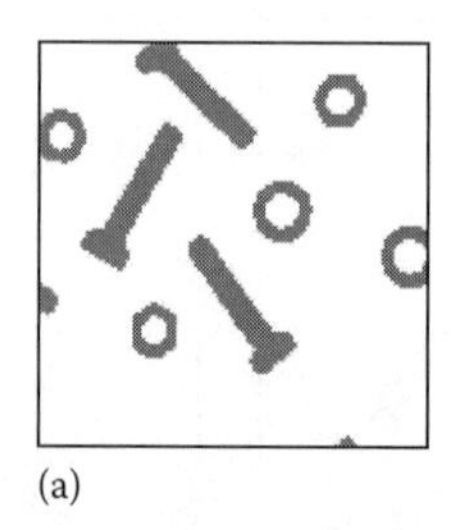
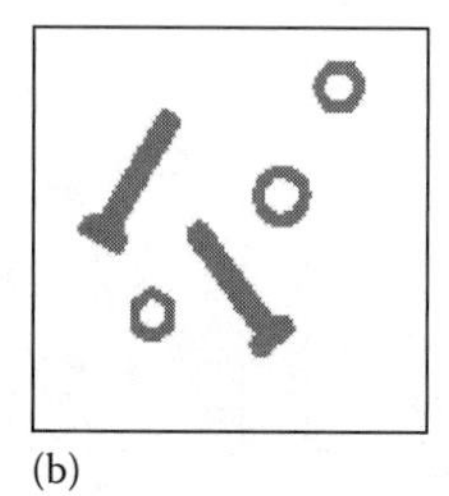

(a) (b)

FIGURE 13.60 Removing objects touching borders. (a) Mask and seed images. (b) Remaining objects.

removed that are 4-connected with the image boundary. If $\boldsymbol{B} = \boldsymbol{N}_8$ is used then objects that 8-connected with the boundary are removed.

- *Exoskeleton*—The exoskeleton of a set of objects is the skeleton of the background that contains the objects. The exoskeleton produces a partition of the image into regions each of which contains one object. The actual skeletonization (Equation 13.169) is performed without the preservation of end pixels and with the border set to "0." The procedure is described below and the result is illustrated in Figure 13.61.

1. Segment image to produce binary image
2. Compute complement of binary image (13.214)
3. Compute skeleton using Equation 13.169 $(1 + 2)$ with border set to "0″

- *Touching objects*—Segmentation procedures frequently have difficulty separating slightly touching, yet distinct, objects. The following procedure provides a mechanism to separate these objects and makes minimal use of "magic numbers." The exoskeleton produces a partition of the image into regions each of which contains one object. The actual skeletonization is performed without the preservation of end pixels and with the border set to "0." The procedure is illustrated in Figure 13.62.

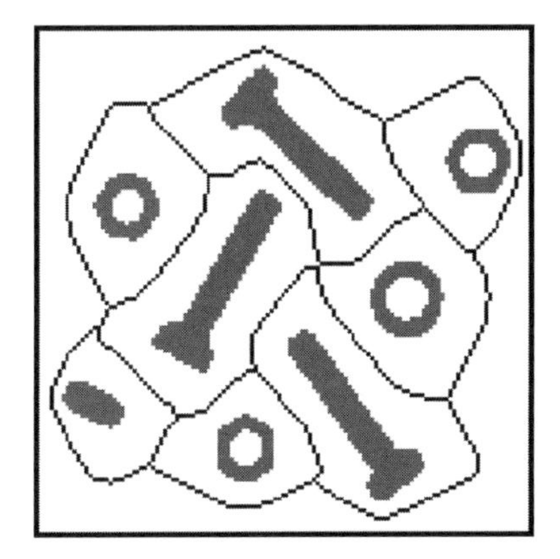

FIGURE 13.61 Exoskeleton.

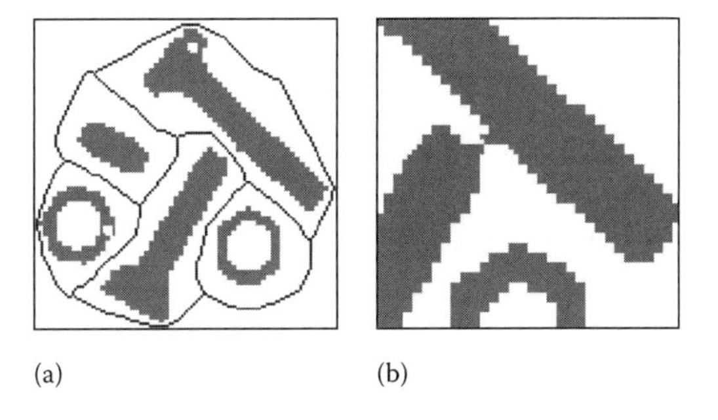

(a) (b)

FIGURE 13.62 Separation of touching objects. (a) Eroded and exo-skeleton images. (b) Objects separated (detail).

1. Segment image to produce binary image
2. Compute a small number of erosions with $B = N_4$
3. Compute exoskeleton of eroded result (13.215)
4. Complement exoskeleton result
5. Compute AND of original binary image and the complemented exoskeleton

The eroded binary image is illustrated in gray in Figure 13.62a and the exoskeleton image is shown in black in that same illustration. An enlarged section of the final result is shown in Figure 13.62b and the separation is easily seen. This procedure involves choosing a small, minimum number of erosions but the number is not critical as long as it initiates a coarse separation of the desired objects. The actual separation is performed by the exoskeleton which, itself, is free of "magic numbers." If the exoskeleton is 8-connected, then the background separating the objects will be 8-connected. The objects, themselves, will be disconnected according to the 4-connected criterion (see Section 13.9.6 and Figure 13.36).

13.10.3.4 Gray-Value Mathematical Morphology

As we have seen in Section 13.10.1.2, gray-value morphological processing techniques can be used for practical problems such as shading correction. In this section several other techniques will be presented.

- *Top-hat transform*—The isolation of gray-value objects that are convex can be accomplished with the top-hat transform as developed by Meyer [43,44]. Depending upon whether we are dealing with light objects on a dark background or dark objects on a light background, the transform is defined as

$$\text{Light objects} - \mathit{Tophat}(\mathbf{A},\mathbf{B}) = \mathbf{A} - (\mathbf{A} \circ \mathbf{B}) = \mathbf{A} - \max_{\mathbf{B}}\left(\min_{\mathbf{B}}(\mathbf{A})\right) \tag{13.216}$$

$$\text{Dark objects} - \mathit{Tophat}(\mathbf{A},\mathbf{B}) = (\mathbf{A} \bullet \mathbf{B}) - \mathbf{A} = \min_{\mathbf{B}}\left(\max_{\mathbf{B}}(\mathbf{A})\right) \tag{13.217}$$

where the structuring element $\mathbf{B}$ is chosen to be bigger than the objects in question and, if possible, to have a convex shape. Because of the properties given in Equations 13.155 and 13.158, Top-hat $(\mathbf{A}, \mathbf{B}) \geq 0$. An example of this technique is shown in Figure 13.63.

The original image including shading is processed by a 15×1 structuring element as described in Equations 13.216 and 13.217 to produce the desired result. Note that the transform for dark objects has been defined in such a way as to yield "positive" objects as opposed to "negative" objects. Other definitions are, of course, possible.

- *Thresholding*—A simple estimate of a locally varying threshold surface can be derived from morphological processing as follows:

$$\text{Threshold surface} - \theta[m,n] = \frac{1}{2}(\max(\mathbf{A}) + \min(\mathbf{A})) \tag{13.218}$$

Once again, we suppress the notation for the structuring element $\mathbf{B}$ under the max and min operations to keep the notation simple. Its use, however, is understood.

- *Local contrast stretching*—Using morphological operations we can implement a technique for local contrast stretching. That is, the amount of stretching that will be applied in a neighborhood will be controlled by the original contrast in that neighborhood. The morphological gradient defined in Equation 13.182 may also be seen as related to a measure of the local contrast in the window defined by the structuring element $\mathbf{B}$:

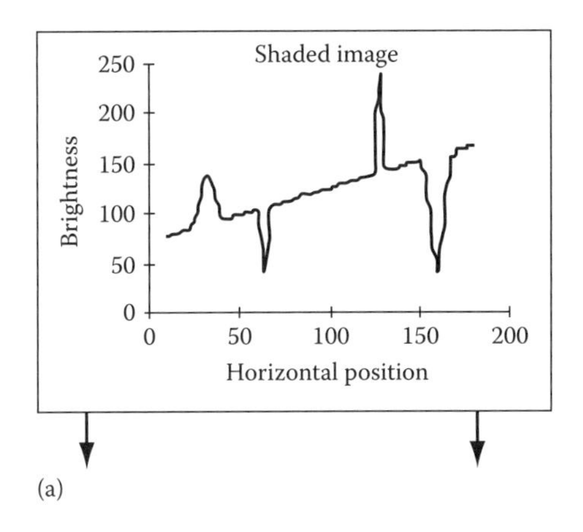

(a)

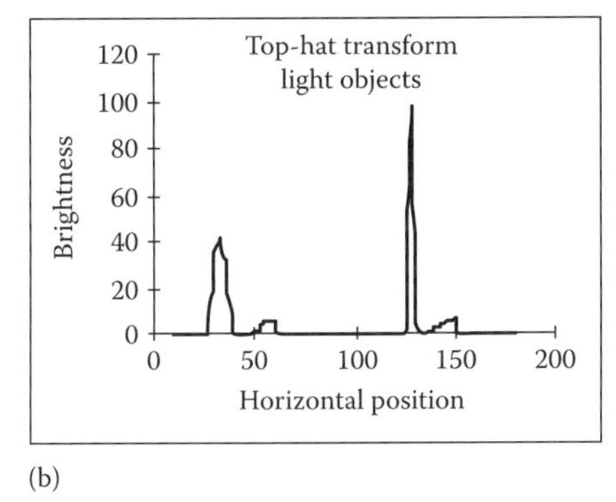

(b)

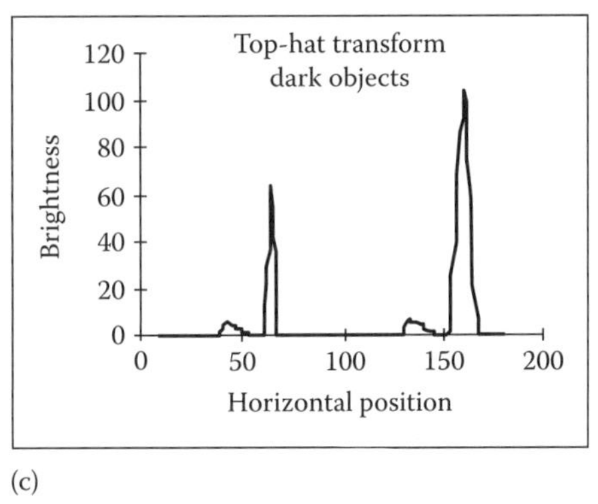

(c)

FIGURE 13.63 Top-hat transforms. (a) Original. (b) Light object transform. (c) Dark object transform.

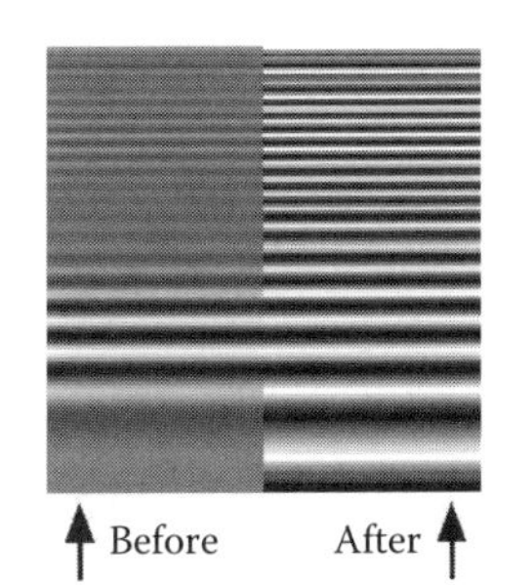

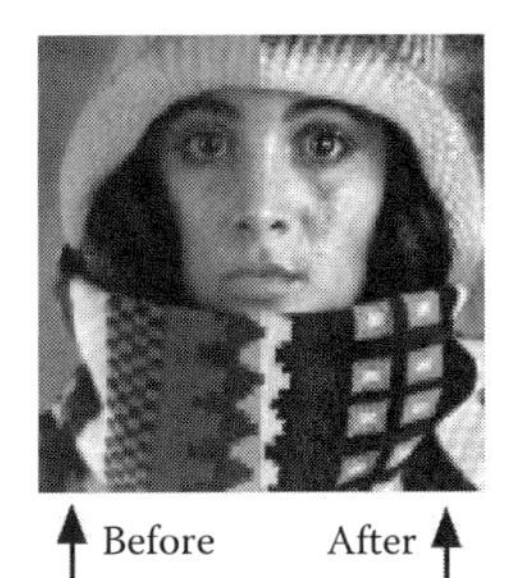

FIGURE 13.64 Local contrast stretching.

$$\text{Local-contrast}\,(\boldsymbol{A}, \boldsymbol{B}) = \max(\boldsymbol{A}) - \min(\boldsymbol{A}) \tag{13.219}$$

The procedure for local contrast stretching is given by

$$c[m, n] = \text{scale} \cdot \frac{\boldsymbol{A} - \min(\boldsymbol{A})}{\max(\boldsymbol{A}) - \min(\boldsymbol{A})} \tag{13.220}$$

The max and min operations are taken over the structuring element $\boldsymbol{B}$. The effect of this procedure is illustrated in Figure 13.64. It is clear that this local operation is an extended version of the point operation for contrast stretching presented in Equation 13.77.

Using standard test images (as we have seen in so many examples) illustrates the power of this local morphological filtering approach.

Acknowledgments

This work was partially supported by the Netherlands Organization for Scientific Research (NWO) Grant 900-538-040, the Foundation for Technical Sciences (STW) Project 2987, the ASCI PostDoc program, and the Rolling Grants program of the Foundation for Fundamental Research in Matter (FOM). Images presented above were processed using TCL–Image and SCIL–Image (both from the TNO-TPD, Stieltjesweg 1, Delft, The Netherlands) and Adobe Photoshop™.

References

1. Dudgeon, D.E. and R.M. Mersereau, *Multidimensional Digital Signal Processing*. 1984, Englewood Cliffs, NJ: Prentice-Hall.
2. Castleman, K.R., *Digital Image Processing*, 2nd edn. 1996, Englewood Cliffs, NJ: Prentice-Hall.
3. Oppenheim, A.V., A.S. Willsky, and I.T. Young, *Systems and Signals*. 1983, Englewood Cliffs, NJ: Prentice-Hall.
4. Papoulis, A., *Systems and Transforms with Applications in Optics*. 1968, New York: McGraw-Hill.
5. Russ, J.C., *The Image Processing Handbook*, 2nd edn. 1995, Boca Raton, FL: CRC Press.
6. Giardina, C.R. and E.R. Dougherty, *Morphological Methods in Image and Signal Processing*. 1988, Englewood Cliffs, NJ: Prentice-Hall, p. 321.
7. Gonzalez, R.C. and R.E. Woods, *Digital Image Processing*. 1992, Reading, MA: Addison-Wesley, p. 716.
8. Goodman, J.W., *Introduction to Fourier Optics. McGraw-Hill Physical and Quantum Electronics Series*. 1968, New York: McGraw-Hill, p. 287.
9. Heijmans, H.J.A.M., *Morphological Image Operators. Advances in Electronics and Electron Physics*. 1994, Boston, MA: Academic Press.
10. Hunt, R.W.G., *The Reproduction of Colour in Photography, Printing & Television*, 4th edn. 1987, Tolworth, England: Fountain Press.
11. Freeman, H., Boundary encoding and processing, in *Picture Processing and Psychopictorics*, B.S. Lipkin and A. Rosenfeld, Eds. 1970, New York: Academic Press, pp. 241–266.
12. Stockham, T.G., Image processing in the context of a visual model. *Proceedings of the IEEE*, 1972. **60**: 828–842.
13. Murch, G.M., *Visual and Auditory Perception*. 1973, New York: Bobbs-Merrill Company, Inc., p. 403.
14. Frisby, J.P., *Seeing: Illusion, Brain and Mind*. 1980, Oxford, England: Oxford University Press, p. 160.
15. Blakemore, C. and F.W.C. Campbell, On the existence of neurons in the human visual system selectively sensitive to the orientation and size of retinal images. *Journal of Physiology*, 1969. **203**: 237–260.
16. Born, M. and E. Wolf, *Principles of Optics*. 6th edn. 1980, Oxford: Pergamon Press.
17. Young, I.T., Quantitative microscopy. *IEEE Engineering in Medicine* and *Biology*, 1996. **15**(1): 59–66.
18. Dorst, L. and A.W.M. Smeulders, *Length Estimators Compared*, in *Pattern Recognition in Practice II*, E.S. Gelsema and L.N. Kanal, Eds. 1986, Amsterdam, the Netherlands: Elsevier Science, pp. 73–80.
19. Young, I.T., Sampling density and quantitative microscopy. *Analytical and Quantitative Cytology and Histology*, 1988. **10**(4): 269–275.
20. Kulpa, Z., Area and perimeter measurement of blobs in discrete binary pictures. *Computer Vision, Graphics and Image Processing*, 1977. **6**: 434–454.

21. Vossepoel, A.M. and A.W.M. Smeulders, Vector code probabilities and metrication error in the representation of straight lines of finite length. *Computer Graphics and Image Processing*, 1982. **20**: 347–364.
22. Photometrics Ltd., Signal processing and noise, in *Series 200 CCD Cameras Manual.* 1990, Tucson, AR.
23. Huang, T.S., G.J. Yang, and G.Y. Tang, A fast two-dimensional median filtering algorithm. *IEEE Transactions on Acoustics, Speech, and Signal Processing*, 1979. **ASSP-27**: 13–18.
24. Groen, F.C.A., R.J. Ekkers, and R. De Vries, Image processing with personal computers. *Signal Processing*, 1988. **15**: 279–291.
25. Verbeek, P.W., H.A. Vrooman, and L.J. Van Vliet, Low-level image processing by max-min filters. *Signal Processing*, 1988. **15**: 249–258.
26. Young, I.T. and L.J. Van Vliet, Recursive implementation of the gaussian filter. *Signal Processing*, 1995. **44**(2): 139–151.
27. Kuwahara, M. et al., Processing of RI-angiocardiographic images, in *Digital Processing of Biomedical Images*, K. Preston and M. Onoe, Eds. 1976, New York: Plenum Press, pp. 187–203.
28. Van Vliet, L.J., Grey-scale measurements in multi-dimensional digitized images, PhD thesis: Delft University of Technology, Delft, the Netherlands, 1993.
29. Serra, J., *Image Analysis and Mathematical Morphology.* 1982, London, U.K.: Academic Press.
30. Vincent, L., Morphological transformations of binary images with arbitrary structuring elements. *Signal Processing*, 1991. **22**(1): 3–23.
31. Van Vliet, L.J. and B.J.H. Verwer, A contour processing method for fast binary neighbourhood operations. *Pattern Recognition Letters*, 1988. **7**(1): 27–36.
32. Young, I.T. et al., A new implementation for the binary and Minkowski operators. *Computer Graphics and Image Processing*, 1981. **17**(3): 189–210.
33. Lantuéjoul, C., Skeletonization in quantitative metallography, in *Issues of Digital Image Processing*, R.M. Haralick and J.C. Simon, Eds. 1980, Groningen, the Netherlands: Sijthoff and Noordhoff.
34. Oppenheim, A.V., R.W. Schafer, and T.G. Stockham, Jr., Non-linear filtering of multiplied and convolved signals. *Proceedings of the IEEE*, 1968. **56**(8): 1264–1291.
35. Ridler, T.W. and S. Calvard, Picture thresholding using an iterative selection method. *IEEE Transactions on Systems, Man, and Cybernetics*, 1978. **SMC-8**(8): 630–632.
36. Zack, G.W., W.E. Rogers, and S.A. Latt, Automatic measurement of sister chromatid exchange frequency. *Journal of Histochemistry and Cytochemistry*, 1977. **25**(7): 741–753.
37. Chow, C.K. and T. Kaneko, Automatic boundary detection of the left ventricle from cineangiograms. *Computers and Biomedical Research*, 1972. **5**: 388–410.
38. Canny, J., A computational approach to edge detection. *IEEE Transactions on Pattern Analysis and Machine Intelligence*, 1986. **PAMI-8**(6): 679–698.
39. Marr, D. and E.C. Hildreth, Theory of edge detection. *Proceedings of the Royal Society of London, Series B*, 1980. **207**: 187–217.
40. Verbeek, P.W. and L.J. Van Vliet, On the location error of curved edges in low-pass filtered 2D and 3D images. *IEEE Transactions on Pattern Analysis and Machine Intelligence*, 1994. **16**(7): 726–733.
41. Lee, J.S.L., R.M. Haralick, and L.S. Shapiro, Morphologic edge detection. In *8th International Conference on Pattern Recognition*. 1986. Paris, France: IEEE Computer Society.
42. Van Vliet, L.J., I.T. Young, and A.L.D. Beckers, A non-linear Laplace operator as edge detector in noisy images. *Computer Vision, Graphics, and Image Processing*, 1989. **45**: 167–195.
43. Meyer, F. and S. Beucher, Morphological segmentation. *Journal of Visual Communication and Image Representation*, 1990. **1**(1): 21–46.
44. Meyer, F., Iterative image transformations for an automatic screening of cervical cancer. *Journal of Histochemistry and Cytochemistry*, 1979. **27**: 128–135.

14

Still Image Compression*

Tor A. Ramstad
Norwegian University of Science and Technology

14.1 Introduction

Digital representation of images is important for digital transmission and storage on different media such as magnetic or laser disks. However, pictorial material requires vast amounts of bits if represented through direct quantization. As an example, an SVGA color image requires $3 \times 600 \times 800$ bytes $= 144$ Mbytes when each color component is quantized using 1 byte per pixel, the amount of bytes that can be stored on one standard 3.5 in. diskette. It is therefore evident that "compression" (often called "coding") is necessary for reducing the amount of data [33].

In this chapter we address three fundamental questions concerning image compression:

- Why is image compression possible?
- What are the theoretical coding limits?
- Which practical compression methods can be devised?

The first two questions concern statistical and structural properties of the image material and human visual perception. Even if we were able to answer these questions accurately, the methodology for image compression (third question) does not follow thereof. That is, the practical coding algorithms must be found otherwise. The bulk of the chapter will review image coding principles and present some of the best proposed still image coding methods.

The prevailing technique for image coding is "transform coding." This is part of the JPEG (Joint Picture Expert Group) standard [14] as well as a part of all the existing video coding standards

* Parts of this chapter are based on Ref. [33].

(H.261, H.263, MPEG-1, MPEG-2) [15–18]. Another closely related technique, "subband coding," is in some respects better, but has not yet been recognized by the standardization bodies. A third technique, "differential coding," has not been successful for still image coding, but is often used to code the low-pass-low-pass band in subband coders, and is an integral part of hybrid video coders for removal of temporal redundancy. "Vector quantization" (VQ) is the ultimate technique if there were no complexity constraints. Because all practical systems must have limited complexity, VQ is usually used as a component in a multicomponent coding scheme. Finally, "fractal or attraclor coding" is based on an idea far from other methods, but it is, nevertheless, strongly related to VQ.

For natural images, no exact digital representation exists because the quantization, which is an integral part of digital representations, is a lossy technique. Lossy techniques will always add noise, but the noise level and its characteristics can be controlled and depend on the number of bits per pixel as well as the performance of the method employed. "Lossless" techniques will be discussed as a component in other coding methods.

14.1.1 Signal Chain

We assume a model where the input signal is properly bandlimited and digitized by an appropriate "analog-to-digital converter." All subsequent processing in the encoder will be digital. The decoder is also digital up to the digital-to-analog converter, which is followed by a low-pass reconstruction filter.

Under idealized conditions, the interconnection of the signal chain excluding the compression unit will be assumed to be noise-free. (In reality, the analog-to-digital conversion will render a noise power which can be approximated by $\Delta^2/12$, where Δ is the quantizer interval. This interval depends on the number of bits, and we assume that it is so high that the contribution to the overall noise from this process is negligible). The performance of the coding chain can then be assessed from the difference between the input and output of the digital compression unit disregarding the analog part.

Still images must be sampled on some two-dimensional grid. Several schemes are viable choices, and there are good reasons for selecting nonrectangular grids. However, to simplify, rectangular sampling will be considered only, and all filtering will be based on separable operations, first performed on the rows and subsequently on the columns of the image. The theory is therefore presented for one-dimensional models, only.

14.1.2 Compressibility of Images

There are two reasons why images can be compressed:

- All meaningful images exhibit some form of internal structure, often expressed through statistical dependencies between pixels. We call this property "signal redundancy."
- The human visual system is not perfect. This means that certain degradations cannot be perceived by human observers. The degree of allowable noise is called "irrelevancy" or "visual redundancy." If we furthermore accept visual degradation, we can exploit what might be termed "tolerance."

In this section, we make some speculations about the compression potential resulting from redundancy and irrelevancy.

The two fundamental concepts in evaluating a coding scheme are "distortion," which measures quality in the compressed signal, and "rate," which measures how costly it is to transmit or store a signal.

Distortion is a measure of the deviation between the encoded/decoded signal and the original signal. Usually, distortion is measured by a single number for a given coder and bit rate. There are numerous ways of mapping an error signal onto a single number. Moreover, it is hard to conceive that a single number could mimic the quality assessment performed by a human observer. An easy-to-use and well-known error measure is the "mean square error" (mse). The visual correctness of this measure is poor. The human visual system is sensitive to errors in shapes and deterministic patterns, but not so much in stochastic textures. The mse defined over the entire image can, therefore, be entirely erroneous in the visual sense. Still, mse is the prevailing error measure, and it can be argued that it reflects well small changes due to optimization in a given coder structure, but poor as for the comparison between different models that create different noise characteristics.

Rate is defined as "bits per pixel" and is connected to the information content in a signal, which can be measured by "entropy."

14.1.2.1 A Lower Bound for Lossless Coding

To define image entropy, we introduce the set **S** containing all possible images of a certain size and call the number of images in the set N_S. To exemplify, assume the image set under consideration has dimension 512×512 pixels and each pixel is represented by 8 bits. The number of different images that exist in this set is $2^{512\times512\times8}$, an overwhelming number!

Given the probability P_i of each image in the set **S**, where $i \in N_S$ is the index pointing to the different images, the source entropy is given by

$$H = -\sum_{i\in N_S} P_i \log_2 P_i. \tag{14.1}$$

The entropy is a lower bound for the rate in lossless coding of the digital images.

14.1.2.2 A Lower Bound for Visually Lossless Coding

In order to incorporate perceptual redundancies, it is observed that all the images in the given set cannot be distinguished visually. We therefore introduce "visual entropy" as an abstract measure which incorporates distortion.

We now partition the image set into disjoint subsets, S_i, in which all the different images have similar appearance. One image from each subset is chosen as the "representation" image. The collection of these N_R representation images constitutes a subset R, that is a set spanning all distinguishable images in the original set.

Assume that image $i \in R$ appears with probability $\hat{P}_i$. Then the "visual entropy" is defined by

$$H_V = -\sum_{i\in N_R} \hat{P}_i \log_2 \hat{P}_i. \tag{14.2}$$

The minimum attainable bit rate is lower bounded by this number for image coders without visual degradation.

14.1.3 The Ideal Coding System

Theoretically, we can approach the visual entropy limit using an unrealistic "vector quantizer" (VQ), in conjunction with an ideal "entropy coder." The principle of such an optimal coding scheme is described next.

The set of representation images is stored in what is usually called a "codebook." The encoder and decoder have similar copies of this codebook. In the encoding process, the image to be coded is compared to all the vectors in the codebook applying the visually correct distortion measure. The codebook member with the closest resemblance to the sample image is used as the "coding approximation." The corresponding codebook index (address) is entropy coded and transmitted to the decoder. The decoder looks up the image located at the address given by the transmitted index.

Obviously, the above method is unrealistic. The complexity is beyond any practical limit both in terms of storage and computational requirement. Also, the correct visual distortion measure is not presently known. We should therefore only view the indicated coding strategy as the limit for any coding scheme.

14.1.4 Coding with Reduced Complexity

In practical coding methods, there are basically two ways of avoiding the extreme complexity of ideal VQ. In the first method, the encoder operates on small image blocks rather than on the complete image. This is obviously suboptimal because the method cannot profit from the redundancy offered by large structures in an image. But the larger the blocks, the better the method. The second strategy is very different and applies some preprocessing on the image prior to quantization. The aim is to remove statistical dependencies among the image pixels, thus avoiding representation of the same information more than once. Both techniques are exploited in practical coders, either separately or in combination.

A typical image encoder incorporating preprocessing is shown in Figure 14.1.

The first block (D) decomposes the signal into a set of coefficients. The coefficients are subsequently quantized in (Q), and are finally coded to a minimum bit representation in (B). This model is correct for frequency domain coders, but in "closed loop differential coders" (DPCM), the decomposition and quantization is performed in the same block, as will be demonstrated later. Usually the decomposition is exact. In "fractal" coding, the decomposition is replaced by approximate modeling.

Let us consider the "decoder" and introduce a series expansion as a unifying description of the different image representation methods:

$$\hat{x}(l) = \sum_{k} \hat{a}_k \phi_k(l). \tag{14.3}$$

The formula represents the recombination of signal components. Here $\{\hat{a}_k\}$ are the coefficients (the parameters in the representation), and $\{\phi_k(l)\}$ are the "basis functions." A major distinction between coding methods is their set of basis functions, as will be demonstrated in the next section.

The complete decoder consists of three major parts as shown in Figure 14.2. The first block (I) receives the bit representation which it partitions into entities representing the different coder parameters and decodes them. The second block (Q^{-1}) is a dequantizer which maps the code to the parametric approximation. The third block (R) reconstructs the signal from the parameters using the series representation.

The second important distinction between compression structures is the coding of the series expansion coefficients in terms of bits. This is dealt with in Section 14.3.

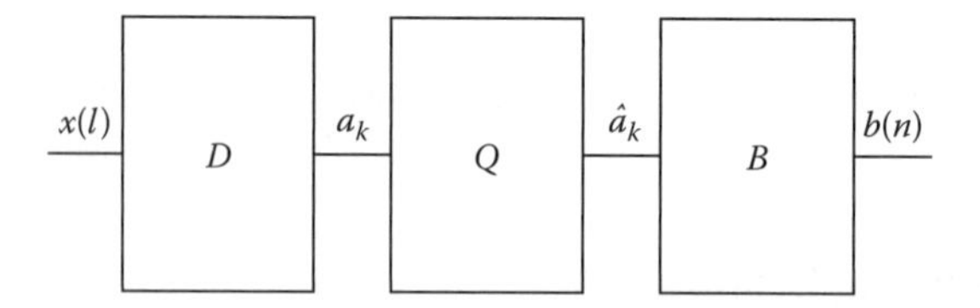

FIGURE 14.1 Generic encoder structure block diagram. D, decomposition unit; Q, quantizer; B, coder for minimum bit representation.

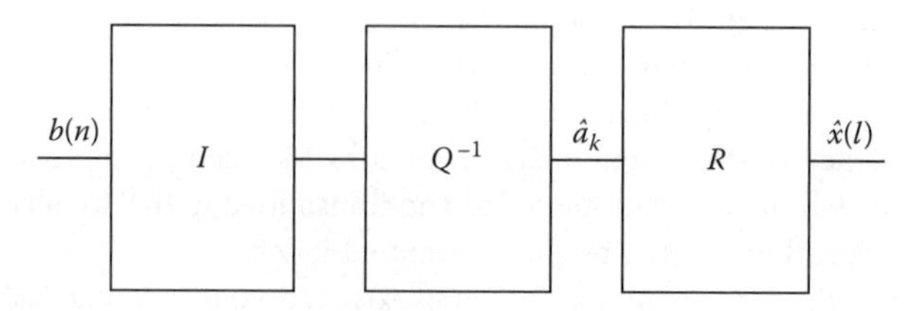

FIGURE 14.2 Block diagram of generic decoder structure. I, bit-representation decoder; Q^{-1}, inverse quantizer; R, signal reconstruction unit.

14.2 Signal Decomposition

As introduced in the previous section, series expansion can be viewed as a common tool to describe signal decomposition. The choice of basis functions will distinguish different coders and influence such features as "coding gain" and the types of distortions present in the decoded image for low bit rate coding. Possible classes of basis functions are

1. Block-oriented basis functions:
   a. The basis functions can cover the whole signal length L. L linearly independent basis functions will make a complete representation.
   b. Blocks of size $N \leq L$ can be decomposed individually. "Transform coders" operate in this way. If the blocks are small, the decomposition can catch fast transients. On the other hand, regions with constant features, such as smooth areas or textures, require long basis functions to fully exploit the correlation.
2. Overlapping basis functions:
   The length of the basis functions and the degree of overlap are important parameters. The issue of reversibility of the system becomes nontrivial.
   a. In "differential coding," one basis function is used over and over again, shifted by one sample relative to the previous function. In this case, the basis function usually varies slowly according to some adaptation criterion with respect to the local signal statistics.
   b. In subband coding using a uniform filter bank, N distinct basis functions are used. These are repeated over and over with a shift between each group by N samples. The length of the basis functions is usually several times larger than the shifts accommodating for handling fast transients as well as long-term correlations if the basis functions taper off at both ends.
   c. The basis functions may be finite (FIR filters) or semi-infinite (IIR filters).

Both time domain and frequency domain properties of the basis functions are indicators of the coder performance. It can be argued that decomposition, whether it is performed by a transform or a filter bank, represents a spectral decomposition. Coding gain is obtained if the different output channels are "decorrelated." It is therefore desirable that the frequency responses of the different basis functions are localized and separate in frequency. At the same time, they must cover the whole frequency band in order to make a complete representation.

The desire to have highly localized basis functions to handle transients, with localized Fourier transforms to obtain good coding gain, are contradictory requirements due to the "Heisenberg uncertainty relation" [33] between a function and its Fourier transform. The selection of the basis functions must be a compromise between these conflicting requirements.

14.2.1 Decomposition by Transforms

When nonoverlapping block transforms are used, the Karhunen–Loève transform decorrelates, in a statistical sense, the signal within each block completely. It is composed of the eigenvectors of the correlation matrix of the signal. This means that one either has to know the signal statistics in advance or estimate the correlation matrix from the image itself.

Mathematically the eigenvalue equation is given by

$$\mathbf{R}_{xx}\mathbf{h}_n = \lambda_n \mathbf{h}_n. \tag{14.4}$$

If the eigenvectors are column vectors, the KLT matrix is composed of the eigenvectors $\mathbf{h}_n$, $n = 0, 1, \ldots, N-1$, as its rows:

$$\mathbf{K} = [\mathbf{h}_0 \mathbf{h}_1 \ldots \mathbf{h}_{N-1}]^{\mathrm{T}}. \tag{14.5}$$

The decomposition is performed as

$$\mathbf{y} = \mathbf{K}\mathbf{x}. \tag{14.6}$$

The eigenvalues are equal to the power of each transform coefficient.

In practice, the so-called cosine transform (of type II) is usually used because it is a fixed transform and it is close to the KLT when the signal can be described as a first-order autoregressive process with correlation coefficient close to 1.

The cosine transform of length N in one dimension is given by

$$y(k) = \sqrt{\frac{2}{N}}\alpha(k)\sum_{n=0}^{N-1} x(n)\cos\frac{(2n+1)k\pi}{2N}, \quad k = 0, 1, \ldots, N-1, \tag{14.7}$$

where

$$\alpha(0) = \frac{1}{\sqrt{2}} \quad \text{and} \quad \alpha(k) = 1 \quad \text{for } k \neq 0. \tag{14.8}$$

The inverse transform is similar except that the scaling factor $\alpha(k)$ is inside the summation.

Many other transforms have been suggested in the literature (discrete Fourier transform [DFT], Hadamard transform, sine transform, etc.), but none of these seem to have any significance today.

14.2.2 Decomposition by Filter Banks

"Uniform" analysis and synthesis filter banks are shown in Figure 14.3.

In the analysis filter bank the input signal is split in contiguous and slightly overlapping frequency bands denoted "subbands." An ideal frequency partitioning is shown in Figure 14.4.

If the analysis filter bank was able to "decorrelate" the signal completely, the output signal would be white. For all practical signals, complete decorrelation requires an infinite number of channels.

In the encoder the symbol $\downarrow N$ indicates decimation by a factor of N. By performing this decimation in each of the N channels, the total number of samples is conserved from the system input to decimator outputs. With the channel arrangement in Figure 14.4, the decimation also serves as a "demodulator." All channels will have a baseband representation in the frequency range $[0, \pi/N]$ after decimation.

The synthesis filter bank, as shown in Figure 14.3, consists of N branches with interpolators indicated by $\uparrow N$ and band-pass filters arranged as the filters in Figure 14.4.

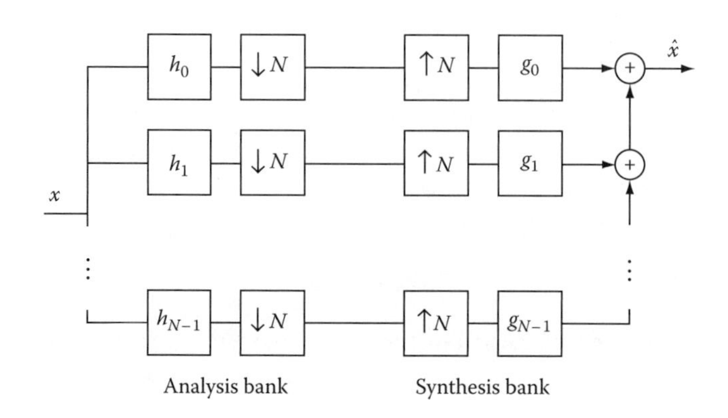

FIGURE 14.3 Subband coder system.

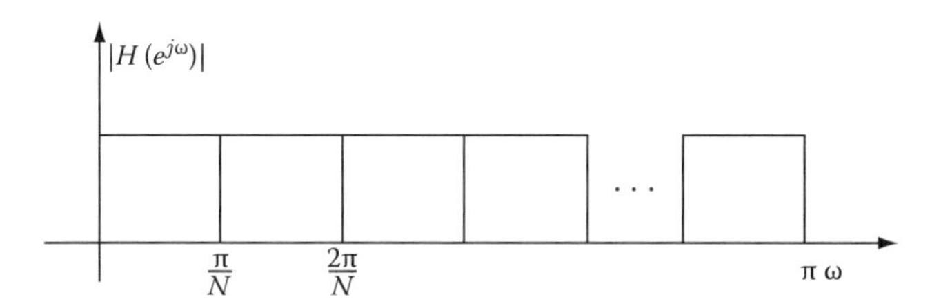

FIGURE 14.4 Ideal frequency partitioning in the analysis channel filters in a subband coder.

The reconstruction formula constitutes the following series expansion of the output signal:

$$\hat{x}(l) = \sum_{n=0}^{N-1} \sum_{k=-\infty}^{\infty} e_n(k)g_n(l - kN), \tag{14.9}$$

where

$\{e_n(k), n = 0, 1, \ldots, N - 1, k = -\infty, \ldots, -1, 0, 1, \ldots, \infty\}$ are the expansion coefficients representing the quantized subband signals

$\{g_n(k),\ n = 0, 1, \ldots, N\}$ are the basis functions, which are implemented as unit sample responses of band-pass filters

14.2.2.1 Filter Bank Structures

Through the last two decades, an extensive literature on filter banks and filter bank structures has evolved. Perfect reconstruction (PR) is often considered desirable in subband coding systems. It is not a trivial task to design such systems due to the downsampling required to maintain a minimum sampling rate. PR filter banks are often called identity systems. Certain filter bank structures inherently guarantee PR.

It is beyond the scope of this chapter to give a comprehensive treatment of filter banks. We shall only present different alternative solutions at an overview level, and in detail discuss an important two-channel system with inherent perfect reconstruction properties.

We can distinguish between different filter banks based on several properties. In the following, five classifications are discussed.

1. FIR vs. IIR filters—Although IIR filters have an attractive complexity, their inherent long unit sample response and nonlinear phase are obstacles in image coding. The unit sample response length influences the "ringing problem," which is a main source of objectionable distortion in subband coders. The nonlinear phase makes the "edge mirroring technique" [30] for efficient coding of images near their borders impossible.
2. Uniform vs. nonuniform filter banks—This issue concerns the spectrum partioning in frequency subbands. Currently it is the general conception that nonuniform filter banks perform better than uniform filter banks. There are two reasons for that. The first reason is that our visual system also performs a nonuniform partioning, and the coder should mimic the type of receptor for which it is designed. The second reason is that the filter bank should be able to cope with slowly varying signals (correlation over a large region) as well as transients that are short and represent high frequency signals. Ideally, the filter banks should be adaptive (and good examples of adaptive filter banks have been demonstrated in the literature [2,11]), but without adaptivity one filter bank has to be a good compromise between the two extreme cases cited above. Nonuniform filter banks can give the best tradeoff in terms of space-frequency resolution.
3. Parallel vs. tree-structured filter banks—The parallel filter banks are the most general, but tree-structured filter banks enjoy a large popularity, especially for octave band (dyadic frequency partitioning) filter banks as they are easily constructed and implemented. The popular subclass

of filter banks denoted "wavelet filter banks" or "wavelet transforms" belong to this class. For octave band partioning, the tree-structured filter banks are as general as the parallel filter banks when perfect reconstruction is required [4].

4. Linear phase vs. nonlinear phase filters—There is no general consensus about the optimality of linear phase. In fact, the traditional wavelet transforms cannot be made linear phase. There are, however, three indications that linear phase should be chosen. (1) The noise in the reconstructed image will be antisymmetrical around edges with nonlinear phase filters. This does not appear to be visually pleasing. (2) The mirror extension technique [30] cannot be used for nonlinear phase filters. (3) Practical coding gain optimizations have given better results for linear than nonlinear phase filters.
5. Unitary vs. nonunitary systems—A unitary filter bank has the same analysis and synthesis filters (except for a reversal of the unit sample responses in the synthesis filters with respect to the analysis filters to make the overall phase linear). Because the analysis and synthesis filters play different roles, it seems plausible that they, in fact, should not be equal. Also, the gain can be larger, as demonstrated in Section 14.2.3, for nonunitary filter banks as long as straightforward scalar quantization is performed on the subbands.

Several other issues could be taken into consideration when optimizing a filter bank. These are, among others, the actual frequency partitioning including the number of bands, the length of the individual filters, and other design criteria than coding gain to alleviate coding artifacts, especially at low rates. As an example of the last requirement, it is important that the different phases in the reconstruction process generate the same noise; in other words, the noise should be stationary rather than cyclostationary. This may be guaranteed through requirements on the norms of the unit sample responses of the polyphase components [4].

14.2.2.2 The Two-Channel Lattice Structure

A versatile perfect reconstruction system can be built from two-channel substructures based on lattice filters [36]. The analysis filter bank is shown in Figure 14.5. It consists of delay-free blocks given in matrix forms as

$$\eta = \begin{bmatrix} a & b \\ c & d \end{bmatrix}, \tag{14.10}$$

and single delays in the lower branch between each block. At the input, the signal is multiplexed into the two branches, which also constitutes the decimation in the analysis system.

A similar synthesis filter structure is shown in Figure 14.6. In this case, the lattices are given by the inverse of the matrix in Equation 14.10:

$$\eta^{-1} = \frac{1}{ad - bc}\begin{bmatrix} d & -b \\ -c & a \end{bmatrix}, \tag{14.11}$$

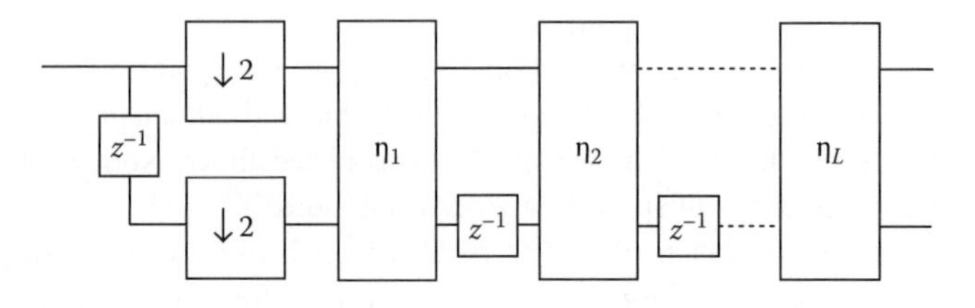

FIGURE 14.5 Multistage two-channel lattice analysis lattice filter bank.

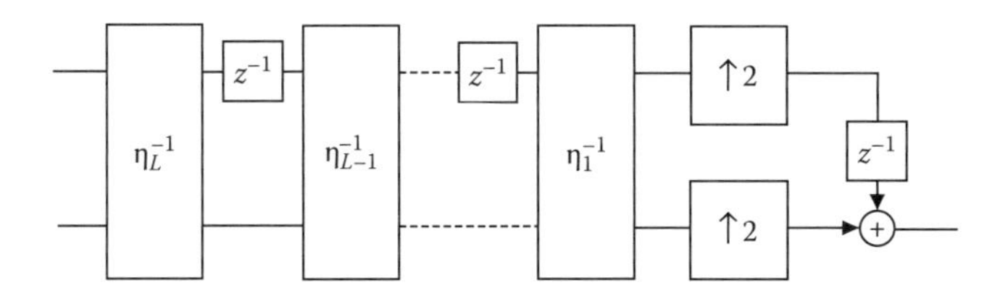

FIGURE 14.6 Multistage two-channel polyphase synthesis lattice filter bank.

and the delays are in the upper branches. It is not hard to realize that the two systems are inverse systems provided $ad - bc \neq 0$, except for a system delay.

As the structure can be extended as much as wanted, the flexibility is good. The filters can be made unitary or they can have a linear phase. In the unitary case, the coefficients are related through $a = d = \cos\phi$ and $b = -c = \sin\phi$, whereas in the linear phase case, the coefficients are $a = d = 1$ and $b = c$. In the linear phase case, the last block (η_L) must be a "Hadamard" transform.

14.2.2.3 Tree-Structured Filter Banks

In tree-structured filter banks, the signal is first split in two channels. The resulting outputs are input to a second stage with further separation. This process can go on as indicated in Figure 14.7 for a system where at every stage the outputs are split further until the required resolution has been obtained.

Tree-structured systems have a rather high flexibility. Nonuniform filter banks are obtained by splitting only some of the outputs at each stage. To guarantee perfect reconstruction, each stage in the synthesis filter bank (Figure 14.7) must reconstruct the input signal to the corresponding analysis filter.

14.2.3 Optimal Transforms/Filter Banks

The "gain" in subband and transform coders depends on the detailed construction of the filter bank as well as the quantization scheme.

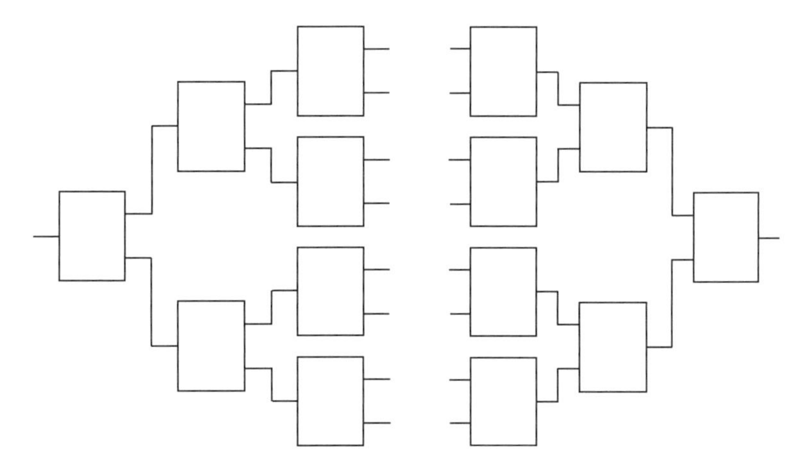

FIGURE 14.7 Left: Tree structured analysis filter bank consisting of filter blocks where the signal is split in two and decimated by a factor of two to obtain critical sampling. Right: Corresponding synthesis filter bank for recombination and interpolation of the signals.

Assume that the analysis filter bank unit sample responses are given by $\{h_n(k), n=0,1,\ldots,N-1\}$. The corresponding unit sample responses of the synthesis filters are required to have unit norm:

$$\sum_{k=0}^{L-1} g_n^2(k) = 1.$$

The coding gain of a subband coder is defined as the ratio between the noise using scalar quantization (pulse code modulation, PCM) and the subband coder noise incorporating optimal bit-allocation as explained in Section 14.3:

$$G_{SBC} = \left[\prod_{n=0}^{N-1} \frac{\sigma_{x_n}^2}{\sigma_x^2}\right]^{-1/N} \tag{14.12}$$

Here σ_x^2 is the variance of the input signal while $\{\sigma_{x_n}^2, n = 0, 1 \ldots, N-1\}$ are the subband variances given by

$$\sigma_{x_n}^2 = \sum_{l=-\infty}^{\infty} R_{xx}(l) \sum_{j=-\infty}^{\infty} h_n(j)h_n(l+j) \tag{14.13}$$

$$= \int_{-\pi}^{\pi} S_{xx}(e^{j\omega})\left|H_n(e^{j\omega})\right|^2 \frac{d\omega}{2\pi} \tag{14.14}$$

The subband variances depend both on the filters and the second order spectral information of the input signal.

For images, the gain is often estimated assuming that the image can be modeled as a first order Markov source (also called an AR(1) process) characterized by

$$R_{xx}(l) = \sigma_x^2 0.95^{|l|}. \tag{14.15}$$

(Strictly speaking, the model is valid only after removal of the image average).

We consider the maximum gain using this model for three special cases. The first is the transform coder performance, which is an important reference as all image and video coding standards are based on transform coding. The second is for unitary filter banks, for which optimality is reached by using ideal brick-wall filters. The third case is for nonunitary filter banks, often denoted "biorthogonal" when the perfect reconstruction property is guaranteed. In the nonunitary case, "halfwhitening" is obtained within each band. Mathematically this can be seen from the optimal magnitude response for the filter in channel n:

$$\left|H_n(e^{j\omega})\right| = \begin{cases} c_2 \left[\frac{S_{xx}(e^{j\omega})}{\sigma_x^2}\right]^{-1/4} & \text{for } \omega \in \pm\left[\frac{\pi n}{N}, \frac{\pi(n+1)}{N}\right] \\ 0 & \text{otherwise,} \end{cases} \tag{14.16}$$

where c_2 is a constant that can be selected for correct gain in each band.

The inverse operation must be performed in the synthesis filter to make completely flat responses within each band.

In Figure 14.8, we give optimal coding gains as a function of the number of channels.

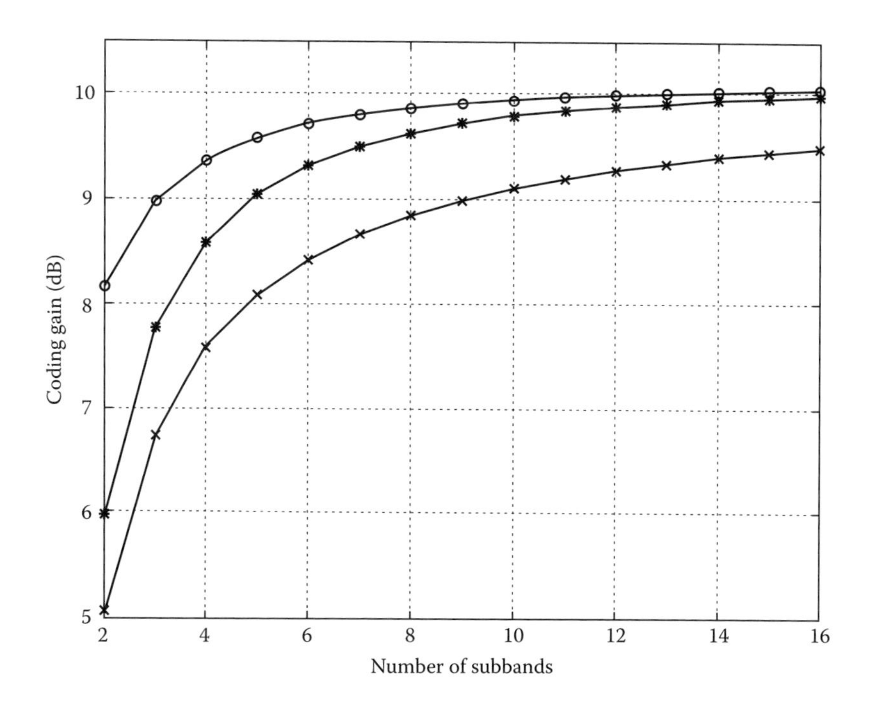

FIGURE 14.8 Maximum coding gain as function of the number of channels for different one-dimensional coders operating on a first order Markov source with one-delay correlation $\rho = 0.95$. Lower curve: Cosine transform. Middle curve: Unitary filter bank. Upper curve: Unconstrained filter bank. Nonunitary case.

14.2.4 Decomposition by Differential Coding

In "closed-loop differential" coding, the generic encoder structure (Figure 14.1) is not valid as the quantizer is placed inside a feedback loop. The decoder, however, behaves according to the generic decoder structure. Basic block diagrams of a closed-loop differential encoder and the corresponding decoder are shown in Figure 14.9a and b, respectively.

In the encoder, the input signal x is represented by the bit-stream b. Q is the quantizer and Q^{-1} the dequantizer, but $QQ^{-1} \neq 1$, except for the case of infinite resolution in the quantizer. The signal d, which is quantized and transmitted by some binary code, is the difference between the input signal and a

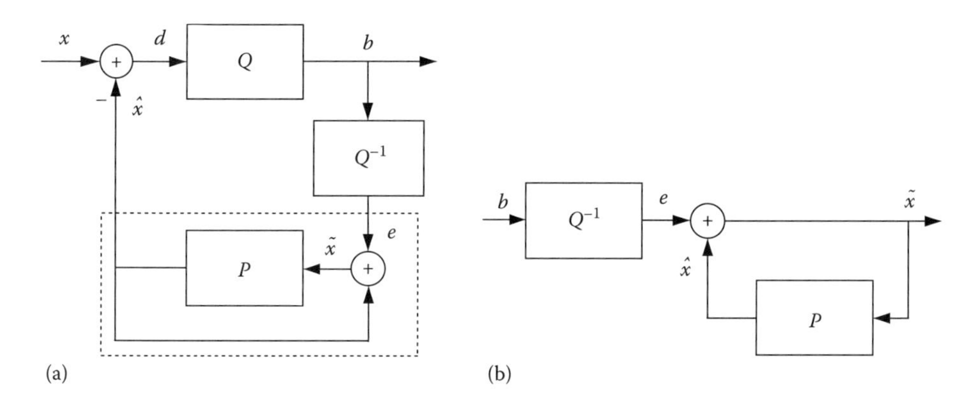

FIGURE 14.9 (a) DPCM encoder. (b) DPCM decoder.

predicted value of the input signal based on previous outputs and a prediction filter with transfer function $G(z) = 1/(1 - P(z))$. Notice that the decoder is a substructure of the encoder, and that $\tilde{x} = x$ in the limiting case of infinite quantizer resolution. The last property guarantees exact representation when discarding quantization.

Introducing the inverse z-transform of $G(z)$ as $g(l)$, the reconstruction is performed on the dequantized values as

$$\tilde{x}(l) = \sum_{k=0}^{\infty} e(k)g(l - k). \tag{14.17}$$

The output is thus a linear combination of unit sample responses excited by the sample amplitudes at different times and, can be viewed as a series expansion of the output signal. In this case, the basis functions are generated by shifts of a single basis function [the unit sample response $g(l)$] and the coefficients represent the coded difference signal $e(n)$.

With an adaptive filter the basis function will vary slowly, depending on some spectral modification derived from the incoming samples.

14.3 Quantization and Coding Strategies

Quantization is the means of providing approximations to signals and signal parameters by a finite number of representation levels. This process is nonreversible and thus always introduces noise. The representation levels constitute a finite alphabet which is usually represented by binary symbols, or "bits." The mapping from symbols in a finite alphabet to bits is not unique. Some important techniques for quantization and coding will be reviewed next.

14.3.1 Scalar Quantization

The simplest quantizer is the "scalar" quantizer. It can be optimized to match the "probability density function" (pdf) of the input signal.

A scalar quantizer maps a continuous variable x to a finite set according to the rule

$$x \in R_i \Rightarrow Q[x] = y_i, \tag{14.18}$$

where $R_i = (x_i, x_{i+1})$, $i = 1, \ldots, L$, are nonoverlapping, contiguous intervals covering the real line, and $(\cdot,\cdot)$ denotes open, half open, or closed intervals. $\{y_i, i = 1, 2, \ldots, L\}$ are referred to as "representation levels" or "reconstruction values." The associated values $\{x_i\}$ defining the partition are referred to as "decision levels" or "decision thresholds." Figure 14.10 depicts the representation and decision levels.

In a "uniform" quantizer, all intervals are of the same length and the representation levels are the midpoints in each interval. Furthermore, in a "uniform threshold" quantizer, the decision levels form a uniform partitioning of the real line, while the representation levels are the "centroids" (see below) in each decision interval. Strictly speaking, uniform quantizers consist of an infinite number of intervals. In practice, the number of intervals is adapted to the dynamic range of the signal. All other quantizers are "nonuniform."

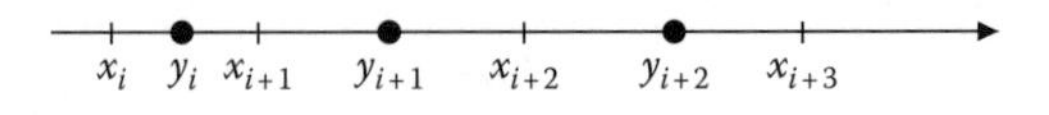

FIGURE 14.10 Quantization notation.

The optimization task is to minimize the average "distortion" between the original samples and the appropriate representation levels given the number of levels. This is the so-called pdf-optimized quantizer. Allowing for variable rate per symbol, the "entropy constrained quantizer" can be used. These schemes are described in the following two subsections.

14.3.1.1 The Lloyd–Max Quantizer

The Lloyd–Max quantizer is a scalar quantizer where the first-order signal pdf is exploited to increase the quantizer performance. It is therefore often referred to as a "pdf-optimized quantizer." Each signal sample is quantized using the same number of bits. The optimization is done by minimizing the total distortion of a quantizer with a given number L of representation levels. For an input signal X with pdf $p_X(x)$, the average mean square distortion is

$$D = \sum_{i=1}^{L} \int_{x_i}^{x_{i+1}} (x - y_i)^2 p_X(x)\mathrm{d}x. \tag{14.19}$$

Minimization of D leads to the following implicit expressions connecting the decision and representation levels:

$$x_{k,\mathrm{opt}} = \frac{1}{2}(y_{k,\mathrm{opt}} + y_{k-1,\mathrm{opt}}), \quad k = 1, \ldots, L-1 \tag{14.20}$$

$$x_{0,\mathrm{opt}} = -\infty \tag{14.21}$$

$$x_{L,\mathrm{opt}} = \infty \tag{14.22}$$

$$y_{k,\mathrm{opt}} = \frac{\int_{x_{k,\mathrm{opt}}}^{x_{k+1,\mathrm{opt}}} x p_X(x)\mathrm{d}x}{\int_{x_{k,\mathrm{opt}}}^{x_{k+1,\mathrm{opt}}} p_X(x)\mathrm{d}x}, \quad k = 0, \ldots, L-1. \tag{14.23}$$

Equation 14.20 indicates that the decision levels should be the midpoints between neighboring representation levels, while Equation 14.23 requires that the optimal representation levels are the "centroids" of the pdf in the appropriate interval.

The equations can be solved iteratively [21]. For high bit rates it is possible to derive approximate formulas assuming that the signal pdf is flat within each quantization interval [21].

In most practical situations the pdf is not known, and the optimization is based on a training set. This will be discussed in Section 14.3.2.

14.3.1.2 Entropy-Constrained Quantization

When minimizing the total distortion for a fixed number of possible representation levels, we have tacitly assumed that every signal sample is coded using the same number of bits: $\log_2 L$ bits/sample. If we allow for a variable number of bits for coding each sample, a further rate-distortion advantage is gained. The Lloyd–Max solution is then no longer optimal. A new optimization is needed, leading to the "entropy constrained quantizer."

At high bit rates, the optimum is reached when using a uniform quantizer with an infinite number of levels. At low bit rates, uniform quantizers perform close to optimum provided the representation levels are selected as the centroids according to Equation 14.23. The performance of the entropy constrained quantizer is significantly better than the performance of the Lloyd–Max quantizer [21].

A standard algorithm for assigning codewords of variable length to the representation levels was given by Huffman [12]. The Huffman code will minimize the average rate for a given set of probabilities and the resulting average bit rate will be close to the entropy bound. Even closer performance to the bound is obtained by "arithmetic coders" [32].

At high bit rates, scalar quantization on "statistically independent" samples renders a bit rate which is at least 0.255 bits/sample higher than the "rate distortion bound" irrespective of the signal pdf. Huffman coding of the quantizer output typically gives a somewhat higher rate.

14.3.2 Vector Quantization

Simultaneous quantization of several samples is referred to as VQ [9], as mentioned in the introductory section. VQ is a generalization of scalar quantization:

A vector quantizer maps a continuous N-dimensional vector $\mathbf{x}$ to a discrete-valued N-dimensional vector according to the rule

$$\mathbf{x} \in C_i \Rightarrow Q[\mathbf{x}] = \mathbf{y}_i, \tag{14.24}$$

where C_i is an N-dimensional cell. The L possible cells are nonoverlapping and contiguous and fill the entire geometric space. The vectors $\{\mathbf{y}_i\}$ correspond to the representation levels in a scalar quantizer. In a VQ setting the collection of representation levels is referred to as the "codebook." The cells C_i, also called "Voronoi regions," correspond to the decision regions, and can be thought of as solid polygons in the N-dimensional space.

In the scalar case, it is trivial to test if a signal sample belongs to a given interval. In VQ an indirect approach is utilized via a "fidelity criterion" or "distortion measure" $d(\cdot,\cdot)$:

$$Q[\mathbf{x}] = \mathbf{y}_i \Longleftrightarrow d(\mathbf{x}, \mathbf{y}_i) \leq d(\mathbf{x}, \mathbf{y}_j), \quad j = 0, \ldots, L-1. \tag{14.25}$$

When the best match, $\mathbf{y}_i$, has been found, the "index i" identifies that vector and is therefore coded as an efficient representation of the vector. The receiver can then reconstruct the vector $\mathbf{y}_i$ by looking up the contents of cell number i in a copy of the codebook. Thus, the bit rate in bits per sample in this scheme is $\log_2 L/N$ when using straightforward bit representation for i. A block diagram of VQ is shown in Figure 14.11.

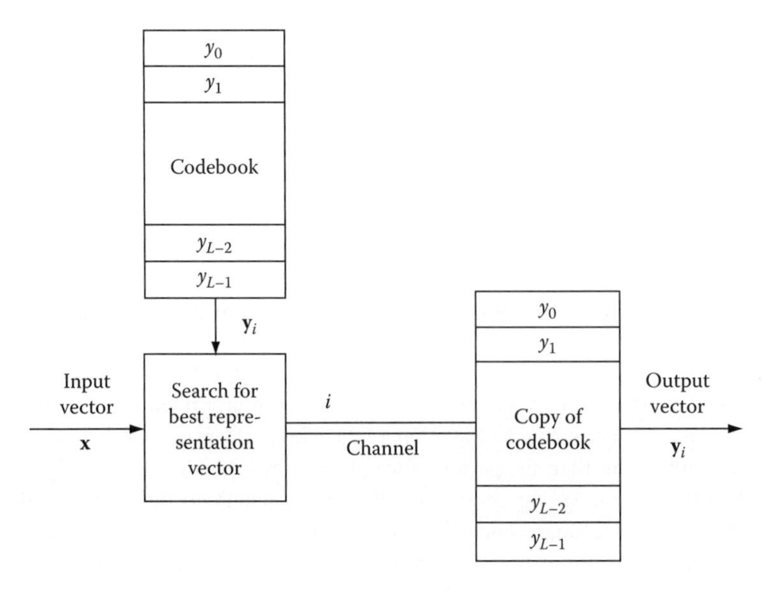

FIGURE 14.11 VQ procedure.

In the previous section we stated that scalar entropy coding was suboptimal, even for sources producing independent samples. The reason for the sub-optimal performance of the entropy constrained quantizer is a phenomenon called "sphere packing." In addition to obtaining good sphere packing, a VQ scheme also exploits both correlation and higher order statistical dependencies of a signal. The higher order statistical dependency can be thought of as "a preference for certain vectors." Excellent examples of sphere packing and higher order statistical dependencies can be found in Ref. [28].

In principle, the codebook design is based on the N-dimensional pdf. But as the pdf is usually not known, the codebook is optimized from a training data set. This set consists of a large number of vectors that are representative for the signal source. A suboptimal codebook can then be designed using an iterative algorithm, for example, the "K-means" or "LBG" algorithm [25].

14.3.2.1 Multistage Vector Quantization

To alleviate the complexity problems of VQ, several methods have been suggested. They all introduce some structure into the codebook which makes fast search possible. Some systems also reduce storage requirements, like the one we present in this subsection. The obtainable performance is always reduced, but the performance in an implementable coder can be improved.

Figure 14.12 illustrates the encoder structure.

The first block in the encoder makes a rough approximation to the input vector by selecting the codebook vector which, upon scaling by e_1, is closest in some distortion measure. Then this approximation is subtracted from the input signal. In the second stage, the difference signal is approximated by a vector from the second codebook scaled by e_2. This procedure continues in K stages, and can be thought of as a successive approximation to the input vector. The indices $\{i(k), k = 1, 2, \ldots, K\}$ are transmitted as part of the code for the particular vector under consideration.

Compared to unstructured VQ, this method is suboptimal but has a much lower complexity than the optimal case due to the small codebooks that can be used.

A special case is the "mean-gain-shape" VQ [9], where one stage only is kept, but in addition the mean is represented separately.

In all multistage VQs, the code consists of the codebook address and codes for the quantized versions of the scaling coefficients.

14.3.3 Efficient Use of Bit Resources

Assume we have a signal that can be split in classes with different statistics. As an example, after applying signal decomposition, the different transform coefficients typically have different variances. Assume also that we have a pool of bits to be used for representing a collection of signal vectors from the different classes, or we try to minimize the number of bits to be used after all signals have been quantized. These two situations are described below.

14.3.3.1 Bit Allocation

Assume that a signal consists of N components $\{x_i, i = 1, 2, \ldots, N\}$ forming a vector $\mathbf{x}$ where the variance of component number i is equal to $\sigma_{x_i}^2$ and all components are zero mean.

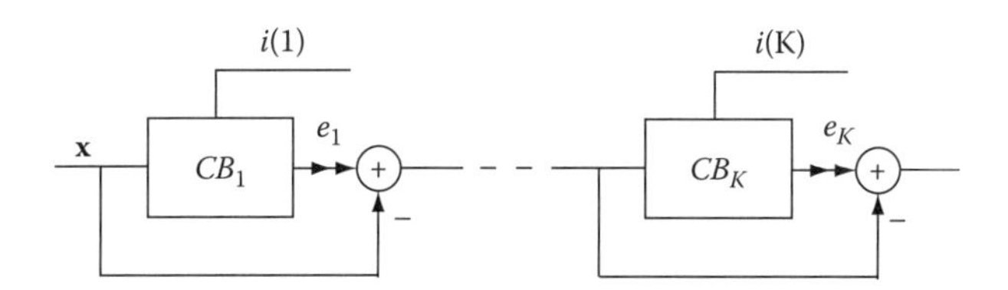

FIGURE 14.12 K-stage VQ encoder structure showing the successive approximation of the signal vector.

We want to quantize the vector **x** using scalar quantization on each of the components and minimize the total distortion with the only constraint that the total number of bits to be used for the whole vector be fixed and equal to B. Denoting the quantized signal components $Q_i(x_i)$, the average distortion per component can be written as

$$D_{\mathrm{DS}} = \frac{1}{N}\sum_{i=1}^{N} E[x_i - Q_i(x_i)]^2 = \frac{1}{N}\sum_{i=1}^{N} D_i, \tag{14.26}$$

where

$E[\cdot]$ is the expectation operator

The subscript DS stands for "decomposed source"

The bit-constraint is given by

$$B = \sum_{i=1}^{N} b_i, \tag{14.27}$$

where b_i is the number of bits used to quantize component number i.

Minimizing D_{DS} with Equation 14.27 as a constraint, we obtain the following bit assignment

$$b_j = \frac{B}{N} + \frac{1}{2}\log_2 \frac{\sigma_{x_j}^2}{\left[\prod_{n=1}^{N} \sigma_{x_n}^2\right]^{1/N}}. \tag{14.28}$$

This formula will in general render noninteger and even negative values of the bit count. So-called "greedy" algorithms can be used to avoid this problem.

To evaluate the coder performance, we use "coding gain." It is defined as the distortion advantage of the component-wise quantization over a direct scalar quantization at the same rate. For the example at hand, the coding gain is found to be

$$G_{\mathrm{DS}} = \frac{\frac{1}{N}\sum_{j=1}^{N} \sigma_{x_j}^2}{\left(\prod_{j=1}^{N} \sigma_{n_j}^2\right)^{1/N}}. \tag{14.29}$$

The gain is equal to the ratio between the arithmetic mean and the geometric mean of the component variances. The minimum value of the variance ratio is equal to 1 when all the component variances are equal. Otherwise, the gain is larger than one. Using the optimal bit allocation, the noise contribution is equal in all components.

If we assume that the different components are obtained by passing the signal through a bank of bandpass filters, then the variance from one band is given by the integral of the power spectral density over that band. If the process is nonwhite, the variances are more different the more colored the original spectrum is. The maximum possible gain is obtained when the number of bands tends to infinity [21]. Then the gain is equal to the maximum gain of a differential coder which again is inversely proportional to the "spectral flatness measure" [21] given by

$$\gamma_x^2 = \frac{\exp\left[\int_{-\pi}^{\pi} \ln S_{xx}(e^{j\omega})\frac{d\omega}{2\pi}\right]}{\int_{-\pi}^{\pi} S_{xx}(e^{j\omega})\frac{d\omega}{2\pi}}, \tag{14.30}$$

where $S_{xx}(e^{j\omega})$ is the spectral density of the input signal. In both subband coding and differential coding, the complexity of the systems must approach infinity to reach the coding gain limit.

To be able to apply bit allocation dynamically to nonstationary sources, the decoder must receive information about the local bit allocation. This can be done either by transmitting the bit allocation table, or the variances from which the bit allocation was derived. For real images where the statistics vary rapidly, the cost of transmitting the side information may become costly, especially for low rate coders.

14.3.3.2 Rate Allocation

Assume we have the same signal collection as above. This time we want to minimize the number of bits to be used after the signal components have been quantized. The first order entropy of the decomposed source will be selected as the measure for the obtainable minimum bit rate when scalar representation is specified.

To simplify, assume all signal components are Gaussian. The entropy of a Gaussian source with zero mean and variance σ_x^2 and statistically independent samples quantized by a uniform quantizer with quantization interval Δ can, for high rates, be approximated by

$$H_G(X) = \frac{1}{2}\log_2\left(2\pi e(\sigma_x/\Delta)^2\right). \tag{14.31}$$

The rate difference [24] between direct scalar quantization of the signal collection using one entropy coder and the rate when using an adapted entropy coder for each component is

$$\Delta H = H_{\text{PCM}} - H_{\text{DS}} = \frac{1}{2}\log_2\frac{\sigma_x^2}{\left[\prod_{i=1}^{N}\sigma_{x_i}^2\right]^{1/N}}, \tag{14.32}$$

provided the decomposition is power conserving, meaning that

$$\sigma_x^2 = \sum_{i=1}^{N}\sigma_{x_i}^2. \tag{14.33}$$

The coding gain in Equation 14.29 and the rate gain in Equation 14.32 are equivalent for Gaussian sources.

In order to exploit this result in conjunction with signal decomposition, we can view each output component as a stationary source, each with different signal statistics. The variances will depend on the spectrum of the input signal. From Equations 14.32 and 14.33 we see that the rate difference is larger the more different the channel variances are.

To obtain the rate gain indicated by Equation 14.32, different "Huffman" or "arithmetic" coders [9] adapted to the rate given in Equation 14.31 must be employed. In practice, a pool of such coders should be generated and stored. During encoding, the closest fitting coder is chosen for each block of components. An index indicating which coder was used is transmitted as side information to enable the decoder to reinterpret the received code.

14.4 Frequency Domain Coders

In this section we present the JPEG standard and some of the best subband coders that have been presented in the literature.

14.4.1 The JPEG Standard

The JPEG coder [37] is the only internationally standardized still image coding method. Presently there is an international effort to bring forth a new, improved standard under the title JPEG2000.

The principle can be sketched as follows: First, the image is decomposed using a two-dimensional cosine transform of size 8×8. Then, the transform coefficients are arranged in an 8×8 matrix as given in Figure 14.13, where i and j are the horizontal and vertical frequency indices, respectively. A vector is formed by a scanning sequence which is chosen to make large amplitudes, on average, appear first, and smaller amplitudes at the end of the scan. In this arrangement, the samples at the end of the scan string approach zero. The scan vector is quantized in a nonuniform scalar quantizer with characteristics as depicted in Figure 14.14.

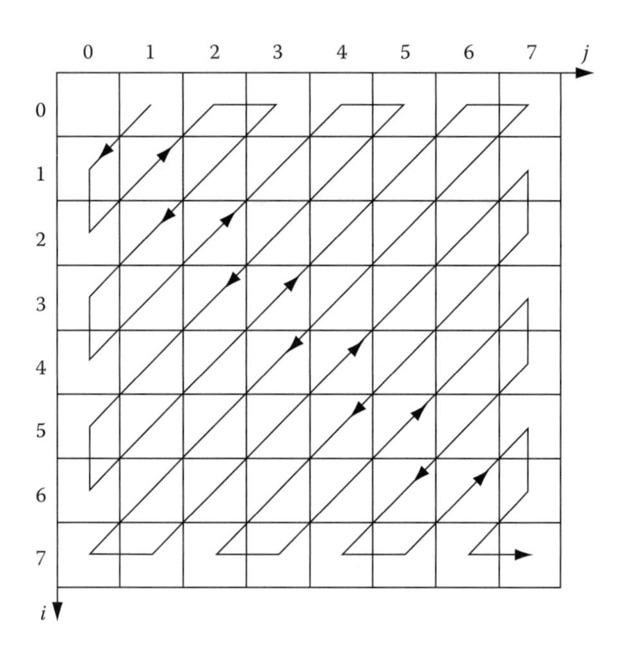

FIGURE 14.13 Zigzag scanning of the coefficient matrix.

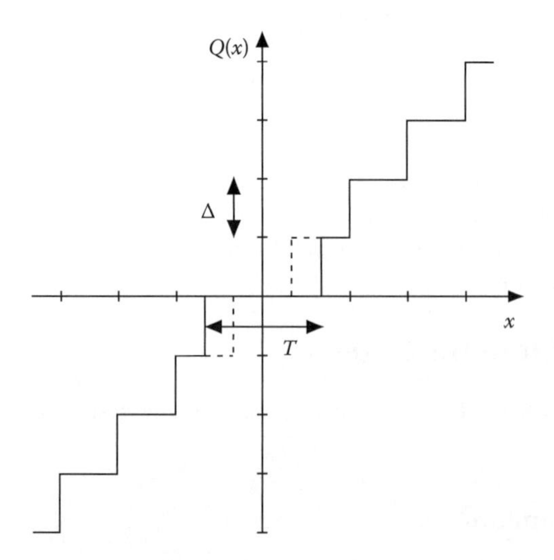

FIGURE 14.14 Nonuniform quantizer characteristic obtained by combining a mid-tread uniform quantizer and a thresholder. Δ is the quantization interval and T is the threshold.

Due to the thresholder, many of the trailing coefficients in the scan vector are set to zero. Often the zero values appear in clusters. This property is exploited by using "runlength coding," which basically amounts to finding zero-runs. After runlength coding, each run is represented by a "number pair" (a, r) where the number a is the amplitude and r is the length of the run. Finally, the number pair is entropy coded using the Huffman method, or arithmetic coding.

The thresholding will increase the distortion and lower the entropy both with and without decomposition, although not necessarily with the same amounts.

As can be observed from Figure 14.13, the coefficient in position $(0, 0)$ is not part of the string. This coefficient represents the block average. After collecting all block averages in one image, this image is coded using a DPCM scheme [37].

Coding results for three images are given in Figure 14.16.

14.4.2 Improved Coders: State-of-the-Art

Many coders that outperform JPEG have been presented in the scientific literature. Most of these are based on subband decomposition (or the special case: wavelet decomposition). Subband coders have a higher potential coding gain by using filter banks rather than transforms, and thus exploiting correlations over larger image areas. Figure 14.8 shows the theoretical gain for a stochastic image model. Visually, subband coders can avoid the blocking-effects experienced in transform coders at low bit rates. This property is due to the overlap in basis functions in subband coders. On the other hand, Gibb's phenomenon is more prevalent in subband coders and can cause severe ringing in homogeneous areas close to edges. The detailed choice and optimization of the filter bank will strongly influence the visual performance of subband coders. The other factor which decides the coding quality is the detailed quantization of the subband signals. The final bit-representation method does not effect the quality, only the rate for a given quality.

Depending on the bit representation, the total rate can be preset for some coders, and will depend on some quality factor specified for other coders. Even though it would be desirable to preset the visual quality in a coder, this is a challenging task, which has not yet been satisfactorily solved.

In the following we present four subband coders with different coding schemes and different filter banks.

14.4.2.1 Subband Coder Based on Entropy Coder Allocation [24]

This coder uses an 8×8 uniform filter bank optimized for reducing blocking and ringing artifacts, plus maximizing the coding gain [1]. The low-pass-low-pass band is quantized using a fixed rate DPCM coder with a third-order two-dimensional predictor. The other subband signals are segmented into blocks of size 4×4, and each block is classified based on the block power. Depending on the block power, each block is allocated a corresponding entropy coder (implemented as an arithmetic coder). The entropy coders have been preoptimized by minimizing the first-order entropy given the number of available entropy coders (see Section 14.3.3). This number is selected to balance the amount of side information necessary in the decoder to identify the correct entropy decoder and the gain by using more entropy coders. Depending on the bit rate, the number of entropy coders is typically 3–5. In the presented results, three arithmetic coders are used. Conditional arithmetic coding has been used to represent the side information efficiently.

Coding results are presented in Figure 14.16 under the name "Lervik."

14.4.2.2 Zero-Tree Coding

Shapiro [35] introduced a method that exploits some dependency between pixels in corresponding location in the bands of an octave band filter bank. The basic assumed dependencies are illustrated in Figure 14.15. The low-pass band is coded separately. Starting in any location in any of the other three bands of same size, any pixel will have an increasing number of "descendants" as one passes down the tree representing information from the same location in the original image. The number of

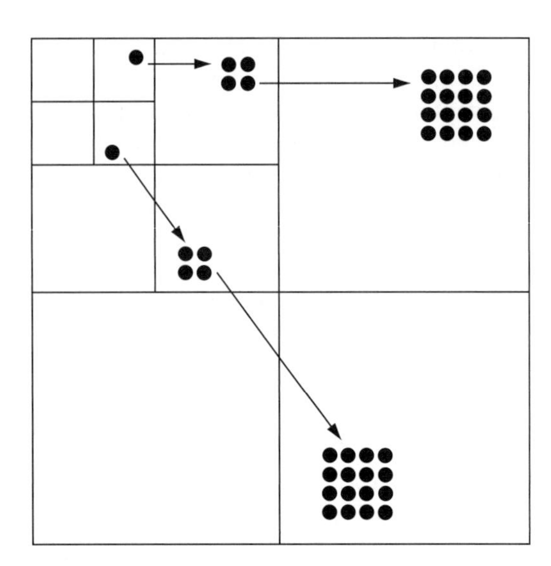

FIGURE 14.15 Zero-tree arrangement in an octave-band decomposed image.

corresponding pixels increases by a factor of four from one level to the next. When used in a coding context, the tree is terminated at any zero-valued pixel (obtained after quantization using some threshold) after which all subsequent pixels are assumed to be zero as well. Due to the growth by a factor of four between levels, many samples can be discarded this way.

What is the underlying mechanism that makes this technique work so well? On one hand, the image spectrum falls off rapidly as a function of frequency for most images. This means that there is a tendency to have many zeros when approaching the leaves of the tree. Our visual system is furthermore more tolerant to high frequency errors. This should be compared to the zigzag scan in the JPEG coder. On the other hand, viewed from a pure statistical angle, the subbands are uncorrelated if the filter bank has done what is required from it! However, the statistical argument is based on the assumption of "local ergodicity," which means that statistical parameters derived locally from the data have the same mean values everywhere. With real images composed of objects with edges, textures, etc. these assumptions do not hold. The "activity" in the subbands tends to appear in the same locations. This is typical at edges. One can look at these connections as energy correlations among the subbands. The zero-tree method will efficiently cope with these types of phenomena.

Shapiro furthermore combined the zero-tree representation with bit-plane coding. Said [34] went one step further and introduced what he calls "set partitioning," The resulting algorithm is simple and fast, and is embedded in the sense that the bit-stream can be cut off at any point in time in the decoder, and the obtained approximation is optimal using that number of bits. The subbands are obtained using the 9/7 biorthogonal spline filters [38].

Coding results from Said's coder are shown in Figure 14.16 and marked "Said."

14.4.2.3 Pyramid VQ and Improved Filter Bank

This coder is based on bit-allocation, or rather, allocation of vector quantizers of different sizes. This implies that the coder is fixed rate, that is, we can preset the total number of bits for an image. It is assumed that the subband signals have a Laplacian distribution, which makes it possible to apply "pyramid" vector quantizers [6]. These are suboptimal compared to trained codebook vector quantizers, but significantly better than scalar quantizers without increasing the complexity too much.

The signal decomposition in the encoder is performed using an 8×8 channel uniform filter bank [1], followed by an octave band filter bank of three stages operating on the resulting low-pass-low-pass band. The uniform filter bank is nonunitary and optimized for coding gain. The building blocks of the octave

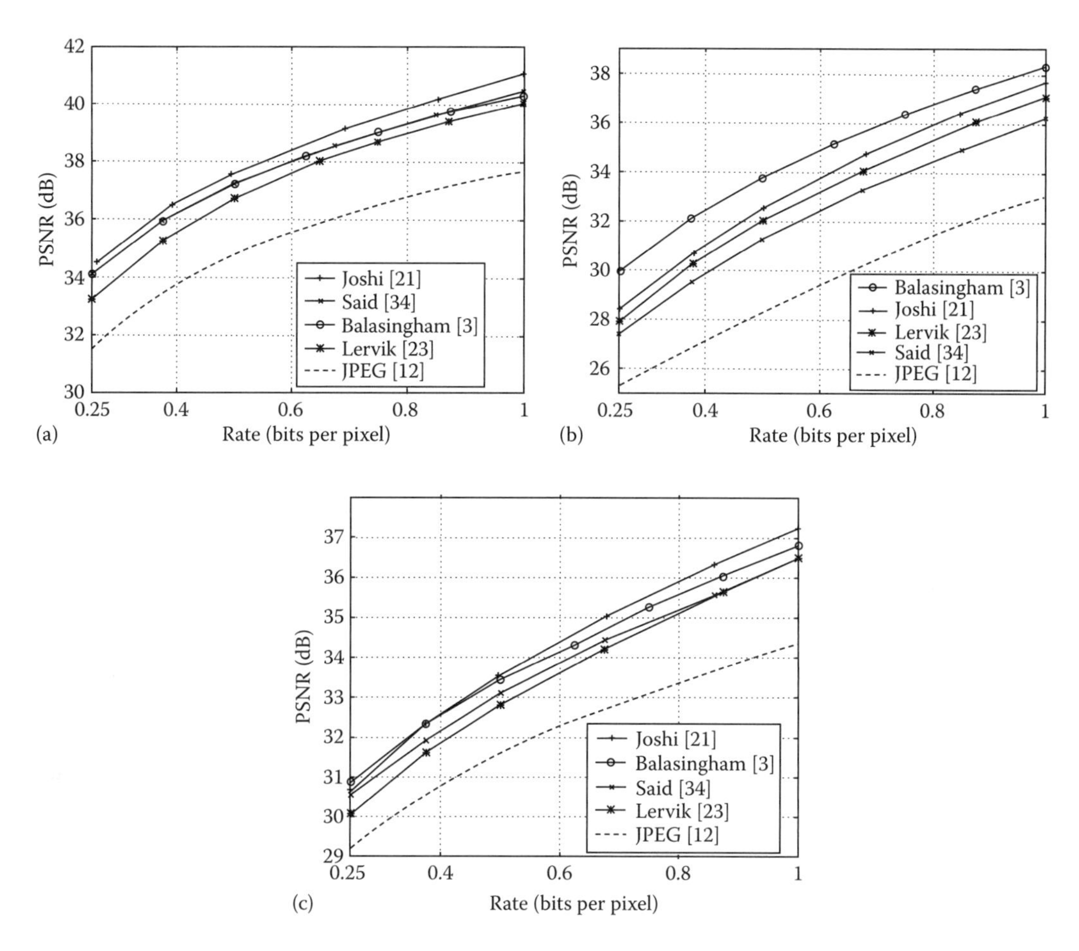

FIGURE 14.16 Coding results. (a) "Lenna," (b) "Barbara," (c) "Goldhill."

band filter bank have been carefully selected from all available perfect reconstruction, two-channel filter systems with limited FIR filter orders.

Coding results from this coder are shown in Figure 14.16 and are marked "Balasingham."

14.4.2.4 Trellis Coded Quantization

Joshi [22] has presented what is presently the "state-of-the-art" coder. Being based on trellis coded quantization [29], the encoder is more complex than the other coders presented. Furthermore, it does not have the embedded character of Said's coder.

The filter bank employed has 22 subbands. This is obtained by first employing a 4×4 uniform filter bank, followed by a further split of the resulting low-pass-low-pass band using a two-stage octave band filter bank. All filters in the curves shown in the next section are 9/7 biorthogonal spline filters [38].

The encoding of the subbands is performed in several stages:

- Separate classification of signal blocks in each band.
- Rate allocation among all blocks.
- Individual arithmetic coding of the trellis-coded quantized signals in each class.

The trellis coded quantization [7] is a method that can reach the rate distortion bound in the same way as VQ. It uses search methods in the encoder, which adds to its complexity. The decoder is much simpler.

Coding results from this coder are shown in Figure 14.16 and are marked "Joshi."

14.4.2.5 Frequency Domain Coding Results

The five coders presented above are compared in this section. All of them are simulated using the three images "Lenna," "Barbara," and "Goldhill" of size 512×512. These three images have quite different contents in terms of spectrum, textures, edges, and so on. Figure 14.16 shows the PSNR as a function of bit rate for the five coders. The PSNR is defined as

$$\text{PSNR} = 10 \log_{10}\left(\frac{255^2}{\frac{1}{NM}\sum_{n=1}^{N}\sum_{m=1}^{M}(x(n,m) - \hat{x}(n,m))^2}\right). \tag{14.34}$$

As is observed, the coding quality among the coders varies when exposed to such different stimuli. The exception is that all subband coders are superior to JPEG, which was expected from the use of better decomposition as well as more clever quantization and coding strategies. Joshi's coder is best for "Lenna" and "Goldhill" at high rates. Balasingham's coder is, however, better for "Barbara" and for "Goldhill" at low rates. These results are interpreted as follows. The Joshi coder uses the most elaborate quantization/coding scheme, but the Balasingham coder applies a better filter bank in two respects. First, it has better high frequency resolution, which explains that the "Barbara" image, with a relatively high frequency content, gives a better result for the latter coder. Second, the improved low frequency resolution of this filter bank also implies better coding at low rates for "Goldhill."

From the results above, it is also observed that the Joshi coder performs well for images with a low-pass character such as the "Lenna" image, especially at low rates. In these cases there are many "zeros" to be represented, and the zero-tree coding can typically cope well with zero-representations. A combination of several of the aforementioned coders, picking up their best components, would probably render an improved system.

14.5 Fractal Coding

This section is placed towards the end of the chapter because fractal coding deviates in many respects from the generic coder on the one hand, but on the other hand can be compared to VQ.

A good overview of the field can be found in Ref. [8].

Fractal coding (also called attractor coding) is based on "Banach's fixed point theorem" and exploits "self-similarity" or "partial self-similarity" among different scales of a given image. A nonlinear transform gives the fractal image representation. Iterative operations using this transform starting from any initial image will converge to the image approximation, called the "attractor." The success of such a scheme will rest upon the compactness, in terms of bits, of the description of the nonlinear transform. A classical example of self-similarity is Michael Barnsley's fern, where each branch is a small copy of the complete fern. Even the branches are composed of small copies of itself. A very compact description can be found for the class of images exhibiting self similarity. In fact, the fern can be described by 24 numbers, according to Barnsley.

Self-similarity is a dependency among image elements (possibly objects) that is not described by correlation, but can be called affine correlation.

There is an enormous potential for image compression if images really have the self-similarity property. However, there seems to be no reason to believe that global self-similarity exists in any complex image created, e.g., by photographing natural or man-made scenes. The less requiring notion of partial self-similarity among image blocks of different scales has proven to be fruitful [19].

In this section we will, in fact, present a practical fractal coder exploiting partial self-similarity among different scales, which can be directly compared to mean-gain-shape vector quantization (MGSVQ). The difference between the two systems is that the vector quantizer uses an optimized codebook based on data from a large collection of different images, whereas the fractal coder uses a "self codebook," in the sense that the codebook is generated from the image itself and implicitly and approximately transmitted to the receiver as part of the image code. The question is then, "Is the 'adaptive' nature of the fractal codebook better than the statistically optimized codebook of standard vector quantization?"

We will also comment on other models and give a brief status of fractal compression techniques.

14.5.1 Mathematical Background

The code of an image in the language of fractal coding is given as the bit representation of a nonlinear transform T. The transform defines what is called the "collage" x_c of the image. The collage is found by

$$x_c = Tx,$$

where x is the original image.

The collage is the object we try to make resemble the image as closely as possible in the encoder through minimization of the distortion function

$$D = d(x, x_c). \tag{14.35}$$

Usually the distortion function is chosen as the Euclidean distance between the two vectors. The decoder cannot reconstruct the collage as it depends on the knowledge of the original image, and not only the transform T. We therefore have to accept reconstruction of the image with less accuracy.

The reconstruction algorithm is based on "Banach's fixed point theorem": If a transform T is "contractive" or "eventually contractive" [26], the fixed point theorem states that the transform then has a unique "attractor" or "fixed point" given by

$$x_T = Tx_T, \tag{14.36}$$

and that the fixed point can be approached by iteration from any starting vector according to

$$x_T = \lim_{n \to \infty} T^n y; \quad \forall y \in X, \tag{14.37}$$

where X is a normed linear space.

The similarity between the collage and the attractor is indicated from an extended version of the collage theorem [27]:

Given an original image x and its collage Tx where $\|x - Tx\| \leq \varepsilon$, then

$$\|x - x_T\| \leq \frac{1 - s_1^K}{(1 - s_1)(1 - s_K)} \varepsilon \tag{14.38}$$

where s_1 and s_K are the Lipschitz constants of T and T^K, respectively, provided $|s_1| < 1$ and $|s_K| < 1$.

Provided the collage is a good approximation of the original image and the Lipschitz constants are small enough, there will also be similarity between the original image and the attractor.

In the special case of "fractal block coding," a given image block (usually called a "domain block") is supposed to resemble another block (usually called a "range block") after some "affine transformation." The transformation that is most commonly used moves the image block to a different position while shrinking the block, rotating it or shuffling the pixels, and adding what we denote a "fixed term," which could be some predefined function with possible parameters to be decided in the encoding process. In most natural images it is not difficult to find affine similarity, e.g., in the form of objects situated at different distances and positions in relation to the camera. In standard block coding methods, only local statistical dependencies can be utilized. The inclusion of "affine redundancies" should therefore offer some extra advantage.

In this formalism we do not see much resemblance with VQ. However, the similarities and differences between fractal coding and VQ were pointed out already in the original work by Jacquin [20]. We shall, in the following section, present a specific model that enforces further similarity to VQ.

14.5.2 Mean-Gain-Shape Attractor Coding

It has been proven [31] that in all cases where each domain block is a union of range blocks, the decoding algorithms for sampled images where the nonlinear part (fixed term) of the transform is orthogonal to the image transformed by the linear part, full convergence is reached after a finite and small number of iterations. In one special case there are no iterations at all [31], and then $x_T = Tx$. We shall discuss only this important case here because it has an important application potential due to its simplicity in the decoder, but, more importantly, we can more clearly demonstrate the similarity to VQ.

14.5.2.1 Codebook Formation

In the encoder two tasks have to be performed, the codebook formation and the codebook search, to find the best representation of the transform T with as few bits as possible.

First the image is split in nonoverlapping blocks of size $L \times L$ so that the complete image is covered. The codebook construction goes as follows:

- Calculate the mean value m in each block.
- Quantize the mean values, resulting in the approximation $\hat{m}$, and transmit their code to the receiver.

These values will serve two purposes:

1. They are the additive, nonlinear terms in the block transform.
2. They are the building elements for the codebook.

All the following steps must be performed both in the encoder and the decoder.

- Organize the quantized mean values as an image so that it becomes a block averaged and down-sampled version of the original image.
- Pick blocks of size $L \times L$ in the obtained image. Overlap between blocks is possible.
- Remove the mean values from each block. The resulting blocks constitute part of the codebook.
- Generate new codebook vectors by a predetermined set of mathematical operations (mainly pixel shuffling).

With the procedure given, the codebook is explicitly known in the decoder, because the mean values also act as the nonlinear part of the affine transforms. The codebook vectors are orthogonal to the nonlinear term due to the mean value removal.

Observe also that the block decimation in the encoder must now be chosen as $L \times L$, which is also the size of the blocks to be coded.

14.5.2.2 The Encoder

The actual encoding is similar to traditional product code VQ.

In our particular case, the image block in position (k, l) is modeled as

$$\hat{x}_{k,l} = \hat{m}_{k,l} + \hat{\alpha}_{k,l}\rho^{(i)}, \tag{14.39}$$

where

$\hat{m}_{k,l}$ is the quantized mean value of the block
$\rho^{(i)}$ is codebook vector number i
$\hat{\alpha}_{k,l}$ is a quantized scaling factor

To optimize the parameters, we minimize the Euclidean distance between the image block and the given approximation,

$$d = \|x_{k,l} - \hat{x}_{k,l}\|. \quad (14.40)$$

This minimization is equivalent to the maximization of

$$P^{(i)} = \frac{\|x_{k,l}, \rho^{(i)}\|^2}{\|\rho^{(i)}\|^2}, \quad (14.41)$$

where $\langle u, v \rangle$ denotes the inner product between u and v over one block. If vector number j maximizes P, then the scaling factor can be calculated as

$$\alpha_{k,l} = \frac{\|x_{k,l}, \rho^{(j)}\|}{\|\rho^{(j)}\|^2}. \quad (14.42)$$

14.5.2.3 The Decoder

In the decoder, the codebook can be regenerated, as previously described, from the mean values. The decoder reconstructs each block according to Equation 14.39 using the transmitted, quantized parameters. In the particular case given above, the following procedure is followed:

Denote by c an image composed of subblocks of size $L \times L$ which contains the correct mean values. The decoding is then performed by

$$x_1 = Tc = Ac + c, \quad (14.43)$$

where A is the linear part of the transform. The operation of A can be described block-wise.

- It takes a block from c of size $L^2 \times L^2$
- Shrinks it to size $L \times L$ after averaging over subblocks of size $L \times L$
- Subtracts from the resulting block its mean value
- Performs the prescribed pixel shuffling
- Multiplies by the scaling coefficient
- Finally inserts the resulting block in the correct position.

Notice that x_1 has the correct mean value due to c, and because Ac does not contribute to the block mean values. Another observation is that each block of size $L \times L$ is mapped to one pixel.

The algorithm just described is equivalent to the VQ decoding given earlier.

The iterative algorithm indicated by Banach's fixed point theorem can be used also in this case. The above described algorithm is the first iteration. In the next iteration we get

$$x_2 = Ax_1 + c = A(Ac) + Ac + c. \quad (14.44)$$

But $A(Ac) = 0$ because A and Ac are orthogonal, therefore $x_2 = x_1$. The iteration can, of course, be continued without changing the result. Note also that $Ac = Ax$, where x is the original image!

We will stress the important fact that as the attractor and the collage are equivalent in the noniterative case, we have direct control of the attractor, unlike any other fractal coding method.

14.5.2.4 Experimental Comparisons with the Performance of MSGVQ

It is difficult to conclude from theory alone as to the performance of the attractor coder model. Experiments indicate, however, that for this particular fractal coder the performance is always worse than for the VQ with optimized codebook for all images tested [23]. The adaptivity of the "self codebook," does not seem to outcompete the VQ codebook which is optimal in a statistical sense.

14.5.3 Discussion

The above model is severely constrained through the required relation between the block size ($L \times L$) and the decimation factor (also $L \times L$). Better coding results are obtained by using smaller decimation factors, typically 2×2.

Even with small decimation factors, no pure fractal coding technique has, in general, been shown to outperform VQ of similar complexity.

However, fractal methods have potential in hybrid block coding. It can efficiently represent edges and other deterministic structures where a shrunken version of another block is likely to resemble the block we are trying to represent. For instance, edges tend to be edges also after decimation. On the other hand, many textures can be hard to represent, as the decimation process requires that another texture with different frequency contents be present in the image to make a good approximation.

Using several block coding methods, where for each block the best method in a distortion-rate sense is selected, has been proven to give good coding performance [5,10].

On the practical side, the fractal encoders have a very high complexity. Several methods have been suggested to alleviate this problem. These methods include limited search regions in the vicinity of the block to be coded, clustering of codebook vectors, and hierarchical search at different resolutions.

The iteration-free decoder is one of the fastest decoders obtainable for any coding method.

14.6 Color Coding

Any color image can be split in three color components and thereafter coded individually for each component. If this is done on the RGB (red, green, blue) components, the bit rate tends to be approximately three times as high as for black and white images.

However, there are many other ways of decomposing the colors. The most used representations split the image in a "luminance" component and two "chrominance" components. Examples are so-called YUV and YIQ representations. One rationale for doing this kind of splitting is that the human visual system has different resolution for luminance and chrominance. The chrominance sampling can therefore be performed at a lower resolution, from two to eight times lower resolution depending on the desired quality and the interpolation method used to reconstruct the image. A second rationale is that the RGB components in most images are strongly correlated and therefore direct coding of the RGB components results in repeated coding of the same information. The luminance/chrominance representations try to decorrelate the components.

The transform between RGB and the luminance and chrominance components (YIQ) used in NTSC is given by

$$\begin{bmatrix} \mathrm{Y} \\ \mathrm{I} \\ \mathrm{Q} \end{bmatrix} = \begin{bmatrix} 0.299 & 0.587 & 0.114 \\ 0.596 & -0.274 & -0.322 \\ 0.058 & -0.523 & 0.896 \end{bmatrix} \begin{bmatrix} \mathrm{R} \\ \mathrm{G} \\ \mathrm{B} \end{bmatrix} \tag{14.45}$$

There are only minor differences between the suggested color transforms. It is also possible to design the optimal decomposition based on the Karhunen–Loève transform. The method could be made adaptive by deriving a new transform for each image based on an estimated color correlation matrix.

We shall not go further into the color coding problem, but state that it is possible to represent color by adding 10%–20% to the luminance component bit rate.

References

1. Aase, S.O., Image subband coding artifacts: Analysis and remedies, PhD thesis, The Norwegian Institute of Technology, Norway, March 1993.
2. Arrowwood, J.L., Jr. and Smith, M.J.T., Exact reconstruction analysis/synthesis filter banks with time varying filters, in *Proceedings of the International Conference on Acoustics, Speech, and Signal Processing (ICASSP)*, Minneapolis, MN, 3, pp. 233–236, April 1993.
3. Balasingham, I., Fuldseth, A., and Ramstad, T.A., On optimal tiling of the spectrum in subband image compression, in *Proceedings of the International Conference on Image Processing (ICIP)*, Santa Barbara, CA, pp. 49–52, 1997.
4. Balasingham, I. and Ramstad, T.A., On the optimality of tree-structured filter banks in subband image compression, *IEEE Trans. Signal Process.* 1997 (submitted).
5. Barthel, K.U., Schüttemeyer, J., Voyé, T., and Noll, P., A new image coding technique unifying fractal and transform coding, in *Proceedings of the International Conference on Image Processing (ICIP)*, Austin, TX, Nov. 1994.
6. Fischer, T.R., A pyramid vector quantizer, *IEEE Trans. Info. Theory*, IT-32:568–583, July 1986.
7. Fischer, T.R. and Mercellin, M.W., Joint trellis coded quantization/modulation, *IEEE Trans. Commun.*, 39(2):172–176, Feb. 1991.
8. Fisher, Y. (Ed.), *Fractal Image Compression. Theory and Applications*, Springer-Verlag, Berlin, 1995.
9. Gersho, A. and Gray, R.M., *Vector Quantization and Signal Compression*, Kluwer Academic Publishers, Boston, MA, 1992.
10. Gharavi-Alkhansari, M., Fractal image coding using rate-distortion optimized matching pursuit, in *Proceedings of the SPIE's Visual Communications and Image Processing*, Orlando, FL, 2727, pp. 1386–1393, March 1996.
11. Herley, C., Kovacevic, J., Ramchandran, K., and Vetterli, M., Tilings of the time-frequency plane: Construction of arbitrary orthogonal bases and fast tiling transforms, *IEEE Trans. Signal Process.*, 41(12):3341–3359, Dec. 1993.
12. Huffman, D.A., A method for the construction of minimum redundancy codes, *Proc. IRE*, 40(9):1098–1101, Sept. 1952.
13. Hung, A.C., *PVRG-JPEG Codec 1.2.1*, Portable Video Research Group, Stanford University, Boston, MA, 1993.
14. ISO/IEC IS 10918-1, *Digital compression and coding of continuous-tone still images, Part 1: Requirements and guidelines*, JPEG.
15. ISO/IEC IS 11172, *Information Technology-Coding of Moving Pictures and Associated Audio for Digital Storage Up to about 1.5 Mbit/s*, MPEG-1.
16. ISO/IEC IS 13818, *Information technology—Generic coding of moving pictures and associated audio information*, MPEG-2.
17. ITU-T (CCITT), Video codec for audiovisual services at p 64 kbit/s, Geneva, Italy, Aug. 1990, Recommendation H.261.
18. ITU-T (CCITT), Video coding for low bitrate communication, May, 1996. Draft Recommendation H.263.
19. Jacquin, A., Fractal image coding: A review, *Proc. IEEE*, 81(10):1451–1465, Oct. 1993.
20. Jacquin, A., Fractal image coding based on a theory of iterated contractive transformations, in *Proceedings of the SPIE's Visual Communications and Image Processing*, Bratislava, Slovakia, pp. 227–239, Oct. 1990.
21. Jayant, N.S. and Noll, P., *Digital Coding of Waveforms, Principles and Applications to Speech and Video*, Prentice-Hall, Englewood Cliffs, NJ, 1984.
22. Joshi, R.L., Subband image coding using classification and Trellis coded quantization, PhD thesis, Washington State University, Pullman, WA, Aug. 1996.

23. Lepsøy, S., Attractor image compression—Fast algorithms and comparisons to related techniques, PhD thesis, The Norwegian Institute of Technology, Norway, June 1993.
24. Lervik, J.M., Subband image communication over digital transparent and analog waveform channels, PhD thesis, Norwegian University of Science and Technology, Norway, Dec. 1996.
25. Linde, Y., Buzo, A., and Gray, R.M., An algorithm for vector quantizer design, *IEEE Trans. Commun.*, COM-28(1):84–95, Jan. 1980.
26. Luenbereger, D.G., *Optimization by Vector Space Methods*, John Wiley & Sons, New York, 1979.
27. Lundheim, L., Fractal signal modelling for source coding, PhD thesis, The Norwegian Institute of Technology, Norway, Sept. 1992.
28. Makhoul, J., Roucos, S., and Gish, H., Vector quantization in speech coding, *Proc. IEEE*, 73(11):1551–1587, Nov. 1985.
29. Marcellin, M.W. and Fischer, T.R., Trellis coded quantization of memoryless and Gauss-Markov sources, *IEEE Trans. Commun.*, 38(1):82–93, Jan. 1990.
30. Martucci, S., Signal extension and noncausal filtering for subband coding of images, *Proc. SPIE Vis. Commun. Image Process.*, 1605:137–148, Nov. 1991.
31. Øien, G.E., L2-optimal attractor image coding with fast decoder convergence, PhD thesis, The Norwegian Institute of Technology, Norway, June 1993.
32. Popat, K., Scalar quantization with arithmetic coding, MSc thesis, Massachusetts Institute of Technology, Cambridge, MA, June 1990.
33. Ramstad, T.A., Aase, S.O., and Husøy, J.H., *Subband Compression of Images—Principles and Examples*, Elsevier Science Publishers BV, North Holland, the Netherlands, 1995.
34. Said, A. and Pearlman, W.A., A new, fast, and efficient image codec based on set partitioning in hierarchical trees, *IEEE Trans. Circuits, Syst. Video Technol.*, 6(3):243–250, June 1996.
35. Shapiro, J.M., Embedded image coding using zerotrees of wavelets coefficients, *IEEE Trans. Signal Process.*, 41:3445–3462, Dec. 1993.
36. Vaidyanathan, P.P., *Multirate Systems and Filter Banks*, Prentice-Hall, Englewood Cliffs, NJ, 1993.
37. Wallace, G.K., Overview of the JPEG (ISO/CCITT) still image compression standard, *Proc. SPIE's Vis. Commun. Image Process.*, 1244:220–223, 1989.
38. Antonini, M., Barland, M., Mathieu, P., and Daubechies, I., Image coding using wavelet transform, *IEEE Trans. Image Process.*, 1:205–220, Apr. 1992.

15

Image and Video Restoration

A. Murat Tekalp
Koç University

15.1 Introduction

Digital images and videos, acquired by still cameras, consumer camcorders, or even broadcast-quality video cameras, are usually degraded by some amount of blur and noise. In addition, most electronic cameras have limited spatial resolution determined by the characteristics of the sensor array. Common causes of blur are out of focus, relative motion, and atmospheric turbulence. Noise sources include film grain, thermal, electronic, and quantization noise. Further, many image sensors and media have known nonlinear input–output characteristics which can be represented as point nonlinearities. The goal of image and video (image sequence) restoration is to estimate each image (frame or field) as it would appear without any degradations, by first modeling the degradation process, and then applying an inverse procedure. This is distinct from the image enhancement techniques which are designed to manipulate an image in order to produce more pleasing results to an observer without making use of particular degradation models. On the other hand, superresolution refers to estimating an image at a resolution higher than that of the imaging sensor. Image sequence filtering (restoration and superresolution) becomes especially important when still images from video are desired. This is because the blur and noise can become rather objectionable when observing a "freeze-frame," although they may not be visible to the human eye at the usual frame rates. Since many video signals encountered in practice are interlaced, we address the cases of both progressive and interlaced video.

The problem of image restoration has sparked widespread interest in the signal processing community over the past 20 or 30 years. Because image restoration is essentially an ill-posed inverse problem which is also frequently encountered in various other disciplines such as geophysics, astronomy, medical imaging, and computer vision, the literature that is related to image restoration is abundant. A concise discussion of early results can be found in the books by Andrews and Hunt [1] and Gonzalez and Woods [2]. More recent developments are summarized in the book by Katsaggelos [3], and review papers by Meinel [4], Demoment [5], Sezan and Tekalp [6], and Kaufman and Tekalp [7]. Most recently, printing high-quality still images from video sources has become an important application for multiframe restoration and superresolution methods. An in-depth coverage of video filtering methods can be found in the book *Digital Video Processing* by Tekalp [8]. This chapter summarizes key results in digital image and video restoration.

15.2 Modeling

Every image restoration/superresolution algorithm is based on an observation model, which relates the observed degraded image(s) to the desired "ideal" image, and possibly a regularization model, which conveys the available a priori information about the ideal image. The success of image restoration and/or superresolution depends on how good the assumed mathematical models fit the actual application.

15.2.1 Intraframe Observation Model

Let the observed and ideal images be sampled on the same 2-D lattice Λ. Then, the observed blurred and noisy image can be modeled as

$$\boldsymbol{g} = s(\boldsymbol{D}\boldsymbol{f}) + \boldsymbol{v} \tag{15.1}$$

where $\boldsymbol{g}$, $\boldsymbol{f}$, and $\boldsymbol{v}$ denote vectors representing lexicographical ordering of the samples of the observed image, ideal image, and a particular realization of the additive (random) noise process, respectively. The operator $\boldsymbol{D}$ is called the blur operator. The response of the image sensor to the light intensity is represented by the memoryless mapping $s(\cdot)$, which is, in general, nonlinear. (This nonlinearity has often been ignored in the literature for algorithm development.)

The blur may be space-invariant or space-variant. For space-invariant blurs, $\boldsymbol{D}$ becomes a convolution operator, which has block-Toeplitz structure; and Equation 15.1 can be expressed, in scalar form, as

$$g(n_1, n_2) = s\left(\sum_{(m_1, m_2) \in S_d} d(m_1, m_2) f(n_1 - m_1, n_2 - m_2) \right) + v(n_1, n_2) \tag{15.2}$$

where $d(m_1, m_2)$ and S_d denote the kernel and support of the operator $\boldsymbol{D}$, respectively. The kernel $d(m_1, m_2)$ is the impulse response of the blurring system, often called the point spread function (PSF). In case of space-variant blurs, the operator $\boldsymbol{D}$ does not have a particular structure; and the observation equation can be expressed as a superposition summation

$$g(n_1, n_2) = s\left(\sum_{(m_1, m_2) \in S_d(n_1, n_2)} d(n_1, n_2; m_1, m_2) f(m_1, m_2) \right) + v(n_1, n_2) \tag{15.3}$$

where $S_d(n_1, n_2)$ denotes the support of the PSF at the pixel location (n_1, n_2).

The noise is usually approximated by a zero-mean, white Gaussian random field which is additive and independent of the image signal. In fact, it has been generally accepted that more sophisticated noise models do not, in general, lead to significantly improved restorations.

15.2.2 Multispectral Observation Model

Multispectral images refer to image data with multiple spectral bands that exhibit interband correlations. An important class of multispectral images is color images with three spectral bands. Suppose we have K spectral bands, each blurred by possibly a different PSF. Then, the vector–matrix model (15.1) can be extended to multispectral modeling as

$$\boldsymbol{g} = \mathcal{D}\boldsymbol{f} + \boldsymbol{v} \tag{15.4}$$

where

$$\boldsymbol{g} \doteq \begin{bmatrix} \boldsymbol{g}_1 \\ \vdots \\ \boldsymbol{g}_K \end{bmatrix}, \quad \boldsymbol{f} \doteq \begin{bmatrix} \boldsymbol{f}_1 \\ \vdots \\ \boldsymbol{f}_K \end{bmatrix}, \quad \boldsymbol{v} \doteq \begin{bmatrix} \boldsymbol{v}_1 \\ \vdots \\ \boldsymbol{v}_K \end{bmatrix}$$

denote $N^2K \times 1$ vectors representing the multispectral observed, ideal, and noise data, respectively, stacked as composite vectors, and

$$\mathcal{D} \doteq \begin{bmatrix} \boldsymbol{D}_{11} & \cdots & \boldsymbol{D}_{1K} \\ \vdots & \ddots & \vdots \\ \boldsymbol{D}_{K1} & \cdots & \boldsymbol{D}_{KK} \end{bmatrix}$$

is an $N^2K \times N^2K$ matrix representing the multispectral blur operator. In most applications, $\mathcal{D}$ is block diagonal, indicating no inter-band blurring.

15.2.3 Multiframe Observation Model

Suppose a sequence of blurred and noisy images $g_k(n_1, n_2)$, $k = 1, \ldots, L$, corresponding to multiple shots (from different angles) of a static scene sampled on a 2-D lattice or frames (fields) of video sampled (at different times) on a 3-D progressive (interlaced) lattice, is available. Then, we may be able to estimate a higher-resolution "ideal" still image $f(m_1, m_2)$ (corresponding to one of the observed frames) sampled on a lattice, which has a higher sampling density than that of the input lattice. The main distinction between the multispectral and multiframe observation models is that here the observed images are subject to subpixel shifts (motion), possibly space-varying, which makes high-resolution reconstruction possible. In the case of video, we may also model blurring due to motion within the aperture time to further sharpen images.

To this effect, each observed image (frame or field) can be related to the desired high-resolution ideal still image through the superposition summation [8]

$$g_k(n_1, n_2) = s\left(\sum_{(m_1, m_2) \in S_d(n_1, n_2; k)} d_k(n_1, n_2; m_1, m_2) f(m_1, m_2) \right) + v_k(n_1, n_2) \tag{15.5}$$

where the support of the summation over the high-resolution grid (m_1, m_2) at a particular observed pixel $(n_1, n_2; k)$ depends on the motion trajectory connecting the pixel $(n_1, n_2; k)$ to the ideal image, the size of the support of the low-resolution sensor PSF $h_a(x_1, x_2)$ with respect to the high-resolution grid, and whether there is additional optical (out of focus, motion, etc.) blur. Because the relative positions of low- and high-resolution pixels in general vary by spatial coordinates, the discrete sensor PSF is space-varying. The support of the space-varying PSF is indicated by the shaded area in Figure 15.1, where the rectangle

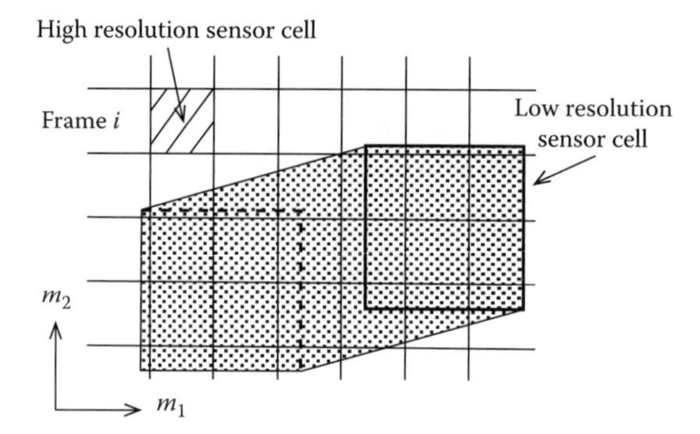

FIGURE 15.1 Illustration of the discrete system PSF.

depicted by solid lines shows the support of a low-resolution pixel over the high-resolution sensor array. The shaded region corresponds to the area swept by the low-resolution pixel due to motion during the aperture time [8].

Note that the model (15.5) is invalid in case of occlusion. That is, each observed pixel $(n_1, n_2; k)$ can be expressed as a linear combination of several desired high-resolution pixels (m_1, m_2), provided that $(n_1, n_2; k)$ is connected to (m_1, m_2) by a motion trajectory. We assume that occlusion regions can be detected a priori using a proper motion estimation/segmentation algorithm.

15.2.4 Regularization Models

Restoration is an ill-posed problem which can be regularized by modeling certain aspects of the desired "ideal" image. Images can be modeled as either 2-D deterministic sequences or random fields. A priori information about the ideal image can then be used to define hard or soft constraints on the solution. In the deterministic case, images are usually assumed to be members of an appropriate Hilbert space, such as a Euclidean space with the usual inner product and norm. For example, in the context of set-theoretic restoration, the solution can be restricted to be a member of a set consisting of all images satisfying a certain smoothness criterion [9]. On the other hand, constrained least squares (CLS) and Tikhonov–Miller (T–M) regularization use quadratic functionals to impose smoothness constraints in an optimization framework.

In the random case, models have been developed for the probability density function (pdf) of the ideal image in the context of maximum a posteriori (MAP) image restoration. For example, Trussell and Hunt [10] have proposed a Gaussian distribution with space-varying mean and stationary covariance as a model for the pdf of the image. Geman and Geman [11] proposed a Gibbs distribution to model the pdf of the image. Alternatively, if the image is assumed to be a realization of a homogeneous Gauss–Markov random process, then it can be statistically modeled through an autoregressive (AR) difference equation [12]

$$f(n_1, n_2) = \sum_{(m_1, m_2) \in \mathcal{S}_c} c(m_1, m_2) f(n_1 - m_1, n_2 - m_2) + w(n_1, n_2) \tag{15.6}$$

where

$\{c(m_2, m_2) : (m_2, m_2) \in \mathcal{S}_c\}$ denote the model coefficients
$\mathcal{S}_c$ is the model support (which may be causal, semicausal, or noncausal)
$w(n_1, n_2)$ represents the modeling error which is Gaussian distributed

The model coefficients can be determined such that the modeling error has minimum variance [12]. Extension of Equation 15.6 to inhomogeneous Gauss–Markov fields was proposed by Jeng and Woods [13].

15.3 Model Parameter Estimation

In this section, we discuss methods for estimating the parameters that are involved in the observation and regularization models for subsequent use in the restoration algorithms.

15.3.1 Blur Identification

Blur identification refers to estimation of both the (size) support and parameters of the PSF $\{d(n_1, n_2): (n_1, n_2) \in S_d\}$. It is a crucial element of image restoration because the quality of restored images is highly sensitive to errors in the PSF [14]. An early approach to blur identification has been based on the assumption that the original scene contains an ideal point source, and that its spread (hence the PSF) can be determined from the observed image. Rosenfeld and Kak [15] show that the PSF can also be determined from an ideal line source. These approaches are of limited use in practice because a scene, in general, does not contain an ideal point or line source and the observation noise may not allow the measurement of a useful spread.

Models for certain types of PSF can be derived using principles of optics, if the source of the blur is known [7]. For example, out of focus and motion blur PSF can be parameterized with a few parameters. Further, they are completely characterized by their zeros in the frequency domain. Power spectrum and cepstrum (Fourier transform of the logarithm of the power spectrum) analysis methods have been successfully applied in many cases to identify the location of these zero-crossings [16,17]. Alternatively, Chang et al. [18] proposed a bispectrum analysis method, which is motivated by the fact that bispectrum is not affected, in principle, by the observation noise. However, the bispectral method requires much more data than the method based on the power spectrum. Note that PSFs, which do not have zero-crossings in the frequency domain (e.g., Gaussian PSF modeling atmospheric turbulence), cannot be identified by these techniques.

Yet another approach for blur identification is the maximum likelihood (ML) estimation approach. The ML approach aims to find those parameter values (including, in principle, the observation noise variance) that have most likely resulted in the observed image(s). Different implementations of the ML image and blur identification are discussed under a unifying framework [19]. Pavlović and Tekalp [20] propose a practical method to find the ML estimates of the parameters of a PSF based on a continuous domain image formation model.

In multiframe image restoration, blur identification using more than one frame at a time becomes possible. For example, the PSF of a possibly space-varying motion blur can be computed at each pixel from an estimate of the frame-to-frame motion vector at that pixel, provided that the shutter speed of the camera is known [21].

15.3.2 Estimation of Regularization Parameters

Regularization model parameters aim to strike a balance between the fidelity of the restored image to the observed data and its smoothness. Various methods exist to identify regularization parameters, such as parametric pdf models, parametric smoothness constraints, and AR image models. Some restoration methods require the knowledge of the power spectrum of the ideal image, which can be estimated, for example, from an AR model of the image. The AR parameters can, in turn, be estimated from the observed image by a least squares (LS) [22] or an ML technique [63]. On the other hand, nonparametric spectral estimation is also possible through the application of periodogram-based methods to a prototype image [23,69]. In the context of MAP methods, the a priori pdf is often modeled by a parametric pdf, such as a Gaussian [10] or a Gibbsian [11]. Standard methods for estimating these parameters do not exist. Methods for estimating the regularization parameter in the CLS, T–M, and related formulations are discussed in Ref. [24].

15.3.3 Estimation of the Noise Variance

Almost all restoration algorithms assume that the observation noise is a zero-mean, white random process that is uncorrelated with the image. Then, the noise field is completely characterized by its variance, which is commonly estimated by the sample variance computed over a low-contrast local region of the observed image. As we will see in the following section, the noise variance plays an important role in defining constraints used in some of the restoration algorithms.

15.4 Intraframe Restoration

We start by first looking at some basic regularized restoration strategies, in the case of a linear space-invariant (LSI) blur model with no pointwise nonlinearity. The effect of the nonlinear mapping $s(\cdot)$ is discussed in Section 15.4.2. Methods that allow PSFs with random components are summarized in Section 15.4.3. Adaptive restoration for ringing suppression and blind restoration are covered in Sections 15.4.4 and 15.4.5, respectively. Restoration of multispectral images and space-varying blurred images are addressed in Sections 15.4.6 and 15.4.7, respectively.

15.4.1 Basic Regularized Restoration Methods

When the mapping $s(\cdot)$ is ignored, it is evident from Equation 15.1 that image restoration reduces to solving a set of simultaneous linear equations. If the matrix $\boldsymbol{D}$ is nonsingular (i.e., $\boldsymbol{D}^{-1}$ exists) and the vector $\boldsymbol{g}$ lies in the column space of $\boldsymbol{D}$ (i.e., there is no observation noise), then there exists a unique solution which can be found by direct inversion (also known as inverse filtering). In practice, however, we almost always have an underdetermined (due to boundary truncation problem [14]) and inconsistent (due to observation noise) set of equations. In this case, we resort to a minimum-norm LS solution. A LS solution (not unique when the columns of $\boldsymbol{D}$ are linearly dependent) minimizes the norm-square of the residual

$$J_{\mathrm{LS}}(\boldsymbol{f}) = \|\boldsymbol{g} - \boldsymbol{D}\boldsymbol{f}\|^2 \tag{15.7}$$

LS solution(s) with the minimum norm (energy) is (are) generally known as pseudo-inverse solution(s) (PIS).

Restoration by pseudo-inversion is often ill-posed owing to the presence of the observation noise [14]. This follows because the pseudo-inverse operator usually has some very large eigenvalues. For example, a typical blur transfer function has zeros; and thus, its pseudo-inverse attains very large magnitudes near these singularities as well as at high frequencies. This results in excessive amplification at these frequencies in the sensor noise. Regularized inversion techniques attempt to roll-off the transfer function of the pseudo-inverse filter at these frequencies to limit noise amplification. It follows that the regularized inverse deviates from the pseudo-inverse at these frequencies which leads to other types of artifacts, generally known as regularization artifacts [14]. Various strategies for regularized inversion (and how to achieve the right amount of regularization) are discussed in the following.

15.4.1.1 Singular-Value Decomposition Method

The pseudo-inverse $\boldsymbol{D}^{+}$ can be computed using the singular-value decomposition (SVD) [1]

$$\boldsymbol{D}^{+} = \sum_{i=0}^{R} \lambda_i^{-1/2} z_i u_i^{\mathrm{T}} \tag{15.8}$$

where

λ_i denote the singular values

z_i and u_i are the eigenvectors of $\boldsymbol{D}^{\mathrm{T}}\boldsymbol{D}$ and $\boldsymbol{D}\,\boldsymbol{D}^{\mathrm{T}}$, respectively

R is the rank of $\boldsymbol{D}$

Clearly, reciprocation of the zero singular values is avoided since the summation runs to R, the rank of $\boldsymbol{D}$. Under the assumption that $\boldsymbol{D}$ is block-circulant (corresponding to a circular convolution), the PIS computed through Equation 15.8 is equivalent to the frequency domain pseudo-inverse filtering

$$D^{+}(u,v) = \begin{cases} 1/D(u,v) & \text{if } D(u,v) \neq 0 \\ 0 & \text{if } D(u,v) = 0 \end{cases} \tag{15.9}$$

where $D(u, v)$ denotes the frequency response of the blur. This is because a block-circulant matrix can be diagonalized by a 2-D discrete Fourier transformation (DFT) [2].

Regularization of the PIS can then be achieved by truncating the singular-value expansion (15.8) to eliminate all terms corresponding to small λ_i (which are responsible for the noise amplification) at the expense of reduced resolution. Truncation strategies are generally ad hoc in the absence of additional information.

15.4.1.2 Iterative Methods (Landweber Iterations)

Several image restoration algorithms are based on variations of the so-called Landweber iterations [25–28,31,32]

$$\boldsymbol{f}_{k=1} = \boldsymbol{f}_k + \boldsymbol{R}\boldsymbol{D}^{\mathrm{T}}(\boldsymbol{g} - \boldsymbol{D}\boldsymbol{f}_k) \tag{15.10}$$

where $\boldsymbol{R}$ is a matrix that controls the rate of convergence of the iterations. There is no general way to select the best $\boldsymbol{C}$ matrix. If the system (15.1) is nonsingular and consistent (hardly ever the case), the iterations (15.10) will converge to the solution. If, on the other hand, (15.1) is underdetermined and/or inconsistent, then Equation 15.10 converges to a minimum-norm LS solution (PIS). The theory of this and other closely related algorithms are discussed by Sanz and Huang [26] and Tom et al. [27]. Kawata and Ichioka [28] are among the first to apply the Landweber-type iterations to image restoration, which they refer to as "reblurring" method.

Landweber-type iterative restoration methods can be regularized by appropriately terminating the iterations before convergence, since the closer we are to the pseudo-inverse, the more noise amplification we have. A termination rule can be defined on the basis of the norm of the residual image signal [29]. Alternatively, soft and/or hard constraints can be incorporated into iterations to achieve regularization. The constrained iterations can be written as [30,31]

$$\boldsymbol{f}_{k=1} = \boldsymbol{C}\left[\boldsymbol{f}_k + \boldsymbol{R}\boldsymbol{D}^{\mathrm{T}}(\boldsymbol{g} - \boldsymbol{D}\boldsymbol{f}_k)\right] \tag{15.11}$$

where $\boldsymbol{C}$ is a nonexpansive constraint operator, i.e., $\|\boldsymbol{C}(f_1) - \boldsymbol{C}(f_2)\| \leq \|f_1 - f_2\|$, to guarantee the convergence of the iterations. Application of Equation 15.11 to image restoration has been extensively studied (see [31,32] and the references therein).

15.4.1.3 Constrained Least-Squares Method

Regularized image restoration can be formulated as a constrained optimization problem, where a functional $\|\boldsymbol{Q}(\boldsymbol{f})\|^2$ of the image is minimized subject to the constraint $\|\boldsymbol{g} - \boldsymbol{D}\boldsymbol{f}\|^2 = \sigma^2$. Here σ^2 is a constant, which is usually set equal to the variance of the observation noise. The CLS estimate minimizes the Lagrangian [34]

$$J_{\mathrm{CLS}}(\boldsymbol{f}) = \|\boldsymbol{Q}(\boldsymbol{f})\|^2 + \alpha(\|\boldsymbol{g} - \boldsymbol{D}\boldsymbol{f}\|^2 - \sigma^2) \tag{15.12}$$

where α is the Lagrange multiplier. The operator $\boldsymbol{Q}$ is chosen such that the minimization of Equation 15.12 enforces some desired property of the ideal image. For instance, if $\boldsymbol{Q}$ is selected as the Laplacian operator, smoothness of the restored image is enforced. The CLS estimate can be expressed, by taking the derivative of Equation 15.12 and setting it equal to zero, as in [1]

$$\hat{f} = (\boldsymbol{D}^{\mathrm{H}}\boldsymbol{D} + \gamma \boldsymbol{Q}^{\mathrm{H}}\boldsymbol{Q})^{-1}\boldsymbol{D}^{\mathrm{H}}\boldsymbol{g} \tag{15.13}$$

where "H" stands for Hermitian (i.e., complex conjugate and transpose). The parameter $\gamma = (1/\alpha)$ (the regularization parameter) must be such that the constraint $\|\boldsymbol{g} - \boldsymbol{D}\boldsymbol{f}\|^2 = \sigma^2$ is satisfied. It is often computed iteratively [2]. A sufficient condition for the uniqueness of the CLS solution is that $\boldsymbol{Q}^{-1}$ exists. For space-invariant blurs, the CLS solution can be expressed in the frequency domain as [34]

$$\hat{F}(u, v) = \frac{D^{*}(u, v)}{|D(u, v)|^2 + \gamma |L(u, v)|^2} G(u, v) \tag{15.14}$$

where * denotes complex conjugation. A closely related regularization method is the T–M regularization [33,35]. T–M regularization has been applied to image restoration [31,32,36]. Recently, neural network structures implementing the CLS or T–M image restoration have also been proposed [37,38].

15.4.1.4 Linear Minimum Mean-Square Error Method

The linear minimum mean-square error (LMMSE) method finds the linear estimate which minimizes the mean-square error between the estimate and ideal image, using up to second-order statistics of the ideal image. Assuming that the ideal image can be modeled by a zero-mean homogeneous random field and the blur is space-invariant, the LMMSE (Wiener) estimate, in the frequency domain, is given by [8]

$$\hat{F}(u, v) = \frac{D^{*}(u, v)}{|D(u, v)|^2 + \sigma_v^2/|P(u, v)|^2} G(u, v) \tag{15.15}$$

where

σ_v^2 is the variance of the observation noise (assumed white)

$|P(u, v)|^2$ stands for the power spectrum of the ideal image

The power spectrum of the ideal image is usually estimated from a prototype. It can be easily seen that the CLS estimate (15.14) reduces to the Wiener estimate by setting $|L(u, v)|^2 = \sigma_v^2/|P(u, v)|^2$ and $\gamma = 1$.

A Kalman filter determines the causal (up to a fixed lag) LMMSE estimate recursively. It is based on a state-space representation of the image and observation models. In the first step of Kalman filtering, a prediction of the present state is formed using an AR image model and the previous state of the system. In the second step, the predictions are updated on the basis of the observed image data to form the estimate of the present state. Woods and Ingle [39] applied 2-D reduced-update Kalman filter (RUKF) to image restoration, where the update is limited to only those state variables in a neighborhood of the present pixel. The main assumption here is that a pixel is insignificantly correlated with pixels outside a certain neighborhood about itself. More recently, a reduced-order model Kalman filtering (ROMKF), where the state vector is truncated to a size that is on the order of the image model support has been proposed [40]. Other Kalman filtering formulations, including higher-dimensional state-space models to reduce the effective size of the state vector, have been reviewed in Ref. [7]. The complexity of higher-dimensional state-space model-based formulations, however, limits their practical use.

15.4.1.5 Maximum A Posteriori Probability Method

The MAP restoration maximizes the a posteriori pdf $p(\boldsymbol{f}\,|\,\boldsymbol{g})$, i.e., the likelihood of a realization of $\boldsymbol{f}$ being the ideal image given the observed data $\boldsymbol{g}$. Through the application of the Bayes rule, we have

$$p(\boldsymbol{f}\,|\,\boldsymbol{g}) \propto p(\boldsymbol{g}|\boldsymbol{f})p(\boldsymbol{f}) \tag{15.16}$$

where

$p(\boldsymbol{g}|\boldsymbol{f})$ is the conditional pdf of $\boldsymbol{g}$ given $\boldsymbol{f}$ (related to the pdf of the noise process)

$p(\boldsymbol{f})$ is the a priori pdf of the ideal image

We usually assume that the observation noise is Gaussian, leading to

$$p(\boldsymbol{g}|\boldsymbol{f}) = \frac{1}{(2\pi)^{N/2}|\boldsymbol{R}_v|^{1/2}} \exp\{-1/2(\boldsymbol{g} - \boldsymbol{D}\boldsymbol{f})^{\mathrm{T}}\boldsymbol{R}_v^{-1}(\boldsymbol{g} - \boldsymbol{D}\boldsymbol{f})\} \tag{15.17}$$

where $\boldsymbol{R}_v$ denotes the covariance matrix of the noise process. Unlike the LMMSE method, the MAP method uses complete pdf information. However, if both the image and noise are assumed to be homogeneous Gaussian random fields, the MAP estimate reduces to the LMMSE estimate, under a linear observation model.

Trussell and Hunt [10] used nonstationary a priori pdf models, and proposed a modified form of the Picard iteration to solve the nonlinear maximization problem. They suggested using the variance of the residual signal as a criterion for convergence. Geman and Geman [11] proposed using a Gibbs random field model for the a priori pdf of the ideal image. They used simulated annealing procedures to maximize Equation 15.16. It should be noted that the MAP procedures usually require significantly more computation compared to, for example, the CLS or Wiener solutions.

15.4.1.6 Maximum Entropy Method

A number of maximum entropy (ME) approaches have been discussed in the literature, which vary in the way that the ME principle is implemented. A common feature of all these approaches, however, is their computational complexity. Maximizing the entropy enforces smoothness of the restored image. (In the absence of constraints, the entropy is highest for a constant-valued image). One important aspect of the ME approach is that the nonnegativity constraint is implicitly imposed on the solution because the entropy is defined in terms of the logarithm of the intensity.

Frieden was the first to apply the ME principle to image restoration [41]. In his formulation, the sum of the entropy of the image and noise, given by

$$J_{\mathrm{ME1}}(\boldsymbol{f}) = -\sum_i f(i) \ln f(i) - \sum_i n(i) \ln n(i) \tag{15.18}$$

is maximized subject to the constraints

$$\boldsymbol{n} = \boldsymbol{g} - \boldsymbol{D}\boldsymbol{f} \tag{15.19}$$

$$\sum_i f(i) = K \doteq \sum_i g(i) \tag{15.20}$$

which enforce fidelity to the data and a constant sum of pixel intensities. This approach requires the solution of a system of nonlinear equations. The number of equations and unknowns are on the order of the number of pixels in the image. The formulation proposed by Gull and Daniell [42] can be viewed as another form of Tikhonov regularization (or CLS formulation), where the entropy of the image

$$J_{\mathrm{ME2}}(\boldsymbol{f}) = -\sum_i f(i) \ln f(i) \tag{15.21}$$

is the regularization functional. It is maximized subject to the following usual constraints

$$\|\boldsymbol{g} - \boldsymbol{D}\boldsymbol{f}\|^2 = \sigma_v^2 \tag{15.22}$$

$$\sum_i f(i) = K \doteq \sum_i g(i) \tag{15.23}$$

on the restored image. The optimization problem is solved using an ascent algorithm. Trussell [43] showed that in the case of a prior distribution defined in terms of the image entropy, the MAP solution is identical to the solution obtained by this ME formulation. Other ME formulations were also proposed [44,45]. Note that all ME methods are nonlinear in nature.

15.4.1.7 Set-Theoretic Methods

In set-theoretic methods, first a number of "constraint sets" are defined such that their members are consistent with the observations and/or some a priori information about the ideal image. A set-theoretic estimate of the ideal image is then defined as a feasible solution satisfying all constraints, i.e., any member of the intersection of the constraint sets. Note that set-theoretic methods are, in general, nonlinear.

Set-theoretic methods vary according to the mathematical properties of the constraint sets. In the method of projections onto convex sets (POCS), the constraint sets C_i are closed and convex in an appropriate Hilbert space $\mathcal{H}$. Given the sets C_i, $i = 1, \ldots, M$, and their respective projection operators $\mathbf{P}_i$, a feasible solution is found by performing successive projections as

$$\boldsymbol{f}_{k+1} = \mathbf{P}_M \mathbf{P}_{M-1} \ldots \mathbf{P}_1 \boldsymbol{f}_k; \quad k = 0, 1, \ldots \tag{15.24}$$

where $\boldsymbol{f}_0$ is the initial estimate (a point in $\mathcal{H}$). The projection operators are usually found by solving constrained optimization problems. In finite-dimensional problems (which is the case for digital image restoration), the iterations converge to a feasible solution in the intersection set [46–48]. It should be noted that the convergence point is affected by the choice of the initialization. However, as the size of the intersection set becomes smaller, the differences between the convergence points obtained by different initializations become smaller. Trussell and Civanlar [49] applied POCS to image restoration. For examples of convex constraint sets that are used in image restoration, see [23]. A relationship between the POCS and Landweber iterations were developed in Ref. [10].

A special case of POCS is the Gerchberg–Papoulis-type algorithms where the constraint sets are either linear subspaces or linear varieties [50]. Extensions of POCS to the case of nonintersecting sets [51] and nonconvex sets [52] have been discussed in the literature. Another extension is the method of fuzzy sets (FS), where the constraints are defined in terms of FS. More precisely, the constraints are reflected in the membership functions defining the FS. In this case, a feasible solution is defined as one that has a high grade of membership (e.g., above a certain threshold) in the intersection set. The method of FS has also been applied to image restoration [53].

15.4.2 Restoration of Images Recorded by Nonlinear Sensors

Image sensors and media may have nonlinear characteristics that can be modeled by a pointwise (memoryless) nonlinearity $s(\cdot)$. Common examples are photographic film and paper, where the nonlinear relationship between the exposure (intensity) and the silver density deposited on the film or paper is specified by a "$d - \log e$" curve. The modeling of sensor nonlinearities was first addressed by Andrews and Hunt [1]. However, it was not generally recognized that results obtained by taking the sensor nonlinearity into account may be far more superior to those obtained by ignoring the sensor nonlinearity, until the experimental work of Tekalp and Pavlović [54,55].

Except for the MAP approach, none of the algorithms discussed above are equipped to handle sensor nonlinearity in a straightforward fashion. A simple approach would be to expand the observation model with $s(\cdot)$ into its Taylor series about the mean of the observed image and obtain an approximate (linearized) model, which can be used with any of the above methods [1]. However, the results do not show significant improvement over those obtained by ignoring the nonlinearity. The MAP method is capable of taking the sensor nonlinearity into account directly. A modified Picard iteration was proposed in Ref. [10], assuming that both the image and noise are Gaussian distributed, which is given by

$$\hat{f}_{k+1} = \bar{f}_k + R_f D^T S_b R_n^{-1}[g - s(Df_k)] \tag{15.25}$$

where

$\bar{f}$ denotes nonstationary image mean
R_f and R_n are the correlation matrices of the ideal image and noise, respectively
S_b is a diagonal matrix consisting of the derivatives of $s(\cdot)$ evaluated at $b = Df$

It is the matrix S_b that R_f maps the difference $[g - s(Df_k)]$ from the observation domain to the intensity domain.

An alternative approach, which is computationally less demanding, transforms the observed density domain image to the exposure domain [54]. There is a convolutional relationship between the ideal and blurred images in the exposure domain. However, the additive noise in the density domain manifests itself as multiplicative noise in the exposure domain. To this effect, Tekalp and Pavlović [54] derive an LMMSE deconvolution filter in the presence of multiplicative noise under certain assumptions. Their results show that accounting for the sensor nonlinearity may dramatically improve restoration results [54,55].

15.4.3 Restoration of Images Degraded by Random Blurs

Basic regularized restoration methods (reviewed in Section 15.4.1) assume that the blur PSF is a deterministic function. A more realistic model may be

$$D = \bar{D} + \Delta D \tag{15.26}$$

where

$\bar{D}$ is the deterministic part (known or estimated) of the blur operator
ΔD stands for the random component

Random component may represent inherent random fluctuations in the PSF, for instance due to atmospheric turbulence or random relative motion, or it may model the PSF estimation error.

A naive approach would be to employ the expected value of the blur operator in one of the restoration algorithms discussed above. The resulting restoration, however, may be unsatisfactory. Slepian [56] derived the LMMSE estimate, which explicitly incorporated the random component of the PSF. The resulting Wiener filter requires the a priori knowledge of the second-order statistics of the blur process. Ward et al. [57,58] also proposed LMMSE estimators. Combettes and Trussell [59] addressed restoration of random blurs within the framework of POCS, where fluctuations in the PSF are reflected in the bounds defining the residual constraint sets. The method of total least squares (TLS) has been used in the mathematics literature to solve a set of linear equations with uncertainties in the system matrix. The TLS method amounts to finding the minimum perturbations on D and g to make the system of equations consistent. A variation of this principle has been applied to image restoration with random PSF by Mesarovic et al. [60]. Various authors have shown that modeling the uncertainty in the PSF (by means of a random component) reduces ringing artifacts that are due to using erroneous PSF estimates.

15.4.4 Adaptive Restoration for Ringing Reduction

LSI restoration methods introduce disturbing ringing artifacts which originate around sharp edges and image borders [36]. A quantitative analysis of the origins and characteristics of ringing and other restoration artifacts was given by Tekalp and Sezan [14]. Suppression of ringing may be possible by means of adaptive filtering, which tracks edges or image statistics such as local mean and variance.

Iterative and set-theoretic methods are well suited for adaptive image restoration with ringing reduction. Lagendijk et al. [36] have extended Miller regularization to adaptive restoration by defining the solution in a weighted Hilbert space, in terms of norms weighted by space-variant weights. Later, Sezan and Tekalp [9] extended the method of POCS to the space-variant case by introducing a region-based bound on the signal energy. In both methods, the weights and/or the regions were identified from the degraded image. Recently, Sezan and Trussell [23] have developed constraints based on prototype images for set-theoretic image restoration with artifact reduction.

Kalman filtering can also be extended to adaptive image restoration. For a typical image, the homogeneity assumption will hold only over small regions. Rajala and de Figueiredo [61] used an off-line visibility function to segment the image according to the local spatial activity of the picture being restored. Later, a rapid edge adaptive filter based on multiple image models to account for edges with various orientations was developed by Tekalp et al. [62]. Jeng and Woods [13] developed inhomogeneous Gauss–Markov field models for adaptive filtering, and ME methods were used for ringing reduction [45]. Results show a significant reduction in ringing artifacts in comparison to LSI restoration.

15.4.5 Blind Restoration (Deconvolution)

Blind restoration refers to methods that do not require prior identification of the blur and regularization model parameters. Two examples are simultaneous identification and restoration of noisy blurred images [63] and image recovery from Fourier phase information [64]. Lagendijk et al. [63] applied the E-M algorithm to blind image restoration, which alternates between ML parameter identification and minimum mean-square error image restoration. Chen et al. [64] employed the POCS method to estimate the Fourier magnitude of the ideal image from the Fourier phase of the observed blurred image by assuming a zero-phase blur PSF so that the Fourier phase of the observed image is undistorted. Both methods require the PSF to be real and symmetric.

15.4.6 Restoration of Multispectral Images

A trivial solution to multispectral image restoration, when there is no interband blurring, may be to ignore the spectral correlations among different bands and restore each band independently, using one of the algorithms discussed above. However, the algorithms that are optimal for single-band imagery may no longer be so when applied to individual spectral bands. For example, restoration of the red, green, and blue bands of a color image independently usually results in objectionable color shift artifacts.

To this effect, Hunt and Kubler [65] proposed employing the Karhunen–Loeve (KL) transform to decorrelate the spectral bands so that an independent-band processing approach can be applied. However, because the KL transform is image dependent, they then recommended using the NTSC YIQ transformation as a suboptimum but easy-to-use alternative. Experimental evidence shows that the visual quality of restorations obtained in the KL, YIQ, or another luminance–chrominance domain are quite similar [65]. In fact, restoration of only the luminance channel suffices in most cases. This method applies only when there is no interband blurring. Further, one should realize that the observation noise becomes correlated with the image under a nonorthogonal transformation. Thus, filtering based on the assumption that the image and noise are uncorrelated is not theoretically founded in the YIQ domain.

Recent efforts in multispectral image restoration are concentrated on making total use of the inherent correlations between the bands [66,67]. Applying the CLS filter expression (15.13) to the observation model (15.4) with $\boldsymbol{Q}^{\mathrm{H}}\boldsymbol{Q} = \mathcal{R}_f^{-1}\mathcal{R}_v$, we obtain the multispectral Wiener estimate $\hat{\boldsymbol{f}}$, given by [68]

$$\hat{\boldsymbol{f}} = \left(\mathcal{D}^{\mathrm{T}}\mathcal{D} + \mathcal{R}_f^{-1}\mathcal{R}_v\right)^{-1}\mathcal{D}^{\mathrm{T}}\boldsymbol{g} \tag{15.27}$$

where

$$\mathcal{R}_f \doteq \begin{bmatrix} \boldsymbol{R}_{f;11} & \cdots & \boldsymbol{R}_{f;1K} \\ \vdots & \ddots & \vdots \\ \boldsymbol{R}_{f;K1} & \cdots & \boldsymbol{R}_{f;KK} \end{bmatrix}, \quad \text{and} \quad \mathcal{R}_v \doteq \begin{bmatrix} \boldsymbol{R}_{v;11} & \cdots & \boldsymbol{R}_{v;1K} \\ \vdots & \ddots & \vdots \\ \boldsymbol{R}_{v;K1} & \cdots & \boldsymbol{R}_{v;KK} \end{bmatrix}$$

Here $\boldsymbol{R}_{f;ij} \doteq \varepsilon\{\boldsymbol{f}_i\boldsymbol{f}_j^{\mathrm{T}}\}$ and $\boldsymbol{R}_{v;ij} \doteq \varepsilon\{\boldsymbol{v}_i\boldsymbol{v}_j^{\mathrm{T}}\}$, $i, j = 1, 2, \ldots, K$ denote the interband, cross-correlation matrices. Note that if $\boldsymbol{R}_{f;ij} = 0$ for $i \neq j$; $i, j = 1, 2, \ldots, K$, then the multiframe estimate becomes equivalent to stacking the K single-frame estimates obtained independently.

Direct computation of $\hat{\boldsymbol{f}}$ through Equation 15.27 requires inversion of a $N^2L \times N^2L$ matrix. Because the blur PSF is not necessarily the same in each band and the interband correlations are not shift-invariant, the matrices $\mathcal{D}$, $\mathcal{R}_f$, and $\mathcal{R}_v$, are not block-Toeplitz; thus, a 3-D DFT would not diagonalize them. However, assuming LSI blurs, each $\boldsymbol{D}_k$ is block Toeplitz. Furthermore, assuming each image and noise band are wide-sense stationary, $\boldsymbol{R}_{f;ij}$ and $\boldsymbol{R}_{v;ij}$ are also block-Toeplitz. Approximating the block-Toeplitz submatrices $\boldsymbol{D}_i$, $\boldsymbol{R}_{f;ij}$, and $\boldsymbol{R}_{v;ij}$ by block-circulant ones, each submatrix can be diagonalized by a separate 2-D DFT operation so that we only need to invert a block matrix with diagonal subblocks. Galatsanos and Chin [66] proposed a method that successively partitions the matrix to be inverted and recursively computes the inverse of these partitions. Later Ozkan et al. [68] have shown that the desired inverse can be computed by inverting N^2 submatrices, each $K \times K$, in parallel. The resulting numerically stable filter was called the cross-correlated multiframe (CCMF) Wiener filter.

The multispectral Wiener filter requires the knowledge of the correlation matrices $\mathcal{R}_f$ and $\mathcal{R}_v$. If we assume that the noise is white and spectrally uncorrelated, the matrix $\mathcal{R}_v$ is diagonal with all diagonal entries equal to σ_v^2. Estimation of the multispectral correlation matrix $\boldsymbol{R}_f$ can be performed by either the periodogram method or 3-D AR modeling [68]. Sezan and Trussell [69] show that the multispectral Wiener filter is highly sensitive to the cross-power spectral estimates, which contain phase information. Other multispectral restoration methods include Kalman filtering approach of Tekalp and Pavlović [67], LS approaches of Ohyama et al. [70] and Galatsanos et al. [71], and set-theoretic approach of Sezan and Trussell [23,69] who proposed multispectral image constraints.

15.4.7 Restoration of Space-Varying Blurred Images

In principle, all basic regularization methods apply to the restoration of space-varying blurred images. However, because Fourier transforms cannot be utilized to simplify large matrix operations (such as inversion or SVD) when the blur is space-varying, implementation of some of these algorithms may be computationally formidable. There exist three distinct approaches to attack the space-variant restoration problem: (1) sectioning, (2) coordinate transformation (CRT), and (3) direct approaches.

The main assumption in sectioning is that the blur is approximately space-invariant over small regions. Therefore, a space-varying blurred image can be restored by applying the well-known space-invariant techniques to local image regions. Trussell and Hunt [73] propose using iterative MAP restoration within rectangular, overlapping regions. Later, Trussell and Fogel proposed using a modified Landweber iteration [21]. A major drawback of sectioning methods is generation of artifacts at the region boundaries. Overlapping the contiguous regions somewhat reduces these artifacts, but does not completely suppress them.

Most space-varying PSF vary continuously from pixel to pixel (e.g., relative motion with acceleration) violating the basic premise of the sectioning methods. To this effect, Robbins and Huang [74] and then Sawchuck [75] proposed a CTR method such that the blur PSF in the transformed coordinates is space-invariant. Then, the transformed image can be restored by a space-invariant filter and then transformed back to obtain the final restored image. However, the statistical properties of the image and noise processes are affected by the CTR, which should be taken into account in restoration filter design. The results reported in Refs. [74,75] have been obtained by inverse filtering; and thus, this statistical issue was of no concern. Also note that the CTR method is applicable to a limited class of space-varying blurs. For instance, blurring due to depth of field is not amenable to CTR.

The lack of generality of sectioning and CTR methods motivates direct approaches. Iterative schemes, Kalman filtering, and set-theoretic methods can be applied to restoration of space-varying blurs in a computationally feasible manner. Angel and Jain [76] propose solving the superposition Equation 15.3 iteratively using a conjugate gradient method. Application of constrained iterative methods was discussed in Ref. [30]. More recently, Ozkan et al. [72] developed a robust POCS algorithm for space-varying image restoration, where they defined a closed, convex constraint set for each observed blurred image pixel, given by

$$\boldsymbol{C}_{n_1,n_2} = \left\{\boldsymbol{y} : |r^{(y)}(n_1, n_2)| \leq \delta_0\right\} \tag{15.28}$$

and

$$r^{(y)}(n_1, n_2) \doteq g(n_1, n_2) - \sum_{(m_1,m_2)\in S_d(n_1,n_2)} d(n_1, n_2;\, m_1, m_2) y(m_1, m_2) \tag{15.29}$$

is the residual at pixel (n_1, n_2) associated with $\boldsymbol{y}$, which denotes an arbitrary member of the set. The quantity δ_0 is an a priori bound reflecting the statistical confidence with which the actual image is a member of the set $\boldsymbol{C}_{n_1,n_2}$. Since $r^{(f)}(n_1, n_2) = v(n_1, n_2)$, the bound δ_0 is determined from the statistics of the noise process so that the ideal image is a member of the set within a certain statistical confidence. The collection of bounded residual constraints over all pixels (n_1, n_2) enforces the estimate to be consistent with the observed image.

The projection of an arbitrary $x(i_1, i_2)$ onto each $\boldsymbol{C}_{n_1,n_2}$ is defined as

$$\mathbf{P}_{n_1,n_2}[x(i_1, i_2)] = \begin{cases} x(i_1, i_2) + \dfrac{r^{(x)}(n_1, n_2) - \delta_0}{\sum_{o_1}\sum_{o_2} h^2(n_1, n_2; o_1, o_2)} h(n_1, n_2; i_1, i_2) & \text{if } r^{(x)}(n_1, n_2) > \delta_0 \\ x(i_1, i_2) & \text{if } -\delta_0 \leq r^{(x)}(n_1, n_2) \leq \delta_0 \\ x(i_1, i_2) + \dfrac{r^{(x)}(n_1, n_2) - \delta_0}{\sum_{o_1}\sum_{o_2} h^2(n_1, n_2; o_1, o_2)} h(n_1, n_2; i_1, i_2) & \text{if } r^{(x)}(n_1, n_2) < -\delta_0 \end{cases} \tag{15.30}$$

The algorithm starts with an arbitrary $x(i_1, i_2)$, and successively projects onto each $\boldsymbol{C}_{n_1,n_2}$. This is repeated until convergence [72]. Additional constraints, such as bounded energy, amplitude, and limited support, can be utilized to improve the results.

15.5 Multiframe Restoration and Superresolution

Multiframe restoration refers to estimating the ideal image on a lattice that is identical with the observation lattice, whereas superresolution refers to estimating it on a lattice that has a higher sampling density than the observation lattice. They both employ the multiframe observation model (15.5), which establishes a relation between the ideal image and observations at more than one instance. Several authors eluded that the sequential nature of video sources can be statistically modeled by means of temporal correlations [68,71]. Multichannel filters similar to those described for multispectral restoration were thus proposed for multiframe restoration. Here, we only review motion-compensated (MC) restoration and superresolution methods, because they are more effective.

15.5.1 Multiframe Restoration

The sequential nature of images in a video source can be used to better estimate the PSF parameters, regularization terms, and the restored image. For example, the extent of a motion blur can be estimated from interframe motion vectors, provided that the aperture time is known. The first MC approach was the MC multiframe Wiener filter (MCMF) proposed by Ozkan et al. [68] who considered the case of frame-to-frame global translations. Then, the auto power spectra of all frames are the same and the cross-spectra are related by a phase factor which can be estimated from the motion information. Given the motion vectors (one for each frame) and the auto power spectrum of the reference frame, they derived a closed-form solution, given by

$$\hat{F}_k(u, v) = \frac{S_{f;k}(u, v) \sum_{i=1}^{N} S_{f;i}^{*}(u, v) D_i^{*}(u, v) G_i(u, v)}{\sum_{i=1}^{N} |S_{f;i}(u, v) D_i(u, v)|^2 + \sigma_v^2}, \tag{15.31}$$

where

k is the index of the ideal frame to be restored

N is the number of available frames

$P_{f;ki}(u, v) = S_{f;k}(u, v) S_{f;i}^{*}(u, v)$ denotes the cross-power spectrum between the frames k and i in factored form

The fact that such a factorization exists was shown in Ref. [68] for the case of global translational motion. The MCMF yields the biggest improvement when the blur PSF changes from frame to frame. This is because the summation in the denominator may not be zero at any frequency, even though each term $D_i(u, v)$ may have zeros at certain frequencies. The case of space-varying blurs may be considered as a special case of the last section which covers superresolution with space-varying restoration.

15.5.2 Superresolution

When the interframe motion is subpixel, each frame, in fact, contains some "new" information that can be utilized to achieve superresolution. Superresolution refers to high-resolution image expansion, which aims to remove aliasing artifacts, blurring due to sensor PSF, and optical blurring given the observation model (15.5). Provided that enough frames with subpixel motion are available, the observation model becomes invertible. It can be easily seen, however, that superresolution from a single observed image is ill-posed because we have more unknowns than equations, and there exist infinitely many expanded images that are consistent with the model (15.5). Therefore, single-frame nonlinear interpolation (also called image expansion and digital zooming) methods for improved definition image expansion employ additional regularization criteria, such as edge-preserving smoothness constraints [77,78]. (It is well-known that no new high-frequency information can be generated by LSI interpolation techniques, including ideal band-limited interpolation, hence the need for nonlinear methods.)

Several early MC methods are in the form of two-stage interpolation-restoration algorithms [79,80]. They are based on the premise that pixels from all observed frames can be mapped back onto a desired frame, based on estimated motion trajectories, to obtain an upsampled reference frame. However, unless we assume global translational motion, the upsampled reference frame is nonuniformly sampled.

In order to obtain a uniformly spaced upsampled image, interpolation onto a uniform sampling grid needs to be performed. Image restoration is subsequently applied to the upsampled image to remove the effect of the sensor blur. However, these methods do not use an accurate image formation model, and cannot remove aliasing artifacts.

MC (multiframe) superresolution methods that are based on the model (15.5) can be classified as those that aim to eliminate: (1) aliasing only, (2) aliasing and LSI blurs, and (3) aliasing and space-varying blurs. In addition, some of these methods are designed for global translational motion only, while others can handle space-varying motion fields with occlusion. Multiframe superresolution was first introduced by Tsai and Huang [81] who exploited the relationship between the continuous and discrete Fourier transforms of the undersampled frames to remove aliasing errors, in the special case of global motion. Their formulation has been extended by Kim et al. [82] to take into account noise and blur in the low-resolution images, by posing the problem in the LS sense. A further refinement by Kim and Su [83] allowed blurs that are different for each frame of low-resolution data, by using a Tikhonov regularization. However, the resulting algorithm did not treat the formation of blur due to motion or sensor size, and suffers from convergence problems.

Inspection of the model (15.5) suggests that the super-resolution problem can be stated in the spatiotemporal domain as the solution of a set of simultaneous linear equations. Suppose that the desired high-resolution frames are $M \times M$, and we have L low-resolution observations, each $N \times N$. Then, from Equation 15.5, we can set up at most $L \times N \times N$ equations in M^2 unknowns to reconstruct a particular high-resolution frame. These equations are linearly independent provided that all displacements between the successive frames are at subpixel amounts. (Clearly, the number of equations will be reduced by the number of occlusion labels encountered along the respective motion trajectories.) In general, it is desirable to set up an overdetermined system of equations, i.e., $L > R^2 = M^2/N^2$, to obtain a more robust solution in the presence of observation noise. Because the impulse response coefficients $h_{ik}(n_1, n_2; m_1, m_2)$ are spatially varying, and hence the system matrix is not block-Toeplitz, fast methods to solve them are not available. Stark and Oskui [86] proposed a POCS method to compute a high-resolution image from observations obtained by translating and/or rotating an image with respect to a CCD array. Irani and Peleg [84,85] employed iterative methods. Patti et al. [87] extended the POCS formulation to include sensor noise and space-varying blurs. Bayesian approaches were also employed for superresolution [88].

15.5.3 Superresolution with Space-Varying Restoration

This section explains extension of the POCS method to space-varying blurs. The POCS method described here addresses the most general form of the superresolution problem based on the model (15.5). The formulation is quite similar to the POCS approach presented for intraframe restoration of space-varying blurred images. In this case, we define a different closed, convex set for each observed low-resolution pixel (n_1, n_2, k) (which can be connected to the desired frame i by a motion trajectory) as

$$C_{n_1,n_2;i,k} = \left\{ x_i(m_1, m_2) : \left| r_k^{(x_i)}(n_1, n_2) \right| \leq \delta_0 \right\}, \quad 0 \leq n_1, n_2 \leq N-1, \quad k = 1, \ldots, L \tag{15.32}$$

where

$$r_k^{(x_i)}(n_1, n_2) \doteq g_k(n_1, n_2) - \sum_{m_1=0}^{M-1} \sum_{m_2=0}^{M-1} x_i(m_1, m_2) h_{ik}(m_1, m_2; n_1, n_2)$$

and δ_0 represents the confidence that we have in the observation and is set equal to $c\sigma_v$, where σ_v is the standard deviation of the noise and $c \geq 0$ is determined by an appropriate statistical confidence bound. These sets define high-resolution images that are consistent with the observed low-resolution frames within a confidence bound that is proportional to the variance of the observation noise. The projection operator which projects onto $C_{n_1,n_2;i,k}$ can be deduced from Equation 15.30 [8]. Additional constraints, such as amplitude and/or finite support constraints, can be utilized to improve the results. Excellent reconstructions have been reported using this procedure [68,87].

A few observations about the POCS method are in order: (1)While certain similarities exist between the POCS iterations and the Landweber-type iterations [79,84,85], the POCS method can adapt to the amount of the observation noise, while the latter generally cannot. (2) The POCS method finds a feasible solution, that is, a solution consistent with all available low-resolution observations. Clearly, the more observations (more frames with reliable motion estimation) we have, the better the high-resolution reconstructed image $\hat{s}_i(m_1, m_2)$ will be. In general, it is desirable that $L > M^2/N^2$. Note, however, that the POCS method generates a reconstructed image with any number L of available frames. The number L is just an indicator of how large the feasible set of solutions will be. Of course, the size of the feasible set can be further reduced by employing other closed, convex constraints in the form of statistical or structural image models.

15.5.4 Special Superresolution Applications

The methods presented in the previous sections have been formulated in the space-domain with the assumption that raw (uncompressed) low-resolution video source is available. More and more frequently video sources are available in compressed formats such as MPEG-2 or MPEG-4/AVC. Recently, Altunbasak et al. [89] proposed an extension of the above methods to MC, transform-domain video data that directly incorporates transform-domain quantization information working given the compressed bitstream.

Another special application of superresolution methods has been on improving the resolution of face images given a training set of high-quality face images [90]. This approach may prove useful if face images obtained from low-quality videos need to be compared with a database of high-quality stills for person identification or recognition.

15.6 Conclusion

At present, factors that limit the success of digital image restoration technology include lack of reliable (1) methods for blur identification, especially identification of space-variant blurs, (2) methods to identify imaging system nonlinearities, and (3) methods to deal with the presence of artifacts in restored images. Our experience with the restoration of real-life blurred images indicates that the choice of a particular regularization strategy (filter) has a small effect on the quality of the restored images as long as the parameters of the degradation model, i.e., the blur PSF and the signal-to-noise ratio (SNR), and any imaging system nonlinearity are properly compensated. Proper compensation of system nonlinearities also plays a significant role in blur identification.

References

1. Andrews, H. C. and Hunt, B. R., *Digital Image Restoration*, Prentice-Hall, Englewood Cliffs, NJ, 1977.
2. Gonzales, R. C. and Woods, R. E., *Digital Image Processing*, Addison-Wesley, Reading, MA, 1992.
3. Katsaggelos, A. K., Ed., *Digital Image Restoration*, Springer-Verlag, Berlin, 1991.
4. Meinel, E. S., Origins of linear and nonlinear recursive restoration algorithms, *J. Opt. Soc. Am.*, A-3(6), 787–799, 1986.
5. Demoment, G., Image reconstruction and restoration: Overview of common estimation structures and problems, *IEEE Trans. Acoust. Speech Signal Process.*, 37, 2024–2036, 1989.
6. Sezan, M. I. and Tekalp, A. M., Survey of recent developments in digital image restoration, *Opt. Eng.*, 29, 393–404, 1990.

7. Kaufman, H. and Tekalp, A. M., Survey of estimation techniques in image restoration, *IEEE Control Syst. Mag.*, 11, 16–24, 1991.
8. Tekalp, A. M., *Digital Video Processing*, Prentice-Hall, Englewood Cliffs, NJ, 1995.
9. Sezan, M. I. and Tekalp, A. M., Adaptive image restoration with artifact suppression using the theory of convex projections, *IEEE Trans. Acoust. Speech Signal Process.*, 38(1), 181–185, 1990.
10. Trussell, H. J. and Hunt, B. R., Improved methods of maximum a posteriori restoration, *IEEE Trans. Comput.*, C-27(1), 57–62, 1979.
11. Geman, S. and Geman, D., Stochastic relaxation, Gibbs distributions, and the Bayesian restoration of images, *IEEE Trans. Pattern Anal. Machine Intell.*, 6(6), 721–741, 1984.
12. Jain, A. K., Advances in mathematical models for image processing, *Proc. IEEE* 69(5), 502–528, 1981.
13. Jeng, F. C. and Woods, J. W., Compound Gauss–Markov random fields for image restoration, *IEEE Trans. Signal Process.*, SP-39(3), 683–697, 1991.
14. Tekalp, A. M. and Sezan, M. I., Quantitative analysis of artifacts in linear space-invariant image restoration, *Multidim. Syst. Signal Process.*, 1(1), 143–177, 1990.
15. Rosenfeld, A. and Kak, A. C., *Digital Picture Processing*, Academic, New York, 1982.
16. Gennery, D. B., Determination of optical transfer function by inspection of frequency-domain plot, *J. Opt. Soc. Am.*, 63(12), 1571–1577, 1973.
17. Cannon, M., Blind deconvolution of spatially invariant image blurs with phase, *IEEE Trans. Acoust. Speech Signal Process.*, ASSP-24(1), 58–63, 1976.
18. Chang, M. M., Tekalp, A. M., and Erdem, A. T., Blur identification using the bispectrum, *IEEE Trans. Signal Process.*, ASSP-39(10), 2323–2325, 1991.
19. Lagendijk, R. L., Tekalp, A. M., and Biemond, J., Maximum likelihood image and blur identification: A unifying approach, *Opt. Eng.*, 29(5), 422–435, 1990.
20. Pavlović, G. and Tekalp, A. M., Maximum likelihood parametric blur identification based on a continuous spatial domain model, *IEEE Trans. Image Process.*, 1(4), 496–504, 1992.
21. Trussell, H. J. and Fogel, S., Identification and restoration of spatially variant motion blurs in sequential images, *IEEE Trans. Image Process.*, 1(1), 123–126, 1992.
22. Kaufman, H., Woods, J. W., Dravida, S., and Tekalp, A. M., Estimation and identification of two-dimensional images, *IEEE Trans. Automatic Control*, 28, 745–756, 1983.
23. Sezan, M. I. and Trussell, H. J., Prototype image constraints for set-theoretic image restoration, *IEEE Trans. Signal Process.*, 39(10), 2275–2285, 1991.
24. Galatasanos, N. P. and Katsaggelos, A. K., Methods for choosing the regularization parameter and estimating the noise variance in image restoration and their relation, *IEEE Trans. Image Process.*, 1(3), 322–336, 1992.
25. Trussell, H. J. and Civanlar, M. R., The Landweber iteration and projection onto convex sets, *IEEE Trans. Acoust. Speech Signal Process.*, ASSP-33(6), 1632–1634, 1985.
26. Sanz, J. L. C. and Huang, T. S., Unified Hilbert space approach to iterative least-squares linear signal restoration, *J. Opt. Soc. Am.*, 73(11), 1455–1465, 1983.
27. Tom, V. T., Quatieri, T. F., Hayes, M. H., and McClellan, J. H., Convergence of iterative nonexpansive signal reconstruction algorithms, *IEEE Trans. Acoust. Speech Signal Process.*, ASSP-29(5), 1052–1058, 1981.
28. Kawata, S. and Ichioka, Y., Iterative image restoration for linearly degraded images. II. Reblurring, *J. Opt. Soc. Am.*, 70, 768–772, 1980.
29. Trussell, H. J., Convergence criteria for iterative restoration methods, *IEEE Trans. Acoust. Speech Signal Process.*, ASSP-31(1), 129–136, 1983.
30. Schafer, R. W., Mersereau, R. M., and Richards, M. A., Constrained iterative restoration algorithms, *Proc. IEEE*, 69(4), 432–450, 1981.
31. Biemond, J., Lagendijk, R. L., and Mersereau, R. M., Iterative methods for image deblurring, *Proc. IEEE*, 78(5), 856–883, 1990.

32. Katsaggelos, A. K., Iterative image restoration algorithms, *Opt. Eng.*, 28(7), 735–748, 1989.
33. Tikhonov, A. N. and Arsenin, V. Y., *Solutions of Ill-Posed Problems*, V. H. Winston and Sons, Washington, D.C., 1977.
34. Hunt, B. R., The application of constrained least squares estimation to image restoration by digital computer, *IEEE Trans. Comput.*, C-22(9), 805–812, 1973.
35. Miller, K., Least squares method for ill-posed problems with a prescribed bound, *SIAM J. Math. Anal.*, 1, 52–74, 1970.
36. Lagendijk, R. L., Biemond, J., and Boekee, D. E., Regularized iterative image restoration with ringing reduction, *IEEE Trans. Acoust. Speech Signal Process.*, 36(12), 1874–1888, 1988.
37. Zhou, Y. T., Chellappa, R., Vaid, A., and Jenkins, B. K., Image restoration using a neural network, *IEEE Trans. Acoust. Speech Signal Process.*, ASSP-36(7), 1141–1151, 1988.
38. Yeh, S. J., Stark H., and Sezan, M. I., Hopfield-type neural networks: their set-theoretic formulations as associative memories, classifiers, and their application to image restoration, in *Digital Image Restoration*, Katsaggelos, A. Ed., Springer-Verlag, Berlin, 1991.
39. Woods, J. W. and Ingle, V. K., Kalman filtering in two-dimensions-further results, *IEEE Trans. Acoust. Speech Signal Process.*, ASSP-29, 188–197, 1981.
40. Angwin, D. L. and Kaufman, H., Image restoration using reduced order models, *Signal Process.*, 16, 21–28, 1988.
41. Frieden, B. R., Restoring with maximum likelihood and maximum entropy, *J. Opt. Soc. Am.*, 62(4), 511–518, 1972.
42. Gull, S. F. and Daniell, G. J., Image reconstruction from incomplete and noisy data, *Nature*, 272, 686–690, 1978.
43. Trussell, H. J., The relationship between image restoration by the maximum *a posteriori* method and a maximum entropy method, *IEEE Trans. Acoust. Speech Signal Process.*, ASSP-28(1), 114–117, 1980.
44. Burch, S. F., Gull, S. F., and Skilling, J., Image restoration by a powerful maximum entropy method, *Comput. Vis. Graph. Image Process.*, 23, 113–128, 1983.
45. Gonsalves, R. A. and Kao, H.-M., Entropy-based algorithm for reducing artifacts in image restoration, *Opt. Eng.*, 26(7), 617–622, 1987.
46. Youla, D. C. and Webb, H., Image restoration by the method of convex projections: Part 1—Theory, *IEEE Trans. Med. Imaging*, MI-1, 81–94, 1982.
47. Sezan, M. I., An overview of convex projections theory and its applications to image recovery problems, *Ultramicroscopy*, 40, 55–67, 1992.
48. Combettes, P. L., The foundations of set-theoretic estimation, *Proc. IEEE*, 81(2), 182–208, 1993.
49. Trussell, H. J. and Civanlar, M. R., Feasible solution in signal restoration, *IEEE Trans. Acoust. Speech Signal Process.*, ASSP-32(4), 201–212, 1984.
50. Youla, D. C., Generalized image restoration by the method of alternating orthogonal projections, *IEEE Trans. Circuits Syst.*, CAS-25(9), 694–702, 1978.
51. Youla, D. C. and Velasco, V., Extensions of a result on the synthesis of signals in the presence of inconsistent constraints, *IEEE Trans. Circuits Syst.*, CAS-33(4), 465–467, 1986.
52. Stark, H., Ed., *Image Recovery: Theory and Application*, Academic, Orlando, FL, 1987.
53. Civanlar, M. R. and Trussell, H. J., Digital image restoration using fuzzy sets, *IEEE Trans. Acoust. Speech Signal Process.*, ASSP-34(8), 919–936, 1986.
54. Tekalp, A. M. and Pavlović, G., Image restoration with multiplicative noise: Incorporating the sensor nonlinearity, *IEEE Trans. Signal Process.*, SP-39, 2132–2136, 1991.
55. Tekalp, A. M. and Pavlović, G., Digital restoration of images scanned from photographic paper, *J. Electron. Imaging*, 2, 19–27, 1993.
56. Slepian, D., Linear least squares filtering of distorted images, *J. Opt. Soc. Am.*, 57(7), 918–922, 1967.
57. Ward, R. K. and Saleh, B. E. A., Deblurring random blur, *IEEE Trans. Acoust. Speech Signal Process.*, ASSP-35(10), 1494–1498, 1987.

58. Quan, L. and Ward, R. K., Restoration of randomly blurred images by the Wiener filter, *IEEE Trans. Acoust. Speech Signal Process.*, ASSP-37(4), 589–592, 1989.
59. Combettes, P. L. and Trussell, H. J., Methods for digital restoration of signals degraded by a stochastic impulse response, *IEEE Trans. Acoust. Speech Signal Process.*, ASSP-37(3), 393–401, 1989.
60. Mesarovic, V. Z., Galatsanos, N. P., and Katsaggelos, A. K., Regularized constrained total least squares image restoration, *IEEE Trans. Image Process.*, 4(8), 1096–1108, 1995.
61. Rajala, S. A. and de Figueiredo, R. P., Adaptive nonlinear image restoration by a modified Kalman filtering approach, *IEEE Trans. Acoust. Speech Signal Process.*, ASSP-29(5), 1033–1042, 1981.
62. Tekalp, A. M., Kaufman, H., and Woods, J., Edge-adaptive Kalman filtering for image restoration with ringing suppression, *IEEE Trans. Acoust. Speech Signal Process.*, ASSP-37(6), 892–899, 1989.
63. Lagendijk, R. L., Biemond, J., and Boekee, D. E., Identification and restoration of noisy blurred images using the expectation-maximization algorithm, *IEEE Trans. Acoust. Speech Signal Process.*, ASSP-38, 1180–1191, 1990.
64. Chen, C. T., Sezan, M. I., and Tekalp, A. M., Effects of constraints, initialization, and finite-word length in blind deblurring of images by convex projections, *Proceedings of the IEEE ICASSP'87*, Dallas, TX, pp. 1201–1204, 1987.
65. Hunt, B. R. and Kubler, O., Karhunen-Loeve multispectral image restoration, Part I: Theory, *IEEE Trans. Acoust. Speech Signal Process.*, ASSP-32(6), 592–599, 1984.
66. Galatsanos, N. P. and Chin, R. T., Digital restoration of multi-channel images, *IEEE Trans. Acoust. Speech Signal Process.*, ASSP-37(3), 415–421, 1989.
67. Tekalp, A. M. and Pavlović, G., Multichannel image modeling and Kalman filtering for multispectral image restoration, *Signal Process.*, 19, 221–232, 1990.
68. Ozkan, M. K., Erdem, A. T., Sezan, M. I., and Tekalp, A. M., Efficient multiframe Wiener restoration of blurred and noisy image sequences, *IEEE Trans. Image Process.*, 1(4), 453–476, 1992.
69. Sezan, M. I. and Trussell, H. J., Use of *a priori* knowledge in multispectral image restoration, *Proceedings of the IEEE ICASSP'89*, Glasgow, Scotland, pp. 1429–1432, 1989.
70. Ohyama, N., Yachida, M., Badique, E., Tsujiuchi, J., and Honda, T., Least-squares filter for color image restoration, *J. Opt. Soc. Am.*, 5, 19–24, 1988.
71. Galatsanos, N. P., Katsaggelos, A. K., Chin, R. T., and Hillery, A. D., Least squares restoration of multichannel images, *IEEE Trans. Signal Process.*, SP-39(10), 2222–2236, 1991.
72. Ozkan, M. K., Tekalp, A. M., and Sezan, M. I., POCS-based restoration of space-varying blurred images, *IEEE Trans. Image Process.*, 3(3), 450–454, 1994.
73. Trussell, H. J. and Hunt, B. R., Image restoration of space-variant blurs by sectioned methods, *IEEE Trans. Acoust. Speech Signal Process.*, ASSP-26(6) 608–609, 1978.
74. Robbins, G. M. and Huang, T. S., Inverse filtering for linear shift-variant imaging systems, *Proc. IEEE*, 60(7), 862–872, 1972.
75. Sawchuck, A. A., Space-variant image restoration by coordinate transformations, *J. Opt. Soc. Am.*, 64(2), 138–144, 1974.
76. Angel, E. S. and Jain, A. K., Restoration of images degraded by spatially varying point spread functions by a conjugate gradient method, *Appl. Opt.*, 17, 2186–2190, 1978.
77. Wang, Y. and Mitra, S. K., Motion/pattern adaptive interpolation of interlaced video sequences, *Proceedings of the IEEE ICASSP'91*, Toronto, Ontario, Canada, pp. 2829–2832, 1991.
78. Schultz, R. R. and Stevenson, R. L., A Bayesian approach to image expansion for improved definition, *IEEE Trans. Image Process.*, 3(3), 233–242, 1994.
79. Komatsu, T., Igarashi, T., Aizawa, K., and Saito, T., Very high-resolution imaging scheme with multiple different aperture cameras, *Signal Process. Image Commun.*, 5, 511–526, 1993.
80. Ur, H. and Gross, D., Improved resolution from subpixel shifted pictures, *CVGIP: Graph. Models Image Process.*, 54(3), 181–186, 1992.

81. Tsai, R. Y. and Huang, T. S., Multiframe image restoration and registration, in *Advances in Computer Vision and Image Processing*, vol. 1, Huang, T. S. Ed., Jai Press, Greenwich, CT, 1984, pp. 317–339.
82. Kim, S. P., Bose, N. K., and Valenzuela, H. M., Recursive reconstruction of high-resolution image from noisy undersampled frames, *IEEE Trans. Acoust., Speech Signal Process.*, ASSP-38(6), 1013–1027, 1990.
83. Kim, S. P. and Su, W.-Y., Recursive high-resolution reconstruction of blurred multiframe images, *IEEE Trans. Image Process.*, 2(4), 534–539, 1993.
84. Irani, M. and Peleg, S., Improving resolution by image registration, *CVGIP: Graph. Models Image Process.*, 53, 231–239, 1991.
85. Irani, M. and Peleg, S., Motion analysis for image enhancement: Resolution, occlusion and transparency, *J. Vis. Comm. Image Rep.*, 4, 324–335, 1993.
86. Stark, H. and Oskoui, P., High-resolution image recovery from image plane arrays using convex projections, *J. Opt. Soc. Am., A* 6, 1715–1726, 1989.
87. Patti, A., Sezan, M. I., and Tekalp, A. M., Superresolution video reconstruction with arbitrary sampling lattices and nonzero aperture time, *IEEE Trans. Image Process.*, 6(8), 1064–1076, 1997.
88. Schultz, R. R. and Stevenson, R. L., Extraction of high-resolution frames from video sequences, *IEEE Trans. Image Process.*, 5(6), 996–1011, 1996.
89. Altunbasak, Y., Patti, A. J., and Mersereau, R. M., Super-resolution still and video reconstruction from MPEG coded video, *IEEE Trans. Circuits Syst. Video Technol.*, 12(4), 217–227, 2002.
90. Gunturk, B., Batur, A. U., Altunbasak, Y., Hayes III, M. H., and Mersereau, R. M., Eigenface-domain super-resolution for face recognition, *IEEE Trans. Image Process.*, 12(5), 597–607, 2003.

16

Video Scanning Format Conversion and Motion Estimation

Gerard de Haan
Philips Research Laboratories

and

Eindhoven University of Technology

Ralph Braspenning
Philips Research Laboratories

Video standards have been designed in the past to strike a particular compromise between quality, cost, transmission, or storage capacity, and compatibility with other standards. The three main formats that were in use until roughly 1980 apply picture rates of 50 or 60 Hz for interlaced video, and 24, 25, or 30 Hz for film material. Conversion between these formats was required for movie broadcast and international programme exchange.

With the arrival of videoconferencing, high-definition television (HDTV), workstations, and personal computers (PCs), many new video formats have appeared. As indicated in Figure 16.1, these include low-end formats such as common intermediate format (CIF) and quarter CIF (QCIF) with smaller picture size and lower picture rates, progressive and interlaced HDTV formats at 50 and 60 Hz, and other video formats used in multimedia PCs, while high-end television displays appeared on the market with high picture rates, 75–240 Hz, to prevent flicker on cathode ray tube (CRT)-displays [11], or to improve the motion portrayal on liquid crystal displays (LCDs) [18]. Despite many attempts to globally standardize video formats, the above observations suggest that video format conversion (VFC) is of a growing importance.

This chapter is organized as follows. In Section 16.1, we shall discuss motion estimation. Next, we will elaborate on de-interlacing in Section 16.2. Both motion estimation and de-interlacing are crucial parts in the more advanced picture-rate converters. Finally, picture-rate conversion methods are discussed in Section 16.3.

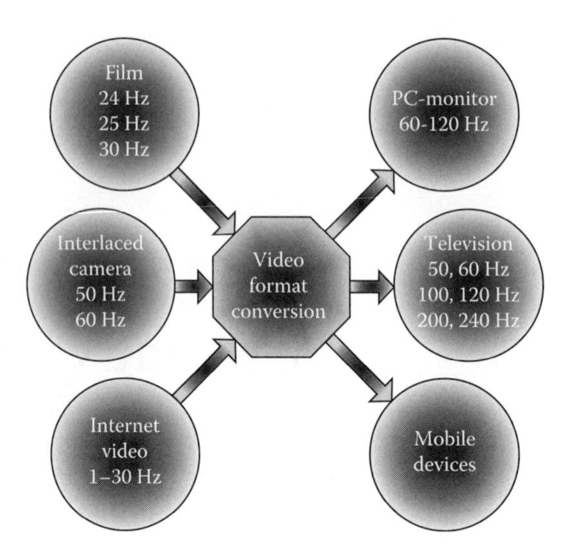

FIGURE 16.1 The wish to show all video material on all displays leads to a growing demand for VFC. Picture-rate conversion is an important element.

16.1 Motion Estimation

Motion estimation (ME) is a powerful means to improve a variety of video processing systems. Many methods have been developed during roughly the last 30 years for different applications, such as motion-compensated (MC) filtering for noise reduction, MC prediction for coding and MC interpolation for video format conversion (MC-VFC, see Section 16.3).

The predictive coding is the least demanding area, and the first that led to widespread use of ME, as it is sufficient for coding that MC decreases the average prediction error. MC-VFC is probably the most demanding application, as it requires estimation of a large vector range picture-rate conversion, and with a subpixel accuracy for de-interlacing.

This section provides an overview of the developments. We start with a short introduction of a commonly used mathematical framework based on Markov random fields (MRFs). In Section 16.1.2, we discuss full search and its efficient derivatives. To improve upon those methods, we will introduce the use of different scales or hierarchy in Section 16.1.4. Further improvements can be achieved by using more complex models, as will be described in Section 16.1.6. Finally in Section 16.1.7, we will present a method that is both computationally efficient and delivers good quality. All categories are illustrated by an example motion field.

16.1.1 Markov Random Field Framework

MRFs are a popular mathematical framework for formulating image and video processing problems, such as ME and image restoration. MRF models provide a framework to better reason about the problem at hand, to come to an explicit model containing all the assumptions made.

MRF models assume that the state of a pixel only depends on a small neighborhood of pixels. Therefore, the probability function for that pixel can be modeled using only information regarding neighboring pixels. The probabilities are often modeled using exponential probability density functions (pdf). Using log likelihoods, the maximum a posteriori probability (MAP) estimation, then becomes an energy minimization task, where the exponents of the likelihoods are the energy functions. Hence, a solution for a specific image processing problem becomes a (MRF) model and an energy minimization algorithm. In the subsequent discussion, we will use the MRF framework to describe and evaluate different ME techniques. For the VFC application, it is of vital importance to obtain a motion field which is close to the true motion in the scene. We will therefore judge the ME techniques according to this requirement.

As mentioned above we want to obtain the motion field that has the highest (a posteriori) probability. The observations that are available are the images, denoted by F. Hence, we are interested in the probability $P(\vec{D}|F_n, F_{n-1})$, where $\vec{D}$ is the motion field and F_n is short for $F(\vec{x}, n)$. The spatial position is indicated by $\vec{x}$ and the frame number by n. Using Bayes' rule, this probability can be rewritten in

$$P(\vec{D}|F_n, F_{n-1}) = \frac{P(F_n|\vec{D}, F_{n-1})P(\vec{D}|F_{n-1})}{P(F_n|F_{n-1})} \tag{16.1}$$

Since $P(F_n|F_{n-1})$ does not depend on the motion field, the MAP estimate for the motion field becomes

$$\vec{D}_{\text{MAP}} = \arg\max_{\vec{D}} P(F_n|\vec{D}, F_{n-1})P(\vec{D}|F_{n-1}) \tag{16.2}$$

Hence, two distinct probabilities can be recognized. The first one, $P(F_n|\vec{D}, F_{n-1})$, models how well a given motion field, in combination with the previous image, explains the observations in the current image. This is often referred to as the likelihood model. A basic assumption is the brightness constancy assumption: the brightness of a point (pixel) on an object does not change along the motion trajectory of that point, i.e.,

$$F(\vec{x}, n) = F(\vec{x} - \vec{D}(\vec{x}, n), n - 1) \tag{16.3}$$

This leads to a model where the probability falls of rapidly with increasing difference in intensity along the motion vector. If we assume a Gaussian pdf, then this boils down to

$$P(F_n(\vec{x})|\vec{D}, F_{n-1}) = \frac{1}{\sqrt{2\pi}\sigma} \exp\left(-\frac{\sum_{\vec{x}_i \in N(\vec{x})}(F_n(\vec{x}_i) - F_{n-1}(\vec{x}_i - \vec{D}))^2}{2\sigma^2}\right) \tag{16.4}$$

where $N(\vec{x})$ is a neighborhood of $\vec{x}$.

The second probability, $P(\vec{D}|F_{n-1})$, expresses the prior knowledge regarding the motion field, and is often referred to as the prior model. Basically any available knowledge or constraints can be used. We expect multiple objects in the scene and the motion of these objects to be continuous within the object and in time; so we expect a piecewise smooth motion field. A common model assumes an exponential distribution, e.g.,

$$P(\vec{D}|F_{n-1}) = \lambda \exp\left(-\lambda \sum_{\vec{x}_i \in N(\vec{x})} ||\vec{D}(\vec{x}) - \vec{D}(\vec{x}_i)||\right) \tag{16.5}$$

Both pdfs from Equations 16.4 and 16.5 involve exponentials. Therefore, in practice, log likelihoods are used. This means that the maximization from Equation 16.2 becomes a minimization of the sum of the exponents of the pdfs:

$$\vec{D}_{\text{MIN}} = \arg\min_{\vec{D}} \lambda_1 E_{\text{data}}(\vec{x}, \vec{D}, n) + \lambda_2 E_{\text{sm}}(\vec{x}, \vec{D}, n) \tag{16.6}$$

where

E_{data} is the exponent of the likelihood or data model

E_{sm} is the exponent of the prior model

These functions are often referred to as energy functions. λ_1 and λ_2 are constants that, in principle, depend on the exact pdf. But because of the minimization, Equation 16.6 can be divided by λ_1, such that there is only one parameter left. In practice, this parameter has to be experimentally determined.

Following Equation 16.4, the E_{data} would become a mean-squared error. However, in practice, a sum of absolute differences (SAD) is most widely used, because it delivers almost equal results, while it has a great computational advantage.

$$E_{\text{data}}(\vec{x}, \vec{D}, n) = \sum_{\vec{x}_i \in B(\vec{x})} |F(\vec{x}_i, n) - F(\vec{x}_i - \vec{D}, n-1)| \tag{16.7}$$

where $B(\vec{x})$ is a block of pixels around $\vec{x}$. To save computational costs, the minimization is often only performed per block instead of for all pixels. This means that Equation 16.7 is only evaluated for block positions, which we will indicate with $\vec{X}$, instead of pixel locations $\vec{x}$. Following Equation 16.5, the other energy function, E_{sm}, becomes

$$E_{\text{sm}}(\vec{x}, \vec{D}, n) = \sum_{\vec{x}_i \in N(\vec{x})} \|\vec{D}(\vec{x}) - \vec{D}(\vec{x}_i)\| \tag{16.8}$$

We will now discuss several minimization techniques. We will start by only considering a simple model consisting of only the data energy function (E_{data}) first.

16.1.2 Full Search and Efficient Derivatives

The most straightforward method for minimizing Equation 16.7 is to calculate the energy for all possible motion vectors within a search range and then select the motion vector that gives the lowest energy. Define CS_{fs} as the set of all motion vectors within a search range of $(2N+1) \times (2M+1)$ pixels, i.e.,

$$CS_{\text{fs}} = \{\vec{v} \mid -N \leq v_x \leq N, -M \leq v_y \leq M\} \tag{16.9}$$

The assigned motion vector for a block position $\vec{X}$ then becomes

$$\vec{D}(\vec{X}, n) = \arg\min_{\vec{v} \in CS_{\text{fs}}} E_{\text{data}}(\vec{X}, \vec{v}, n) \tag{16.10}$$

Since this method considers all possible motion vectors within a search range, it is often referred to as full search or exhaustive search. As one can see immediately, this minimization requires $(2N+1)$ $(2M+1)$ times the computation of the SAD, which is a costly operation. Therefore, researchers have been looking into more efficient methods for the minimization. Especially for video coding purposes, this has lead to numerous efficient search strategies, i.e., subsampled full search, N-step search or logarithmic search, one-at-a-time search, diamond search, etc. We will briefly discuss the first two here, but all these methods have the same objective, namely approaching the minimum from Equation 16.10 with as few operations as possible.

Also the approach is the same for all these methods, namely to reduce the number of motion vectors to be tested, i.e., taking a subset of the set from Equation 16.9. The "subsampled full search" starts by considering only half of the possible motion vectors

$$CS_1 = \{\vec{v} \in CS_{\text{fs}} | (v_x + v_y) \bmod 2 = 0\} \tag{16.11}$$

Figure 16.2 illustrates this option, further showing the candidate vectors in the second step of the algorithm

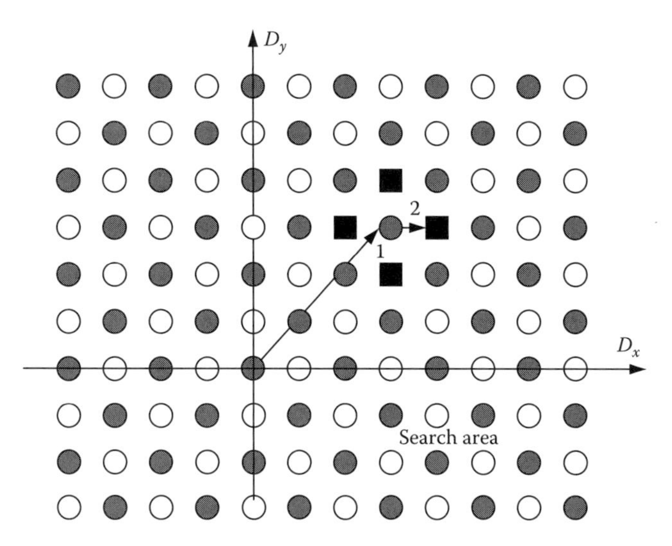

FIGURE 16.2 Candidate vectors tested in the second step (black squares), around the best matching vector in the first step, $\vec{D}_1(\vec{X}, n)$, for subsampled full search block-matching.

$$CS_2 = \{\vec{D}_1, \vec{D}_1 + \vec{u}_x, \vec{D}_1 + \vec{u}_y, \vec{D}_1 - \vec{u}_x, \vec{D}_1 - \vec{u}_y\} \tag{16.12}$$

which is centered around the result vector from the first step $\vec{D}_1$. The vectors $\vec{u}_x$ and $\vec{u}_y$ are unit vectors in the horizontal and vertical direction, respectively.

Another idea is to use a coarse to fine strategy. This is used by the "*N*-step search" or "logarithmic search algorithms." As illustrated in Figure 16.3, the first step, of a three-step block-matcher [30], performs a search on a coarse grid consisting of only nine candidate vectors in the entire search area. The second step includes a finer search, with eight candidate vectors around the best matching vector of the first step, and finally in the third step, a search on the full resolution grid is performed, with another

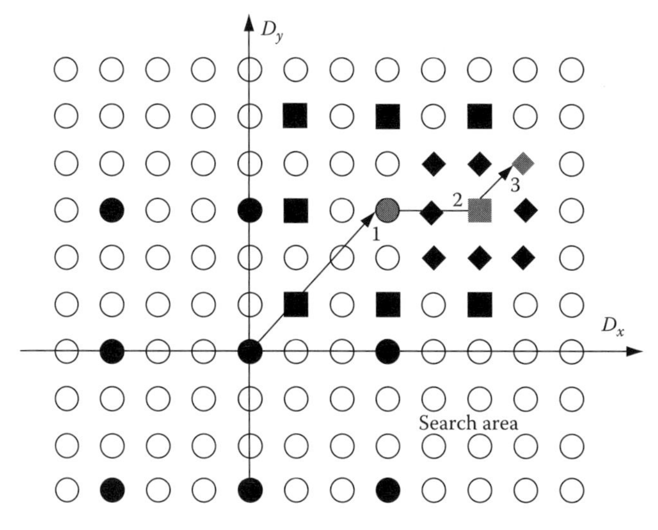

FIGURE 16.3 Illustration of the three-step search block-matching algorithm. The candidates of successive search steps are on increasing grid densities.

eight candidates around the best vector of the second step. Formally, the candidate set from which the best vector is selected is adapted in the following consecutive steps:

$$CS_i(\vec{X}, n) = \{\vec{v} \in CS_{fs} | \vec{v} = \vec{D}_{i-1}(\vec{X}, n) + \vec{U}, U_x = \{0, \pm(4 - i)\}, U_y = \{0, \pm(4 - i)\}\} \quad (16.13)$$

where $\vec{D}_{i-1}$ is the result from the previous iteration, or equals the zero vector initially, while the iteration index i varies from 1 to 3 in the original algorithm of Koga [30].

Note that a search range of ± 6 pixels is assumed. Other search areas require modifications, either resulting in less accurate vectors, or in more consecutive steps. Generalizations to N-steps block-matching are obvious. Related is the "2-D logarithmic search," or "cross search," method [37]:

$$CS_i(\vec{X}, n) = \left\{\vec{v} \in CS_{fs} | \vec{v} = \vec{D}_{i-1}(\vec{X}, n) + \vec{U}, U_x = \{0, \pm 2^{N-i}\}, U_y = \{0, \pm 2^{N-i}\}\right\} \quad (16.14)$$

where i varies from 1 to N.

Surprisingly enough neither full search nor the efficient variants is the ultimate solution, because true ME is an ill-posed problem [45]. Therefore, it suffers from three problems, respectively called the aperture problem, sensitivity to noise, and the occlusion problem. The first two problems will be discussed in this section, while occlusion will be briefly discussed in Section 16.3.5.

The aperture problem states that the motion can only be measured orthogonal to the local spatial image gradient [45]. In practice, for instance this means that locally for a vertical edge it is impossible to determine the vertical motion, since a lot of different values for the vertical component of the motion vector will yield very similar SAD values. Even more problematic are areas with uniform luminance values, since here both components of the motion vector are unreliable. Hence, to overcome this problem, reliable information regarding the motion vectors must come from looking at a scale that contains sufficient gradient information or from propagating reliable information to neighboring sites.

The aperture problem is illustrated in Figure 16.4. It shows three SAD profiles for three different places in the image in the top left of Figure 16.4. The locations are indicated with black squares. The SAD profiles show the SAD value (in gray values) as a function of the tested motion vector for a specific location. For the bottom left profile, it is clear that multiple motion vectors all lead to low SAD values (black areas) and that there is an uncertainty in the motion vector along the edge. For homogeneous areas, the situation is even worse. For those areas, there is an uncertainty in motion vector in all directions. The bottom right profile shows the aperture problem for a repeating texture. Here also multiple minima exist that repeat according to the repetition pattern of the texture. The global minimum is indicated with the crosshair combined with its position. This illustrates the other problem with ME, which is the sensitivity to noise. Because of the noise in the images, one local minimum will be slightly lower or higher than another. Therefore, there is no guarantee that the global minimum is the correct minimum, i.e., the minimum that corresponds to the true motion vector. In areas with sufficient texture, the minimum is well defined and it does correspond to the true motion vector. This is illustrated in the top right of Figure 16.4.

Concluding, searching for the global minimum within a search range, as is the objective of full search, will not result in the true motion field. Neither will the efficient search strategies, because they have the same objective, although these latter methods are more likely to get stuck in a local minimum, instead of finding the global minimum. An example of a motion field obtained with full search is shown in the middle of Figure 16.5. It shows per block in gray values, the horizontal component of the estimated motion field. One can clearly see the noisiness of the estimated motion field, which clearly deviates significantly from the true motion field.

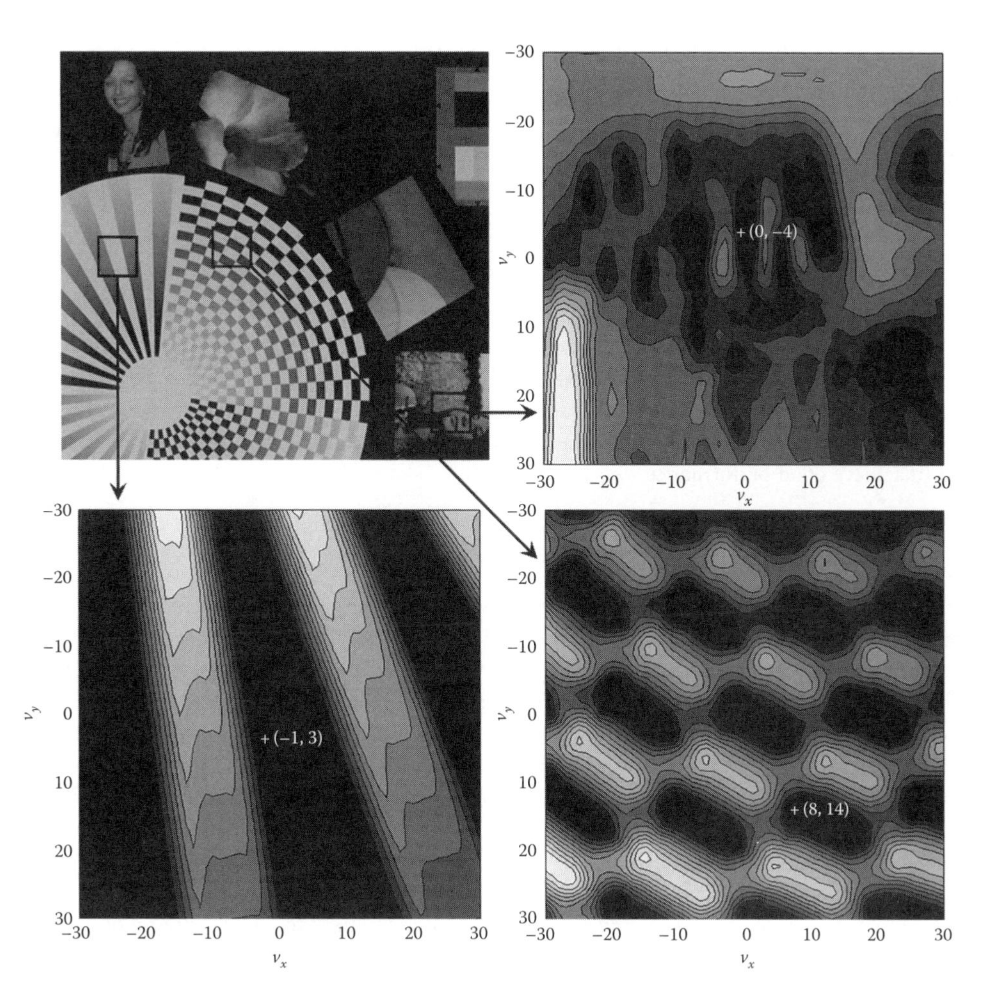

FIGURE 16.4 The SAD value of an 8 × 8 block as function of the motion vector for three different locations in the wheel sequence. A darker value means a lower SAD value. The global minimum (over the search range) is indicated with the crosshair.

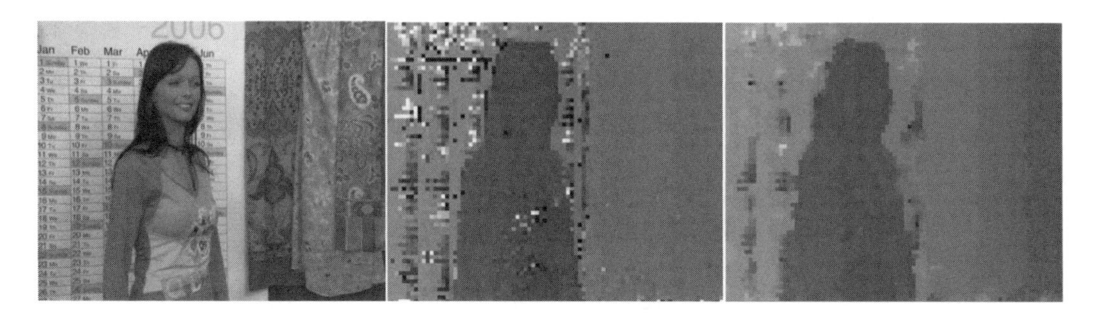

FIGURE 16.5 The horizontal component of the motion field estimated with full search for the Yvonne sequence (middle) and the result after postprocessing that motion field with a 3 × 3 vector median (right).

16.1.3 Improvements through Postprocessing

To improve the correlation with the true motion, postprocessing can be applied to reduce the noisiness. Low-pass filtering can be applied, but nonlinear filtering is more effective in this case, since it preserves the object discontinuities better and is better suited for removing shot noise. An example of such a nonlinear filter is a median filter, i.e.,

$$\vec{D}_{\mathrm{filt}}(\vec{X}, n) = \mathrm{med}_{\vec{X}_i \in N(\vec{X})}(\vec{D}(\vec{X}_i)) \tag{16.15}$$

where $N(\vec{X})$ is a neighborhood of block locations around $\vec{X}$. An example result is shown in the right part of Figure 16.5. Here a 3×3 neighborhood was used. The median operator can either be a component-wise median or a vector median. The latter was used in the example. As can be seen from Figure 16.5, the motion field greatly improves. However, the result is still not satisfactory and other approaches are needed to achieve good performance.

16.1.4 Considering Different Scales

As mentioned in Section 16.1.2, one way to attempt to overcome the aperture problem is to look at a scale that contains sufficient gradient information. The granularity of the required scale (coarse to fine) depends on the granularity of the local texture in the image, which can be seen in Figure 16.4. Unfortunately this granularity varies within the image which complicates the choice of a fixed scale. To overcome this, one could choose to work at a coarse scale, i.e., with large blocks, such that there is always sufficient gradient information within the blocks. However, this greatly compromises the spatial accuracy of the estimated motion field. Of course numerous methods have been proposed to deal with this problem, here we will discuss two.

The first one uses a "hierarchical approach." The method first builds a resolution pyramid. At the top of the pyramid, the image is strongly prefiltered and subsampled. The bandwidth of the filter increases and the subsampling factor decreases going down in the hierarchy until the full resolution of the original image. Often factors of 2 are used. Motion vectors are estimated at one scale, starting at the top, and propagated to a lower scale as a prediction for a more accurate estimate.

Remember the candidate set for an N-step logarithmic search is defined as (see Equation 16.14)

$$CS_i(\vec{X}, n) = \left\{\vec{v} \in CS_{\mathrm{fs}} | \vec{v} = \vec{D}_{i-1}(\vec{X}, n) + \vec{U}, U_x = \{0, \pm 2^{N-i}\}, U_y = \{0, \pm 2^{N-i}\}\right\} \tag{16.16}$$

where i varies from 1 to N. A hierarchical variant of this logarithmic search would perform a regular logarithmic search at each level, however taking the estimated value of the previous level as a starting point. Normally the search starts at zero, i.e.,

$\vec{D}_0(\vec{X}, n) = 0$. For the hierarchical variant, we introduce a motion field at each level l, $\vec{D}(\vec{X}, n, l)$. This motion field is found as:

$$\vec{D}_i(\vec{X}, n, l) = \begin{cases} \vec{D}_N(\vec{X}, n, l-1) & i = 0 \\ \arg\min_{\vec{v} \in CS_i(\vec{X}, n, l)} E_{\mathrm{data}}(\vec{X}, \vec{v}, n) & i > 0 \end{cases} \tag{16.17}$$

where $CS_i(\vec{X}, n, l)$ is defined similar to Equation 16.16, but with $\vec{D}_{i-1}(\vec{X}, n)$ replaced by $\vec{D}_{i-1}(\vec{X}, n, l)$ (the motion vector of the previous iteration i at that level l). At the first level, $l = 1$, the starting point is zero, i.e., $\vec{D}_0(\vec{X}, n, 1) = 0$. The final motion field is the estimated motion field at level L, $\vec{D}_N(\vec{X}, n, L)$. Combinations with other than the logarithmic search strategy are possible, and the hierarchical method is not limited to block-matching algorithms either.

Another method called "phase plane correlation" uses a two-level hierarchy. This method was devised for television-studio applications at BBC research [46]. In the first level, on fairly large blocks (typically 64×64), a limited number of candidate vectors, usually less than 10, are generated, which are fed to the second level. Here, one of these candidate vectors is assigned as the resulting vector to a much smaller area (typically 1×1 up to 8×8 is reported) inside the large initial block. The name of the method refers to the procedure used for generating the candidate vectors in the first step. It makes use of the shift property of the Fourier transform, which states that a shift in the spatial domain corresponds to a phase shift in the frequency domain. If $G(\vec{f}, n)$ is the Fourier transform of image $F(\vec{x}, n)$, then using the shift property Equation 16.3 becomes

$$G(\vec{f}, n) = G(\vec{f}, n-1)e^{-j2\pi \vec{f}\vec{D}} \tag{16.18}$$

Now consider the cross power spectrum $\mathrm{CPS}(\vec{f}, n) = G(\vec{f}, n) \cdot G^*(\vec{f}, n-1)$. Substituting Equation 16.18 and rewriting the complex values using amplitude ($|\cdot|$) and phase ($\vec{\phi}$) leads to

$$\begin{aligned}\mathrm{CPS}(\vec{f}, n) &= G(\vec{f}, n) \cdot G^*(\vec{f}, n)e^{-j2\pi \vec{f}\vec{D}} \\ &= |G(\vec{f}, n)|e^{-j2\pi\vec{\phi}} \cdot |G(\vec{f}, n)|e^{j2\pi\vec{\phi}}e^{-j2\pi \vec{f}\vec{D}} = |G(\vec{f}, n)|^2 e^{-j2\pi \vec{f}\vec{D}}\end{aligned} \tag{16.19}$$

Hence, the phase of the cross power spectrum yields the displacement. This phase can be computed directly as

$$e^{\vec{\phi}_c} = \frac{\mathrm{CPS}(\vec{f}, n)}{|\mathrm{CPS}(\vec{f}, n)|} = \frac{G(\vec{f}, n) \cdot G^*(\vec{f}, n-1)}{|G(\vec{f}, n) \cdot G^*(\vec{f}, n-1)|} \tag{16.20}$$

The inverse Fourier transform of Equation 16.20, called the phase difference matrix, yields a delta function located at the displacement $\vec{D}$ (see Equation 16.19).

A "peak hunting" algorithm is applied to find the p largest peaks in the phase difference matrix, which correspond to the p best matching candidate vectors. Subpixel accuracy better than a tenth of a pixel can be achieved fitting a quadratic curve through the elements in this matrix [46]. The second step is a simple block-matching on the smaller subblocks, trying only the p candidate vectors resulting from the actual phase correlation step and the SAD as an error metric.

An example of a motion field obtained with both methods described above is shown in Figure 16.6. The middle parts show the horizontal component of the motion vector obtained with the hierarchical logarithmic search and the right parts show the results obtained with phase plane correlation. As one can see, these motion fields are much smoother than the one obtained with full search (see Figure 16.5) and are therefore much better suited for use in VFC. However, there are still inaccuracies present and also the smoothness could be improved.

FIGURE 16.6 The horizontal component of the motion field estimated with the hierarchical logarithmic search (middle) and phase plane correlation (right) for the Yvonne sequence.

16.1.5 Dense Motion Fields through Postprocessing

For VFC, a dense motion field is required. Block-based methods only estimate one motion vector per block of pixels. This block-based motion field must then be up-sampled to pixel level. A simple zeroth-order up-sampling (repetition) can suffice; however, this results in a misalignment of the discontinuities in the estimated motion field and the object boundaries and can lead to blocking artifacts. Higher-order linear up-sampling is not recommended, since it introduces new (interpolated) motion vectors that were not estimated. Furthermore, it removes the discontinuities in the motion field, while at moving object boundaries these discontinuities should be present. To circumvent this, a linear up-sampled motion vector could be quantized to the most similar, originally estimated, motion vector in the neighborhood.

Another approach is to use a nonlinear up-sampling method that preserves motion discontinuities. A median operator is an exquisite candidate for this nonlinear up-sampling. A proven method, called block erosion [14], uses a hierarchical median-based approach. In each step, it subdivides the current block into four subblocks and calculates a new motion vector for each subblock. For each subblock, the new motion vector is computed based on the current motion vector (of the whole block) and its neighboring blocks. This is depicted in Figure 16.7. The current block is indicated with a gray color. The three motion vectors used for each subblock are indicated using the lines with dotted endpoints. If we number the subblocks from left to right, top to bottom, then the motion vector of the subblocks become

$$\begin{aligned}\vec{D}_1 &= \text{med}(\vec{D}_c, \vec{D}_n, \vec{D}_w)\\ \vec{D}_2 &= \text{med}(\vec{D}_c, \vec{D}_n, \vec{D}_e)\\ \vec{D}_3 &= \text{med}(\vec{D}_c, \vec{D}_s, \vec{D}_w)\\ \vec{D}_4 &= \text{med}(\vec{D}_c, \vec{D}_s, \vec{D}_e)\end{aligned} \tag{16.21}$$

where

$\vec{D}_c$ is the motion vector of the center block

$\vec{D}_n, \vec{D}_s, \vec{D}_w$, and $\vec{D}_e$ are the motion vectors of the neighboring blocks north, south, west, and east, respectively, of the current block

Furthermore, the median operator can be either a component-wise median or a vector median. This process is repeated until the subblocks become pixels.

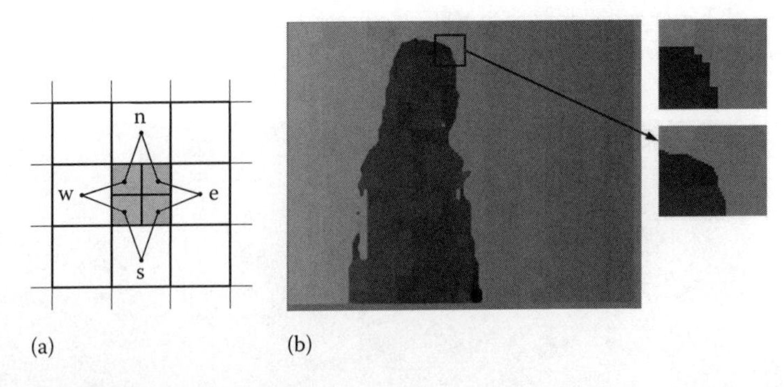

FIGURE 16.7 (a) Motion vector selection for one step of block erosion. The current block (gray) is split into four equal parts. A new vector for each part is computed based on the motion vectors obtained from the positions indicated by the dots. (b) Result of block erosion applied to the motion field obtained with phase plane correlation shown in Figure 16.6. To the right a part is zoomed in (bottom) and shown together with the original input (top).

The result of block erosion is also shown in Figure 16.7. The block-based motion field resulting from the phase plane correlation method presented in Section 16.1.4 is used as input (see Figure 16.6). As can be seen, block erosion preserves the orientation of the edges in the motion field, but slightly changes the position to come to smooth contours. The results might be improved by involving the underlying image data in the process.

Another option is to follow the multiscale approach as was discussed in Section 16.1.4 up to the pixel level. The example shown in Figure 16.6 for the hierarchical logarithmic search already used block sizes ranging from 8×8 until 2×2. The problem is however that at these very small block sizes (2×2 and 1×1) the aperture problem is huge. Therefore, reliable ME at those levels is problematic and the motion fields tend to become noisy, as can be seen in Figure 16.6 when comparing the middle and the right result. If we do want to estimate the motion for these very small blocks, but want to reduce the noisiness, we would need to go to more complex models instead of using only the SAD.

16.1.6 Going to More Complex Models

In Section 16.1.2, we have seen that minimizing only the data energy function does not lead to the correct result, because of the problems associated with ME. Therefore, additional measures are needed to produce a smooth vector field, as is attempted by the methods described in Section 16.1.4, which resulted in clear improvements.

We will now expand the model to include the prior information as was outlined in Section 16.1.1. We will use the energy functions as described there, but others have been reported as well [22]. To minimize the energy function from Equation 16.6, a "simple" full search cannot be used anymore and therefore even more complex minimization algorithms are required. To make matters worse, it has been shown that this minimization for MRF models is a NP-hard problem. However, some good approximation techniques exist, for instance "Graph Cuts" [31] and "Belief Propagation" [22].

The effect of adding a smoothness term like the one in Equation 16.8 to the energy function is that in areas where several vectors yield similar SAD values the smoothness term will dominate the energy function. However, in an area with a distinct minimum in the SAD for a particular motion vector, the data term will dominate over the smoothness term. Hence, from locations where a proper motion estimate is possible (areas with sufficient gradient information), motion vectors are propagated to areas with less reliable information. This is another strategy for coping with the aperture problem, as was mentioned in Section 16.1.2. Some methods use the strategy of communicating information regarding the motion vectors to neighboring sites explicitly for minimizing the energy function, as is the case for belief propagation and 3-D recursive search (3DRS) which will be explained later.

The graph cuts and belief propagation methods are too complex to explain here in detail, but the interested reader is referred to [31] and [22], respectively. Both methods compute a new state for each site, where this state consists of the likelihood for each possible motion vector. As can be seen in Equation 16.6, this likelihood depends on the states of the neighboring sites as well. After this new state has been computed for all sites, the states for the neighboring sites have changed, and the process is repeated until convergence is reached, i.e., the states of the sites change less than a threshold value. Hence, these methods are global iterative optimization methods. As mentioned they require several iterations, typically in the order of ten. Furthermore they consider all possible motion vectors within a fixed search range. For broadcast resolutions (720×576 and higher) and associated sensible search ranges ($\pm 24 \times \pm 12$ and larger), this becomes a massive computational undertaking. Therefore, today these methods are far too complex to be of use for real-time VFC systems, although they have shown to deliver good results. A resulting motion field from graph cuts and belief propagation is shown in Figure 16.8.*

* The code for graph cuts was obtained online from the authors of Ref. [31] and the code for belief propagation is based on code from the authors of Ref. [22] but extended for motion estimation.

FIGURE 16.8 The horizontal component of the motion field estimated with Graph Cuts (middle) and Belief Propagation (right) for the Yvonne sequence.

16.1.7 3-D Recursive Search

Now we will discuss an algorithm that is suited for use in VFC both from a complexity and a quality point of view. It is called 3-D recursive search (3DRS) and has been successfully implemented in ICs [11] and brought to the market. It follows a similar strategy as discussed in the previous section, where reliable information is propagated to areas with less reliable information. But a major difference is that 3DRS does not test all possible motion vectors within a search range. Instead, it constructs a candidate set with a limited number of possible motion vectors. This greatly reduces the computational cost and promotes smoothness. Although the original algorithm [13] has been mainly applied to ME, it can be regarded as a more general minimization technique, and we will explain it first as such. Following that, the details of each part will be discussed one by one.

Say the unknown quantity to be estimated is denoted by $\vec{q}$ (for ME $\vec{q}$ equals the motion field $\vec{D}$).

1. For each location (pixel or block) $\vec{x}$, construct a candidate set C, e.g.,

$$C = \left\{ \begin{array}{c} \vec{q}(\vec{x} + k \cdot \vec{u}_x - \vec{u}_y, n), \\ \vec{q}(\vec{x} - \vec{u}_x, n), \vec{q}(\vec{x}, n-1), \vec{q}(\vec{x} + \vec{u}_x, n-1), \\ \vec{q}(\vec{x} + l \cdot \vec{u}_x + \vec{u}_y,\ n-1), \\ \vec{q}(\vec{x} - \vec{u}_x, n) + \underline{\vec{\eta}}, \vec{q}(\vec{x} - \vec{u}_y, n) + \underline{\vec{\eta}} \end{array} \right\} \quad k = -1, 0, 1, \quad l = -1, 0, 1 \qquad (16.22)$$

   where $\underline{\vec{\eta}}$ is a random value. Usually this random value is drawn from a fixed update set [13].

2. The estimated value for $\vec{q}(\vec{x})$ then is

$$\vec{q}(\vec{x}) = \arg\min_{\vec{q}_c \in C} \left(E_{\text{tot}}(\vec{x}, \vec{q}_c) + E_{\text{p}}(\vec{q}_c^{\,*})\right) \qquad (16.23)$$

   where E_{p} is the penalty depending on the type of the candidate $\vec{q}_c$ which is denoted by $\vec{q}_c^{\,*}$.

Important is that steps 1 and 2 are performed sequentially for each location. Hence, the newly estimated value is assigned to the location before moving to the next location. Therefore, this new value becomes part of the candidate set of the next location, directly influencing the estimate for that next location, in contrast to e.g., belief propagation where this happens after a complete iteration over all locations.

Since we are expecting a solution that is spatiotemporally smooth (see Section 16.1.6), the underlying idea of 3DRS is that already estimated neighboring values are good predictions for the current value to be estimated. These neighboring values are called "spatial" candidates ($\vec{q}(\cdot,\ n)$ in Equation 16.22). Unfortunately, because of the sequential nature of the algorithm, there is a causality problem. Not all neighboring values have already been estimated. However, previous estimates both in time and iteration are also good predictions, but less reliable then spatial candidates, because of the motion of the objects and the change in motion. These predictions from previous estimates are called "temporal" candidates ($\vec{q}(\cdot, n-1)$ in Equation 16.22). The reliability of the different types of predictors is taken into account by the penalty in Equation 16.23.

Which neighboring locations already have been estimated in the current pass and which have not depends on the order in which the locations are processed. This is dependent on the scanning direction. Figure 16.9 shows the configuration of spatial and temporal candidates when processing line by line from top to bottom and location by location from left to right. This is the scanning order used in Equation 16.22. All locations can be processed in this order. However, this means that good motion vector estimates can only propagate in one direction, namely along the scanning direction. If a good estimate is found at the bottom of the image, it can take a long time before it is propagated to the top of the image. Therefore, to improve the propagation of good estimates, it is beneficial to change the scanning direction at certain times. In practice, two mechanisms are used for this. The first one is quite obvious. After a scan from top to bottom, the next scan over all locations is run from bottom to top. This is alternated continuously. The next mechanism is called meandering and that means that after a line is scanned from left to right, the next line is scanned from right to left. If both mechanisms are used, good estimates can propagate in four different directions, which ensure a quick spreading of good estimates all over the image. The example candidate set from Equation 16.22 is correct for scanning from top to bottom and from left to right, i.e., along the unit vectors defining the axes of the image. If the scanning direction is changed, the relative position of the spatial candidates changes accordingly. Hence, the unit vectors $\vec{u}_x$ and $\vec{u}_y$ in Equation 16.22 should be changed in unit vectors defining the current scanning direction, $\vec{s}_x$ and $\vec{s}_y$.

Another important aspect in the construction of the candidate set is the "update" candidates ($\vec{q}(\cdot, n) + \underline{\vec{\eta}}$ in Equation 16.22). Both spatial and temporal candidates contain values that already have been estimated. But of course new values need to be introduced as well to find new motion vectors (for appearing objects) and accommodate for changes in the motion of objects. These new values are introduced by adding small random values to existing values (spatial candidates). These random values can be drawn from a random distribution, e.g., a Gaussian distribution ($\mathcal{N}$). But typically these random values are taken sequentially from a fixed set (US) [13].

$$\underline{\vec{\eta}} \sim \mathcal{N}(0, \sigma) \quad \text{or} \quad \underline{\vec{\eta}} \in US \tag{16.24}$$

These values are small to promote smoothness ($\sigma \leq 2$). Furthermore, they can be small because of the assumption that objects have inertia. Hence, the motion of objects will change gradually from image to

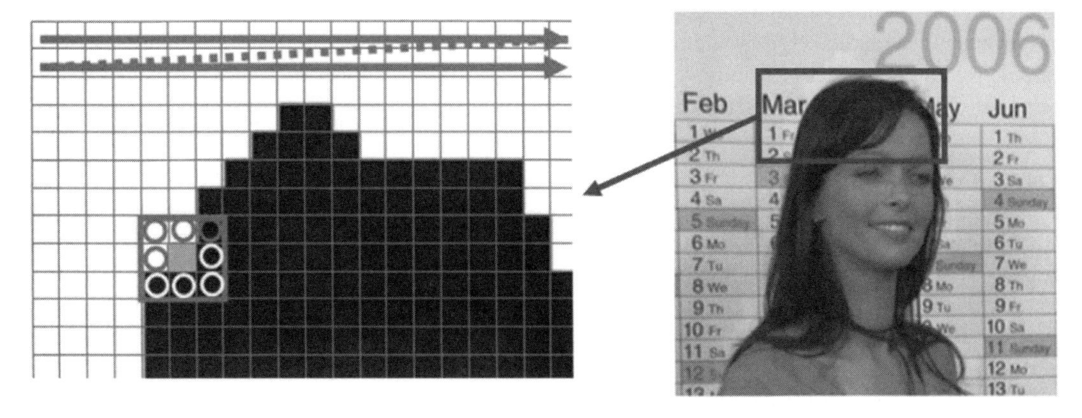

FIGURE 16.9 Configuration of the spatial and temporal candidates for the scanning direction indicated by the gray arrows in the block grid. The light gray block is the current block. The spatial candidates are indicated with the gray circles, the temporal candidates with the white circles (see Equation 16.22).

FIGURE 16.10 The horizontal component of the motion field estimated with 3DRS for the Yvonne sequence.

image. To find a new motion vector that differs significantly from the previously estimated vectors takes several consecutive updates. This process is called convergence. Since updated values have not been found to be good estimates before, they are less reliable predictors than both spatial and temporal candidates.

3DRS finds a spatiotemporal smooth estimate of the motion field that is a local minimum in the energy function. Typically a simple model is used, hence, only a data term (SAD) is used. This is sufficient to produce good results. But more complex models like the one in Equation 16.6 (including a smoothness term) can also be used. When using only the SAD as energy function the penalties can be related to the SAD values. Typically the penalty for spatial candidates is fixed to zero. Research has shown that ½ per pixel (i.e., $E_p(\vec{q}_c^{\,*}) = 32$ for an 8×8 block) as penalty for temporal candidates and 2 per pixel for update candidates (i.e., $E_p(\vec{q}_c^{\,*}) = 128$) produce good results. An example motion field for the Yvonne sequence is shown in Figure 16.10. These results were obtained by applying block erosion (see Section 16.1.5) to the motion field estimated with 3DRS using blocks of 8×8 pixels.

The evaluation of the energy function is typically the most expensive part of an ME algorithm. In the case of 3DRS, the energy function only needs to be evaluated for each candidate. Therefore, the complexity of 3DRS is linear in the number of candidates as opposed to the complexity of many other methods (like the ones presented in Sections 16.1.2 and 16.1.6) that is linear in the size of the search range. The size of the candidate set can be tuned to achieve good quality with a minimum number of candidates. The candidate set from Equation 16.22 contains 11 candidates. However, because of the smoothness of the motion field, often neighboring candidate locations result in the same prediction. Therefore, the number of candidates can be reduced to six without a loss in quality. Further quality improvements can be achieved by adding candidate vectors from "outside" sources, for instance, a global parametric motion model or feature correspondences. The limited number of candidates makes 3DRS very well suited for real-time implementation both in hardware [11] and in software [3].

16.2 De-Interlacing

The human visual system is less sensitive to flickering details than to large-area flicker [21]. Television displays apply interlacing to profit from this fact, while broadcast formats were originally defined to match the display scanning format. As a consequence, interlace is found throughout the video chain. If we describe interlacing as a form of spatiotemporal subsampling, then de-interlacing, the topic of this section, is the reverse operation aiming at the removal of the subsampling artifacts.

Figure 16.11 illustrates the de-interlacing task. The input video fields, containing samples of either the odd or the even vertical grid positions (lines) of an image, have to be converted to frames. These frames represent the same image as the corresponding input field but contain the samples of all lines. Formally, we shall define the output frame $F_o(\vec{x}, n)$ as:

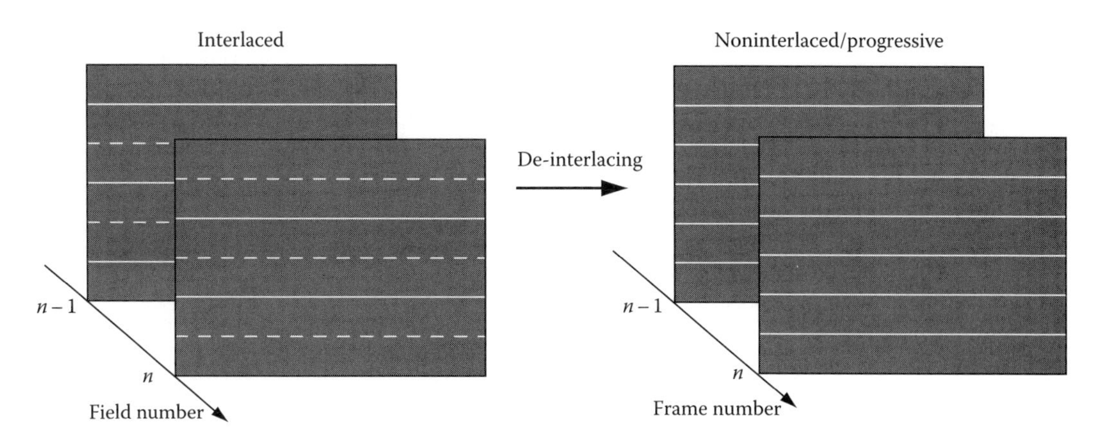

FIGURE 16.11 The de-interlacing task.

$$F_o(\vec{x}, n) = \begin{cases} F(\vec{x}, n), & ((y + n) \bmod 2 = 0) \\ F_i(\vec{x}, n), & \text{(otherwise)} \end{cases} \tag{16.25}$$

with $\vec{x} = \begin{pmatrix} x \\ y \end{pmatrix}$ designating the spatial position, field number n, $F(\vec{x}, n)$ the input field defined for $(y + n)$ mod $2 = 0$ only, and $F_i(\vec{x}, n)$ the interpolated pixels.

De-interlacing doubles the vertical-temporal (VT) sampling density, and aims at removing the first repeat spectrum caused by the interlaced sampling of the video (Figure 16.12). It is not, however, a straightforward linear sampling rate up-conversion problem [48], as television signals do not fulfill the demands of the sampling theorem: the prefiltering prior to sampling, required to suppress frequencies outside the chosen unit cell of the reciprocal sampling lattice, is missing. As in a television system the pickup device in the camera samples the scene (vertically and temporally), the prefilter should be in the optical path. This is hardly feasible, or at least absent in practical systems.

Due to these practical and fundamental problems, researchers have proposed many de-interlacing algorithms. Some neglected the problems with linear theory, and showed that acceptable results could nevertheless be achieved. Until the end of the 1970s, this was the common approach for television applications. From roughly the early 1980s onwards, others suggested that with nonlinear means, linear methods can sometimes be outperformed. Next, motion compensation has been suggested to escape from problems in scenes with motion, but was considered to be too expensive for nonprofessional applications until the beginning of the 1990s [17]. We shall discuss the relevant categories in the Sections 16.3 and 16.4.

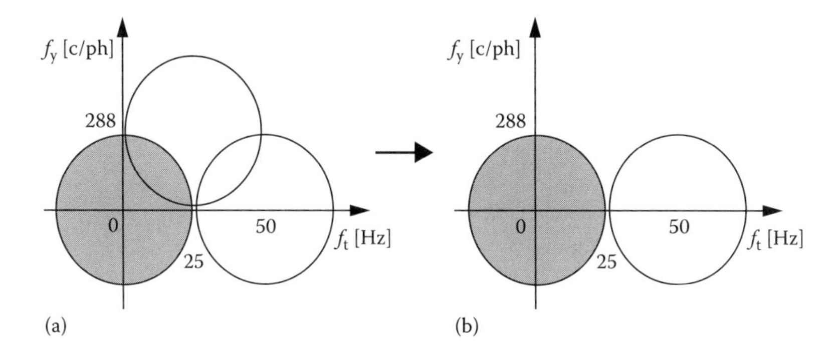

FIGURE 16.12 (a) Spectrum of the interlaced input. (b) The target spectrum of the de-interlacing.

16.2.1 Linear Techniques

The spatial and temporal filters are no longer popular in television products. For multimedia PCs, however these techniques, under the name "Bob" and "Weave" [35], are still used together with the spatiotemporal linear filters and therefore deserve our attention. All linear methods are defined by:

$$F_i(\vec{x}, n) = \sum\nolimits_k F(\vec{x} + k\vec{u}_y, n + m)h(k, m), \quad k, m \in \mathcal{Z}, \ (k + m) \bmod 2 = 1 \tag{16.26}$$

with $h(k, m)$ as the impulse response of the filter in the VT domain

$$\vec{u}_y = \begin{pmatrix} 0 \\ 1 \end{pmatrix} \quad \text{and} \quad \vec{u}_x = \begin{pmatrix} 1 \\ 0 \end{pmatrix}$$

The actual choice of $h(k, m)$ determines whether it is a spatial, a temporal, or a spatiotemporal filter.

Spatial filtering: Spatial de-interlacing techniques exploit the correlation between vertically neighboring samples in a field when interpolating intermediate pixels. Their all-pass temporal frequency response guarantees the absence of motion artifacts. Defects occur with high vertical frequencies only. The strength of the spatial or intrafield methods is their low implementation cost as no field memories are required. The simplest form is line repetition, which results by selecting $h(k, 0) = 1$ for $k = -1$, and $h(k, m) = 0$ otherwise. The frequency response of this interpolator is given by

$$H_y(f_y) = |\cos(\pi f_y)| \tag{16.27}$$

with

f_y as the vertical frequency (normalized to the vertical sampling frequency)
$H_y(f_y)$ as the frequency response in the vertical direction

This frequency characteristic has no steep roll-off. As a consequence, the first spectral replica is not much suppressed, while the base-band is partly suppressed. This causes alias and blur in the output signal.

The alias suppression can be improved by increasing the order of the interpolator. Line averaging, or "Bob" as it is called by the PC community, is one of the most popular methods, for which $h(k, 0) = 0.5$ for $k = -1, 1$, and $h(k, m) = 0$ otherwise. Its response:

$$H_y(f_y) = \frac{1}{2} + \frac{1}{2}\cos(2\pi f_y) \tag{16.28}$$

indicates a higher alias suppression. However, this suppresses the higher part of the base-band spectrum as well. Generally, purely spatial filters cannot discriminate between base-band and repeat spectrum regardless of their length. They always balance between alias and resolution loss as illustrated for a 50 Hz format in Figure 16.13. The gray-shaded area indicates the passband, that either suppresses vertical detail, or passes the alias.

Temporal filtering: Temporal de-interlacing techniques exploit the correlation in the time domain. Pure temporal interpolation implies a spatial all-pass. Consequently, there is no degradation of stationary images.

The analogy in the temporal domain of the earlier line repetition method of the previous subsection is field repetition or field insertion. It results from selecting $h(0, -1) = 1$, and $h(k, m) = 0$ otherwise. The frequency characteristic of field repetition too is the analogy of line repetition. It is defined by replacing f_y by f_t in Equation 16.27.

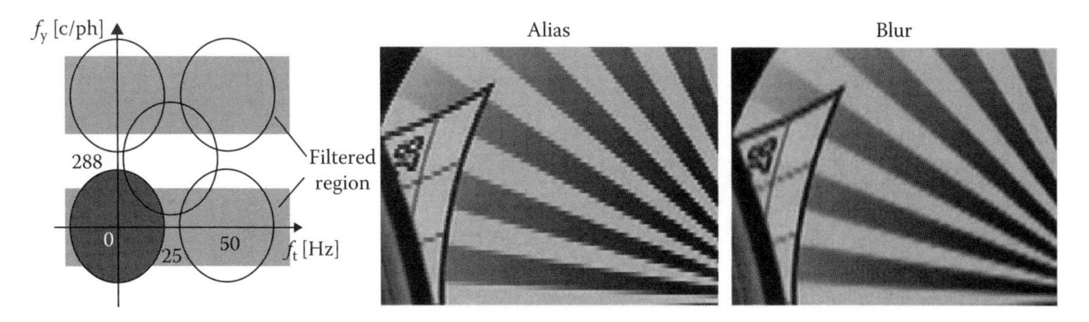

FIGURE 16.13 The VT frequency response with a spatial filter only allows a balance between alias and blur.

Field insertion, also called "Weave" in the PC world, provides an all-pass characteristic in the vertical frequency domain. It is the best solution in case of still images, as all vertical frequencies are preserved. However, moving objects are not shown at the same position for odd and even lines of a single output frame. This causes serration of moving edges, which is a very annoying artifact illustrated in Figure 16.14.

Longer temporal Finite Impulse Response (FIR) filters require multiple-field storage. They are, therefore, economically unattractive, particularly as they also cannot discriminate between base-band and repeat spectra, as shown in Figure 16.14.

VT filtering: A VT interpolation filter would theoretically solve the de-interlacing problem if the signal were bandwidth-limited prior to interlacing. The required prefilter would be similar to the up-conversion filter. The required frequency characteristic is shown in Figure 16.15.

Although the prefilter is missing, and there are problems with motion-tracking viewers [47], Figure 16.15 illustrates that the VT filter is certainly the best linear approach, in that it prevents both alias and blur in stationary images. The vertical detail is gradually reduced with increasing temporal frequencies. Such a loss of resolution with motion is not unnatural.

The filter is usually designed such that the contribution from the neighboring fields is limited to the higher vertical frequencies. As a consequence, motion artifacts are absent for objects without vertical detail that move horizontally. This can be achieved using

$$18h(k,m) = \begin{cases} 1,8,8,1, & (k=-3,-1,1,3) \wedge (m=0) \\ -5,10,-5, & (k=-2,0,2) \wedge (m=-1) \\ 0, & \text{(otherwise)} \end{cases} \tag{16.29}$$

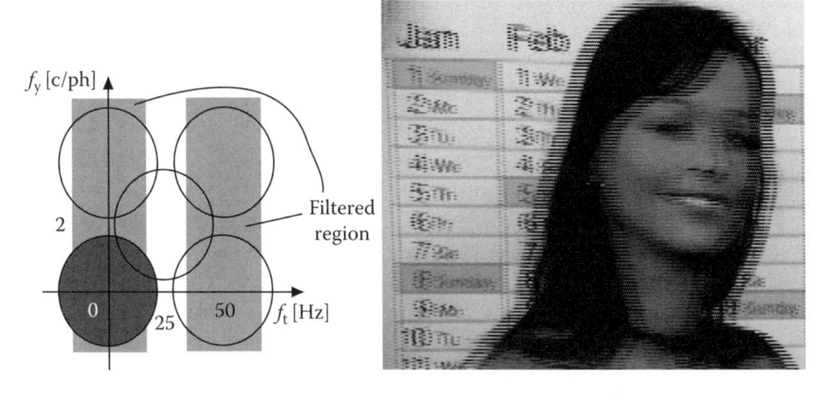

FIGURE 16.14 Frequency response of the temporal filter (left) and artifacts of field insertion with motion (right).

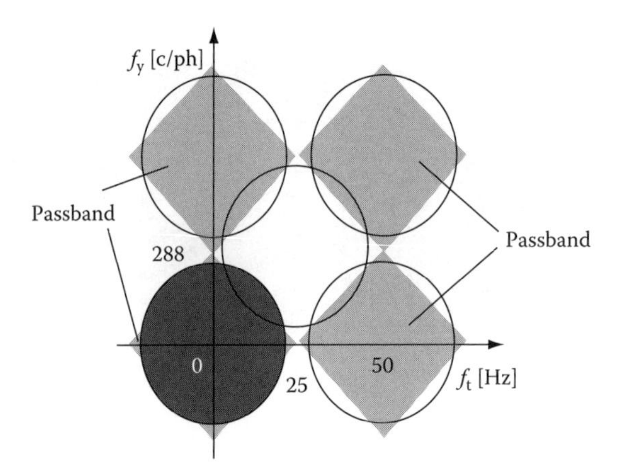

FIGURE 16.15 Video spectrum and the stop- and passbands of a VT filter.

16.2.2 Nonlinear Techniques

Linear temporal interpolators are perfect in the absence of motion. Linear spatial methods have no artifacts in case no vertical detail occurs. It seems logical, therefore, to adapt the interpolation strategy to motion and/or vertical detail. Many such systems have been proposed, mainly in the eighties, and the detection of motion/detail can be explicit, or implicit.

16.2.2.1 Motion-Adaptive Algorithms

A motion detector (MD) can be applied to switch or preferably fade between two processing modes, the one optimal for stationary and the other for moving image parts. Achiha et al. [1] and Prodan [40] mention that temporal and vertical filters may be combined to reject alias components and preserve true frequency components in the two-dimensional VT frequency domain by applying motion-adaptive fading. Bock [7] also mentioned the possibility to fade between an interpolator optimized for static image parts and one for moving image parts according to

$$F_i(\vec{x}, n) = \alpha F_{st}(\vec{x}, n) + (1 - \alpha) F_{mot}(\vec{x}, n) \tag{16.30}$$

with

F_{st} as the result of interpolation for static image parts
F_{mot} as the result for moving image parts

A MD determines the mix factor α.

Seth-Smith and Walker [43] suggested that a well-defined VT filter can perform as well as the best motion-adaptive filter, at a lower price. Their argument is that, in order to prevent switching artifacts, the fading results in something very similar to VT filtering that needs no expensive MD to realize that. Their case seems rather strong, but requires the (subjective) weighting of entirely different artifacts.

Filliman et al. [23] propose to fade between more than two interpolators. The high-frequency information for the interpolated line is extracted from the previous line. The low-frequency information is determined by a motion-adaptive interpolator.

$$F_i(\vec{x}, n) = F_{HF}(\vec{x} + \vec{u}_y, n) + \alpha F_{av}(\vec{x}, n) + (1 - \alpha) F_{LF}(\vec{x}, n - 1) \tag{16.31}$$

with F_{HF} and F_{LF} as the high-pass and low-pass filtered version of input signal F, respectively, F_{av} defined by

$$F_{av} = \frac{1}{2}(F_{LF}(\vec{x} - \vec{u}_y, n) + F_{LF}(\vec{x} + \vec{u}_y, n)) \quad (16.32)$$

with α controlled by the MD. The MD of Filliman et al. uses the frame difference. Field insertion results for the lower frequencies in the absence of motion, and line averaging in case of significant motion. Small frame differences yield an intermediate output.

Hentschel [26,27] proposed to detect vertical edges, rather than motion, within a field. The edge detector output signal ED is defined by

$$\text{ED}(\vec{x}, n) = g(F(\vec{x} - \vec{u}_y, n) - F(\vec{x} + \vec{u}_y, n)), \quad ((y + n) \bmod 2 = 1) \quad (16.33)$$

with $g()$ being a nonlinear function that determines the presence of an edge. The output of $g()$ is either 0 or 1. Note that this detector does not discriminate between still and moving areas, but merely shows where temporal interpolation could be advantageous.

16.2.2.2 Implicit Motion-Adaptive Methods

Next to the adaptive linear filters for de-interlacing, nonlinear filters have been described that implicitly adapt to motion or edges. Median filtering [2] is by far the most popular example. The simplest version is the three-tap VT median filter, illustrated in Figure 16.16. The interpolated samples are found as the median luminance value of the vertical neighbors (A and B), and the temporal neighbor in the previous field (C):

$$F_i(\vec{x}, n) = \text{med}\{F(\vec{x} - \vec{u}_y, n), F(\vec{x} + \vec{u}_y, n), F(\vec{x}, n - 1)\} \quad (16.34)$$

where the "med"-function ranks a set of values and then returns the value on the middle rank.

The underlying assumption is that, in case of stationarity, $F(\vec{x}, n - 1)$ is likely to have a value between that of its vertical neighbors in the current field. This results in temporal interpolation. However, in the case of motion, intrafield interpolation often results, since then the correlation between the samples in the current field is likely to be the highest. Median filtering automatically realizes this "intra/inter" switch on pixel basis.

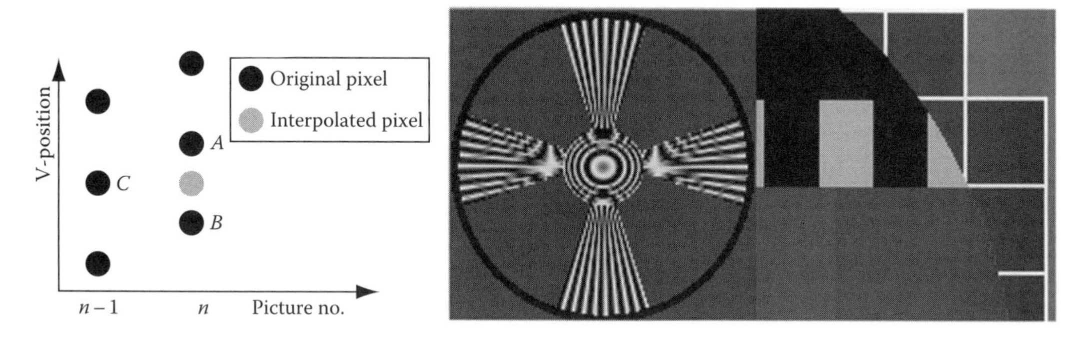

FIGURE 16.16 VT median filtering. The interpolated output pixel results as the median of its three spatiotemporal neighboring pixels *A*, *B*, and *C*. The output (middle and right image) show favorable performance on stationary edges and a weakness (alias) in textures with high vertical frequencies.

The major drawback of median filtering is that it distorts vertical details and introduces alias as shown in Figure 16.16. However, its superior properties at vertical edges and its low hardware cost have made it very successful [39].

16.2.2.3 Directional Interpolation

The algorithms reviewed so far can only give perfect results for stationary image parts. For moving image parts, alias and blur can be balanced but never disappear. Since the alias is most objectionable on long straight edges, Doyle et al. [20] use a 2-D spatial neighborhood of samples to include information of the edge orientation. If intrafield interpolation is necessary because of motion, then the interpolation filter should preferably preserve the base-band spectrum. After determining the direction in which the signal has the highest correlation, the signal is interpolated in that direction. As shown in Figure 16.17, the interpolated sample X is determined by a luminance gradient indication calculated from its direct neighborhood:

$$F_{\mathrm{i}}(\vec{x}, n) = X = \begin{cases} X_{\mathrm{A}}, & ((|A - F| < |C - D|) \wedge (|A - F| < |B - E|)) \\ X_{\mathrm{C}}, & ((|C - D| < |A - F|) \wedge (|C - D| < |B - E|)) \\ X_{\mathrm{B}}, & (\text{otherwise}) \end{cases} \tag{16.35}$$

where X_{A}, X_{B}, and X_{C} are defined by

$$X_{\mathrm{A}} = \frac{A + F}{2}, \quad X_{\mathrm{B}} = \frac{B + E}{2}, \quad X_{\mathrm{C}} = \frac{C + D}{2} \tag{16.36}$$

and the pixels A, B, C, D, E, and F are the ones indicated in Figure 16.17, and defined by

$$\begin{aligned} &A = F(\vec{x} - \vec{u}_x - \vec{u}_y, n) \quad B = F(\vec{x} - \vec{u}_y, n) \\ &C = F(\vec{x} + \vec{u}_x - \vec{u}_y, n) \quad D = F(\vec{x} - \vec{u}_x + \vec{u}_y, n) \\ &E = F(\vec{x} + \vec{u}_y, n) \quad F = F(\vec{x} + \vec{u}_y + \vec{u}_x, n) \\ &G = F(\vec{x} - 3\vec{u}_y, n) \quad H = F(\vec{x} + 3\vec{u}_y, n) \end{aligned} \tag{16.37}$$

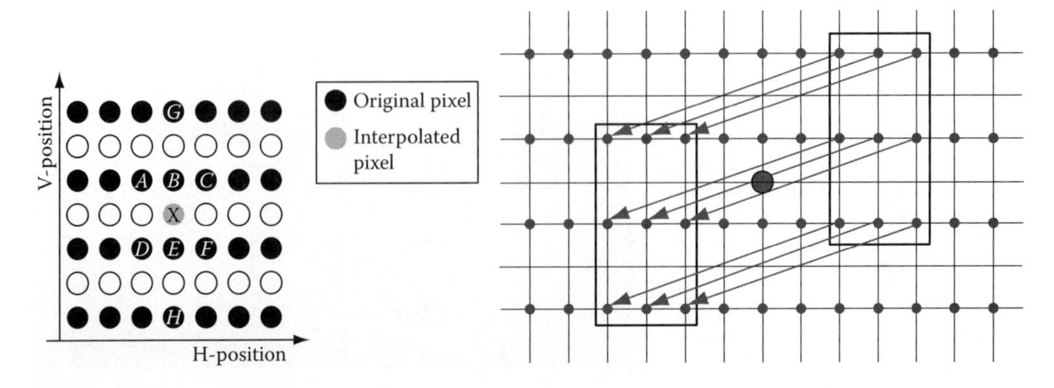

FIGURE 16.17 Identification of pixels in the aperture of edge-dependent interpolators (left). Instead of using the absolute difference between pixel pairs, the sum of absolute pixel differences gives a more robust edge orientation detection (right).

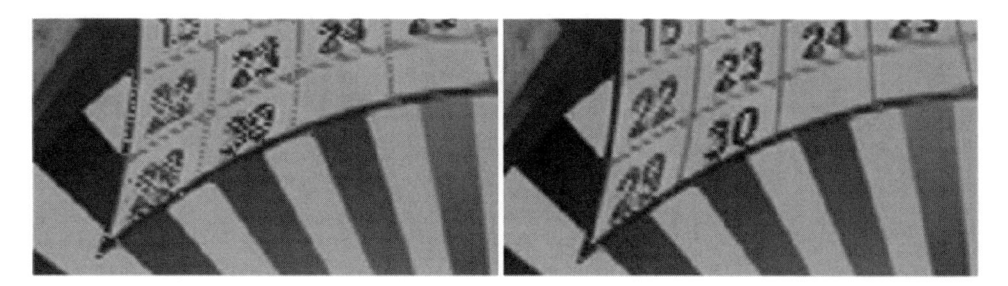

FIGURE 16.18 Directional interpolation, here simple ELA (left), gives favorable results on long straight edges and is weak in detailed areas. Protection of the interpolated sample using a median filter (right) gives a considerable improvement.

Many modifications to this so-called edge-dependent line averaging (ELA) have been proposed in the de-interlacing literature.

It is possible to increase the ED consistency [42], by checking also the edge orientation at the neighboring pixel. In Ref. [42] directional-ED operators are defined. For example, the error measure for a vertical orientation is defined by

$$|B - E| + |C - F|, \text{ alternative} : \text{Error (angle 90)} = |\text{B} - \text{E}| + |\text{C} - \text{F}| \tag{16.38}$$

and for an edge under 116°:

$$|A - E| + |B - F|, \text{ alternative} : \text{Error (angle 116)} = |\text{A} - \text{E}| + |\text{B} - \text{F}| \tag{16.39}$$

Consistency of edge information is further increased by looking for a dominating main direction in a 1-D or even 2-D neighborhood. Figure 16.17 illustrates this for a 3×3 block. Figure 16.18 illustrates that this indeed increases the robustness, but also shows remaining artifacts. Further analysis of the SADs as a function of the orientation has been proposed as an improvement [9]. As a final alternative, we mention postfiltering the interpolation result, e.g., with a median filter, which is quite effective as shown in Figure 16.18.

16.2.2.4 Hybrid Methods

In the literature, many combinations of the earlier described methods have been proposed. Lehtonen and Renfors [34] combine a VT filter with a five-point median. The output of the VT filter is one of the inputs of a five-point median. The remaining four inputs are nearest neighbors on the VT sampling grid.

Salo et al. [41] extend the aperture of the median filter in the horizontal domain to enable implicit edge adaptation.

Haavisto et al. [24] extend this concept with a MD. They propose a seven-point spatiotemporal window as a basis for weighted median filtering. The MD controls the importance or "weight" of these individual pixels at the input of the median filter.

Simonetti et al. [44] describe yet another combination of implicit/explicit edge and motion adaptivity. Their de-interlacing algorithm uses a hierarchical three-level MD which provides indications of static, slow, and fast motion. Based on this analysis, one of the three different interpolators is selected.

Kim et al. [29] detect motion by comparing an environment within the previous field with the same environment in the next field. Motion is detected if the (weighted) SAD between corresponding pixels in the two environments exceeds a motion threshold value. Furthermore, vertical edges are detected by comparing the absolute difference of vertically neighboring samples with a threshold value.

Depending on the ED and MD, their output at interpolated lines switches between temporal averaging and edge-dependent interpolation where the interpolation directions are determined using SAD between groups of pixels on the top and bottom neighboring lines.

16.2.3 Motion-Compensated Methods

The most advanced de-interlacing algorithms use motion compensation. It is only since the mid-1990s that motion estimators became feasible at consumer price level. Motion estimators are currently available in studio scan rate converters, in the more advanced television receivers [17], and in single-chip consumer MPEG2 encoders [8].

Similar to many previous algorithms, MC methods try to interpolate in the direction with the highest correlation. With motion vectors available,* this is an interpolation along the motion trajectory. Motion compensation allows us to virtually convert a moving sequence into a stationary one. Methods that perform better for stationary than for moving image parts will profit from motion compensation. Replacing the pixels $F(\vec{x}, n+m)$ with $F(\vec{x} + m\vec{D}(\vec{x}, n), n+m)$ converts a non-MC method in a MC version. Indeed, MC field insertion, MC field averaging, MC VT filtering, MC median filtering, and combinations with edge adaptivity have been proposed.

In this section, we shall focus on methods that cannot readily be deduced from the non-MC algorithms. The common feature of these methods is that they provide a solution to the fundamental problem of motion-compensating subsampled data. This problem arises if the motion vector used to modify coordinates of pixels in a neighboring field does not point to a pixel on the interlaced sampling grid. In the horizontal domain, this causes no problem, as sampling rate conversion theory is applicable. In the vertical domain, however, the demands for applying the sampling theorem are not satisfied, prohibiting correct interpolation.

16.2.3.1 Temporal Backward Projection

A first approximation to cope with this fundamental problem is to nevertheless perform a spatial interpolation whenever the motion vector points at a nonexisting sample, or even round to the nearest pixel.

Woods et al. [28] depart from this approximation. However, before actually performing an intrafield interpolation, the motion vector is extended into the preprevious field to check whether this extended vector arrives in the vicinity of an existing pixel. Figure 16.19 illustrates the procedure. Only if this is not the case spatial interpolation in the previous field is proposed:

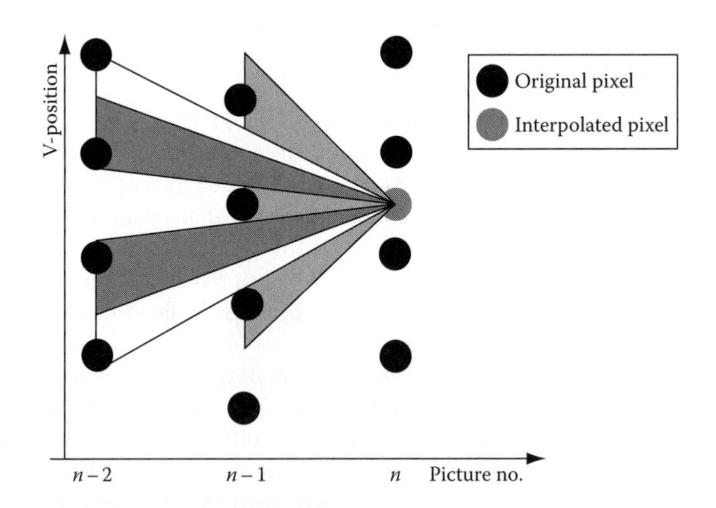

FIGURE 16.19 Temporal backward projection. Pixels from the previous, or the preprevious field are inserted, whichever is closest to the desired position after motion compensation.

* Motion estimation is described later in this chapter.

$$F_i(\vec{x}, n) = \begin{cases} F(\vec{x} - \vec{D}(\vec{x}, n) - \vec{\varepsilon}, n - 1), \\ \quad ((y - D_y - \varepsilon_y) \mod 2 = (n - 1) \mod 2) \\ F(\vec{x} - 2\vec{D}(\vec{x}, n) - 2\vec{\varepsilon}, n - 2), \\ \quad ((y - D_y - 2\varepsilon_y) \mod 2 = (n - 2) \mod 2) \\ F(\vec{x} - \vec{D}(\vec{x}, n), n - 1), \quad \text{(otherwise)} \end{cases} \tag{16.40}$$

where $\vec{\varepsilon} = \begin{pmatrix} 0 \\ \varepsilon_y \end{pmatrix}$, and ε_y is the small error resulting from rounding to the nearest grid position. This ε_y has to be smaller than a threshold. If no MC pixel appears in the vicinity of the required position, it would be possible to find one even further backwards in time. This, however, is not recommended as the motion vector loses validity by extending it too much.

The algorithm implicitly assumes uniform motion over a two-field period, which is a drawback. Furthermore, the robustness to incorrect motion vectors is poor, since no protection is proposed.

16.2.3.2 Time-Recursive De-Interlacing

The MC time-recursive (TR) de-interlacing of Wang et al. [50] uses the previously de-interlaced field (frame) instead of the previous field in a "field"-insertion algorithm. Once a perfectly de-interlaced image is available, and the motion vectors are accurate, sampling rate conversion theory can be used to interpolate the samples required to de-interlace the current field:

$$F_i(\vec{x}, n) = F_O(\vec{x} - \vec{D}(\vec{x}, n), n - 1) \tag{16.41}$$

As can be seen in Figure 16.20, the interpolated samples generally depend on previous original samples as well as previously interpolated samples. Thus, errors originating from an output frame can propagate into subsequent output frames. This is inherent to the recursive approach, and is the most important drawback of this method.

To prevent serious errors from propagating, solutions have been described in Ref. [50]. Particularly, the median filter is recommended for protection. As a consequence, the TR de-interlacing becomes

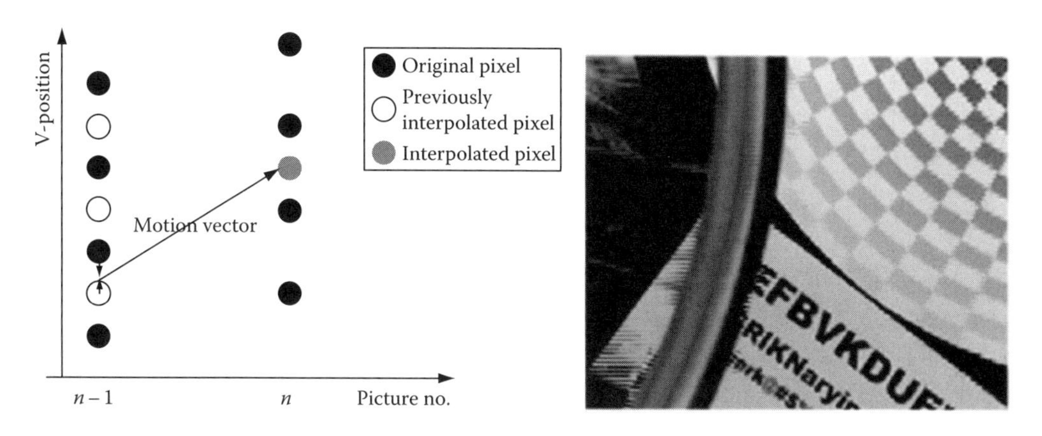

FIGURE 16.20 Time-recursive de-interlacing. The interpolated pixel is interpolated from the previously de-interlaced image, using both original and interpolated pixels from that image (left). Protection is necessary to prevent error propagation (right).

similar to the MC median filter approach, albeit that the previous image consists of a previously de-interlaced field instead of the previous field. The output is defined by

$$F_i(\vec{x}, n) = \text{med}\left\{\begin{array}{c} F_O(\vec{x} - \vec{D}(\vec{x}, n), n - 1), \\ F(\vec{x} - \vec{u}_y, n), \\ F(\vec{x} + \vec{u}_y, n), \end{array}\right\} \tag{16.42}$$

This is a very effective method, although the median filter can introduce aliasing in the de-interlaced image as illustrated in Figure 16.16.

To improve on this, de Haan et al. [12] proposed adaptive-recursive (AR) de-interlacing:

$$F_o(\vec{x}, n)\begin{cases} kF(\vec{x}, n) + (1 - k)F_O(\vec{x} - \vec{D}(\vec{x}, n), n - 1), & ((y + n) \bmod 2 = 0) \\ pF_i(\vec{x}, n) + (1 - p)F_O(\vec{x} - \vec{D}(\vec{x}, n), n - 1), & (\text{otherwise}) \end{cases} \tag{16.43}$$

where

k and p are adaptive parameters

F_i is the output of any initial de-interlacing algorithm

Preferably, a simple method is used, e.g., line averaging. The derivation of k is comparable to what we see in edge-preserving recursive filters, e.g., for motion-adaptive noise reduction. A similar derivation for p is not obvious, since the difference would heavily depend upon the quality of the initial de-interlacing method.

To calculate a value for p, we define

$$\begin{array}{ll} U = F(\vec{x} - \vec{u}_y, n) & U_{mc} = F_O(\vec{x} - \vec{D} - \vec{u}_y, n - 1) \\ C = F_O(\vec{x}, n) & C_{mc} = F_O(\vec{x} - \vec{D}, n - 1) \\ L = F(\vec{x} + \vec{u}_y, n) & L_{mc} = F_O(\vec{x} - \vec{D} + \vec{u}_y, n - 1) \end{array} \tag{16.44}$$

We can now rewrite Equation 16.43 for the interpolated lines as

$$C = p\frac{(U + L)}{2} + (1 - p)C_{mc} \tag{16.45}$$

The essence of the AR de-interlacing method is to select the factor p such that the nonstationarity along the motion trajectory of the resulting output for the interpolated pixels equals that of the vertically neighboring original pixels, i.e.,

$$|C_{mc} - C| = \left|\frac{U_{mc} - U}{2} + \frac{L_{mc} - L}{2}\right| \tag{16.46}$$

Substituting Equation 16.45 in Equation 16.46 and some rewriting lead to the value of the recursive filter coefficient p:

$$p = \frac{|(U_{mc} + L_{mc}) - (U + L)| + \delta}{|(2C_{mc} - (U + L)| + \delta} \tag{16.47}$$

where δ, a small constant, prevents division by zero.

16.2.3.3 Interlace and Generalized Sampling

The sampling theorem states that a bandwidth-limited signal with maximum frequency $0.5f_s$ can exactly be reconstructed, if this signal is sampled with a frequency of at least f_s. In 1956, Yen [51] showed a generalization of this theorem. Yen proved that any signal that is limited to a frequency of $0.5f_s$ can be exactly reconstructed from N independent sets of samples, representing the same signal with a sampling frequency f_s/N. This theorem can effectively be used to solve the problem of interpolation on a subsampled signal, as first presented by Delogne [19] and Vandendorpe [49]. We shall call this method the generalized sampling theorem (GST) de-interlacing method.

Figure 16.21 shows the calculation of the samples to be interpolated. Samples from the previous field are shifted over the motion vector towards the current field in order to create two independent sets of samples valid at the same temporal instant. A filter calculates the output sample.

Filter coefficients, based on a sync interpolator, are derived in the papers of Delogne [19] and Vandendorpe [49]. In Ref. [4], the simpler linear interpolator is used to derive coefficients. The de-interlaced output is defined by

$$F_i(\vec{x},n) = \sum\nolimits_k F(\vec{x} - (2k+1)\vec{u}_y, n)h_1(k,\delta_y) + \sum\nolimits_m F(\vec{x} - \vec{e}(\vec{x},n) - 2m\vec{u}_y, n-1)h_2(m,\delta_y) \tag{16.48}$$

with h_1 and h_2 defining the GST filter, and the modified motion vector $\vec{e}(\vec{x},n) = (e_x(\vec{x},n), e_y(\vec{x},n))^T$ defined as

$$\vec{e}(\vec{x},n) = \begin{pmatrix} d_x(\vec{x},n) \\ 2\text{Round}\left(\frac{d_y(\vec{x},n)}{2}\right) \end{pmatrix} \tag{16.49}$$

with Round( ) rounding to the nearest integer value and the vertical motion fraction δ_y defined by

$$\delta_y(\vec{x},n) = \left| d_y(\vec{x},n) - 2\text{Round}\left(\frac{d_y(\vec{x},n)}{2}\right)\right| \tag{16.50}$$

The GST filter, composed of h_1 and h_2, depends on the vertical motion fraction $\delta_y(\vec{x},n)$ and the subpixel interpolator type. The equation shows that output samples are completely determined by the original

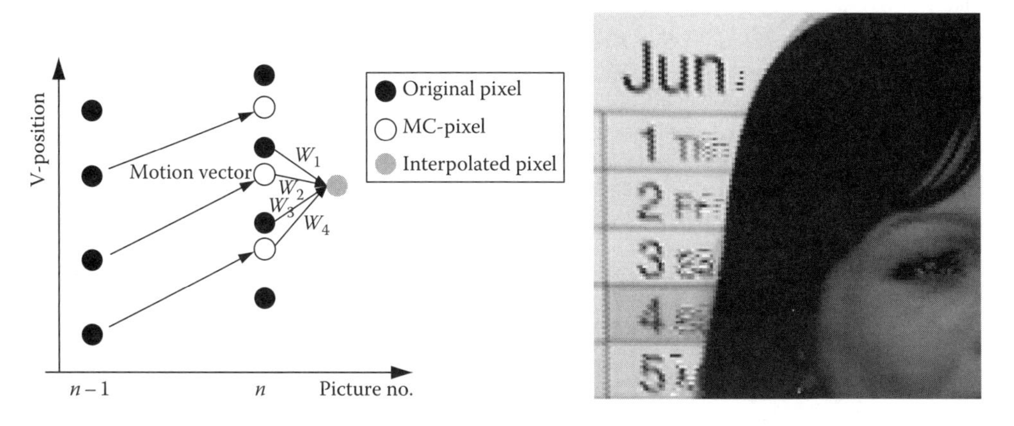

FIGURE 16.21 De-interlacing and generalized sampling. The interpolated pixel results as a weighted sum of pixels from the current field an MC pixels from the previous field (left). As the right-hand image shows, motion vector inaccuracies still lead to artifacts.

samples of the current and the previous field. No previously interpolated samples are used. Therefore, errors will not propagate, which is a clear advantage over the time-recursive and the adaptive-recursive algorithms.

To increase the robustness for incorrect motion vectors, it is conceivable to use a median filter that eliminates outliers in the output signal produced by the GST de-interlacing method. Bellers et al. [5,6] found that the median degraded the performance too much in areas with correct motion vectors. To improve on that, they proposed to selectively apply protection, as it was found that the GST de-interlacing produces artifacts particularly in areas with near critical velocities. Consequently, the proposed median protector is applied to the interpolated pixels, $F_i(\vec{x}, n)$, for near critical velocities only.

A bigger improvement resulted from the later observation [10] that GST cannot only be applied to de-interlace a video signal using samples from the current and the previous field, but that equally well the samples could be taken from the current and the next field. Using both options, two output samples result theoretically the same. If not, the motion vector is, at least locally, unreliable and the difference between the two options provides a quality indicator for every interpolated pixel. This indicator enables discrimination between areas that need protection and image portions where GST yields (near) perfect result that should not be modified.

16.2.4 Hybrids with Motion Compensation

Although clearly, on the average, the MC methods are superior over the non-MC methods, there are good reasons to consider hybrids. A closer study reveals that MC algorithms and methods based on direction interpolation (DI) have orthogonal strengths. MC is poor on long edges, due to the so-called aperture problem, but is strong on details. On the other hand, DI is strong on long edges, but poor in detailed areas. Figure 16.22 illustrates the above.

To combine MC and non-MC de-interlacing methods, Nguyen [36] and Kovacevic [32] describe de-interlacing methods that mix four methods: line averaging ($F_1(\vec{x}, n)$), edge-dependent interpolation ($F_2(\vec{x}, n)$), field averaging ($F_3(\vec{x}, n)$) and MC field averaging ($F_4(\vec{x}, n)$).

The interpolated lines of the output frame are defined by

$$F_i(\vec{x}, n) = \sum_{j=1}^{4} k_j F_j(\vec{x}, n) \tag{16.51}$$

The weights k_j associated with the corresponding interpolation methods are determined by calculating the "likely correctness" of the corresponding filter. To this end, they are calculated from the absolute difference of the corresponding method within a small region around the current position.

Kwon et al. [33] advocate switching instead of fading, and propose a decision on block basis. They include no edge adaptivity, but extend the number of MC-interpolators by distinguishing forward and backward field insertion, as well as MC field averaging.

A final proposal in this category was proposed by Zhao et al. [52], who use a classification-based mixing with coefficients obtained with an least mean squars (LMS) optimization on a set of training sequences.

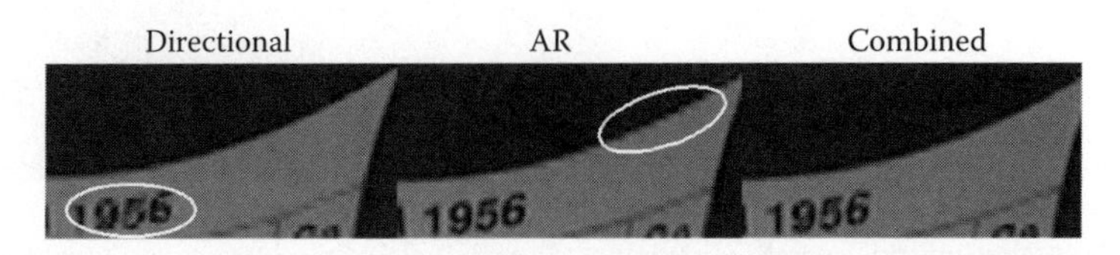

FIGURE 16.22 Motion compensation and directional interpolation have orthogonal strengths, which makes a combination logical.

16.3 Picture-Rate Conversion

Temporal interpolation, essential for picture-rate conversion, has fundamental problems because the object-tracking of the human visual system creates a difference between the temporal frequencies in the video signal and the temporal frequencies on the retina. This severely complicates the bandwidth requirements of the interpolation filter, as we shall see in this section.

The simple conversion methods neglect these problems and, consequently, negatively influence the resolution and the motion portrayal. More advanced algorithms apply motion vectors to predict the position of the moving objects at the unregistered temporal instances to improve the quality of the picture at the output format. A so-called motion estimator extracts these vectors from the input signal.

16.3.1 First-Order Linear Picture Interpolation

The simplest method to deal with a picture-rate mismatch exists in writing the input images into a picture memory and reading them from this memory at the, different, output rate. The procedure is cost-effective, but has severe flaws. Particularly, the read- and write-addresses have to be identical at some points in time because the rates are unequal. As a consequence, the output image may be built from parts of different images. Figure 16.23 shows the simple circuit, the timing diagram, and a resulting output image to illustrate the above.

An improved version of this simple processing results, if the reading of an output image from a memory starts only if this image has been completely written into that memory to prevent tearing. This means that the next input image has to be written into a different memory bank, which leads to increased cost as now (at least) two picture memories are required. In case the input picture-rate is higher than the output picture-rate, this processing implies that some input images will be dropped from the sequence, while we (repeatedly) display an image until a more recent one becomes available in case the output rate is higher than the input rate.

A possible implementation, which is quite common in video/graphics-cards of PCs, consists of two or more image buffers, also referred to as "back buffers," arranged such that when one is busy writing an input image at the input rate, another outputs its image content at the required output rate. As soon as both the writing of an input image and the reading of an output image is completed, the memory banks swap function. Figure 16.24 illustrates the result and shows the block and timing diagrams of this circuit for an integer up-sampling, where two back buffers suffice.

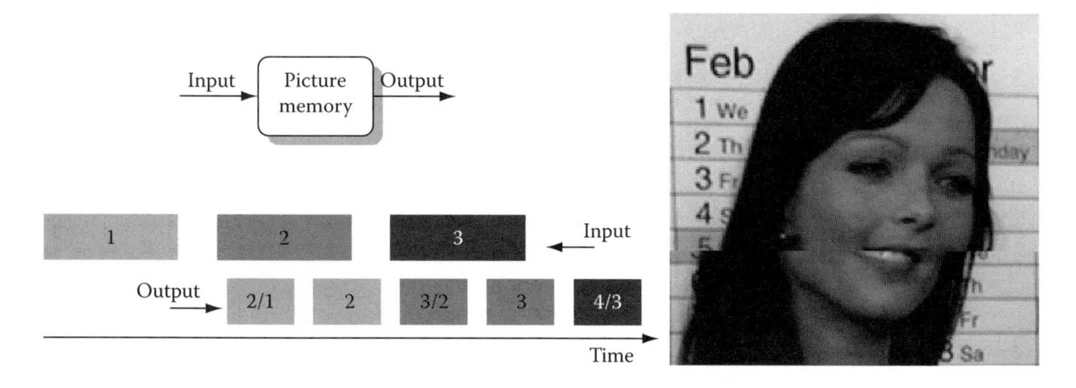

FIGURE 16.23 Picture-rate conversion with a single picture memory. The timing diagram shows that the images are written at the input rate and read at the output rate. Consequently, output images may be composed of different input pictures. This so-called em tearing artifact is illustrated in the screen shot.

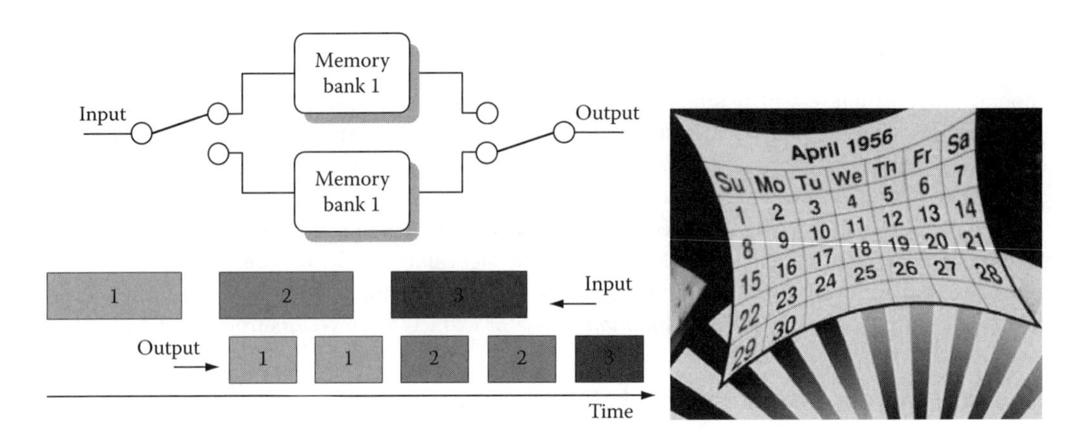

FIGURE 16.24 Picture repetition with two memory banks. When the first bank is writes an input image at the input rate, the second (repeatedly) outputs its content at the required output rate. After the writing of an input image is completed, the memory banks swap function at the moment a new output image is required, which eliminates the tearing.

Although the tearing artifact has disappeared with the use of two memory banks, the resulting motion portrayal obtained with this solution is still rather poor, as we shall see next.

16.3.1.1 Motion Portrayal with First-Order Interpolation

To understand the impact of picture repetition on the perceived picture quality, we analyze what happens when an input sequence shows a moving ball, as indicated in Figure 16.25, and the processing increases the picture-rate with a factor of 2.

If only the original images are shown on the screen, as in Figure 16.25, one could expect that, due to the integration of the eye that causes the roll-off in the frequency domain shown in Figure 16.25, a blurred ball would be seen by the observer (also shown in Figure 16.25). The expectation is correct, but only if we assume that the viewer focuses on a fixed image part.

In practice, the blur is not perceived. This results from the very sophisticated human motion-tracking capability, which makes that, when confronted with a moving ball on a screen, the ball is automatically projected on a fixed location of the retina. The object-tracking compensates for the motion of the ball and makes it stationary at the retina. Consequently, the temporal frequency response of the viewer is irrelevant as the retinal image is stationary in the temporal domain. Figure 16.26 shows the position–time

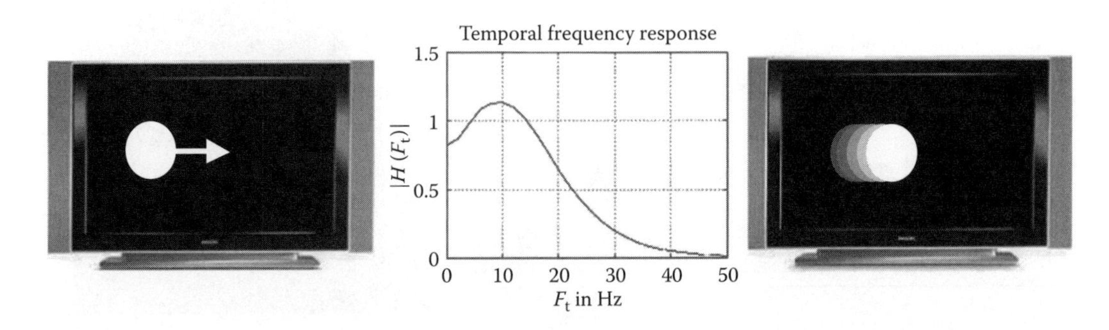

FIGURE 16.25 A ball moving over the screen (left). The integration of light by the retina of the eye causes a low-pass frequency behavior of the human visual system (middle). If the viewer focuses on a fixed location of the screen, a moving ball would seem blurred (right).

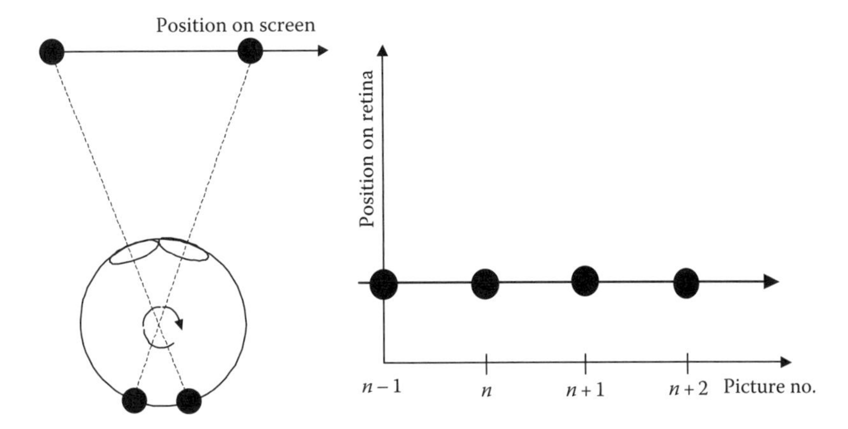

FIGURE 16.26 In practice, a viewer will track the moving ball, due to which the position on the retina is stationary, and no blurring results from the integration of light in the retina.

plot of the ball on the retina of the viewer due to which the result, even after integration by the retina, is a perfectly sharp ball. The sensation of motion remains as the human visual system is aware of the motion-tracking of the eye.

This brief analysis helps us understand the effect of picture repetition as it results from the simple picture-rate up-conversion circuit of Figure 16.24. If the input is 50 Hz and we show 100 images per second at the output, every image will be shown twice on the screen as illustrated in Figure 16.27. The consequence is also shown in the same figure; at the retina, the ball will now be projected onto two different places and the separation between these two positions is determined by the distance the ball travels in a picture period.

It is assumed here that the irregular motion cannot be tracked, but rather the average speed on the screen is determined by the angular velocity of the eye. With an input rate higher than 50 Hz this is reasonable, since 50 Hz flicker is hardly noticeable and both objects seem to be continuously present at the retina. If the input rate is much lower than 50 Hz, the ball will not seem to be continuously present at both positions of the retina and the motion will appear jerky. This jerky motion artifact is usually referred to as motion judder, and appears, e.g., when 25 Hz* film is shown on 50 Hz television.

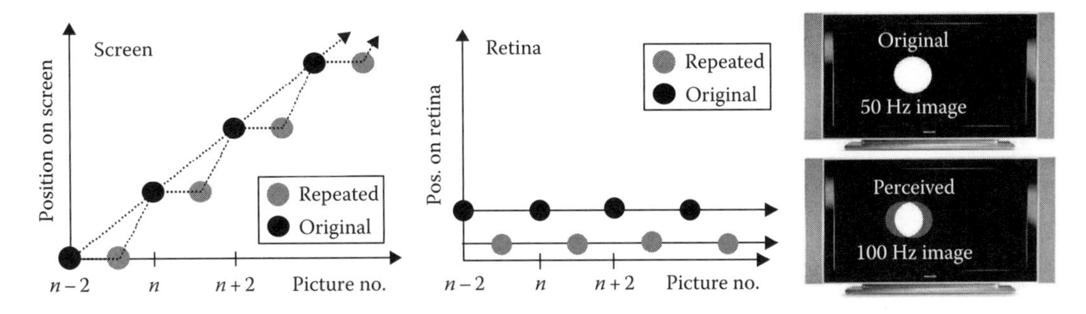

FIGURE 16.27 If every image is displayed twice to arrive at 100 Hz picture rate with a 50 Hz input sequence, the ball will no longer be projected on a single location of the retina, but rather at two different locations. This distance between these locations depends on the displacement of the ball in a picture period, and leads to a blurred ball in our perception.

* Actually 25 Hz film is normal 24 pictures per second film material as shown in the cinema, but for 50 Hz it is run 4% faster to arrive at 25 Hz which is more convenient for picture-rate conversion.

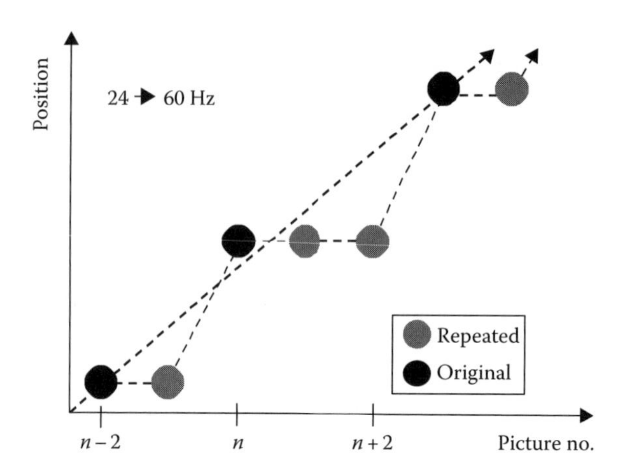

FIGURE 16.28 The motion quality of film on television is worse in 60 Hz countries, where images are repeated alternatingly two and three times.

This 25 Hz motion judder can be quite objectionable, and the situation is even worse when using the simple picture-rate conversion technique to up-convert 24 Hz film to a 60 Hz television display. This situation is illustrated in Figure 16.28 and results in the so-called 2–3 pull-down artifact, as images are alternatingly repeated two and three times on the television screen.

16.3.1.2 What Film-Directors Do

It may seem strange that the simple picture-repetition procedure is common practice for film on television and yet rarely leads to complaints. This is because 24 images per second cannot be watched without heavy screen flicker. Therefore, from the very start, film has been repeatedly projected onto the screen, also in the cinema, and art-directors take this into account when shooting the material. To prevent the judder from becoming unacceptable, film directors avoid rapid motion, by using slow zooms and pans and particularly use tracking shots. To prevent the background, which is not tracked, from juddering excessively, they also use a shallow depth of focus as out-of-focus parts of image cause less high temporal frequencies. Finally, it is common procedure, to use a large temporal aperture, i.e., insensitive film material that requires long shutter times, to blur objects moving relative to the camera.

16.3.1.3 Formal Description of First-Order Interpolation

All linear methods for a conversion factor of k/l, with integers k and l, can formally be defined using an up-conversion with a factor k that consists of an intermediate step of zero stuffing:

$$F_z\left(\vec{x}, \frac{a}{k}\right) = \begin{cases} F(\vec{x}, n), & \left(\frac{a}{k} = 1, 2, 3 \ldots\right) \\ 0, & \text{(otherwise)} \end{cases} \tag{16.52}$$

and a low-pass filtering step:

$$F_{\text{lp}}(\vec{x}, a) = \sum_m F_z(\vec{x}, a + m)h(m), \quad (m = \ldots, -1, 0, 1, 2 \ldots) \tag{16.53}$$

with $h(m)$ the impulse response of the temporal filter, followed by a down-conversion, or decimation, step with a factor l:

$$F_{\text{o}}(\vec{x}, n') = F_{\text{lp}}(\vec{x}, la) \tag{16.54}$$

Changing k, l, and $h(m)$ in the above equations leads to the different techniques including picture repetition that we know from the previous section, weighted picture averaging, and higher-order temporal filtering. We shall elucidate the effect of these linear methods on image quality with some practical examples.

First, we discuss the old problem of displaying film, which registers 24 pictures per second, on a 60 Hz television screen. Figure 16.27 illustrates the consequence of picture repetition for our moving ball sequence. As Figure 16.27 shows, the film input pictures are alternatingly repeated two or three times at the 60 Hz display. The eye will track with approximately the average speed, i.e., the original speed of the ball. Consequently, the ball will not always be projected at the same position of the retina, but on five different locations depending on the output picture number.

More formally, an up-conversion with a factor of 2.5 is required, i.e., in Equations 16.52 and 16.54 $k=5$, $l=2$, and the positions of the ball on the retina can be understood from the impulse response of the interpolation filter of Equation 16.53:

$$h(m) = \begin{cases} 1, & (-2 \leq m \leq 2) \\ 0, & \text{(otherwise)} \end{cases} \tag{16.55}$$

which shows five nonzero coefficients.

The time interval between images showing the ball at the same relative position, 83.3 ms, is too large to result in complete integration, i.e., blurring. Instead, an irregular, jerky, motion portrayal results.

Our second example concerns 50 Hz video displayed at a 100 Hz screen, using picture repetition. Televisions with this processing, to eliminate flicker on large bright screens, arrived on the consumer market in 1988 [39]. Again, the motion trajectory is irregular. This time, however, the ball is projected at only two different positions in alternate output images. The positions of the ball on the retina can again be understood from the impulse response of the interpolation filter of Equation 16.53:

$$h(m) = \begin{cases} 1, & (m = -1, 0) \\ 0, & \text{(otherwise)} \end{cases} \tag{16.56}$$

which shows two nonzero coefficients.

In this second example, the time interval between images showing the ball at the same relative position is much shorter, only 10 ms, and complete integration results. Therefore, the ball is constantly seen in the two positions. This results in a constant blurring for lower velocities, or becomes visible as a double image at higher object velocities.

16.3.2 Higher-Order Linear Picture Interpolation

Since picture repetition can be seen as a first-order linear interpolation, the question may arise whether perhaps a higher-order interpolation could give a much improved motion portrayal. Weighted picture averaging is the time domain equivalent of what is usually referred to as the (bi)-linear interpolation in the spatial domain, i.e., a two-tap temporal interpolation filter with coefficients inversely related to the distance between the output picture and the original pictures. Somewhat loosely stated, the output image, $F_{avg}(\vec{x}, n+\alpha)$, at a time distance α times the picture period from the current picture, n, is calculated from its neighboring pictures n and $n+1$ according to

$$F_{avg}(\vec{x}, n+\alpha) = (1-\alpha)F(\vec{x}, n) + \alpha F(\vec{x}, n+1); \quad 0 \leq \alpha \leq 1 \tag{16.57}$$

FIGURE 16.29 (a) Picture showing the perceived moving image in case of 50–100 Hz up-conversion applying picture repetition. (b) Picture averaging. (c) Higher-order filtering.

If we stick to the previously introduced practical examples, the up-conversion filter is defined by

$$h(m) = \begin{cases} \frac{4}{5}, \frac{3}{5}, & (m = \pm 1, \pm 2) \\ \frac{2}{5}, \frac{1}{5}, & (m = \pm 3, \pm 4) \\ 1, 0, & (m = 0, \text{otherwise}) \end{cases} \quad (16.58)$$

in case of 24 Hz film to 60 Hz television conversion, and by

$$h(m) = \begin{cases} 1, & (m = 0) \\ \frac{1}{2}, & (m = \pm 1) \\ 0, & (\text{otherwise}) \end{cases} \quad (16.59)$$

for the up-conversion of 50 Hz video to 100 Hz flicker-free television display.

Since the number of nonzero coefficients in the interpolation filters has increased, it shall be clear that the number of echoes in the image has also increased. Figure 16.29b shows the resulting image for the 100 Hz conversion. Although the blurring is somewhat worse, since the bandwidth of the temporal filter is less, the legibility of the text may be somewhat better, because the echoes have a reduced intensity.

By now, it should be clear that a further increase of the order of the temporal interpolation filter, to better approximate the sinc-interpolation filter, cannot eliminate the artifacts, as is the case for spatial interpolation. Figure 16.29c shows the perceived image in case an interpolation filter with an improved frequency response would have been used. The effect of the negative coefficients, necessary to arrive at a steeper roll-off, is clearly visible.

We conclude that for the object-tracking observer the temporal low pass filter (LPF) is transformed into a spatial low-pass filter (LPF) . For an object velocity of 1 pixel per picture period (1 pel/pp), its frequency characteristic equals the temporal frequency characteristic of the interpolating LPF. A speed of 1 pel/pp is a slow motion, as in broadcast picture material velocities in a range exceeding 20 pel/pp do occur. Thus, the spatial blur caused by the linear interpolation is unacceptable already for moderate object velocities.

Linear interpolation can only yield a perfect dynamic resolution for moving images, if the interpolation is performed along the motion trajectory rather than along the time axis. This more advanced technique is the topic of the following section.

16.3.3 Motion-Compensated Picture Interpolation

The integration of light in the retina of the eye causes a temporal bandwidth limitation from which we could profit when interpolation is necessary. However, since the viewer tracks moving objects with his or

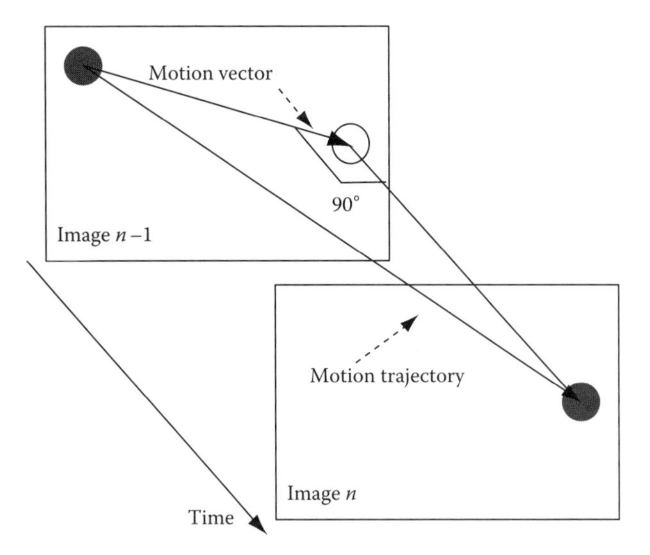

FIGURE 16.30 Definition of motion trajectory and motion vector.

her eyes, we can only profit from the integration of the eye along the motion trajectory. Consequently, our temporal interpolation filter should be a low-pass filter not in the pure temporal dimension, but rather in a spatiotemporal direction that coincides with this motion trajectory.

Figure 16.30 defines the motion trajectory as the line that connects identical picture parts in a sequence of pictures. The projection of this motion trajectory between two successive pictures on the image plane, called the motion vector, is also shown in Figure 16.30. Knowledge of object motion allows us to interpolate image data at any temporal instance between two successive pictures.

16.3.3.1 Motion-Compensated First-Order Linear Interpolation

The most straightforward thing to do is to calculate motion vectors for every location of the image and shift the local image data over the estimated local motion vector. Figure 16.31 shows the required circuit diagram and the effect this processing would have on our moving ball sequence.

Although in this simple example, the processing seems effective, the processing is totally unacceptable in practice, as illustrated in Figure 16.32. To understand the demonstrated artifacts, it is sufficient to note that a moving object, in this case a person, causes discontinuities in the motion vector field at the

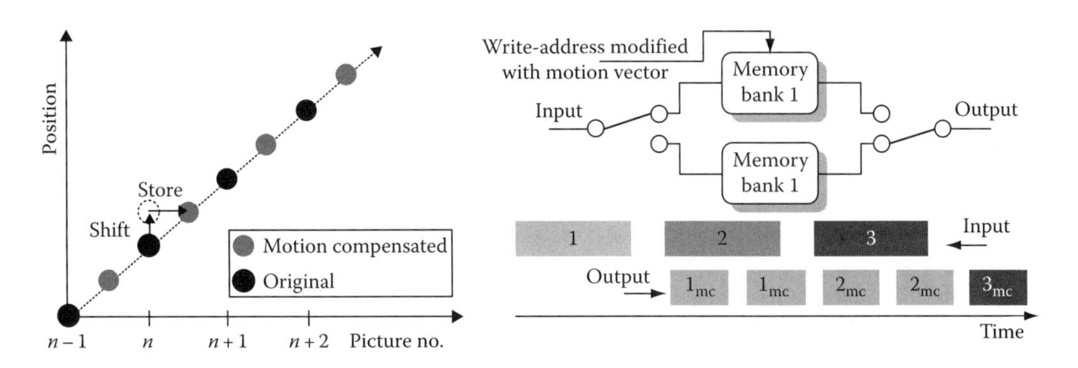

FIGURE 16.31 Straightforward motion compensation. By modifying the write address of the image memory, the object is shifted depending on the motion vector.

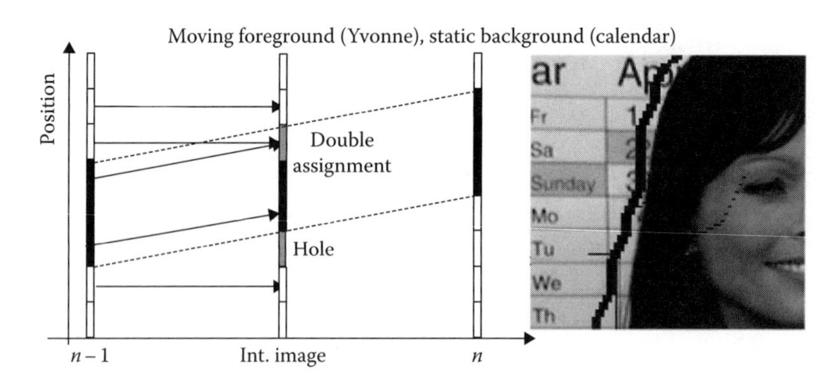

FIGURE 16.32 MC shifting of original images leads to interpolated images with holes and double assignments. The analysis is shown left, while the right-hand image shows the result. Yvonne moves to the right, so the holes (black) show in the uncovered areas.

boundary of the object and the background. Such discontinuities imply that there will be areas in the MC image to which no motion vector points and there will be areas to which more than one motion vector points. The first problem leads to holes in the interpolated image, while the second leads to double assignments. The holes are visible in Figure 16.32 as black pixels. The double assignments will not show, as they cause image data that has been shifted to this location of the memory to be overwritten by image data that has been shifted to that same location at a later point in the processing.

Formally, the interpolated image is defined by

$$F_{\text{MCshift}}(\vec{x} + \alpha\vec{D}(\vec{x}, n),\, n + \alpha) = F(\vec{x}, n), \quad (0 \leq \alpha \leq 1) \tag{16.60}$$

and the holes and double assignments are understood from this equation as only the interpolated image pixels at position $\vec{x} + \alpha\vec{D}(\vec{x}, n)$ are, sometimes multiply, defined.

The solution to this problem is simple, as far as the actual picture-rate conversion is concerned, and exists in changing the read address of the picture memory rather than the write address. Since there will be a modified read address for all (interpolated) picture parts and there is (original) image data at every address that can be generated,* holes are effectively prevented and double assignments cannot occur as always just a single read address is generated (Figure 16.33). In this case, the interpolated image is defined by

$$F_{\text{MCfetch}}(\vec{x}, n + \alpha) = F(\vec{x} - \alpha\vec{D}(\vec{x},\, n + \alpha), n), \quad (0 \leq \alpha \leq 1) \tag{16.61}$$

As can be seen from this equation, the problem now has been moved to the motion estimator, which has to estimate motion vectors that are valid at the time the interpolated image is displayed and not at the time instance of an existing image.

This is not a trivial demand. In any current image part, it is usually possible to find a strong correlation with an image part in the previous, or in the next image of a sequence and thus define a motion vector. For a temporal instance between original images, it may often be possible to find a correlation between symmetrically located image parts in the previous and next image, see Figure 16.34, but in occlusion areas no good correlation is to be expected.

* Apart from image boundaries, which are neglected here.

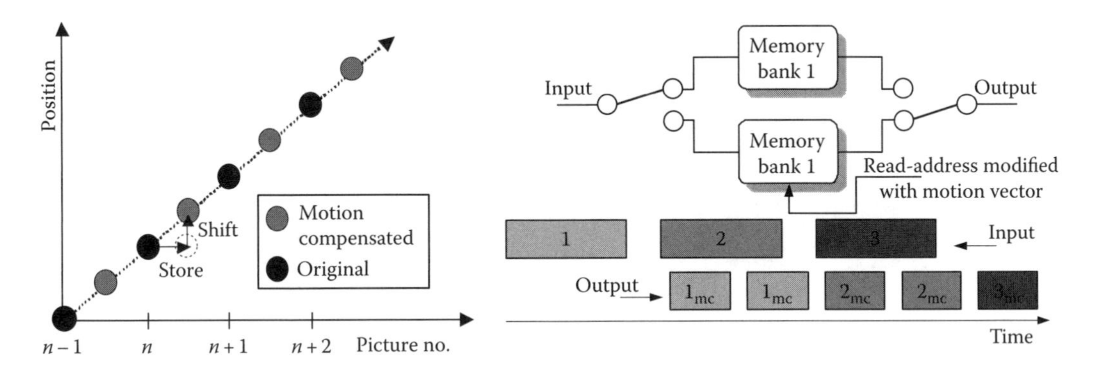

FIGURE 16.33 Improved motion compensation. By modifying the read address of the image memory, the object is fetched from the previous original image depending on the motion vector. The motion vector has to be valid at the time the interpolated image is displayed.

16.3.3.2 Higher-Order MC Picture Interpolation

The linear methods described in the section on non-MC picture-rate conversion methods all have their MC variant. Replacing the pixels $F(\vec{x}, n + m)$ with $F(\vec{x} + m\vec{D}, n + m)$, where $\vec{D}$ indicates the displacement or motion vector, converts a non-MC method to a MC-method. Indeed, MC picture repetition, MC picture averaging, MC higher-order filtering, and also some MC variants of nonlinear methods have been proposed. We shall limit the description of the linear higher-order methods, here, to the most popular MC picture averaging, demonstrate why we are unhappy with linear interpolation, and then move to the nonlinear MC algorithms.

The MC picture averaging algorithm combines two adjacent pictures linearly. From each picture, pixels are fetched from a shifted location that depends on the required temporal position of the interpolated image and the estimated object velocities:

$$F_{\text{mca}}(\vec{x}, n + \alpha) = \frac{1}{2}(F(\vec{x} - \alpha\vec{D}(\vec{x}, n + \alpha), n) + F(\vec{x} + (1 - \alpha)\vec{D}(\vec{x}, n + \alpha), n + 1)), \quad (0 \leq \alpha \leq 1) \tag{16.62}$$

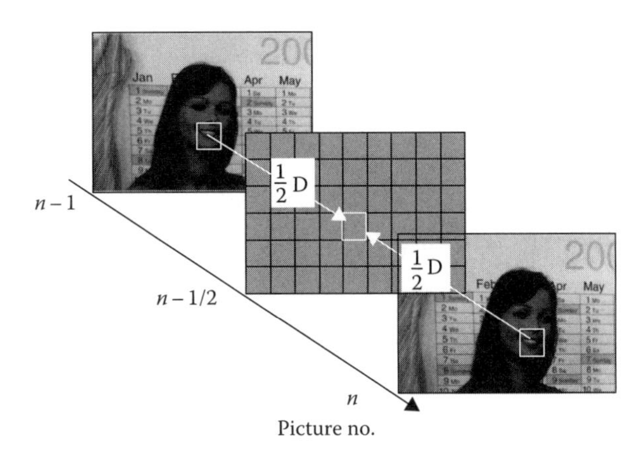

FIGURE 16.34 Finding a motion vector, valid for an intermediate image, by establishing correspondences between image parts symmetrically located around the intermediate image part in the next and previous image. These correspondences can usually be found for foreground objects, but the background may be covered differently in the two images.

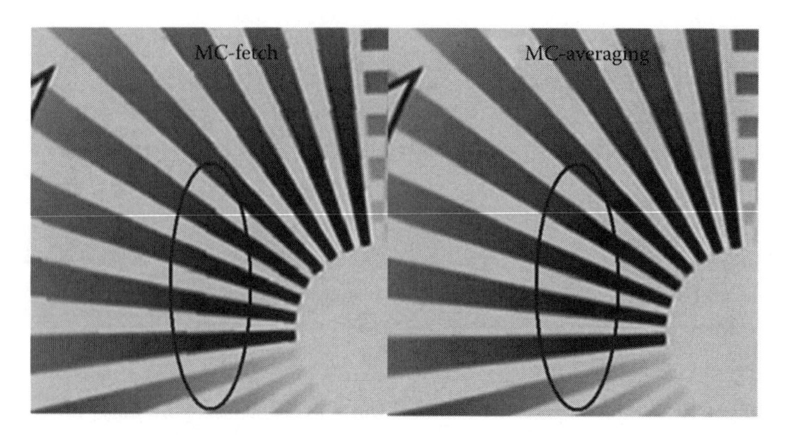

FIGURE 16.35 MC picture averaging is better than MC picture fetching. This is because inaccuracies of motion vectors result in blurring rather than miss-positioning.

The improvement realized by using MC averaging is significant, as shown in Figure 16.35. However, if we compare the quality with non-MC picture averaging for a picture of our video sequence to which a stationary subtitle has been added, as in Figure 16.36, we also notice drawbacks of motion compensation. The blur in moving image parts disappeared, but artifacts are visible in the subtitles. This is because due to the moving background not all motion vectors at the stationary text are zero vectors. Clearly, a higher-order interpolation can add little to solve this problem and we have to switch to the nonlinear approaches to arrive at more significant improvements.

16.3.4 Nonlinear MC Picture Interpolation

Linear techniques applying motion vectors can at best decrease the visibility of artifacts due to erroneous vectors by linearly combining different predictions. This leads to blurring, but never to elimination of the artifacts. In the nonlinear algorithms described in this section, a number of pixels from MC pictures are ranked based on their magnitude. These pixels may correspond to different candidate motion vectors. One could consider them as candidates for the calculation of the current output pixel. In an rank-order filter, it is now possible to combine these pixels, where their contribution to the output value depends on their rank. This way, it can be prevented that unlikely candidates, outliers, contribute to the output. The principle was introduced and patented in Ref. [16], while a more elaborate evaluation was given in Ref. [38].

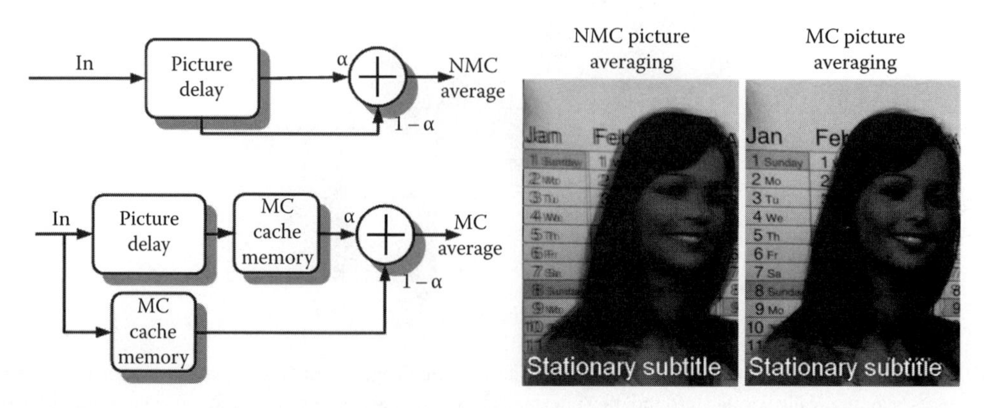

FIGURE 16.36 MC picture averaging gives a great improvement for moving image parts. However, there is also a risk that stationary parts, e.g., the subtitles in the above images, are damaged due to vector inaccuracies.

16.3.4.1 MC Static Median Filtering

MC averaging is especially vulnerable when incorrect vectors are used to interpolate stationary text, or any other fine stationary structures, superimposed on moving objects. We can effectively preserve such structures by using a simple implementation of the rank-order filter—the static median filter.

Let F_{sta} denote the output of the static median filter, such that:

$$F_{sta}(\vec{x}, n + \alpha) = \text{med}\{F(\vec{x}, n),\ F(\vec{x}, n + 1),\ F_{mca}(\vec{x}, n + \alpha)\} \tag{16.63}$$

where F_{mca} is as defined in Equation 16.62. This filter can produce a MC output only if the corresponding pixels in the neighboring original images are different. Consequently, it is robust against spurious vectors in stationary areas.

Figure 16.37 shows the improvement that this filtering brings in the stationary text area. Also in moving areas with little detail, such as the head of the lady, the interpolation result is correct, as the output is determined by the MC average. However, a rather serious problem occurs in detailed textures. In Figure 16.37, this shows as a significant blur of the moving calendar text.

16.3.4.2 MC Dynamic Median Filtering

On the average a much improved output results from a slightly modified median filter. Similar to the static median, this filter uses four pixels for the processing: two with motion compensation, and two without [15,16], but its output is defined by

$$F_{dyn}(\vec{x}, n + \alpha) = \text{med}\{F(\vec{x} - \alpha\vec{D}(\vec{x}, n + \alpha), n),\ F(\vec{x} + (1 - \alpha)\vec{D}(\vec{x}, n + \alpha),\ n + 1),\ F_{avg}(\vec{x}, n + \alpha)\} \tag{16.64}$$

where F_{avg} is the non-MC picture average, and conforms the definition in Equation 16.57.

If the motion vector is accurate, the compensated pixels will have about the same luminance amplitude, and thus the median filter will select either of them. However, if the motion vector is unreliable, then it is more likely that the uncompensated input will be in the middle in the ranking. This filter successfully strikes a compromise between MC averaging and the static median filter, as can be seen in Figure 16.38. Because the MC inputs are in the majority, we shall call this algorithm the "dynamic median" filter.

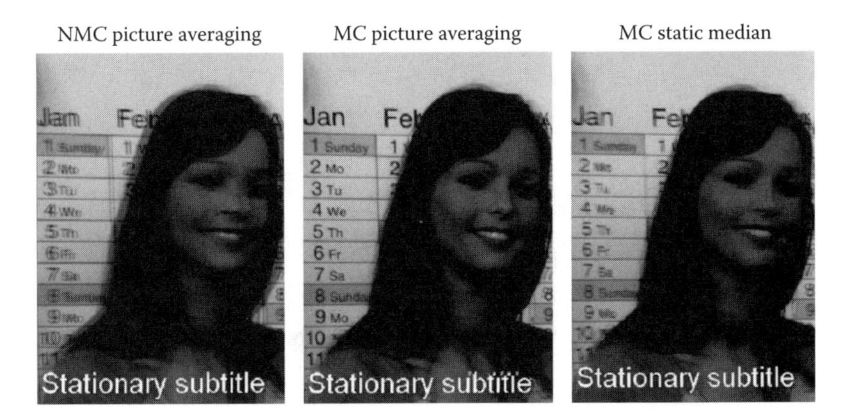

FIGURE 16.37 The static median gives a great improvement for stationary image parts, e.g., the subtitles, but there is a loss of resolution in texture areas.

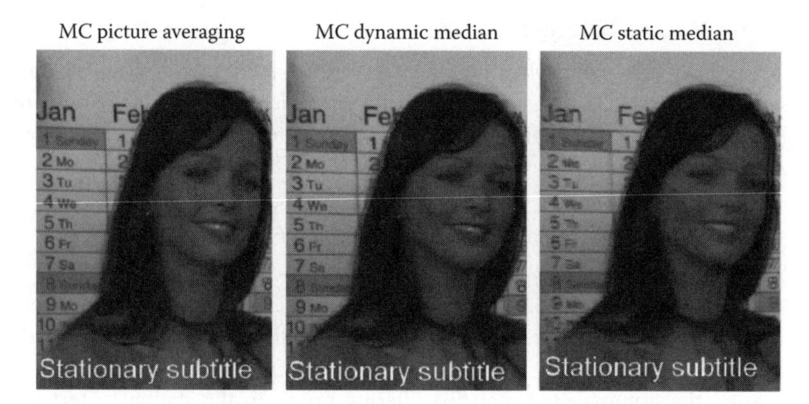

FIGURE 16.38 The performance of the dynamic median balances between the good resolution in moving areas of MC averaging and the good stationary text of the static median.

16.3.4.3 Reciprocal Mixing

A possible alternative to the MC median filtering is mixing between MC averaging and temporal averaging [25].

Consider again Equations 16.57 (F_{avg}) and 16.62 (F_{mca}). We can define a soft switch between these two outputs according to

$$F_{mix}(\vec{x}, n+\alpha) = (1-k)F_{avg} + kF_{mca}; \quad 0 \leq k \leq 1 \tag{16.65}$$

where the mixing factor, k, varies with the local reliability of the motion vectors. Low values of k indicate very reliable motion vectors, and vice versa. In Ref. [25], two differences, a first one between the MC pixels, and a second one between the non-MC pixels, are used to control k. If we define:

$$\begin{aligned} &F(\vec{x}, n) = A, F(\vec{x}, n+1) = B, \\ &F(\vec{x} - \alpha\vec{D}(\vec{x}, n+\alpha), n) = C \quad \text{and} \\ &F(\vec{x} + \alpha\vec{D}(\vec{x}, n+\alpha), n) = D \end{aligned} \tag{16.66}$$

then in Ref. [25], k is defined as

$$k = \frac{|B-C|}{\beta|B+C-A-D| + \delta} \tag{16.67}$$

where δ is a small number to prevent division by zero, while β is a gain parameter. The effect of β on the resulting picture quality is shown in Figure 16.39.

16.3.4.4 Cascaded Median Filtering

The static median preserves stationary structures in homogeneous areas, but is not suitable for detailed image parts. The dynamic median reasonably preserves stationary structures, but it can serrate edges slightly. A careful combination of the strengths of these filters could provide a robust up-conversion algorithm.

A practical implementation [38] to approach this ideal uses again a median filter to remove outliers:

$$F_{out} = \text{med}\{F_{dyn}, F_{sta}, F_{mix}\} \tag{16.68}$$

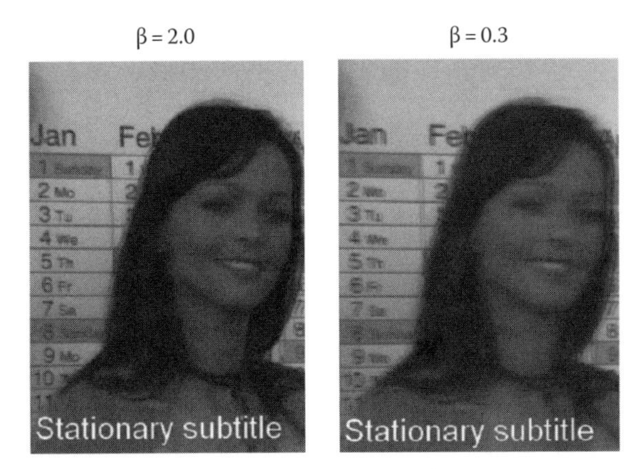

FIGURE 16.39 The influence of parameter β in reciprocal mixing. The performance ranges from something close to the static median to something similar to the dynamic median.

where the output of the dynamic median, static median, and reciprocal mix filters are represented as F_{dyn}, F_{sta}, and F_{mix}, respectively. In contrast with the description under reciprocal mix, it has been proposed in Ref. [38] to control the mix factor k with a local vector consistency measure.

The block diagram of this "cascaded median" algorithm, along with an example of the performance, is shown in Figure 16.40.

16.3.5 Occlusion-Adapted Picture Interpolation

Unfortunately, motion vectors cannot always be correct. It is particularly problematic to establish correspondences between two successive images in a sequence for background areas that are covered in one of the images by a foreground moving object. Commonly, a motion estimator that estimates displacements between these two images, as illustrated in Figure 16.34, tends to extend the foreground motion vector into the background area that is covered in one of the images. The explanation for this behavior is that, generally, the background better resembles an incorrectly shifted piece of the background than the foreground object, as illustrated in Figure 16.41.

In picture-rate conversion, the consequence is illustrated in Figure 16.42 and exists in so-called halos, visible around moving objects and due to distortions in the background that is interpolated with foreground motion vectors.

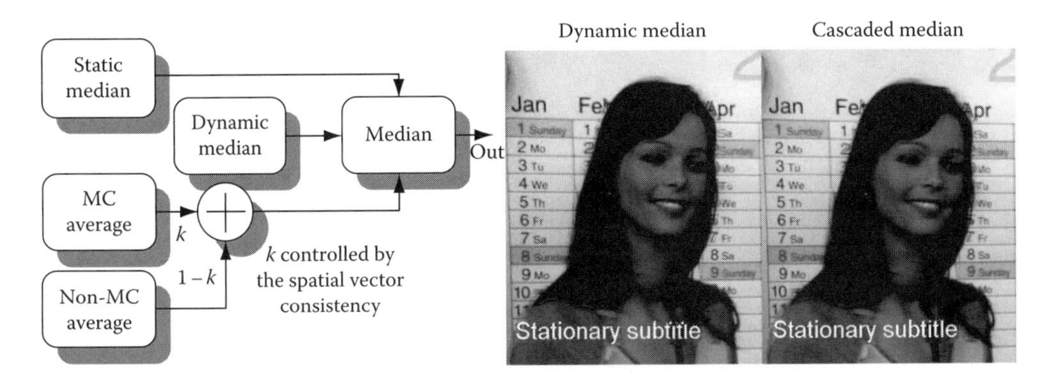

FIGURE 16.40 Block diagram of the cascaded median algorithm, and an illustration of its performance. Due to the inconsistent motion vectors near the subtitles, the static median dominates the performance there.

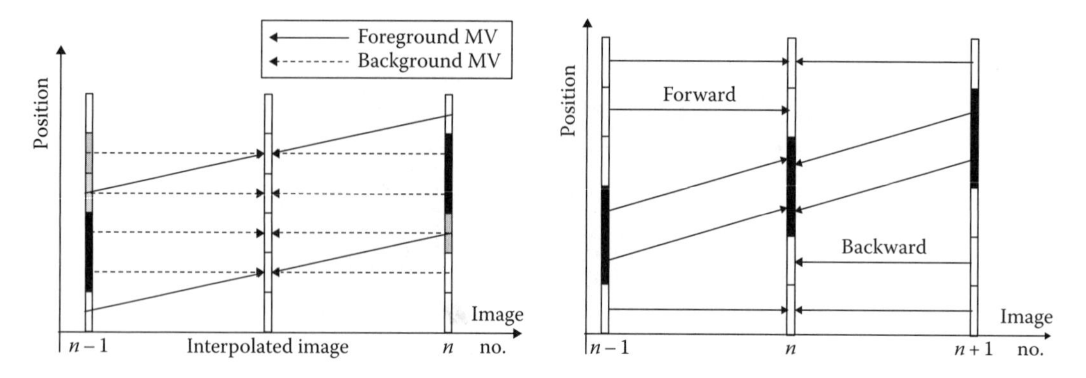

FIGURE 16.41 When matching previous and next image, the foreground vector often gives a better match in the occluded background area than the, correct, background vector (left). This is because an erroneously shifted background portion often better resembles the background than the foreground. When both previous and next image are available, it is possible to find correspondences even for occlusion areas (right). The best match occurs, either in the previous, or in the next image, depending on the kind of occlusion, i.e., covering or uncovering.

The problem is fundamental if only two images are used to estimate motion. Correct motion vectors in occlusion areas can be established easily, if in addition to a previous image a next image is also available. In this case, the occluded background in the current image can be found, either in the previous image for covering image parts, or in the next image for uncovering areas. Figure 16.41 illustrates a three-picture ME procedure, while Figure 16.42 shows the improved motion vector alignment.

The motion vector fields from this motion estimator, however, are valid for the current original image and not for the image to be interpolated at an arbitrary temporal position between two original images. Motion compensation of this motion vector field seems the obvious thing to do, but we would need again a vector field valid at the intermediate temporal instance. It seems that we end with a chicken-and-egg-problem here.

FIGURE 16.42 The extension of foreground motion vectors into background areas (left) leads to the so-called halo-artifact around moving foreground objects (middle). A motion estimator using three successive pictures leads to a more accurate vector field (right), but this is valid for the original image and not for the image to be interpolated.

References

1. M. Achiha, K. Ishikura, and T. Fukinuki. A motion-adaptive high-definition converter for NTSC color TV signals. *SMPTE Journal*, 93(5), 470–476, May 1984.
2. M. J. J. C. Annegarn, T. Doyle, P. H. Frencken, and D. A. van Hees. Video signal processing circuit for processing an interlaced video signal, April 1988. U.S. Patent 4,740,842.
3. E. B. Bellers and J. Janssen. Architectural implications for high-quality video format conversion. *Proceeding of the SPIE*, 5685(2), 672–682, 2005.
4. E. B. Bellers and G. de Haan. Advanced de-interlacing techniques. In *Proceedings of the ProRISC/IEEE Workshop on Circuits Systems and Signal Processing*, pp. 7–17, November 1996. Mierlo, the Netherlands.
5. E. B. Bellers and G. de Haan. Advanced motion estimation and motion compensated de-interlacing. In *Proceedings of the International Workshop on HDTV*, October 1996. Los Angeles, CA, Session A2, 3rd paper.
6. E. B. Bellers and G. de Haan. Advanced motion estimation and motion compensated de-interlacing. *SMPTE Journal*, 106(11), 777–786, November 1997.
7. A. M. Bock. Motion-adaptive standards conversion between formats of similar field rates. *Signal Processing: Image Communication*, 6, 275–280, June 1994.
8. W. Brüls, A. van der Werf, R. Kleihorst, T. Friedrich, E. Salomons, and F. Jorritsma. A single-chip MPEG2 encoder for consumer storage applications. In *Digest of the ICCE*, pp. 262–263, Chicago, IL, June 1997.
9. J. J. Campbell. Adaptive diagonal interpolation for image resolution enhancement, October 2000. U.S. Patent 6,133,957.
10. C. Ciuhu and G. de Haan. A two-dimensional generalized sampling theory and application to de-interlacing. In *SPIE, Proceedings of VCIP*, pp. 700–711, January 2004.
11. G. de Haan. IC for motion compensated de-interlacing, noise reduction and picture rate conversion. *IEEE Transactions on Consumer Electronics*, 45(3), 617–624, August 1999.
12. G. de Haan and E. B. Bellers. De-interlacing of video data. *IEEE Transaction on Consumer Electronics*, 43(3), 819–825, August 1997.
13. G. de Haan and P. W. A. C. Biezen. Sub-pixel motion estimation with 3-D recursive search blockmatching. *Signal Processing: Image Communications*, 6, 229–239, 1994.
14. G. de Haan, P. W. A. C. Biezen, H. Huijgen, and O. A. Ojo. True motion estimation with 3-D recursive search block-matching. *IEEE Transactions on Circuits and Systems for Video Technology*, 3(5), 368–388, October 1993.
15. G. de Haan, P. W. A. C. Biezen, H. Huijgen, and O. A. Ojo. Graceful degradation in motion compensated field-rate conversion. In *Signal Processing of HDTV*, vol. V, pp. 249–256, 1994. Elsevier, Amsterdam, the Netherlands.
16. G. de Haan, P. W. A. C. Biezen, and O. A. Ojo. Apparatus for performing motion-compensated picture signal interpolation, July 1996. U.S. Patent 5,534,946.
17. G. de Haan, J. Kettenis, and B. Deloore. IC for motion compensated 100Hz TV, with a smooth motion movie-mode. In *Digest of the ICCE*, pp. 40–41, Chicago, IL, June 1995.
18. G. de Haan and M. A. Klompenhouwer. An overview of flaws in emerging television displays and remedial video processing. *IEEE Transactions on Consumer Electronics*, 47(3), 326–334, August 2001.
19. P. Delogne, L. Cuvelier, B. Maison, B. Van Caillie, and L. Vandendorpe. Improved interpolation, motion estimation and compensation for interlaced pictures. *IEEE Transactions on Image Processing*, 3(5), 482–491, September 1994.
20. T. Doyle and M. Looymans. Progressive scan conversion using edge information. In *Signal Processing of HDTV*, vol. II, pp. 711–721, 1990. Elsevier Science Publishers, Amsterdam, the Netherlands.

21. E. W. Engstrom. A study of television image characteristics. Part two: Determination of frame frequency for television in terms of flicker characteristics. *Proceedings of the IRE*, 23(4), 295–310, April 1935.
22. P. Felzenszwalb and D. Huttenlocher. Efficient belief propagation for early vision. *International Journal of Computer Vision*, 70(1), 41–54, October 2006.
23. P. D. Filliman, T. J. Christopher, and R. T. Keen. Interlace to progressive scan converter for IDTV. *IEEE Transactions on Consumer Electronics*, 38(3), 135–144, August 1992.
24. P. Haavisto, J. Juhola, and Y. Neuvo. Scan rate up-conversion using adaptive weighted median filtering. In *Signal Processing of HDTV*, vol. II, pp. 703–710, 1990. Elsevier Science Publishers, Amsterdam, the Netherlands.
25. M. Hahn, G. Scheffler, P. Rieder, and C. Tuschen. Motion-compensated frame interpolation, March 2004. EU Patent 1,397,003.
26. C. Hentschel. Comparison between median filtering and vertical edge controlled interpolation for flicker reduction. *IEEE Transactions on Consumer Electronics*, 35(3), 279–289, August 1989.
27. C. Hentschel. Television with increased image quality; flicker reduction through an increased picture rate in the receiver, Februari 1990. *Institut fur Nachrichtentechnik, Tech. Univ. Braunschweig*, Germany, (in German).
28. J. W. Woods and Soo-Chul Han. Hierarchical motion compensated de-interlacing. *Proceedings of the SPIE*, 1605, 805–810, 1991.
29. Y. Kim and Y. Cho. Motion adaptive deinterlacing algorithm based on wide vector correlations and edge dependent motion switching. In *Proceedings of HDTV Workshop*, Taipei, Taiwan, pp. 8B9–8B16, 1995.
30. T. Koga, K. Iinuma, A. Hirano, Y. Iilima, and T. Ishiguro. Motion-compensated interframe coding for video conferencing. In *Proceedings of the National Telecommunications Conference*, vol. 4, pp. 531–534, December 1981. New Orleans LA.
31. V. Kolmogorov and R. Zabih. Computing visual correspondence with occlusions using graph cuts. In *Proceeding of the International Conference on Computer Vision*, pp. 508–515, July 2001. Vancouver, British Columbia, Canada.
32. J. Kovacevic, R. J. Safranek, and E. M. Yeh. Deinterlacing by successive approximation. *IEEE Transactions on Image Processing*, 6(2), 339–344, February 1997.
33. J. S. Kwon, K. Seo, J. Kim, and Y. Kim. A motion-adaptive de-interlacing method. *IEEE Transactions on Consumer Electronics*, 38(3), 145–150, August 1992.
34. A. Lethonen and M. Renfors. Non-linear quincunx interpolation filtering. In *Proceedings of the SPIE Visual Communication and Image Processing*, pp. 135–142, October 1990. Lausanne, Switzerland.
35. Microsoft Corp. Broadcast-enabled computer hardware requirements, 1997. WinHEC, *Broadcast Technologies White Paper*, pp. 11–12, 1997.
36. A. Nguyen and E. Dubois. Spatio-temporal adaptive interlaced-to-progressive conversion. In *Signal Processing of HDTV*, vol. IV, pp. 749–756, 1993. Elsevier Science Publishers, Amsterdam, the Netherlands.
37. Y. Ninomiya and Y. Ohtsuka. A motion compensated interframe coding scheme for television pictures. *IEEE Transactions on Communications*, 30(1), 201–211, January 1982.
38. O. A. Ojo and G. de Haan. Robust motion-compensated video up-conversion. *IEEE Transactions on Consumer Electronics*, 43(4), 1045–1055, November 1997.
39. Datasheet SAA4990, PROZONIC, 1995. Philips Semiconductors.
40. R. S. Prodan. Multidimensional digital signal processing for television scan conversion. *Philips Journal of Research*, 41(6), 576–603, 1986.
41. J. Salo, Y. Nuevo, and V. Hameenaho. Improving TV picture quality with linear-median type operations. *IEEE Transactions on Consumer Electronics*, 34(3), 373–379, August 1988.

42. J. Salonen and S. Kalli. Edge adaptive interpolation for scanning rate conversion. In *Signal Processing of HDTV*, vol. IV, pp. 757–764, 1993. Elsevier Science Publishers, Amsterdam, the Netherlands.
43. N. Seth-Smith and G. Walker. Flexible upconversion for high quality TV and multimedia displays. In *Digest of the ICCE*, pp. 338–339, June 1996. Chicago, IL.
44. R. Simonetti, S. Carrato, G. Ramponi, and A. Polo Filisan. Deinterlacing of HDTV images for multimedia applications. In *Signal Processing of HDTV*, vol. IV, pp. 765–772, 1993. Elsevier Science Publishers, Amsterdam, the Netherlands.
45. A. Tekalp. *Digital Video Processing*, 1995. Prentice-Hall, Upper Saddle River, NJ. ISBN: 0-13-190075-7.
46. G. A. Thomas. Television motion measurement for DATV and other applications. Technical report, BBC Research Report, November 1987. BBC RD 1987/11.
47. G. J. Tonge. Television motion portrayal, September 1985. *Presented at Les. Assises des Jeunes Chercheurs*, Rennes, France.
48. A. W. M. van den Enden and N. A. M. Verhoeckx. *Discrete-Time Signal Processing*, 1989. Prentice Hall, Upper Saddle River, NJ.
49. L. Vandendorpe, L. Cuvelier, B. Maison, P. Quelez, and P. Delogne. Motion-compensated conversion from interlaced to progressive formats. In *Signal Processing: Image Communication*, vol. 6, pp. 193–211, 1994. Elsevier, Amsterdam, the Netherlands.
50. F. Wang, D. Anastassiou, and A. Netravali. Time-recursive deinterlacing for IDTV and pyramid coding. *Signal Processing: Image Communications*, 2, 365–374, 1990.
51. J. L. Yen. On nonuniform sampling of bandwidth-limited signals. *IRE Transactions on Circuit Theory*, 3, 251–257, December 1956.
52. M. Zhao, C. Ciuhu, and G. de Haan. Classification based data mixing for hybrid de-interlacing techniques. In *Proceedings of the 13th European Signal Processing Conference (EUSIPCO)*, September 2005. Antalya, Turkey.

17

Document Modeling and Source Representation in Content-Based Image Retrieval

Soo Hyun Bae
Sony US Research Center

Biing-Hwang Juang
Georgia Institute of Technology

17.1 Introduction

With the rapid growth of digitized images available, efficient and robust techniques for managing the images have become increasingly more important. A particular challenge is that it is considerably difficult to organize the raw digital form of images in a semantically sensible structure for easy indexing and retrieval. There has been a plethora of research that tackles this problem by analyzing multidimensional data and extracting features from them for the purpose of retrieval.

Generally, there are primarily three groups of image retrieval techniques according to the query modalities. Initial attempts at organizing the stored images in a semantically relevant manner led to the use of precompiled metadata associated with them, e.g., filename, caption, and even geographical data. This group of systems is called *text-based image retrieval*. They first try to limit or reduce semantic ambiguity in the given query, then search only for a match between the keywords associated with the query and each image in the database. Many of the current commercial image search engines, e.g., Google and Yahoo!, employ this type of systems. The need for a rapid search and retrieval of desired images by analyzing the inferred semantic notions of the stored images continues to receive great attention in both research and commercial domains. The second group is *content-based image retrieval* (CBIR) that accepts various forms of imagery queries, such as imagery objects, synthetic scene, and hand drawing, which convey the full or a part of the querying concept. To our best knowledge, among the first CBIR systems are IBM QBIC [1] in the commercial domain and MIT Photobook [2] in a research domain. Since then, various types of feature extraction, image modeling, and machine learning techniques have been proposed. Finally, *interaction-based image retrieval* is capable of interacting with users in order to process the queries and improve the indexing results.

Before the study of CBIR became active, text document retrieval has been extensively investigated in many research sectors, and various aspects of the algorithms have been well studied and successfully applied to commercial WEB document search engines as well as prototype systems. The success has later stimulated CBIR research to apply such techniques to image database modeling, and significant achievements have been made since then. However, one important problem, the issue of source representation, has not been given sufficient attention. In many CBIR systems as well as image analysis applications, more than frequently a given image is partitioned into a number of fixed-shape blocks which are fed into further processes. This is likely due to the important role of block transforms such as discrete cosine transform (DCT) in conventional image processing systems. Or in some other systems, a few objects from a given image are extracted by image segmentation techniques, and the semantic meanings are estimated from the objects. However, it is still unclear that those image representation techniques are the most efficient ones for image analysis applications. In recent research [3], a new class of source representation based on a universal source coding has been proposed, and in a subsequent study it was applied to the design of CBIR systems. In this chapter, we set our focus on the two particular dimensions of CBIR systems: machine learning techniques originated from natural language processing (NLP) and recent advances of source representations. Note that a comprehensive survey of image retrieval techniques can be found in Refs. [4,5].

17.2 Document Modeling in Content-Based Image Retrieval

Historically there have been two main issues in linguistic information processing: *synonymy* and *polysemy*. Recent advances in document retrieval have successfully dealt with these two problems and have been implemented into many commercial products. The success has become a strong motivation for employing the techniques in CBIR systems. In this section, we study two classes of document modeling techniques and their applications to CBIR.

17.2.1 Latent Semantic Analysis

One fundamental observation behind the latent semantic analysis (LSA) technique [6] is that similarities among documents may be measured by the occurrences of conceptual words in the documents. Basically what LSA does is to project a given set of document onto a lower-dimensional conceptual space and provide a framework for measuring the similarity among documents in that space. The main assumption of LSA is that "a bag of words" is capable of retaining enough information for semantic association for document retrieval applications.

Let D be a collection of N documents $D = \{d_1, \ldots, d_N\}$ and W be a lexicon with M words $W = \{w_1, \ldots, w_M\}$. The vector space model [7] represents each document in D as an M-dimensional vector. Thus, for the given N documents, we generate a co-occurrence matrix $L = \{l_{i,j}\}$ where the jth column corresponds to the occurrence vector of the jth document. Then, by the standard singular value decomposition (SVD) the co-occurrence matrix is decomposed into $L = U\Sigma V^{\mathrm{T}}$, where U and V are unitary matrices, and the superscript "T" denotes matrix transpose. By taking K largest singular values in Σ, L is approximated as $\hat{L}_K = \hat{U}\hat{\Sigma}_K\hat{V}^{\mathrm{T}}$, where each column vector $\hat{\mathbf{d}}_j$ corresponds to the document vector projected onto K-reduced space, called "latent semantic space." If a query is given in a form of a document, similarly it is represented as an M-dimensional vector $\mathbf{q}$, and it is mapped onto the K-reduced space by

$$\hat{\mathbf{q}} = \hat{\Sigma}_K^{-1} U^{\mathrm{T}} \mathbf{q}. \tag{17.1}$$

Then, the similarity between the jth document in D and the query document can be computed as

$$s(\hat{\mathbf{d}}_j, \hat{\mathbf{q}}) = \frac{\hat{\mathbf{d}}_j^{\mathrm{T}} \hat{\mathbf{q}}}{\|\hat{\mathbf{d}}_j\| \|\hat{\mathbf{q}}\|}. \tag{17.2}$$

If $s(\hat{\mathbf{d}}_j, \hat{\mathbf{q}})$ is 1.0, the two documents are semantically equivalent while decreasing values of $s(\hat{\mathbf{d}}_j, \hat{\mathbf{q}}) < 1.0$ mean decreasing similarity between them.

In addition to the basic formulation of LSA above, by taking term frequency (TF) and interdocument frequency (IDF) into account, one can achieve better separability and higher retrieval precision [8]. According to document size and word entropy, the co-occurrence matrix L can be weighted by a TF-normalized entropy. The normalized entropy ε_i for ith word is

$$\varepsilon_i = -\frac{1}{\log N}\sum_{j=1}^{N}\frac{l_{i,j}}{t_i}\log\frac{l_{i,j}}{t_i}, \tag{17.3}$$

where $t_i = \sum_{j=1}^{N} l_{i,j}$ is the number of occurrences of ith word in D. A normalized co-occurrence matrix $L_n = \{\tilde{l}_{i,j}\}$ is computed as

$$\tilde{l}_{i,j} = (1-\varepsilon_i)\frac{l_{i,j}}{\Sigma_i l_{i,j}}. \tag{17.4}$$

In image retrieval, different from document retrieval, the definition of the *visual word* is still nebulous; thus, each image retrieval system generates the co-occurrence in a different way depending on the chosen definition of lexicon. To our best knowledge, the LSA-based CBIR research in the literature was first reported by Pečenović [9]. The CBIR system, called "Circus," generates a feature vector of a given image by exploiting color histogram, angular second moment, inverse difference moment, and so on. By collating all the feature vectors from the given image database, the co-occurrence matrix is generated and decomposed by SVD, and then the similarities between images are measured by Equation 17.2. Cascia et al. [10] also employed the LSA paradigm in image database indexing. However, the system implemented used manually annotated metadata as a word lexicon; hence, it is not a CBIR indeed. Similar to [9], Zhao et al. used a histogram of hue and saturation components of images or subimages, and generated the co-occurrence matrix [11]. In addition to the LSA-based image retrieval framework, they also attempted to combine the image features with manually annotated data in order to improve the retrieval performance. Also, some different types of visual features have been used under the same paradigm, for example, luminance components [12], text image patches [13], or patch clustering labels [14]. In Refs. [15,16], variable-size image patches generated by a universal source coding technique, called the parsed representations, are employed for CBIR systems and the performance of the proposed system is compared to that of a benchmark system based on a fixed-shape block representation. The details of the parsed representation will be discussed in Section 17.3.3.

17.2.2 Probabilistic Latent Semantic Analysis

Although the LSA paradigm has been widely implemented in various types of information retrieval systems, several fundamental limitations of LSA have been of concern for a decade. First, the approximate co-occurrence matrix in LSA forms a joint Gaussian distribution while a Poisson is observed to be more relevant to count data [17, Section 15.4.3]. Second, since the approximate co-occurrence matrix is a Gaussian distribution, elements in the matrix may not be larger than, or equal to, zero for term counts, which is not sensible. One of the notable advances in text document retrieval is the graphical modeling of linguistic words and documents in a probabilistic setup with an "aspect modeling" technique trained by probabilistic latent semantic analysis (pLSA) [18], which overcomes the aforementioned drawbacks.

Similar to the setup in Section 17.2.1, we are given a collection of documents $D = \{d_1, \ldots, d_N\}$ and a lexicon $W = \{w_1, \ldots, w_M\}$. From the co-occurrence observation, pLSA estimates the term–document joint distribution $P(w, d)$ that minimizes the Kullback–Leibler (KL) divergence [19] with respect to the empirical distribution $p(w, d)$ subject to K latent aspects. Basically, aspect modeling posits that each document is a mixture of K latent aspects $z_k \in Z = \{z_1, \ldots, z_K\}$. Also, the modeling has two independence assumptions: observations pairs (w_i, d_j) are generated independently and the pairs of random variables (w_i, d_j) are conditionally independent given the hidden aspect z_k, i.e.,

$$P(w_i, d_j|z_k) = P(w_i|z_k)P(d_j|z_k). \tag{17.5}$$

By marginalizing Equation 17.5 over z_k, the joint distribution of the observation pairs is

$$\begin{aligned} P(w_i, d_j) &= P(d_j)P(w_i|d_j) \\ &= P(d_j) \sum_{z_k \in Z} P(w_i|z_k)P(z_k|d_j). \end{aligned} \tag{17.6}$$

The graphical model representation is shown in Figure 17.1 corresponding to a joint distribution $P(w, d, z)$. A document d_j is selected with the probability $P(d_j)$, and for each aspect distribution $P(z_k|d_j)$, each word w_i is selected with probability $P(w_i|z_k)$.

Following the maximum likelihood principle, one determines $P(w_i|z_k)$ and $P(z_k|d_j)$ by maximization of the likelihood, given observed data. The likelihood is defined by

$$\mathcal{L} = \prod_{j=1}^{N} \prod_{i=1}^{M} P(d_i) \sum_{k=1}^{K} P(z_k|d_j)P(w_i|z_k)^{p(w_i,d_j)}, \tag{17.7}$$

equivalently

$$\log \mathcal{L} = \sum_{j=1}^{N} \sum_{i=1}^{M} p(w_i, d_j) \log P(w_i, d_j). \tag{17.8}$$

This means that the pLSA attempts to minimize the cross entropy of $p(w, d)$ and $P(w, d)$:

$$H(p(w, d), P(w, d)) = -\sum_{j=1}^{N} \sum_{i=1}^{M} p(w_i, d_j) \log P(w_i, d_j), \tag{17.9}$$

Also it can be interpreted as minimization of KL divergence of the two distributions, one empirical and the other taking the form of a mixture. The model parameters are trained by the expectation-maximization (EM) algorithm [20]. In the expectation step, the conditional probability distribution

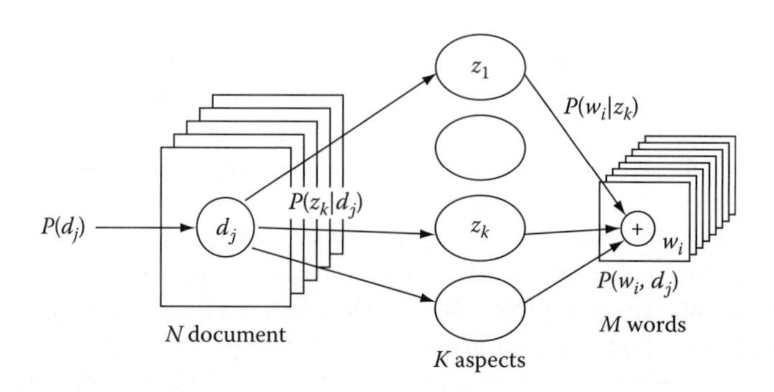

FIGURE 17.1 Graphical model of pLSA for a corpus with N documents and an M-words lexicon subject to K latent aspects. The documents and the words are observed.

of the kth latent aspect z_k given the observation pair (w_i, d_j) is computed based on the parameters estimated at the previous iteration:

$$P(z_k|w_i, d_j) = \frac{P(w_i|z_k)P(z_k|d_j)}{\sum_{k'=1}^{K} P(w_i|z_{k'})P(z_{k'}|d_j)}. \tag{17.10}$$

In the maximization step, the parameters of the two multinomial distributions $P(w_i|z_k)$ and $P(z_k|d_j)$ are updated with the conditional probability distribution $P(z_k|w_i, d_j)$:

$$P(w_i|z_k) = \frac{\sum_{j=1}^{N} p(w_i, d_j)P(z_k|w_i, d_j)}{\sum_{i=1}^{M}\sum_{j=1}^{N} p(w_i, d_j)P(z_k|w_i, d_j)}, \tag{17.11}$$

$$P(z_k|d_j) = \frac{\sum_{i=1}^{M} p(w_i, d_j)P(z_k|w_i, d_j)}{p(d_j)}. \tag{17.12}$$

From this EM procedure, the pLSA estimates the two multinomial distributions $P(w|z)$ and $P(z|d)$ and finally the term–count joint distribution $P(w, d)$.

In practical applications of the aspect modeling technique, the observed pairs may not be sufficiently reliable for dealing with unseen test data. Thus, it is necessary to control the prediction trade-off between the training data (the empirical distribution) and the unseen data by a regularization term (on top of the implicit smoothing by the aspect mixture). To deal with this problem, in [18], a tempered EM (TEM) replaces Equation 17.10 with

$$P_\beta(z_k|w_i, d_j) = \frac{P(z_k)[P(w_i|z_k)P(d_j|z_k)]^\beta}{\sum_{z'_k=1}^{K} P(z')[P(w_i|z'_k)P(d_j|z'_k)]^\beta}, \tag{17.13}$$

where β is a control parameter that scales the likelihood function.

In the image retrieval domain, three research groups simultaneously reported three different applications of the aspect modeling technique in the same conference, *IEEE Conference on Computer Vision* (ICCV) 2005. In Ref. [21], Quelhas et al. proposed a scene categorization technique that learns from a bag-of-visterm model. From a given set of images, the technique generates a visual vocabulary with difference of Gaussians (DoG) features and scale-invariant feature transform (SIFT) descriptors [22], which are invariant to various scene perturbations, e.g., translation, rotation, and illumination. Then, it classifies the extracted features into a finite set of visterms by the k-means algorithm. The latent aspects of the bag-of-visterm co-occurrence are learned by pLSA and three-class classification systems were tested. Sivic et al. proposed an unsupervised object discovery technique based on the aspect modeling [23]. Image features are extracted by SIFT descriptors on affine covariant regions and then quantized by the k-means algorithm to be registered into a visual vocabulary. Again, once the feature-image co-occurrence is generated, its set of aspects is learned. The performance of the proposed technique is evaluated in three different dimensions: topic discovery, image classification, and image segmentation. Finally, Fergus et al. incorporated spatial information and scale-invariant features into the two modified aspect modeling frameworks for category recognition [24]. In absolute position pLSA (ABS-pLSA), instead of observing the word–document distribution $p(w, d)$, the word–location–document joint distribution $p(w, x, d)$ is learned in order to embed the location information of features into the framework. Thus, Equation 17.6 is replaced with

$$P(w, x, d) = P(d)\sum_{z\in\mathcal{Z}} P(w, x|z)P(z|d). \tag{17.14}$$

In translation and scale-invariant pLSA (TSI-pLSA), by employing the centroid of the object within each image, represented as a latent variable c, the model becomes invariant to scale and translation as follows:

$$P(w, x|z) = \sum_{c \in C} P(w, x, c|z) = \sum_{c \in C} P(w, x|c, z)P(c). \tag{17.15}$$

Since then, there has been a number of techniques based on pLSA for image retrieval as well as some homogeneous applications. Zhang et al. used image segments as visual words and trained the aspect models by pLSA [25]. Monay and Gatica-Perez [26] used annotated data and textual modality for visual words and Bosch et al. [27] attempted various visual patches, e.g., gray/color patches or SIFT descriptors, then applied the pLSA paradigm for image retrieval.

17.3 Source Representation in Content-Based Image Retrieval

In any CBIR, or more widely in any visual information analysis, "feature extraction" from a given set of imagery data is one of the key components in the design of the system. In the literature, a sizable number of feature extraction techniques has been proposed, the primary process of those techniques is "source representation." A given image is represented by a number of small chunks of pixels with typically two types of techniques: fixed-shape partition and image segmentation. In addition, some techniques are based on global pixel statistics or on salient regions (e.g., by SIFT descriptors and its variants), but not many. Recently, a new representation technique, called parsed representation, was proposed in Ref. [3] and has been applied to image compression and CBIR. In this section, we discuss the details of the three source representation techniques.

17.3.1 Fixed-Shape Block Representation

One of the traditional and prevalent source representations is a fixed-shape block representation due to its computational simplicity. Not just in CBIR, the fixed-shape block representations have also been used in standard image and video compression techniques for a few decades.

The simplest way of generating the fixed-shape block representation is to partition a given image into a number of $k_1 \times k_2$ block-shape pixel patches. The size $k_1 \times k_2$ may be either constant or variable in the given image domain. The primary purpose of the use of variable-size blocks is to capture the global and the local statistics on the set of blocks. In large blocks, one can estimate wide-range color or texture statistics that are more or less sufficient for representing global semantic information. On the other hand, in small blocks, local texture or edge information can be more sensibly obtained. The size of block at the current pixels of interest can be determined by the amount of resources (in compression applications), a measurement of detailedness (e.g., pixel entropy, second-order statistics), or the size of the given image. One example of fixed-shape block representation is given in Figure 17.2a.

In CBIR, this representation has been employed in accordance with the design of feature extraction and its corresponding similarity measures. Table 17.1 provides a comparison of CBIR systems based on the fixed-shape block representations. Note that the systems in the table are only a small portion of the entire CBIR systems based on the representation. Most of the systems in the table determine the size of the blocks for their specific applications except the two systems by Smith and Chang [28] and Teng and Lu [29]. The first system uses quadtree indexing for single-region query, while the second system attempted to design a set of vector quantization (VQ) codebooks according to the block size in order to evaluate the performance variation of the block size. Although most of the systems partition the given images into a number of nonoverlapping blocks, the two systems proposed by Carneiro et al. [30] and Natsev et al. [31] allow overlaps by a small number of pixels. As shown in Table 17.1, VQ has been popularly used for clustering the

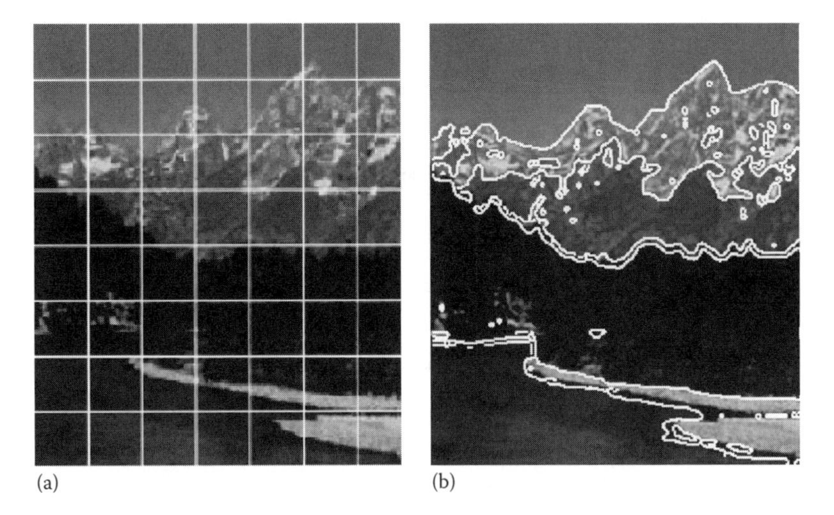

FIGURE 17.2 (a) Image is partitioned into 32 × 32 square pixel regions. (b) image is segmented into six perceptually homogeneous regions.

TABLE 17.1 CBIR Systems Based on Fixed-Shape Block Representations

CBIR System	Block Size	Image Space	Clustering	Feature Size	Overlap
Monay and Gatica-Perez [26]	32 × 32	Hue-saturation (HS)	*k*-means	96	None
Smith and Chang [28]	Quadtree	Hue-saturation-value (HSV)	VQ	166	None
Teng and Lu [29]	2 × 2 ~ 16 × 16	RGB, HSV, L*u*v	VQ	128 ~ 4096	None
Carneiro et al. [30]	8 × 8	YCbCr	Gaussian mixture model	64	Overlap
Natsev et al. [31]	64 × 64	Wavelet	Preclustering phase	Variable	Overlap
Vogel and Schiele [32]	10 × 10	HSV	Support vector machine	180	None
Fei-Fei and Perona [33]	11 × 11	Gray value	*k*-means	174	None
Qiu [34]	4 × 4	RGB	VQ	256	None
Yeh and Kuo [35]	4 × 4	Wavelet	VQ	256	None
Jeong et al. [36]	2 × 2	HSV	Gaussian mixture VQ	256	None
Vailaya et al. [37]	10 × 10	L*u*v	VQ	10, 15, 30	None

extracted features in order to generate a compact representation of a given image. In Section 17.3.4, we will compare the performance of the fixed-shape block representation trained by VQ with that of other source representations in terms of retrieval precisions of some CBIR systems.

17.3.2 Segmentation-Based Representation

Image segmentation is generally defined as following: For a given $m \times n$ image with c components,

$$X = \left\{ \vec{X}_{x_1,x_2} : 0 \le x_1 < m, 0 \le x_2 < n, \vec{X} \in \mathbb{R}^c, m, n \in Z_+ \right\}, \tag{17.16}$$

TABLE 17.2 CBIR Systems Based on Segmentation-Based Representations

CBIR System	Segmentation Technique	Image Space	# of Objects
Zhang and Zhang [25]	*k*-means	L*a*b*, Haar wavelet	$5\sim6$
Wang et al. [39]	*k*-means	L*u*v*	$2\sim16$
Pappas et al. [40]	Adaptive perceptual segmentation [38]	L*a*b*	$1\sim4$
Jeon et al. [41]	Normalized cuts [42]	RGB	$1\sim10$
Liu and Zhou [43]	Texture segmentation	Wavelet	Varying
Sural et al. [44]	*k*-means	HSV	$2\sim12$

segmentation is a process of partitioning a given image into a finite number of connected regions:

$$\bigcup_{l=1}^{k} S_l = X \quad \text{with } S_i \bigcap S_j = \emptyset, \quad \text{for } i \neq j. \tag{17.17}$$

One example of image segmentation is given in Figure 17.2b, in which the given image is partitioned into six regions; sky, mountain with snow, mountain, forest, beach, and river. Image segmentation has been widely used in many image processing and computer vision applications. Also, it is one of the important and challenging components of many CBIR systems, since it partitions a given image into perceptually homogeneous regions. From each segmented region, one can extract scene or object features from each segment so as to estimate its semantic meaning. Once the given image's semantic meaning is estimated, whether it is in the notion of linguistic expressions or numeric indices, CBIR systems then compare the semantic similarity between the query image and the images from the database.

Generally speaking, the segmentation-based representation is less favored than the fixed-shape block representation in CBIR. It is noted in Refs. [38,39] that a complete and detailed description of every object in the given image is a very hard task; even humans cannot easily produce consistent and homogeneous results. However, since it can group such pixels that have similar perceptual significance into one specific region, it can better contribute to the perceptual and semantic classification of images toward an ideal CBIR system than the fixed-shape block representation. In that sense, there have been various types of CBIR systems based on the representation in the literature. Table 17.2 compares some of the CBIR systems based on the representation. Apparently most of the systems in the table are based on *k*-means clustering, even including an adaptive perceptual segmentation by Pappas et al. [40]. One interesting point is that the color spaces of the CBIR systems are different from each other, since there is no color space that is most prominent in image segmentation for extracting high-level semantics. Once a given image is partitioned into a small number of object regions, the procedures for estimating high-level notions from the low-level features extracted from the segmented regions are usually followed. These tasks are primarily accomplished with pattern recognition/classification as well as machine learning techniques.

17.3.3 Parsed Representation

In Sections 17.3.1 and 17.3.2, we studied two popular source representations primarily for CBIR applications. In the literature, either of the representations has been chosen without paying strong attention to how well it is relevant to capturing high-level semantics from primitive pixel information. However, many natural objects are not tessellations of fixed-shape blocks and machine segmentation of images is still away from mimicking the way a human would identify real-world objects. Let us consider another case. When a human is given a linguistic sequence of symbols, say the letters, she or he tries to parse the sequence in unit of linguistic words which are the smallest unit for conveying a semantic meaning. Then, one can associate the words with other words toward phrases, sentences, paragraphs, and finally documents by applying syntax and semantics. These structural rules have played a key role in the success of the recent document retrieval systems. However, if the sequence of symbols is parsed in unit of four, it becomes very difficult for machines as well as humans to interpret its semantic meaning. This example applies not only to linguistic documents but also to two-dimensional images. Thus, it is unclear that a conventional fixed-block or segmentation-based representation is the best practice for semantic higher-level processing. To tackle this source representation problem, a new class of source representation generated by a universal source coding was recently proposed in Ref. [3]. In the following section, we study the multidimensional incremental parsing technique and the application to CBIR.

17.3.3.1 Lempel–Ziv Incremental Parsing

Recently some universal source coding techniques, i.e., the Lempel–Ziv incremental parsing scheme (LZ78) [45] have been found to have a similar nature to the linguistic parsing. The scheme is originally designed for encoding a given sequence of finite symbols with a minimum number of bit resources by reducing the redundancy of the given source. Also, it is theoretically proved that for a stationary ergodic source, the coding scheme is asymptotically optimal and achieves a source rate approaching the entropy of the source without any prior knowledge of the statistical distribution of the given source.

Let $\mathbf{X} = \{X_t\}_{t=1}^{n}$ be a stationary ergodic sequence taking values from a finite alphabet $\mathcal{A}$ with cardinality $|\mathcal{A}| < \infty$. For a given dictionary $\mathbb{D}$, which is initially empty, at each coding epoch LZ78 searches the longest match from the dictionary and gives the dictionary index j, called "pattern matching." Since at the decoder the given symbols are reconstructed without information loss, the searching procedure finds the precise match with the source symbols at the current coding location. Then, the dictionary is augmented with the last parsed phrases $\mathbb{D}_j$ appended with the next source symbol X_i, denoted by $\mathbb{D}_j \circ X_i$. It then transmits the codeword $\{j, X_i\}$ to the decoder by spending $\lceil \log_2 \Gamma \rceil + \lceil \log_2 |\mathcal{A}| \rceil$ bits, where $\lceil x \rceil$ denotes the least integer not smaller than x, and Γ corresponds to the number of dictionary entries. Accordingly, the total length of the code generated by LZ78 is

$$\mathcal{L}(\mathbf{X}) = \Gamma \cdot \left(\lceil \log_2 \Gamma \rceil + \lceil \log_2 |\mathcal{A}| \rceil \right). \tag{17.18}$$

17.3.3.2 Multidimensional Incremental Parsing

Because of the aforementioned rich properties, many research attempts have been made to apply the coding algorithm to higher-dimensional compression applications (e.g., image or video compression). To the best of our knowledge, the first trial was made by Lempel and Ziv, reported in Ref. [46]. The technique was designed to linearize a given two-dimensional discrete source by a Hilbert–Peano space-filling curve in order to fit for the use of the one-dimensional coding scheme as many following research efforts do. However, it is not clearly verified that any of the linearization methods is capable of analyzing the local property of a given higher-dimensional source so as to allow the one-dimensional coding scheme to achieve the optimal coding rate.

In Ref. [3], a new source coding technique has been proposed, which accounts for a generalized incremental parsing scheme for a multidimensional discrete source. Since we are aiming at the application of the coding scheme to CBIR, here we limit our focus on two-dimensional universal source coding. As LZ78 does, the two-dimensional incremental parsing scheme starts with an empty dictionary entry or some preregistered entries. At each coding epoch, it first searches the dictionary for the maximum decimation or the maximal patch that matches the two-dimensional source at the current coding point according to a given distortion criterion $\rho(\cdot)$. Then, it augments the dictionary with two new entries, which are incrementally expanded along the horizontal and the vertical axes. For example, if the match found is of size $n_1 \times n_2$, the new entries then are of $(n_1 + 1) \times n_2$ and $n_1 \times (n_2 + 1)$ in size. Note that for an m-dimensional incremental parsing, m augmentative patches of $m - 1$ dimensions generated along all the m-axes are appended into the dictionary at each coding epoch. Let $\mathbf{X}$ be a two-dimensional vector field taking values from a set of three-dimensional finite vectors. Each element of the vector represents each component of the pixel. In this implementation, each element corresponds to red, green, and blue color channel. $\mathbf{X}(\vec{x})$ denotes the vector at the location $\vec{x} \in \mathbb{Z}^2$. Also, a subset of $\mathbf{X}$ for an area vector $\vec{a} \in \mathbb{Z}^2$ is defined as

$$\mathbf{X}(\vec{x};\vec{a}) = \{\mathbf{X}(\bar{x}_1,\bar{x}_2) : x_i \leq \bar{x}_i < x_i + a_i,\ i = 1,2\}. \tag{17.19}$$

Given a dictionary $\mathbb{D}$, $|\mathbb{D}|$ and $[\mathbb{D}_j]$ denote the number of elements of $\mathbb{D}$ and an area vector whose element represents the number of pixels of the jth patch along each axis, respectively. Also, $|\mathbb{D}_j|$ corresponds to the number of symbols of the jth patch.

In the CBIR system called IPSILON [15,16] for a lossy color image compression, a minimax distortion function is employed:

$$\rho_{\text{m}}(\mathbf{X},\hat{\mathbf{X}}) = \max_{\vec{x}\in\mathbf{X}} \left\{ \max_{c\in\{r,g,b\}} \left(0, \frac{|\mathbf{X}_c(\vec{x}) - \hat{\mathbf{X}}_c(\vec{x})| - \mathcal{T}_c(\vec{x})}{\mathcal{T}_c(\vec{x})} \right) \right\}, \tag{17.20}$$

where $\hat{\mathbf{X}}$ denotes an approximation of $\mathbf{X}$ and $\mathcal{T}_c(\vec{x})$ is the distortion threshold of the corresponding color component at the location $\vec{x}$.

In their implementation, the threshold values are computed with a color just-noticeable distortion (JND) model. The JND values represent the threshold levels of distortion visibility below which human visual system cannot perceive the distortion. A suprathreshold color image compression can be achieved by a minimally noticeable distortion (MND) model, which is simply multiplying every element of JND by a constant scale factor:

$$\mathcal{T}_{\text{mnd}} = \mathcal{T}_{\text{jnd}} \times \theta_{\text{jnd}}. \tag{17.21}$$

In the implementation, $\mathcal{T}_{\text{mnd}}$ is used for the threshold value $\mathcal{T}_c$ in Equation 17.20. In the CBIR system called AMPARS [47], instead of the minimax distortion function, a color structural texture similarity measure for color image patches is proposed as following:

$$\rho_s(\mathbf{X},\hat{\mathbf{X}}) = w_{\text{Y}}\text{STSIM}(\mathbf{X}_{\text{Y}},\hat{\mathbf{X}}_{\text{Y}}) + w_{\text{C}_b}\text{SSIM}(\mathbf{X}_{\text{C}_b},\hat{\mathbf{X}}_{\text{C}_b}) + w_{\text{C}_r}\text{SSIM}(\mathbf{X}_{\text{C}_r},\hat{\mathbf{X}}_{\text{C}_r}), \tag{17.22}$$

where STSIM($\cdot$) and SSIM($\cdot$) are the structural texture similarity metric proposed by Zhao et al. [48] and the structural similarity metric proposed by Wang et al. [49]. Also, w_{Y}, w_{C_b}, and w_{C_r} are the weights for each color component.

At each coding epoch, the parsing scheme finds the match that covers a maximum previously uncoded area, called "maximum decimation matching." If the source vector $\mathbf{X}(\vec{x})$ is already encoded, then $\vec{x} \in \mathbf{E}$, otherwise it is in $\mathbf{E}^c$. And the decimation field $\mathbf{F}(\vec{x})$ is defined as

$$\mathbf{F}(\vec{x}) = \begin{cases} 0, & \vec{x} \in \mathbf{E} \\ 1, & \vec{x} \in \mathbf{E}^c. \end{cases} \tag{17.23}$$

As in Equation 17.19, the decimation field area for an area vector $\vec{a}$ is

$$\mathbf{F}(\vec{x};\vec{a}) = \{F(\bar{x}_1,\bar{x}_2) : x_i \leq \bar{x}_i < x_i + a_i,\ i = 1,2\}. \tag{17.24}$$

Then, the set of indices by ε_{m}-bounded distortion at the current coding epoch is

$$\mathbf{H} = \{j | \rho(\mathbb{D}_j, \mathbf{X}(\Delta; [\mathbb{D}_j])) \leq \varepsilon_{\text{m}}, 0 \leq j < |\mathbb{D}|, \varepsilon_{\text{m}} \in \mathbb{R}_+\}, \tag{17.25}$$

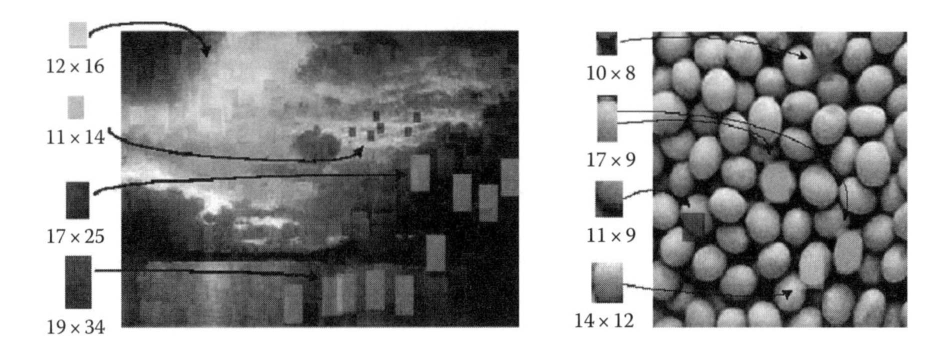

FIGURE 17.3 Example of the two-dimensional incremental parsing for image compression.

where Δ corresponds to the current coding location. Typically, ε_m is set to 0. Now, the set **H** contains the dictionary indices whose patches are matched with the pixels at the current coding location. In IPSILON [15,16], the encoder measures the decimation level of each patches in **H** and finally selects the match of the index k_m that gives the highest level of decimation by

$$k_m = \underset{k \in \mathbf{H}}{\operatorname{argmax}} \left\{ \sum_{\vec{x} \in \mathbf{F}(\Delta; [\mathbb{D}_k])} \mathbf{F}(\vec{x}) \right\}. \tag{17.26}$$

On the other hand, in AMPARS [47], rather than computing the decimation levels, the encoder just selects the maximal match index k_s by

$$k_s = \underset{k \in \mathbf{H}}{\operatorname{argmax}} \{ |\mathbb{D}_k| \}. \tag{17.27}$$

Figure 17.3 provides the two reconstructed images by the incremental parsing scheme introduced above. Figure 17.3a shows the four largest patches that were used for the reconstruction. It is observed that if a patch is observed at a certain location of the image, the patch may appear in the vicinity of the current location. In NLP, the distribution of time interval between successive occurrences of words in documents has been a long research topic and it is observed that the distribution obeys a Poisson distribution or a stretched exponential distribution [50,51]. Figure 17.3a suggests that the spatial statistics of visual patches may be an interesting issue and be investigated for visual linguistics, as done in NLP for the distribution of time interval. Figure 17.3b shows a result of a texture image with the four largest patches captured by the incremental parsing. All the four patches shown in this figure contain the boundary region of the texture objects, which may be considered an essential property for measuring semantic similarity of texture images as well as natural scenes.

17.3.4 Retrieval Precision

In this section, we compare the retrieval precision of five CBIR systems; AMPARS [47], IPSILON [15], two benchmark systems, and the system called SIMPLIcity proposed by Wang et al. [39]. The AMPARS system analyzes the hidden semantic concepts by the incremental parsing implemented with Equations 17.22 and 17.27 under the pLSA paradigm studied in Section 17.2.2. The IPSILON system also characterizes the given images by the incremental parsing implemented with Equations 17.20 and 17.26 under the LSA paradigm presented in Section 17.2.1. Note that the threshold $\theta_{jnd} = 1.6$ is chosen for Equation 17.21 in IPSILON. The two benchmark systems, denoted as "VQ + pLSA" and "VQ + LSA," are implemented with the fixed-shape block representations under the corresponding paradigms in order to estimate the benefit of the use of the incremental parsing in CBIR systems. By a k-means technique, the

TABLE 17.3 Total Average Precisions of the Five Image Retrieval Systems

	AMPARS	IPSILON	VQ + pLSA	VQ + LSA	SIMPLIcity [39]
$r = 20$	0.602	0.502	0.420	0.322	0.381
$r = 40$	0.550	0.435	0.380	0.294	0.323
$r = 100$	0.457	0.353	0.324	0.250	0.264

SIMPLIcity system partitions a given image into a few object regions; then, by an integrated region matching (IRM), it tries to arrange the correspondences among the segmented regions and measures the semantic similarity between the two images.

A common practice in CBIR is to use precision/recall as performance indicators. However, in this evaluation, a weighted average precision, called "total average precision," which takes a nonuniform prior of images into account, is computed. Also, rather than examining the entire images in the database, we only evaluate the r retrieved images. The five systems are tested with 20,000 images of natural scenes taken from the Corel photo stock library and 600 images among them are chosen as query images. Table 17.3 compares the five CBIR systems at $r = \{20, 40, 100\}$. For the three r values, the total average precision of the AMPARS system is the highest among the five followed by the IPSILON system. Specifically, the precision of AMPARS is higher than that of SIMPLIcity by as much as 0.2 for $r = \{20, 40\}$. Also, we can see the superiority of the parsed representations generated by the incremental parsing when compared with the benchmark systems. Again, note that the only difference between the benchmark systems and the AMPARS/IPSILON system is the source representations: the parsed representation and the fixed-shape block representation.

As an illustrative example, Figure 17.4 provides the first two dimensions of the latent semantic spaces of the IPSILON and the corresponding benchmark systems. The two axes in each plot represent the first- and the second-orthogonal dimensions that capture the first- and the second-largest eigenvalues of the given co-occurrence matrices. Figure 17.4a shows the superimposed images at the corresponding image locations in the space, and the images have three primary concepts, e.g., "underwater," "blue sky," and "bright sky," although the three concepts contain similar color components. On the other hand, Figure 17.4b presents the images on the latent semantic spaces formed by the fixed-shape block representations. It is observed that the images in Figure 17.4b have dominant black regions regardless of their different semantic concepts, for example, flowers, fireworks, birds, and so on.

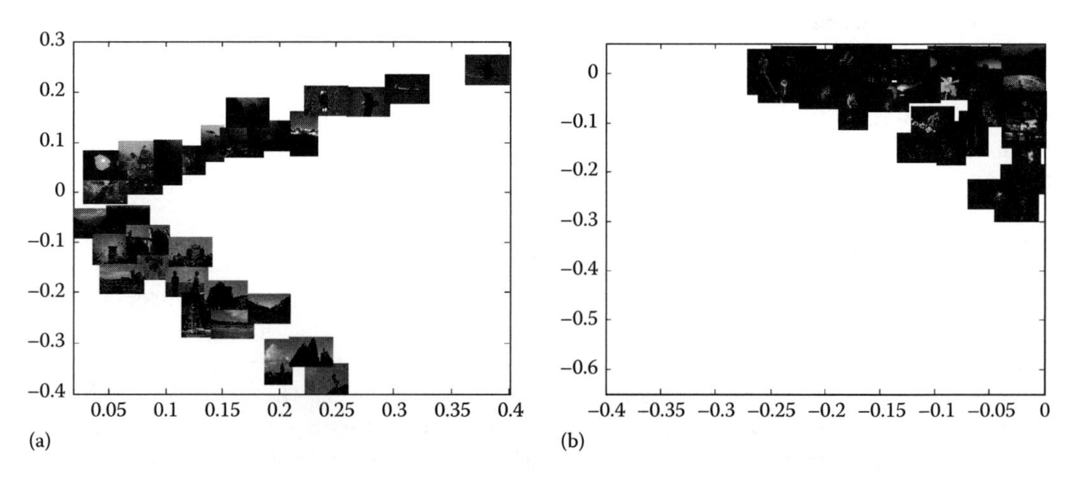

FIGURE 17.4 Two-dimensional illustrations of latent semantic dimensions of images. (a) Parsed representation in IPSILON (b) Fixed-shape block representation in VQ + LSA. (Reproduced from Bae S.H. and Juang B.-H., IPSILON: Incremental parsing for semantic indexing of latent concepts [15].)

17.4 Summary

Notwithstanding the intensive study of CBIR systems for the last two decades, the issues of document modeling and source representation have not been addressed with much attention. The success of the commercial search engines motivated the CBIR research to employ some of the relevant document retrieval techniques, LSA, pLSA, or latent Dirichlet allocation, and they have started playing an important role in CBIR. Since they are based on the vector space model [7] and the given document (or image) is represented as a single vector, they ignore the spatial occurrence pattern of words (or visual patches). As noted in Section 17.3.3.2, the successive occurrences of words in linguistic documents obey a Poisson distribution [50]. Here, we conjecture that the occurrence patterns of visual patches in natural scenes follow a different model with a low second-order statistics. Also, a formal study of the occurrence patterns will provide rich information for the design of high-performance CBIR systems.

In Section 17.3, we studied three primary source representations; fixed-shape block representation, segmentation-based representation, and parsed representation, and compared the retrieval precisions of the five CBIR systems based on the aforementioned representations. The first two have embodied a number of recent CBIR systems with different color models, pattern recognition, and machine learning techniques. However, as mentioned in Section 17.3.3, either of the representations has been selected without a strong consideration of its relevancy to the design of CBIR systems. Thus, we studied a new class of source representation generated by a universal source coding technique, called parsed representation in Section 17.3.3.2. The Lempel–Ziv incremental parsing has become one of the most prevalent lossless data compression techniques, and recently its extension for multidimensional discrete source without the aid of a scanning scheme has been proposed in Ref. [3].

The coding technique parses a given multidimensional source into a number of variable-size patches, each of which asymptotically contains the same amount of information. As a subsequent application of the source coding technique, there have been CBIR systems based on the parsed representation with different machine learning techniques [15,16,47]. In Section 17.3.4, we compared the performance of the systems with that of other existing systems in terms of retrieval precision.

In this chapter, two particular dimensions of recent CBIR systems are discussed. We hope to bring up the two issues that are not addressed here but deserve much attention: "robustness" of CBIR systems and "image database" for evaluation. In Section 17.3.4, we took retrieval precision into account, but frequently the query image may contain some perturbations that hinder a machine from understanding its semantic meaning because of visual synonymy, visual polysemy, or visual ambiguity. Formal studies on these visual perturbations and the techniques for lessening the adverse effect of the perturbations are necessary. Regarding the image database, the performance of a large portion of CBIR systems proposed in the literature have been evaluated on different sets of image databases, most popularly subsets of Corel photo stock library as done in Section 17.3.4. In Ref. [52], Müller et al. have criticized the use of the Corel library for the performance evaluation of CBIR systems, which are generally acknowledged. Because there is no relevant image database in a public domain for an intensive evaluation of CBIR systems to date, we have not been able to avoid the use of it. Thus, it is expected that an organization of the image database for this specific purpose would contribute to the further advances of CBIR systems.

References

1. M. Flickner, H. Sawhney, W. Niblack, J. Ashley, Q. Huang, B. Dom, M. Gorkani, J. Hafner, D. Lee, D. Petkovic, D. Steele, and P. Yanker, Query by image and video content: The QBIC system, *IEEE Trans. Comput.*, 28(9), 23–32, Sept. 1995.
2. T. P. Minka and R. W. Picard, Interactive learning using a society of models, *Pattern Recogn.*, 30(3), 565–581, Apr. 1997.
3. S. H. Bae and B.-H. Juang, Multidimensional incremental parsing for universal source coding, *IEEE Trans. Image Process.*, 17(10), 1837–1848, Oct. 2008.

4. A. W. M. Smeulders, M. Worring, S. Santini, A. Gupta, and R. Jain, Content-based image retrieval at the end of the early years, *IEEE Trans. Pattern Anal. Machine Intell.*, 22(12), 1349–1380, Dec. 2000.
5. R. Datta, D. Joshi, J. Li, and J. Z. Wang, Image retrieval: Ideas, influences, and trends of the new age, *ACM Comput. Surv.*, 40(2), 5:1–5:60, Apr. 2008.
6. S. Deerwester, S. T. Dumais, G. W. Furnas, T. K. Landauer, and R. Hershman, Indexing by latent semantic analysis, *J. Am. Soc. Info. Sci.*, 41(6), 391–407, Sept. 1990.
7. G. Salton, A. Wong, and C. S. Yang, A vector space model for automatic indexing, *Commun. ACM*, 18(11), 613–620, Nov. 1975.
8. J. R. Bellegarda, J. W. Butzberger, Y.-L. Chow, N. B. Coccaro, and D. Naik, A novel word clustering algorithm based on latent semantic analysis, in *Proceedings of the IEEE International Conference on Acoustics, Speech, and Signal Processing*, Atlanta, GA, May 1996, vol. 1, pp. 172–175.
9. Z. Pečenović, Image retrieval using latent semantic indexing, Master's thesis, École Polytechnique Fédérale De Lausanne (EPFL), Lausanne, Switzerland, June 1997.
10. M. L. Cascia, S. Sethi, and S. Sclaroff, Combining textual and visual cues for content-based image retrieval on the world wide web, in *Proceedings of the IEEE Workshop on Content-Based Access of Image and Video Libraries*, Santa Barbara, CA, June 1998, pp. 24–28.
11. R. Zhao and W. I. Grosky, From features to semantics: Some preliminary results, in *Proceedings of the IEEE International Conference on Multimedia and Expo*, New York, July–Aug. 2000, pp. 679–682.
12. P. Praks, J. Dvorsky, and V. Snasel, Latent semantic indexing for image retrieval systems, in *Proceedings of the SIAM Conference on Applied Linear Algebra*, New York, July 2003.
13. S. Banerjee, G. Harit, and S. Chaudhury, Word image based latent semantic indexing for conceptual querying in document image databases, in *Proceedings of the Ninth International Conference on Document Analysis and Recognition*, Parana, Brazil, Sept. 2007, pp. 640–644.
14. L. Hohl, F. Souvannavong, B. Merialdo, and B. Huet, Using structure for video object retrieval, in *Proceedings of the Third International Conference on Image and Video Retrieval*, Dublin, Ireland, July 2004, pp. 564–572.
15. S. H. Bae and B.-H. Juang, IPSILON: Incremental parsing for semantic indexing of latent concepts, *IEEE Trans. Image Process.*, submitted.
16. S. H. Bae and B.-H. Juang, Incremental parsing for latent semantic indexing of images, in *Proceedings of the IEEE International Conference on Image Processing*, San Diego, CA, Oct. 2008, pp. 925–928.
17. C. D. Manning and H. Schütze, *Foundations of Statistical Natural Language Processing*, Cambridge, MA: The MIT Press, 1999.
18. T. Hofmann, Unsupervised learning by probabilistic latent semantic analysis, *Mach. Learn.*, 42, 177–196, Jan. 2001.
19. S. Kullback and R. A. Leibler, On information and sufficiency, *Ann. Math. Stat.*, 22, 79–87, 1951.
20. A. Dempster, N. Laird, and D. Rubin, Maximum likelihood from incomplete data via the EM algorithm, *J. R. Stat. Soc.*, 39(1), 1–38, 1977.
21. P. Quelhas, F. Monay, J.-M. Odobez, D. Gatica-Perez, T. Tuytelaars, and L. V. Gool, Modeling scenes with local descriptors and latent aspects, in *Proceedings of the IEEE International Conference on Computer Vision (ICCV)*, Beijing, China, Oct. 2005, pp. 883–890.
22. D. G. Lowe, Object recognition from local scale-invariant features, in *Proceedings of the IEEE International Conference on Computer Vision (ICCV)*, Kerkyra, Greece, Sept. 1999, pp. 1150–1157.
23. J. Sivic, B. C. Russell, A. A. Efros, A. Zisserman, and W. T. Freeman, Discovering objects and their locations in images, in *Proceedings of the IEEE International Conference on Computer Vision (ICCV)*, Beijing, China, Oct. 2005, pp. 370–307.
24. R. Fergus, L. Fei-Fei, P. Perona, and A. Zisserman, Learning object categories from Google's image search, in *Proceedings of the IEEE International Conference on Computer Vision (ICCV)*, Beijing, China, Oct. 2005, pp. 1816–1823.

25. R. Zhang and Z. Zhang, Effective image retrieval based on hidden concepts discovery in image database, *IEEE Trans. Image Process.*, 16(2), 562–572, Feb. 2007.
26. F. Monay and D. Gatica-Perez, Modeling semantic aspects for cross-media image indexing, *IEEE Trans. Pattern Anal. Machine Intell.*, 29(10), 1802–1817, Oct. 2007.
27. A. Bosch, A. Zisserman, and X. Munoz, Scene classification using a hybrid generative discriminative approach, *IEEE Trans. Pattern Anal. Machine Intell.*, 30(4), 712–727, Apr. 2008.
28. J. R. Smith and S.-F. Chang, VisualSEEk: A fully automated content-based image query system, in *Proceedings of the ACM Multimedia*, Boston, MA, Nov. 1996, pp. 87–98.
29. S. W. Teng and G. Lu, Image indexing and retrieval based on vector quantization, *Pattern Recogn.*, 40(11), 3299–3316, Nov. 2007.
30. G. Carneiro, A. B. Chan, P. J. Moreno, and N. Vasconcelos, Supervised learning of semantic classes for image annotation and retrieval, *IEEE Trans. Pattern Anal. Machine Intell.*, 29(3), 394–410, Mar. 2007.
31. A. Natsev, R. Rastogi, and K. Shim, WALRUS: A similarity retrieval algorithm for image databases, *IEEE Trans. Knowl. Data Eng.*, 15(5), 1–16, Sept./Oct. 2003.
32. J. Vogel and B. Schiele, Semantic modeling of natural scenes for content-based image retrieval, *Int. J. Comput. Vis.*, 72(2), 133–157, Jan. 2007.
33. L. Fei-Fei and P. Perona, A Bayesian hierarchical model for learning natural scene categories, in *Proceedings of the IEEE Conference on Computer Vision and Pattern Recognition (CVPR)*, 2, San Diego, CA, June 2005, pp. 524–531.
34. G. Qiu, Color image indexing using BTC, *IEEE Trans. Image Process.*, 12(1), 93–101, Jan. 2003.
35. C.-H. Yeh and C.-J. Kuo, Content-based image retrieval through compressed indices based on vector quantized images, *Opt. Eng.*, 45(1), 43–50, Jan. 2006.
36. S. Jeong, C. S. Won, and R. M. Gray, Image retrieval using color histograms generated by Gauss mixture vector quantization, *Comput. Vis. Image Underst.*, 94(1–3), 44–66, Apr.–Jun. 2004.
37. A. Vailaya, M. A. T. Figueiredo, A. K. Jain, and H.-J. Zhang, Image classification for content-based indexing, *IEEE Trans. Image Process.*, 10(1), 117–130, Jan. 2001.
38. J. Chen, T. N. Pappas, A. Mojsilović, and B. E. Rogowitz, Adaptive perceptual color-texture image segmentation, *IEEE Trans. Image Process.*, 14(10), 1524–1536, Oct. 2005.
39. J. Z. Wang, J. Li, and G. Wiederhold, SIMPLIcity: Semantics-sensitive integrated matching for picture libraries, *IEEE Trans. Pattern Anal. Machine Intell.*, 23(9), 947–963, Sept. 2001.
40. T. N. Pappas, J. Chen, and D. Depalov, Perceptually based techniques for image segmentation and semantic classification, *IEEE Commun. Mag.*, 45(1), 44–51, Jan. 2007.
41. J. Jeon, V. Lavrenko, and R. Manmatha, Automatic image annotation and retrieval using cross-media relevance models, in *ACM SIGIR*, Toronto, Canada, Jul.–Aug. 2003, pp. 119–126.
42. J. Shi and J. Malik, Normalized cuts and image segmentation, *IEEE Trans. Pattern Anal. Machine Intell.*, 22(8), 888–905, Aug. 2000.
43. Y. Liu and X. Zhou, Automatic texture segmentation for texture-based image retrieval, in *Proceedings of the International Multimedia Modeling Conference*, Brisbane, Australia, Jan. 2004, pp. 285–290.
44. S. Sural, G. Qian, and S. Pramanik, Segmentation and histogram generation using the HSV color space for image retrieval, in *Proceedings of the IEEE International Conference on Image Processing*, Rochester, NY, Sept. 2002, vol. 2, pp. 589–592.
45. J. Ziv and A. Lempel, Compression of individual sequences via variable rate coding, *IEEE Trans. Info. Theory*, IT-24(5), 530–536, Sept. 1978.
46. A. Lempel and J. Ziv, Compression of two-dimensional data, *IEEE Trans. Infor. Theory*, IT-32(1), 2–8, Jan. 1986.
47. S.H. Bae and B.-H. Juang, Aspect modeling of parsed representation for image retrieval, in *Proceedings of the IEEE Conference on Acoustics, Speech and Signal Processing*, Apr. 2009, pp. 1137–1140.
48. X. Zhao, M. G. Reyes, T. N. Pappas, and D. L. Neuhoff, Structural texture similarity metrics for retrieval applications, in *Proceedings of the IEEE International Conference on Image Processing*, San Diego, CA, Oct. 2008, pp. 1196–1199.

49. Z. Wang, A. C. Bovik, H. R. Sheikh, and E. P. Simoncelli, Image quality assessment: From error visibility to structural similarity, *IEEE Trans. Image Process.*, 13(8), 600–612, Apr. 2004.
50. G. K. Zipf, *The Psycho-biology of Language: An Introduction to Dynamic Philology*, Boston, MA: Houghton Mifflin, 1935.
51. E. G. Altmann, J. B. Pierrehumbert, and A. E. Motter, Beyond word frequency: Bursts, lulls, and scaling in the temporal distributions of words, arXiv:0901.2349v1 [cs.CL], Jan. 2009.
52. H. Müller, S. Marchand-Maillet, and T. Pun, The truth about Corel—evaluation in image retrieval, in *Proceedings of the International Conference on Image and Video Retrieval (CIVR)*, London, U.K. July 2002, pp. 38–49.

18

Technologies for Context-Based Video Search over the World Wide Web

Arshdeep Bahga
Georgia Institute of Technology

Vijay K. Madisetti
Georgia Institute of Technology

We present methods and a system for video search over the Internet or the intranet. Our objective is to design a real time and automated video clustering and search system that provides users of the search engine the most relevant videos available that are responsive to a query at a particular moment in time, and supplementary information that may also be useful. The chapter highlights methods to mitigate the effect of the semantic gap faced by current content-based video search approaches. A context-sensitive video ranking scheme is used, wherein the context is generated in an automated manner.

18.1 Introduction

Videos have become a regular part of our lives due to the recent advances in video compression technologies, availability of affordable digital cameras, high-capacity digital storage media and systems, as well as growing accessibility to high speed communication networks and computers. Thousands of new videos are being uploaded over the Internet every second. However, without a fast and reliable video search engine it is difficult to retrieve videos. There are many different approaches to video search, as discussed below:

1. Content-based video search: There are two categories of content-based video search approaches which either use the low-level visual content or high-level semantic content, as described below:
   a. *Low-level visual content-based search*: The low-level content-based approach uses low-level visual content characterized by visual features such as color, shapes, textures, motion, edges, etc. for video search. These low level features can be extracted automatically to represent the video content. There

are several different classification schemes for video content. For example, MPEG-7 is a multimedia content description scheme, which has standardized more than 140 classification schemes that describe properties of multimedia content. MPEG-7 provides different descriptors for color, motion, shapes, textures, etc., to store the features extracted from video in a fully standards-based searchable representation. Other multimedia description schemes used in the past are Thesaurus of Graphical Material (TGM-I), TV-Anytime, SMPTE Metadata Registry from Society of Motion Picture and Television Engineers and P/Meta Metadata Scheme from European Broadcasting Union. A limitation of low-level visual content-based approach is the semantic gap between users' queries and the low-level features that can be automatically extracted. Virage video engine [26], CueVideo [27] and VideoQ [28] are some of the low-level content-based video search engines.

b. *High-level semantic content-based search:* The high-level semantic content-based approach uses high-level semantic content characterized by high-level concepts like objects and events for video search. Unlike low-level features which can be automatically extracted, the high-level semantic content is difficult to characterize from raw video data. The reason being that at physical level, a video is nothing but a temporal sequence of pixel regions without a direct relation to its semantic content. There are two different types of high-level semantic content-based approaches:

   i. *Concept-based video search:* The concept-based video search approaches use concept detectors (like building, car, etc.) to extract semantics from low level features [16–20]. These use shared knowledge ontology such as WordNet or external information from Internet to bridge the semantic gap between the user queries and raw video data. For example, LSCOM (Large-Scale Concept Ontology for Multimedia) [21] includes 834 semantic concepts. Media-Mill [22] extended the LSCOM-lite set by adding more high level semantic features (HLFs) amounting to a total of 101 features. Informedia [23] is another well-known system which uses HLFs for video search. Though semantic concepts are useful in retrieving shots which cannot be retrieved by textual features alone, the search accuracy is low. To overcome these limitations, event-based approaches have been used.

   ii. *Event/topic-based video search:* The event-/topic-based approaches use event/topic structures from video for providing additional partial semantics for search. Text annotations, closed captions and keywords are used to detect events and topics in a video. The concept of text-based topic detection and tracking for news videos, in which the news clusters are generated based on lexical similarity of news texts was introduced in Ref. [14]. Semantics extracted from news clusters for video story boundary detection and search were utilized in Ref. [15]. The importance of event text for video search was demonstrated in Ref. [24].

   The approaches discussed so far belong to a broad category that we define as content-based video search, which either uses low-level visual content or high-level semantic content (or both). While the process of extraction of visual features is usually automatic and domain independent, extracting the semantic content is more complex, because it requires domain knowledge or user interaction or both. The high-level content-based approach does not have the limitation of "semantic gap." It is mainly based on attribute information like text annotations and closed captions, which is associated to video manually by human. The process of manual annotation is not only time consuming but also subjective. Moreover, multiple semantic meanings such as metaphorical, hidden or suppressed meanings can be associated with the same video content which makes the process of content description even more complex. For example, a HLF like "fire" in a video sequence could have different semantic meanings like "explosion," "forest fire," etc. To overcome the limitations of both the low-level and high-level content-based approaches, hybrid video search approaches have been proposed which provide an automatic mapping from low-level features to high-level concepts [25].

2. Context-based video search: Context-based video search approach uses contextual cues to improve search precision. This approach differs from content-based approach as it uses story-level contextual cues, instead of (or supplementing) the shot-level visual or semantic content, for video search.

The contextual cues intuitively broaden query coverage and facilitate multimodal search. For example, [29] used story-level matching against a parallel news corpus obtained from Google News™, for news video search. The user query is searched on Google News and the text from resulting articles is matched to the text obtained from the recognized speech in the video, for video search. Multimodal event information extracted from news video, Web news articles, and news blogs was combined in Refs. [30] for an event-based analysis.

Commercial video search engines like Google Video™ and YouTube™ use text annotations and captions for video search. Repositories of large number of videos are searched using the keywords extracted from captions and text annotations. However, neither the keywords are effectively linked to the video content, nor are they sufficient to make an effective video search engine. Moreover, the process of building such repositories is user dependant and relies on the textual content provided by the user while uploading the video. The search process in these search engines is off-line as it is dependent on user-generated repositories. In the case of news videos, these search engines result in a very disappointing performance as the news videos retrieved are usually old and are presented in an unorganized manner. Moreover, for videos which are uploaded with non-English captions and annotations, the search performance is poor as it relies on translations of the non-English content which often changes the context. All these limitations make these search engines unsuitable for real time and automated video repository generation and search.

A recent product developed by EveryZing (www.everyzing.com) appears to allow the ability to extract and index the full text from any video file, using speech recognition to find spoken words inside videos. Google appears to be experimenting indexing based on audio (www.labs.google.com/gaudi). However, relying completely on speech recognition text may lead to incorrect results, as this technology is still not perfect.

An important aspect of any search engine is the ranking of the search results. Commercial search engines like Google use a "page ranking" scheme that measures the citation importance of a page. Pages with higher ranks are the ones to which a larger number of other pages link with [31]. An outcome of this page ranking scheme is that for a particular query, the same search results are produced, irrespective of the user's context of search. However, the context underlying the search for each user may be entirely different. For example, a graduate student working on a thesis related to video processing is more likely to search for research papers and articles related to this topic. A query term like "video" by such a user is more likely to be related to video processing material. On the other hand the same query for a user who watches a lot of music videos is more likely to be related to music videos rather than research papers on video processing. The context of the user query in the above two cases is very different. The user's context of search is also based on the user's geographical location. For example, a query term like "fire" from a user in California is more likely to be related to forest fires in California rather than volcanic fires in Japan. The current search engines provide search results which are same for all the users. Moreover, the search results are fairly static and a query may return the same result for a number of days, until the crawlers update the indexes and the pages are reranked. The user's objectives for video search on the other hand are dynamic. A system which does a context sensitive ranking of search results can provide much more meaningful results for a user (Figure 18.1).

We propose a novel context-based video clustering and search approach which attempts to make the generation of automated real time video repositories efficient and also tries to make the process of video browsing and search more meaningful. The system is very effective particularly for news video search. Our system is different from the existing context-based video search systems for the following reasons: (1) The "news context" is derived from a cluster of similar text articles and is used to crawl and cluster videos from different video broadcast sources. (2) A dynamic mapping from the generated context to the video content is done using the automatic speech recognition (ASR) text, text annotations, and closed captions. (3) A context-sensitive ranking scheme called VideoRank™, is used to assign ranks to videos with respect to different contexts. (4) A query expansion technique is used to enhance the search precision as the user's objective of the search may not be clear with the short and imprecise query terms provided. The above differences are significant in terms of generating good results for video search as the video clustering, ranking and search processes are guided by comprehensive contexts generated from cluster of similar text articles.

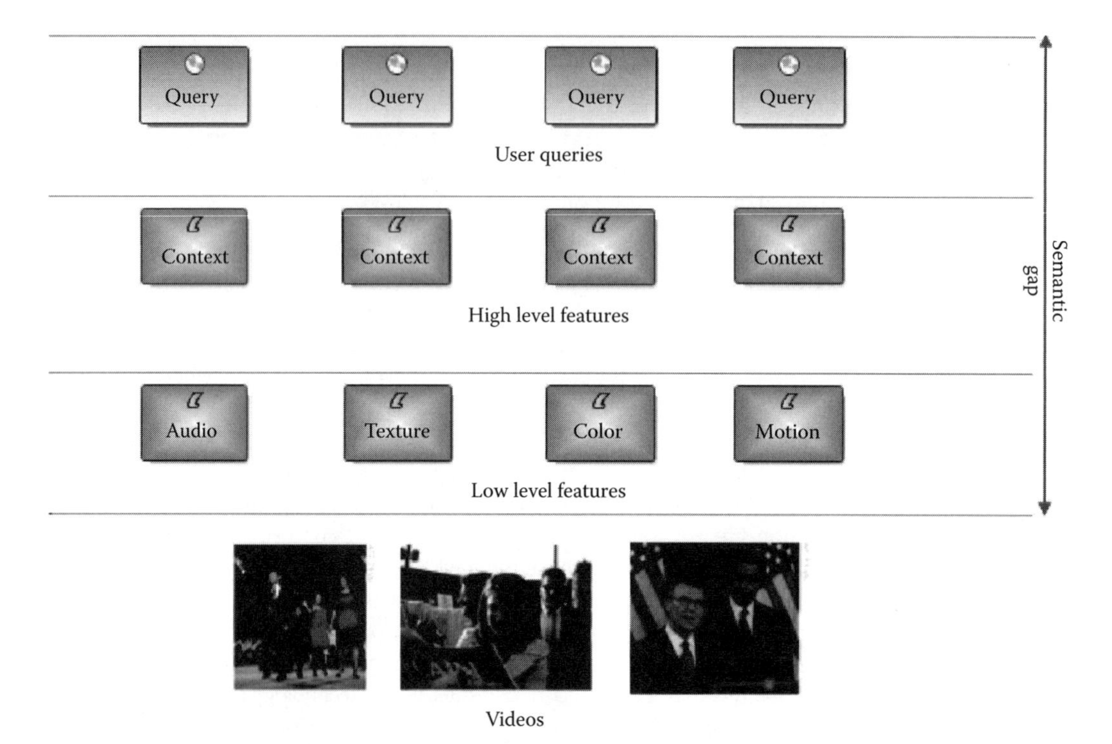

FIGURE 18.1 Semantic gap between the low level features and user queries is bridged by the video context.

There are at least five factors that affect the quality of our search, particularly for news videos: (1) The context of a news item is dynamic in nature and needs to be updated regularly, (2) a context derived from multiple news sources is more meaningful than the one from just a single news source, (3) clustering of news videos from multiple sources can be made more meaningful based on a context derived from multiple textual news articles, (4) searching news videos clustered automatically in this manner makes the search process more accurate, and (5) similar news topics and events tend to yield similar videos and thus the clustering of videos can be guided by a comprehensive context generated from several news sources.

18.2 News Clustering

Our system crawls various news sources and Internet news clustering services (e.g., Google News or Samachar.com) and extracts all the news items, along with the links to the news sources. A local cache of the news text and links extracted from different news sources is made. The clustered news items are then classified into different "context classes" (e.g., political, business, weather, sports, entertainment, technology, etc.). The idea is that the online news sources broadcasting videos may not have enough text-based information to derive a complete context of the news story. Similarly, the online text-based news services may not be directly linked to the news video broadcast Web sites.

Our system tries to bridge the gap between these two categories of news services. The ASR-text obtained from the video is not always accurate. Reference [13] showed that incorrectly recognized speech can often change the context of the news. Therefore, approaches which rely only on extracting the video context from the ASR-text are not accurate. Figure 18.2 shows the steps involved in news clustering using Google News.

18.3 Context Generation

Our system then crawls to various news sources, whose links were extracted from Internet news clustering services like Google News and extracts the news text from the Web pages. For every news item, the "news text" from several sources is fetched and cached. The summarization module then analyses the news text

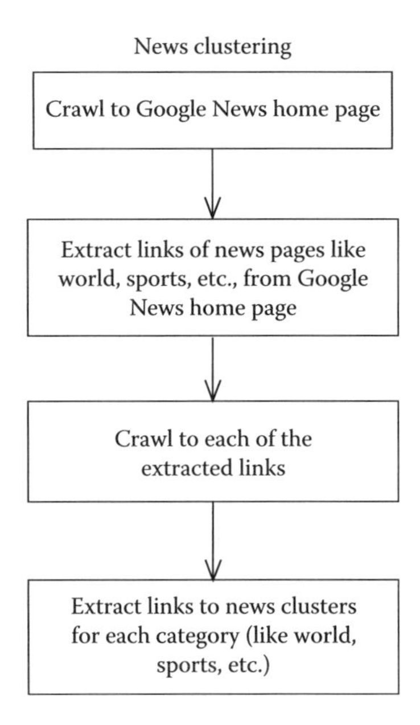

FIGURE 18.2 News clustering flow chart.

from different news sources and creates news summaries. The idea here is to use the news summaries to generate a comprehensive context, which will guide the video clustering and search process. The context is dynamic in nature and as newer news items are clustered, the context is updated.

A news context can be divided into two categories, (1) News topic, and (2) News event. For example, "Presidential elections in the United States," may qualify as a news topic, which gives a broad categorization of news. On the other hand, news like "Obama elected as U.S. President," is a news event. The essential difference between these two categories is the lifetime. While a news topic may have a lifetime as long as a year, a news event on the other hand may have a lifetime of only a day. On a further higher level, each news context can be classified into a context class. For example, the above news context, along with news topic and event qualify for the political news context class. Such a hierarchical context classification scheme makes the search process more precise. This hierarchical classification scheme is shown in Figure 18.3.

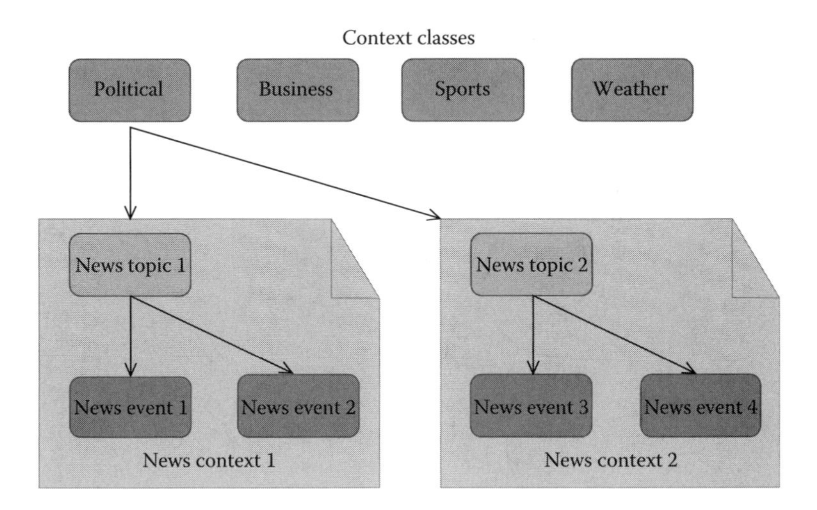

FIGURE 18.3 Hierarchical context classification scheme.

At the highest level are the context classes like political, business, weather, sports, etc., represented as the parent nodes. The next level has news topic contexts and the third level has the news event contexts. Each news event context has a number of child nodes which are basically the news items clustered in Step II. Every node at each of these four levels is represented by a set of keywords. While the highest level may be represented by a few hundred keywords like rain, temperature, storm, etc., for the weather context class, the lowest level node may have only a few keywords which are more specific, e.g., Florida, storm, etc. Thus in this hierarchical structure, a parent node has all the characteristics of a child node.

To create summaries, the system first extracts all the keywords from the news text extracted from different sources. Then a sentence ranking algorithm is used to assign ranks to sentences in the text of different news articles. The following criteria are used for sentence ranking:

1. Location of the sentence in the news article: Generally, sentences which appear in the beginning of the news article contain important content as compared to the sentences toward the end of the article.
2. Number of title keywords in the sentence: This is obtained by the number of matching keywords between the title and sentence.
3. Ratio of number of words in a sentence to the number of keywords: A high ratio generally means that the sentence is important.
4. Sentence length: Shorter sentences having small number of keywords are generally less important.
5. News source ranking: A sentence which comes from a news source with higher rank is given more importance than sentences from lower ranked news sources. The ranking of news sources is done based on criteria like the number of hits, popularity of the news source and the geographical proximity of the news source to the place of origin of a news story.

After ranking the sentences, the system finds the union and intersection sets of sentences in news articles from different sources. To generate a short summary, the intersection set is used. The intersection set is formed such that among two similar sentences from different news sources, the one which provides more information (in terms of keywords), and has a higher rank is chosen (Figure 18.4).

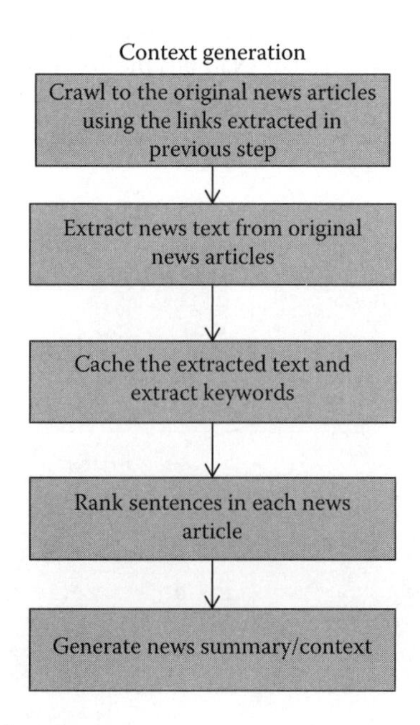

FIGURE 18.4 Context generation flow chart.

18.4 Video Clustering

Our system crawls different news sources which broadcast news videos and extracts videos. The video extraction and clustering process is guided by the news contexts generated in the previous step. The video clustering is done in two steps (Figure 18.5). In the first step, the textual content surrounding these videos, including the captions and annotations is mapped to the contexts generated in the previous step. Based on this mapping the system associates the extracted news videos to the news contexts. In the second step ASR is performed on the clustered videos and a more precise mapping is done to the news contexts. The two-step approach makes the system efficient for real-time video clustering, as time consuming process of speech recognition is not involved in the initial clustering of the videos. The system does not rely completely on the ASR-text, as it is not always accurate. However, the ASR-text is useful to extract some keywords from the video speech, which are used for a more precise mapping to the news contexts. This precise mapping is used to perform a "reclustering" of videos to different contexts, in case there is an error in the initial clustering which relies only on the text annotations and captions. Here an MPEG-7-based framework is used for video description. MPEG-7 descriptors like AudioVisualSegment, MediaTime, MediaUri, MediaLocator, TextAnnotation, KeywordAnnotation, etc. are used to capture the metadata in an XML-based format. The videos are classified into "context classes,"

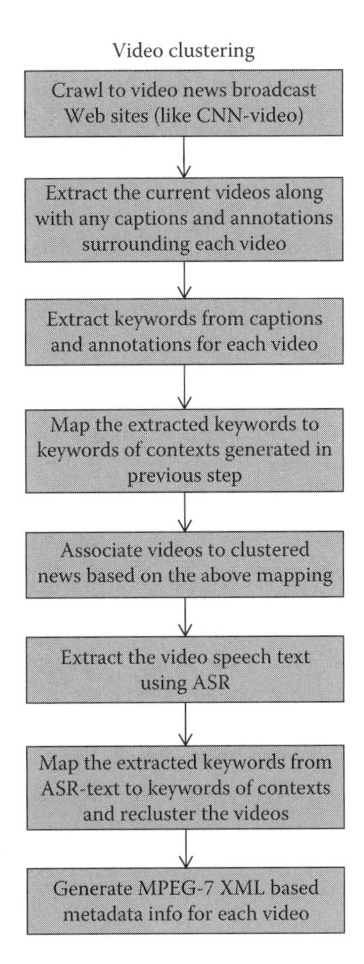

FIGURE 18.5 Video clustering flow chart.

like political, business, weather, sports, entertainment, technology, etc. This classification process is guided by context class lexicons. For each context class there is a separate lexicon which contains the frequently used keywords. For example, in weather news, keywords like rain, storm, and temperature are common. Other information like the video broadcast time, source, and news event date is also tagged with the video to help in the search process.

The system makes a news video repository, which allows users to search news videos. The popularity of video formats like FLV and MP4 has made embedding and extraction of videos from the original sources easier. Our system differs from video sharing Web sites like YouTube, in the sense that such Web sites rely on the users to upload videos manually and attach captions and annotation for efficient search. Our system is completely automated and the videos are clustered and tagged without any user intervention.

18.5 Video Indexing and Ranking

Video indexing and ranking is an important step for efficient browsing and search of videos. After clustering the videos our system assigns relevance weights to the videos with respect to a context and calculates the VideoRank. VideoRank indicates the likelihood of the presence of a context in a video. This ranking scheme is different from the page ranking schemes used by search engines like Google, where pages with more citations are ranked higher. Such a ranking scheme does not work well for video search in a particular context as a page may have a number of embedded videos, which need to be ranked individually according to their relevance to a context. Our approach to video ranking is based on the context. A number of criteria are used for calculating the VideoRank, like the number of matching keywords between the news context and the video metadata information contained in the MPEG-7 XML format. The news source ranking is also taken into account for the process of video ranking. The news source ranking is based on criteria like the number of hits, popularity of the news source, and the geographical proximity of the news source to the place of origin of a news story. As in the case of context generation, the video ranking process is also dynamic in nature. As the system constantly clusters videos, the rankings also keep changing. Other criteria like the time of broadcast of video and news event date can also be taken into account. Thus newer videos matching a particular context may be ranked higher than the older videos.

The VideoRank of a video v is defined as

$$\mathrm{VR}(v) = R(c) + R(n)$$

where

$R(c)$ is the relevance rank of the video with respect to a context c

$R(n)$ is the rank of news source n

Let C be a set of N contexts $C = \{c_1, c_2, c_3, \ldots, c_N\}$, where context c_i is characterized by a set of M keywords $K = \{k_1, k_2, k_3, \ldots, k_M\}$. Every context may have a number of keywords which may be common with other contexts. A keyword which occurs in many contexts is not a good discriminator, and should be given less weight than one which occurs in few contexts. To find the discriminating power of a keyword for a context, we calculate the inverse document frequency,

$$\mathrm{IDF}(k_i) = \log(N)/n_i$$

where

N is the total number of contexts

n_i is the number of contexts in which the keyword k_i occurs

The relevance rank of the video with respect to a context c is defined as follows,

$$R(c) = \sum \mathrm{IDF}(y_i)\mathrm{TF}(y_i)$$

where y_i are the matching keywords between the context c and the video metadata information contained in the MPEG-7 XML format (which includes the ASR text, captions and annotations). $\mathrm{TF}(y_i)$ is the term frequency of the keyword y_i, i.e., the number of times the keyword occurs in the video metadata information.

A user query may have different contexts with respect to different domains. For example, if the user has sports in mind while searching for a term like "videos," then he is clearly interested in sports videos and not music videos. Our system not only ranks the results based on the context but also provides a clear separation of different domains of search like sports, politics, weather, etc. A domain classification module classifies different search results into different domains. A domain relevance rank is computed for each search result and based on this rank the search result is classified to a particular domain. There may be a case where the relevance rank for a search result is almost the same for two or more domains. In this case the result is classified to all the domains for which the relevance rank is greater than a threshold. Ranking of the search result is then done separately in each domain as compared to the other results.

A probabilistic support vector machine (SVM) model is used for classification of a search result into different domains. Let R be a set of n results $R = \{r_1, r_2, r_3, \ldots, r_n\}$ and D be a set of k domains $D = \{d_1, d_2, d_3, \ldots, d_k\}$. A probabilistic model for classification of results into domains will select a domain $d \in D$ for result $r \in R$ with probability $p(d|r)$. The $p(d|r)$ can be used to score domain d among possible domains D for result r. The conditional probability $p(d|r)$ is estimated using probabilistic support vector machine.

18.6 Video Search

Queries for video search that are short may not contain enough information to map it to one of the contexts generated in Step III. Our system uses a query expansion technique to enhance contextual information of the query which can then be mapped more efficiently to the contexts generated in Step III, thus enhancing the recall and improving the search precision. References [11,12] have shown the usefulness of this query expansion technique.

The idea is to analyze the query terms and find other similar terms or keywords which have a high correlation with the query terms. For example, consider a query which has only one term "elections." This single term cannot give any contextual information. However based on the current news clusters and the generated contexts, the query can be expanded such that keywords among all the generated contexts which have a high correlation with the query term are added to the query. So if there is a current news cluster on "Presidential elections in United States," then the system will attach keywords like "President," "United States," etc. to the original query. Our approach to query expansion is novel in the sense that it takes into consideration the contexts generated from the current news clusters. As queries are sensitive to time, the query expansion process is dynamic in nature and depends on the current contexts.

18.7 User Interface

Commercial news services like Google News provide links to the news stories clustered from several sources. Links to the original news sources are presented and the user has to visit different news articles to get a comprehensive view of the news story. Such an interface may present links to hundreds of news articles for a particular news story which is generally overwhelming for most users. Video search engines like YouTube, on the other hand, provide lists of videos arranged in order of relevance. This interface again overwhelms the users with hundreds of videos, most of which may not be relevant to the user in a particular context. Our system overcomes the limitations of both the commercial news services like Google News and video search engines like YouTube. A novel user interface is provided which not only gives the links to the original news articles but also provides the news summaries, related news videos, and images all at one place. This makes the process of news search more interesting as a user can read a brief summary of the news and watch related news videos at the same time.

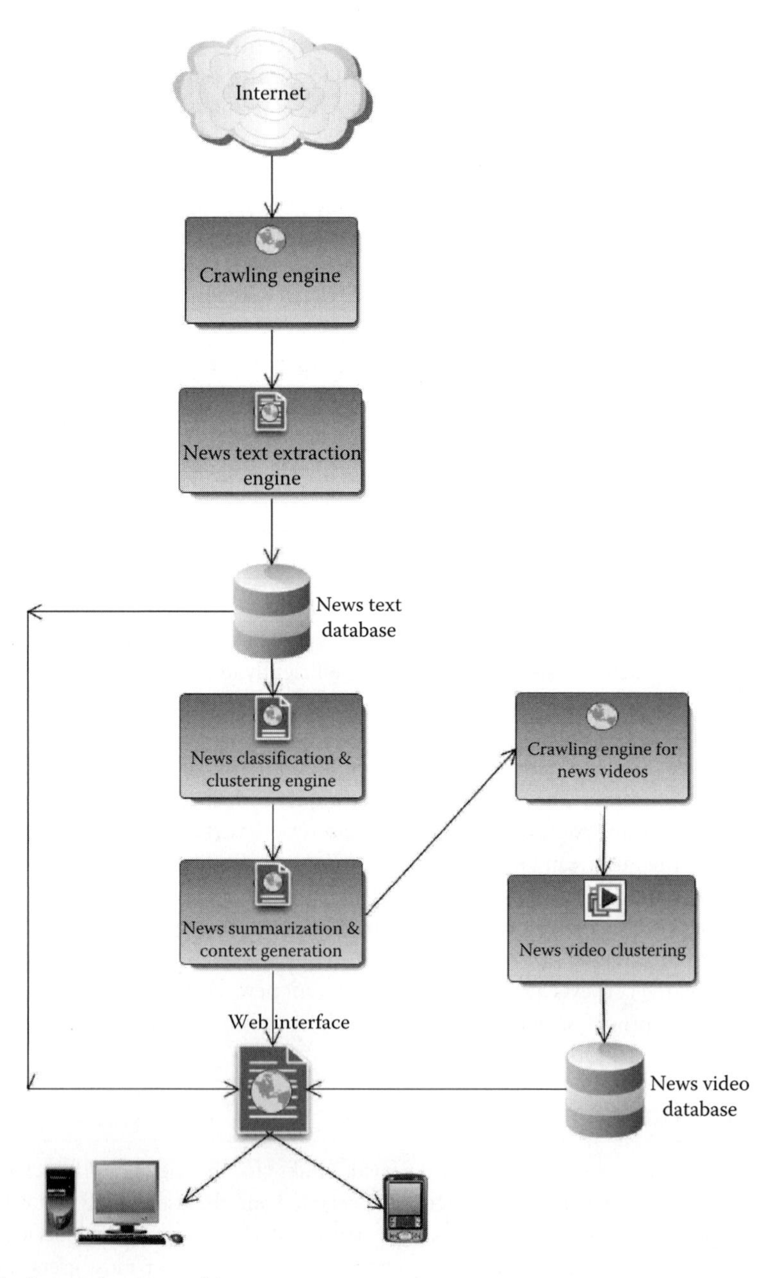

FIGURE 18.6 Proposed system architecture.

Due to the dynamic nature of the content on the Web, a user may be interested in getting automated updates for a query. For example, a user who is interested in videos of a sports tournament or weather related videos may be interested in getting automated updates whenever new videos are available. Our system provides this feature by the creation of custom feeds for a particular query. As new videos are crawled, the system can send updates to an interested user (Figures 18.6 and 18.7).

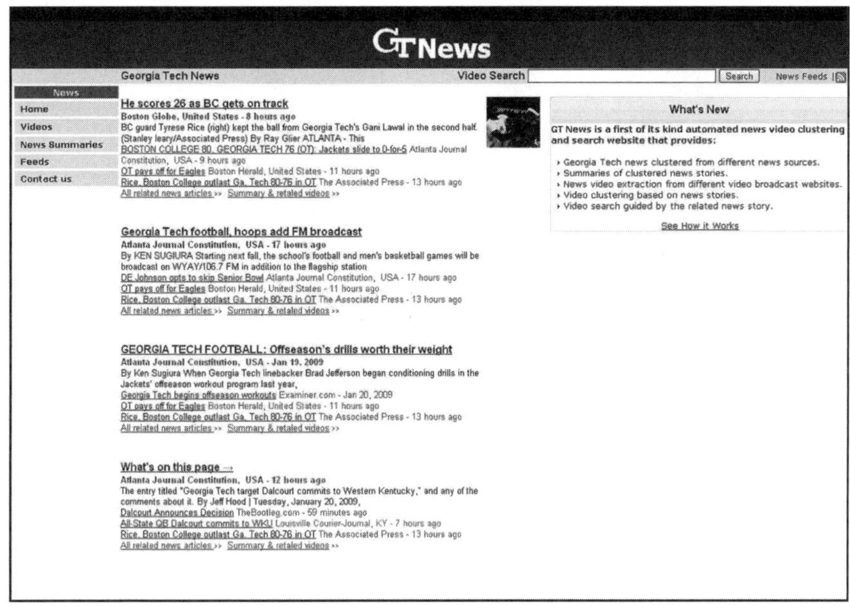

(a)

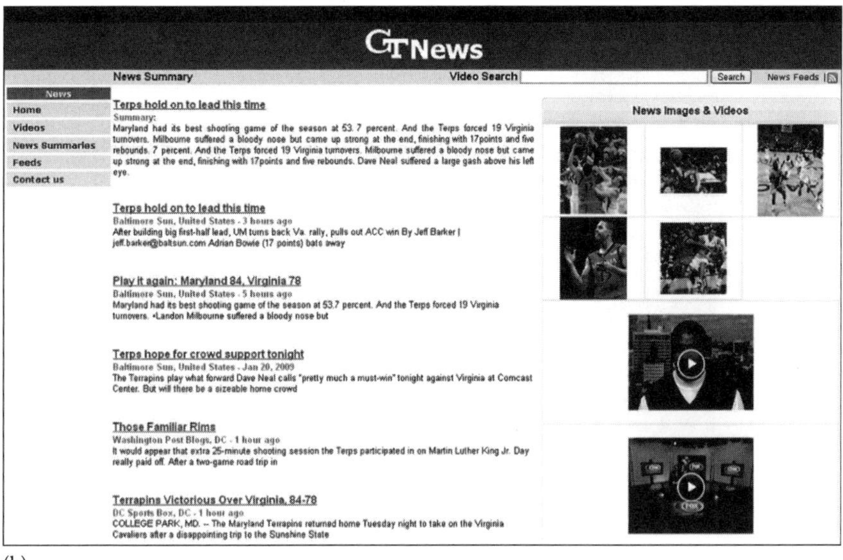

(b)

FIGURE 18.7 (a) Screen shot of home page. (b) News summary page with related videos and images.

(continued)

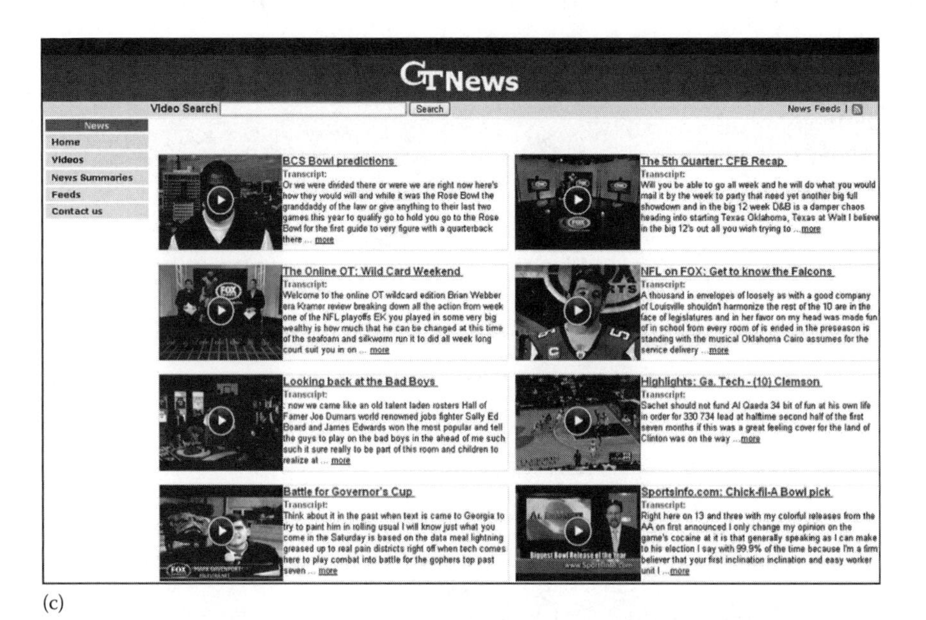

(c)

(d)

FIGURE 18.7 (continued) (c) News videos page. (d) Video search results.

TABLE 18.1 Feature Comparison Chart

Feature	YouTube	Our System
Video search	Yes	Yes
Automated video clustering	No	Yes
Dynamic updates for videos	No	Yes
Video context information	No	Yes
RSS feeds	No	Yes

TABLE 18.2 Feature Comparison Chart

Feature	Google News	Our System
News clustering by topic	Yes	Yes
News clustering by category	Yes	Yes
News summaries	No	Yes
Related news images	Yes	Yes
Related news videos	No	Yes
News text search	Yes	Yes
News videos search	No	Yes
RSS feeds	Yes	Yes

18.8 Feature Comparison

Tables 18.1 and 18.2 show feature comparison charts of our proposed system with YouTube and Google News.

18.9 Demand Analysis

Context-based video search finds applications in many areas, including digital video broadcasts, video on demand, and video surveillance. Most applications require the ability to search videos based on the semantic and contextual information available. A limitation of content-based video search approaches is that it is difficult to relate the low level features with semantics. The commercial video search engines available today depend on the users to upload videos manually and the search is again dependent on the captions and annotations provided by the user. Thus they are not able to keep up with the rapid rate at which new videos are being uploaded by various video broadcasts Web sites as there is a complete lack of automation. There is an increasing demand for online video broadcast services, driven primarily by the ease with which users can access videos through mobile phones, PDA's, and other hand held devices. Thus a real time and automated video clustering and search system which not only provide the users the most relevant videos available at a particular moment but also the related contexts of the videos, summarized from a number of different sources, will become indispensable for users in future.

References

1. D. Radev, J. Otterbacher, A. Winkel, and S. Blair-Goldensohn. NewsInEssence: Summarizing online news topics, *Commun. AeM*, 48(10), 95–98, 2005.
2. D. K. Evans, J. L. Klavans, and K. R. McKeown. Columbia newsblaster: Multilingual news summarization on the web. *Human Language Technology* (*HLT*), Boston, MA, May, 2004.

3. V. Thapar, A.A. Mohamed, and S. Rajasekaran. A consensus text summarizer based on meta-search algorithms. *IEEE International Symposium on Signal Processing and Information Technology*, Vancouver, British Columbia, Canada, 2006.
4. H. Geng, P. Zhao, E. Chen, and Q. Cai. A novel automatic text summarization study based on term co-occurrence. *5th IEEE International Conference on Cognitive Informatics*, Beijing, China, 2006.
5. O. Sornil and K. Gree-ut. An automatic text summarization approach using content-based and graph-based characteristics. *IEEE Conference on Cybernetics and Intelligent Systems*, Bangkok, Thailand, 2006.
6. L. Yu, J. Ma, F. Ren, and S. Kuroiwa. Automatic text summarization based on lexical chains and structural features, *Proceedings of the Eighth International IEEE ACIS Conference*, pp. 574–578, 2007.
7. K. Kaikhah. Automatic text summarization with neural networks. *Proceedings of the Second IEEE International Conference on Intelligent Systems*, pp. 40–44, 2004.
8. M. T. Maybury and A. E. Merlino. Multimedia summaries of broadcast news, *Proceedings on Intelligent Information Systems*, Grand Bahama Island, Bahamas, pp. 442–449, 1997.
9. A. Chongsuntornsri and O. Sornil. An automatic thai text summarization using topic sensitive PageRank, *International Symposium on Communications and Information Technologies*, Bangkok, Thailand, pp. 574–552, 2006.
10. C. Li and S. Wang. Study of automatic text summarization based on natural language understanding, *IEEE International Conference on Industrial Informatics*, Singapore, pp. 712–714, 2006.
11. S. Y. Neo, J. Zhao, M. Y. Kan, and T. S. Chua, Video search using high-level features: Exploiting query-matching and confidence-based weighting, *CIVR 2006*, Tempe, AZ, pp. 143–152, July 2006.
12. H. Yang, L. Chaisorn, Y. Zhao, S.-Y. Neo, and T. S. Chua, VideoQA: Question answering on news video *ACM Multimedia 2003*, Berkeley, CA, pp. 632–641, Nov. 2003.
13. J. McCarley and M. Franz. Influence of speech recognition errors on topic detection. *SIGIR 2000*, pp. 342–344, New York, 2000.
14. J. Allan, R. Papka, and V. Lavrenko, On-line new event detection and tracking *SIGIR 1998*, Melbourne, Australia, pp. 37–45, 1998.
15. W. H. Hsu, L. Kennedy, S. F. Chang, M. Franz, and J. Smith, Columbia-IBM news video story segmentation in TRECVID 2004, Columbia ADVENT Technical Report, New York 2005.
16. A. P. Natsev, M. R. Naphade, and J. R. Smith, Semantic representation, search and mining of multimedia content, *ACM International Conference on Knowledge Discovery and Datamining (SIGKDD)*, Seattle, WA, 2004.
17. S.-Y. Neo, J. Zhao, M.-Y. Kan, and T.-S. Chua, Video search using high level features: Exploiting query matching and confidence-based weighting, *International Conference on Image and Video Search (CIVR)*, Tempe, AZ, 2006.
18. M. Campbell, S. Ebadollahi, D. Joshi, M. Naphade, A. P. Natsev, J. Seidl, J. R. Smith, K. Scheinberg, J. Tešić, L. Xie, and A. Haubold, IBM research TRECVID-2006 video search system, *TRECVID*, Gaithersburg, MD, 2006.
19. C. G. M. Snoek, B. Huurnink, L. Hollink, M. de Rijke, G. Schreiber, and M. Worring, Adding semantics to detectors for video search, *IEEE Trans. Multimedia*, 9(5), 975–986, 2007.
20. S. F. Chang, W. Hsu, W. Jiang, L. Kennedy, D. Xu, A. Yanagawa, and E. Zavesky, Columbia university TRECVID-2006 video search and high-level feature extraction, in *TRECVID*, Gaithersburg, MD, 2006.
21. M. Naphade, J. R. Smith, J. Tešić, S.-F. Chang, W. Hsu, L. Kennedy, A. Hauptmann, and J. Curtis, Large-scale concept ontology for multimedia, *IEEE Multimedia*, 13(3), 86–91, 2006.
22. C. G. M. Snoek, J. C. van Gemert, J. M. Geusebroek, B. Huurnink, D. C. Koelma, G. P. Nguyen, O. De Rooij, F. J. Seinstra., A. W. M. Smeulders, C. J. Veenman., and M. Worring, The MediaMill TRECVID 2005 semantic video search engine, *TRECVID Workshop*, NIST, Gaithersburg, MD, Nov. 2005.

23. A. Hauptmann., M. Christel, R. Concescu, J. Gao, Q. Jin, W. H. Lin, J. Y. Pan, S. M. Stevens, R. Yan, J. Yang, and Y. Zhang, CMU Informedia's TRECVID 2005 skirmishes, *TRECVID Workshop*, NIST, Gaithersburg, MD, Nov. 2005.
24. H. Yang, T.-S. Chua, S. Wang, and C.-K. Koh, Structured use of external knowledge for event-based open-domain question answering, *SIGIR 2003*, Toronto, Ontario, Canada, pp. 33–40, July 2003.
25. M. Petković, Content-based Video Search.
26. A. Hampapur, A. Gupta, B. Horowitz, C.-F. Shu, C. Fuller, J. Bach, M. Gorkani, and R. Jain, Virage video engine, *SPIE*, 3022, 188–198, 1997.
27. D. Ponceleon, S. Srinivasan, A. Amir, D. Petkovic, and D. Diklic, Key to effective video search: effective cataloging and browsing, *ACM Multimedia'98*, Bristol, U.K., pp. 99–107, 1998.
28. S.-F. Chang, W. Chen, H. Meng, H. Sundaram, and D. Zhong, A fully automated content based video search engine supporting spatio-temporal queries, *IEEE Trans. Circuit Syst. Video Technol.*, 8(5), 302–615, Sept., 1998.
29. K. Wan, Exploiting story-level context to improve video search, *IEEE International Conference on Multimedia and Expo*, Hannover, Germany, pp. 298–292, 2008.
30. S.-Y. Neo, Y. Ran, H.-K. Goh, Y. Zheng, T.-S. Chua, and J. Li, The use of topic evolution to help users browse and find answers in news video corpus, *Proceedings of the 15th International Conference on Multimedia*, Augsburg, Germany, pp. 198–207, 2008.
31. S. Brin and L. Page, The anatomy of a large-scale hypertextual web search engine, *Comput. Netw. ISDN Syst.*, 30(1–7), 107–117.

19

Image Interpolation*

Yucel Altunbasak
Georgia Institute of Technology

The spatial image interpolation problem can be stated in its most general form as follows: Given a digital image obtained by sampling an ideal continuous image on a certain spatial grid, obtain the pixel values sampled on a different spatial grid. If the new sampling grid is denser than the input sampling grid, then the interpolated image has more samples than the input image. This case is referred to as zoom-in and the main challenge is to compute the values of the new pixels with high precision without creating artifacts. If the new sampling grid is sparser than the input sampling grid, then the interpolated image has fewer samples. This is referred to as zoom-out. Since zoom-out effectively performs down-sampling, the challenge is to obtain a smaller image free of aliasing artifacts with minimal blur.

Many digital image processing tasks require spatial interpolation at some step. As an example, consider rotating a digital image around its center by an arbitrary angle. This may appear as a trivial task at first. We can simply derive the mapping between the input and the rotated pixel coordinates, using basic trigonometry, and map the input pixels one-by-one. But there is a technical detail that complicates this task: both the input and the output are digital images represented on fixed rectangular sample grids. Since the coordinates of the rotated pixels cannot be guaranteed to fall on a rectangular sample grid, we are not done by just mapping the input pixel coordinates. These rotated pixel locations are simply not meaningful. To complete the rotation, we have to resample the rotated image at the rectangular grid points, and this resampling operation is performed by spatial interpolation. Other important applications that require spatial interpolation are video/image standards conversion, frame freeze for interlaced video, zoom-in, zoom-out and subpixel image registration. Considering the variety of applications that require spatial interpolation, it should not surprise us to find a considerable amount of work devoted

* The content of this chapter substantially overlaps with a digital image processing textbook that is currently being authored by Prof. Yucel Altunbasak and Prof. Russell M. Mersereau.

to interpolation in the digital image processing literature. A comprehensive list of different approaches to the spatial interpolation problem may be given as

- Linear spatially invariant interpolation
- Edge-adaptive interpolation
- Statistical learning-based interpolation
- Optimal recovery-based image interpolation
- Transform domain interpolation

Although each of these approaches has its own (dis)advantages, in terms of the visual quality, the computational complexity, the ease of hardware implementation, and the integration into existing image/video standards, there is a common trade-off that applies to all interpolation methods. Basically, all these interpolation methods are spatial domain filters that operate on the input image pixels. The number of pixels considered for interpolation affects the complexity of the computation and the quality of the result. The greater the number of pixels considered, the greater the computational load is and the more accurate the result gets. Hence, there is a trade-off between computational resources and quality. When choosing an interpolation method for a specific purpose, we may need to trade-off computational load and/or memory to achieve the desired quality in the result or vice versa.

19.1 Linear Spatially Invariant Filters

From a signal processing point of view, any discrete signal can be interpreted as a sampled version of some continuous signal. In this context, we can model the interpolation as a resampling of the continuous image underlying the input discrete image. This model is shown in Figure 19.1, where $g(x, y)$ is the continuous image from which the input discrete image $g[m, n]$ is assumed to be obtained through sampling. C/D and D/C stand for continuous to discrete and discrete to continuous conversion, respectively. $\hat{g}(x, y)$ is the continuous version of $g[m, n]$ that we can obtain with a practically feasible D/C converter* and $f[m, n]$ is the interpolated discrete image obtained from $\hat{g}(x, y)$ through sampling.

For equally spaced data, D/C conversion can be represented in the form of a convolution

$$\begin{aligned}\hat{g}(x, y) &= \left(\sum_{m}\sum_{n} g[m, n]\delta[x - mT_x, y - nT_y]\right) * u\left(\frac{x}{T_x}, \frac{y}{T_y}\right),\\ &= \sum_{m}\sum_{n} g[m, n]u\left(\frac{x - mT_x}{T_x}, \frac{y - nT_y}{T_y}\right),\end{aligned} \tag{19.1}$$

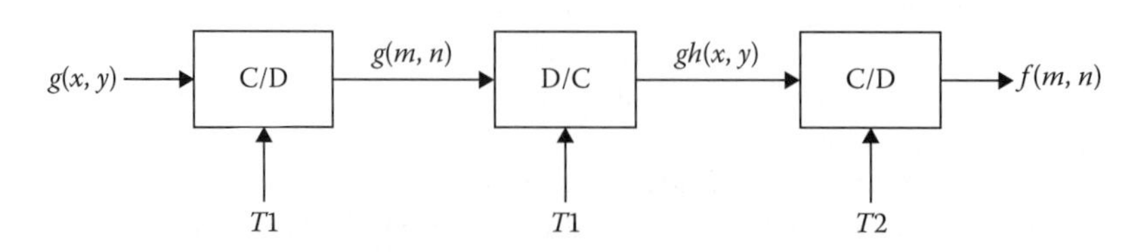

FIGURE 19.1 Resampling interpretation.

* Due to the impossibility of designing perfect low-pass filters we can never achieve ideal D/C converters, hence reconstruction of $g(x, y)$ from $g[m, n]$ is not feasible. We can only hope to get very close to $g(x, y)$ assuming we design our reconstruction filters properly and negligible aliasing occurred during sampling to obtain $g[m, n]$.

where

T_x is the sampling interval in x-direction
T_y is the sampling interval in y-direction
$u(x, y)$ is the reconstruction kernel

It is worth noting that a band-limited signal can be perfectly reconstructed by choosing

$$u(x,y) = \text{sinc}(x,y) \triangleq \frac{\sin\left(\pi\sqrt{x^2+y^2}\right)}{\pi\sqrt{x^2+y^2}}. \tag{19.2}$$

Unfortunately, sinc function is band-limited, which implies infinite spatial support. Hence, a practical implementation of Equation 19.2 is not possible. One way to approximate Equation 19.2 is to apply a finite length window to the sinc function to get a finite support. The shape and the length of the window can be optimized to satisfy certain properties. Note that regardless of the specific interpolation kernel used in Equation 19.1, we require $\hat{g}(mT_x, nT_y) = g[m, n]$ for $\forall[m, n]$, i.e., the reconstructed signal must match the input values on the original discrete grid. Among the interpolation methods that can be characterized in this manner are the nearest neighborhood interpolation, the bilinear interpolation, and the cubic splines.

19.1.1 Pixel Replication

Pixel replication corresponds to using a zero-order hold to obtain $\hat{g}(x, y)$. Hence, the reconstruction kernel is given by

$$u(x,y) = \begin{cases} 1, & \text{if } 0 \leq x \leq 1 \quad \text{and} \quad 0 \leq y \leq 1 \\ 0, & \text{otherwise.} \end{cases} \tag{19.3}$$

An imaginary one-dimensional reconstruction (considering x-direction only) is shown in Figure 19.2. Pixel replication is the simplest and the fastest of all the interpolation methods. Every input pixel is repeated L times in both the spatial directions to obtain the output image. No other input pixels are considered during this operation. Pixel replication provides the worst results in terms of smoothness. Figure 19.3 demonstrates the result of the pixel replication for a scaling ratio of $L = 3$. The pixels assigned the value of G_1 are marked by arrows.

For an arbitrary scaling ratio L (which may be noninteger), pixel replication is performed as follows:

$$f[m,n] = g\left[\left\lfloor \frac{m}{L} \right\rfloor, \left\lfloor \frac{n}{L} \right\rfloor\right], \tag{19.4}$$

where $\lfloor x \rfloor$ denotes floor operation, which rounds x to the nearest integer less than or equal to x.

19.1.2 Bilinear Interpolation

Bilinear interpolation corresponds to using the first-order hold to obtain $\hat{g}(x, y)$. Hence, the reconstruction kernel is given as

$$u(x,y) = u(x)u(y), \tag{19.5}$$

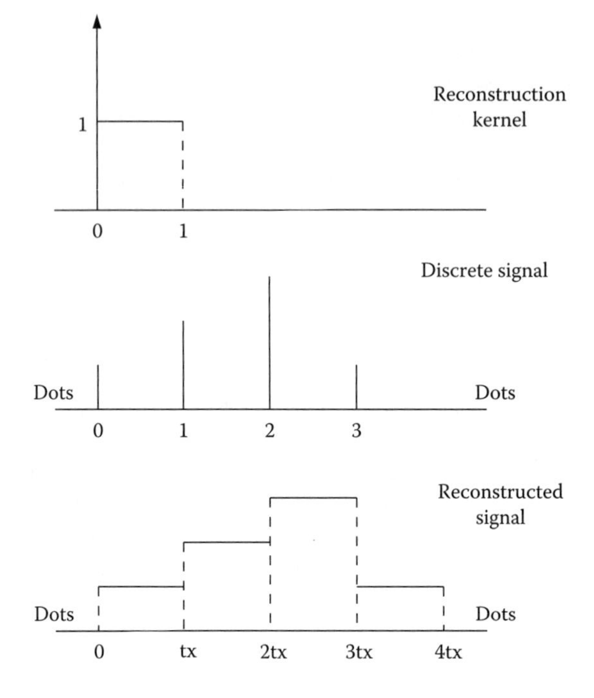

FIGURE 19.2 Reconstruction using zero-order hold.

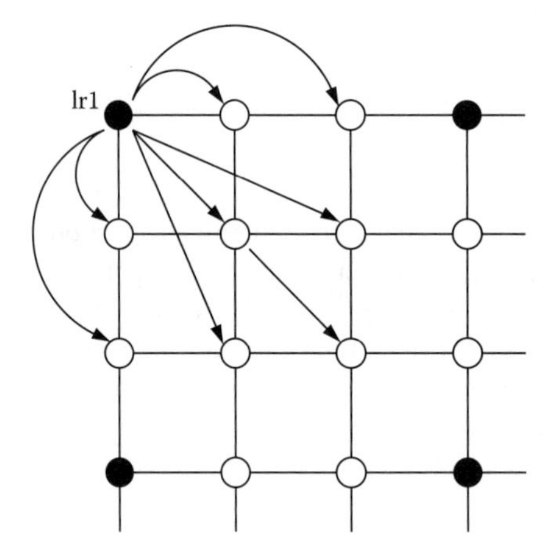

FIGURE 19.3 Pixel replication for $L = 3$.

where $u(x)$ and $u(y)$ are given as

$$u(x) = \begin{cases} 1 - |x|, & \text{if } -1 \leq x \leq 1 \\ 0, & \text{otherwise.} \end{cases} \tag{19.6}$$

$$u(y) = \begin{cases} 1 - |y|, & \text{if } -1 \leq y \leq 1 \\ 0, & \text{otherwise.} \end{cases} \tag{19.7}$$

An imaginary one-dimensional reconstruction (considering x-direction only) is shown in Figure 19.4. The output pixel value is the weighted average of the pixels in the nearest two-by-two neighborhood.

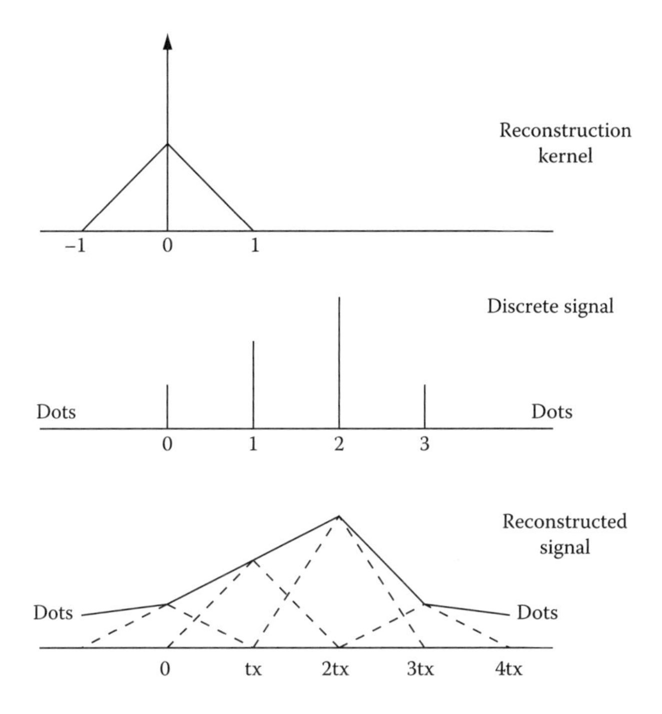

FIGURE 19.4 Reconstruction using first-order hold.

The weights are obtained by sampling the reconstruction kernel given in Equation 19.5 at respective locations. Figure 19.5 demonstrates bilinear interpolation for a scaling ratio of $L = 3$. The interpolated pixel marked with $\times$ symbol is given as

$$(1 - \delta_m)(1 - \delta_n)G_1 + (1 - \delta_m)\delta_n G_2 + \delta_m(1 - \delta_n)G_3 + \delta_m\delta_n G_4 \tag{19.8}$$

Compared to the nearest neighborhood interpolation, bilinear interpolation uses more memory and requires more computations. Unlike the nearest neighbor interpolation, its output has continuous pixel values. But continuity of the first derivative is not guaranteed.

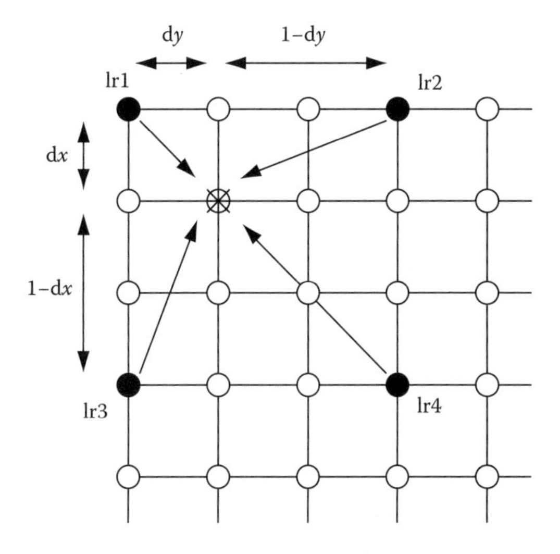

FIGURE 19.5 Bilinear interpolation for $L = 3$.

For an arbitrary scaling ratio L (which may be noninteger), bilinear interpolation is performed as follows:

$$\hat{f}[m,n] = \sum_{k=0}^{k=1} \sum_{l=0}^{l=1} w_{m,k}\, w_{n,l}\, g[\lfloor m/L \rfloor + k, \lfloor n/L \rfloor + l], \tag{19.9}$$

where the weights are given as

$$w_{m,0} = (1 - \delta_m), \quad w_{m,1} = \delta_m \tag{19.10}$$

$$w_{n,0} = (1 - \delta_n), \quad w_{n,1} = \delta_n, \tag{19.11}$$

and the distances δ_m and δ_n are defined as

$$\begin{aligned} \delta_m &= \frac{m}{L} - \left\lfloor \frac{m}{L} \right\rfloor \\ \delta_n &= \frac{n}{L} - \left\lfloor \frac{n}{L} \right\rfloor. \end{aligned} \tag{19.12}$$

19.1.3 Bicubic Interpolation

The reconstruction kernels used for the nearest neighborhood and the bilinear interpolation methods are quite straightforward. The bicubic reconstruction kernel, however, is derived from a set of conditions imposed to maximize the accuracy of the interpolated image. We require the reconstruction kernel to be continuous with a continuous first derivative. To obtain the reconstruction kernel that satisfies these requirements, we start with a parametric piecewise cubic function given as

$$u(x) = \begin{cases} A_1|x|^3 + B_1|x|^2 + C_1|x| + D_1, & \text{if } 0 \leqslant |x| \leqslant 1 \\ A_2|x|^3 + B_2|x|^2 + C_2|x| + D_2, & \text{if } 1 \leqslant |x| \leqslant 2 \\ 0 & \text{if } 2 < |x|. \end{cases} \tag{19.13}$$

As we mentioned previously, regardless of the complexity of the reconstruction kernel, the reconstructed continuous signal is required to agree with the discrete samples. This implies that $u(0) = 1$ and $u(n) = 0$ for $\forall n \in \mathbb{Z} - \{0\}$. These conditions combined with the continuity requirement $u(1^+) = u(1^-)$ imply

$$1 = u(0) = D_1 \tag{19.14}$$

$$0 = u(1^-) = A_1 + B_1 + C_1 + D_1 \tag{19.15}$$

$$0 = u(1^+) = A_2 + B_2 + C_2 + D_2 \tag{19.16}$$

$$0 = u(2^-) = 8A_2 + 4B_2 + 2C_2 + D_2. \tag{19.17}$$

Imposing the continuity of the first derivative we obtain

$$-C_1 = u'(0^-) = u'(0^+) = C_1 \tag{19.18}$$

$$3A_1 + 2B_1 + C_1 = u'(1^-) = u'(1^+) = 3A_2 + 2B_2 + C_2 \tag{19.19}$$

$$12A_2 + 4B_2 + C_2 = u'(2^-) = u'(2^+) = 0. \tag{19.20}$$

In all, the constraints imposed on the reconstruction kernel result in seven equations. But since there are eight unknown coefficients, one more equation is needed to obtain a unique solution. The final equation is user-specified, and it often comes from setting $A_2 = -1/2$.* Solving these equations for the unknowns $A_1, B_1, C_1, D_1, A_2, B_2, C_2$, and D_2, we obtain the reconstruction kernel for the 1D case as

$$u(x) = \begin{cases} \frac{3}{2}|x|^3 - \frac{5}{2}|x|^2 + 1, & \text{if } 0 \leqslant x \leqslant 1 \\ -\frac{1}{2}|x|^3 + \frac{5}{2}|x|^2 - 4|x| + 2, & \text{if } 1 \leqslant x \leqslant 2 \\ 0 & \text{if } 2 < |x|. \end{cases} \tag{19.21}$$

An imaginary 1D reconstruction (considering x-direction only) is shown in Figure 19.6. The separable 2D reconstruction kernel can be obtained from the 1D version as

$$u(x, y) = u(x)u(y), \tag{19.22}$$

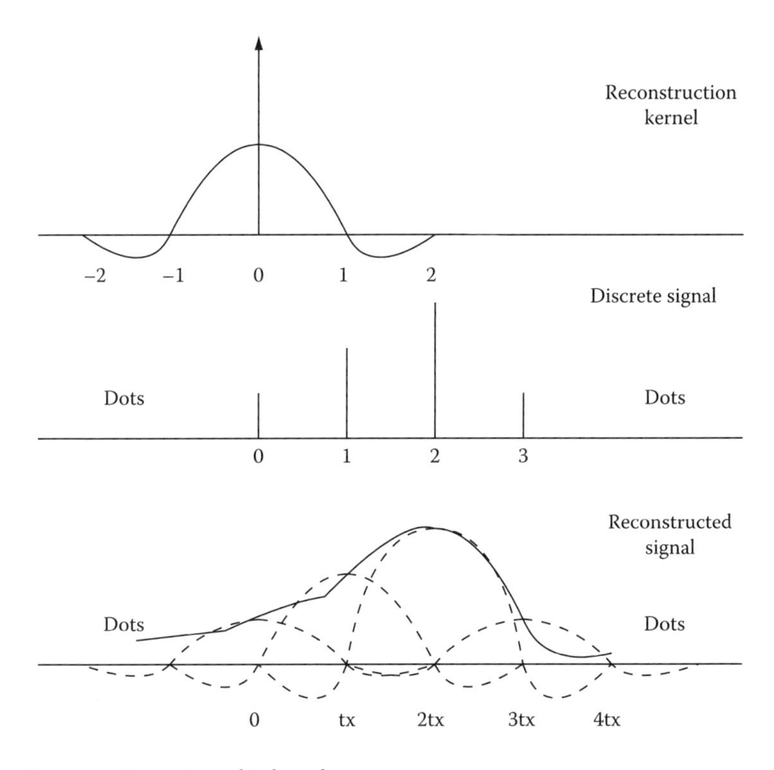

FIGURE 19.6 Reconstruction using cubic kernel.

* $A_2 = 1$ is often used as well.

where $u(x)$ and $u(y)$ are given as

$$u(x) = \begin{cases} \frac{3}{2}|x|^3 - \frac{5}{2}|x|^2 + 1, & \text{if } 0 \leqslant |x| \leqslant 1 \\ -\frac{1}{2}|x|^3 + \frac{5}{2}|x|^2 - 4|x| + 2, & \text{if } 1 \leqslant x \leqslant 2 \\ 0 & \text{if } 2 < |x|. \end{cases} \tag{19.23}$$

$$u(y) = \begin{cases} \frac{3}{2}|y|^3 - \frac{5}{2}|y|^2 + 1, & \text{if } 0 \leqslant |y| \leqslant 1 \\ -\frac{1}{2}|y|^3 + \frac{5}{2}|y|^2 - 4|y| + 2, & \text{if } 1 \leqslant |y| \leqslant 2 \\ 0 & \text{if } 2 < |y|. \end{cases} \tag{19.24}$$

Figure 19.7 demonstrates bicubic interpolation for a scaling ratio of $L = 3$. The interpolated pixels are obtained as weighted linear combinations of G_i, $i = 1, \ldots, 16$. The weights are obtained by sampling the reconstruction kernel given in Equation 19.22 at the respective locations. Bicubic interpolation (for a detailed derivation see [1]) requires more memory and execution time than either the nearest neighbor or the linear methods. However, both the interpolated image and its derivative are continuous. The continuity of the second derivative is not guaranteed.

For an arbitrary scaling ratio L (which may be noninteger), bicubic interpolation is performed as follows

$$\hat{f}[m,n] = \sum_{k=-1}^{k=2} \sum_{l=-1}^{l=2} w_{m,k}\, w_{n,l}\, g[\lfloor m/L \rfloor + k, \lfloor n/L \rfloor + l], \tag{19.25}$$

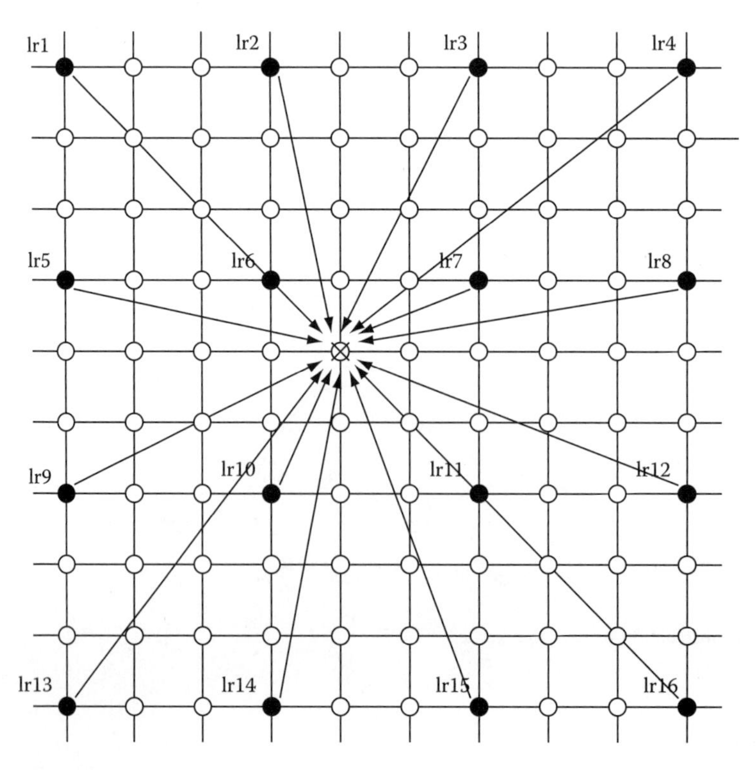

FIGURE 19.7 Bicubic interpolation for $L = 3$.

where the weights are given as

$$
\begin{aligned}
w_{m,-1} &= \left(-\delta_m^3 + 2\delta_m^2 - \delta_m\right)/2 \\
w_{m,0} &= \left(3\delta_m^3 - 5\delta_m^2 + 2\right)/2 \\
w_{m,1} &= \left(-3\delta_m^3 + 4\delta_m^2 + \delta_m\right)/2 \\
w_{m,2} &= \left(\delta_m^3 - \delta_m^2\right)/2 \\
w_{n,-1} &= \left(-\delta_n^3 + 2\delta_n^2 - \delta_n\right)/2 \\
w_{n,0} &= \left(3\delta_n^3 - 5\delta_n^2 + 2\right)/2 \\
w_{n,1} &= \left(-3\delta_n^3 + 4\delta_n^2 + \delta_n\right)/2 \\
w_{n,2} &= \left(\delta_n^3 - \delta_n^2\right)/2,
\end{aligned}
$$

and the distances δ_m and δ_n are defined as

$$
\begin{aligned}
\delta_m &= \frac{m}{L} - \left\lfloor \frac{m}{L} \right\rfloor \\
\delta_n &= \frac{n}{L} - \left\lfloor \frac{n}{L} \right\rfloor.
\end{aligned}
$$

A closely related interpolation method is the cubic spline interpolation. By achieving continuity in the second derivative, it produces the smoothest results of all the linear shift-invariant interpolation methods presented so far. Compared to bicubic interpolation, cubic spline interpolation requires more execution time but less memory.

The main advantage of the linear interpolation methods presented so far is their low computational cost. Hence, these methods are preferred for applications where keeping computational load at a minimum is the main concern. Unless the computational concern is critical enough to justify the use of the nearest neighborhood interpolation, bicubic is usually the method of choice. When image quality is the main concern, these methods are rarely used. The main drawback of these methods is their lack of spatial adaptivity. Local characteristics of an image can change depending on the spatial content. Linear spatially shift-invariant filters cannot handle these changes since they use the same interpolation strategy regardless of the spatial content. This results in serious visual artifacts in the interpolated images, namely jaggy and blurred edges. Jaggy edges are most easily observed in the nearest neighborhood interpolation. Blurred edges are most evident when using bilinear and bicubic interpolation. This is a serious shortcoming of these methods, since it is a well-established fact that edges are among the most important image features for human observers. There are many spatially adaptive interpolation algorithms designed to handle edge pixels in a spatially adaptive way to avoid or reduce these artifacts. The next section is devoted to these edge-adaptive algorithms.

19.2 Edge-Adaptive Methods

Edge information can be utilized to locally adapt interpolation filter coefficients. There are numerous edge-adaptive interpolation schemes in digital image processing literature. These edge-adaptive methods can be roughly classified under two main branches: methods that require explicit edge detection prior to interpolation and methods that extract edge characteristics locally through some processing. The explicit edge-adaptive methods generally have higher computational complexity due to additional edge processing. Robust edge-detectors require filtering with multiple filters corresponding to different

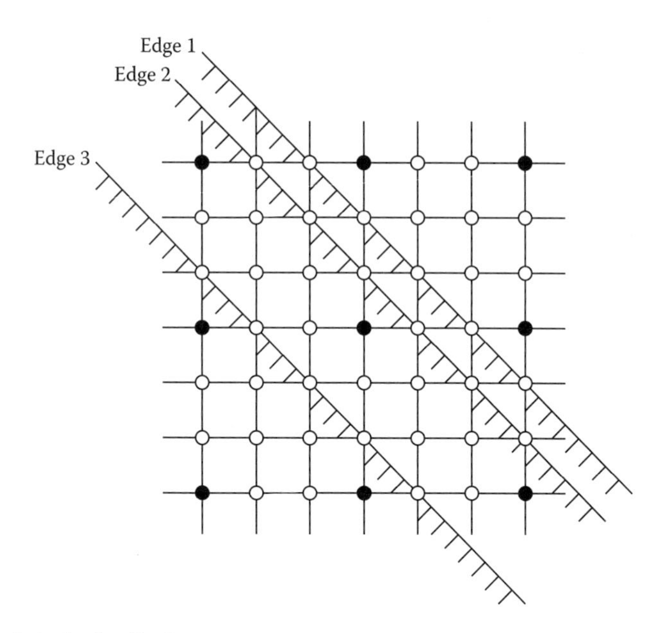

FIGURE 19.8 Subpixel edge localization.

edge orientations and have high computational burden. Note that for spatial interpolation, edge location by itself is not good enough. In addition to the location of an edge, we require its orientation.

Before moving on, we would like to briefly discuss a problem that plagues all edge-directed methods, namely the subpixel edge localization. This problem can most easily be explained with the help of a figure. Figure 19.8 demonstrates three subpixel resolution edges. The black dots represent low-resolution pixels, and the empty circles represent high-resolution pixels. A careful observation of the figure should make it clear that assuming ideal step edges, it is not possible to differentiate between edge 1 and edge 2 using only the low-resolution samples. Edge 3, on the other hand, can be differentiated from edges 1 and 2. Of course, in natural images we rarely have ideal step edges, and the problem becomes how to differentiate between edges 1 and 2, assuming smoother edges with finite gradients. Not all discrete edge (and edge orientation) detection filters can be used for this purpose. Most of the time Laplacian-of Gaussian (LoG) filters are employed in a two-step fashion. Although these filters offer fairly good performance, subpixel edge localization is a difficult task due to its noise sensitivity and high computational requirements. As a careful inspection of Figure 19.8 should make it clear, failing to correctly localize a subpixel edge shifts the edge location in the interpolated image. Quite detailed subpixel edge localization methods have been derived to precisely locate subpixel edges, but their noise sensitivity, excessive computational complexity and the extent of their assumptions about the edge structure to be detected limit their utility.

19.2.1 Explicit Edge Detection-Based Methods

Perhaps the most straightforward approach to edge-adaptive interpolation is to use an explicit edge map to guide the interpolation process. The edge map of the interpolated (output) image can be obtained in two different ways. We can first apply a simple interpolation technique like bicubic interpolation to obtain an initial interpolated image and extract an estimate edge map from this initial interpolated image using conventional edge detection masks. Alternatively, we can first extract the edge map of the input image and interpolate this edge map to obtain the edge map of the output image. Regardless of the way we obtain it, the edge map is only an approximation and some form of preprocessing is required to refine it. The preprocessing is usually in the form of assuring connectedness so that isolated false edge locations are eliminated. Once we have the edge map, we can use it to guide the interpolation process. Since the edge detection phase is computationally intense, to keep the overall computational load at an acceptable level, a simple interpolation technique may be preferred. One way to achieve this is to slightly modify bilinear interpolation to make it edge-adaptive. Bilinear interpolation computes the output pixel as a weighted linear combination of its closest four neighbors where the weights are chosen as inversely proportional to the distance between the neighbors and the pixel to be interpolated. In Figure 19.9, we illustrate the modified bilinear interpolation for a scaling ratio of $L = 3$. Consider the output pixel marked by $\times$ sign. Bilinear interpolation would compute this pixel as

$$\frac{1}{3}\times\frac{1}{3}G_1+\frac{1}{3}\times\frac{2}{3}G_2+\frac{2}{3}\times\frac{1}{3}G_3+\frac{2}{3}\times\frac{2}{3}G_4=\frac{1}{9}G_1+\frac{2}{9}G_2+\frac{2}{9}G_3+\frac{4}{9}G_4. \tag{19.26}$$

Note that for both of the edge scenarios shown in Figure 19.9, this formula simply combines pixel values from each side of the edges resulting in a smeared edge in the output image. Modified bilinear interpolation avoids (or reduce) this effect. The first step is to determine whether or not any of the input pixels G_1, G_2, G_3, or G_4 are separated from the target output pixel by edges. If an input pixel is separated from the target output pixel by an edge, instead of using this pixel value directly, we compute a replacement value according to a heuristic procedure. Assume that our estimated edge map revealed that the edge structure is as given by edge 1 in Figure 19.9. Clearly, G_1 is separated from the target output pixel. To calculate a replacement value for G_1 we linearly interpolate the midpoint of the line G_2–G_3 and call this point as G_{23}. Then we extrapolate along the line G_4–G_{23} to obtain the replacement value of G_1. This replacement rule can be generalized for other cases. For example, if both pixels G_1 and G_3 were separated from the target output pixel by the edge (corresponding to edge 2), we would check to see if any edges crossed the lines G_2–G_5 and G_4–G_6. If no edges were detected, we would linearly extrapolate along the lines G_2–G_5 and G_4–G_6 to generate the replacement values for G_1 and G_3, respectively. To eliminate possible discontinuities, we clamp the result of any extrapolation to lie between the values of the input pixels on either side of the target output pixel whose value is being interpolated.

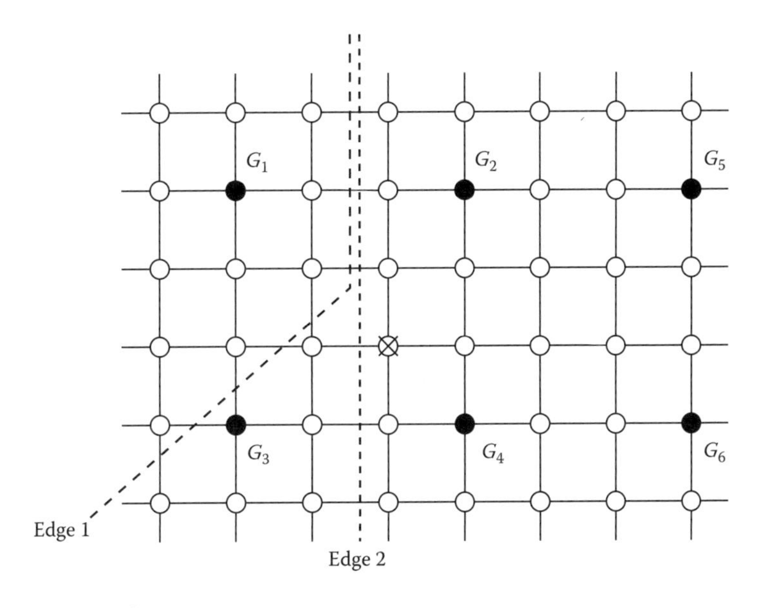

FIGURE 19.9 Edge-directed.

We can further improve this approach by exploiting a model of the imaging system used to capture the original low-resolution data in an iterative fashion. The fidelity of the initial edge-directed interpolation output can be checked by passing it through the imaging model. Ideally, the output of the imaging model should be equal to the initial low-resolution input image, but because of the imperfect interpolation filtering and the modeling errors this can never be the case. The error between the output of the imaging model and the initial low-resolution image can be used to correct the interpolated image in an iterative fashion. One of the most serious drawbacks of this scheme is that its precision is limited with the precision of the edge detection method and the imaging model. Even the most advanced edge detection algorithms like the Canny edge detector are prone to mistakes, and the subpixel edge localization problem is always present. Furthermore, edge detection algorithms generally require thresholding at some step and finding a universal threshold that would perform well for all input images is quite challenging.

19.2.2 Warped-Distance Space-Variant Interpolation

As mentioned before, the problem with linear shift-invariant (LSI) methods like bicubic interpolation is their lack of spatial adaptivity. These methods can be made spatially adaptive by making the interpolation weights dependent on the local pixel values. One such method referred to as warped-distance space-variant interpolation is capable of putting more weight on either side of an edge by utilizing a scoring function to decide on the side that the high-resolution pixel came from. The interpolation method we present in this section operates on rows and columns separately. Such interpolation methods are referred to as separable. We first interpolate all the rows through 1D processing. Then we process the columns in a similar fashion to obtain the final interpolated image. Figure 19.10 illustrates 1D row-wise bicubic* interpolation. 1D bicubic interpolation can be interpreted as resampling a continuous 1D signal reconstructed from the input discrete signal using the following reconstruction kernel:

$$u(x) = \begin{cases} \frac{3}{2}|x|^3 - \frac{5}{2}|x|^2 + 1, & \text{if } 0 \leqslant x \leqslant T_x \\ -\frac{1}{2}|x|^3 + \frac{5}{2}|x|^2 - 4x + 2, & \text{if } T_x \leqslant x \leqslant 2T_x \\ 0, & \text{if } 2T_x < |x|. \end{cases} \tag{19.27}$$

For the scaling ratio of $L = 3$, the target output image would be computed as

$$\begin{aligned} &\left[(-\delta^3 + 2\delta^2 - \delta)/2\right]g[n-1] + \left[(3\delta^3 - 5\delta^2 + 2)/2\right]g[n] \\ &\quad + \left[(-3\delta^3 + 4\delta^2 + \delta_m)/2\right]g[n+1] + \left[(\delta^3 - \delta^2)/2\right]g[n+2], \end{aligned}$$

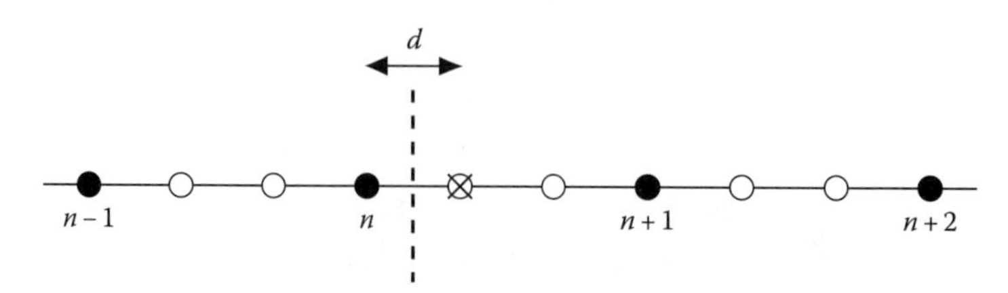

FIGURE 19.10 Warped distance.

* Strictly speaking, in 1-D, this method should no longer be called "bicubic," but for the sake of continuity we will do so.

where δ is defined as the distance between the location of the output pixel to be interpolated and the input pixel right before this location. As illustrated in Figure 19.10, assume that there is a vertical edge between $g[n]$ and $g[n+1]$. Similar to the standard bilinear interpolation, bicubic interpolation computes the output pixel as a weighted linear combination of pixels from both sides of the edge resulting in a blurred edge in the output image. The idea behind warped-distance space-variant interpolation is to modify, δ, the distance used in the interpolation equation, in such a way that the pixels on the different sides on the edge are given different weights. This method is applicable to any LSI filter, including bilinear and bicubic interpolation. To modify any LSI interpolation filter, a warped distance δ' is used in place of δ. The net effect of this warped distance is to move the pixel itself toward those neighbors which are able to yield a visually better estimate for it. To decide on the warping, we need to evaluate to which side of an edge the output pixel should belong. A simple heuristics that gives acceptable results is to associate the output pixel to the most homogeneous side of the edge. To achieve this effect, we first estimate the asymmetry of the data in the neighborhood with the operator

$$A = \frac{|g[n+1] - g[n-1]| - |g[n+2] - g[n]|}{R-1}, \tag{19.28}$$

where
  $R = 255$ for 8 bit luminance images
  A is in the range $[-1, 1]$

It is easy to validate that if the input image is sufficiently smooth and the luminance changes at object edges can be approximated by sigmoidal functions; $A = 0$ indicates symmetry. For positive values of A, the edge is more homogeneous on the right side, and we conclude that the pixel to be interpolated belongs to the right side. Hence, pixels on the right side of the edge should be given more weight in the interpolated value. Hence, we require $\delta' > \delta$ for $A > 0$. The opposite holds, if $A < 0$. This suggests that the desired effect can be obtained by mapping δ as

$$\delta + f(A) \tag{19.29}$$

where $f(A)$ is a function that is increasing with A. To avoid artifacts, we must assure that δ' stays in the same range as δ, i.e., $[0, 1]$. Also observe that for $\delta \approx 0$ and $\delta \approx 1$, the output location is already very close to one side and assigning it to the other side does not make much sense. Hence, we use a nonlinear mapping for the function $f(A)$ that allows a strong warping when δ is around 0.5 but suppresses changes when δ is close to 0 or 1. The simplest mapping that has these properties is

$$f(A) = -kA\delta(1-\delta). \tag{19.30}$$

Then the warped distance is given as

$$\delta' = \delta - kA\delta(1-\delta) \tag{19.31}$$

where the positive parameter k controls the intensity of the warping. It can be shown that we require $k < 1$ to keep δ in $[0, 1]$ for any possible configuration of the input. We should mention that the symmetry measure A given in Equation 19.28 and the mapping $f(A)$ given in Equation 19.30 are not the only possibilities. Any measure or mapping with the desired properties as explained above can be utilized. Also note that using a scoring function to measure symmetry based on local pixel values is not the only option to introduce spatial adaptivity. Conventional methods like bicubic filter can be made adaptive to the local edge structure by enforcing certain concavity constraints.

19.2.3 New Edge-Directed Interpolation

The new* edge-directed interpolation (NEDI) method is a linear spatially adaptive interpolation filter. NEDI is based on exploiting local covariances to adapt the interpolation coefficients to the local edge structure without explicit edge detection. To explain the role of local covariance in adaptation, we point out that the spatial interpolation using linear filters can be viewed as a linear estimation problem. In a linear estimation problem, we are given a vector of samples of a random variable **x**, and our goal is to estimate the best value of a correlated random variable **y**. Both random variables are assumed to have zero mean. The form of the linear minimum mean-squared interpolation estimator is known to be

$$\mathbf{w}_o = R_{xx}^{-1} R_{xy}. \tag{19.32}$$

This results in the optimal estimate

$$\hat{\mathbf{y}}_o = \mathbf{w}_o^t \mathbf{x} = R_{xy}^t R_{xx}^{-1} \mathbf{x} \tag{19.33}$$

where R_{xx} is the covariance matrix of **x** defined as

$$R_{xx} = \mathcal{E}\{\mathbf{x}\mathbf{x}^{\sqcup}\} \tag{19.34}$$

and R_{xy} is the cross-correlation matrix of **x** and **y** defined as

$$R_{xy} = \mathcal{E}\{\mathbf{x}\mathbf{y}\}. \tag{19.35}$$

For the image interpolation problems, the vector of the samples is simply the pixel values that will be used in the interpolation, and the random variable we are trying to estimate is the high-resolution pixel to be interpolated. If we can estimate R_{xx} and R_{xy}, we can simply use Equation 19.33 to obtain the best linear filter. One way to achieve this is to use the local correlation matrices extracted from low-resolution image to estimate the correlation matrices in the high-resolution image. These estimated correlations are then used to capture the local edge orientations in the high-resolution image.

For the sake of clarity we will consider the case of $L = 2$. Consider Figure 19.11, which shows the low- and high-resolution grids superimposed on top of each other. As usual, filled dots represent the low-resolution grid and empty dots represent the high-resolution grid. A careful observation of this figure tells us that the spatial interpolation problem is not the same for all high-resolution pixels. For resolution enhancement by a factor of $L = 2$, there are three types of high-resolution grid pixels to be interpolated. The first type, represented by letter A, has its two closest low-resolution neighbors in the vertical direction. The second type, represented by letter C, has its two closest neighbors in the horizontal direction. On the other hand, the third type, represented by letter B, has its four closest neighbors in diagonal directions. Due to this asymmetry, the interpolation of type B pixels is different from the interpolation of pixel types A and C. Type B pixels have four low-resolution neighbors immediately available for interpolation; on the other hand, type A and C pixels have only two. For this reason, NEDI first operates on type B pixels interpolating their values from the four available closest neighbors. Once type B pixel values are obtained, they can be used for interpolating type A and C pixels. Again a careful observation should make it clear that once type B pixels are available, the interpolation problem for types A and C (which now have four closest neighbors in a symmetric configuration) becomes similar to type B pixels. Hence, we can apply the algorithm used to obtain type B pixels to get type A and C pixels without any changes.

* The word "new" is not an adjective; it is part of the name of the algorithm proposed in Ref. [1].

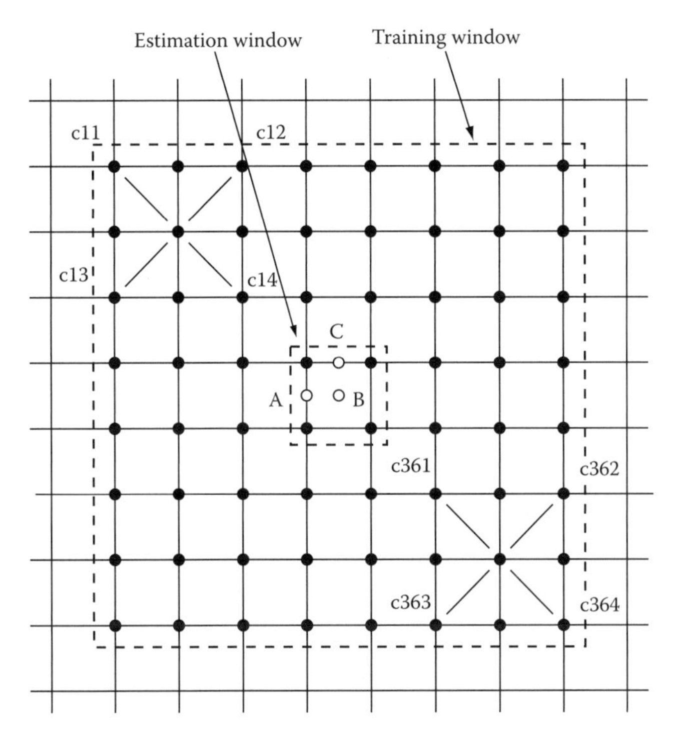

FIGURE 19.11 NEDI.

NEDI algorithm uses an $M \times M$ training window to adapt the filter coefficients to the local edge characteristics. Figure 19.11 depicts such a window for $M = 6$. Note that due to boundary issues, the effective training window size is 8×8. Let us define **C** matrix and **y** vector are as

$$\mathbf{C} = \begin{bmatrix} c_{1,1} & c_{1,2} & c_{1,3} & c_{1,4} \\ c_{2,1} & c_{2,2} & c_{2,3} & c_{2,4} \\ \vdots & \vdots & \vdots & \vdots \\ c_{M^2,1} & c_{M^2,2} & c_{M^2,3} & c_{M^2,4} \end{bmatrix} \quad \mathbf{y} = \begin{bmatrix} c_1 \\ c_2 \\ \vdots \\ c_{M^2} \end{bmatrix}. \tag{19.36}$$

Hence, **y** is an $M^2 \times 1$ vector that consists of the pixels within the training window, and **C** is an $M^2 \times 4$ matrix whose rows are the diagonal neighbors of the pixels within the training window in raster-scan order. Our aim is to find four filter coefficients $\alpha = [\alpha_1, \alpha_2, \alpha_3, \alpha_4]^t$ such that the squared error defined as

$$\text{Error} = \|\mathbf{e}\|^2 \quad \text{where } \mathbf{e} = \begin{bmatrix} c_1 \\ c_2 \\ \vdots \\ c_{M^2} \end{bmatrix} - \begin{bmatrix} c_{1,1} & c_{1,2} & c_{1,3} & c_{1,4} \\ c_{2,1} & c_{2,2} & c_{2,3} & c_{2,4} \\ \vdots & \vdots & \vdots & \vdots \\ c_{M^2,1} & c_{M^2,2} & c_{M^2,3} & c_{M^2,4} \end{bmatrix} \begin{bmatrix} \alpha_1 \\ \alpha_2 \\ \alpha_3 \\ \alpha_4 \end{bmatrix},$$

is minimized. Assuming **C** is full column-rank (i.e., its columns are linearly independent) this problem becomes the classic least-squares (LS) problem whose solution is given by

$$\boldsymbol{\alpha} = \mathbf{R}^{-1}\mathbf{r} \quad \text{where } \mathbf{R} = \mathbf{C}^t\mathbf{C} \quad \text{and} \quad \mathbf{r} = \mathbf{C}^t\mathbf{y}. \tag{19.37}$$

Comparing $\boldsymbol{\alpha}$ with Equation 19.32, we see that $\boldsymbol{\alpha}$ is an approximate LMS estimator with

$$\begin{aligned} \mathbf{R} &\approx R_{xx} \\ \mathbf{r} &\approx R_{xy}. \end{aligned} \quad (19.38)$$

The resulting interpolation coefficients given by $\boldsymbol{\alpha}$ can effectively adapt to the local edge structure assuming that the local window has a valid edge structure with arbitrary orientation. The most serious shortcoming of NEDI is its high computational complexity. The optimal estimator requires matrix multiplication and inversion of fairly large matrices. Both of these operations are quite intensive. Another drawback of NEDI is the lack of a constraint to force the coefficients to sum up to 1. The net effect of this shortcoming is a possible change in the DC component of the frame. Fortunately, for natural images, the sum of coefficients was observed to stay within a close neighborhood of 1.

19.3 Statistical Learning-Based Methods

An advanced approach to spatially adaptive interpolation is context-based pixel classification in a statistical framework. In image processing literature, this method is referred to as resolution synthesis (RS). RS is based on the observation that pixels can be classified as belonging to different spatial context classes such as edges of different orientation and smooth textures. This is similar to the approach mentioned at the beginning of the section, but has two important improvements that make this approach one of the most effective interpolation methods ever. First, the pixel classes are derived in the statistical framework and found to be superior when compared to the previously mentioned methods. Second, in RS, optimal filters for every class are derived separately to take the full advantage of the pixel classification process.

Figure 19.12 illustrates building blocks of RS. The first block is the feature extractor. For every input pixel, a 5×5 window centered at the current pixel is used to extract a feature vector denoted by $\mathbf{y}$. Let us raster-scan this 5×5 window into a 25×1 vector and call it $\mathbf{z}$. Then feature vector extraction can be defined as a mapping from $\mathbf{z}$ to $\mathbf{y}$. The extracted feature vector $\mathbf{y}$ is usually of lower dimensionality. The main goal of feature extraction is to reduce the computational load and the number of parameters required to model different contexts. Instead of using the entire neighborhood, the feature extraction process is designed to summarize the useful local contextual information in a lower-dimensional feature vector. Note that feature vector selection requires a good understanding of the image interpolation

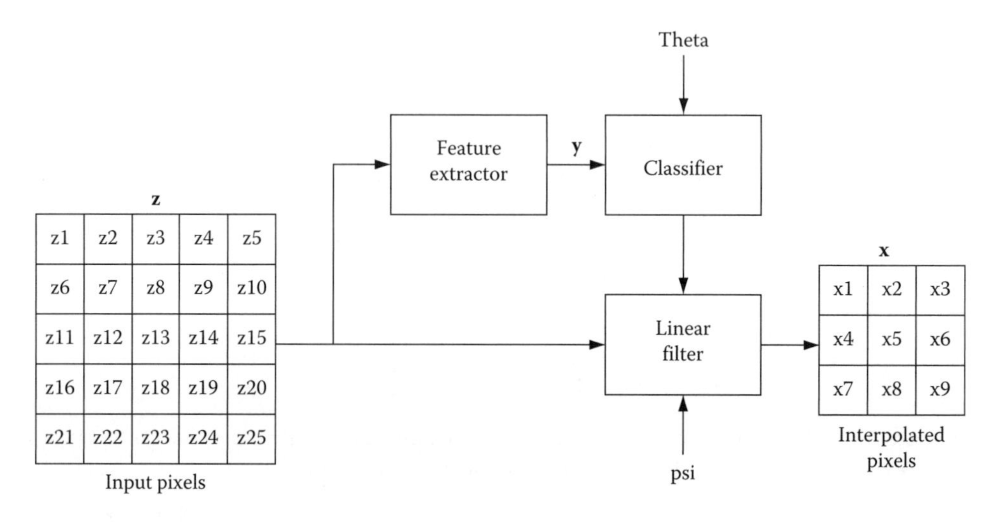

FIGURE 19.12 Resolution synthesis.

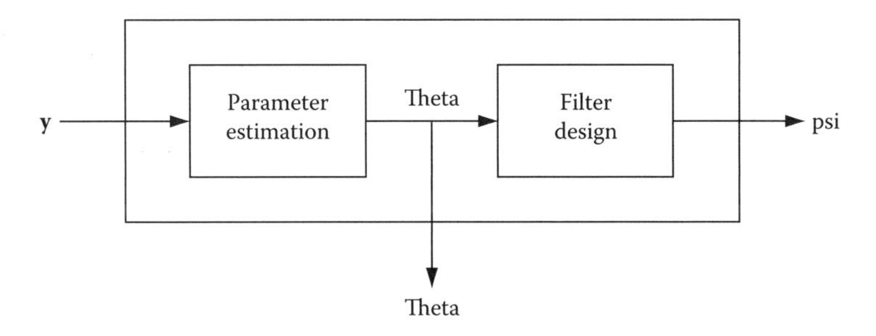

FIGURE 19.13 EM-based off-line training.

process, and the extracted feature vector affects the interpolation quality to a great extent. The extracted feature vector is passed to the context classifier. Context classifier uses the context parameters to classify the input pixel as a member of a certain number of context classes. These classes represent different image regions like edges, smooth, and textured regions. Each class is represented by its feature vector. The feature vectors are modeled as a multivariate Gaussian mixture. There is a different interpolation filter for every context optimized for the spatial characteristics of that context. The parameters that specify different context classes and the corresponding interpolation filters are estimated beforehand in a computationally demanding training phase shown in Figure 19.13. Training procedure uses the well-known expectation maximization method to obtain the maximum-likelihood estimates of the required parameters. Then, these model parameters are used to interpolate arbitrary input images. Finally, the 5×5 window centered at the input pixel is input to the detected context's interpolation filter, and the output pixels are obtained as an $L \times L$ patch. If this $L \times L$ patch of output pixels is raster-scanned into a vector denoted by $\mathbf{x}$, the interpolation filtering can be stated in matrix-vector form as

$$\mathbf{x} = \mathbf{A}\mathbf{z} + \boldsymbol{\beta}, \tag{19.39}$$

where

$\mathbf{A}$ is the interpolation matrix

$\boldsymbol{\beta}$ is the bias vector

Together $\mathbf{A}$ and $\boldsymbol{\beta}$ are called as the interpolation filter coefficients and they are obtained during the off-line training phase.

19.4 Adaptively Quadratic Image Interpolation

An advanced interpolation technique that uses spatial correlation to adapt to the local image characteristics is the adaptively quadratic (AQUA) image interpolation [2]. AQUA is based on a mathematical tool called optimal recovery (OR) theory. In very rough terms, OR theory can be used to recover a signal from the partial information provided in the form of linear functionals given that the signal belongs to a certain class. To apply OR theory to an image interpolation problem, we first specify the quadratic signal class as

$$\mathcal{K} = \{\mathbf{x} \in \mathbb{R}^n : \mathbf{x}^t\mathbf{Q}\mathbf{x} < \varepsilon\}, \tag{19.40}$$

where the $\mathbf{Q}$ matrix is obtained from training data. To obtain spatial adaptivity the training data can be extracted from the local neighborhoods. Note that given $\mathbf{Q}$ is symmetric, $\mathbf{x}^t\mathbf{Q}\mathbf{x}$ is a quadratic form and $\mathcal{K}$ as defined in Equation 19.40 has a specific shape dictated by the properties of $\mathbf{Q}$. For the case at hand, $\mathbf{Q}$ will be a positive definite matrix and the resulting $\mathcal{K}$ will be an ellipsoid in $\mathbb{R}^n$.

For now, assume that we are given a training set $\mathcal{T} = \{\mathbf{x}_1, \ldots, \mathbf{x}_m\}$ to determine the local quadratic signal class, i.e., to estimate the **Q** matrix. The vectors $\mathbf{x}_i$ in the training set are simply raster-scanned image patches extracted from a local region. The **Q** matrix that satisfies

$$\mathbf{x}^t\mathbf{Q}\mathbf{x} < \varepsilon \tag{19.41}$$

for all $\mathbf{x}_i \in \mathcal{T}$ and for a fixed ε is a representative of the local signal class. Hence, if a new vector **z** satisfies Equation 19.41, then it is similar to the vectors on the training set $\mathcal{T}$. To see this, assume or consider a matrix **S** whose columns are training vectors $\mathbf{x}_i$. Consider the following matrix-vector equation that relates the training set to the new vector **z** (which is again a raster-scanned image patch):

$$\mathbf{S}\mathbf{a} = \mathbf{z}, \tag{19.42}$$

where **a** is a column vector of weights that specify a weighted linear combination of the training vectors to obtain **z**. If we have enough training vectors and if **z** is similar to the training vectors, then Equation 19.42 can be satisfied with very small weights, i.e., $\|\mathbf{a}\|^2 < \delta$, for some small constant δ. This can easily be demonstrated using the singular-value decomposition of $\mathcal{S}$. We start with writing Equation 19.42 as

$$\begin{aligned} \mathbf{S}\mathbf{a} &= \mathbf{z} \\ \mathbf{U}\Lambda\mathbf{V}^t\mathbf{a} &= \mathbf{z} \\ \mathbf{a} &= \mathbf{V}\Lambda^{-1}\mathbf{U}^t\mathbf{z}. \end{aligned} \tag{19.43}$$

Then we have

$$\begin{aligned} \|\mathbf{a}\|^2 &= \mathbf{a}^t\mathbf{a} \\ &= \mathbf{z}^t\mathbf{U}\Lambda^{-1}\mathbf{V}^t\mathbf{V}\Lambda^{-1}\mathbf{U}^t\mathbf{z} \\ &= \mathbf{z}^t\mathbf{U}\Lambda^{-2}\mathbf{U}^t\mathbf{z}. \end{aligned} \tag{19.44}$$

Noting that

$$\begin{aligned} \mathbf{S}\mathbf{S}^t &= \mathbf{U}\Lambda\mathbf{V}^t\mathbf{V}\Lambda\mathbf{U}^t \\ &= \mathbf{U}\Lambda^2\mathbf{U}^t, \end{aligned} \tag{19.45}$$

we can see

$$\begin{aligned} \mathbf{a}^t\mathbf{a} &= \mathbf{z}^t\mathbf{U}\Lambda^{-2}\mathbf{U}^t\mathbf{z} \\ &= \mathbf{z}^t(\mathbf{S}\mathbf{S}^t)^{-1}\mathbf{z}. \end{aligned} \tag{19.46}$$

If we define $\mathbf{Q} \triangleq (\mathbf{S}\mathbf{S}^t)^{-1}$, then if **z** is similar to the training set, i.e., $\mathbf{z}^t\mathbf{Q}\mathbf{z} < \varepsilon$, we have

$$\mathbf{a}^t\mathbf{a} = \mathbf{z}^t\mathbf{Q}\mathbf{z} < \varepsilon = \delta. \tag{19.47}$$

Since **Q** effectively defines the quadratic signal class, the training samples used to estimate **Q** are of utmost importance. The most straightforward way of selecting the training vectors is based on the proximity of their locations to the position of the location being modeled. Ideally this would require pixels from the image to be interpolated. Since we do not have the output image, patches can simply be extracted from the input image. While this approach works well for small interpolation factors, the approach quickly deteriorates when interpolating by larger factors. For larger interpolation factors, the size of training patches is larger and the distances of the pixels in the extracted patches to the current input pixel quickly increase. Hence, the patches cannot effectively represent the local image characteristics. An alternative approach is to interpolate the input image by using a simple interpolation method like bicubic to obtain an estimate output image. Then we can extract training patches from the estimate output image to determine the quadratic signal class. Once we estimate **Q**, the final interpolated image can be obtained by AQUA.

After obtaining **Q**, the next step is to decide on a crude imaging model. Assume that the scaling ratio is $L = 2$. Then for every input pixel, there are four output pixels. The imaging model hypothesizes that the low-resolution input image $g[m, n]$ is obtained from a high-resolution image $f_o[m, n]$ through some observation process. The interpolated output image $f[m, n]$ is then interpreted as an estimate of $f_o[m, n]$. For example, we can choose to hypothesize that the low-resolution image is a down-sampled version of the high-resolution image,

$$g[m, n] = f_o[2m, 2n]. \tag{19.48}$$

A second alternative is to assume that the original high-density image has been low-pass filtered first, before decimation by two. (This is quite realistic for digital cameras, where every pixel in the CCD sensor array corresponds to the average light energy detected in the area of the pixel.) In this case, we have

$$\begin{aligned} g[m, n] = {} & 0.25f_o[2m, 2n] + 0.25f_o[2m, 2n + 1] \\ & + 0.25f_o[2m + 1, 2n] + 0.25f_o[2m + 1, 2n + 1], \end{aligned} \tag{19.49}$$

where we assumed a simple 2×2 averaging filter. Once we decide on the imaging model, we relate the model to OR theory as follows. Recall that we modeled the output image as a member of a quadratic signal class. The model is specified as

$$\mathbf{x}^t \mathbf{Q} \mathbf{x} < \varepsilon, \tag{19.50}$$

where **x** is a vector obtained by raster-scanning a patch from the output image. Our imaging model tells us what kind of constraints we can introduce on the output image as linear functionals. For example, consider the upper left corner of the input image $g[m, n]$ to be interpolated. In case of the model given in Equation 19.48, the functionals $F_i(\mathbf{x})$ are given as

$$\begin{aligned} F_1(\mathbf{x}) &= g[0, 0] = f_o[0, 0] \\ F_2(\mathbf{x}) &= g[0, 1] = f_o[0, 2] \\ F_3(\mathbf{x}) &= g[1, 0] = f_o[2, 0] \\ F_4(\mathbf{x}) &= g[1, 1] = f_o[2, 2]. \end{aligned}$$

whereas for the model given in Equation 19.49 we have

$$\begin{aligned} F_1(\mathbf{x}) &= g[0, 0] = 0.25f_o[0, 0] + 0.25f_o[0, 1] + 0.25f_o[1, 0] + 0.25f_o[1, 1] \\ F_2(\mathbf{x}) &= g[0, 1] = 0.25f_o[0, 2] + 0.25f_o[0, 3] + 0.25f_o[1, 2] + 0.25f_o[1, 3] \\ F_3(\mathbf{x}) &= g[1, 0] = 0.25f_o[2, 0] + 0.25f_o[3, 0] + 0.25f_o[2, 1] + 0.25f_o[3, 1] \\ F_4(\mathbf{x}) &= g[1, 1] = 0.25f_o[2, 2] + 0.25f_o[2, 3] + 0.25f_o[3, 2] + 0.25f_o[3, 3]. \end{aligned}$$

Note that these linear functionals can be written as

$$\begin{aligned} F_1(\mathbf{x}) &= \mathbf{x}^t\mathbf{e}_1 \\ F_2(\mathbf{x}) &= \mathbf{x}^t\mathbf{e}_2 \\ F_3(\mathbf{x}) &= \mathbf{x}^t\mathbf{e}_3 \\ F_4(\mathbf{x}) &= \mathbf{x}^t\mathbf{e}_4. \end{aligned} \qquad (19.51)$$

where

$$\begin{aligned} \mathbf{e}_1 &= [1\,0\,0\,0\,0\,0\,0\,0\,0\,0\,0\,0\,0\,0\,0\,0]^t, \\ \mathbf{e}_2 &= [0\,0\,1\,0\,0\,0\,0\,0\,0\,0\,0\,0\,0\,0\,0\,0]^t, \\ \mathbf{e}_3 &= [0\,0\,0\,0\,0\,0\,0\,0\,1\,0\,0\,0\,0\,0\,0\,0]^t, \\ \mathbf{e}_4 &= [0\,0\,0\,0\,0\,0\,0\,0\,0\,0\,1\,0\,0\,0\,0\,0]^t, \end{aligned}$$

and $\mathbf{x}$ is a 16×1 vector obtained by raster-scanning the top left patch of the output image $f[m, n]$, i.e.,

$$\mathbf{x} = [\,f[0,0]f[0,1]f[0,2]f[0,3]f[1,0]\ldots f[3,3]]^t. \qquad (19.52)$$

For the model of Equation 19.49, Equation 19.51 applies with

$$\begin{aligned} \mathbf{e}_1 &= \left[\frac{1}{4}\frac{1}{4}\,0\,0\,\frac{1}{4}\frac{1}{4}\,0\,0\,0\,0\,0\,0\,0\,0\,0\,0\right]^t, \\ \mathbf{e}_2 &= \left[0\,0\,\frac{1}{4}\frac{1}{4}\,0\,0\,\frac{1}{4}\frac{1}{4}\,0\,0\,0\,0\,0\,0\,0\,0\right]^t, \\ \mathbf{e}_3 &= \left[0\,0\,0\,0\,0\,0\,0\,0\,\frac{1}{4}\frac{1}{4}\,0\,0\,\frac{1}{4}\frac{1}{4}\,0\,0\right]^t, \\ \mathbf{e}_4 &= \left[0\,0\,0\,0\,0\,0\,0\,0\,0\,0\,\frac{1}{4}\frac{1}{4}\,0\,0\,\frac{1}{4}\frac{1}{4}\right]^t, \end{aligned}$$

and the same $\mathbf{x}$. To obtain the interpolated patch, we define the known representors as the vectors given by

$$\phi_i \stackrel{\Delta}{=} \mathbf{Q}^{-1}\mathbf{e}_i. \qquad (19.53)$$

If we denote the interpolated patch in the raster-scanned order as $\mathbf{u}$, then by applying OR theory $\mathbf{u}$ can be computed as

$$\mathbf{u} = \sum_i c_i\phi_i. \qquad (19.54)$$

The weighting coefficients c_i are computed as the solution of

$$\begin{bmatrix} F_1(\mathbf{x}) \\ F_2(\mathbf{x}) \\ \vdots \\ F_K(\mathbf{x}) \end{bmatrix} = \begin{bmatrix} \langle\phi_1,\phi_1\rangle_{\mathbf{Q}} & \langle\phi_1,\phi_2\rangle_{\mathbf{Q}} & \cdots & \langle\phi_i,\phi_i\rangle_{\mathbf{Q}} \\ \langle\phi_2,\phi_1\rangle_{\mathbf{Q}} & \langle\phi_2,\phi_2\rangle_{\mathbf{Q}} & \cdots & \langle\phi_i,\phi_i\rangle_{\mathbf{Q}} \\ \vdots & \vdots & \ddots & \vdots \\ \langle\phi_K,\phi_1\rangle_{\mathbf{Q}} & \langle\phi_K,\phi_2\rangle_{\mathbf{Q}} & \cdots & \langle\phi_K,\phi_K\rangle_{\mathbf{Q}} \end{bmatrix} \begin{bmatrix} c_1 \\ c_2 \\ \vdots \\ c_K \end{bmatrix}, \qquad (19.55)$$

where

K is the number of the linear functionals imposed by the imaging system model

$\langle \cdot,\cdot \rangle_{\mathbf{Q}}$ denotes weighted inner product where the weighting matrix is $\mathbf{Q}$

For a detailed explanation of the AQUA method the interested reader is referred to Ref. [2].

19.5 Transform Domain Methods

Currently the most popular transform domain methods are wavelet transform (WT) and discrete cosine transform (DCT)-based methods. These transforms offer the advantage of decomposing the image into specific frequency bands, and work on these bands separately for the obvious cost of largely increased computational complexity. Also wavelet and block-DCT transform-based approaches can adapt to the local frequency characteristics of the input image. The usual properties of increased space and frequency domain resolution transforms are observed. Specifically, as the quality of the frequency partitioning increases (as the band-pass filters get sharper, or yet in another way, as the neighboring frequency bands are better separated), it becomes possible to filter (interpolate) separate frequency bands with less modification to the neighboring bands, which in turn gives better interpolation results. On the other hand, as the spatial resolution increases, better spatial adaptation can be achieved.

19.6 Summary

In this chapter, we have considered the spatial interpolation techniques. We classified the spatial interpolation methods under five main categories, namely, linear spatially invariant methods, edge-adaptive methods, statistical learning-based methods, model-based methods, and transform domain methods.

We started our discussion with linear spatially invariant interpolation filters. We discussed that LSI interpolation can be interpreted as resampling a continuous image reconstructed from the input discrete image. Depending on the complexity of the reconstruction kernel used to obtain the continuous image, different LSI interpolation filters resulted. For the simplest case of zero-order hold, we obtained the nearest neighborhood interpolation. Nearest neighborhood provided the worst interpolation results, failing to guarantee even the continuity of the interpolated image. Nearest neighborhood interpolation was leveraged by its simplicity. Since the interpolated pixel values are simply copies from their neighbors, nearest neighborhood interpolation does not require any arithmetic operations. Bilinear interpolation resulted by using first-order hold reconstruction. We mentioned that bilinear interpolation guaranteed continuity of the interpolated image (but not any of its gradients). In terms of quality and computational complexity, bilinear interpolation provided us with a middle ground between the nearest neighborhood and the bicubic interpolation. The most complicated LSI interpolation method we presented was the bicubic interpolation and its variant cubic splines.

Edge-adaptive interpolation methods required spatial domain filtering to detect edge locations and orientations. Once the edge pixels are identified, they are specially handled to avoid averaging across edges. We discussed three edge-adaptive methods. The first edge-adaptive method we discussed was based on explicit edge detection. We first extracted an estimate edge map of the interpolated image. This estimate edge map was used to guide the interpolation process such that interpolation across edges was avoided. The second and more advanced edge-directed method was the warped-distance space-variant interpolation. The idea behind warped-distance space-variant interpolation was to modify, δ, the distance used in the interpolation equation, in such a way that pixels on different sides on an edge are given different weights. Finally, we presented the NEDI method. NEDI did not require explicit edge detection. Instead, NEDI method was based on using LS estimation with a locally estimated correlation matrix.

We discussed RS as a statistical learning-based interpolation method. RS used a statistical framework to classify pixels into different context classes such as edges of different orientation and smooth textures.

Using an intensive off-line training procedure, the pixel classes and the corresponding optimal interpolation filters were extracted from natural image sets.

We presented AQUA as a model-based interpolation method. AQUA interpolation technique was based on a quadratic signal model. The quadratic signal model was shown to be represented by a positive definite matrix **Q**. Spatial adaptivity was achieved by locally estimating the **Q** matrix.

We pointed out that the transform domain methods offer the advantage of decomposing the image into specific frequency bands, and working on these bands separately for the obvious cost of largely increased computational complexity.

19.7 Bibliography and Historical Remarks

For a detailed treatment of LSI interpolation techniques see [3]. For a detailed treatment of bicubic interpolation see [1]. For reading on spline-based interpolation see [4]. For the 2D generalization of bicubic interpolation see [5].

For further reading on sub-pixel edge localization methods see [6]. For a detailed explanation of the warped-distance interpolation method see [7]. For an alternative way of modifying standard LSI interpolation filters based on local concavity constraints see [8]. For further reading on edge-adaptive interpolation methods see [9–14]. For a detailed treatment of NEDI refer to [15].

For a detailed treatment of the RS methods see [16].

For a detailed treatment of AQUA methods see [2].

For further reading on transform domain methods see [17–22].

For a color image interpolation method with edge enhancement see [21].

References

1. R. G. Keys, Cubic convolution interpolation for digital image processing, *IEEE Transactions on Acoustics, Speech, and Signal Processing*, 29, 1153–1160, Dec. 1981.
2. D. Muresan and T. W. Parks, Adaptively quadratic (aqua) image interpolation, *IEEE Transactions on Image Processing*, 13, 690–698, May 2004.
3. R. Schafer and L. Rabiner, A digital signal processing approach to interpolation, *Proceedings of the IEEE*, 61, 692–702, 1973.
4. A. Gotchev, J. Vesna, T. Saramaki, and K. Egiazarian, Digital image resampling by modified b-spline functions, in *IEEE Nordic Signal Processing Symposium*, Kolmarden, Sweden, pp. 259–262, June 2000.
5. S. E. Reichenbach and F. Geng, Two-dimensional cubic convolution, *IEEE Transactions on Image Processing*, vol. 12, pp. 854–865, Aug. 2003.
6. K. Jensen and D. Anastassiou, Subpixel edge localization and the interpolation of still images, *IEEE Transactions on Image Processing*, 4, 285–295, Mar. 1995.
7. G. Ramponi, Warped distance for space-variant linear image interpolation, *IEEE Transactions on Image Processing*, 8, 629–639, May 1999.
8. J. K. Han and S. U. Baek, Parametric cubic convolution scaler for enlargement and reduction of image, *IEEE Transactions on Consumer Electronics*, 46, 247–256, May 2000.
9. J. Allebach and P. W. Wong, Edge-directed interpolation, in *Proceedings of the IEEE International Conference on Image Processing*, Lausanne, Switzerland, pp. 707–710, Sept. 1996.
10. H. Shi and R. Ward, Canny edge based image expansion, in *Proceedings of the IEEE International Symposium on Circuits and Systems*, Scottsdale, AZ, pp. 785–788, May 2002.
11. Q. Wang and R. Ward, A new edge-directed image expansion scheme, in *Proceedings of the IEEE International Conference on Image Processing*, Thessaloniki, Greece, pp. 899–902, Oct. 2001.
12. A. M. Darwish, M. S. Bedair, and S. I. Shaheen, Adaptive re-sampling algorithm for image zooming, in *IEE Proceedings of Visual Image and Processing*, 144, pp. 207–212, Aug. 1997.

13. H. Jiang and C. Moloney, A new direction adaptive scheme for image interpolation, in *Proceedings of the IEEE International Conference on Image Processing*, Rochester, NY, 3, pp. 369–372, June 2002.
14. S. Dube and L. Hong, An adaptive algorithm for image resolution enhancement, in *Proceedings of the IEEE Conference on Signals Systems and Computers*, Pacific Grove, CA, pp. 1731–1734, Oct. 2000.
15. X. Li and M. T. Orchard, New edge-directed interpolation, *IEEE Transactions on Image Processing*, 10, 1521–1527, Oct. 2001.
16. C. B. Atkins, C. A. Bouman, and J. P. Allebach, Optimal image scaling using pixel classification, in *Proceedings of the IEEE International Conference on Image Processing*, Thessaloniki, Greece, pp. 864–867, Oct. 2001.
17. Y. Zhu, S. C. Schwartz, and M. T. Orchard, Wavelet domain image interpolation via statistical estimation, in *Proceedings of the IEEE International Conference on Image Processing*, Thessaloniki, Greece, pp. 840–843, Oct. 2001.
18. K. Kinebuchi, D. D. Muresan, and T. W. Parks, Image interpolation using wavelet-based hidden markov trees, in *Proceedings of the IEEE International Conference on Acoustics Speech and Signal Processing*, Salt Lake City, UT, pp. 1957–1960, May 2001.
19. D. D. Muresan and T. W. Parks, Prediction of image detail, in *Proceedings of the IEEE International Conference on Image Processing*, Vancouver, British Columbia, Canada, pp. 323–326, Sept. 2000.
20. W. K. Carey, D. B. Chuang, and S. S. Hemami, Regularity-preserving image interpolation, *IEEE Transactions on Image Processing*, 8, 1293–1297, Sept. 1999.
21. L. S. DeBrunner and V. DeBrunner, Color image interpolation with edge-enhancement, in *Proceedings of the IEEE International Conference on Signals, Systems and Computers*, Pacific Grove, CA, pp. 901–905, Oct. 2000.
22. S. A. Martucci, Image resizing in the discrete cosine transform domain, in *Proceedings of the IEEE International Conference on Image Processing*, Washington, D.C., vol. 2, pp. 244–247, Oct. 1995.

20

Video Sequence Compression

Osama Al-Shaykh
Packet Video

Ralph Neff
University of California

David Taubman
Hewlett Packard

Avideh Zakhor
University of California

The image and video processing literature is rich with video compression algorithms. This chapter overviews the basic blocks of most video compression systems, discusses some important features required by many applications, e.g., scalability and error resilience, and reviews the existing video compression standards such as H.261, H.263, MPEG-1, MPEG-2, and MPEG-4.

20.1 Introduction

Video sources produce data at very high bit rates. In many applications, the available bandwidth is usually very limited. For example, the bit rate produced by a 30 fps color common intermediate format (CIF) (352×288) video source is 73 Mbps. In order to transmit such a sequence over a 64 kbps channel (e.g., ISDN line), we need to compress the video sequence by a factor of 1140. A simple approach is to subsample the sequence in time and space. For example, if we subsample both chroma components by 2 in each dimension, i.e., 4 : 2 : 0 format, and the whole sequence temporally by 4, the bit rate becomes 9.1 Mbps. However, to transmit the video over a 64 kbps channel, it is necessary to compress the subsampled sequence by another factor of 143. To achieve such high compression ratios, we must tolerate some distortion in the subsampled frames.

Compression can be either lossless (reversible) or lossy (irreversible). A compression algorithm is lossless if the signal can be reconstructed from the compressed information; otherwise it is lossy. The compression performance of any lossy algorithm is usually described in terms of its rate-distortion curve, which represents the potential trade-off between the bit rate and the distortion associated with the lossy representation. The primary goal of any lossy compression algorithm is to optimize the rate-distortion curve over some range of rates or levels of distortion. For video applications, rate is usually expressed in terms of bits per second. The distortion is usually expressed in terms of the peak-signal-to-noise ratio (PSNR) per frame or, in some cases, measures that try to quantify the subjective nature of the distortion.

In addition to good compression performance, many other properties may be important or even critical to the applicability of a given compression algorithm. Such properties include robustness to errors in the compressed bit-stream, low complexity encoders and decoders, low latency requirements, and scalability. Developing scalable video compression algorithms has attracted considerable attention in recent years. Generally speaking, scalability refers to the potential to effectively decompress subsets of the compressed bit-stream in order to satisfy some practical constraint, e.g., display resolution, decoder computational complexity, and bit rate limitations.

The demand for compatible video encoders and decoders has resulted in the development of different video compression standards. The international standards organization (ISO) has developed MPEG-1 to store video on compact discs, MPEG-2 for digital television, and MPEG-4 for a wide range of applications including multimedia. The international telecommunication union (ITU) has developed H.261 for video conferencing and H.263 for video telephony.

All existing video compression standards are hybrid systems. That is, the compression is achieved in two main stages. The first stage, motion compensation and estimation, predicts each frame from its neighboring frames, compresses the prediction parameters, and produces the prediction error frame. The second stage codes the prediction error. All existing standards use block-based discrete cosine transform (DCT) to code the residual error. In addition to DCT, others nonblock-based coders, e.g., wavelets and matching pursuit, can be used.

In this chapter, we will provide an overview of hybrid video coding systems. In Section 20.2, we discuss the main parts of a hybrid video coder. This includes motion compensation, signal decompositions and transformations, quantization, and entropy coding. We compare various transformations such as DCT, subband, and matching pursuit. In Section 20.3, we discuss scalability and error resilience in video compression systems. We also describe a nonhybrid video coder that provides scalable bit-streams [28]. Finally, in Section 20.4, we review the key video compression standards: H.261, H.263, MPEG-1, MPEG-2, and MPEG-4.

20.2 Motion-Compensated Video Coding

Virtually all video compression systems identify and reduce four basic types of video data redundancy: interframe (temporal) redundancy, interpixel redundancy, psychovisual redundancy, and coding redundancy. Figure 20.1 shows a typical diagram of a hybrid video compression system. First the current frame is predicted from previously decoded frames by estimating the motion of blocks or objects, thus reducing the interframe redundancy. Afterwards to reduce the interpixel redundancy, the

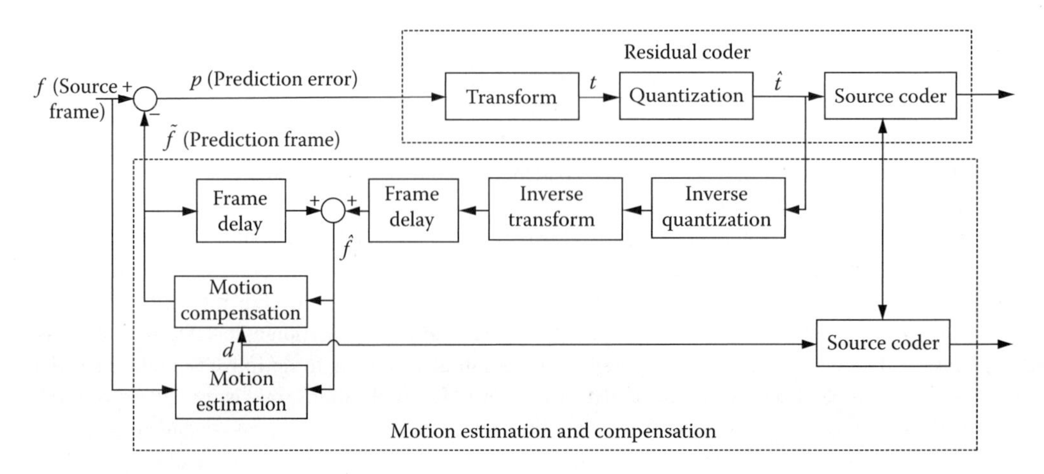

FIGURE 20.1 Motion-compensated coding of video.

residual error after frame prediction is transformed to another format or domain such that the energy of the new signal is concentrated in few components and these components are as uncorrelated as possible. The transformed signal is then quantized according to the desired compression performance (subjective or objective). The quantized transform coefficients are then mapped to codewords that reduce the coding redundancy. The rest of this section will discuss the blocks of the hybrid system in more detail.

20.2.1 Motion Estimation and Compensation

Neighboring frames in typical video sequences are highly correlated. This interframe (temporal) redundancy can be significantly reduced to produce a more compressible sequence by predicting each frame from its neighbors. Motion compensation is a nonlinear predictive technique in which the feedback loop contains both the inverse transformation and the inverse quantization blocks, as shown in Figure 20.1.

Most motion compensation techniques divide the frame into regions, e.g., blocks. Each region is then predicted from the neighboring frames. The displacement of the block or region, d, is not fixed and must be encoded as side information in the bit-stream. In some cases, different prediction models are used to predict regions, e.g., affine transformations. These prediction parameters should also be encoded in the bit-stream.

To minimize the amount of side information, which must be included in the bit-stream, and to simplify the encoding process, motion estimation is usually block based. That is, every pixel $\vec{i}$ in a given rectangular block is assigned the same motion vector, d. Block-based motion estimation is an integral part of all existing video compression standards.

20.2.2 Transformations

Most image and video compression schemes apply a transformation to the raw pixels or to the residual error resulting from motion compensation before quantizing and coding the resulting coefficients. The function of the transformation is to represent the signal in a few uncorrelated components. The most common transformations are linear transformations, i.e., the multidimensional sequence of input pixel values, $f[\vec{i}]$, is represented in terms of the transform coefficients, $t[\vec{k}]$, via

$$f[\vec{i}] = \sum_{\vec{k}} t[\vec{k}]w_{\vec{k}}[\vec{i}] \tag{20.1}$$

for some $w_{\vec{k}}[\vec{i}]$. The input image is thus represented as a linear combination of basis vectors, $w_{\vec{k}}$. It is important to note that the basis vectors need not be orthogonal. They only need to form an over-complete set (matching pursuits), a complete set (DCT and some subband decompositions), or very close to complete (some subband decompositions). This is important since the coder should be able to code a variety of signals. The remainder of the section discusses and compares DCT, subband decompositions, and matching pursuits.

20.2.2.1 The DCT

There are two properties desirable in a unitary transform for image compression: the energy should be packed into a few transform coefficients, and the coefficients should be as uncorrelated as possible. The optimum transform under these two constraints is the Karhunen–Loéve transform (KLT) where the eigenvectors of the covariance matrix of the image are the vectors of the transform [10]. Although the KLT is optimal under these two constraints, it is data-dependent, and is expensive to compute. The DCT performs very close to KLT especially when the input is a first order Markov process [10].

The DCT is a block-based transform. That is, the signal is divided into blocks, which are independently transformed using orthonormal discrete cosines. The DCT coefficients of a one-dimensional signal, f, are computed via

$$t^{\mathrm{DCT}}[Nb+k] = \frac{1}{\sqrt{N}} \begin{cases} \sum\limits_{i=0}^{N-1} f[Nb+i], & k=0 \\ \sum\limits_{i=0}^{N-1} \sqrt{2} f[Nb+i] \cos\dfrac{(2i+1)k\pi}{2N}, & 1 \le k < N \end{cases} \quad \forall b \tag{20.2}$$

where
N is the size of the block
b denotes the block number

The orthonormal basis vectors associated with the one-dimensional DCT transformation of Equation 20.2 are

$$w_k^{\mathrm{DCT}}[i] = \frac{1}{\sqrt{N}} \begin{cases} 1, & k=0, 0 \le i < N \\ \sqrt{2}\cos\dfrac{(2i+1)k\pi}{2N}, & 1 \le k < N, 0 \le i < N \end{cases} \tag{20.3}$$

Figure 20.2a shows these basis vectors for $N = 8$.

The one-dimensional DCT described above is usually separably extended to two dimensions for image compression applications. In this case, the two-dimensional basis vectors are formed by the tensor product of one-dimensional DCT basis vectors and are given by

$$w_{\vec{k}}^{\mathrm{DCT}}[\vec{i}] = w_{k_1,k_2}^{\mathrm{DCT}}[i_1, i_2] w_{k_1}^{\mathrm{DCT}}[i_1] \cdot w_{k_2}^{\mathrm{DCT}}[i_2]; \quad 0 \le k_1, k_2, i_1, i_2 < N$$

Figure 20.2b shows the two-dimensional basis vectors for $N = 8$.

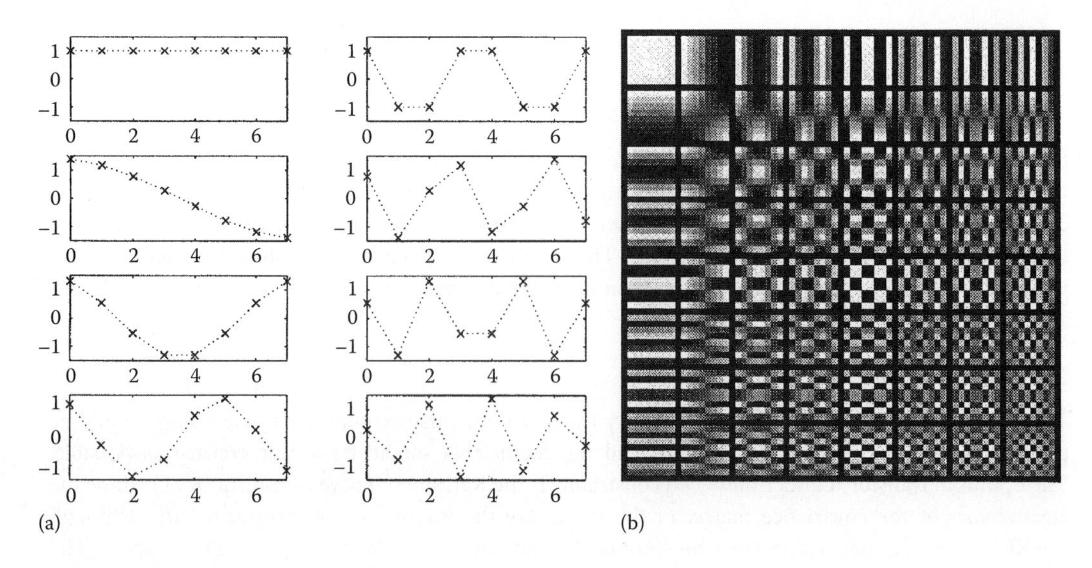

FIGURE 20.2 DCT basis vectors ($N = 8$): (a) one-dimensional and (b) separable two-dimensional.

The DCT is the most common transform in video compression. It is used in the JPEG still image compression standard, and all existing video compression standards. This is because it performs reasonably well at different bit rates. Moreover, there are fast algorithms and special hardware chips to compute the DCT efficiently.

The major objection to the DCT in image or video compression applications is that the nonoverlapping blocks of basis vectors, $w_{\vec{k}}$, are responsible for distinctly "blocky" artifacts in the decompressed frames, especially at low bit rates. This is due to the quantization of the transform coefficients of a block independent from neighboring blocks. Overlapped DCT representation addresses this problem [15]; however, the common solution is to post-process the frame by smoothing the block boundaries [18,22].

Due to bit rate restrictions, some blocks are only represented by one or a small number of coarsely quantized transform coefficients, hence the decompressed block will only consist of these basis vectors. This will cause artifacts commonly known as ringing and mosquito noise.

Figure 20.3b shows frame 250 of the 15 fps CIF COAST-GUARD sequence coded at 112 kbps using a DCT hybrid video coder.* This figure provides a good illustration of the "blocking" artifacts.

FIGURE 20.3 Frame 250 of COAST-GUARD sequence, original shown in (a), coded at 112 kbps using (b) DCT-based coder (H.263) [3], (c) ZTS coder [16], and (d) matching pursuit coder [20]. Blocking artifacts can be noticed on the DCT-coded frame. Ringing artifacts can be noticed on the subband-coded frame.

* It is coded using H.263 [3], which is an ITU standard.

20.2.2.2 Subband Decomposition

The basic idea of subband decomposition is to split the frequency spectrum of the image into (disjoint) subbands. This is efficient when the image spectrum is not flat and is concentrated in a few subbands, which is usually the case. Moreover, we can quantize the subbands differently according to their visual importance.

As for the DCT, we begin our discussion of subband decomposition by considering only a one-dimensional source sequence, $f[i]$. Figure 20.4 provides a general illustration of an N-band one-dimensional subband system. We refer to the subband decomposition itself as analysis and to the inverse transformation as synthesis. The transformation coefficients of bands $1, 2, \ldots, N$ are denoted by the sequences $u_1[k], u_2[k], \ldots, u_N[k]$, respectively. For notational convenience and consistency with the DCT formulation above, we write $t^{SB}[\cdot]$ for the sequence of all subband coefficients, arranged according to $t^{SB}[(\beta - 1) + Nk] = u_\beta[k]$, where $1 \leq \beta \leq N$ is the subband number. These coefficients are generated by filtering the input sequence with filters $H_1, \ldots, H_N$ and downsampling the filtered sequences by a factor of N, as depicted in Figure 20.4. In subband synthesis, the coefficients for each band are upsampled, interpolated with the synthesis filters, $G_1, \ldots, G_N$, and the results summed to form a reconstructed sequence, $\tilde{f}[i]$, as depicted in Figure 20.4.

If the reconstructed sequence, $\tilde{f}[i]$, and the source sequence, $f[i]$, are identical, then the subband system is referred to as perfect reconstruction (PR) and the corresponding basis set is a complete basis set. Although PR is a desirable property, near perfect reconstruction (NPR), for which subband synthesis is only approximately the inverse of subband analysis, is often sufficient in practice. This is because distortion introduced by quantization of the subband coefficients, $t^{SB}[k]$, usually dwarfs that introduced by an imperfect synthesis system.

The filters, $H_1, \ldots, H_N$, are usually designed to have band-pass frequency responses, as indicated in Figure 20.5, so that the coefficients $u_\beta[k]$ for each subband, $1 \leq \beta \leq N$, represent different spectral components of the source sequence.

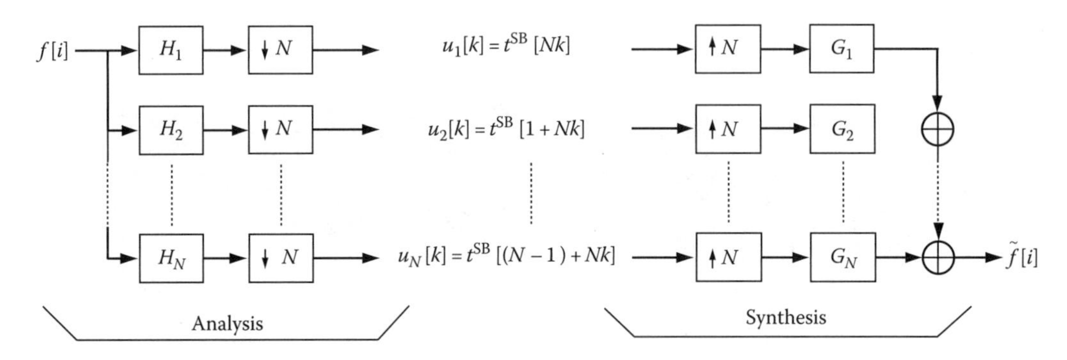

FIGURE 20.4 One-dimensional, N-band subband analysis and synthesis block diagrams. (From Taubman, D. et al., Directionality and scalability in subband image and video compression, in *Image Technology: Advances in Image Processing, Multimedia, and Machine Vision*, Jorge L.C. Sanz, Ed., Springer-Verlag, New York, 1996.)

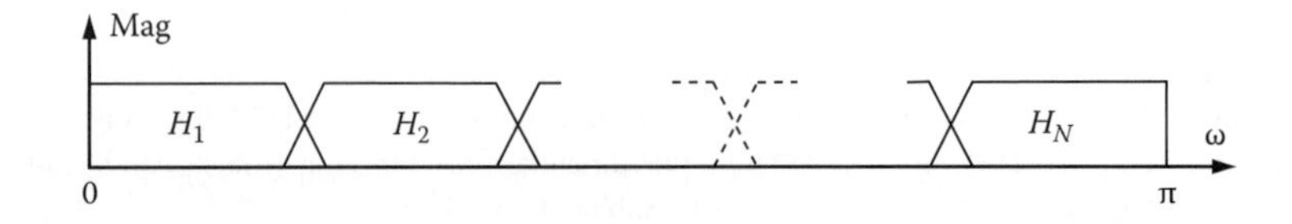

FIGURE 20.5 Typical analysis filter magnitude responses. (From Taubman, D. et al., Directionality and scalability in subband image and video compression, in *Image Technology: Advances in Image Processing, Multimedia, and Machine Vision*, Jorge L.C. Sanz, Ed., Springer-Verlag, New York, 1996.)

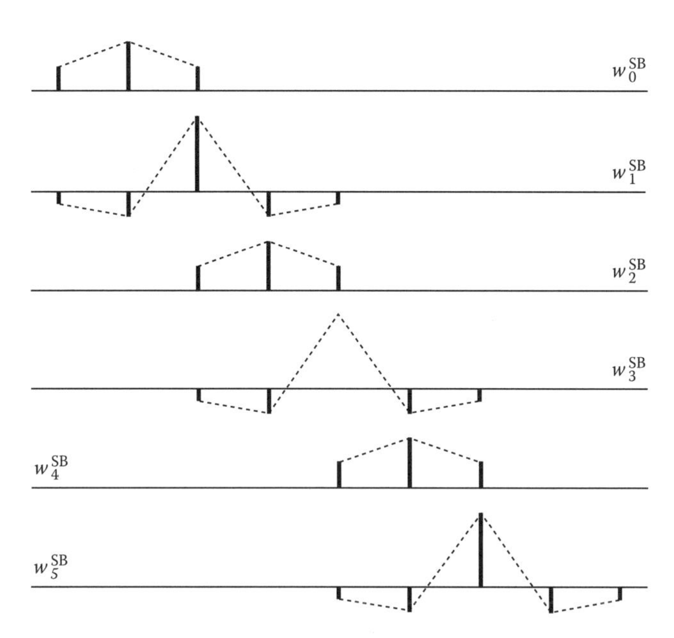

FIGURE 20.6 Subband basis vectors with $N = 2, h_1[-2\ldots2] = \sqrt{2}\cdot(-\frac{1}{8},\frac{1}{4},\frac{3}{4},\frac{1}{4},-\frac{1}{8}), h_2[-2\ldots0] = \sqrt{2}\cdot(-\frac{1}{4},\frac{1}{2},-\frac{1}{4}), g_1[-1\ldots1] = \sqrt{2}\cdot(\frac{1}{4},\frac{1}{2},\frac{1}{4})$, and $g_2[-1\ldots3] = \sqrt{2}\cdot(-\frac{1}{8},-\frac{1}{4},\frac{3}{4},-\frac{1}{4},-\frac{1}{8}) h_i$ and g_i are the impulse responses of the H_i (analysis) and G_i (synthesis) filters, respectively. (From Taubman, D. et al., Directionality and scalability in subband image and video compression, in *Image Technology: Advances in Image Processing, Multimedia, and Machine Vision*, Jorge L.C. Sanz, Ed., Springer-Verlag, New York, 1996.)

The basis vectors for subband decomposition are the N-translates of the impulse responses, $g_1[i],\ldots,g_N[i]$, of synthesis filters $G_1,\ldots,G_N$. Specifically, denoting the kth basis vector associated with subband β by $w_{Nk+\beta-1}^{SB}$, we have

$$w_{Nk+\beta-1}^{SB}[i] = g_\beta[i - Nk] \tag{20.4}$$

Figure 20.6 illustrates five of the basis vectors for a particularly simple, yet useful, two-band PR subband decomposition, with symmetric FIR analysis and synthesis impulse responses. As shown in Figure 20.6 and in contrast with the DCT basis vectors, the subband basis vectors overlap.

As for the DCT, one-dimensional subband decompositions may be separably extended to higher dimensions. By this we mean that a one-dimensional subband decomposition is first applied along one dimension of an image or video sequence. Any or all of the resulting subbands are then further decomposed into subbands along another dimension and so on. Figure 20.7 depicts a separable two-dimensional subband system. For video compression applications, the prediction error is sometimes decomposed into subbands of equal size.

Two-dimensional subband decompositions have the advantage that they do not suffer from the disturbing blocking artifacts exhibited by the DCT at high compression ratios. Instead, the most noticeable quantization-induced distortion tends to be "ringing" or "rippling" artifacts, which become most bothersome in the vicinity of image edges. Figures 20.3c and 20.11c clearly show this effect. Figure 20.8 shows frame 210 of the PING-PONG sequence compressed using a scalable, three-dimensional subband coder [28] at 1.5 Mbps, 300 kbps, and 60 kbps. As the bit rate decreases, we notice loss of detail and introduction of more ringing noise. Figure 20.3c shows frame 250 of the COAST-GUARD sequence compressed at 112 kbps using a zero-tree scalable coder [16]. The edges of the trees and the boat are affected by ringing noise.

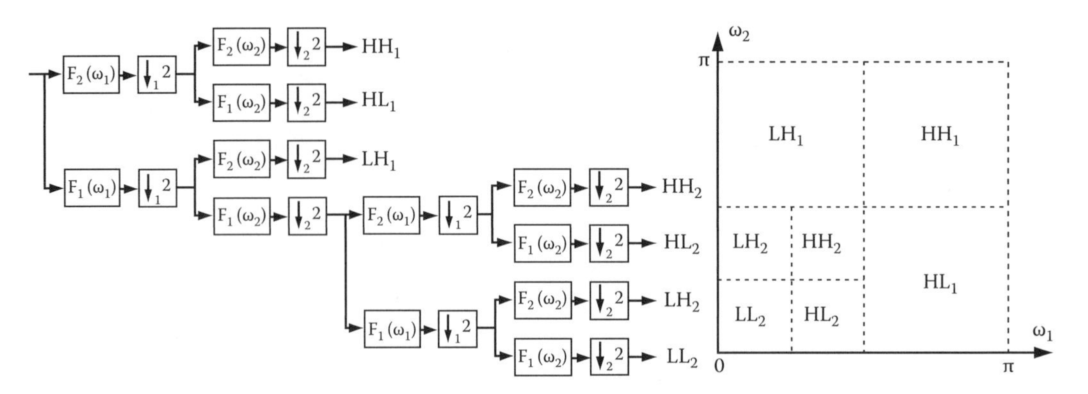

FIGURE 20.7 Separable spatial subband pyramid. Two level analysis system configuration and subband passbands shown. (From Taubman, D. et al., Directionality and scalability in subband image and video compression, in *Image Technology: Advances in Image Processing, Multimedia, and Machine Vision*, Jorge L.C. Sanz, Ed., Springer-Verlag, New York, 1996.)

FIGURE 20.8 Frame 210 of PING-PONG sequence decoded from scalable bit-stream at (a) 1.5 Mbps, (b) 300 kbps, and (c) 60 kbps [28]. (From Taubman, D. et al., Directionality and scalability in subband image and video compression, in *Image Technology: Advances in Image Processing, Multimedia, and Machine Vision*, Jorge L.C. Sanz, Ed., Springer-Verlag, New York, 1996.)

20.2.2.3 Matching Pursuit

Representing a signal using an over-complete basis set implies that there is more than one representation for the signal. For coding purposes, we are interested in representing the signal with the fewest basis vectors. This is an NP-complete problem [14]. Different approaches have been investigated to find or approximate the solution. Matching pursuits is a multistage algorithm, which in each stage finds the basis vector that minimizes the mean-squared-error [14].

Suppose we want to represent a signal $f[i]$ using basis vectors from an over-complete dictionary (basis set) $\mathcal{G}$. Individual dictionary vectors can be denoted as

$$w_\gamma[i] \in \mathcal{G}. \tag{20.5}$$

Here γ is an indexing parameter associated with a particular dictionary element. The decomposition begins by choosing γ to maximize the absolute value of the following inner product:

$$t = \langle f[i], w_\gamma[i] \rangle \tag{20.6}$$

where t is the transform (expansion) coefficient. A residual signal is computed as

$$R[i] = f[i] - tw_{\gamma}[i] \tag{20.7}$$

This residual signal is then expanded in the same way as the original signal. The procedure continues iteratively until either a set number of expansion coefficients are generated or some energy threshold for the residual is reached. Each stage k yields a dictionary structure specified by γ_k, an expansion coefficient $t[k]$, and a residual R_k, which is passed on to the next stage. After a total of M stages, the signal can be approximated by a linear function of the dictionary elements:

$$\hat{f}[i] = \sum_{k=1}^{M} t[k]w_{\gamma_k}[i] \tag{20.8}$$

The above technique has useful signal representation properties. For example, the dictionary element chosen at each stage is the element that provides the greatest reduction in mean square error between the true signal $f[i]$ and the coded signal $\hat{f}[i]$. In this sense, the signal structures are coded in order of importance, which is desirable in situations where the bit budget is limited. For image and video coding applications, this means that the most visible features tend to be coded first. Weaker image features are coded later, if at all. It is even possible to control which types of image features are coded well by choosing dictionary functions to match the shape, scale, or frequency of the desired features.

An interesting feature of the matching pursuit technique is that it places very few restrictions on the dictionary set. The original Mallat and Zhang paper considers both Gabor and wave packet function dictionaries, but such structure is not required by the algorithm itself [14]. Mallat and Zhang showed that if the dictionary set is at least complete, then $\hat{f}[i]$ will eventually converge to $f[i]$, though the rate of convergence is not guaranteed [14]. Convergence speed and thus coding efficiency are strongly related to the choice of dictionary set. However, true dictionary optimization can be difficult because there are so few restrictions. Any collection of arbitrarily sized and shaped functions can be used with matching pursuits, as long as completeness is satisfied.

Bergeaud and Mallat used the matching pursuit technique to represent and process images [1]. Neff and Zakhor have used the matching pursuit technique to code the motion prediction error signal [20]. Their coder divides each motion residual into blocks and measures the energy of each block. The center of the block with the largest energy value is adopted as an initial estimate for the inner product search. A dictionary of Gabor basis vectors, shown in Figure 20.9, is then exhaustively matched to an $S \times S$ window around the initial estimate. The exhaustive search can be thought of as follows. Each $N \times N$ dictionary structure is centered at each location in the search window, and the inner product between the structure and the corresponding $N \times N$ region of image data is computed. The largest inner product is then quantized. The location, basis vector index, and quantized inner product are then coded together.

Video sequences coded using matching pursuit do not suffer from either blocking or ringing artifacts, because the basis vectors are only coded when they are well-matched to the residual signal. As bit rate decreases, the distortion introduced by matching pursuit coding takes the form of a gradually increasing blurriness (or loss of detail). Since matching pursuits involves exhaustive search, it is more complex than DCT approaches, especially at high bit rates.

Figure 20.3d shows frame 250 of the 15 fps CIF COAST-GUARD sequence coded at 112 kbps using the matching pursuit video coder described by Neff and Zakhor [20]. This frame does not suffer from the blocky artifacts, which affect the DCT coders as shown in Figure 20.3b. Moreover, it does not suffer from the ringing noise, which affects the subband coders as shown in Figures 20.3c and 20.8c.

20.2.3 Discussion

Figure 20.3 shows frame 250 of the 15 fps CIF COAST-GUARD sequence coded at 112 kbps using DCT, subband, and matching pursuit coders. The DCT coded frame suffers from blocking artifacts. The subband coded frame suffers from ringing artifact.

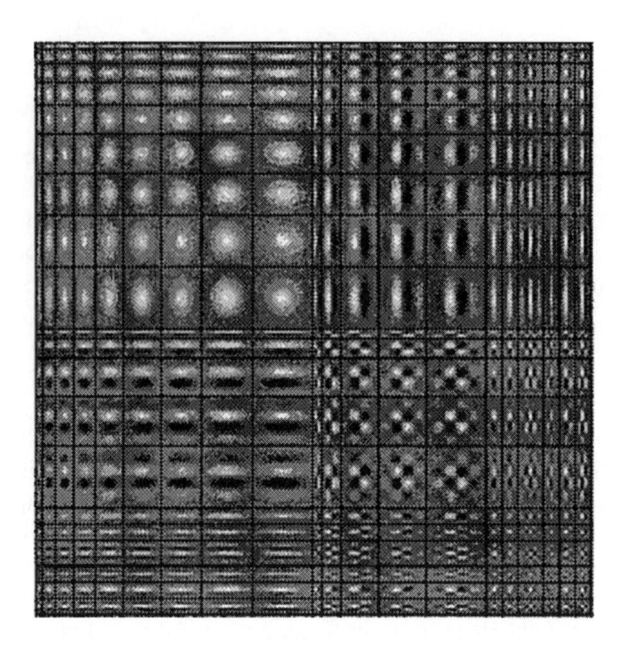

FIGURE 20.9 Separable two-dimensional 20 × 20 Gabor dictionary.

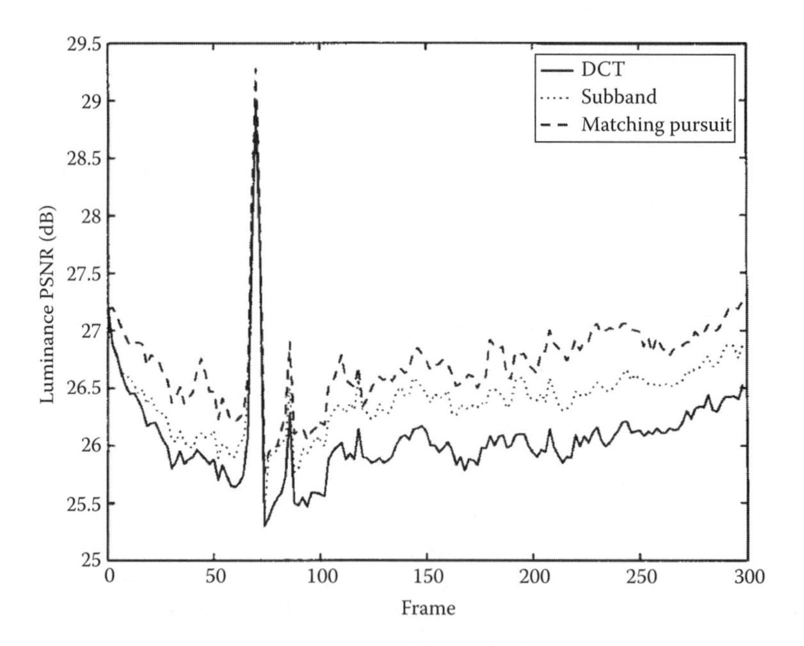

FIGURE 20.10 Frame-by-frame distortion of the luminance component of the COAST-GUARD sequence, reconstructed from 112 kbps H.263 bit-stream (solid line) [3], a ZTS bit-stream (dotted line) [16], and from a matching pursuit bit-stream (dashed line) [20]. Consistently, the matching pursuit coder had the highest PSNR while the DCT coder had the lowest PSNR.

Figure 20.10 compares the PSNR performance of the matching pursuit coder [20] to a DCT (H.263) coder [3] and a zero-tree subband (ZTS) coder [16] when coding the COAST-GUARD sequence at 112 kbps. The matching pursuit coder [20] in this example has consistently higher PSNR than the H.263 [3] and the ZTS [16] coders. Table 20.1 shows the average luminance PSNRs for different sequences at different

TABLE 20.1 The Average Luminance PSNR of Different Sequences at Different Bit Rates When Coding Using a DCT Coder (H.263) [3], ZTS Coder [16], and Matching Pursuit Coder (MP) [20]

		Rate		PSNR (dB)		
Sequence	Format	Bit	Frame	DCT	ZTS	MP
Container-ship	QCIF	10K	7.5	29.43	28.01	31.10
Hall-monitor	QCIF	10K	7.5	30.04	28.44	31.27
Mother-daughter	QCIF	10K	7.5	32.50	31.07	32.78
Container-ship	QCIF	24K	10.0	32.77	30.44	34.26
Silent-voice	QCIF	24K	10.0	30.89	29.41	31.71
Mother-daughter	QCIF	24K	10.0	35.17	33.77	35.55
COAST-GUARD	QCIF	48K	10.0	29.00	27.65	29.82
News	QCIF	48K	7.5	30.95	29.97	31.96

bit rates. In all examples mentioned in Table 20.1, the matching pursuit coder has higher average PSNR than the DCT coder. The subband coder has the lowest average PSNR.

20.2.4 Quantization

Motion compensation and residual error decomposition reduce the redundancy in the video signal. However, to achieve low bit rates, we must tolerate some distortion in the video sequence. This is because we need to map the residual and motion information to a fewer collection of codewords to meet the bit rate requirements.

Quantization, in a general sense, is the mapping of vectors (or scalars) of an information source into a finite collection of codewords for storage or transmission [8]. This involves two processes: encoding and decoding. The encoder blocks the source $\{t[i]\}$ into vectors of length n, and maps each vector $\mathcal{T}^n \in \mathcal{T}^n$ into a codeword c taken from a finite set of codewords $\mathcal{C}$. The decoder maps the codeword c into a reproduction vector $\mathcal{Y}^n \in \mathcal{Y}^n$ where $\mathcal{Y}$ is a reproduction alphabet. If $n = 1$, it is called scalar quantization. Otherwise, it is called vector quantization.

The problem of optimum mean-squared scalar quantization for a given reproduction alphabet size was independently solved by Lloyd [13] and Max [17]. They found that if t is a real scalar random variable with continuous probability density function $p_t(t)$, then the quantization thresholds are

$$\hat{t}_k = \frac{r_k + r_{k-1}}{2} \tag{20.9}$$

which is the geometric mean of the interval $(r^{k-1}, r^k]$, where

$$r_k = \frac{\int_{\hat{t}_k}^{\hat{t}_{k+1}} x p_x(x) \mathrm{d}x}{\int_{\hat{t}_k}^{\hat{t}_{k+1}} p_x(x) \mathrm{d}x} \tag{20.10}$$

are the reconstruction levels. Iterative numerical methods are required to solve for the reconstruction and quantization levels.

The simplest scalar quantizer is the uniform quantizer for which the reconstruction intervals are of equal length. The uniform quantizer is optimal when the coefficients have a uniform distribution. Moreover, due to its simplicity and good general performance, it is commonly used in coding systems.

A fundamental result of Shannon's rate distortion theory is that better performance can be achieved by coding vectors instead of scalars, even if the source is memoryless [8,19]. Linde et al. [12] generalized the

Lloyd–Max algorithm to vector quantization. Vector quantization exploits spatial redundancy in images, a function also served by the transformation block of Figure 20.1, so it is sometimes applied directly to the image or video pixels [19].

Memory can be incorporated into scalar quantization by predicting the current sample from the previous samples and quantizing the residual error, e.g., linear predictive coding.

The human visual system is sensitive to some frequency bands more than others. So, humans tolerate more losses in some bands and less in others. In practice, the DCT coefficients corresponding to a particular frequency are grouped together to form a band, or in the case of subband decomposition, the bands are simply the subband channels. Different quantizers are then applied to each band according to its visual importance.

20.2.5 Coding of Quantized Symbols

The simplest method to code quantized symbols is to assign a fixed number of bits per symbol. For an alphabet of L symbols, this approach requires $\lceil \log_2 L \rceil$ bits per symbol. This method, however, does not exploit the coding redundancy in the symbols. Coding redundancy is eliminated by minimizing the average number of bits per symbol. This is achieved by giving fewer bits to more frequent symbols and more bits to less frequent symbols. Huffman [9] or arithmetic coding [21] schemes are usually used for this purpose.

In image and video coding, a significant number of the transform coefficients are zeros. Moreover, the "significant" DCT transform coefficients (low-frequency coefficients) of a block can be predicted from the neighboring blocks resulting in a larger number of zero coefficients. To code the zero coefficients, run-length is performed on a reordered version of the transform coefficients. Figure 20.11a shows a commonly used zigzag scan to code 8×8 block DCT coefficients. Figure 20.11b shows a scan used to code subband coefficients commonly known as zero-tree coding [24]. The basic idea behind zero-tree coding is that if a coefficient in a lower frequency band (coarse scale) is zero or insignificant, then all the coefficients of the same orientation at higher frequencies (finer scales) are very likely to be zero or insignificant [16,24]. Thus, the subband coefficients are organized in a data structure design based on this observation.

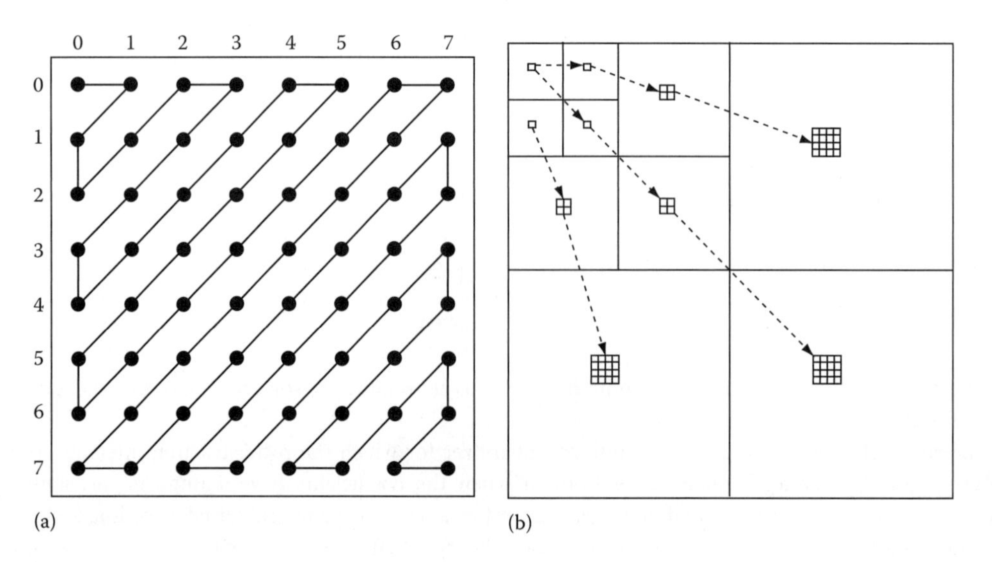

FIGURE 20.11 A common scan for (a) an 8 × 8 block DCT and (b) subband decompositions (zero-tree).

20.3 Desirable Features

Some video applications require the encoder to provide more than good compression performance. For example, it is desirable to have scalable video compression schemes so that different users with different bandwidth, resolution, or computational capabilities can decode from the same bit-stream. Cellular applications require the coder to provide a bit-stream that is robust when transmission errors occur. Other features include object-based manipulation of the bit-stream and the ability to perform content search. This section addresses two important desired features, namely scalability and error resilience.

20.3.1 Scalability

Developing scalable video compression algorithms has attracted considerable attention in recent years. Scalable compression refers to encoding a sequence in such a way so that subsets of the encoded bit-stream correspond to compressed versions of the sequence at different rates and resolutions. Scalable compression is useful in today's heterogeneous networking environment in which different users have different rate, resolution, display, and computational capabilities.

In rate scalability, appropriate subsets are extracted in order to trade distortion for bit rate at a fixed display resolution. Resolution-scalability, on the other hand, means that extracted subsets represent the image or video sequence at different resolutions. Rate- and resolution-scalability usually also provide a means of scaling the computational demands of the decoder. Resolution-scalability is best thought of as a property of the transformation block of Figure 20.1. Both the DCT and subband transformations may be used to provide resolution-scalability. Rate-scalability, however, is best thought of as a property of the quantization and coding blocks.

Hybrid video coders can achieve scalability using multilayer schemes. For example, in a two layer rate-scalable coder, the first layer codes the video at a low bit rate, while the second layer codes the residual error based on the source material and what has been coded thus far. These layers are usually called the base and enhancement layers. Such schemes, however, do not support fully scalable video, i.e., they can only provide a few levels of scalability, e.g., a few rates. The bottleneck is motion compensation, which is a nonlinear feedback predictor. To understand this, observe that the storage block of Figure 20.1 is a memory element, storing values $\tilde{f}[\vec{i}]$ or $\tilde{t}[\vec{k}]$, recovered during decoding, until they are required for prediction. In scalable compression algorithms, the value of $\tilde{f}[\vec{i}]$ or $\tilde{t}[\vec{k}]$, obtained during decoding, depends on constraints, which may be imposed after the bit-stream has been generated. For example, if the algorithm is to permit rate scalability, then the value of $\tilde{f}[\vec{i}]$ or $\tilde{t}[\vec{k}]$ obtained by decoding a low rate subset of the bit-stream can be expected to be a poorer approximation to $f[\vec{i}]$ or $t[\vec{k}]$, respectively, than the value obtained by decoding from a higher rate subset of the bit-stream. This ambiguity presents a difficulty for the compression algorithm, which must select a particular value for $\tilde{f}[\vec{i}]$ or $\tilde{t}[\vec{k}]$ to serve as a prediction reference.

This inherent nonscalability of motion compensation is particularly problematic for video compression where scalability and motion compensation are both highly desirable features. As a solution, Taubman and Zakhor [28,29] used three-dimensional subband decompositions to code video. They first compensated for the camera pan motion, then used three-dimensional subband decomposition. The coefficients in each subband are then quantized by a layered quantizer in order to generate a fully scalable video with fine granularity of bit rates. Temporal filtering, however, introduces significant overall latency, a critical parameter for interactive video compression applications. To reduce this effect, it is possible to use a 2-tap temporal filter, which results in one frame of delay.

As a visual demonstration of the quality trade-off inherent to rate-scalable video compression, Figure 20.8 shows frame 210 of the PING-PONG video sequence, decompressed at bit rates of 1.5 Mbps, 300 kbps, and 60 kbps for monochrome display using the scalable coder developed by Taubman and Zakhor [28]. As the bit rate decreases, the frame is less detailed and suffers more from ringing noise, i.e., the visual quality decreases. Figure 20.12 shows the PSNR characteristics of the scalable coder and MPEG-1 coder as a function of bit rate. The curve corresponding to the scalable coder corresponds to

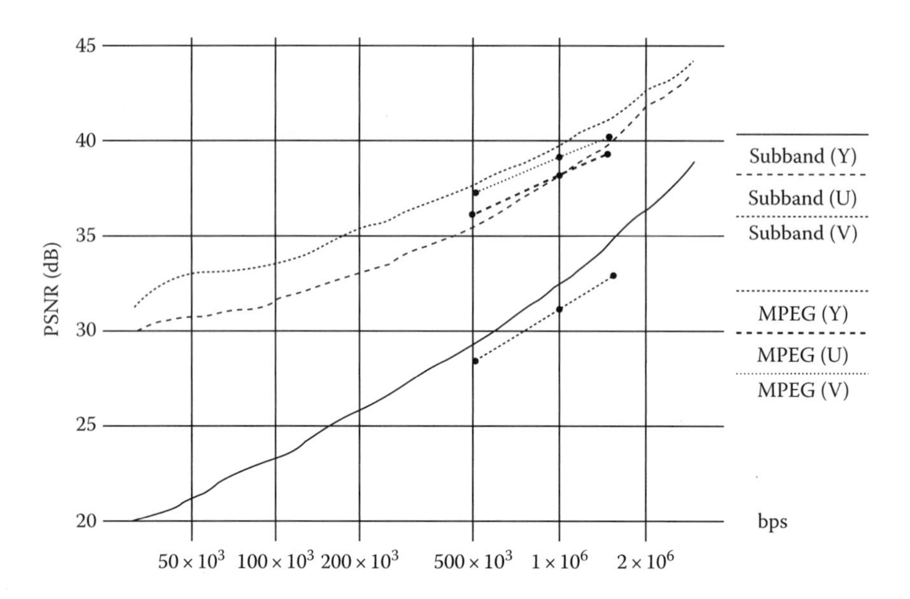

FIGURE 20.12 Rate-distortion curves for PING-PONG sequence. Overall PSNR values for Y, U, and V components for the codec in Ref. [28] are plotted against the bit rate limit imposed on the rate-scalable bit-stream prior to decompression. MPEG-1 distortion values are also plotted as connected dots for reference. (From Taubman, D. et al., Directionality and scalability in subband image and video compression, in *Image Technology: Advances in Image Processing, Multimedia, and Machine Vision*, Jorge L.C. Sanz, Ed., Springer-Verlag, New York, 1996.)

one encoded bit-stream decoded at arbitrary bit rates, while the three points for the MPEG-1 coder correspond to three different encoded bit-streams encoded and decoded at these different rates. As seen the scalable codec offers a fine granularity of available bit rates with little or no loss in PSNR as compared to MPEG-1 codec.

Real-time software only implementation of scalable video codec has also received a great deal of attention over the past few years. Tan et al. [27] have recently proposed a real-time software only implementation of the modified version of the algorithm in [28] by replacing the arithmetic coding with block coding. The resulting scalable coder is symmetric in encoding and decoding complexity and can encode up to 17 fps for rates as high as 1 Mbps on a 171 MHz Ultra-Sparc workstation.

20.3.2 Error Resilience

When transmitting video over noisy channels, it is important for bit-streams to be robust to transmission errors. It is also important, in case of errors, for the error to be limited to a small region and not to propagate to other areas. If the coder is using fixed-length codes, the error will be limited to the region of the bit-stream where it occurred and the rest of the bit-stream will not be affected. Unfortunately, fixed-length codes do not provide good compression performance, especially since the histogram of the transform coefficients has a significant peak around low frequency.

In order to achieve such features when using variable length codes, the bit-stream is usually partitioned into segments that can be independently decoded. Thus, if a segment is lost, only that region of the video is affected. A segment is usually a small part of a frame. If an error occurs, the decoder should have enough information to know the beginning and the end of a segment. Therefore, synchronization codes are added to the beginning and end of each segment. Moreover, to limit the error to a smaller part of the segment, reversible variable length codes may be used [26]. So, if an error occurs, the decoder will advance to the next synchronization code and can decode in the backward direction till the error is reached.

As is evident, there is a trade-off between good compression performance and error resilience. In order to reduce the cost of error resilient codes, some approaches jointly optimize the source and channel codes [6,23].

20.4 Standards

In this section we review the major video compression standards. Essentially, these schemes are based on the building blocks introduced in Section 20.2. All these standards use the DCT. Table 20.2 summarizes the basic characteristics and functionalities supported by existing standards. Sections 20.4.2, 20.4.3, and 20.4.5 outline the Motion Picture Experts Group (MPEG) standards for video compression. Sections 20.4.1 and 20.4.4 review the CCITT H.261 and H.263 standards for digital video communications. This section lists the standards according to their chronological order in order to provide an understanding of the progress of the video compression standardization process.

20.4.1 H.261

Recommendation H.261 of the CCITT Study Group XV was adopted in December 1990 [2] as a video compression standard to be used for video conferencing applications. The bit rates supported by H.261 are $p \times 64$ kbps, where p is in the range 1–30. H.261 supports two source formats: CIF (352×288 luminance and 176×144 chrominance) and QCIF (176×144 luminance and 88×72 chrominance). The chrominance components are subsampled by two in both the vertical and horizontal directions.

TABLE 20.2 Summary of the Functionalities and Characteristics of the Existing Standards

ITU Attribute	H.261	H.263	ISO MPEG-1	MPEG-2	MPEG-4
Applications	Videoconferencing	Videophone	CD storage	Broadcast	Wide range (multimedia)
Bit rate	64K–1 M	<64 K	1.0–1.5 M	2–10 M	5K–4 M
Material	Progressive	Progressive	Progressive, interlaced	Progressive, interlaced	Progressive, interlaced
Object shape	Rectangular	Arbitrary (simple)	Rectangular	Rectangular	Arbitrary
Residual coding					
Transform	8×8 DCT	8×8 DCT	8×8 DCT	8×8 DCT	8×8 DCT
Quantizer	Uniform	Uniform	Weighted uniform	Weighted uniform	Weighted uniform
Motion compensation					
Type	Block	Block	Block	Block	Block, sprites
Block size	16×16	16×16, 8×8	16×16	16×16	16×16, 8×8
Prediction type	Forward	Forward, backward	Forward, backward	Forward, backward	Forward, backward
Accuracy	One pixel	Half pixel	Half pixel	Half pixel	Half pixel
Loop filter	Yes	No	No	No	No
Scalability					
Temporal	No	Yes	Yes	Yes	Yes
Spatial	No	Yes	No	Yes	Yes
Bit rate	No	Yes	No	Yes	Yes
Object	No	No	No	No	Yes

The transformation used in H.261 is the 8×8 block-DCT. Thus, there are four luminance (Y) DCT blocks for each pair of U and V chrominance DCT blocks. These six DCT blocks are collectively referred to as a "macroblock." The macroblocks are grouped together to construct a group of blocks (GOB), which relates to 11×3 region of macroblocks. Each macroblock may individually be specified as intracoded or intercoded. The intracoded blocks are coded independently of the previous frame and so do not conform to the model of Figure 20.1. They are used when successive frames are not related, such as during scene changes, and to avoid excessive propagation of the effects of communication errors. Intercoded blocks use the motion compensation predictive feedback loop of Figure 20.1 to improve compression performance. The motion estimation scheme is based on 16×16 pixel blocks. Each macroblock is predicted from the previous frame and is assigned exactly one motion vector with one pixel accuracy.

The data for each frame consists of a picture header that includes a start code, a temporal reference for the current coded picture, and the source format. The picture header is followed by the GOB layer. The data of each GOB has a header that includes a start code to indicate the beginning of a GOB, the GOB number to indicate the position of the GOB, and all information necessary to code each GOB independently. This will limit the loss if an error occurs during the transmission of a GOB. The header of the GOB is followed by the motion data, and then followed by the block information.

20.4.2 MPEG-1

The first (MPEG) video compression standard [7], MPEG-1, is intended primarily for progressive video at 30 fps. The targeted bit rate is in the range 1.0–1.5 Mbps. MPEG-1 was designed to store video on compact discs. Such applications require MPEG-1 to support random access to the material on the disc, fast forward and backward searches, reverse playback, and audio visual synchronization. MPEG-1 is also a hybrid coder that is based on the 8×8 block DCT and 16×16 motion-compensated macroblocks with half pixel accuracy.

The most significant departure from H.261 in MPEG-1 is the introduction of the concept of bidirectional prediction, together with that of group of pictures (GOP). These concepts may be understood with the aid of Figure 20.13. Each GOP commences with an intracoded picture (frame), denoted I in the figure. The motion-compensated predictive feedback loop of Figure 20.1 is used to compress the subsequent intercoded frames, marked P. Finally, the bidirectionally predicted frames, marked B in Figure 20.13, are coded using motion compensated prediction based on both previous and successive I or P frames. Bidirectional prediction conforms essentially to the model of Figure 20.1, except that the prediction signal is given by

$$a\tilde{f}\left[\vec{i} - d_{\vec{i}}^{\vec{f}}\right] + b\tilde{f}\left[\vec{i} - d_{\vec{i}}^{\vec{b}}\right]$$

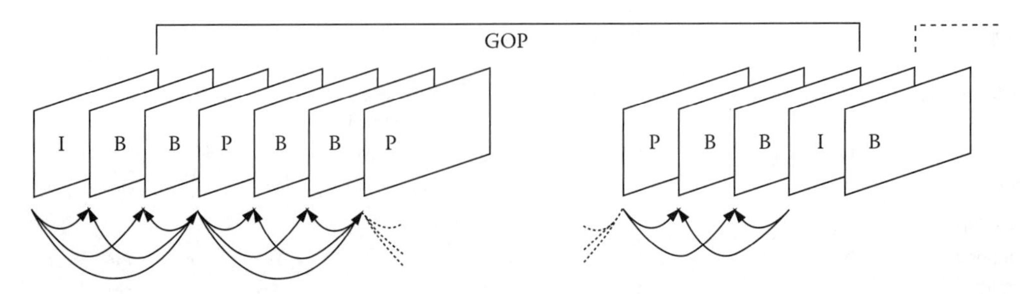

FIGURE 20.13 MPEG's GOP. Arrows represent direction of prediction. (From Taubman, D. et al., Directionality and scalability in subband image and video compression, in *Image Technology: Advances in Image Processing, Multimedia, and Machine Vision*, Jorge L.C. Sanz, Ed., Springer-Verlag, New York, 1996.)

In this notation, $\tilde{f}$ is a reconstructed frame, $d^{\tilde{f}}_{\vec{i}}\left(h^{f}_{\vec{i}}, v^{f}_{\vec{i}}, n^{f}\right)$, where $\left(h^{f}_{\vec{i}}, v^{f}_{\vec{i}}\right)$ is a forward motion vector describing the motion from the previous I or P frame, and n^{f} is the frame distance to this previous I or P frame. Similarly, $d^{\tilde{b}}_{\vec{i}} = \left(h^{b}_{\vec{i}}, v^{b}_{\vec{i}}, -n^{b}\right)$, where $\left(h^{b}_{\vec{i}}, v^{b}_{\vec{i}}\right)$ is a backward motion vector describing the motion to the next I or P frame, and n_b is the temporal distance to that frame. The weights a and b are given either by

$$\begin{matrix} a = 1 \\ b = 0 \end{matrix}, \quad \begin{matrix} a = 0 \\ b = 1 \end{matrix}, \quad \text{or} \quad \begin{matrix} a = n_b/(n_f + n_b) \\ b = n_f/(n_f + n_b) \end{matrix}$$

corresponding to forward, backward, and average prediction, respectively. Each bidirectionally predicted macroblock is independently assigned one of these three prediction strategies.

An MPEG-1 decoder can reconstruct the I and P frames without the need to decode the B frames. This is a form of temporal scalability and is the only form of scalability supported by MPEG-1.

20.4.3 MPEG-2

The second MPEG standard, MPEG-2, targets 60 fields/s interlaced television; however, it also supports progressive video. The targeted bit rate is between 2 and 10 Mbps. MPEG supports frames sizes up to $2^{14} - 1$ in each direction; however, the most popular formats are CCIR 601 (720×480), CIF (352×288), and SIF (352×240). The chrominance can be sampled in either the 4:2:0 (half as many samples in the horizontal and vertical directions), 4:2:2 (half as many samples in the horizontal direction only), or 4:4:4 (full chrominance size) formats.

MPEG-2 supports scalability by offering four tools: data partitioning, signal-to-noise ratio (SNR) scalability, spatial scalability, and temporal scalability. Data partitioning can be used when two channels are available. The bit-stream is partitioned into two streams according to their importance. The most important stream is transmitted in the more reliable channel for better error resilience performance. SNR (rate), spatial, and temporal scalable bit-streams are achieved through the definition of a two-layer coder. The sequence is encoded into two bit-streams called lower and enhancement layer bit-streams. The lower bit-stream can be encoded independently from the enhancement layer using an MPEG-2 basic encoder. The enhancement layer is combined with the lower layer to get a higher quality sequence. The MPEG-2 standard supports hybrid scalabilities by combining these tools.

20.4.4 H.263

The ITU recommended H.263 standard to be used for video telephony (video coding for narrow telecommunications channels) [3]. Although, the bit rates specified are smaller than 64 kbps, H.263 is also suitable for higher bit rates. H.263 supports three source formats: CIF (352×288 luminance and 176×144 chrominance), QCIF (176×144 luminance and 88×72 chrominance), and sub-QCIF (128×96 luminance and 64×48 chrominance).

The transformation used in H.263 is the 8×8 block-DCT. As in H.261, a macroblock consists of four luminance and two chrominance blocks. The motion estimation scheme is based on 16×16 and 8×8 pixel blocks. It alternates between them according to the residual error in order to achieve better performance. Each intercoded macroblock is assigned one or four motion vectors with half pixel accuracy. Motion estimation is done in both forward and backward directions.

H.263 provides a scalable bit-stream in the same fashion MPEG-2 does. This includes temporal, spatial, and rate (SNR) scalabilities. Moreover, H.263 has been extended to support coding of video objects of arbitrary shape. The objects are segmented and then coded the same way rectangular objects are coded with slight modification at the boundaries of the object. The shape information is embedded in the chrominance part of the stream by assigning the least used color to the parts outside the object in the rectangular frame. The decoder uses the color information to detect the object in the decoded stream.

20.4.5 MPEG-4

The moving picture expert group is developing a video standard that targets a wide range of applications including Internet multimedia, interactive video games, videoconferencing, videophones, multimedia storage, wireless multimedia, and broadcasting applications. Such a wide range of applications needs a large range of bit rates, thus MPEG-4 supports a bit rate range of 5 kbps–4 Mbps. In order to support multimedia applications effectively, MPEG-4 supports synthetic and natural image and video in both progressive and interlaced formats. It is also required to provide object-based scalabilities (temporal, spatial, and rate) and object-based bit-stream manipulation, editing, and access [5,25]. Since it is also intended to be used in wireless communications, it should be robust to high error rates. The standard is expected to be finalized in 1998.

Acknowledgment

The authors would like to acknowledge support from AFOSR grants F49620–93–1–0370 and F49620-94-1-0359, ONR grant N00014-92-J-1732, Tektronix, HP, SUN Microsystems, Philips, and Rockwell. Thanks to Iraj Sodagar of David Sarnoff Research Center for providing the zero-tree-coded video sequence.

References

1. Bergeaud, F. and Mallat, S., Matching pursuit of images, *Proceedings of the IEEE-SP International Symposium on Time-Frequency and Time-Scale Analysis*, Philadelphia, PA, pp. 330–333, Oct. 1994.
2. *CCITT Recommendation H.261, Video Codec for Audio Visual Services at p × 64 kbit/s*, Geneva, Switzerland, 1990.
3. *CCITT Recommendation H.263, Video codec for Audio Visual Services at p × 64 kbit/s*, Seattle, WA, 1995.
4. Chao, T.-H., Lau, B., and Miceli, W. J., Optical implementation of a matching pursuit for image representation, *Opt. Eng.*, 33(2), 2303–2309, July 1994.
5. Chiarilione, L., MPEG and multimedia communications, *IEEE Trans. Circuits Syst. Video Technol.*, 7(1), 5–18, Feb. 1997.
6. Cheung, G. and Zakhor, A., Joint source/channel coding of scalable video over noisy channels, *Proceedings of the IEEE International Conference on Image Processing*, Lausanne, Switzerland, vol. 3, pp. 767–770, 1996.
7. *Committee Draft of Standard ISO11172, Coding of Moving Pictures and Associated Audio*, ISO/MPEG 90/176, Dec. 1990.
8. Gray, R., Vector quantization, *IEEE Acoust. Speech Signal Process. Mag.*, 1, 4–29, April 1984.
9. Huffman, D., A method for the construction of minimal redundancy codes, *Proc. IRE*, 40(9), 1098–1101, Sept. 1952.
10. Jain, A. K., *Fundamentals of Digital Image Processing*, Prentice-Hall, Englewood Cliffs, NJ, 1989.
11. Jayant, N. and Noll, P., *Digital Coding of Waveforms*, Prentice-Hall, Englewood Cliffs, NJ, 1984.
12. Linde, Y., Buzo, A., and Gray, R. M., An algorithm for vector quantizer design, *IEEE Trans. Commun.*, COM-28(1), 84–95, Jan. 1980.
13. Lloyd, S. P., Least squares optimization in PCM, *IEEE Trans. Info. Theory* (reproduction of a paper presented at the Institute of Mathematical Statistics meeting in Atlantic City, NJ, September 10–13, 1957), IT-28(2), 129–137, Mar. 1982.
14. Mallat, S. and Zhang, Z., Matching pursuits with time-frequency dictionaries, *IEEE Trans. Signal Process.*, 41(12), 3397–3415, Dec. 1993.
15. Malvar, H. S., *Signal Processing with Lapped Transforms*, Artech House, Norwood, MA, 1992.

16. Martucci, S. A., Sodagar, I., Chiang, T., and Zhang, Y.-Q., A zerotree wavelet coder, *IEEE Trans. Circuits Syst. Video Technol.*, 7(1), 109–118, Feb. 1997.
17. Max, J., Quantization for minimum distortion, *IRE Trans. Info. Theory*, IT-16(2), 7–12, Mar. 1960.
18. Minami, S. and Zakhor, A., An optimization approach for removing blocking effects in transform coding, *IEEE Trans. Circuits Syst. Video Technol.*, 5(2), 74–82, April 1995.
19. Nasrabadi, N. M. and King, R. A., Image coding using vector quantization: A review, *IEEE Trans. Commun.*, 36(8), 957–971, Aug. 1988.
20. Neff, R. and Zakhor, A., Very low bit rate video coding based on matching pursuits, *IEEE Trans. Circuits Syst. Video Technol.*, 7(1), 158–171, Feb. 1997.
21. Rissanen, J. and Langdon, G., Arithmetic coding, *IBM J. Res. Dev.*, 23(2), 149–162, Mar. 1979.
22. Rosenholtz, R. and Zakhor, A., Iterative procedures for reduction of blocking effects in transform image coding, *IEEE Trans. Circuits Syst. Video Technol.*, 2, 91–95, Mar. 1992.
23. Ruf, M. J. and Modestino, J. W., Rate-distortion performance for joint source channel coding of images, *Proceedings of the IEEE International Conference on Image Processing*, vol. 2, pp. 77–80, 1995.
24. Shapiro, J. M., Embedded image coding using zerotrees of wavelet coefficients, *IEEE Trans. Signal Process.*, 41(12), 3445–3462, Dec. 1993.
25. Sikora, T., The MPEG-4 video standard verification model, *IEEE Trans. Circuits Syst. Video Technol.*, 7(1), 19–31, Feb. 1997.
26. Takishima, Y., Wada, M., and Murakami, H., Reversible variable length codes, *IEEE Trans. Commun.*, 43(2-4), 158–162, Feb.–Apr. 1995.
27. Tan, W., Chang, E., and Zakhor, A., Real time software implementation of scalable video codec, *IEEE International Conference on Image Processing*, Lausanne, Switzerland, vol. 1, pp. 17–20, 1996.
28. Taubman, D. and Zakhor, A., Multirate 3-D subband coding of video, *IEEE Trans. Image Process.*, 3(5), 572–588, Sept. 1994.
29. Taubman, D. and Zakhor, A., A common framework for rate and distortion based scaling of highly scalable compressed video, *IEEE Trans. Circuits Syst. Video Technol.*, 6(4), 329–354, Aug. 1996.
30. Vetterli, M. and Kalker, T., Matching pursuit for compression and application to motion compensated video coding, *Proceedings of the IEEE International Conference on Image Process.*, Austin, TX, vol. 1, pp. 725–729, Nov. 1994.
31. Woods, J., Ed., *Subband Image Coding*, Kluwer Academic Publishers, Norwell, MA, 1991.
32. Taubman, D., Chang, E., and Zakhor, A., Directionality and scalability in subband image and video compression, in *Image Technology: Advances in Image Processing, Multimedia, and Machine Vision*, Jorge L. C. Sanz, Ed., Springer-Verlag, New York, 1996.

21

Digital Television

Kou-Hu Tzou
Hyundai Network Systems

21.1 Introduction

Digital television is being widely adopted for various applications ranging from high-end applications, such as studio recording, to consumer applications, such as digital cable TV and digital DBS (Direct Broadcasting Satellite) TV. For example, several digital video tape recording standards, using component format (D1 and D5), composite format (D2 and D3), or compressed component formats (Digital Betacam) are commonly used by broadcasters and TV studios [1]. These standards preserve the best possible picture quality at the expense of high data rates, ranging from approximately 150 to 300 Mbps. When captured in a digital format, the picture quality can be free from degradation during multiple generations of recording and playback, which is extremely attractive to studio editing. However, transmission of these high data-rate signals may be hindered due to lack of transmission media with an adequate bandwidth. Although it is possible, the associated transmission cost will be very high. The bit rate requirement for high definition television (HDTV) is even more demanding, which may exceed 1 Gbps in an uncompressed form. Therefore, data compression is essential for economical transmission of digital TV/HDTV.

Before motion-compensated discrete cosine transform (DCT) coding technology became mature in recent years, transmission of high-quality digital television used to be carried out at 45 Mbps using differential pulse code modulation (DPCM) techniques. Today, by incorporating advanced motion-compensated DCT coding, comparable picture quality can be achieved at about one-third of the rate required by DPCM-coded video. For entertainment applications, the requirement on picture quality can be relaxed a little bit to allow more TV channels to fit into the same bandwidth. It is generally agreed that 3–4 Mbps for movie-originated or low-activity interlaced video (talk shows, etc.) materials is acceptable, and 6–8 Mbps for high-activity interlaced video (sports, etc.) is acceptable. The targeted bit rate for

HDTV transmission is usually around 20 Mbps, which is chosen to match the available digital bandwidth of terrestrial broadcast channels allocated for conventional TV signals.

21.2 EDTV/HDTV Standards

The concept of HDTV system and efficient transmission format was originally explored by researches at NHK (Japan Broadcasting Corp.) more than 20 years ago [2] in order to offer superior picture quality while conserving bandwidth. Main HDTV features, including more scan lines, higher horizontal resolution, wider aspect ratio, better color representation, and higher frame rate, were identified. With these new features, HDTV is geared to offer picture quality close to that of 35 mm prints. However, the transmission of such a signal will require a very wide bandwidth. During the last 20 years, intensive research efforts have been engaged toward video coding to reduce bandwidth.

Currently there are two dominant HDTV production formats being used worldwide; one is the 1125-line/60 Hz system primarily used in Japan and the United States and the other is the 1250-line/50 Hz system primarily used in Europe. The main scanned raster characteristics of these two formats are listed in Table 21.1. The nominal bandwidth of the luminance component is about 30 MHz (in some cases, 20 MHz was quoted). Roughly speaking, the HDTV signal can carry about six times as much information as a conventional TV signal.

Development of HDTV transmission techniques in the early days was focused on bandwidth-compatible approaches that use the same analog bandwidth as a conventional TV signal. In some cases, in order to conserve bandwidth or to offer compatibility with an existing conventional signal or display, a compromised system—Enhanced or Extended Definition TV—was developed instead. The EDTV signal does not offer the picture quality and resolution required for an HDTV signal; however, it enhances the picture quality/resolution of conventional TV.

21.2.1 MUSE System

The most well-known early development in HDTV coding is the MUSE (Multiple Sub-Nyquist Sampling Encoding) system at NHK [3,4]. The main concept of the MUSE system is adaptive spatial–temporal subsampling. Since human eyes have better spatial sensitivity for stationary or slow-moving scenes, the full spatial resolution is preserved while the temporal resolution is reduced for these scenes in the MUSE system. For fast moving scenes, the spatial sensitivity of human eyes declines so that reducing the spatial resolution will not significantly affect perceived picture quality. The MUSE signal is intended for analog transmission with a baseband bandwidth of 8.1 MHz, which can be fitted into a satellite transponder for a conventional analog TV signal. However, it should be noted that most signal processing employed in the MUSE system is in the digital domain. The MUSE coding technique was later modified to reduce bandwidth requirement for transmission over 6 MHz terrestrial broadcasting channels (Narrow-MUSE) [5]. Currently, MUSE-based HDTV programming is being broadcast regularly through a DBS in Japan.

TABLE 21.1 Main Scanned Raster Characteristics of the 1125-Line/60 Hz System and the 1250-Line/50 Hz System

Format	Total Scan Lines per Frame	Active Lines per Frame	Scanning Format	Aspect Ratio	Field Rate
1	1125	1035	2 : 1 interlaced	16 : 9	60.00/59.94
2	1250	1152	2 : 1 interlaced	16 : 9	50.00

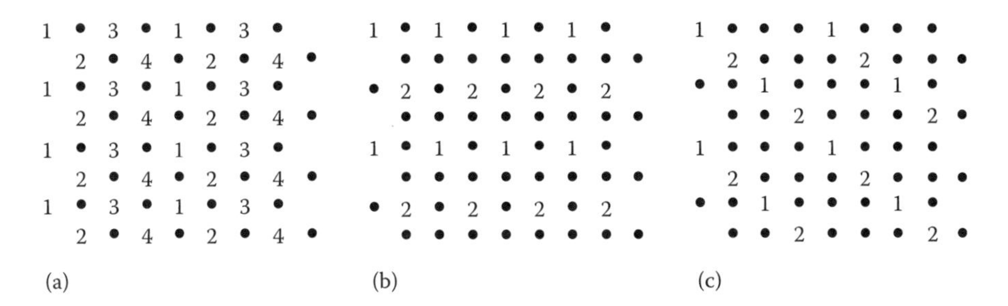

FIGURE 21.1 Adaptive spatial–temporal subsampling of the HD-MAC system. (a) The 80 ms mode for stationary to very-slow moving scenes, (b) the 40 ms mode for medium-speed moving scenes, and (c) the 20 ms mode for fast moving scene.

21.2.2 HD-MAC System

A development similar to the MUSE was initiated in Europe as well. The system, HD-MAC (high-definition multiplexed analog component), is also based on the concept of adaptive spatial–temporal subsampling. Depending on the amount of motion, each block, consisting of 8×8 pixels, is classified into either the 20, 40, or 80 ms mode [6]. For a fast-moving block (the 20 ms mode), it is transmitted at the full temporal resolution, but at 1/4 spatial resolution. For a stationary or slow-moving block (the 80 ms mode), it is transmitted at full spatial resolution, but at 1/4 temporal resolution (25/4 fps). For the 40 ms block, it is transmitted at half spatial and half temporal resolutions. The mode associated with each block is transmitted as side information through a digital channel at a bit rate nearly 1 Mbps. The subsampling process of the HD-MAC system is illustrated in Figure 21.1, where the numbers indicate the corresponding fields of transmitted pixels and the "•" indicates a pixel not transmitted.

21.2.3 HDTV in North America

HDTV development in North America started much later than that in Japan and Europe. The Advisory Committee on Advanced Television Services (ACATS) was formed in 1987 to advise Federal Communications Commission (FCC) on the facts and circumstances regarding advanced television systems for terrestrial broadcasting. The proposed systems in early days were all intended for analog transmission [7]. However, the direction of U.S. HDTV development took a 180° turn in 1990 since General Instrument (GI) entered the U.S. HDTV race by submitting an all-digital HDTV system proposal to the FCC. The final contender in the U.S. HDTV race consisted of one analog system (Narrow-MUSE) and four digital systems, which all employed motion compensated DCT coding. Extensive testings on the five proposed systems were conducted in 1991 and 1992 and the testing concluded that there are major advantages in the performance of the digital HDTV systems and only the digital system shall be considered as the standard. However, none of these four digital systems was ready to be selected as the standard without implementing improvements.

With the encouragement from ACATS, the four U.S. HDTV proponents formed the Grand Alliance (GA) to combine their efforts for developing a better system. Two HDTV scan formats were adopted by the GA. The main parameters are shown in Table 21.2. The lower-resolution format, 1280×720, is only used for progressive source materials while the high-resolution format, 1920×1080, can be used for both progressive and interlaced source materials. The digital formats of GA HDTV are carefully designed to accommodate the square-pixel feature, which provides better interoperability with digital video/graphics in the computer environment. Since the main structure of MPEG-2 system and video coding standards were settled at that time and the MPEG-2 video coding standard provides extension to accommodate HDTV formats, the GA adopted MPEG-2 system and video coding (Main Profile [MP] at High Level [HL]) standards for the U.S. HDTV, instead of creating another

TABLE 21.2 Main Scanned Raster Characteristics of the GA HDTV Input Signals

Active Samples/Line	Active Lines per Frame	Scanning Format	Aspect Ratio	Frame Rate
1280	720	1 : 1 progressive	16 : 9 square pixels	60.00/59.94 30/29.97 24/23.976
1920	1080	1 : 1 progressive	16 : 9 square pixels	30/29.97 24/23.976
1920	1080	2 : 1 interlaced	16 : 9 square pixels	30/29.97

standard [8]. However, the GA HDTV adopted the AC-3 audio compression standard [9] instead of the MPEG-2 Layer 1 and Layer 2 audio coding.

21.2.4 EDTV

EDTV refers to the TV signal that offers quality between the conventional TV and HDTV. Usually, EDTV has the same number of scan lines as the conventional TV, but offers better horizontal resolution. Though it is not a required feature, most EDTV systems offer a wide aspect ratio. When the compatibility with a conventional TV signal is of concern, the additional information (more horizontal details, side panels, etc.) required by the EDTV signal is embedded in the unused spatial–temporal spectrum (called "spectrum holes") of the conventional TV signal and can be transmitted in either an analog or digital form [10,11]. When the compatibility with the conventional TV is not required, EDTV can use the component format to avoid the artifacts caused by mixing of chrominance and luminance signals in the composite format. For example, several multiplexed analog component (MAC) systems for analog transmission were adopted in Europe for DBS and cable TV applications [12,13]. Usually, these signals offer better horizontal resolution and better color fidelity. There were many fully digital TV systems developed in the past. These systems that used adequate spatial resolution and higher bit rates were likely to achieve superior quality to the conventional TV and were qualified as EDTV [14]. Nevertheless, an efficient EDTV system is already embedded in the MPEG-2 video coding standard. Within the context of the standard, the 16:9 aspect ratio and horizontal and vertical resolutions exceeding the conventional TV can be specified in the "Sequence Header." When coded with adequate bit rates, the resulting signal can be qualified as EDTV.

21.3 Hybrid Analog/Digital Systems

Today, existing conventional TV sets and other home video equipment represent a massive investment by consumers. The introduction of any new video system that is not compatible with the existing system may face strong resistance in initial acceptance and may take a long time to penetrate households. One way to circumvent this problem during the transition period is to "simulcast" a program in both formats. The redundant conventional TV, being simulcast in a separate channel, can be phased out gradually when most households are able to receive the EDTV or HDTV signal. Intuitively, a more bandwidth efficient approach may be achieved if the transmitted conventional TV signal can be incorporated as a baseline signal and only the enhancement signal is transmitted in an additional channel (called "augmentation channel"). In order to facilitate the compatibility, an analog conventional TV signal has to be transmitted to allow conventional TV sets to receive the signal. On the other hand, digital video compression techniques may be employed to code the enhancement signals in order to accomplish the best compression efficiency. Such systems belong to the category of hybrid analog/digital system. A generic system structure for the hybrid analog/digital approach is shown in Figure 21.2. Due to the

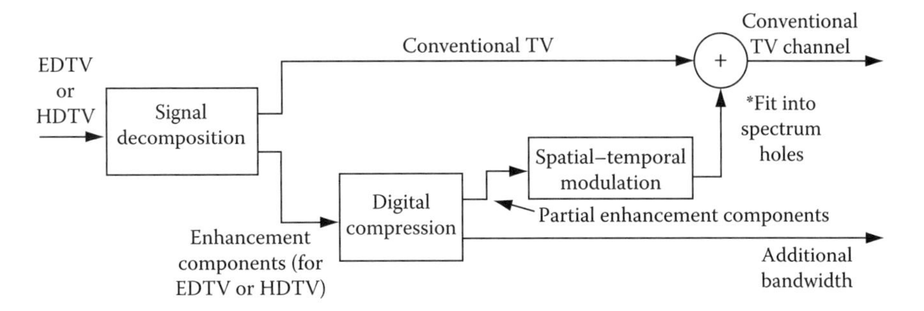

FIGURE 21.2 A generic hybrid analog digital HDTV coding system.

interlacing processing used in TV standards, there are some unused holes in the spatial–temporal spectrum [15], which can be used to carry partial enhancement components as shown in Figure 21.2.

The Advanced Compatible Television System II (ACTV-II), developed by the consortium of NBC, RCA, and the David Sarnoff Research Center during the U.S. ATV standardization process, is an example of a hybrid system. The ACTV-II signal uses a 6 MHz channel to carry an NTSC compatible ACTV-I signal and uses an additional 6 MHz channel to carry the enhancement signal. The ACTV-I consists of a main signal, which is fully compatible with the conventional NTSC signal, and enhancement components (luminance horizontal details, luminance vertical-temporal details, and side-panel details of the wide-screen signal), which are transmitted in 3-D spectrum holes of the NTSC signal. The differences between the input HDTV signal and the ACTV-I signal are digitally coded using 4-band subband coding. The digitally coded video difference signal and digital audio signal require a total bandwidth of 20 Mbps and are expected to fit into the 6 MHz bandwidth by using the 16-QAM modulation. The enhancement components of the ACTV-I signal are digitally processed (time expansion and compression) and transmitted in an analog format. Nevertheless, they could be digitally compressed and transmitted, which would result in a hybrid analog/digital ACTV-I signal. For users with conventional TV sets, conventional TV pictures (4 : 3 aspect ratio) will be displayed. For users with an ACTV-I decoder and a wide screen (16 : 9) TV monitor, the wide-screen EDTV can be viewed by receiving the signal from the main channel. For those who have an ACTV-II decoder and an HDTV monitor, the HDTV picture can be received by using signals from both the main channel and the associated augmentation channel.

The HDS/NA system developed by Philips Laboratories is another example of hybrid analog/digital system where the augmentation signal is carried in a 3 MHz channel [16]. The augmentation signal consists of side panels to convert the aspect ratio from 4 : 3 to 16 : 9, and high-resolution spatial components. The side panels from two consecutive frames are combined into one frame of panels and are intraframe compressed by using DCT coding with a block size of 16×16 pixels. Both the horizontal and vertical high-resolution components are also compressed by intraframe DCT coding with some modifications to take into account the characteristics of these signals. The augmentation signals result in a total bit rate of 6 Mbps, which is expected to fit into a 3 MHz channel using modulation schemes with efficiency of 2 bits/Hz. However, the HDS/NA system was later modified into an analog simulcast system, HDS/NA-6, which occupies only a 6 MHz bandwidth and is intended to be transmitted simultaneously with a conventional TV in a "taboo" channel.

The augmentation-based hybrid analog/digital approach may be more efficient than the simulcast approach when both conventional TV and HDTV receivers have to be accommodated at the same time. However, for the augmentation-based approach, the reconstruction of the HDTV signal relies on the availability of the conventional TV signal, which implies that the main channel carrying the conventional TV signal can never be eliminated. Due to the inefficient use of bandwidth by the conventional analog TV signal, the overall bandwidth efficiency of the hybrid analog/digital approach is inferior to that of the fully digital-based simulcast approach. Furthermore, the system complexity of the hybrid approach is likely to be higher than that of the fully digital approach because it requires both analog and digital types of processing.

21.4 Error Protection and Concealment

Video coding results in a very compact representation of digital video by removing its redundancy, which leaves the compressed data very vulnerable to transmission errors. Usually, a single transmission error will only affect a single pixel for uncompressed data. However, due to the coding process employed, such as DCT transform and motion-compensated interfield/frame prediction, a single transmission error may affect a whole block or blocks in consecutive frames. Furthermore, variable length coding is extensively used in most video coding systems, which is even more susceptible to transmission errors. For variable-length coded data, a single bit error may cause the decoder to lose track of codeword boundaries and results in decoding errors in subsequent data. Generally speaking, a single transmission error may result in noticeable picture impairment if no error concealment is applied.

21.4.1 FEC

The first effort to protect the compressed digital video in an environment susceptible to transmission errors should be to reduce transmission errors by employing forward error correction (FEC) coding. FEC adds redundancy, just opposite to data compression, in order to protect the underlying data from transmission errors. One trivial FEC example is to transmit each bit repeatedly, say three times. A single bit error in each three transmitted bits can be easily corrected by a majority-vote circuit. There are many known FEC techniques which can achieve much better protection without devoting too much bandwidth to redundancy. Today, two types of FEC codes are popularly used for digital transmission over various media. One is Reed-Solomon (RS) code, which belongs to the class of block codes. The other is the convolutional code, which usually operates on continuous data.

The RS code appends a number of redundant bytes to a block of data to achieve error correction. Usually $2n$ redundant bytes can correct up to n byte errors. When a higher level protection is required, more redundant bytes can be attached or alternatively the redundant bytes can be added to shorter data blocks. For digital transmission using the MPEG-2 transport format, in order to maintain the structure of the MPEG-2 transport packets, the (204,188) RS code has been particularly chosen by many standards, which appends 16 redundant bytes to each MPEG-2 transport packet. On the other hand, the U.S. GA-HDTV chose the (207,187) RS code, where the RS redundancy computation is based on the 187 byte data block with the sync byte excluded.

The convolutional code is a powerful FEC code, which generates m output bits for every n input bits. The code rate, r, is defined as $r = n/m$. The output bits are not only determined by the current input bits, but also depend on previous input bits. The depth of the previous input data affecting the output is called the constraint length, k. The output stream of the convolutional code is the result of a generator function convolved with the input stream. Viterbi decoding is an efficient algorithm to decode convolutionally coded data. The complexity of the Viterbi algorithm is proportional to 2^k. Therefore, longer constraint length results in higher decoding complexity. However, longer constraint length also improves FEC performance. A lower rate convolutional code provides more protection at the expense of higher redundancy. For a $r = 1/2$ and $k = 7$ convolutional code, a BER of 10^{-2} can be reduced to below 10^{-5}.

In order to maintain nearly error-free transmission, a very low BER has to be achieved. For example, if an average error-free interval of two hours needs to be achieved for a 6 Mbps compressed bit stream, the required BER is 2.3×10^{-11}. For some transmission media that have limited carrier-to-noise ratio (CNR), such a low BER may not be achievable using the RS code or convolutional code alone. However, an extremely powerful coding can be accomplished by concatenating the RS code and the convolutional code, where the RS code (called outer code) is used toward the source or sink side and the convolutional

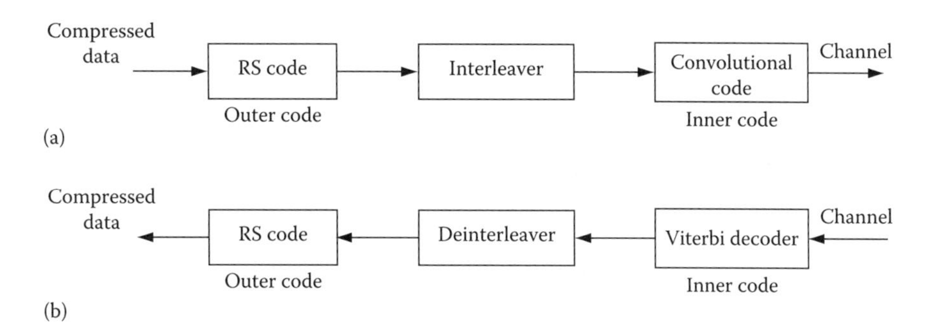

FIGURE 21.3 Block diagram of concatenated RS code and convolutional code. (a) Transmit side and (b) receive side.

code (called inner code) is used toward the channel side. An interleaver to spread bursts of errors is usually used between the inner and outer code in order to improve error correction capability. The interleaver needs to be carefully designed so that the locations of the sync byte in the ATM packets remain unchanged through the interleaver. A block diagram of the concatenated RS code and convolutional code is shown in Figure 21.3. Some simulations showed that satisfactory performance can be achieved by using the concatenate codes for digital video transmission over the satellite link [17]. In Ref. [17], the overall BER is about 2^{-11}, which corresponds to a BER of about 2×10^{-4} using the convolutional code only.

21.4.2 Error Detection and Confinement

While FEC techniques can improve BER significantly, there are still chances that errors may occur. As mentioned earlier, a single bit error may cause catastrophic effects on compressed digital video if precaution is not exercised. To avoid the infinite error propagation, one needs to identify the occurrences of errors and to confine the errors during decoding. Due to the use of variable length coding, a single bit error in the compressed bit stream may cause the decoder to lose track of codeword boundaries. Even though the decoder may regain code synchronization later, the number of decoded data may be more or less than the actual number of samples transmitted, which will affect proper display of the remaining samples. To avoid error propagation, compressed data need to be organized into smaller self-contained data units with "unique words" to identify the beginning or boundaries of the data unit. In case transmission errors occur in preceding data units, the current data unit can still be properly decoded. In the MPEG-2 video coding standard, the "slice" is the smallest self-contained data unit, which has a unique 32-bit "slice_start_code" and information regarding its location within a picture [18]. Therefore, a transmission error in one "slice" will not affect the proper decoding of subsequent "slices." However, for interfield/frame coded pictures, the artifacts in the error-contaminated slice will still propagate to subsequent pictures, which use this slice as reference. Error concealment is a technique to mitigate artifacts caused by transmission errors in the reconstructed picture.

21.4.3 Error Concealment

For DCT-based video coding, some analytic work was conducted in Ref. [19] to derive an optimal reconstruction method based on received blocks with missing DCT coefficients. The solution consists of three linear interpolation in the spatial, temporal, and frequency domains from the boundary data, reconstructed reference block, and received DCT block, respectively. When the complete block is missing, the optimal solution becomes a linear combination of a block replaced by the corresponding block in the previous frame and a spatially interpolated block from boundary pixels. This method needs to go through an iterative process to restore damaged data when consecutive blocks are corrupted by errors. The above concealment technique was further improved in Refs. [20,21] by incorporating an adaptive spatial–temporal interpolation scheme and a multidirectional spatial interpolation scheme.

When a temporal concealment scheme is used, the picture quality in the moving area can be improved by incorporating a motion compensation technique The motion vector for a missing or corrupted macroblock can be estimated from the motion vectors of surrounding macroblocks. For example, the motion vector can be estimated based on the averaged motion vector from the macroblocks above and below the underlying block, as suggested in the MPEG-2 video standard. However, when the neighboring reference macroblocks are intracoded, there are no motion vectors associated with these macroblocks. The MPEG-2 video coding standard allows transmission of the "concealment_motion_vectors" associated with intracoded macroblocks, which can be used to estimate the motion vector for the missing or corrupted macroblock.

21.4.4 Scalable Coding for Error Concealment

When the requirement of error-free transmission cannot be met, it may be useful to provide different protection of underlying data according to the visual importance of the compressed data. This will be useful for transmission media which have different delivery priorities or provide different levels of FEC protection for underlying data. The data that can be used to reconstruct basic pictures are usually treated as high-priority data while the data used to enhance the pictures are treated as low-priority data. For these visually important data, high redundancy is used to offer more protection (or high priority in a cell-based transport system). Therefore, the high-priority data can always be reliably delivered. On the other hand, any errors in the low-priority data will only result in minor degradation. Therefore, if any error is detected in the low-priority data, the affected data can be discarded without significantly degrading the picture quality. Nevertheless, if concealment techniques by spatial–temporal interpolation as described above can be applied to affected areas, this will further improve picture quality. The scalable source coding processes the underlying signal in a hierarchical fashion according to the spatial resolution, temporal resolution, or picture signal-to-noise ratio (SNR), and organizes the compressed data into layers so that a lower-level data set can be used to reconstruct a basic video sequence and the quality can be improved by adding higher levels. Many coding systems can offer the scalable coding feature if the underlying data is carefully partitioned [22,23]. The MPEG-2 video coding standard also offers scalable extension to accommodate spatial, temporal, and SNR scalability.

21.5 Terrestrial Broadcasting

In conventional analog TV standards, in order to allow low-cost TV receivers to acquire the carrier and subcarrier frequencies easily, the transmitted analog signals always contain these two frequencies in high strength, which are the potential cause for cochannel and adjacent-channel interferences. This problem becomes more prominent in the terrestrial broadcasting environment, where the transmitter of an undesired signal (adjacent channel) may be much closer than that of a desired signal. The strong undesired signal may interfere with the desired weak signal. Therefore, some of the terrestrial broadcasting channels (taboo channels) are prohibited in the same coverage area in order to reduce the potential interference. In digital TV transmission, the power spectrum of the signal is widespread over the allocated spectrum, which substantially reduces the potential interference. On the other hand, the bandwidth efficiency of digital coding may significantly increase the capacity of terrestrial broadcasting. Therefore, digital video coding is a very attractive alternative to solving the channel congestion problem in major cities.

21.5.1 Multipath Interference

One notorious impairment of the terrestrial broadcasting channel is the multipath interference, which manifests as the ghost effect in received pictures. For digital transmission, the multipath interference will cause signal distortion and degrade system performance. An effective way to cope with multipath interference is to use adaptive equalization, which can restore the impaired signal by using a known training data sequence. The GA-HDTV system for terrestrial broadcasting adopted this method to overcome the multipath problem [9]. A very different approach—coded orthogonal frequency division multiplexing (COFDM)—has been advocated in Europe for terrestrial broadcasting [10]. The COFDM technology employs multiple carriers to transport parallel data so that the data rate for each carrier is very low. The COFDM system is carefully designed to ensure that the symbol duration for each carrier is longer than the multipath delay. Consequently, the effect of multipath interference will be significantly reduced. The carrier spacing of the COFDM system is carefully arranged so that each subcarrier is orthogonal to the other subcarriers, which achieves high spectrum efficiency. A performance simulation of COFDM for terrestrial broadcasting was reported in Ref. [24], which indicated that COFDM is a viable alternative to digital transmission of 20 Mbps in a 6 MHz terrestrial channel.

21.5.2 Multiresolution Transmission

In terrestrial broadcasting, the CNR of the received signal decreases gradually when the distance between a receiver and the transmitter increases. In an analog transmission system, the picture quality usually degrades gracefully when the CNR decreases. In a digital transmission system, a lower CNR will result in a higher BER and the decoded picture contaminated by errors may become unusable when the BER exceeds a certain threshold. A technique to extend the coverage area of terrestrial broadcasting is to use scalable source coding in conjunction with multiresolution (MR) channel coding [25,26]. In MR modulation, the constellation of the modulated signal is carefully organized in a hierarchical fashion so that a low-density modulation can be derived from the constellation with high protection while a high-density modulation can be achieved by further demodulation of the received signal. An example of MR modulation using quadrature amplitude modulation (QAM) is shown in Figure 21.4, where the nonuniform constellation represents 4-QAM/16-QAM MR modulation. The scalable source coding processes the underlying signal in a hierarchical fashion according to the spatial resolution, temporal resolution, or picture SNR, and organizes the compressed data in layers so that a lower-level data set can be used to reconstruct a basic video sequence and the quality can be improved by adding more levels. The MPEG-2 video coding standard also offers scalable extension to accommodate spatial, temporal, and SNR scalability. In light of the fact that MPEG-2-based systems are being widely used for digital satellite TV broadcasting and being adopted by the Digital Audio-Visual Interactive Council (DAVIC) for the set-top box standard, MPEG-2-based scalable coding in conjunction with the MR modulation likely will be used to offer graceful degradation in terrestrial broadcasting if it is desired.

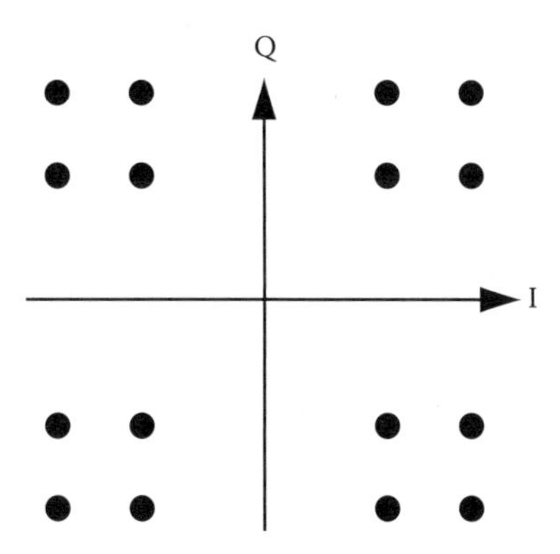

FIGURE 21.4 The constellation of a MR modulation for a 4-QAM/16-QAM.

21.6 Satellite Transmission

Satellite video broadcasting provides an effective way for point-to-multipoint video distribution. It has been widely used in video distribution to cable headends and to satellite TVRO (TV receive only) users for years. Due to recent development in high-powered Ku-band satellite transponders, satellite video broadcasting to small home antennas becomes feasible. The cost of consumer satellite receive systems, including receive dish antenna/LNB and integrated receiver/decoder (IRD) falls below $600 today and is expected to decline gradually. Furthermore, due to advances in digital video compression technology, the capacity of the satellite transponder has been increased substantially. Today, digital TV with 100 or more channels per satellite is being broadcast in North America.

In an analog satellite transmission system, the baseband video signal is FM modulated and transmitted from an uplink site to geostationary satellite. The signal is received by the satellite and retransmitted downward at a different frequency. At a receive site, the signal is received by the receive antenna, block frequency converted to a lower frequency band, and carried through a coax cable to an indoor IRD unit. A simplified system is shown in Figure 21.5.

Due to constraint on the power limit, the satellite transmitters are normally operated in the saturated mode, which introduces system nonlinearity and causes waveform distortion. The available SNR for satellite channels is usually much lower than that for cable channels. In order to overcome the nonlinearity as well as to improve the SNR, the FM technique is always used for analog TV transmission over satellites. For Ku-band applications, a 27 MHz bandwidth is normally allocated to carry one analog TV signal.

For digital transmission over satellite, the quadrature phase shift keying (QPSK) modulation is the most popular technique. The QAM technique, which requires a linear system response, is not suitable for satellite applications. In the North American region, the Ku-band DBS uses the 12/14 GHz frequency band (14 GHz for uplink and 12 GHz for downlink), which allows the subscribers to use a smaller dish antenna. However, the Ku-band link is more susceptible to rain fading than the C-band link and, therefore, more margin for rain fading is required for the Ku-band link. Due to the typical low SNR available for satellite links, powerful coding techniques are required in order to achieve a high-quality link. For an MPEG-2 video stream at 10 Mbps, an average of 1 h error-free transmission will require a BER of 2.778×10^{-11}.

Satellite link has been notorious for its nonlinearity and relatively low CNR. Due to the nonlinearity, any amplitude modulation technique is discouraged in satellite environment. Without forward error

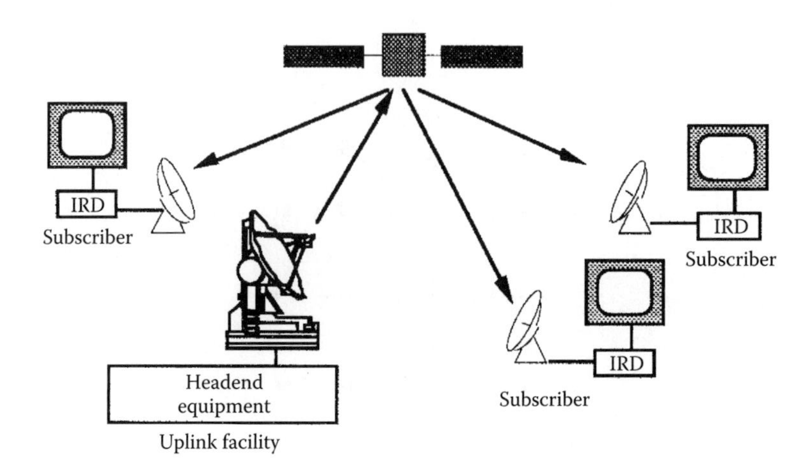

FIGURE 21.5 A satellite video transmission system.

correction coding, typical satellite links can only achieve a BER around 10^{-2}–10^{-5}. This BER is far from the targeted quality of service for compressed digital video. As discussed earlier, concatenated inner and outer codes are very effective for satellite communications, which can reduce the BER to below 10^{-12} from 10^{-4}.

Recently, European Broadcasting Union (EBU) launched a project intended to set a standard for digital video transmission over satellitc, cable, and satellite master antenna TV (SMATV) channels. A draft standard [27] was published by EBU/European Telecommunications Standards Institute (ETSI). This draft specifies a powerful error correction scheme based on concatenation of convolutional and RS codes as shown in Figure 21.3. The convolutional code can be configured to operate at different rates, including 1/2, 2/3, 3/4, 5/6, 7/8, and 1 to optimize the performance for transponder power and bandwidth. At the receive end, Viterbi decoding with soft-decision is often used to decode the convolutional code. By using the convolutional code alone, a BER between 10^{-3} and 10^{-8} may be achieved for typical satellite links. However, this is still not adequate for real-time digital video applications.

In order to further improve the BER performance, an outer code using the RS code is applied to correct errors remaining uncorrected by the convolutional code. Channel errors generated at the output of Viterbi decoder tend to occur in bursts. The RS code operates on byte-oriented data and is effective in correcting burst errors. To improve the effectiveness of the RS code, an interleaver is usually used between the convolutional code and the RS code. By using the (204,188) RS code and a convolutional interleaver of depth 12, the BER of 2×10^{-4} for the convolutional code can be improved to around 10^{-11}. A recent report [28] showed that a BER around 10^{-11} can be achieved for typical high-powered DBS with bit rates ranging from 23 to 41 Mbps by using concatenate convolutional and RS codes.

21.7 ATM Transmission of Video

ATM is a cell-based transport technology that multiplexes fixed-length cells from a variety of sources to a variety of remote locations. Each ATM cell consists of a 5 byte header and 48 byte payload. The routing, flow control and payload type information is carried in the header, which is then protected by a 1 byte error correction code. However, unlike the packet data communication, ATM is a connection-oriented protocol. Connections, either permanent, semitemporary, or permanent, between ATM users are established before data exchanges commence. The header information in each cell determines to which port at an ATM switch the cell should be routed. This substantially reduces the processing complexity required in a switching equipment. The flexibility of ATM technology allows both constant rate and variable rate services to be easily offered through the network. Also, it allows multimedia services, such as video, voice, and data of different characteristics to be multiplexed into a single stream and delivered to customers.

21.7.1 ATM Adaptation Layer for Digital Video

In order to carry data units other than the 48-octets payload size in ATM cells, an adaptation layer is needed. The ATM adaptation layer (AAL) provides for segmentation and reassembly (SAR) of higher-layer data units and detection of errors in transmission. Five AALs are specified in ITU-T Recommendation, I.363. AAL1 is intended for constant bit rate services while AAL2 is intended for variable bit rate services with a required timing relationship between the source and destination. AAL3/4 is intended for variable bit rate services that require bursty bandwidth. AAL5 is a simple and efficient adaptation layer intended to reduce the complexity and overhead of AAL3/4. Both AAL1 and AAL5 have been seriously considered as a candidate for real-time digital video applications. However, the AAL5 was adopted by the ATM Forum as the standard for Audiovisual Multimedia Services (AMS) [29].

The standard process of ATM is undertaken by several international standard bodies such as ATM Forum, and International Telecommunication Union-Transmission (ITU-T) Study Groups (SG) 9, 13, and 15. For digital television transmission, the MPEG-2 transport standard seems to be the sole format

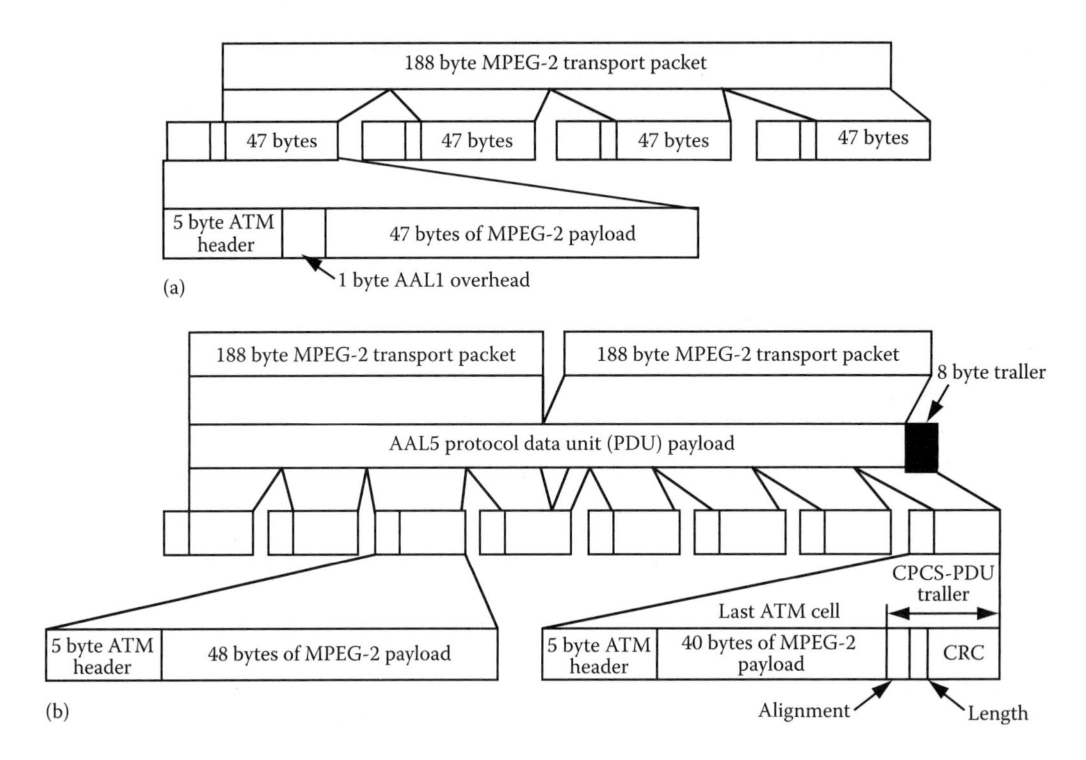

FIGURE 21.6 Mapping MPEG-2 TS packets into AAL PDU. (a) AAL-1 and (b) AAL5.

being considered. MPEG-2 transport standard relies on frequent and low-jitter delivery of transport stream (TS) packets containing presentation clock references (PCR) to recover the 27 MHz clock at the receiving end. There are several key parameters in designing an AAL for digital video, which include packaging efficiency, complexity, error handling capability and performance, and PCR jitter. When AAL1 is employed, each MPEG-2 TS packet is mapped into 4 ATM cells as shown in Figure 21.6a. A 1 byte AAL1 header is inserted into the first payload byte of each ATM cell. The AAL1 header contains a sequence number field and a sequence number protection field. The AAL1 uses the synchronous residual time stamp (SRTS) method to support source clock recovery.

The AAL5 specified in [29] maps *N* MPEG-2 single program TS (SPTS) packets into an AAL5-SDU (service data unit) unless there are fewer than *N* TS packets left in the sequence. In the case when there are fewer than *N* packets left in the SPTS, the last AAL5-SDU contains all the remaining packets. The default value for *N* is 2, which results in a default SDU size of 376 bytes. This default SDU along with an 8 byte trail fits nicely into the payloads of 8 ATM cells, as shown in Figure 21.6b. The trailer contains a 2 byte alignment field, a 2 byte length indicator field, and a 4 byte CRC field. For constant bit rate transmission, the MPEG-2 SPTS is considered as a constant packet rate (CPR) stream of information, which implies that the interarrival time between packets of the MPEG-2 TS is constant. In order to ensure satisfactory timing recovery, the time interval of the last byte containing the PCR should be constant.

The AAL5 is meant for both constant bit rate and variable bit rate applications while the AAL1 is mainly intended for constant bit rate applications. The AAL5 contains a 4 byte CRC field and a 2 byte length indicator field to check the payload integrity. On the other hand, the AAL1 only offers sequence integrity to detect lost cells. The most attractive factor of AAL5 is the wide support of major service and equipment vendors.

21.7.2 Cell Loss Protection

In the ATM environment, cells may be corrupted due to transmission errors or lost due to traffic congestion. The transmission bit error rate usually is very small for fiber-based systems. However, the cell loss due to congestion seems to be unavoidable in order to increase link utilization efficiency. Depending on how compressed data is mapped into ATM cells, the loss of a single cell may corrupt a number of cells. In the ATM header, there is 1 bit information to indicate the delivery priority of the underlying payload. This priority bit can be used to cope with the cell loss issue. To take advantage of the priority bit, the coding systems will have to separate the compressed data into high- and low-priority layers and to pack the data into cells with a corresponding priority indicator. When network congestion occurs, these cells labeled with low priority are subject to discarding at the switch. Since the low-priority cells carry visually less important information, the impairments in the reconstructed low-priority data will be less objectionable. Some two-layer coding techniques were proposed for MPEG-2 video and have shown significant improvement over a single-layer coding under cell loss circumstance [20,21].

References

1. Strachan, D. and Conrad, R., Serial video basics, *SMPTE J.*, 254–257, Aug. 1994.
2. Fujii, T. et al., Film simulation for high definition TV picture and subjective test of picture quality, NHK Technical Report, 18, No. 11, 1975. Some papers related to HDTV camera and display also appeared in the same issue.
3. Nonomiya, Y., MUSE coding system for HDTV broadcast, *Proceedings of the 1st International HDTV Signal Processing Workshop*, Torino, Italy, 1986.
4. Nonomiya, Y. et al., HDTV broadcasting and transmission system—MUSE, *Proceedings of the 2nd International HDTV Signal Processing Workshop*, Torino, Italy, 1986.
5. Nishizawa, et al., HDTV and transmission system—MUSE and its family, *Proceedings of the 1988 International Broadcasting Conference*, Brighton, U.K., pp. 37–40, 1988.
6. Vreeswijk, F.W.P. et al., An HD-MAC coding system, *Proceedings of the 2nd International HDTV Signal Processing Workshop*, Torino, Italy, 1988.
7. Hopkins, R., Advanced televisions systems, *IEEE Trans. Consum. Electron.*, 34(1), 1–15, Feb. 1988.
8. United States Advanced Television Systems Committee, *Digital Television Standard for HDTV Transmission*, Doc. A/53, Apr. 12, 1995.
9. United States Advanced Television Systems Committee, *Digital Audio Compression* (AC-3), Doc. A/52, 1994.
10. Isnardi, M. et al., Decoding issues in the ACTV system, *IEEE Trans. Consum. Electron.*, 34(1), 111–120, Feb. 1988.
11. Kawai, K. et al., A wide screen EDTV, *IEEE Trans. Consum. Electron.*, 35(3), 133–141, Aug. 1989.
12. Gardiner, P.N., The UK D-MAC/packet standard for DBS, *IEEE Trans. Consum. Electron.*, 34(1), 128–136, Feb. 1988.
13. Garault, T. et al., A digital MAC decoder for the display of a 16/9 aspect ratio picture on a conventional TV receiver, *IEEE Trans. Consum. Electron.*, 34(1), 137–146, Feb. 1988.
14. Jalali, A. et al., A component CODEC and line multiplexer, *IEEE Trans. Consum. Electron.*, 34(1), 156–165, Feb. 1988.
15. Fukinuki, T. and Hirano, Y., Extended definition TV fully compatible with existing standards, *IEEE Trans. Commun.*, COM-32, 948–953, Aug. 1984.
16. Tsinberg, M., Compatible introduction of HDTV: The HDS/NA system, *Proceedings of the 3rd International HDTV Signal Processing Workshop*, Torino, Italy, 1989.
17. Cominetti, M. and Morello, A., Direct-to-home digital multi-programme television by satellite, *Proceedings of the International Broadcasting Convention*, pp. 358–365, Sept. 16–20, 1994.

18. ISO/IEC IS 13818-2/ITU-T Recommendation H.262, Information technology—generic coding of moving picture and associated audio—Part 2: Video, ISO/IEC, May 10, 1994.
19. Zhu, Q.-F., Wang, Y., and Shaw, L., Coding and cell-loss recovery in DCT-based packet video, *IEEE Trans. Circuits Syst. Video Technol.*, 3(3), 238–247, June 1993.
20. Sun, H. and Zdepski, J., Adaptive error concealment algorithm for MPEG compressed video, *Proceedings of the SPIE, Visual Communication and Image Processing*, vol. 1818, pp. 814–824, Nov. 1992.
21. Kwok, W. and Sun, H., Multi-directional interpolation for spatial error concealment, *IEEE Trans. Consum. Electron.*, 39(3), 455–460, Aug. 1993.
22. Yu, Y. and Anastassiou, D., High quality two layer video coding using MPEG-2 syntax, *Proceedings of the 6th International Workshop on Packet Video*, A4.1–4, Portland, OR, Sept. 26–27, 1994.
23. Chan, S.K. et al., Layer transmission of MPEG-2 video in ATM environment, *Proceedings of the 6th International Workshop on Packet Video*, D1.1–4, Portland, OR, Sept. 26–27, 1994.
24. Wu, Y. and Zou, W., Performance simulation of COFDM for TV broadcasting application, *SMPTE J.*, 104(5), 258–265, May 1995.
25. Schreiber, W.F., Advanced television systems for terrestrial broadcasting: Some problems and some proposed solutions, *Proc. IEEE*, 83(6), 958–981, June 1995.
26. deBot, P.G.M., Multiresolution transmission over the AWGN channel, Technical reports, Philips Labs., Eindhoven, the Netherlands, June 1992.
27. EBU/ETSI JTC, *Draft Digital Broadcasting System for Television, Sound and Data Services; Framing Structure, Channel Coding and Modulation for 11/12 GHz Satellite Services*, Draft prETS 300 421, June 1994.
28. Cominetti, M. and Morello, A., Direct-to-home digital multi-programme television by satellite, *Proceedings of the International Broadcasting Convention*, 358–365, June 1994.
29. ATM Forum, *Audiovisual Multimedia Services: Video on Demand Implementation Agreement 1.0*, ATMF/95-0012R6, Oct. 1995.

22

Stereoscopic Image Processing*

Reginald L. Lagendijk
Delft University of Technology

Ruggero E. H. Franich
AEA Technology

Emile A. Hendriks
Delft University of Technology

22.1 Introduction

Static images and dynamic image sequences are the projection of time-varying three-dimensional (3-D) real world scenes onto a two-dimensional (2-D) plane. As a result of this planar projection, depth information of objects in the scene is generally lost. Only by cues such as shadow, relative size and sharpness, interposition, perspective factors, and object motion, can we form an impression of the depth organization of the real world scene.

In a wide variety of image processing applications, explicit depth information is required in addition to the scene's gray value information (representing intensities, color, densities, etc.) [2,4,7]. Examples of such applications are found in 3-D vision (robot vision, photogrammetry, remote sensing systems); in medical imaging (computer tomography, magnetic resonance imaging, microsurgery); in remote handling of objects, for instance in inaccessible industrial plants or in space exploration; and in visual communications aiming at virtual presence (conferencing, education, virtual travel and shopping, virtual reality). In each of these cases, depth information is essential for accurate image analysis or for enhancing the realism. In remote sensing the terrain's elevation needs to be accurately determined for map production, in remote handling an operator needs to have precise knowledge of the 3-D organization of the area to avoid collisions and misplacements, and in visual communications the quality and ease of information exchange significantly benefits from the high degree of realism provided by scenes with depth.

Depth in real-world scenes can be explicitly measured by a number of range sensing devices such as by laser range sensors, structured light, or ultrasound. Often it is, however, undesirable or unnecessary to have separate systems for acquiring the intensity and the depth information because of the relative low

* This work was supported in part by the European Union under the RACE-II projects DISTIMA and the ACTS project PANORAMA.

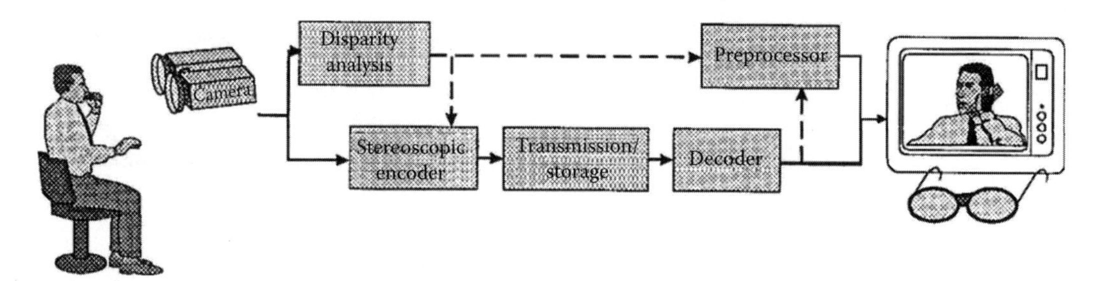

FIGURE 22.1 Illustration of system for stereoscopic image (sequence) recording, processing, transmission, and display.

resolution of the range sensing devices and because of the question of how to fuse information from different types of sensors.

An often used alternative to acquire depth information is to record the real-world scene from different perspective viewpoints. In this way, multiple images or (preferably time-synchronized) image sequences are obtained that implicitly contain the scene's depth information. In the case that multiple views of a single scene are taken without any specific relation between the spatial positions of the viewpoints, such recordings are called multiview images. Generally speaking, when recordings are obtained from an increasing number of different viewpoints, the 3-D surfaces and/or interior structures of the real-world scene can be reconstructed more accurately. The terms stereoscopic image and stereoscopic image sequence are reserved for the special case that two perspective viewpoints are recorded or computed such that they can be viewed by a human observer to produce the effect of natural depth perception (see Figure 22.1). Therefore, the two views are required to be recorded under specific constraints such as the cameras' separation, convergence angle, and alignment [8]. Stereoscopic images are not truly 3-D images since they merely contain information about the 2-D projected real-world surfaces plus the depth information at the perspective viewpoints. They are, therefore, sometimes called 2.5-D images.

In the broadest meaning of the word, a digital stereoscopic system contains the following components: stereoscopic camera setup, depth analysis of the digitized and recorded views, compression, transmission or storage, decompression, preprocessing prior to display, and, finally, the stereoscopic display system. The emphasis here is on the image processing components of this stereoscopic system; that is, depth analysis, compression, and preprocessing prior to the stereoscopic display. Nonetheless, we first briefly review the perceptual basis for stereoscopic systems and techniques for stereoscopic recording and display in Section 22.2. The issue of depth or disparity analysis of stereoscopic images is discussed in Section 22.3, followed by the application of compression techniques to stereoscopic images in Section 22.4. Finally, Section 22.5 considers the issue of stereoscopic image interpolation as a preprocessing step required for multiviewpoint stereoscopic display systems.

22.2 Acquisition and Display of Stereoscopic Images

The human perception of depth is brought about by the hardly understood brain process of fusing two planar images obtained from slightly different perspective viewpoints. Due to the different viewpoint of each eye, a small horizontal shift exists, called disparity, between corresponding image points in the left and right view images on the retinas. In stereoscopic vision, the objects to which the eyes are focused and accommodated have zero disparity, while objects to the front and to the back have negative and positive disparity, respectively, as is illustrated in Figure 22.2. The differences in disparity are interpreted by the brain as differences in depth ΔZ.

In order to be able to perceive depth using recorded images, a stereoscopic camera is required which consists of two cameras that capture two different, horizontally shifted perspective viewpoints. This

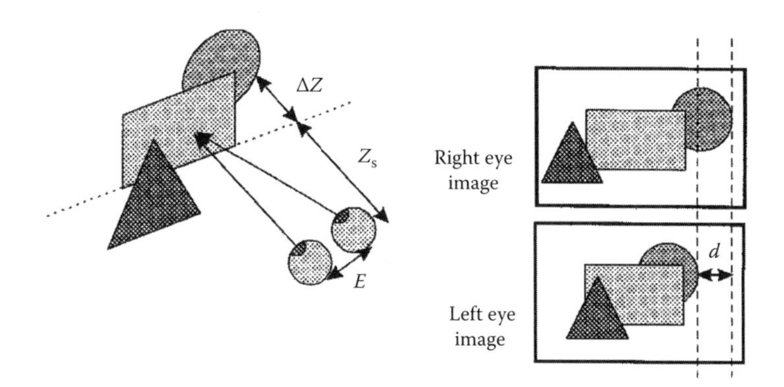

FIGURE 22.2 Stereoscopic vision, resulting in different disparities depending on depth.

results in a shift (or disparity) of objects in the recorded scene between the left and the right view depending on their depth. In most cases, the interaxial separation or baseline B between the two lenses of the stereoscopic camera is in the same order as the eye distance E (6–8 cm). In a simple camera model, the optical axes are assumed to be parallel. The depth Z and disparity d are then related as follows:

$$d = \lambda \frac{B}{\lambda - Z} \tag{22.1}$$

where λ is the focal length of the cameras. Figure 22.3a illustrates this relation for a camera with $B = 0.1$ m and $\lambda = 0.05$ m. A more complicated camera model takes into account the convergence of the camera axes with angle β. The resulting relation between depth and disparity, which is a much more elaborate expression in this case, is illustrated in Figure 22.3b for the same camera parameters and $\beta = 1°$. It shows that, in this case, the disparity is not only dependent on the depth Z of an object, but also on the horizontal object position X. Furthermore, a converging camera configuration also leads to small vertical disparity components, which are, however, often ignored in subsequent processing of the stereoscopic data. Figure 22.4a and b show as an example a pair of stereoscopic images encountered in video communications.

When recording stereoscopic image sequences, the camera setup should be such that, when displaying the stereoscopic images, the resulting shifts between corresponding points in the left and right view images on the display screen allow for comfortable viewing. If the observer is at a distance Z_s from the screen, then the observed depth Z_{obs} and displayed disparity d are related as

$$Z_{obs} = Z_s \frac{E}{E - d} \tag{22.2}$$

In the case that the camera position and focusing are changing dynamically, as is the case, for instance, in stereoscopic television production where the stereoscopic camera may be zooming, the camera geometry is controlled by a set of production rules. If the recorded images are to be used for multi-viewpoint stereoscopic display, a larger interaxial lens separation needs to be used, sometimes even up to 1 m. In any case, the camera setup should be geometrically calibrated such that the two cameras capture the same part of the real world scene. Furthermore, the two cameras and A/D converters need to be electronically calibrated to avoid unbalances in gray value of corresponding points in the left and right view image.

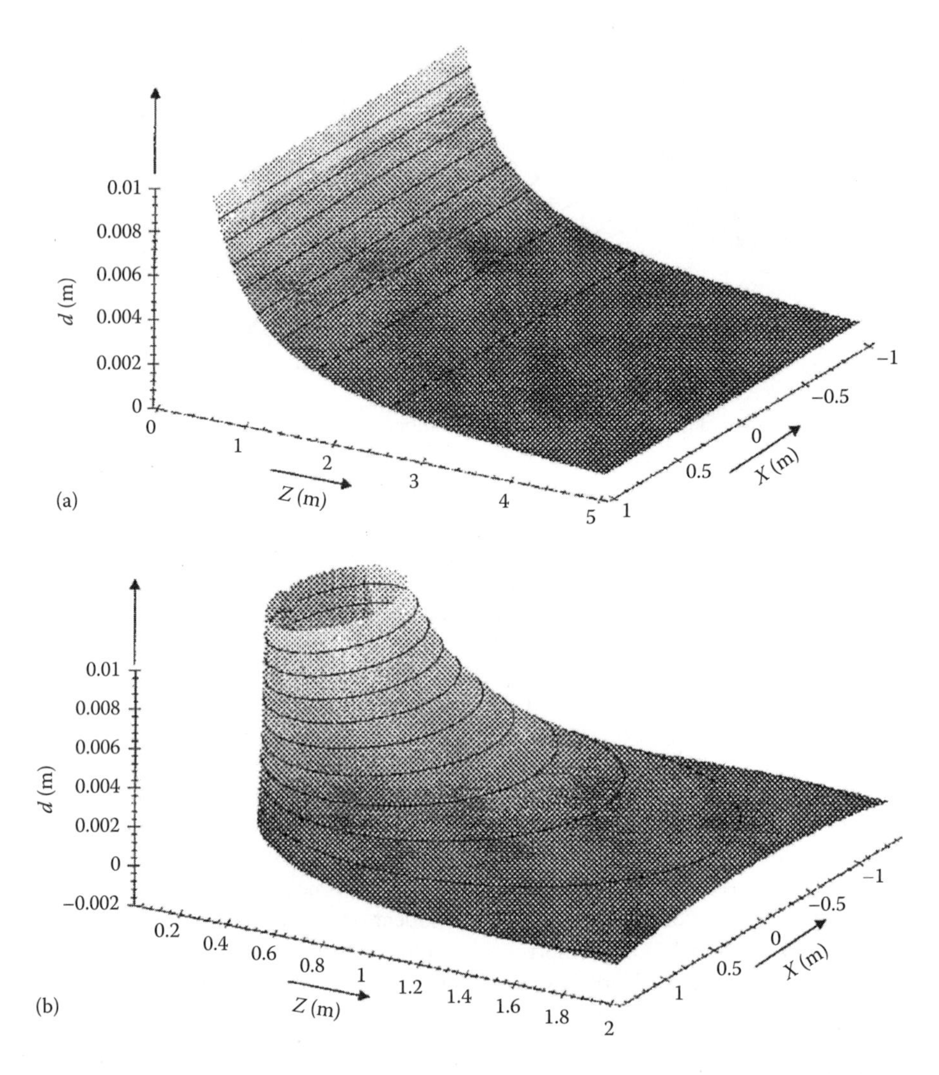

FIGURE 22.3 (a) Disparity as a function of depth for a sample parallel camera configuration; (b) disparity for a sample converging camera configuration.

The stereoscopic image pair should be presented such that each perspective viewpoint is seen only by one of the eyes. Most practical state-of-the-art systems require viewers to wear special viewing glasses [6]. In a time-parallel display system, the left and right view images are presented simultaneously to the viewer. The views are separated by passive viewing glasses such as red–green viewing glasses requiring the left and right view to be displayed in red and green, respectively, or polarized viewing glasses requiring different polarization of the two views. In a time-sequential stereoscopic display, the left and right view images are multiplexed in time and displayed at a double field rate, for instance 100 or 120 Hz. The views are separated by means of the active synchronized shuttered glasses that open and close the left and right eyeglasses depending on the viewpoint being shown. Alternatively, lenticular display screens can be used to create spatial interference patterns such that the left and right view images are projected directly into the viewer's eyes. This avoids the need of wearing viewing glasses.

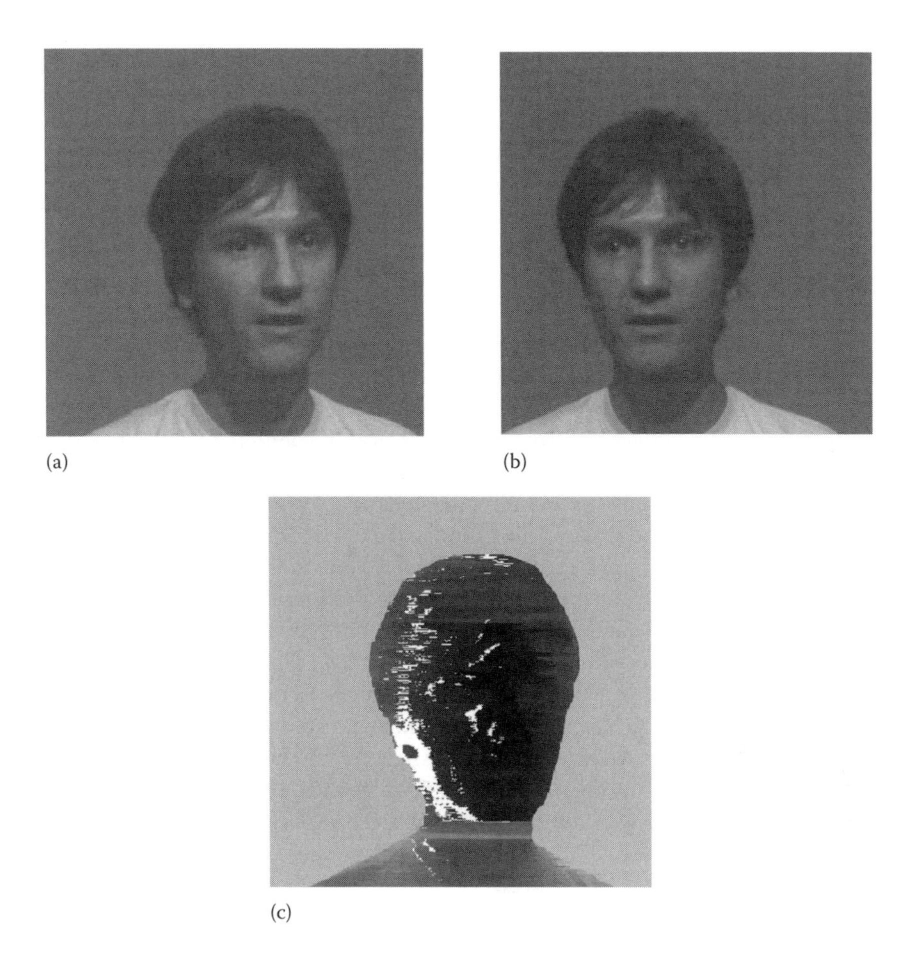

FIGURE 22.4 The left (a) and right (b) view image from a stereoscopic image pair. (c) Disparity field in the stereoscopic image pair represented as gray values (black is foreground, gray is background, white is occlusion).

22.3 Disparity Estimation

The key difference between planar and stereoscopic images and image sequences is that the latter implicitly contains depth information in the form of disparity between the left and right view images. Not only is the presence of disparity information essential to the ability of humans to perceive depth, disparity can also be exploited for automated depth segmentation of real world scenes, and for compression and interpolation of stereoscopic images or image sequences [1].

To be able to exploit disparity information in a stereoscopic pair in image processing applications, the relation between the contents of the left view image and the right view image has to be established, yielding the disparity (vector) field. The disparity field indicates for each point in the left view image the relative shift of the corresponding point in the right view image and vice versa. Since some parts of one view image may not be visible in the alternate view image due to occlusion, not all points in the image pair can be assigned a disparity vector.

Disparity estimation is essentially a correspondence problem. The correspondence between the two images can be determined by either matching features or by operating on or matching of small patches of gray values. Feature matching requires as a preprocessing step the extraction of appropriate features from the images, such as object edges and corners. After obtaining the features, the correspondence problem is first solved for the spatial locations at which the features occur, from which next the full disparity field can be deduced by, for instance, interpolation or segmentation procedures. Feature-based disparity estimation is especially useful in the analysis of scenes for robot vision applications [4,11].

Disparity field estimation by operating directly on the image gray value information is not unlike the problem of motion estimation [11,12]. The first difference is that disparity vectors are approximately horizontally oriented. Deviations from the horizontal orientation are caused by the convergence of the camera axes and by differences between the camera optics. Usually vertical disparity components are either ignored or rectified. A second difference is that disparity vectors can take on a much larger range of values within a single image pair. Furthermore, the disparity field may have large discontinuities associated with objects neighboring in the planar projection but having a very much different depth. In those regions of the stereoscopic image pair where one finds large discontinuities in the disparity field due to abrupt depth changes, large regions of occlusion will be present. Estimation methods for disparity fields must therefore be able not only to find the correspondence between information in the left and right view images, but must also be able to detect and handle discontinuities and occlusions [1].

Most disparity estimation algorithms used in stereoscopic communications rely on matching small patches of gray values from one view to the gray values in the alternate view. The matching of this small patch is not carried out in the entire alternate image, but only within a relatively small search region to limit the computational complexity. Standard methods typically use a rectangular match block of relatively small size (e.g., 8×8 pixels), as illustrated in Figure 22.5. The relative horizontal shift between a match block and the block within the search region of the alternate image that results in the smallest value of a criterion function used is then assigned as disparity vector to the center of that match block. Often used criterion functions are the sum of squares and the sum of the absolutes values of the differences between the gray values in the match block and the block being considered in the search region [3,12].

The above procedure is carried out for all pixels, first matching the blocks from the left view image to the right view image, then vice versa. From the combination of the two resulting disparity fields and the values of the criterion function, the final disparity field is computed, and occluding areas in the stereoscopic image pair are detected. For instance, one way of detecting occlusions is a local abrupt increase of the criterion function, indicating that no acceptable correspondence between the two image pairs could be found locally. Figure 22.4c illustrates the result of a disparity estimation process as an image in which different gray values correspond to different disparities (and thus depth), and in which "white" indicates occluding regions that can be seen in the left view image but that cannot be seen in the right view image.

More advanced versions of the above block matching disparity estimator use hierarchical or recursive approaches to improve the consistency or smoothness of the resulting disparity field, or are based on the

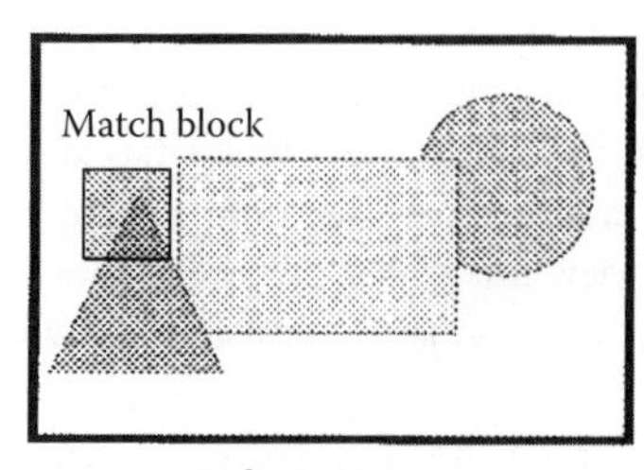

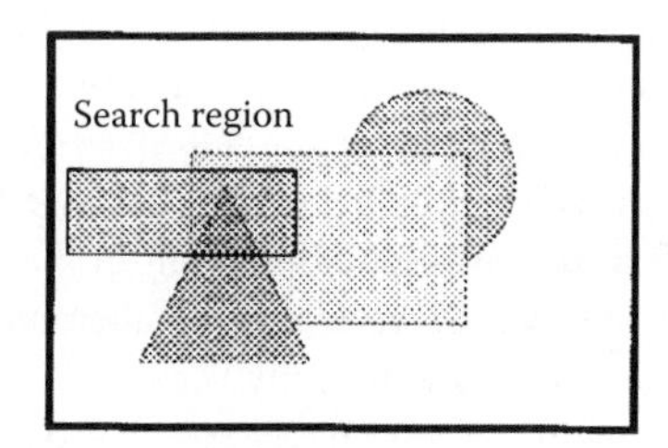

FIGURE 22.5 Block matching disparity estimation procedure by comparing a match block from the left image to the blocks within a horizontally oriented search region in the right image.

optical flow model often used in motion estimation. Other approaches use preprocessing steps to determine the dominant disparity values that are then used as candidate solutions during the actual estimation procedure. Finally, most recent approaches use advanced Markov random field models for the disparity field and/or they make use of more complicated cost functions such as the disparity space image. These approaches typically require exhaustive optimization procedures but they have the potential of accurately estimating large discontinuities and of precisely detecting the presence of occluding regions [1].

In image analysis problems, disparity estimation is often considered in combination with the segmentation of the stereoscopic image pair. Joint disparity estimation and texture segmentation methods partition the image pair into spatially homogeneous regions of approximately equal depth. Disparity estimation in image sequences is typically carried out independently on successive frame pairs. Nevertheless, the need for temporal consistency of successive disparity fields often requires temporal dependencies to be exploited by postprocessing of the disparity fields. If an image sequence is recorded as an interlaced video signal, disparity estimation should be carried out on the individual fields instead of frames to avoid confusion between motion displacements and disparity.

22.4 Compression of Stereoscopic Images

Compression of digital images and image sequences is necessary to limit the required transmission bandwidth or storage capacity [3,5]. One of the compression principles underlying the JPEG and MPEG standards is to avoid transmitting or storing gray value information that is predictable from the signal's spatial or temporal past, i.e., information that is redundant. In both JPEG and MPEG, this principle is exploited by a spatial DPCM system, while in MPEG motion-compensated temporal prediction is also used to exploit temporal redundancies.

When dealing with stereoscopic image pairs, a third dimension of redundancy appears, namely the mutual predictability of the two perspective views [9]. Although the left and right view images are not identical, gray value information in, for instance, the left view image is highly predictable from the right view image if the horizontal shift of corresponding points, i.e., the disparity, is taken into account. Thus, instead of transmitting or storing both views of a stereoscopic image pair, only the right view image is retained, together with the disparity field. Since the construction of the left view image from the right view is not perfect due to errors in the estimated disparity field and due to presence of occluding areas and perspective differences, some information of the disparity-compensated prediction error of the left view (i.e., the difference between the predicted gray values and the actual gray values in the left view image) also needs to be retained. Figure 22.6

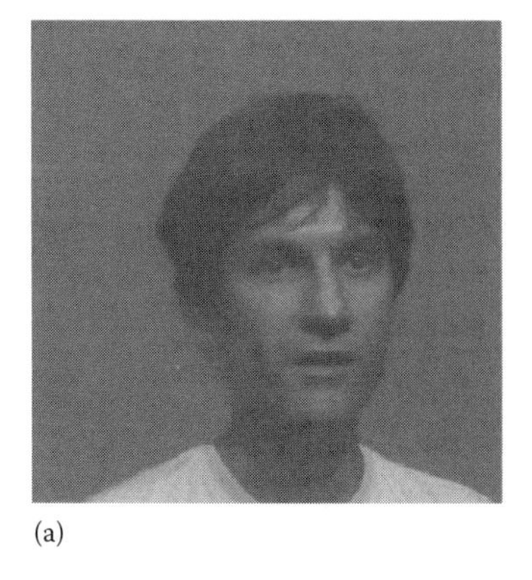

(a)

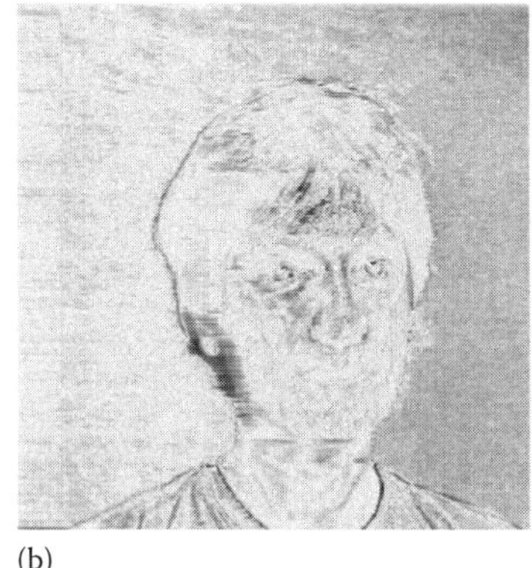

(b)

FIGURE 22.6 (a) Disparity-compensated prediction and (b) disparity-compensated prediction error of the left view image (scaled for maximal visibility) in Figure. 22.4. Black areas indicate a large error.

shows the disparity-compensated prediction and the disparity-compensated prediction error of the left view image from Figure 22.4a using the right view image in Figure 22.4b and the disparity field in Figure 22.4c. In most cases, the sum of the bit rates needed for coding the disparity vector field and the disparity-compensated prediction error is much smaller than the bit rate needed for the left view image when compressed without disparity compensation.

In image sequence, left view images can be compressed efficiently by carrying out motion-compensated prediction from previous left view images, by disparity-compensated prediction from the corresponding right view image, or by a combination of the two by choosing for motion-compensation or disparity-compensation on a block-by-block basis, as illustrated in Figure 22.7a. Basically this is a direct extension of the MPEG compression standard with an additional prediction mode for the left view image sequence. The effect of this additional (disparity-compensated) prediction mode is that the variance of the prediction error of the left view image sequence is further decreased (see Figure 22.7b), meaning that

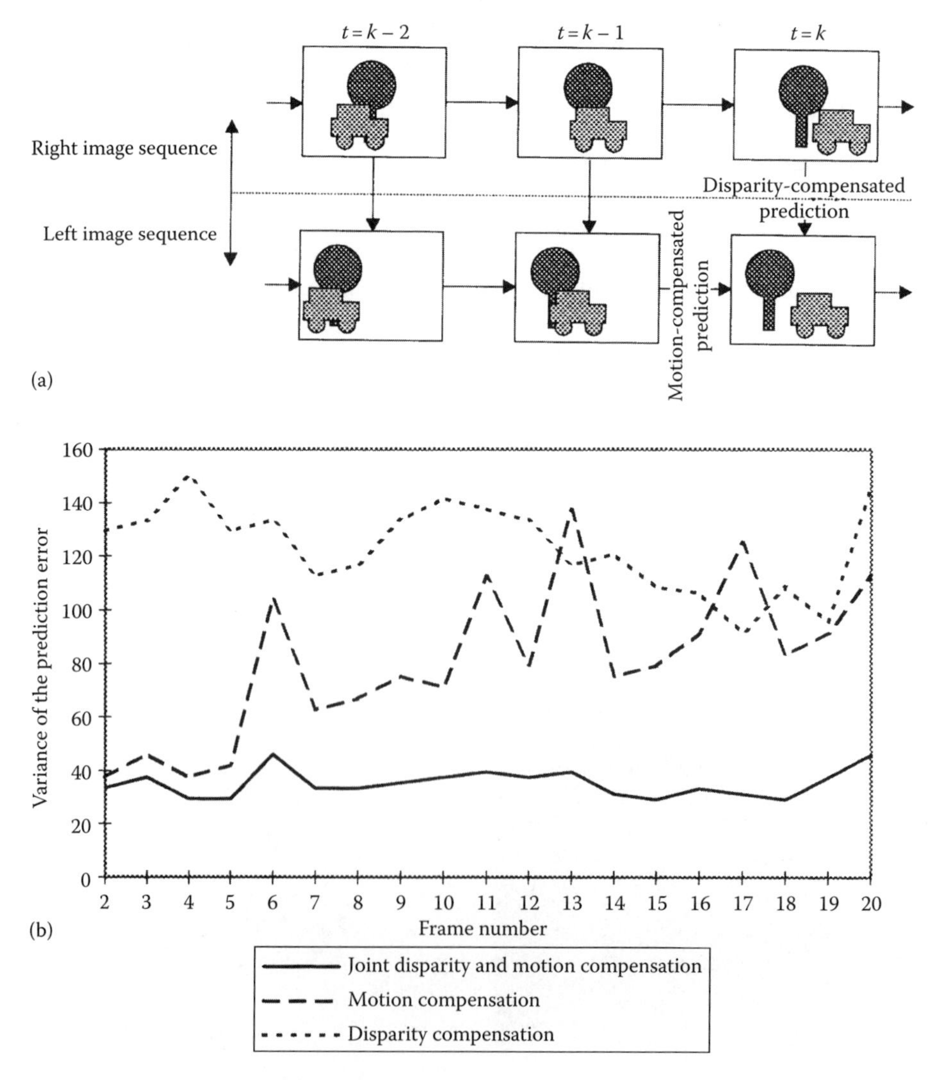

FIGURE 22.7 (a) Principle of joint disparity- and motion-compensated prediction for the left view of a stereoscopic image sequence; (b) variance of the prediction error of the left view image sequence when using motion-compensation, disparity-compensation, or joint motion–disparity compensation on a block-by-block basis.

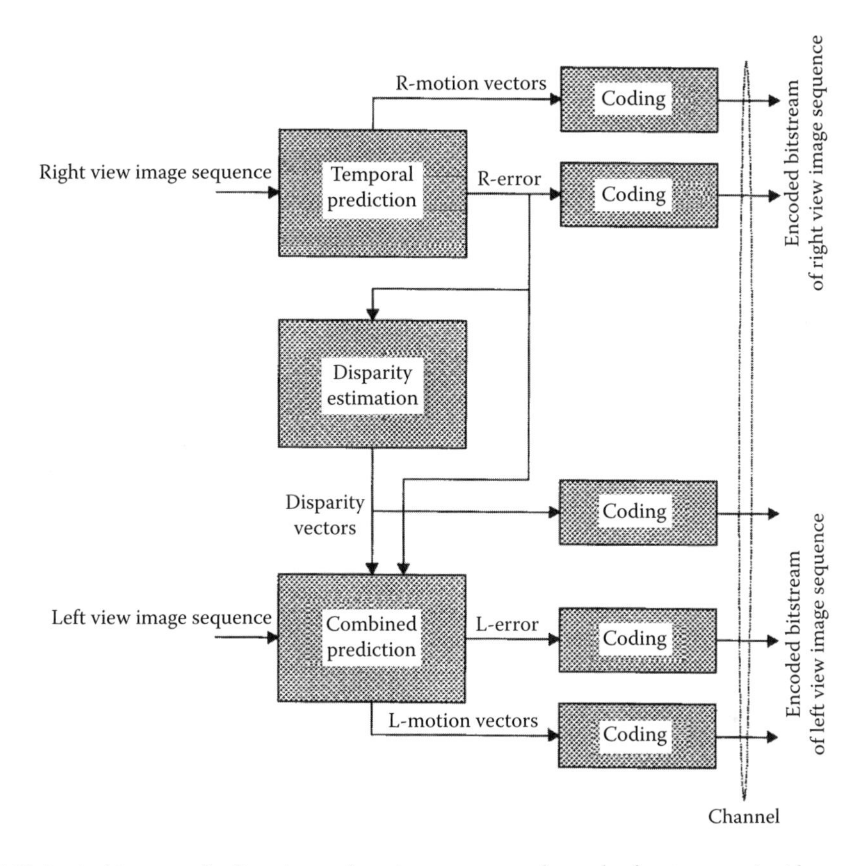

FIGURE 22.8 Architecture of a disparity- and motion-compensated encoder for stereoscopic video.

more compression of the left view sequence is possible than when independently compressing the two views of the stereoscopic sequence. Figure 22.8 schematically shows the architecture of a disparity- and motion-compensated encoder for stereoscopic video.

22.5 Intermediate Viewpoint Interpolation

The system illustrated in Figure 22.1 assumes that the stereoscopic image captured by the cameras is directly displayed at the receiver's end. One of the shortcomings of such a two-channel stereoscopic system is that shape and depth distortion occur when the stereoscopic images are viewed from an off-center position. Furthermore, since the cameras are in a fixed position, the viewer's (horizontal) movements do not provide additional information about, for instance, objects that are partly occluded. The lack of this "look around" capability especially is a limiting factor in the truly realistic visualization of a recorded real world scene.

In a multichannel or multiview stereoscopic system, multiple viewpoints of the same real world scene are available. The stereoscopic display then shows only those two perspective views which correspond as well as possible with the viewer's position. To this end some form of tracking the viewer's position is necessary. The additional viewpoints could be obtained by installing more cameras at a wide range of possible viewpoints. On grounds of complexity and costs the number of cameras will typically be limited to three to five, meaning that not all possible positions of the viewer are covered in this way. If, because of the viewer's position, a view of the scene is needed from an unavailable camera position, a virtual camera or intermediate viewpoint must be constructed from the available camera viewpoints (see Figure 22.9).

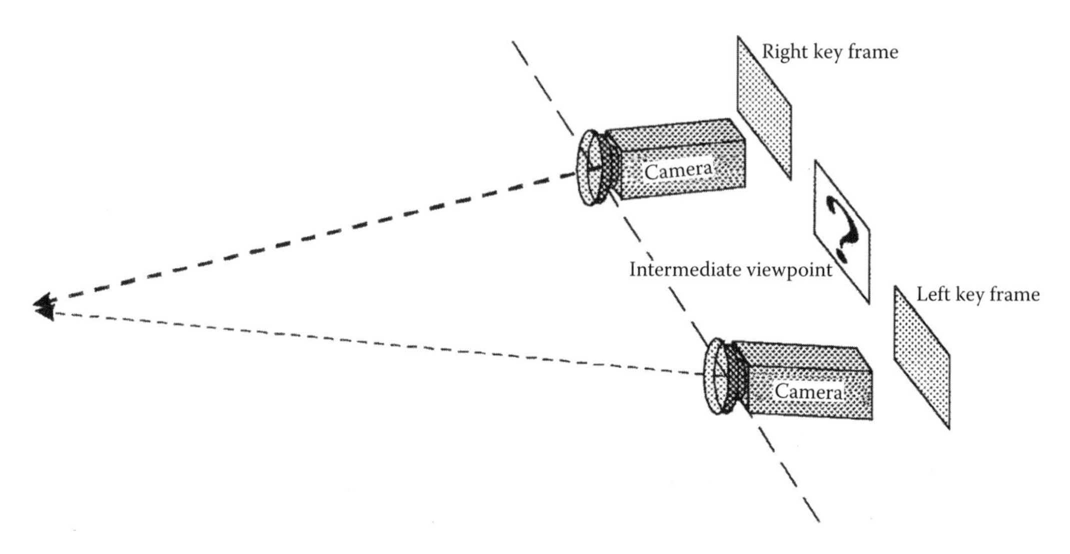

FIGURE 22.9 Multiview stereoscopic system with interpolated intermediate viewpoint (virtual camera).

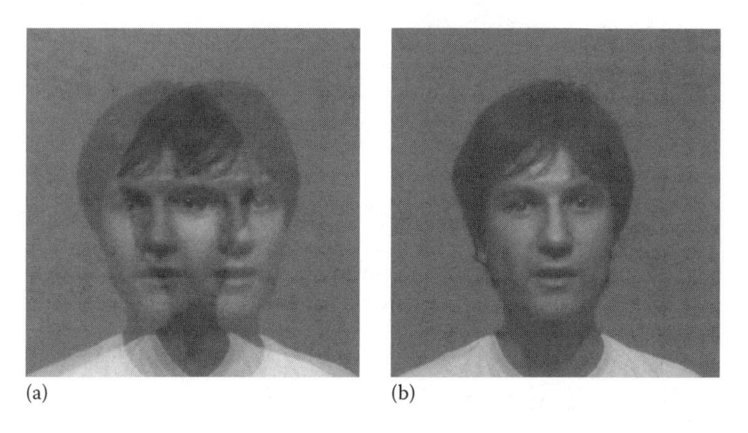

FIGURE 22.10 Interpolation of an intermediate viewpoint image of the stereoscopic pair in Figure 22.4: (a) without and (b) with taking into account the disparity information between the key frames.

The construction of intermediate viewpoints is an interpolation problem, which has much in common with the problem of video standards conversion [11]. In its most simple form, the interpolated viewpoint is merely a weighted average between the images from the nearest two camera viewpoints, which are called the key images. Such a straightforward averaging ignores the presence of disparity between the key images, yielding a highly blurred and essentially useless result (see Figure 22.10a). If, however, the disparity vector field between the two key images has been estimated and the areas of occlusions are known, the interpolation can be carried out along the disparity axis, such that the disparity information in the interpolated image corresponds exactly to the virtual camera position. For the points where a correspondence exists between the two key images, this construction process is called disparity-compensated interpolation, while for the occluding regions extrapolation has to be carried out from the key images [10]. Figure 22.10b illustrates the result of intermediate viewpoint interpolation on the stereoscopic image pair in Figure 22.4a and b.

References

1. Siegel, M., Grinberg, V., Jordan, A., McVeigh, J., Podnar, G., Safier, S., and Sriram, S., *Software for 3D TV and 3D-stereoscopic workstations, Proceedings of the International Workshop on Stereoscopic and Three-Dimensional Imaging,* S.N. Efstratiadis et al., Eds., pp. 251–260, Santorini, Greece, Sep. 1995.
2. Dhond, U.R. and Aggerwal, J.K., Structure from stereo, *IEEE Trans. Sys., Man Cybern.*, 19(6), 1489–1509, 1989.
3. Hang, H.-M and Woods, J.W., *Handbook of Visual Communications*, Academic Press, San Diego, CA, 1995.
4. Horn, B.K.P., *Robot Vision*, MIT Press, Cambridge, MA, 1986.
5. Jayant, N.S. and Noll, P., *Digital Coding of Waveforms*, Prentice-Hall, Englewood Cliffs, NJ, 1984.
6. Lipton, L., *The Crystal Eyes Handbook,* StereoGraphics Corporation, San Rafael, CA, 1991.
7. Marr, D., *Vision*, Freeman, San Francisco, CA, 1982.
8. Pastoor, S., 3-D television: A survey of recent research results on subjective requirements, *Signal Process.: Image Commun.*, 4(1), 21–32, 1991.
9. Perkins, M.G., Data compression of stereopairs, *IEEE Trans. Commun.*, 40(4), 684–696, 1992.
10. Skerjanc, R. and Liu, J., A three camera approach for calculating disparity and synthesizing intermediate pictures, *Signal Processing: Image Communications*, 4(1), 55–64, 1991.
11. Tekalp, A.M., *Digital Video Processing,* Prentice-Hall, Upper Saddle River, NJ, 1995.
12. Tziritas, G. and Labit, C., *Motion Analysis for Image Sequence Coding*, Elsevier, Amsterdam, the Netherlands, 1994.

23

A Survey of Image Processing Software and Image Databases

Stanley J. Reeves
Auburn University

Image processing has moved into the mainstream, not only of the engineering world, but of society in general. Personal computers are now capable of handling large graphics and images with ease, and fast networks and modems transfer images in a fraction of the time required just a few years ago. Image manipulation software is a common item on PCs, and CD-ROMs filled with images and multimedia databases are standard fare in the realm of electronic publishing. Furthermore, the development of areas such as data compression, neural networks and pattern recognition, computer vision, and multimedia systems have all contributed to the use of and interest in image processing. Likewise, the growth of image processing as an engineering discipline has fueled interest in these other areas. As a result of this symbiotic growth, image processing has increasingly become a standard tool in the repertoire of the engineer.

Because of the popularity of image processing, a large array of tools has emerged for accomplishing various image processing tasks. In addition, a variety of image databases has been created to address the needs of various specialty areas. In this chapter, we will survey some of the tools available for accomplishing basic image processing tasks and indicate where they may be obtained. Furthermore, we will describe and provide pointers to some of the most generally useful images and image databases. The goal is to identify a basic collection of images and software that will be of use to the nonspecialist. It should also be of use to the specialist who needs a general tool in an area outside his or her specialty.

23.1 Image Processing Software

Image processing has become such a broad area that it is sometimes difficult to distinguish what might be considered an image processing package from other software systems. The boundaries among the areas of computer graphics, data visualization, and image processing have become blurred. Furthermore, to discuss or even to list all the image processing software available would require many pages and would not be particularly useful to the nonspecialist. Therefore, we emphasize a representative set of image processing software packages that embody core capabilities in scientific image processing applications. Core capabilities, in our view, include the following:

- *Image utilities*: These include display, manipulation, and file conversion. Images come in such a variety of formats that a package for converting images from one format to another is essential. Furthermore, basic display and manipulation (cropping, rotating, etc.) are essential for almost any image processing task. The ability to edit images using cut-and-paste, draw, and annotate operations is also useful in many cases.
- *Image filtering and transformation*: These are necessary capabilities for most scientific applications of image processing. Convolution, median filtering, fast Fourier Transforms (FFTs), morphological operations, scaling, and other image functions form the core of many scientific image processing algorithms.
- *Image compression*: Anyone who works with images long enough will learn that they require a large amount of storage space. A number of standard image compression utilities are available for storing images in compressed form and for retrieving compressed images from image databases.
- *Image analysis*: Scientific image processing applications often have the goal of deriving information from an image. Simple image analysis tools such as edge detection and segmentation are powerful methods for gleaning important visual information.
- *Programming and data analysis environment*: While many image processing packages have a wide variety of functions, a whole new level of utility and flexibility arises when the image processing functions are built around a programming and/or data analysis environment. Programming environments allow for tailoring image processing techniques to the specific task, developing new algorithms, and interfacing image processing tasks with other scientific data analysis and numerical computational techniques.

Other capabilities include higher-level object recognition and other computer vision tasks, visualization and rendering techniques, computed imaging such as medical image reconstruction, and morphing and other special effects of the digital darkroom and the film industry. These areas require highly specialized software and/or very specialized skills to apply the methods and are not likely to be part of the image processing world of the nonspecialist.

The packages to be discussed here encompass as a group all of the core image processing capabilities mentioned above. Because these packages offer such a wide variety and mix of functions, they defy simple categorization. We have chosen to group the packages into three categories: general image utilities, specialized utilities, and programming/analysis environments. Keep in mind, however, that the distinctions among these groups is blurry at best. We have chosen to emphasize packages that are freely distributable and available on the Internet because these can be obtained and used with a minimum of expense and hassle.

23.1.1 General Image Utilities

23.1.1.1 netpbm

pbmplus is a set of tools that allows the user to convert to and from a large number of common image formats. The package has its own intermediate formats so that the conversion routines can be written to convert to or from one of these formats. The user can then convert to and from any combination of formats by going through one of the intermediate formats. Functions are also provided to convert from different color resolutions, such as from color to grayscale. Several other functions do basic image manipulation such as cropping, rotating, and smoothing. The source is available from ftp://ftp.wustl.edu/graphics/graphics/packages/NetPBM/.

23.1.1.2 xv

xv is an X11 utility that combines several important image handling functions. It can display images in a wide variety of display formats, including binary, 8 bit, and 24 bit. It allows the user to manipulate the colormap both in RGB and HSV space. It crops, resizes, smooths, rotates, detects edges, and produces other special effects. In addition, it reads and writes a large variety of image formats, so it can serve as a format conversion utility. Until recently, xv has been freely distributable. The latest version, however, is shareware and requires a small fee to become a registered user. The source is available from http://www.trilon.com/xv.

23.1.1.3 NCSA Image

NCSA Image is available in versions for the Mac, DOS, and Unix (X11). The Unix version is called ximage. ximage allows the user to display color images. It can also display the actual data in the form of a spreadsheet. A number of other display options are available. Like xv, it allows for manipulation of the colormap in a variety of ways. In addition, the user may display multiple images as an animated sequence, either from disk or server memory. The functionality of NCSA Image is augmented by other programs available from NCSA, including DataSlice for visualization tasks and Reformat for converting image formats. The source is available from NCSA by ftp at ftp://ftp.ncsa.uiuc.edu/Visualization/Image/.

23.1.1.4 ImageMagick

ImageMagick is an X11 package for display and interactive image manipulation. It reads and writes a large number of standard formats, does standard operations such as cropping and rotating as well as more specialized editing operations such as cutting, pasting, color filling, annotating, and drawing. Separate utilities are provided for grabbing images from a display, for converting, combining, resizing, blurring, adding borders, and doing many other operations. The source is available by ftp from ftp://ftp.x.org/contrib/applications/ImageMagick/.

23.1.1.5 NIH Image

NIH Image is available only in a Macintosh version. However, the popularity of NIH Image among Mac users and the breadth of features justify inclusion of the package in this survey. It reads/writes a small number of image formats, acquires images using compatible frame grabbers, and displays. It allows image manipulation such as flipping, rotating, and resizing; and editing such as drawing and annotating. It has a number of built-in enhancement and filtering functions: contrast enhancement, smoothing, sharpening, median filtering, and convolution. It supports a number of analysis operations such as edge detection and measurement of area, mean, centroid, and perimeter of user-defined regions of interest. It also performs automated particle analysis. In addition, the user can animate a set of images. NIH Image has a Pascal-like macro capability and the ability to add precompiled plug-in modules. The source is available from NIH by ftp at ftp://zippy.nimh.nih.gov/pub/nih-image/.

23.1.1.6 LaboImage

LaboImage is an X11 package for mouse- and menu-driven interactive image processing. It reads/writes a special format as well as Sun raster format and displays grayscale and RGB and provides dithering. Basic filtering operations are possible, as well as enhancement tasks such as background subtraction and histogram equalization. It computes various measures such as histograms, image statistics, and image power. Region outlining and object counting can be done as well. Images can be modified interactively at the pixel level, and an expert system is available for region segmentation. LaboImage has a macro capability for combining operations. LaboImage can be obtained from http://cuiwww.unige.ch/ftp/sgaico/research/geneve/vision/labo.html.

23.1.1.7 Paint Shop Pro

Paint Shop Pro is a Windows-based package for creating, displaying, and manipulating images. It has a large number of image editing features, including painting, photo retouching, and color enhancement. It reads and writes a large number of formats. It includes several standard image processing filters and geometrical transformations. It can be obtained from http://www.jasc.com/psp.html. It is shareware and costs $69.

23.1.2 Specialized Image Utilities

23.1.2.1 Compression

JPEG is a standard for image compression developed by the Joint Photographic Experts Group. Free, portable C code that implements JPEG compression and decompression has been developed by the Independent JPEG Group, a volunteer organization. It is available from ftp://ftp.uu.net/graphics/jpeg. The downloadable package contains source and documentation. The code converts between JPEG and several other common image formats. A lossless JPEG implementation can be obtained from ftp://ftp.cs.cornell.edu/pub/multimed/.

A fractal image compression program is available from ftp://inls.ucsd.edu/pub/young-fractal/. The package contains source for both compression and decompression. A number of other fractal compression programs are also available and can be found in the sci.fractal FAQ at ftp://rtfm.mit.edu/pub/usenet/news.answers/sci/fractals-faq.

JBIG is a standard for binary image compression developed by the Joint Binary Images Group. A JBIG coder/decoder can be obtained from ftp://nic.funet.fi/pub/graphics/misc/test-images/.

MPEG is a standard for video/audio compression developed by the Moving Pictures Experts Group. A set of MPEG tools is available from ftp://mm-ftp.cs.berkeley.edu/pub/multimedia/mpeg/. These tools allow for encoding, decoding (playing), and analyzing the MPEG data.

H.261 and H.263 are standards for video compression for videophone applications. An H.261 coder/decoder is available from ftp://havefun.stanford.edu/pub/p64/. An H.263 video coder/decoder is available from http://www.fou.telenor.no/brukere/DVC/h263_software/.

23.1.2.2 Computer Vision

Vista is an X11-based image processing environment specifically designed for computer vision applications. It allows a variety of display and manipulation options. It has a library that lets the user easily create applications with menus, mouse interaction, and display options. Vista defines a very flexible data format that represents a variety of images, collections of images, or other objects. It also has the ability to add new objects or new image attributes without changing existing software or data files. It does edge detection and linking, optical flow estimation and camera calibration, and viewing of images and edge vectors. Vista includes routines for common image processing operations such as convolution, FFTs, simple enhancement tasks, scaling, cropping, and rotating. Vista is available from http://www.cs.ubc.ca/nest/lci/vista/vista.html.

23.1.3 Programming/Analysis Environments

23.1.3.1 Khoros

Khoros is a comprehensive software development and data analysis environment. It allows the user to perform a large variety of image and signal processing and visualization tasks. A graphical programming environment called Cantata allows the user to construct programs visually using a data flowgraph approach. It has a user interface design tool with automatic code generation for writing customized applications. Software objects (programs) are accessible from the command line, from within Cantata, and in libraries.

A large set of standard numerical and statistical algorithms are available within Khoros. Common image processing operations such as FFTs, convolution, median filtering, and morphological operators are available. In addition, a variety of image display and geometrical manipulation programs, animation, and colormap editing are included. Khoros has a very general data model that allows for images of up to five dimensions.

Khoros is free-access software—it is available for downloading free of charge but cannot be distributed without a license. It can be obtained from Khoral Research, Inc., at ftp://ftp.khoral.com/pub/. Note that the Khoros distribution is quite large and requires significant disk space.

23.1.3.2 MATLAB®

MATLAB is a general numerical analysis and visualization environment. Matrices are the underlying data structure in MATLAB, and this structure lends itself well to image processing applications. All data in MATLAB is represented as double-precision, which makes the calculations more precise and interaction more convenient. However, it may also mean that MATLAB uses more memory and processing time than necessary.

A large number of numerical algorithms and visualization options are available with the standard package. The Image Processing Toolbox provides a great deal of added functionality for image processing applications. It reads/writes several of the most common image formats; does convolution, FFTs, median filtering, histogram equalization, morphological operations, two-dimensional filter design, general nonlinear filtering, colormap manipulation, and basic geometrical manipulation. It also allows for a variety of display options, including surface warping and movies.

MATLAB is an interactive environment, which makes interactive image processing and manipulation convenient. One can also add functionality by creating scripts or functions that use MATLAB's functions and other user-added functions. Additionally, one can add functions that have been written in C or Fortran. Conversely, C or Fortran programs can call MATLAB and MATLAB library functions. MATLAB is commercial software. More information on MATLAB and how to obtain it can be found through the homepage of The Mathworks, Inc., at http://www.mathworks.com/.

23.1.3.3 PV-Wave

PV-Wave is a general graphical/visualization and numerical analysis environment. It can handle images of arbitrary dimensionality—2-D, 3-D, and so on. The user can specify the data type of each data structure, which allows for flexibility but may be inconvenient for interactive work.

PV-Wave contains a large collection of visualization and rendering options, including colormap manipulation, volume rendering, and animation. In addition, the IMSL library is available through PV-Wave. Basic image processing operations such as convolution, FFTs, median filtering, morphological operations, and contrast enhancement are included.

Like MATLAB, PV-Wave is an interactive environment. One can create scripts or functions from the PV-Wave language to add functionality. It can also call C or Fortran functions. PV-Wave can be invoked from within C or Fortran too. PV-Wave is commercial software. More information on PV-Wave can be found through the homepage of Visual Numerics, Inc., at http://www.vni.com/.

23.2 Image Databases

A huge number of image databases and archives are available on the Internet now, and more are continually being added. These databases serve various purposes. For the practicing engineer, the primary value of an image database is for developing, testing, evaluating, or comparing image processing and manipulation algorithms. Standard images provide a benchmark for comparing various algorithms. Furthermore, standard test images can be selected so that their characteristics are particularly suited to demonstrating the strengths and weaknesses of particular types of image processing techniques. In some areas of image processing no real standards exist, although de facto standards have arisen. In the discussion that follows, we provide pointers to some standard images, some de facto standards, and a few other databases that might provide images of value to algorithm work in image processing. We have deliberately steered away from images whose copyright is known to prohibit use for research purposes. However, some of the images in the list have certain copyright restrictions. Be sure to check any auxiliary information provided with the images before assuming that they are public domain.

The images listed are in a variety of formats and may require conversion using one of the packages discussed previously such as netpbm. We list the databases according to two categories: (1) form and (2) content. By form, we mean that the images are organized according to the form of the image—color, stereo, sequence, etc. By content, we mean that the images are grouped according to the image content—faces, fingerprints, etc.

23.2.1 Images by Form

23.2.1.1 Binary Images

A set of standard CCITT fax test images has been made available for testing compression schemes. These are binary images that have come from scanning actual documents. They can be found at ftp://nic.funet.fi/pub/graphics/misc/test-images/ under ccitt[1-8].pbm.gz.

23.2.1.2 Grayscale Images

A collection of grayscale images can be obtained from ftp://ipl.rpi.edu/pub/image/still/canon/gray/. A compilation of de facto standard images can be found at http://www.sys.uea.ac.uk/Research/ResGroups/SIP/images_ftp/index.html.

Note that the Lena image is copyrighted and should not be used in publications.

23.2.1.3 Color Images

A set of test images that were used by the JPEG committee in the development of the JPEG algorithm are available from ftp://ipl.rpi.edu/pub/image/still/jpeg/bgr/.

These are 24 bit RGB images. Other 24 bit color images can be found at ftp://ipl.rpi.edu/pub/image/still/canon/bgr/.

A set of miscellaneous images in JPEG and Kodak CD format can be found at http://www.kodak.com/digitalImages/samples/samples.shtml.

23.2.1.4 Image Sequences

Image sequences may be intended for study of computer vision applications or video coding. A huge set of sequences for computer vision applications are archived at http://www.ius.cs.cmu.edu/idb/.

A set of sequences commonly used for video coding applications can be found at ftp://ipl.rpi.edu/pub/image/sequence/.

23.2.1.5 Stereo Image Pairs

Stereo image pairs are available from http://www.ius.cs.cmu.edu/idb/.

23.2.1.6 Texture Images

A large set of texture images can be found at http://www-white.media.mit.edu/vismod/imagery/VisionTexture/vistex.html.

These images include textures from various angles and under different lighting conditions.

23.2.1.7 Face Images

The USENIX FACES database contains hundreds of face images in various formats. The database is archived at ftp://ftp.uu.net/published/usenix/faces/.

23.2.1.8 Fingerprint Images

Fingerprint images can be obtained from ftp://sequoyah.ncsl.nist.gov/pub/databases/data/.

23.2.1.9 Medical Images

A variety of medical images are available over the Internet. An excellent collection of CT, MRI, and cryosection images of the human body has been made available by the National Library of Medicine's

The Visual Human Project. Samples of these images can be acquired at http://www.nlm.nih.gov/research/visible/visible_human.html.

A collection of over 3500 images that cover an entire human body is available via ftp and on tape by signing a license agreement.

MRI and CT volume images are available from ftp://omicron.cs.unc.edu/pub/projects/softlab.v/CHVRTD/.

PET images and other modalities can be found in gopher://gopher.austin.unimelb.edu.au/11/images/petimages.

23.2.1.10 Astronomical Images

A collection of astronomical images can be found at https//www.univ-rennesl.fr/ASTRO/astro.english.html.

Hubble telescope imagery can be obtained from http://archive.stsci.edu/archive.html.

23.2.1.11 Range Images

Range images are available from http://www.eecs.wsu.edu/?rl/3DDB/RID/, along with a list of other sources of range imagery, and also from http://marathon.csee.usf.edu/range/DataBase.html.

The tools and databases discussed here should provide a convenient set of capabilities for the nonspecialist. The capabilities that are readily available are not static, however. Image processing will continue to become more and more mainstream, so we expect to see the development of image processing tools representing greater variety and sophistication. The advent of the World Wide Web will also stimulate further development and publishing of image databases on the Internet. Therefore, image processing capabilities will continue to grow and will be more readily available. The items discussed here are only a small sample of what will be available as time goes on.

24

VLSI Architectures for Image Communications

P. Pirsch
University of Hannover

W. Gehrke
Philips Semiconductors

24.1 Introduction

Video processing has been a rapidly evolving field for telecommunications, computer, and media industries. In particular, for real-time video compression applications a growing economical significance is expected for the next years. Besides digital TV broadcasting and videophone, services such as multimedia education, teleshopping, or video mail will become audiovisual mass applications.

To facilitate worldwide interchange of digitally encoded audiovisual data, there is a demand for international standards, defining coding methods, and transmission formats. International standardization committees have been working on the specification of several compression schemes. The Joint Photographic Experts Group (JPEG) of the International Standards Organization (ISO) has specified an algorithm for compression of still Images [4]. The ITU proposed the H.261 standard for video telephony and videoconference [1]. The Motion Pictures Experts Group (MPEG) of ISO has completed its first standard MPEG-1, which will be used for interactive video and provides a picture quality comparable to VCR quality [2]. MPEG made substantial progress for the second phase of standards MPEG-2, which will provide audiovisual quality of both broadcast TV and HDTV [3]. Besides the availability of international standards, the successful introduction of the named services depends on the availability of VLSI components, supporting a cost-efficient implementation of video compression applications. In the following, we give a short overview of recent coding schemes and discuss implementation alternatives. Furthermore, the efficiency estimation of architectural alternatives is discussed and implementation examples of dedicated and programmable architectures are presented.

24.2 Recent Coding Schemes

Recent video coding standards are based on a hybrid coding scheme that combines transform coding and predictive coding techniques. An overview of these hybrid encoding schemes is depicted in Figure 24.1.

The encoding scheme consists of the tasks motion estimation, typically based on block matching algorithms, computation of the prediction error, discrete cosine transform (DCT), quantization (Q), variable length coding (VLC), inverse quantization (Q^{-1}), and inverse discrete cosine transform (IDCT or DCT^{-1}). The reconstructed image data are stored in an image memory for further predictions. The decoder performs the tasks variable length decoding (VLC^{-1}), Q^{-1}, and motion-compensated reconstruction.

Generally, video processing algorithms can be classified in terms of regularity of computation and data access. This classification leads to three classes of algorithms:

- *Low-level algorithms*—These algorithms are based on a predefined sequence of operations and a predefined amount of data at the input and output. The processing sequence of low-level algorithms is predefined and does not depend on the values of data processed. Typical examples of low-level algorithms are block matching or transforms such as the DCT.
- *Medium-level algorithms*—The sequence and number of operations of medium-level algorithms depend on the data. Typically, the amount of input data is predefined, whereas the amount of output data varies according to the input data values. With respect to hybrid coding schemes, examples for these algorithms are Q, Q^{-1}, or VLC.
- *High-level algorithms*—High-level algorithms are associated with a variable amount of input and output data and a data-dependent sequence of operations. As for medium-level algorithms, the sequence of operations is highly data dependent. Control tasks of the hybrid coding scheme can be assigned to this class.

Since hybrid coding schemes are applied for different video source rates, the required absolute processing power varies in the range from a few hundred mega operations per second (MOPS) for video signals

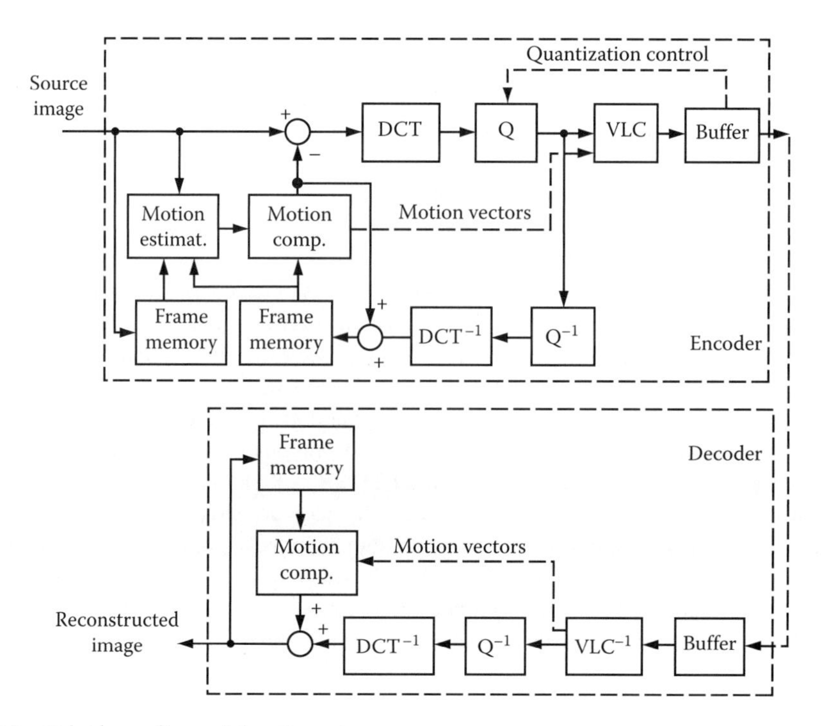

FIGURE 24.1 Hybrid encoding and decoding scheme.

in QCIF format to several giga operations per second (GOPS) for processing of TV or HDTV signals. Nevertheless, the relative computational power of each algorithmic class is nearly independent of the processed video format. In case of hybrid coding applications, approximately 90% of the overall processing power is required for low-level algorithms. The amount of medium-level tasks is about 7% and nearly 3% is required for high-level algorithms.

24.3 Architectural Alternatives

In terms of a VLSI implementation of hybrid coding applications, two major requirements can be identified. First, the high computational power requirements have to be provided by the hardware. Second, low manufacturing cost of video processing components is essential for the economic success of an architecture. Additionally, implementation size and architectural flexibility have to be taken into account.

Implementations of video processing applications can either be based on standard processors from workstations or PCs or on specialized video signal processors. The major advantage of standard processors is their availability. Application of these architectures for implementation of video processing hardware does not require the time consuming design of new VLSI components. The disadvantage of this implementation strategy is the insufficient processing power of recent standard processors. Video processing applications would still require the implementation of cost-intensive multiprocessor systems to meet the computational requirements. To achieve compact implementations, video processing hardware has to be based on video signal processors, adapted to the requirements of the envisaged application field.

Basically, two architectural approaches for the implementations of specialized video processing components can be distinguished. "Dedicated architectures" aim at an efficient implementation of one specific algorithm or application. Due to the restriction of the application field, the architecture of dedicated components can be optimized by an intensive adaptation of the architecture to the requirements of the envisaged application, e.g., arithmetic operations that have to be supported, processing power, or communication bandwidth. Thus, this strategy will generally lead to compact implementations. The major disadvantage of dedicated architecture is the associated low flexibility. Dedicated components can only be applied for one or a few applications. In contrast to dedicated approaches with limited functionality, "programmable architectures" enable the processing of different algorithms under software control. The particular advantage of programmable architectures is the increased flexibility. Changes of architectural requirements, e.g., due to changes of algorithms or an extension of the aimed application field, can be handled by software changes. Thus, a generally cost-intensive redesign of the hardware can be avoided. Moreover, since programmable architectures cover a wider range of applications, they can be used for low-volume applications, where the design of function specific VLSI chips is not an economical solution.

For both architectural approaches, the computational requirements of video processing applications demand for the exploitation of the algorithm-inherent independence of basic arithmetic operations to be performed. Independent operations can be processed concurrently, which enables the decrease of processing time and thus an increased throughput rate. For the architectural implementation of concurrency, two basic strategies can be distinguished: pipelining and parallel processing.

In case of pipelining several tasks, operations or parts of operations are processed in subsequent steps in different hardware modules. Depending on the selected granularity level for the implementation of pipelining, intermediate data of each step are stored in registers, register chains, FIFOs, or dual-port memories. Assuming a processing time of T_P for a nonpipelined processor module and $T_{D,IM}$ for the delay of intermediate memories, we get in the ideal case the following estimation for the throughput rate $R_{T,Pipe}$ of a pipelined architecture applying N_{Pipe} pipeline stages:

$$R_{T,Pipe} = \frac{1}{\frac{T_P}{N_{Pipe}} + T_{D,IM}} = \frac{N_{Pipe}}{T_P + N_{Pipe} \cdot T_{D,IM}} \tag{24.1}$$

From this follows that the major limiting factor for the maximum applicable degree of pipelining is the access delay of these intermediate memories.

The alternative to pipelining is the implementation of parallel units, processing independent data concurrently. Parallel processing can be applied on operation level as well as on task level. Assuming the ideal case, this strategy leads to a linear increase of processing power and we get

$$R_{T,Par} = \frac{N_{Par}}{T_P} \tag{24.2}$$

where N_{Par} is the number of parallel units.

Generally, both alternatives are applied for the implementation of high-performance video processing components. In the following sections, the exploitation of algorithmic properties and the application of architectural concurrency are discussed considering the hybrid coding schemes.

24.4 Efficiency Estimation of Alternative VLSI Implementations

Basically, architectural efficiency can be defined by the ratio of performance over cost. To achieve a figure of merit for architectural efficiency we assume in the following that performance of a VLSI architecture can be expressed by the achieved throughput rate R_T and the cost is equivalent to the required silicon area A_{Si} for the implementation of the architecture:

$$E = \frac{R_T}{A_{Si}} \tag{24.3}$$

Besides the architecture, efficiency mainly depends on the applied semiconductor technology and the design-style (semi-custom, full-custom). Therefore, a realistic efficiency estimation has to consider the gains provided by the progress in semiconductor technology. A sensible way is the normalization of the architectural parameters according to a reference technology. In the following we assume a reference process with a grid length $\lambda_0 = 1.0$ μm. For normalization of silicon area, the following equation can be applied:

$$A_{Si,0} = A_{Si}\left(\frac{\lambda_0}{\lambda}\right)^2 \tag{24.4}$$

where the index 0 is used for the system with reference gate length λ_0.

According to [7] the normalization of throughput can be performed by

$$R_{T,0} = R_T\left(\frac{\lambda}{\lambda_0}\right)^{1.6} \tag{24.5}$$

From Equations 24.3 through 24.5, the normalization for the architectural efficiency can be derived:

$$E_0 = \frac{R_{T,0}}{A_{Si,0}} = \frac{R_T}{A_{Si}}\left(\frac{\lambda}{\lambda_0}\right)^{3.6} \tag{24.6}$$

E can be used for the selection of the best architectural approach out of several alternatives. Moreover, assuming a constant efficiency for a specific architectural approach leads to a linear relationship of throughput rate and silicon area and this relationship can be applied for the estimation of the required silicon area for a specific application. Due to the power of 3.6 in Equation 24.6, the chosen semiconductor technology for implementation of a specific application has a significant impact on the architectural efficiency.

In the following, examples of dedicated and programmable architectures for video processing applications are presented. Additionally, the discussed efficiency measure is applied to achieve a figure of merit for silicon area estimation.

24.5 Dedicated Architectures

Due to their algorithmic regularity and the high processing power required for the DCT and motion estimation, these algorithms are the first candidates for a dedicated implementation. As typical examples, alternatives for a dedicated implementation of these algorithms are discussed in the following.

The DCT is a real-valued frequency transform similar to the discrete Fourier transform (DFT). When applied to an image block of size $L \times L$, the two-dimensional DCT (2D-DCT) can be expressed as follows:

$$Y_{k,l} = \sum_{i=0}^{L-1} \sum_{j=0}^{L-1} x_{i,j} \cdot C_{i,k} \cdot C_{j,l} \tag{24.7}$$

where

$$C_{n,m} = \begin{cases} \dfrac{1}{\sqrt{2}} & \text{for } m = 0 \\ \cos\left[\dfrac{(2n+1)m\pi}{2L}\right] & \text{otherwise} \end{cases}$$

with

(i, j) as coordinates of the pixels in the initial block
(k, l) as coordinates of the coefficients in the transformed block
$x_{i,j}$ as the value of the pixel in the initial block

Computing a 2D-DCT of size $L \times L$ directly according to Equation 24.7 requires L^4 multiplications and L^4 additions.

The required processing power for the implementation of the DCT can be reduced by the exploitation of the arithmetic properties of the algorithm. The 2D-DCT can be separated into two one-dimensional DCTs (1D-DCTs) according to Equation 24.8.

$$Y_{k,l} = \sum_{i=0}^{L-1} C_{i,k} \cdot \left[\sum_{j=0}^{L-1} x_{i,j} \cdot C_{j,l}\right] \tag{24.8}$$

The implementation of the separated DCT requires $2L^3$ multiplications and $2L^3$ additions. As an example, the DCT implementation according to Ref. [9] is depicted in Figure 24.2. This architecture is based on two 1D-DCT processing arrays. Since this architecture is based on a pipelined multiplier/accumulator implementation in carry–save technique, vector merging adders are located at the output of each array. The results of the 1D-DCT have to be reordered for the second 1D-DCT stage. For this purpose, a transposition memory is used. Since both one-dimensional processor arrays require identical DCT coefficients, these coefficients are stored in a common ROM.

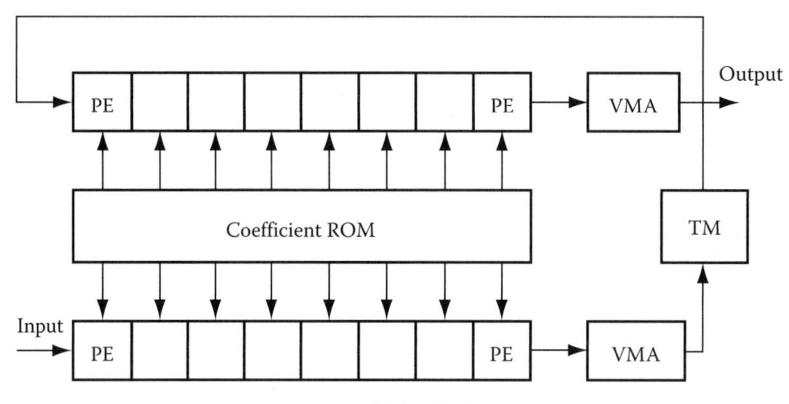

FIGURE 24.2 Separated DCT implementation. (Totzek, U. et al., CMOS VLSI implementation of the 2D-DCT with linear processor arrays, *Proceedings of the International Conference on Acoustics Speech and Signal Processing*, Albuquerque, NM, vol. 2, pp. 937–940, Apr. 1990.)

Moving from a mathematical definition to an algorithm that can minimize the number of calculations required is a problem of particular interest in the case of transforms such as the DCT. The 1D-DCT can also be expressed by the matrix–vector product:

$$[\mathbf{Y}] = [\mathbf{C}][\mathbf{X}] \tag{24.9}$$

where

[**C**] is and $L \times L$ matrix
[**X**] and [**Y**] 8-point input and output vectors

As an example, with $\theta = p/16$, the 8-points DCT matrix can be computed as denoted in Equation 24.10.

$$\begin{bmatrix} Y_0 \\ Y_1 \\ Y_2 \\ Y_3 \\ Y_4 \\ Y_5 \\ Y_6 \\ Y_7 \end{bmatrix} = \begin{bmatrix} \cos 4\theta & \cos 4\theta & \cos 4\theta & \cos 4\theta & \cos 4\theta & \cos 4\theta & \cos 4\theta & \cos 4\theta \\ \cos\theta & \cos 3\theta & \cos 5\theta & \cos 7\theta & -\cos 7\theta & -\cos 5\theta & -\cos 3\theta & -\cos\theta \\ \cos 2\theta & \cos 6\theta & -\cos 6\theta & -\cos 2\theta & -\cos 2\theta & -\cos 6\theta & \cos 6\theta & \cos 2\theta \\ \cos 3\theta & -\cos 7\theta & -\cos\theta & -\cos 5\theta & \cos 5\theta & \cos\theta & \cos 7\theta & -\cos 3\theta \\ \cos 4\theta & -\cos 4\theta & -\cos 4\theta & \cos 4\theta & \cos 4\theta & -\cos 4\theta & -\cos 4\theta & \cos 4\theta \\ \cos 5\theta & -\cos\theta & \cos 7\theta & \cos 3\theta & -\cos 3\theta & -\cos 7\theta & \cos\theta & -\cos 5\theta \\ \cos 6\theta & -\cos 2\theta & \cos 2\theta & -\cos 6\theta & -\cos 6\theta & \cos 2\theta & -\cos 2\theta & \cos 6\theta \\ \cos 7\theta & -\cos 5\theta & \cos 3\theta & -\cos\theta & \cos\theta & -\cos 3\theta & \cos 5\theta & -\cos 7\theta \end{bmatrix} \begin{bmatrix} x_0 \\ x_1 \\ x_2 \\ x_3 \\ x_4 \\ x_5 \\ x_6 \\ x_7 \end{bmatrix} \tag{24.10}$$

$$\begin{bmatrix} Y_0 \\ Y_2 \\ Y_4 \\ Y_6 \end{bmatrix} = \begin{bmatrix} \cos 4\theta & \cos 4\theta & \cos 4\theta & \cos 4\theta \\ \cos 2\theta & \cos 6\theta & -\cos 6\theta & -\cos 2\theta \\ \cos 4\theta & -\cos 4\theta & -\cos 4\theta & \cos 4\theta \\ \cos 6\theta & -\cos 2\theta & \cos 2\theta & -\cos 6\theta \end{bmatrix} \begin{bmatrix} x_0 + x_7 \\ x_1 + x_6 \\ x_2 + x_5 \\ x_3 + x_4 \end{bmatrix} \tag{24.11}$$

$$\begin{bmatrix} Y_1 \\ Y_3 \\ Y_5 \\ Y_7 \end{bmatrix} = \begin{bmatrix} \cos\theta & \cos 3\theta & \cos 5\theta & \cos 7\theta \\ \cos 3\theta & -\cos 7\theta & -\cos\theta & -\cos 5\theta \\ \cos 5\theta & -\cos\theta & \cos 7\theta & \cos 3\theta \\ \cos 7\theta & -\cos 5\theta & \cos 3\theta & -\cos\theta \end{bmatrix} \begin{bmatrix} x_0 + x_7 \\ x_1 + x_6 \\ x_2 + x_5 \\ x_3 + x_4 \end{bmatrix} \tag{24.12}$$

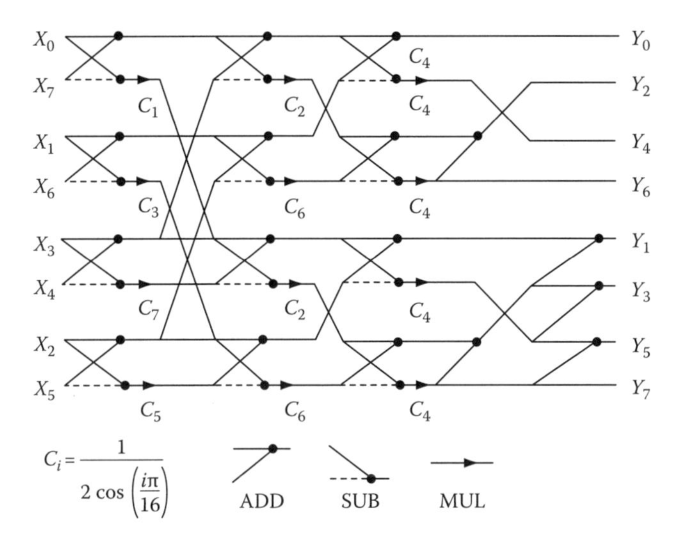

FIGURE 24.3 Lee FDCT flowgraph for the one-dimensional 8-points DCT. (Lee, B. G., *IEEE Trans. Acoust. Speech Signal Process.*, 32, 1243, Dec. 1984.)

More generally, the matrices in Equations 24.11 and 24.12 can be decomposed in a number of simpler matrices, the composition of which can be expressed as a flowgraph. Many fast algorithms have been proposed. Figure 24.3 illustrates the flowgraph of the Lee's algorithms, which is commonly used [10]. Several implementations using fast flowgraphs have been reported [11,12].

Another approach that has been extensively used is based on the technique of distributed arithmetic. Distributed arithmetic is an efficient way to compute the DCT totally or partially as scalar products. To illustrate the approach, let us compute a scalar product between two length-M vectors C and X:

$$Y = \sum_{i=0}^{M-1} c_i \cdot x_i \quad \text{with} \quad x_i = -x_{i,0} + \sum_{j=1}^{B-1} x_{i,j} \cdot 2^{-j} \tag{24.13}$$

where

$\{c_i\}$ are N-bit constants
$\{x_i\}$ are coded in B bits in 2s complement

Then Equation 24.13 can be rewritten as

$$Y = \sum_{j=0}^{B-1} C_j \cdot 2^{-j} \quad \text{with} \quad C_{j\neq 0} = \sum_{i=0}^{M-1} c_i x_{i,j} \quad \text{and} \quad C_0 = -\sum_{i=0}^{M-1} c_i x_{i,0} \tag{24.14}$$

The change of summing order in i and j characterizes the distributed arithmetic scheme in which the initial multiplications are distributed to another computation pattern. Since the term C_j has only 2^M possible values (which depend on the $x_{i,j}$ values), it is possible to store these 2^M possible values in a ROM. An input set of M bits $\{x_{0,j}, x_{1,j}, x_{2,j}, \ldots, x_{M-1,j}\}$ is used as an address, allowing retrieval of the C_j value. These intermediate results are accumulated in B clock cycles, for producing one Y value. Figure 24.4 shows a typical architecture for the computation of a M input inner product. The inverter and the MUX are used for inverting the final output of the ROM in order to compute C_0.

Figure 24.5 illustrates two typical uses of distributed arithmetic for computing a DCT. Figure 24.5a implements the scalar products described by the matrix of Equation 24.10. Figure 24.5b takes advantage

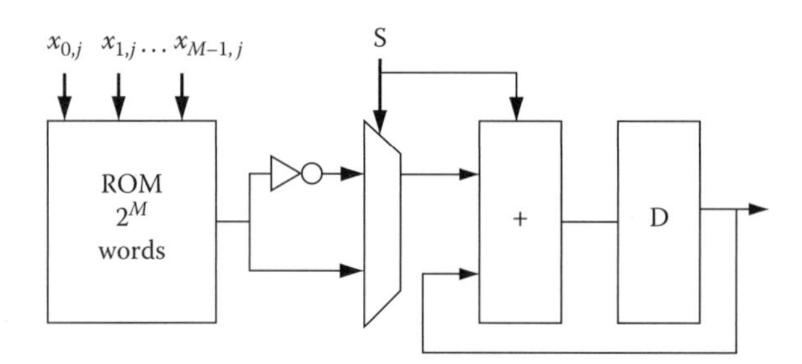

FIGURE 24.4 Architecture of a M input inner product using distributed arithmetic.

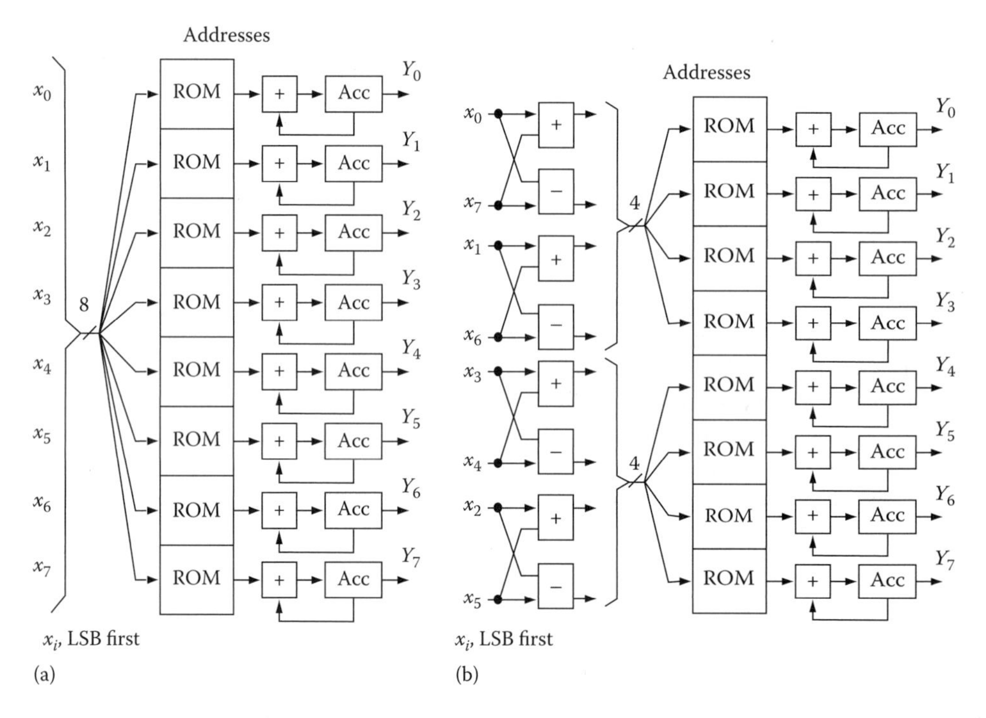

FIGURE 24.5 Architecture of an 8-point one-dimensional DCT using distributed arithmetic. (a) Pure distributed arithmetic. (b) Mixed D.A.: first stage of flowgraph decomposition products of 8 points followed by 2 times 4 scalar products of 4 points.

of a first stage of additions and subtractions and the scalar products described by the matrices of Equations 24.11 and 24.12.

Properties of several dedicated DCT implementations have been reported in [6]. Figure 24.6 shows the silicon area as a function of the throughput rate for selected design examples. The design parameters are normalized to a fictive 1.0 μm CMOS process according to the discussed normalization strategy. As a figure of merit, a linear relationship of throughput rate and required silicon area can be derived:

$$\alpha_{T,0} \approx 0.5\,\text{mm}^2/\text{Mpel/s} \tag{24.15}$$

Equation 24.15 can be applied for the silicon area estimation of DCT circuits. For example, assuming TV signals according to the CCIR-601 format and a frame rate of 25 Hz, the source rate equals 20.7 Mpel/s. As a figure of merit from Equation 24.15 a normalized silicon area of about 10.4 mm^2 can be derived. For HDTV signals the video source rate equals 110.6 Mpel/s and approximately 55.3 mm^2 silicon area is

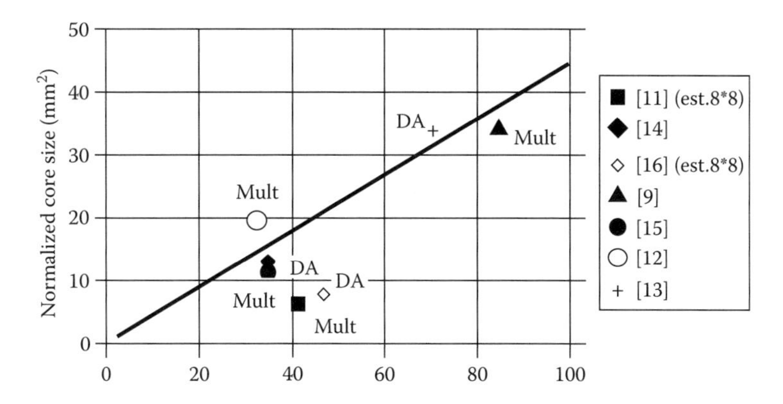

FIGURE 24.6 Normalized silicon area and throughput for dedicated DCT circuits.

required for the implementation of the DCT. Assuming an economically sensible maximum chip size of about 100 mm^2 to 150 mm^2, we can conclude that the implementation of the DCT does not necessarily require the realization of a dedicated DCT chip and the DCT core can be combined with several other on-chip modules that perform additional tasks of the video coding scheme.

For motion estimation several techniques have been proposed in the past. Today, the most important technique for motion estimation is block matching, introduced by [21]. Block matching is based on the matching of blocks between the current and a reference image. This can be done by a full (or exhaustive) search within a search window, but several other approaches have been reported in order to reduce the computation requirements by using an "intelligent" or "directed" search [17–19,23,25–27].

In case of an exhaustive search block matching algorithm, a block of size $N \times N$ pels of the current image (reference block, denoted X) is matched with all the blocks located within a search window (candidate blocks, denoted Y) The maximum displacement will be denoted by w. The matching criterium generally consists in computing the mean absolute difference (MAD) between the blocks. Let $x(i, j)$ be the pixels of the reference block and $y(i, j)$ the pixels of the candidate block. The matching distance (or distortion) D is computed according to Equation 24.16. The indexes m and n indicate the position of the candidate block within the search window. The distortion D is computed for all the $(2w+1)^2$ possible positions of the candidate block within the search window (Equation 24.16) and the block corresponding to the minimum distortion is used for prediction. The position of this block within the search window is represented by the motion vector $\mathbf{v}$ (Equation 24.17).

$$D(m, n) = \sum_{i=0}^{N-1} \sum_{j=0}^{N-1} |x(i, j) - y(i+m, j+n)| \tag{24.16}$$

$$\mathbf{v} = \left[\begin{matrix} m(56) \\ n \end{matrix} \right] \Bigg|_{D_{\min}} \tag{24.17}$$

The operations involved for computing $D(m, n)$ and $D_{\min}$ are associative. Thus, the order for exploring the index spaces (i, j) and (m, n) are arbitrary and the block matching algorithm can be described by several different dependence graphs (DGs). As an example, Figure 24.7 shows a possible DG for $w=1$ and $N=4$. In this figure, AD denotes an absolute difference and an addition, M denotes a minimum value computation.

The DG for computing $D(m, n)$ is directly mapped into a 2-D array of processing elements (PE), while the DG for computing $\mathbf{v}(X, Y)$ is mapped into time Figure 24.8. In other words, block matching is performed by a sequential exploration of the search area, while the computation of each distortion is performed in parallel. Each of the AD nodes of the DG is implemented by an AD processing element

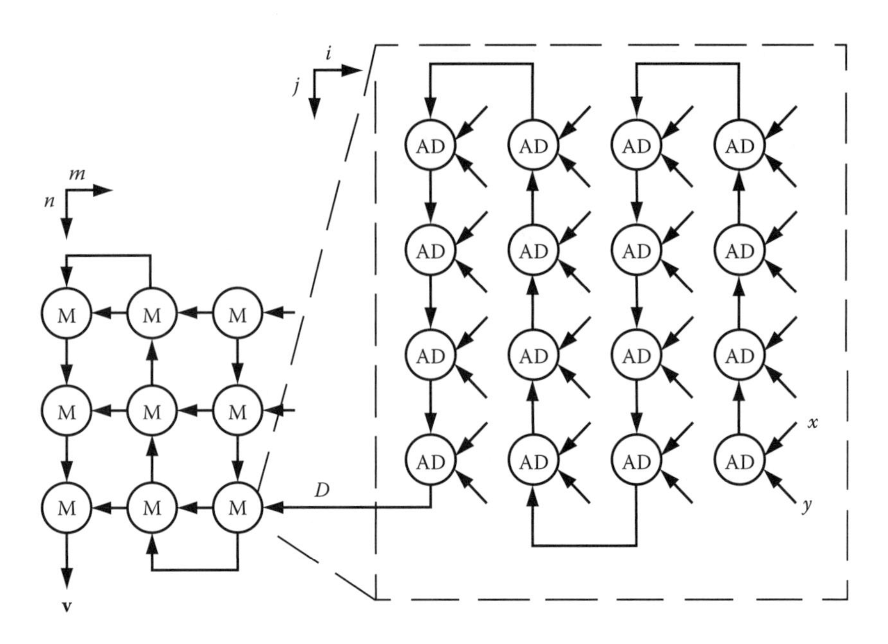

FIGURE 24.7 Dependence graphs of the block matching algorithm. The computation of $v(X, Y)$ and $D(m, n)$ are performed by 2-D linear DGs.

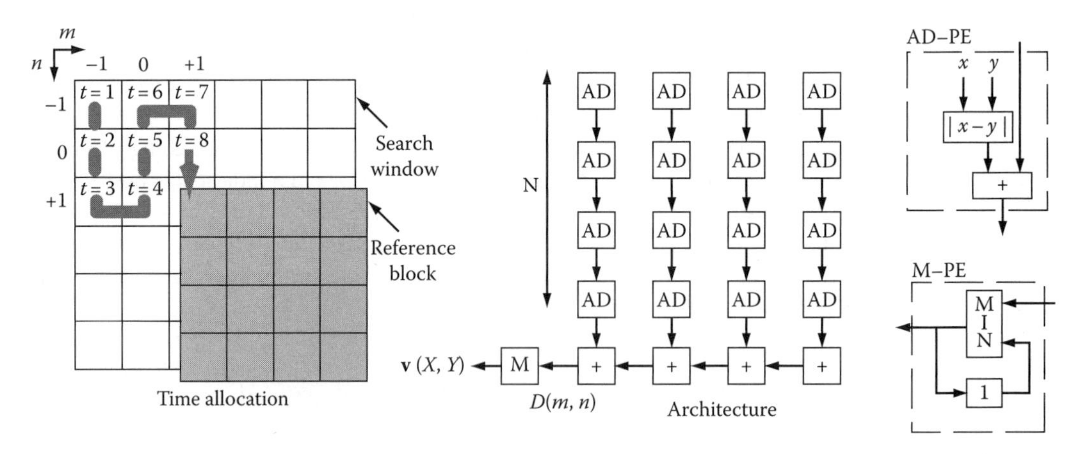

FIGURE 24.8 Principle of the 2-D block-based architecture.

(AD-PE). The AD-PE stores the value of $x(i, j)$ and receives the value of $y(m + i, n + j)$ corresponding to the current position of the reference block in the search window. It performs the subtraction and the absolute value computation, and adds the result to the partial result coming from the upper PE. The partial results are added on columns and a linear array of adders performs the horizontal summation of the row sums, and computes $D(m, n)$. For each position (n, m) of the reference block, the M-PE checks if the distortion $D(m, n)$ is smaller than the previous smaller distortion value, and, in this case, updates the register which keeps the previous smaller distortion value.

To transform this naive architecture into a realistic implementation, two problems must be solved: (1) a reduction of the cycle time and (2) the I/O management.

1. The architecture of Figure 24.8 implicitly supposes that the computation of $D(m, n)$ can be done combinatorially in one cycle time. While this is theoretically possible, the resulting cycle time would be very large and would increase as $2N$. Thus, a pipeline scheme is generally added.
2. This architecture also supposes that each of the AD-PE receives a new value of $y(m+i, n+j)$ at each clock cycle.

Since transmitting the N^2 values from an external memory is clearly impossible, advantage must be taken from the fact that these values belong to the search window. A portion of the search window of size $N^*(2w+N)$ is stored in the circuit, in a 2-D bank of shift registers able to shift in the up, down, and right direction. Each of the AD-PEs has one of these registers and can, at each cycle, obtain the value of $y(m+i, n+j)$ that it needs. To update this register bank, a new column of $2w+N$ pixels of the search area is serially entered in the circuit and is inserted in the bank of registers. A mechanism must also be provided for loading a new reference with a low I/O overhead: a double buffering of $x(i, j)$ is required, with the pixels $x'(i, j)$ of a new reference block serially loaded during the computation of the current reference block (Figure 24.9).

Figure 24.10 shows the normalized computational rate vs. normalized chip area for block matching circuits. Since one MAD operation consists of three basic ALU operations (SUB, ABS, ADD), for a 1.0 μm CMOS process, we can derive from this figure that:

$$\alpha_{T,0} \approx 30\,\text{mm}^2 + 1.9\,\text{mm}^2/\text{GOPS} \tag{24.18}$$

The first term of this expression indicates that the block matching algorithm requires a large storage area (storage of parts of the actual and previous frame), which cannot be reduced even when the throughput is reduced. The second term corresponds to the linear dependency on computation throughput. The second term has the same amount as that determined for the DCT for GADDS because the three types of operations for the matching require approximately the same expense of additions.

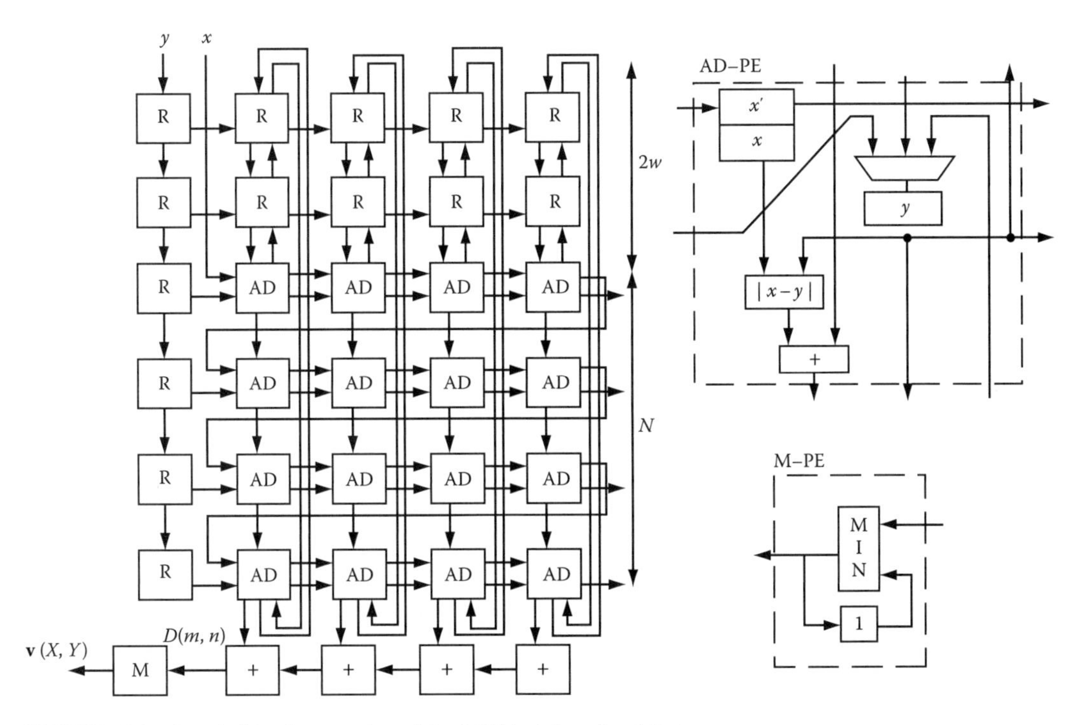

FIGURE 24.9 Practical implementation of the 2-D block-based architecture.

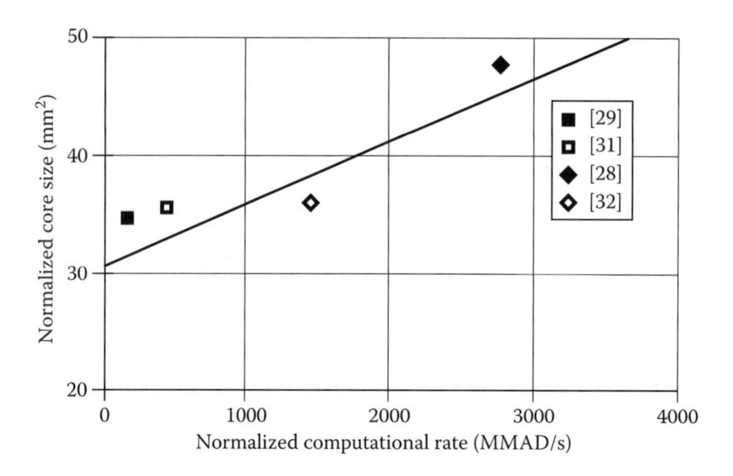

FIGURE 24.10 Normalized silicon area and computational rate for dedicated motion estimation architectures.

From Equation 24.19, the silicon area required for the dedicated implementation of the exhaustive search block matching strategy for a displacement of $\pm w$ pels can be derived by

$$\alpha_{T,0} \approx 0.0057 \cdot (2w+1)^2 \cdot R_S + 30\,\text{mm}^2 \qquad (24.19)$$

According to Equation 24.19, a dedicated implementation of exhaustive search block matching for telecommunication applications based on a source rate of $R_S = 1.01$ Mpel/s (CIF format, 10 Hz frame rate) and a maximum displacement of $w = 15$, the required silicon area can be estimated to 35.5 mm^2. For TV ($R_S = 10.4$ Mpel/s) the silicon area for $w = 31$ can be estimated to 265 mm^2. Estimating the required silicon area for HDTV signals and $w = 31$ leads to 1280 mm^2 for the fictive 1.0 μm CMOS process. From this follows that the implementation for TV and HDTV applications will require the realization of a dedicated block matching chip. Assuming a recent 0.5 μ m semiconductor processes the core size estimation leads to about 22 mm^2 for TV signals and 106 mm^2 for HDTV signals.

To reduce the high computational complexity required for exhaustive search block matching, two strategies can be applied:

1. Decrease of the number of candidate blocks
2. Decrease of the pels per block by subsampling of the image data

Typically, (1) is implemented by search strategies in successive steps. As an example, a modified scheme according to the original proposal of [25] will be discussed. In this scheme, the best match $\mathbf{v}_{s-1}$ in the previous step $s-1$ is improved in the present step s by comparison with displacements $\pm\Delta_s$. The displacement vector $\mathbf{v}_s$ for each step s is calculated according to

$$D_s(m_s, n_s) = \sum_{i=0}^{N-1}\sum_{j=0}^{N-1} |x(i,j) - y(i + m_s + q\cdot\Delta_s, j + n_s + q\cdot\Delta_s)|$$

with $q \in \{-1, 0, 1\}$

$$\begin{bmatrix} m_s \\ n_s \end{bmatrix} = \mathbf{v}_{s-1} \quad \text{for}\, s > 0$$

$$\begin{bmatrix} m_s \\ n_s \end{bmatrix} = \begin{bmatrix} 0 \\ 0 \end{bmatrix} \quad \text{for}\, s = 0$$

and

$$\mathbf{v}_s = \begin{bmatrix} m_s \\ n_s \end{bmatrix} \Bigg|_{D_{s,\min}} \tag{24.20}$$

Δ_s depends on the maximum displacement w and the number of search steps N_s. Typically, when $w = 2^k - 1$, N_s is set to $k = \log_2(w+1)$ and $\Delta_s = 2^{k-s+1}$. For example, for $w = 15$, four steps with $\Delta_s = 8, 4, 2, 1$ are performed. This strategy reduces the number of candidate blocks from $(2w+1)^2$ in case of exhaustive search to $1 + 8 * \log_2(w+1)$, e.g., for $w = 15$ the number of candidate blocks is reduced from 961 to 33 which leads to a reduction of processing power by a factor of 29. For large block sizes N, the number of operations for the match can be further reduced by combining the search strategy with subsampling in the first steps. Architectures for block matching based on hierarchical search strategies are presented in Refs. [20,22,24,30].

24.6 Programmable Architectures

According to the three ways for architectural optimization, adaptation, pipelining, and parallel processing, three architectural classes for the implementation of video signal processors can be distinguished:

- *Intensive pipelined architectures*—These architectures are typically scalar architectures that achieve high clock frequencies of several hundreds of MHz due to the exploitation of pipelining.
- *Parallel data paths*—These architectures exploit data distribution for the increase of computational power. Several parallel data paths are implemented on one processor die, which leads in the ideal case to a linear increase of supported computational power. The number of parallel data paths is limited by the semiconductor process, since an increase of silicon area leads to a decrease of hardware yield.
- *Coprocessor architectures*—Coprocessors are known from general processor designs and are often used for specific tasks, e.g., floating-point operations. The idea of the adaptation to specific tasks and increase of computational power without an increase of the required semiconductor area has been applied by several designs. Due to their high regularity and the high processing power requirements, low-level tasks are the most promising candidates for an adapted implementation. The main disadvantage of this architectural approach is the decrease of flexibility by an increase of adaptation.

24.6.1 Intensive Pipelined Architectures

Applying pipelining for the increase of clock frequency leads to an increased latency of the circuit. For algorithms that require a data-dependent control flow, this fact might limit the performance gain. Additionally, increasing arithmetic processing power leads to an increase of data access rate. Generally, the required data access rate cannot be provided by external memories. The gap between provided external and required internal data access rate increases for processor architectures with high clock frequency. To provide the high data access rate, the amount of internal memory which provides a low access time has to be increased for high-performance signal processors. Moreover, it is unfeasible to apply pipelining to speedup on-chip memory. Thus, the minimum memory access time is another limiting factor for the maximum degree of pipelining. At least speed optimization is a time-consuming task of the design process, which has to be performed for every new technology generation.

Examples for video processors with high clock frequency are the S-VSP [39] and the VSP3 [40]. Due to intensive pipelining, an internal clock frequency of up to 300 MHz can be achieved. The VSP3 consists of two parallel data paths, the pipelined arithmetic logic unit (PAU) and pipelined convolution unit (PCU) (Figure 24.11). The relatively large on-chip data memory of size 114 kb is split into seven blocks, six data

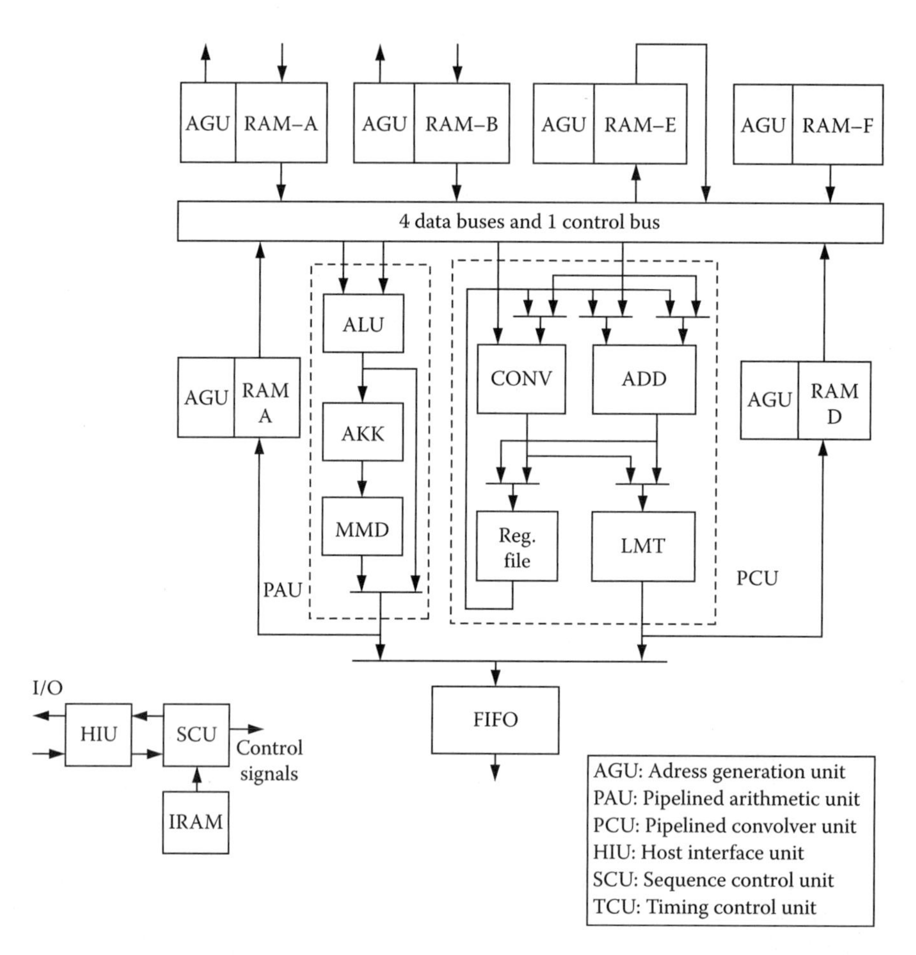

FIGURE 24.11 VSP3 architecture. (Inoue, T. et al., *IEEE J. Solid-State Circuits*, 28, 1321, Dec. 1993.)

memories and one FIFO memory for external data exchange. Each of the six data memories is provided with an address generation unit (AGU), which provides the addressing modes "block," "DCT," and "zigzag." Controlling is performed by a sequence control unit (SCU) which involves a 1024×32 bit instruction memory. A host interface unit (HIU) and a timing control unit (TCU) for the derivation of the internal clock frequency are integrated onto the VSP3 core.

The entire VSP3 core consists of 1.27 million transistors, implemented based on a 0.5 μm BiCMOS technology on a 16.5×17.0 mm^2 die. The VSP3 performs the processing of the CCITT-H.261 tasks (neglecting Huffman coding) for one macroblock in 45 μs. Since real-time processing of 30 Hz-CIF signals requires a processing time of less than 85 μs for one macroblock, a H.261 coder can be implemented based on one VSP3.

24.6.2 Parallel Data Paths

In the previous section, pipelining was presented as a strategy for processing power enhancement. Applying pipelining leads to a subdivision of a logic operation into suboperations, which are processed in parallel with increased processing speed. An alternative to pipelining is the distribution of data among functional units. Applying this strategy leads to an implementation of parallel data paths.

Typically, each data path is connected to an on-chip memory which provides the access distributed image segments.

Generally, two types of controlling strategies for parallel data paths can be distinguished. An MIMD concept provides a private control unit for each data path, whereas SIMD-based controlling provides a single common controller for parallel data paths. Compared to SIMD, the advantage of MIMD is a greater flexibility and a higher performance for complex algorithms with highly data dependent control flow. On the other hand, MIMD requires a significantly increased silicon area. Additionally, the access rate to the program memory is increased, since several controllers have to be provided with program data. Moreover, a software-based synchronization of the data paths is more complex. In case of an SIMD concept synchronization is performed implicitly by the hardware.

Since actual hybrid coding schemes require a large amount of processing power for tasks that require a data independent control flow, a single control unit for the parallel data path provides sufficient processor performance. The controlling strategy has to provide the execution of algorithms that require a data dependent control flow, e.g., quantization. A simple concept for the implementation of a data dependent control flow is to disable the execution of instruction in dependence of the local data path status. In this case, the data path utilization might be significantly decreased, since several of the parallel data path idle while others perform the processing of image data. An alternative is a hierarchical controlling concept. In this case, each data path is provided with a small local control unit with limited functionality and the global controller (GC) initiates the execution of control sequences of the local data path controllers. To reduce the required chip area for this controlling concept, the local controller can be reduced to a small instruction memory. Addressing of this memory is performed by the global control unit.

An example of a video processor based on parallel identical data path with a hierarchical controlling concept is the IDSP [42] (Figure 24.12). The IDSP processor includes four pipelined data processing units (DPU0–DPU3), three parallel I/O ports (PIO0–PIO2), one 8×16 bit register file, five dual-ported memory blocks of size 512×16 bit each, an AGU for the data memories, and a program sequencer with 512×32 bit instruction memory and 32×32 bit boot ROM.

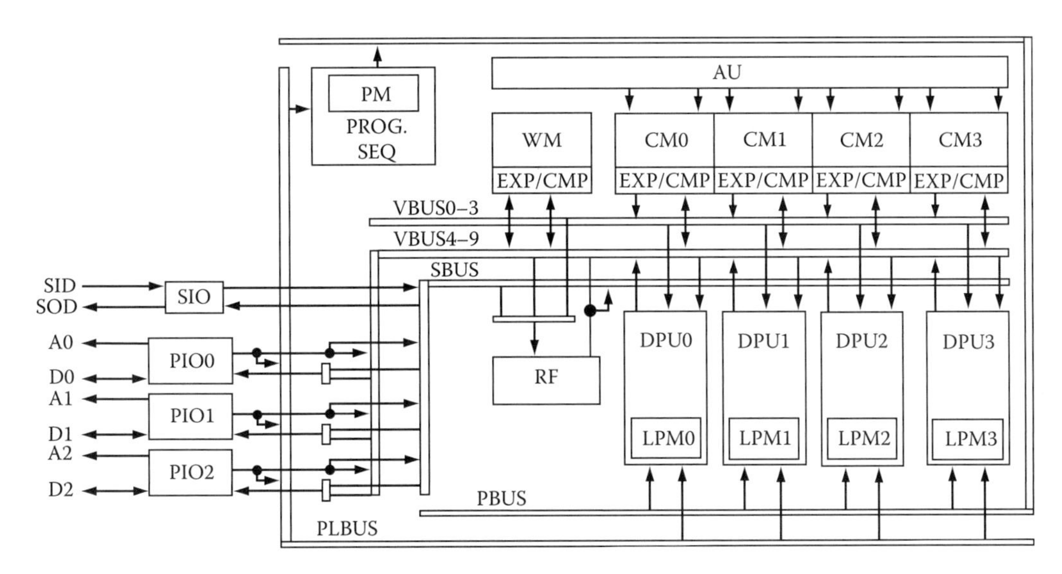

FIGURE 24.12 IDSP architecture. (Yamauchi, H. et al., *IEEE Trans. Circuits Syst. Videotechnol.*, 2, 207, June 1992.)

The data processing units consist of a three-stage pipeline structure based on a ALU, multiplier, and an accumulator. This data path structure is well suited for L1 and L2 norm calculations and convolution-like algorithms. The four parallel data paths support a peak computational power of 300 MOPS at a typical clock frequency of 25 MHz. The data required for parallel processing are supplied by four cache memories (CM0–CM3) and a work memory (WM). Address generation for these memories is performed by an AGU which supports address sequences such as block scan, bit reverse, and butterfly. The three parallel I/O units contain a data I/O port, an AGU, and a DMA control processor (DMAC).

The IDSP integrates 910,000 transistors in $15.2 \times 15.2\ \text{mm}^2$ using an 0.8 μm BiCMOS technology. For a full-CIF H.261 video codec four IDSP are required.

Another example of an SIMD-based video signal processor architecture based on identical parallel data paths is the HiPAR-DSP [44] (Figure 24.13). The processor core consists of 16 RISC data paths, controlled by a common VLIW instruction word. The data paths contain a multiplier/accumulator unit, a shift/round unit, an ALU, and a 16×16 bit register file. Each data path is connected to a private data cache. To support the characteristic data access pattern of several image processing tasks efficiently, a shared memory with parallel data access is integrated on-chip and provides parallel and conflict-free access to the data stored in this memory. The supported access patterns are "matrix," "vector," and "scalar," Data exchange with external devices is supported by an on-chip DMA unit and a hypercube interface.

At present, a prototype of the HiPAR-DSP, based on four parallel data paths, is implemented. This chip will be manufactured in a 0.6 μm CMOS technology and will require a silicon area of about $180\ \text{mm}^2$. One processor chip is sufficient for real-time decoding of video signals, according to MPEG-2 main profile at main level. For encoding an external motion estimator is required.

In contrast to SIMD-based HiPAR-DSP architecture, the TMS320C80 (MVP) is based on an MIMD approach [43]. The MVP consists of four parallel processors (PP) and one master processor (Figure 24.14). The processors are connected to 50 kB on-chip data memory via a global crossbar interconnection network. A DMA controller provides the data transfer to an external data memory and video I/O is supported by an on-chip video interface.

The master processor is a general-purpose RISC processor with an integral IEEE-compatible floating-point unit (FPU). The processor has a 32 bit instruction word and can load or store 8, 16, 32, and 64 bit data sizes. The master processor includes a 32×32 bit general-purpose register file. The master processor

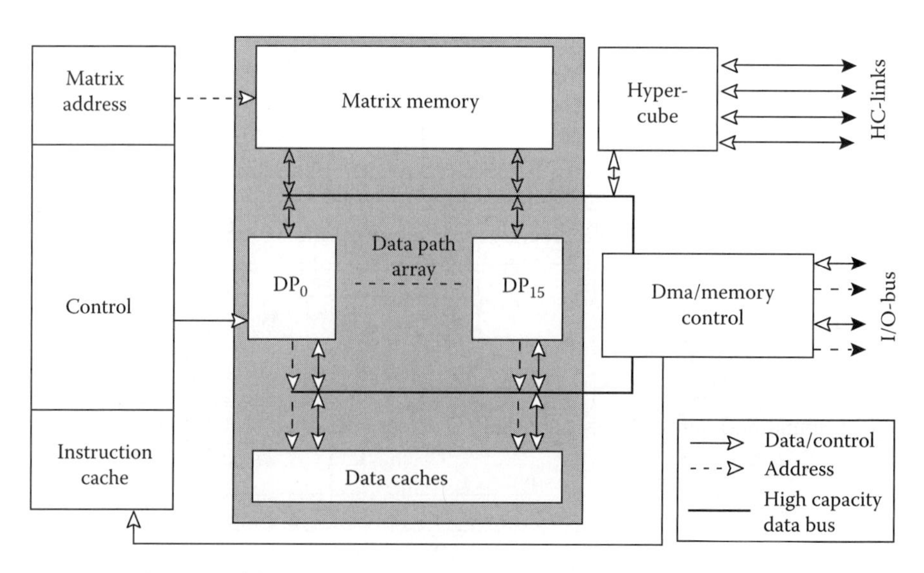

FIGURE 24.13 Architecture of the HiPAR-DSP. (Kneip, J. et al., *Proceedings of the SPIE Visual Communications Image Process.*, 2308, 1753, Sept. 1994.)

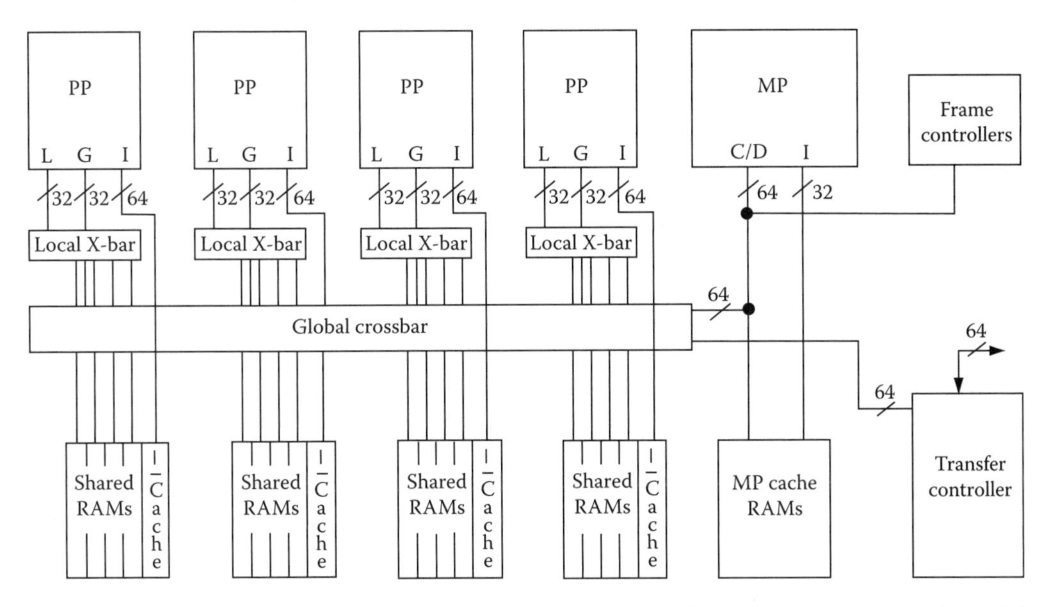

FIGURE 24.14 TMS320C80 (MVP). (Guttag, K., The multiprocessor video processor, MVP, *Proceedings of the IEEE Hot Chips V*, Stanford, CA, Aug. 1993.)

is intended to operate as the main supervisor and distributor of tasks within the chip and is also responsible for the communication with external processors. Due to the integrated FPU, the master processor will perform tasks such as audio signal processing and 3-D graphics transformation.

The PP architecture has been designed to perform typical DSP algorithms, e.g., filtering, DCT, and to support bit and pixel manipulations for graphics applications. The PP contain two address units, a program flow control unit, and a data unit with 32 bit ALU, 16×16 bit multiplier, and a barrel rotator.

The MVP has been designed using a 0.5 μm CMOS technology. Due to the supported flexibility, about four million transistors on a chip area of 324 mm^2 are required. A computational power of 2 GOPS is supported. A single MVP is able to encode CIF-30 Hz video signals according to the MPEG-1 standard.

24.6.3 Coprocessor Concept

Most programmable architectures for video processing applications achieve an increase of processing power by an adaptation of the architecture to the algorithmic requirements. A feasible approach is the combination of a flexible programmable processor module with one or more adapted modules. This approach leads to an increase of processing power for specific algorithms and leads a significant decrease of required silicon area. The decrease of silicon area is caused by two effects. At first, the implementation of the required arithmetic operations can be optimized, which leads to an area reduction. Second, dedicated modules require significantly less hardware expense for module controlling, e.g., for program memory.

Typically, computation-intensive tasks, such as DCT, block matching, or VLC, are candidates for an adapted or even dedicated implementation. Besides the adaptation to one specific task, mapping of several different tasks onto one adapted processor module might be advantageous. For example, mapping successive tasks, such as DCT, Q, Q^{-1}, IDCT, onto the same module reduces the internal communication overhead.

Coprocessor architectures that are based on highly adapted coprocessors achieve high computational power on a small chip area. The main disadvantage of these architectures is the limited flexibility.

Changes of the envisaged applications might lead to an unbalanced utilization of the processor modules and therefore to a limitation of the effective processing power of the chip.

Applying the coprocessor concept opens up a variety of feasible architecture approaches, which differ in achievable processing power and flexibility of the architecture. In the following several architectures are presented, which clarify the wide variety of sensible approaches for video compression based on a coprocessor concept. Most of these architectures aim at an efficient implementation of hybrid coding schemes. As a consequence, these architectures are based on highly adapted coprocessors.

A chip set for video coding has been proposed in Ref. [8]. This chip set consists of four devices: two encoder options (the AVP1300E and AVP1400E), the AVP1400D decoder, and the AVP1400C system controller. The AVP1300E has been designed for H.261 and MPEG-1 frame-based encoding. Full MPEG-1 encoding (I-frame, P-frame, and B-frame) is supported by the AVP1400E. In the following, the architecture of the encoder chips is presented in more detail.

The AVP1300E combines function-oriented modules, mask programmable modules, and user programmable modules (Figure 24.15). It consists of a dedicated motion estimator for exhaustive search block matching with a search area of ± 15 pels. The variable length encoder unit contains an ALU, a register array, a coefficient RAM, and a table ROM. Instructions for the VLE unit are stored in a program ROM. Special instructions for conditional switching, run-length coding, and variable-to-fixed-length conversion are supported. The remaining tasks of the encoder loop, i.e., DCT/IDCT, Q, and Q^{-1}, are performed in two modules called SIMD processor and quantization processor (QP). The SIMD processor consists of six PP each with ALU, multiplier-accumulator units. Program information for this module is again stored in a ROM memory. The QP's instructions are stored in a 1024×28-bit RAM. This module contains 16-bit ALU, a multiplier, and a register file of size 144×16-bit. Data communication with external DRAMs is supported by a memory management unit (MMAFC). Additionally, the processor scheduling is performed by a GC.

Due to the adaptation of the architecture to specific tasks of the hybrid coding scheme, a single chip of size 132 mm^2 (at 0.9 μm CMOS technology) supports the encoding of CIF-30 Hz video signals according to the H.261 standard, including the computation intensive exhaustive search motion estimation strategy. An overview of the complete chipset is given in Ref. [33].

The AxPe640V [37] is another typical example of the coprocessor approach (Figure 24.16). To provide high flexibility for a broad range of video processing algorithms, the two processor modules are fully user programmable. A scalar RISC core supports the processing of tasks with data dependent control flow, whereas the typically more computation intensive low level tasks with data independent control flow can be executed by a parallel SIMD module.

The RISC core functions as a master processor for global control and for processing of tasks such as variable length encoding and Q. To improve the performance for typical video coding schemes, the data

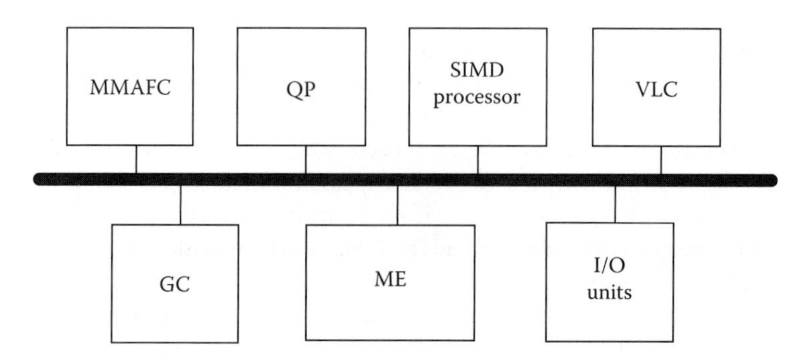

FIGURE 24.15 AVP encoder architecture. (Rao, S.K. et al., *Proceedings of the IEEE International Solid State Circuits Conference*, San Francisco, CA, pp. 32–35, 1993.)

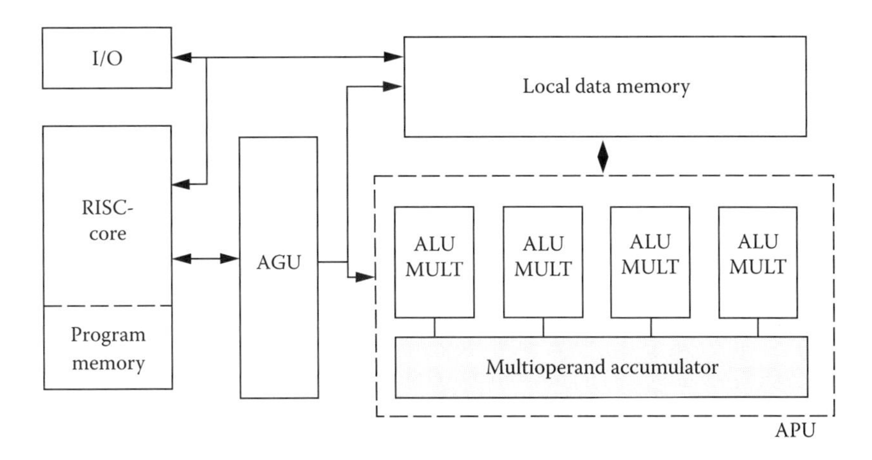

FIGURE 24.16 AxPe640V architecture. (Gaedke, K. et al., *J. VLSI Signal Process.*, 5, 159, April 1993.)

path of the RISC core has been adapted to the requirements of Q and VLC, by an extension of the basic instruction set. A program RAM of size is placed on-chip and can be loaded from an external PROM during start-up. The SIMD-oriented arithmetic processing unit (APU) contains four parallel data paths with a subtracter–complementer–multiplier pipeline. The intermediate results of the arithmetic pipelines are fed into a multioperand accumulator with shift/limit circuitry. The results of the APU can be stored in the internal local memory or read out to the external data output bus.

Since both RISC core and APU include a private program RAM and AGU, these processor modules are able to work in parallel on different tasks. This MIMD-like concept enables an execution of two tasks in parallel, e.g., DCT and Q.

The AxPe640V is currently available in a 66 MHz version, designed in a 0.8 μm CMOS technology. A QCIF-10 Hz H.261 codec can be realized with a single chip. To achieve higher computation power several AxPe640V can be combined to a multiprocessor system. For example, three AxPe640V are required for an implementation of a CIF-10 Hz codec.

The examples presented above clarify the wide range of architectural approaches for the VLSI implementation of video coding schemes. The applied strategies are influenced by several demands, especially the desired flexibility of the architecture and maximum cost for realization and manufacturing. Due to the high computational requirements of real-time video coding, most of the presented architectures apply a coprocessor concept with flexible programmable modules in combination with modules that are more or less adapted to specific tasks of the hybrid coding scheme. An overview of programmable architectures for video coding applications is given in Ref. [6].

Equations 24.4 and 24.5 can be applied for the comparison of programmable architectures. The result of this comparison is shown in Figure 24.17, using the coding scheme according to ITU recommendation H.261 as a benchmark. Assuming a linear dependency between throughput rate and silicon area, a linear relationship corresponds to constant architectural efficiency, indicated by the two gray lines in Figure 24.17. According to these lines, two groups of architectural classes can be identified. The first group consists of adapted architectures, optimized for hybrid coding applications. The architectures contain one or more adapted modules for computation intensive tasks, such as DCT or block matching. It is obvious that the application field of these architectures is limited to a small range of applications. This limitation is avoided by the members of the second group of architectures. Most of these architectures do not contain function specific circuitry for specific tasks of the hybrid coding scheme. Thus, they can be applied for wider variety of applications without a significant loss of sustained computational power. On the other hand, these architectures are associated with a decreased

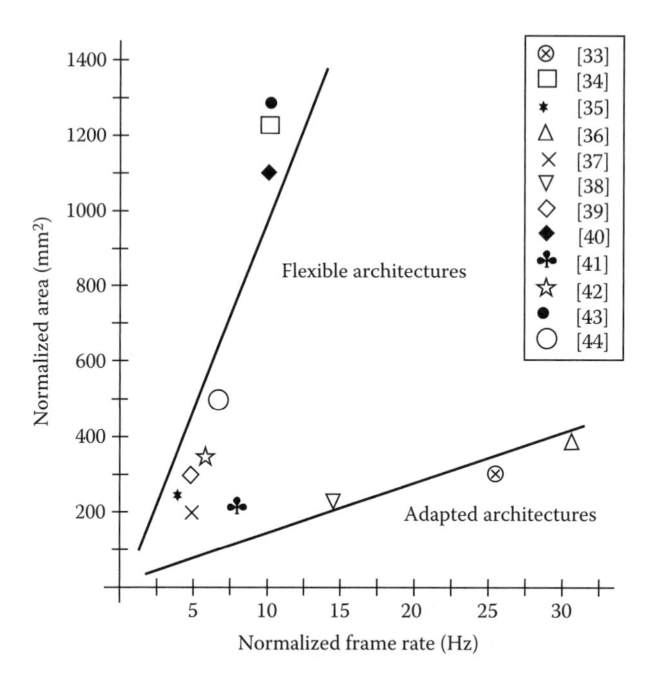

FIGURE 24.17 Normalized silicon area and throughput (frame rate) for adapted and flexible programmable architectures for a H.261 codec.

architectural efficiency compared to the first group of proposed architectures: Adapted architectures achieve an efficiency gain of about 6–7.

For a typical videophone application a frame rate of 10 Hz can be assumed. For this application the required normalized silicon area of about 130 mm^2 is required for adapted programmable approaches and approximately 950 mm^2 are required for flexible programmable architectures. For a rough estimation of the required silicon area for an MPEG-2 decoder, we assume that the algorithmic complexity of an MPEG-2 decoder for CCIR-601 signals is about half the complexity of an H.261 codec. Additionally, it has to be taken into account that the number of pixels per frame is about 5.3 times larger for CCIR signals than for CIF signals. From this the normalized implementation size of an MPEG-2 decoder for CCIR-601 signals and a frame rate of 25 Hz can be estimated to 870 mm^2 for an adapted architecture and 6333 mm^2 for flexible programmable architecture. Scaling these figures according to the defined scaling rules, a silicon area of about 71 and 520 mm^2 can be estimated for an implementation based on an 0.5 μm CMOS process. Thus, the realization of video coding hardware for TV or HDTV based on flexible programmable processors still requires several monolithic components.

24.7 Conclusion

The properties of recent hybrid coding schemes in terms of VLSI implementation have been presented. Architectural alternatives for the dedicated realization of the DCT and block matching have been discussed. Architectures of programmable video signal processors have been presented and compared in terms of architectural efficiency. It has been shown that adapted circuits achieve a six to seven times higher efficiency than flexible programmable circuits. This efficiency gap might decrease for future coding schemes associated with a higher amount of medium- and high-level algorithms. Due to their flexibility, programmable architectures will become more and more attractive for future VLSI implementations of video compression schemes.

Acknowledgment

Figures 24.1 through 24.12 and 24.14 through 24.17 are reprinted from Pirsch, P., Demassieux, N., and Gehrke, W., VLSI architectures for video compression—a survey, *Proc. IEEE*, 83(2), 220–246, Feb., 1995, and used with permission from IEEE.

References

1. ITU-T Recommendation H.261, Video codec for audiovisual services at px64 kbit/s, 1990.
2. ISO-IEC IS 11172, Coding of moving pictures and associated audio for digital storage media at up to about 1.5 Mbit/s, 1993.
3. ISO-IEC IS 13818, Generic coding of moving pictures and associated audio, 1994.
4. ISO-IEC IS 10918, Digital compression and coding of continuous-tone still images, 1992.
5. ISO/IEC JTC1/SC29/WG11, MPEG-4 functionalities, Nov. 1994.
6. Pirsch, P., Demassieux, N., and Gehrke, W., VLSI architectures for video compression—a survey, *Proc. IEEE*, 83(2), 220–246, Feb. 1995.
7. Bakoglu, H. B., *Circuits Interconnections and Packaging for VLSI*, Addison Wesley, Reading, MA, 1987.
8. Rao, S. K., Matthew, M. H. et al., A real-time P*64/MPEG video encoder chip, *Proceedings of the IEEE International Solid State Circuits Conference*, San Francisco, CA, pp. 32–35, 1993.
9. Totzek, U., Matthiesen, F., Wohlleben, S., and Noll, T.G., CMOS VLSI implementation of the 2D-DCT with linear processor arrays, *Proceedings of the International Conference on Acoustics Speech and Signal Processing*, Albuquerque, NM, vol. 2, pp. 937–940, Apr. 1990.
10. Lee, B. G., A new algorithm to compute the discrete cosine transform, *IEEE Trans. Acoust. Speech Signal Process.*, 32(6), 1243–1245, Dec. 1984.
11. Artieri, A., Macoviak, E., Jutand, F., and Demassieux, N., A VLSI one chip for real time two-dimensional discrete cosine transform, *Proceedings of the IEEE International Symposium on Circuits and Systems*, Helsinki, 1988.
12. Jain, P. C., Schlenk, W., and Riegel, M., VLSI implementation of two-dimensional DCT processor in real-time for video codec, *IEEE Trans. Consum. Electron.*, 38(3), 537–545, Aug. 1992.
13. Chau, K. K., Wang, I. F., and Eldridge, C. K., VLSI implementation of 2D-DCT in a compiler, *Proceedings of the IEEE ICASSP*, Toronto, Canada, pp. 1233–1236, 1991.
14. Carlach, J. C., Penard, P., and Sicre, J. L., TCAD: a 27 MHz 8x8 discrete cosine transform chip, *Proceedings of the International Conference on Acoustics Speech and Signal Processing*, Glasgow, U.K., vol. 4, pp. 2429–2432, May 1989.
15. Kim, S. P. and Pan, D. K., Highly modular and concurrent 2-D DCT chip, *Proceedings of the IEEE International Symposium on Circuits and Systems*, San Diego, CA, vol. 3, pp. 1081–1084, May 1992.
16. Sun, M. T., Chen, T. C., and Gottlieb, A. M., VLSI implementation of a 16×16 discrete cosine transform, *IEEE Trans. Circuits Syst.*, 36(4), 610–617, April 1989.
17. Bierling, M., Displacement estimation by hierarchical block-matching, *Proc. SPIE Vis. Commun. Image Process.*, 1001, 942–951, 1988.
18. Chow, K. H. and Liou, M. L., Genetic motion search for video compression, *Proceedings of the IEEE Visual Signal Processing and Communication*, Melbourne, Australia, pp. 167–170, Sept. 1993.
19. Ghanbari, M., The cross-search algorithm for motion estimation, *IEEE Trans. Commun.*, COM 38(7), 950–953, July 1990.
20. Gupta, G. et al., VLSI architecture for hierarchical block matching, *Proceedings of the IEEE International Symposium on Circuits and Systems*, London, U.K., vol. 4, pp. 215–218, 1994.
21. Jain, J. R. and Jain, A. K., Displacement measurement and its application in interframe image coding, *IEEE Trans. Commun.*, COM 29(12), 1799–1808, Dec. 1981.

22. Jong, H. M. et al., Parallel architectures of 3-step search block-matching algorithms for video coding, *Proceedings of the IEEE International Symposium on Circuits and Systems*, vol. 3, pp. 209–212, 1994.
23. Kappagantula, S. and Rao, K. R., Motion compensated interframe image prediction, *IEEE Trans. Commun.*, COM 33(9), 1011–1015, Sept. 1985.
24. Kim, H. C. et al., A pipelined systolic array architecture for the hierarchical block-matching algorithm, *Proceedings of the IEEE International Symposium on Circuits and Systems*, London, U.K., vol. 3, pp. 221–224, 1994.
25. Koga, T., Iinuma, K., Hirano, A., Iijima, Y., and Ishiguro, T., Motion compensated interframe coding for video conferencing, *Proceedings of the National Telecommunications Conference*, New Orleans, LA, G5.3.1–5.3.5, Nov. 29–Dec. 3, 1981.
26. Puri, A., Hang, H. M., and Schilling, D. L., An efficient block-matching algorithm for motion compensated coding, *Proceedings of the IEEE ICASSP*, San Diego, CA, pp. 25.4.1–25.4.4, 1987.
27. Srinivasan, R. and Rao, K. R., Predictive coding based on efficient motion estimation, *IEEE Trans. Commun.*, COM 33(8), 888–896, Aug. 85.
28. Colavin, O., Artieri, A., Naviner, J. F., and Pacalet, R., A dedicated circuit for real-time motion estimation, *EuroASIC*, Paris, France, pp. 96–99, 1991.
29. Dianysian, R. et al., Bit-serial architecture for real-time motion compensation, *Proc. SPIE Visual Communications and Image Processing*, 1988.
30. Komarek, T. et al., Array architectures for block-matching algorithms, *IEEE Trans. Circuits Syst.*, 36(10), 1301–1308, Oct. 1989.
31. Yang, K. M. et al., A family of VLSI designs for the motion compensation block-matching algorithms, *IEEE Trans. Circuits Syst.*, 36(10), 1317–1325, Oct. 1989.
32. Ruetz, P., Tong, P., Bailey, D., Luthi, D. A., and Ang, P. H., A high-performance full-motion video compression chip set, *IEEE Trans. Circuits Syst. Video Technol.*, 2(2), 111–122, June 1992.
33. Ackland, B., The role of VLSI in multimedia, *IEEE J. Solid-State Circuits*, 29(4), 1886–1893, April 1994.
34. Akari, T. et al., Video DSP architecture for MPEG2 codec, *Proceedings of the ICASSP'94*, Adelaide, SA, vol. 2, pp. 417–420, 1994, IEEE Press.
35. Aono, K. et al., A video digital signal processor with a vector-pipeline architecture, *IEEE J. Solid-State Circuits*, 27(12), 1886–1893, Dec. 1992.
36. Bailey, D. et al., Programmable vision processor/controller, *IEEE MICRO*, 12(5), 33–39, Oct. 1992.
37. Gaedke, K., Jeschke, H., and Pirsch, P., A VLSI based MIMD architecture of a multiprocessor system of real-time video processing applications, J. *VLSI Signal Process.*, 5, 159–169, April 1993.
38. Gehrke, W., Hoffer, R., and Pirsch, P., A hierarchical multiprocessor architecture based on heterogeneous processors for video coding applications, *Proceedings of the ICASSP'94*, Adelaide, SA, vol. 2, 1994.
39. Goto, J. et al., 250-MHz BiCMOS super-high-speed video signal processor (S-VSP) ULSI, *IEEE J. Solid-State Circuits*, 26(12), 1876–1884, 1991.
40. Inoue, T. et al., A 300-MHz BiCMOS video signal processor, *IEEE J. Solid-State Circuits*, 28(12), 1321–1329, Dec. 1993.
41. Micke, T., Müller, D., and Heiß, R., ISDN-bildtelefon auf der grundlage eines array-prozessor-IC, mikroelektronik, vde-verlag, 5(3), 116–119, May/June 1991. (In German.)
42. Yamauchi, H. et al., Architecture and implementation of a highly parallel single chip video DSP, *IEEE Trans. Circuits Syst. Videotechnol.*, 2(2), 207–220, June 1992.
43. Guttag, K., The multiprocessor video processor, MVP, *Proceedings of the IEEE Hot Chips V*, Stanford, CA, Aug. 1993.
44. Kneip, J. Rönner, K., and Pirsch, P., A single chip highly parallel architecture for image processing applications, *Proc. SPIE Vis. Commun. Image Process.*, 2308(3), 1753–1764, Sept. 1994.

Index

A

I

K

L

M

N

O

P

Q

R

S

T

V

W

Z